Chebyshev an Spectral Methods

Second Edition (Revised)

JOHN P. BOYD

DOVER PUBLICATIONS, INC.
Mineola, New York

Bibliographical Note

This Dover edition, first published in 2001, is a revised and enlarged second edition of the work first published in 1969 by Springer-Verlag, Berlin, in the series "Lecture Notes in Engineering."

Library of Congress Cataloging-in-Publication Data

Boyd, J. P. (John Philip), 1941–
    Chebyshev and Fourier spectral methods / John P. Boyd.—2nd ed., rev.
       p. cm.
    Includes bibliographical references and index.
    ISBN-13: 978-0-486-41183-5
    ISBN-10: 0-486-41183-4 (pbk.)
    1. Chebyshev polynomials. 2. Fourier analysis. 3. Spectral theory (Mathematics) I. Title.

QA404.5 .B69 2001
515'.55—dc21

                                                        00-027475

Manufactured in the United States by LSC Communications

www.doverpublications.com

41183409   2023

Dedication

To Marilyn, Ian, and Emma

"A computation is a temptation that should be resisted as long as possible."
— J. P. Boyd, paraphrasing T. S. Eliot

Contents

Preface

[Preface to the First Edition (1988)]

The goal of this book is to teach spectral methods for solving boundary value, eigenvalue and time-dependent problems. Although the title speaks only of Chebyshev polynomials and trigonometric functions, the book also discusses Hermite, Laguerre, rational Chebyshev, sinc, and spherical harmonic functions.

These notes evolved from a course I have taught the past five years to an audience drawn from half a dozen different disciplines at the University of Michigan: aerospace engineering, meteorology, physical oceanography, mechanical engineering, naval architecture, and nuclear engineering. With such a diverse audience, this book is not focused on a particular discipline, but rather upon solving differential equations in general. The style is not lemma-theorem-Sobolev space, but algorithm-guidelines-rules-of-thumb.

Although the course is aimed at graduate students, the required background is limited. It helps if the reader has taken an elementary course in computer methods and also has been exposed to Fourier series and complex variables at the undergraduate level. However, even this background is not absolutely necessary. Chapters 2 to 5 are a self-contained treatment of basic convergence and interpolation theory.

Undergraduates who have been overawed by my course have suffered not from a lack of knowledge, but a lack of sophistication. This volume is not an almanac of unrelated facts, even though many sections and especially the appendices can be used to look up things, but rather is a travel guide to the Chebyshev City where the individual algorithms and identities interact to form a community. In this mathematical village, the special functions are special friends. A differential equation is a pseudospectral matrix in drag. The program structure of grids point/basisset/collocation matrix is as basic to life as cloud/rain/river/sea.

It is not that spectral concepts are difficult, but rather that they link together as the components of an intellectual and computational ecology. Those who come to the course with no previous adventures in numerical analysis will be like urban children abandoned in the wilderness. Such innocents will learn far more than hardened veterans of the arithmurgical wars, but emerge from the forests with a lot more bruises.

In contrast, those who have had a couple of courses in numerical analysis should find this book comfortable: an elaboration of familiar ideas about basis sets and grid point representations. Spectral algorithms are a new worldview of the same computational landscape.

These notes are structured so that each chapter is largely self-contained. Because of this and also the length of this volume, the reader is strongly encouraged to skip-and-choose.

The course on which this book is based is only one semester. However, I have found it necessary to omit seven chapters or appendices each term, so the book should serve equally well as the text for a two-semester course.

Although these notes were written for a graduate course, this book should also be useful to researchers. Indeed, half a dozen faculty colleagues have audited the course.

The writing style is an uneasy mixture of two influences. In private life, the author has written fourteen published science fiction and mystery short stories. When one has described zeppelins jousting in the heavy atmosphere of another world or a stranded explorer alone on an artificial toroidal planet, it is difficult to write with the expected scientific dullness.

Nonetheless, I have not been too proud to forget most of the wise precepts I learned in college English: the book makes heavy use of both the passive voice and the editorial "we". When I was still a postdoc, a kindly journal editor took me in hand, and circled every single "I" in red. The scientific abhorrence of the personal pronoun, the active voice, and lively writing is as hypocritical as the Victorian horror of "breast" and "pregnant". Nevertheless, most readers are so used to the anti-literature of science that what would pass for good writing elsewhere would be too distracting. So I have done my best to write a book that is not about its style but about its message.

Like any work, this volume reflects the particular interests and biases of the author. While a Harvard undergraduate, I imagined that I would grow up in the image of my professors: a pillar of the A. M. S., an editorial board member for a dozen learned journals, and captain and chief executive officer of a large company of graduate students and postdocs. My actual worldline has been amusingly different.

I was once elected to a national committee, but only after my interest had shifted. I said nothing and was not a nuisance. I have never had any connection with a journal except as a reviewer. In twelve years at Michigan, I have supervised a single Ph. D. thesis. And more than three-quarters of my 65 papers to date have had but a single author.

This freedom from the usual entanglements has allowed me to follow my interests: chemical physics as an undergraduate, dynamic meteorology as a graduate student, hydrodynamic stability and equatorial fluid mechanics as an assistant professor, nonlinear waves and a stronger interest in numerical algorithms after I was tenured. This book reflects these interests: broad, but with a bias towards fluid mechanics, geophysics and waves.

I have also tried, not as successfully as I would have wished, to stress the importance of analyzing the physics of the problem before, during, and after computation. This is partly a reflection of my own scientific style: like a sort of mathematical guerrilla, I have ambushed problems with Padé approximants and perturbative derivations of the Korteweg-deVries equation as well as with Chebyshev polynomials; numerical papers are only half my published articles.

However, there is a deeper reason: the numerical agenda is always set by the physics. The geometry, the boundary layers and fronts, and the symmetries are the topography of the computation. He or she who would scale Mt. Everest is well-advised to scout the passes before beginning the climb.

When I was an undergraduate — ah, follies of youth — I had a quasi-mystical belief in the power of brute force computation.

Fortunately, I learned better before I could do too much damage. Joel Primack (to him be thanks) taught me John Wheeler's First Moral Principle: Never do a calculation until you already know the answer.

The point of the paradox is that one can usually deduce much about the solution — orders-of-magnitude, symmetries, and so on — before writing a single line of code. A thousand errors have been published because the authors had no idea what the solution ought to look like. For the scientist, as for Sherlock Holmes, it is the small anomalies that are the clues to the great pattern. One cannot appreciate the profound significance of the unexpected without first knowing the expected.

The during-and-after theory is important, too. My thesis advisor, Richard Lindzen, never had much interest in computation *per se*, and yet he taught me better than anyone else the art of good scientific number-crunching. When he was faced with a stiff boundary value problem, he was not too proud to run up and down the halls, knocking on doors, until he finally learned of a good algorithm: centered differences combined with the tridiagonal elimination described in Appendix B. This combination had been known for twenty years, but was only rarely mentioned in texts because it was hard to prove convergence theorems.[1] He then badgered the programming staff at the National Center for Atmospheric Research to help him code the algorithm for the most powerful computer then available, the CDC 7600, with explicit data swaps to and from the core.

A scientist who is merely good would have stopped there, but Lindzen saw from the numerical output that equatorial waves in vertical shear satisfied the separation-of-scales requirement of singular perturbation theory. He then wrote two purely analytical papers to derive the perturbative approximation, and showed it agreed with his numerical calculations. The analysis was very complicated — a member of the National Academy of Sciences once described it to me, laughing, as "the most complicated damn thing I've ever seen" — but the final answers fits on one line.

In sad contrast, I see far too many students who sit at their workstation, month after month, trying to batter a problem into submission. They never ask for help, though Michigan has one of the finest and broadest collections of arithmurgists on the planet. Nor will they retreat to perturbation theory, asymptotic estimates, or even a little time alone in the corner.

It is all too easy to equate multiple windows with hard work, and multiple contour plots with progress. Nevertheless, a scientist by definition is one who listens for the voice of God. It is part of the fallen state of man that He whispers.

In order that this book may help to amplify those whispers, I have been uninhibited in expressing my opinions. Some will be wrong; some will be soon outdated.[2] Nevertheless, I hope I may be forgiven for choosing to stick my neck out rather than drown the reader in a sea of uninformative blandness. The worst sin of a thesis advisor or a textbook writer is to have no opinions.

[Preface to the Second Edition, January, 1999]

In revising this book ten years after, I deleted the old Chapter 11 (case studies of fluid computations) and Appendix G (least squares) and added four new chapters on eigenvalue problems, aliasing and spectral blocking, the slow manifold and Nonlinear Galerkin theory, and semi-Lagrangian spectral methods. All of the chapters have been updated and most have been rewritten. Chapter 18 has several new sections on polar coordinates. Appendix E contains a new table giving the transformations of first and second derivatives for a two-dimensional map. Appendix F has new analytical formulas for the Legendre-Lobatto grid points up to nine-point grids, which is sufficient for most spectral element applications.

My second book, *Weakly Nonlocal Solitary Waves and Beyond-All-Orders-Asymptotics* (Kluwer, 1998) has two chapters that amplify on themes in this volume. Chapter 8 is an expanded version of Appendices C and D here, describing a much wider range of strategies for nonlinear algebraic equations and for initializing interations. Chapter 9 explains how a standard infinite interval basis can be extended to approximate functions that oscillate rather than decay-to-zero at infinity.

Other good books on spectral methods have appeared in recent years. These and a selection of review articles are catalogued in Chapter 23.

---

[1] Alas, numerical analysis is still more proof-driven than accomplishment-driven even today.

[2] Surely, too, the book has typographical errors, and the reader is warned to check formulas and tables before using them.

My original plan was to build a bibliographical database on spectral methods and applications of spectral algorithms that could be printed in full here. Alas, this dream was overtaken by events: as the database grew past 2000 items, I was forced to limit the bibliography to 1025 references. Even so, this partial bibliography and the Science Citation Index should provide the reader with ample entry points into any desired topic. The complete database is available online at the author's homepage, currently at **http://www-personal.engin.umich.edu/~jpboyd**. To paraphrase Newton, it is better to stand on the shoulders of giants than to try to recreate what others have already done better.

Spectral elements have become an increasingly important part of the spectral world in the last decade. However, the first edition, with but a single chapter on spectral elements, was almost 800 pages long. (Students irrevently dubbed it the "Encyclopedia Boydica".) So, I have reluctantly included only the original chapter on domain decomposition in this edition. A good treatment of spectral elements in the lowbrow spirit of this book will have to await another volume.

Perhaps it is just as well. The bibliographic explosion is merely a symptom of a field that is still rapidly evolving. The reader is invited to use this book as a base camp for his or her own expeditions.

The Heart of Africa has lost its mystery; the planets of Tau Ceti are currently unknown and unreachable. Nevertheless, the rise of digital computers has given this generation its galleons and astrolabes. The undiscovered lands exist, in one sense, only as intermittent electric rivers in dendritic networks of copper and silicon, invisible as the soul. And yet the mystery of scientific computing is that its new worlds over the water, wrought only of numbers and video images, are as real as the furrowed brow of the first Cro-Magnon who was mystified by the stars, and looked for a story.

Acknowledgments

The author's work has been supported by the National Science Foundation through the Physical Oceanography, Meteorology, Computational Engineering and Computational Mathematics programs via grants OCE7909191, OCE8108530, OCE8305648, OCE8509923, OCE812300, DMS8716766 and by the Department of Energy. My leave of absence at Harvard in 1980 was supported through grant NASA NGL-22-007-228 and the hospitality of Richard Lindzen. My sabbatical at Rutgers was supported by the Institute for Marine and Coastal Sciences and the hospitality of Dale Haidvogel.

I am grateful for the comments and suggestions of William Schultz, George Delic, and the students of the course on which this book is based, especially Ahmet Selamet, Mark Storz, Sue Haupt, Mark Schumack, Hong Ma, Beth Wingate, Laila Guessous, Natasha Flyer and Jeff Hittinger. I thank Andreas Chaniotis for correcting a formula

I am also appreciative of the following publishers and authors for permission to reproduce figures or tables. Fig. 3.3: C. A. Coulson, *Valence* (1973), Oxford University Press. Fig. 7.3: H. Weyl, *Symmetry* (1952) [copyright renewed, 1980], Princeton University Press. Tables 9.1 and Figs. 9.1 and 9.2: D. Gottlieb and S. A. Orszag, *Numerical Analysis of Spectral Methods* (1977), Society for Industrial and Applied Mathematics. Fig. 12-4: C. Canuto and A. Quarteroni, *Journal of Computational Physics* (1985), Academic Press. Tables 12.2 and 12.3: T. Z. Zang, Y. S. Wong and M. Y. Hussaini, *Journal of Computational Physics* (1984), Academic Press. Fig. 13.1 and Table 13.2: J. P. Boyd, *Journal of Computational Physics* (1985), Academic Press. Fig. 14.3: E. Merzbacher, *Quantum Mechanics* (1970), John Wiley and Sons. Figs. 14.4, 14.5, 14.7, 14.8, 14.9, 14.10, and 14.11: J. P. Boyd *Journal of Computational Physics* (1987), Academic Press. Fig. 15.1: W. D'Arcy Thompson, *Growth and Form* (1917), Cambridge University Press. Fig. D.1 (wth changes): J. P. Boyd, *Physica D* (1986), Elsevier. Fig. D.2: E. Wasserstrom, *SIAM Review* (1973), Society for Industrial and Applied Mathematics.

I thank Gene, Dale, Dave and Terry of the Technical Illustration Dept., DRDA [now disbanded], for turning my rough graphs and schematics into camera-ready drawings.

I also would like to acknowledge a debt to Paul Bamberg of the Harvard Physics department. His lecturing style strongly influenced mine, especially his heavy reliance on class notes both as text and transparencies.

I thank Joel Primack, who directed my undergraduate research, for his many lessons. One is the importance of preceding calculation with estimation. Another is the need to write quick-and-rough reports, summary sheets and annotations for even the most preliminary results. It is only too true that "otherwise, in six months all your computer output and all your algebra will seem the work of a stranger."

I am also thankful for Richard Goody's willingness to humour an undergraduate by teaching him in a reading course. Our joint venture on tides in the Martian atmosphere was scooped, but I found my calling.

I am grateful for Richard Lindzen's patient tolerance of my first experiments with Chebyshev polynomials. His running commentary on science, scientists, and the interplay of numerics and analysis was a treasured part of my education.

I thank Steven Orszag for accepting this manuscript for the Lecture Notes in Engineering series (Springer-Verlag) where the first edition appeared. The treatment of timestepping methods in Chapter 10 is heavily influenced by his MIT lectures of many years ago, and the whole book is strongly shaped by his many contributions to the field.

I am appreciative of John Grafton and the staff of Dover Press for bringing this book back into print in an expanded and corrected form.

Lastly, I am grateful for the support of the colleagues and staff of the University of Michigan, particularly Stan Jacobs for sharing his knowledge of nonlinear waves and perturbation theory, Bill Schultz for many fruitful collaborations in applying spectral methods to mechanical engineering, and Bill Kuhn for allowing me to introduce the course on which this book is based.

Errata and Extended-Bibliography

These may be found on author's homepage, currently at

http://www-personal.engin.umich.edu/~jpboyd

Errata and comments may be sent to the author at the following:

jpboyd@umich.edu

Thank you!

Chapter 1

Introduction

"I have no satisfaction in formulas unless I feel their numerical magnitude."
  –Sir William Thomson, 1st Lord Kelvin (1824–1907)

"It is the increasingly pronounced tendency of modern analysis to substitute ideas for calculation; nevertheless, there are certain branches of mathematics where calculation conserves its rights."
  –P. G. L. Dirichlet (1805–1859)

1.1 Series expansions

Our topic is a family of methods for solving differential and integral equations. The basic idea is to assume that the unknown $u(x)$ can be approximated by a sum of $N + 1$ "basis functions" $\phi_n(x)$:

$$u(x) \approx u_N(x) = \sum_{n=0}^{N} a_n \, \phi_n(x) \qquad (1.1)$$

When this series is substituted into the equation

$$Lu = f(x) \qquad (1.2)$$

where L is the operator of the differential or integral equation, the result is the so-called "residual function" defined by

$$R(x; a_0, a_1, \dots, a_N) = Lu_N - f \qquad (1.3)$$

Since the residual function $R(x; a_n)$ is identically equal to zero for the exact solution, the challenge is to choose the series coefficients $\{a_n\}$ so that the residual function is minimized. The different spectral and pseudospectral methods differ mainly in their minimization strategies.

1

1.2 First Example

These abstract ideas can be made concrete by a simple problem. Although large problems are usually programmed in FORTRAN and C, it is very educational to use an algebraic manipulation language like Maple, Mathematica, Macsyma or Reduce. In what follows, Maple statements are shown in bold face. The machine's answers have been converted into standard mathematical notation.

The example is the linear, one-dimensional boundary value problem:

$$u_{xx} - (x^6 + 3x^2)u = 0 \qquad (1.4)$$

$$u(-1) = u(1) = 1 \qquad (1.5)$$

The exact solution is (Scraton, 1965)

$$u(x) = \exp([x^4 - 1]/4) \qquad (1.6)$$

Polynomial approximations are recommended for most problems, so we shall choose a spectral solution of this form. In order to satisfy the boundary conditions independently of the unknown spectral coefficients, it is convenient to write the approximation as **u2:=1 + (1-x*x)*(a0 + a1*x + a2*x*x);**

$$u_2 = 1 + (1 - x^2)(a_0 + a_1 x + a_2 x^2) \qquad (1.7)$$

where the decision to keep only three degrees of freedom is arbitrary.

The residual for this approximation is **Resid:= diff(u,x,x) - (x**6 + 3*x**2)*u;**

$$R(x; a_0, a_1, a_2) = u_{2,xx} - (x^6 + 3x^2)u_2 \qquad (1.8)$$

$$
\begin{aligned}
R = {} & (2a_2 + 2a_0) - 6a_1 x - (3 + 3a_0 + 12a_2)x^2 - 3a_1 x^3 + 3(a_0 - a_2)x^4 \qquad (1.9) \\
& + 3a_1 x^5 + (-1 - a_0 + 3a_2)x^6 - a_1 x^7 + (a_0 - a2)x^8 + a_1 x^9 + 10a_2 x^{10}
\end{aligned}
$$

As error minimization conditions, we choose to make the residual zero at a set of points equal in number to the undetermined coefficients in $u_2(x)$. This is called the "collocation" or "pseudospectral" method. If we arbitrarily choose the points $x_i = (-1/2, 0, 1/2)$, this gives the three equations: **eq1:=subs(x=-1/2,Resid); eq2:=subs(x=0,Resid); eq3:=subs(x=1/2,Resid);**

$$
\begin{aligned}
eq1 &= -\frac{659}{256}a_0 + \frac{1683}{512}a_1 - \frac{1171}{1024}a_2 - \frac{49}{64} \\
eq2 &= -2(a_0 - a_2) \\
eq3 &= -\frac{659}{256}a_0 - \frac{1683}{512}a_1 - \frac{1171}{1024}a_2 - \frac{49}{64}
\end{aligned}
\qquad (1.10)
$$

The coefficients are then determined by solving $eq1 = eq2 = eq3 = 0$; **solutionarray:= solve({eq1,eq2,eq3}, {a0,a1,a2});** yields

$$a_0 = -\frac{784}{3807}, \qquad a_1 = 0, \qquad a_2 = a_0 \qquad (1.11)$$

Figure 1.1 shows that this low order approximation is quite accurate.

However, the example raises a whole host of questions including:

1. What is an optimum choice of basis functions?

2. Why choose "collocation" as the residual-minimizing condition?

3. What are the optimum collocation points?

4. Why is a_1 zero? Could we have anticipated this, and used a trial solution with just two degrees of freedom for the same answer?

5. How do we solve the algebraic problem for the coefficients when the Maple "solve" function isn't available?

The answer to the first question is that choosing powers of x as a basis is actually rather dangerous unless N, the number of degrees-of-freedom, is small or the calculations are being done in exact arithmetic, as true for the Maple solution here. In the next section, we describe the good choices. In an algebraic manipulation language, different rules apply as explained in Chapter 20.

The second answer is: Collocation is the simplest choice which is guaranteed to work, and if done right, nothing else is superior. To understand why, however, we shall have to understand both the standard theory of Fourier and Chebyshev series and Galerkin methods (Chapters 2 and 3) and the theory of interpolation and cardinal functions (Chapters 4 and 5).

The third answer is: once the basis set has been chosen, there are only two optimal sets of interpolation points for each basis (the Gauss-Chebyshev points and the Gauss-Lobatto points); both are given by elementary formulas in Appendices A and F, and which one is used is strictly a matter of convenience.

The fourth answer is: Yes, the irrelevance of a_1 could have been anticipated. Indeed, one can show that for this problem, all the odd powers of x have zero coefficients. Symmetries of various kinds are extremely important in practical applications (Chapter 8).

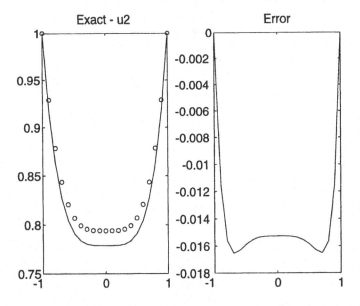

Figure 1.1: Left panel: Exact solution $u = \exp([x^4 - 1]/4)$ (solid) is compared with the three-coefficient numerical approximation (circles). Right panel: $u - u_2$.

Table 1.1: Maple program to solve linear boundary-value problem

```
u2:=1 + (1-x*x)*(a0 + a1*x + a2*x*x);
Resid:= diff(u,x,x) - (x**6 + 3*x**2)*u;
eq1:=subs(x=-1/2,Resid); eq2:=subs(x=0,Resid); eq3:=subs(x=1/2,Resid);
solutionarray:=solve({eq1,eq2,eq3},{a0,a1,a2});
```

The fifth answer is: the algebraic equations can be written (for a linear differential equation) as a matrix equation, which can then be solved by library software in FORTRAN or C.

Many other questions will be asked and answered in later chapters. However, some things are already clear.

First, the method is not necessarily harder to program than finite difference or finite element algorithms. In Maple, the complete solution of the ODE/BVP takes just five lines (Table 1.1)!

Second, spectral methods are not purely numerical. When N is sufficiently small, Chebyshev and Fourier methods yield an *analytic* answer.

1.3   Comparison with finite element methods

Finite element methods are similar in philosophy to spectral algorithms; the major difference is that finite elements chop the interval in x into a number of sub-intervals, and choose the $\phi_n(x)$ to be *local* functions which are polynomials of *fixed* degree which are non-zero only over a couple of sub-intervals. In contrast, spectral methods use *global* basis functions in which $\phi_n(x)$ is a polynomial (or trigonometric polynomial) of *high* degree which is non-zero, except at isolated points, over the entire computational domain.

When more accuracy is needed, the finite element method has three different strategies. The first is to subdivide each element so as to improve resolution uniformly over the whole domain. This is usually called "h-refinement" because h is the common symbol for the size or average size of a subdomain. (Figure 1.2).

The second alternative is to subdivide only in regions of steep gradients where high resolution is needed. This is "r-refinement".

The third option is to keep the subdomains fixed while increasing p, the degree of the polynomials in each subdomain. This strategy of "p-refinement" is precisely that employed by spectral methods. Finite element codes which can quickly change p are far from universal, but those that can are some called "p-type" finite elements.

Finite elements have two advantages. First, they convert differential equations into matrix equations that are *sparse* because only a handful of basis functions are non-zero in a given sub-interval. ("Sparse" matrices are discussed in Appendix B; suffice it to say that "sparse" matrix equations can be solved in a fraction of the cost of problems of similar size with "full" matrices.) Second, in multi-dimensional problems, the little sub-intervals become little triangles or tetrahedra which can be fitted to irregularly-shaped bodies like the shell of an automobile. Their disadvantage is low accuracy (for a given number of degrees of freedom N) because each basis function is a polynomial of low degree.

Spectral methods generate algebraic equations with full matrices, but in compensation, the high order of the basis functions gives high accuracy for a given N. When fast iterative matrix–solvers are used, spectral methods can be much more efficient than finite element or finite difference methods for many classes of problems. However, they are most useful when the *geometry* of the problem is fairly smooth and regular.

So-called "spectral element" methods gain the best of both worlds by hybridizing spectral and finite element methods. The domain is subdivided into elements, just as in finite elements, to gain the flexibility and matrix sparsity of finite elements. At the same time, the degree of the polynomial p in each subdomain is sufficiently high to retain the high accuracy and low storage of spectral methods. (Typically, $p = 6$ to 8, but spectral element codes are almost always written so that p is an arbitrary, user-choosable parameter.)

It turns out that most of the theory for spectral elements is the *same* as for global spectral methods, that is, algorithms in which a single expansion is used everywhere. Consequently, we shall concentrate on spectral methods in the early going. The final chapter will describe how to match expansions in multiple subdomains.

Low order finite elements can be derived, justified and implemented without knowledge of Fourier or Chebyshev convergence theory. However, as the order is increased, it turns out that *ad hoc* schemes become increasingly ill-conditioned and ill-behaved. The only practical way to implement "nice" high order finite elements, where "high order" generally means sixth or higher order, is to use the technology of spectral methods.

Similarly, it turns out that the easiest way to match spectral expansions across subdomain walls is to use the variational formalism of finite elements.

Thus, it really doesn't make much sense to ask: Are finite elements or spectral methods better? For sixth or higher order, they are essentially the same. The big issue is: Does one need high order, or is second or fourth order sufficient?

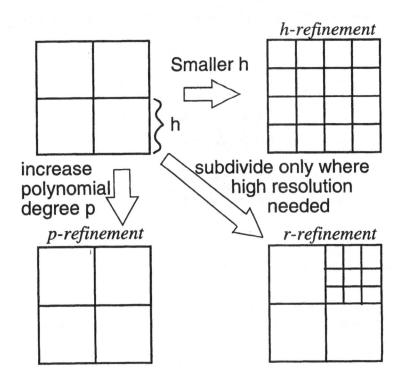

Figure 1.2: Schematic of three types of finite elements

1.4   Comparisons with Finite Difference Method: Why Spectral Methods are Accurate and Memory-Minimizing

Finite difference methods approximate the unknown $u(x)$ by a sequence of overlapping polynomials which interpolate $u(x)$ at a set of grid points. The derivative of the local interpolant is used to approximate the derivative of $u(x)$. The result takes the form of a weighted sum of the values of $u(x)$ at the interpolation points.

Spectral
One high-order polynomial for WHOLE domain

Finite Difference
Multiple Overlapping Low-Order Polynomials

Finite Element/Spectral Element
Non-Overlapping Polynomials,
One per Subdomain

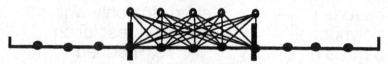

Figure 1.3: Three types of numerical algorithms. The thin, slanting lines illustrate all the grid points (black circles) that *directly* affect the estimates of derivatives at the points shown above the lines by open circles. The thick black vertical lines in the bottom grid are the subdomain walls.

The most accurate scheme is to center the interpolating polynomial on the grid point where the derivative is needed. Quadratic, three-point interpolation and quartic, five-point interpolation give

$$df/dx \approx [f(x+h) - f(x-h)]/(2h) + O(h^2) \tag{1.12}$$

$$df/dx \approx [-f(x+2h) + 8f(x+h) - 8f(x-h) + f(x-2h)]/(12h) + O(h^4) \tag{1.13}$$

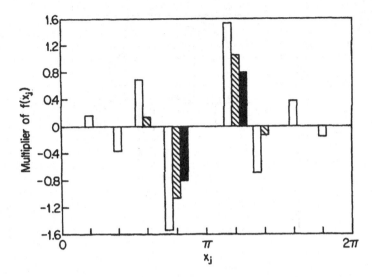

Figure 1.4: Weights w_j in the approximation $df/dx \mid_{x=x_0} \approx \sum_j w_j f(x_0 + jh)$ where $x_0 = \pi$ and $h = \pi/5$. In each group, the Fourier weights are the open, leftmost bars. Middle, cross-hatched bars ($j = \pm 1, \pm 2$ only): Fourth-order differences. Rightmost, solid bars ($j = \pm 1$ only): weights for second order differences.

The function O( ), the "Landau gauge symbol", denotes that in order-of-magnitude, the errors are proportional to h^2 and h^4, respectively.

Since the pseudospectral method is based on evaluating the residual function only at the "selected points", $\{x_i\}$, we can take the grid point values of the approximate solution, the set $\{u_N(x_i)\}$, as the unknowns instead of the series coefficients. Given the value of a function at (N+1) points, we can compute the $(N + 1)$ series coefficients $\{a_n\}$ through polynomial or trigonometric interpolation. Indeed, this symbolic equation

$$\text{series coefficients}\{a_n\} \Longleftrightarrow \text{grid point values}\{u_N(x_i)\} \qquad (1.14)$$

is one of the most important themes we will develop in this course, though the mechanics of interpolation will be deferred to Chapters 4 and 5.

Similarly, the finite element and spectral element algorithms approximate derivatives as a weighted sum of grid point values. However, only those points which lie within a given subdomain contribute directly to the derivative approximations in that subdomain. (Because the solution in one subdomain is matched to that in the other subdomain, there is an *indirect* connection between derivatives at a point and the whole solution, as true of finite differences, too.) Figure 1.3 compares the regions of direct dependency in derivative formulas for the three families of algorithms.

Figs.1.4 and 1.5 compare the weights of each point in the second and fourth-order finite difference approximations with the $N = 10$ Fourier pseudospectral weights. Since the basis functions can be differentiated analytically and since each spectral coefficient a_n is determined by *all* the grid point values of $u(x)$, it follows that the pseudospectral differentiation rules are not 3-point formulas, like second-order finite differences, or even 5-point formulas, like the fourth-order expressions; rather, the pseudospectral rules are N-point formulas. To equal the accuracy of the pseudospectral procedure for $N = 10$, one would need a tenth-order finite difference or finite element method with an error of $O(h^{10})$.

As N is increased, the pseudospectral method benefits in two ways. First, the interval h

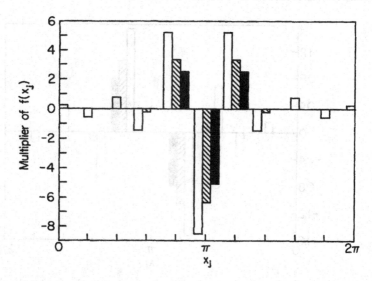

Figure 1.5: Same as previous figure except for the second derivative. Hollow bars: pseudospectral. Cross-hatched bars: Fourth-order differences. Solid bars: Second-order differences.

between grid points becomes smaller – this would cause the error to rapidly decrease even if the order of the method were fixed. Unlike finite difference and finite element methods, however, the order is *not fixed*. When N increases from 10 to 20, the error becomes $O(h^{20})$ in terms of the new, smaller h. Since h is $O(1/N)$, we have

$$\text{Pseudospectral error} \approx O[(1/N)^N] \tag{1.15}$$

The error is decreasing faster than *any* finite power of N because the power in the error formula is always increasing, too. This is *"infinite order"* or *"exponential"* convergence.[1]

This is the magic of pseudospectral methods. When many decimal places of accuracy are needed, the contest between pseudospectral algorithms and finite difference and finite element methods is not an even battle but a rout: pseudospectral methods win hands-down. This is part of the reason that physicists and quantum chemists, who must judge their calculations against experiments accurate to as many as fourteen decimal places (atomic hydrogen maser), have always preferred spectral methods.

However, even when only a crude accuracy of perhaps 5% is needed, the high order of pseudospectral methods makes it possible to obtain this modest error with about half as many degrees of freedom – in *each dimension* – as needed by a fourth order method. In other words, spectral methods, because of their high accuracy, are *memory-minimizing*. Problems that require high resolution can often be done satisfactorily by spectral methods when a three-dimensional second order finite difference code would *fail* because the need for eight or ten times as many grid points would exceed the core memory of the available computer.

'Tis true that virtual memory gives almost limitless memory capacity *in theory*. In practice, however, swapping multi-megabyte blocks of data to and from the hard disk is very *slow*. Thus, in a practical (as opposed to theoretical) sense, virtual storage is not an option when core memory is exhausted. The Nobel Laureate Ken Wilson has observed that because of this, memory is a more severe constraint on computational problem-solving than

---

[1]Chapter 2 shows show that the convergence is *always* exponential for well-behaved functions, but (1.15) is usually too optimistic. The error in an N-point method is $O(M[n]h^n)$ where $M(n)$ is a proportionality constant; we ignored the (slow) growth of this constant with n to derive (1.15).

CPU time. It is easy to beg a little more time on a supercomputer, or to continue a job on your own workstation for another night, but if one runs out of memory, one is simply stuck – unless one switches to an algorithm that uses a lot less memory, such as a spectral method.

For this reason, pseudospectral methods have triumphed in metereology, which is most emphatically an area where high precision is impossible!

The drawbacks of spectral methods are three-fold. First, they are usually more difficult to program than finite difference algorithms. Second, they are more costly per degree of freedom than finite difference procedures. Third, irregular domains inflict heavier losses of accuracy and efficiency on spectral algorithms than on lower-order alternatives. Over the past fifteen years, however, numerical modellers have learned the "right" way to implement pseudospectral methods so as to minimize these drawbacks.

1.5 Parallel Computers

The current generation of massively parallel machines is *communications-limited*. That is to say, each processor is a workstation-class chip capable of tens of megaflops or faster, but the rate of interprocessor transfers is considerably slower.

Spectral elements function very well on massively parallel machines. One can assign a single large element with a high order polynomial approximation within it to a single processor. A three-dimensional element of degree p has roughly p^3 internal degrees of freedom, but the number of grid points on its six walls is $O(6p^2)$. It is these wall values that must be shared with other elements – i. e., other processors – so that the numerical solution is everywhere continuous. As p increases, the ratio of internal grid points to boundary grid points increases, implying that more and more of the computations are internal to the element, and the shared boundary values become smaller and smaller compared to the total number of unknowns. Spectral elements generally require more computation per unknown than low order methods, but this is *irrelevant* when the slowness of *interprocessor data transfers*, rather than CPU time, is the limiting factor.

To do the same calculation with low order methods, one would need roughly eight times as many degrees of freedom in three dimensions. That would increase the interprocessor communication load by at least a factor of four. The processors would likely have a lot of idle time: After applying low order finite difference formulas quickly throughout its assigned block of unknowns, each processor is then idled while boundary values from neighboring elements are communicated to it.

Successful applications of spectral elements to complicated fluid flows on massively parallel machines have been given by Fischer(1990, 1994a,b, 1997) Iskandarani, Haidvogel and Boyd (1994), Taylor, Tribbia and Iskandarani(1997) and Curchitser, Iskandarani and Haidvogel(1998), among others.

1.6 Choice of basis functions

Now that we have compared spectral methods with other algorithms, we can return to some fundamental issues in understanding spectral methods themselves.

An important question is: What sets of "basis functions" $\phi_n(x)$ will work? It is obvious that we would like our basis sets to have a number of properties: (i) easy to compute (ii) rapid convergence and (iii) completeness, which means that any solution can be represented to arbitrarily high accuracy by taking the truncation N to be sufficiently large.

Although we shall discuss many types of basis functions, the best choice for 95% of all applications is an ordinary Fourier series, or a Fourier series in disguise. By "disguise" we mean a change of variable which turns the sines and cosines of a Fourier series into different functions. The most important disguise is the one worn by the Chebyshev polynomials, which are defined by

$$T_n(cos\theta) \equiv \cos(n\theta) \tag{1.16}$$

Although the $T_n(x)$ are polynomials in x, and are therefore usually considered a separate and distinct species of basis functions, a Chebyshev series is really just a Fourier cosine expansion with a change of variable. This brings us to the first of our proverbial sayings:

MORAL PRINCIPLE 1:

(i) When in doubt, use Chebyshev polynomials unless the solution is spatially periodic, in which case an ordinary Fourier series is better.

(ii) Unless you're sure another set of basis functions is better, use Chebyshev polynomials.

(iii) Unless you're really, really sure that another set of basis functions is better, use Chebyshev polynomials.

There are exceptions: on the surface of a sphere, it is more efficient to use spherical harmonics than Chebyshev polynomials. Similarly, if the domain is infinite or semi-infinite, it is better to use basis sets tailored to those domains than Chebyshev polynomials, which in theory and practice are associated with a finite interval.

The general rule is: *Geometry chooses the basis set*. The engineer never has to make a choice. Table A-1 in Appendix A and Figure 1.6 summarize the main cases. When multiple basis sets are listed for a single geometry or type of domain, there is little to choose between them.

It must be noted, however, that the non-Chebyshev cases in the table only strengthen the case for our first Moral Principle. Though not quite as good as spherical harmonics, Chebyshev polynomials in latitude and longtitude work just fine on the sphere (Boyd, 1978b). The rational Chebyshev basis sets are actually just the images of the usual Chebyshev polynomials under a change of coordinate that stretches the interval [-1, 1] into an infinite or semi-infinite domain. Chebyshev polynomials are, as it were, almost idiot-proof.

Consequently, our analysis will concentrate almost exclusively upon Fourier series and Chebyshev polynomials. Because these two basis sets are the same except for a change of variable, the theorems for one are usually trivial generalizations of those for the other. The formal convergence theory for Legendre polynomials is essentially the same as for Chebyshev polynomials except for a couple of minor items noted in Chapter 2. Thus, understanding Fourier series is the key to all spectral methods.

1.7 Boundary conditions

Normally, boundary and initial conditions are not a major complication for spectral methods. For example, when the boundary conditions require the solution to be spatially periodic, the sines and cosines of a Fourier series (which are the natural basis functions for all periodic problems) *automatically* and *individually* satisfy the boundary conditions. Consequently, our only remaining task is to choose the coefficients of the Fourier series to minimize the residual function.

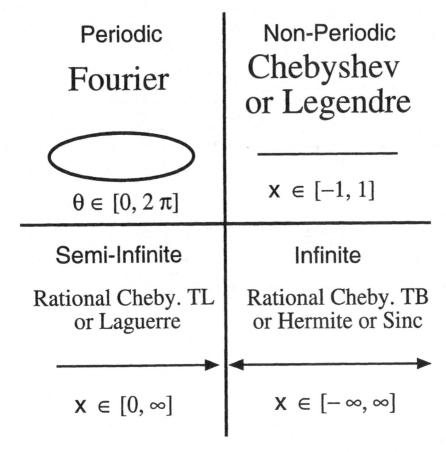

Figure 1.6: Choice of basis functions. Upper left: on a periodic interval, use sines and cosines. This case is symbolized by a ring because the dependence on an angular coordinate, such as longitude, is always periodic. Upper right: a finite interval, which can always be rescaled and translated to $x \in [-1, 1]$. Chebyshev or Legendre polynomials are optimal. Lower left: semi-infinite interval $x \in [0, \infty]$, symbolized by a one-sided arrow. Rational Chebyshev functions $TL_n(x)$ are the generic choice, but Laguerre functions are sometimes more convenient for particular problems. Lower right: $x \in [-\infty, \infty]$ (double-ended arrow). Rational Chebyshev functions $TB_n(x)$ are the most general, but sinc and Hermite functions are widely used, and have similar convergence properties.

For non-periodic problems, Chebyshev polynomials are the natural choice as explained in the next chapter. They do not satisfy the appropriate boundary conditions, but it is easy to add explicit constraints such as

$$\sum_{n=0}^{N} a_n \, \phi_n(1) = \alpha \qquad (1.17)$$

to the algebraic equations obtained from minimizing $R(x; a_0, a_1, \ldots, a_N)$ so that $u(1) = \alpha$ is satisfied by the approximate solution.

Alternatively, one may avoid explicit constraints like (1.17) by writing the solution as

$$u(x) \equiv v(x) + w(x) \qquad (1.18)$$

where $w(x)$ is a *known* function chosen to satisfy the inhomogeneous boundary conditions. The new unknown, $v(x)$, satisfies homogeneous boundary conditions. For (1.17), for example, $w(1) = \alpha$ and $v(1) = 0$. The advantage of homogenizing the boundary conditions is that we may combine functions of the original basis, such as the Chebyshev polynomials, into new basis functions that *individually* satisfy the homogeneous boundary conditions. This is surprisingly easy to do; for example, to satisfy $v(-1) = v(1) = 0$, we expand $v(x)$ in terms of the basis functions

$$\begin{aligned} \phi_{2n}(x) &\equiv T_{2n}(x) - 1, & n &= 1, 2, \dots \\ \phi_{2n+1}(x) &\equiv T_{2n+1}(x) - x, & n &= 1, 2, \dots \end{aligned} \qquad (1.19)$$

where the $T_n(x)$ are the usual Chebyshev polynomials whose properties (including boundary values) are listed in Appendix A. This basis is complete for functions which vanish at the ends of the interval. The reward for the switch of basis set is that it is unnecessary, when using basis recombination, to waste rows of the discretization matrix on the boundary conditions: All algebraic equations come from minimizing the residual of the differential equation.

1.8   The Two Kingdoms: Non-Interpolating and Pseudospectral Families of Methods

Spectral methods fall into two broad categories. In the same way that all of life was once divided into the "plant" and "animal" kingdoms[2], most spectral methods may be classed as either "interpolating" or "non–interpolating". Of course, the biological classification may be ambiguous – is a virus a plant or animal? How about a sulfur-eating bacteria? The mathematical classification may be ambiguous, too, because some algorithms mix ideas from both the interpolating and non-interpolating kingdoms. Nonetheless, the notion of two exclusive "kingdoms" is a useful taxonomical starting point for both biology and numerical analysis.

The "interpolating" or "pseudospectral" methods associate a grid of points with each basis set. The coefficients of a known function $f(x)$ are found by requiring that the truncated series agree with $f(x)$ at each point of the grid. Similarly, the coefficients a_n of a pseudospectral approximation to the solution of a differential equation are found by requiring that the residual function interpolate $f \equiv 0$:

$$R(x_i; a_0, a_1, \dots, a_N) = 0, \qquad i = 0, 1, \dots, N \qquad (1.20)$$

In words, the pseudospectral method demands that the differential equation be exactly satisfied at a set of points known as the "collocation" or "interpolation" points. Presumably, as $R(x; a_n)$ is forced to vanish at an increasingly large number of discrete points, it will be smaller and smaller in the gaps between the collocation points so that $R \approx x$ everywhere in the domain, and therefore $u_N(x)$ will converge to $u(x)$ as N increases. Methods in this kingdom of algorithms are also called "orthogonal collocation" or "method of selected points".

The "non–interpolating" kingdom of algorithms includes Galerkin's method and the Lanczos tau-method. There is no grid of interpolation points. Instead, the coefficients of

---

[2]Modern classification schemes use three to five kingdoms, but this doesn't change the argument.

a known function $f(x)$ are computed by multiplying $f(x)$ by a given basis function and *integrating*. It is tempting to describe the difference between the two algorithmic kingdoms as "integration" versus "interpolation", but unfortunately this is a little simplistic. Many older books, such as Fox and Parker (1968), show how one can use the properties of the basis functions – recurrence relations, trigonometric identities, and such – to calculate coefficients without explicitly performing any integrations. Even though the end product is identical with that obtained by integration, it is a little confusing to label a calculation as an "integration-type" spectral method when there is not an integral sign in sight! Therefore, we shall use the blander label of "non-interpolating".

Historically, the "non–interpolating" methods were developed first. For this reason, the label "spectral" is sometimes used in a narrow sense as a collective tag for the "non–interpolating" methods. In these notes, we shall use "spectral" only as catch–all for global expansion methods in general, but the reader should be aware of its other, narrower usage. (Actually, there are *several* other uses because "spectral" has other meanings in time series analysis and functional analysis – ugh!).

Many spectral models of time-dependent hydrodynamics split the calculation into several subproblems and apply different techniques to different subproblems. To continue with our biological metaphor, the computer code becomes a little ecology of interacting "interpolating" and "non–interpolating" parts. Each algorithm (or algorithmic kingdom) has ecological niches where it is superior to all competition, so we must master both non–interpolating and pseudospectral methods.

At first glance, there is no obvious relation between the pseudospectral method and the alternatives that use weighted integrals of $R(x; a_n)$ to choose the $\{a_n\}$. Worse still, we now have the further burden of choosing the interpolation points, $\{x_i\}$. Fortunately, there is a natural choice of interpolation points for each of the common basis sets. These points are the Gaussian quadrature points for the integrals of Galerkin's method. The pseudospectral method is therefore *equivalent* to the spectral if we evaluate the integrals of the latter by numerical quadrature with $(N + 1)$ points. This is the reason why the interpolation-based methods are now commonly called "pseudospectral".

Better yet, we shall show in later chapters that the accuracy of pseudospectral methods is only a little bit poorer than that of the non-interpolating kingdom – too little to outweigh the much greater simplicity and computational efficiency of the pseudospectral algorithms. Consequently, we shall emphasize pseudospectral methods in this book. Nonetheless, the justification for the pseudospectral kingdom is derived from that for the non-interpolating methods, and the latter are still superior to interpolation for specialized but important applications. We cannot understand the high efficiency of either kingdom of spectral algorithms without first reviewing the theory of Fourier series (Chapter 2).

1.9 Nonlinearity

Nonlinearity is not a major complication for spectral methods *per se*. For expository simplicity, we shall usually concentrate on linear algorithms, especially in explaining the basic ideas. The extension to nonlinear problems usually only requires minor modifications.

To illustrate this, we shall do a very simple nonlinear boundary value problem here. Such equations normally are solved by Newton's iteration. If we set up the iteration by first linearizing the differential equation about the current iterate (the "Newton-Kantorovich" method of Appendix C), then we solve a *linear* differential equation at each step.

In this example, we shall instead apply the spectral method first. The only difference from a linear problem is that the system of algebraic equations for the coefficients is nonlinear. It is usually irrelevant whether the Newton iteration is created before or after applying the spectral method; take your choice!

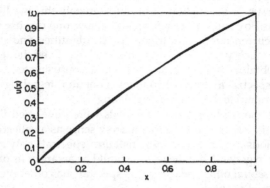

Figure 1.7: Comparison of exact and approximate solutions: Nonlinear diffusion equation

The *nonlinear* boundary value problem is

$$u_{xx} + \alpha(u_x)^2 + \alpha u u_{xx} = 0 \tag{1.21}$$

subject to the boundary conditions that

$$u(0) = 0; \qquad u(1) = 1 \tag{1.22}$$

We will take the approximate solution $u_2(x)$ to be a quadratic polynomial. The most general quadratic polynomial which satisfies the boundary conditions is

$$u_2 = x + a_2(x^2 - x) \tag{1.23}$$

Since there is only one undetermined coefficient a_2, only a single collocation point is needed. The obvious choice, the midpoint of the interval, is best.

The residual function is

$$R(x; a_2) = a_2^2 \, \alpha \, [6x^2 - 6x + 1] + 2a_2 \, [3\alpha x + 1 - \alpha] + \alpha \tag{1.24}$$

The condition that $R(x = 1/2; a_2) = 0$ then gives the quadratic equation

$$a_2^2 \, \alpha \, [-1/2] + 2a_2 \, [\alpha/2 + 1] + \alpha = 0 \tag{1.25}$$

We note an amusing fact: although pseudospectral methods are usually considered only as numerical techniques, we have in fact obtained an *analytical* solution to this *nonlinear* problem. To see how accurate it is, let us specialize to $\alpha = 1$ for which the exact solution is

$$u(x; \alpha = 1) = -1 + (1 + 3x)^{1/2} \tag{1.26}$$

There are two roots to the quadratic, of course, but one gives an unphysical heat flux towards the boundary source at $x = 1$, so it can be rejected.[3]

---

[3]The ambiguity of multiple solutions is a difficulty raised by the nonlinearity of the differential equation, not by the method used to solve it. All algorithms for solving nonlinear boundary value problems have the drawback that the algebraic equations that are the discretization of the differential equation have multiple solutions. Most are unphysical and must be rejected on various grounds including (i) imaginary parts (ii) unrealistic behavior such as the heat flux for this example or (iii) failure to converge as N is varied.

The other gives the approximate solution

$$u_2(x; \alpha = 1) = x - 0.317(x^2 - x) \tag{1.27}$$

Fig. 1.7 compares the exact and approximate solutions. The maximum of $u(x)$ is 1.00; the maximum absolute error of the 1-point pseudospectral solution is only 0.014. The figure shows that even though the functional forms of (1.26) and (1.27) bear no obvious resemblance, the two graphs differ so little that it is hard to tell them apart.

In real-life problems, of course, the exact solution is not known, but the accuracy of an approximate solution can be tested by repeating the calculation with higher N. This problem is particularly difficult because it is nonlinear, so for all N we will invariably be left with a nonlinear algebraic equation or set of equations to determine the solution. However, these can be easily solved by Newton's method since the lowest approximation, obtained analytically, is sufficiently close to the exact solution to furnish a good first guess for the iteration. One of the great virtues of the pseudospectral method is the ease with which it can be applied to nonlinear differential equations.

1.10 Time-dependent problems

Although it is possible to treat the time coordinate spectrally, and we shall describe some special cases and special algorithms where this has been done, it is generally most efficient to apply spectral methods only to the spatial dependence. The reason is that the time-dependence can be marched forward, from one time level to another. Marching is much cheaper than computing the solution *simultaneously* over all space-time.

A space-only spectral discretization reduces the original partial differential equation to a set of *ordinary* differential equations in time, which can then be integrated by one's favorite Runge-Kutta or other ODE time-marching scheme. (This approach, of discretizing one or more coordinates to generate a system of ODEs in the remaining coordinate, is sometimes called the "method of lines", especially in the Russian literature.)

As an illustration, consider the following generalized diffusion problem:

$$u_t = u_{xx} - 2q \cos(2x) u \tag{1.28}$$

with the boundary conditions that the solution must be periodic with a period of 2π. The exact general solution is

$$u(x, t) = \sum_{n=0}^{\infty} a_n(0) \exp(-\lambda_n t) ce_n(x) + \sum_{n=1}^{\infty} b_n(0) \exp(-\mu_n t) se_n(x) \tag{1.29}$$

where the $ce_n(x)$ and $se_n(x)$ are transcendental functions known as Mathieu functions and the λ_n and μ_n are the corresponding eigenvalues. The coefficients $a_n(0)$ and $b_n(0)$ are the values of the coefficients of the Mathieu function series for $u(x)$ at $t = 0$. As for a Fourier series, they can be calculated via

$$a_n(0) = (u[x, t = 0], ce_n)/(ce_n, ce_n) \tag{1.30}$$

$$b_n(0) = (u[x, t = 0], se_n)/(se_n, se_n) \tag{1.31}$$

where

$$(f, g) \equiv \int_0^{2\pi} f(x) g(x) dx \qquad [\text{"inner product"}] \tag{1.32}$$

In the next chapter, we will discuss "inner products"; the $ce_n(x)$ and $se_n(x)$ are computed using "sideband truncation" in Chapter 19.

As a numerical example, take

$$u(x, t = 0) \equiv 1 \tag{1.33}$$

and employ two-point collocation with the basis functions

$$u_2(x) = a_0(t) + a_2(t) \cos(2x) \tag{1.34}$$

and the collocation or interpolation points

$$x_0 = 0; x_1 = \pi/3 \tag{1.35}$$

The reasons for omitting $\cos(x)$ and any and all sine functions are discussed in the chapter on parity and symmetry (Chapter 8). The choice of collocation points is standard for a periodic interval as explained in Chapter 4.

The residual function $R(x; a_0, a_2)$ is

$$R(x; a_0, a_2) = -\{\ [2q \cos(2x)\, a_0 + a_{0,t}] + \cos(2x)[\ (4 + 2q \cos(2x))\, a_2 + a_{2,t}]\ \} \tag{1.36}$$

The collocation conditions that (i) $R(x = 0; a_0, a_2) = 0$ and (ii) $R(x = \pi/3; a_0, a_2) = 0$ give two coupled, ordinary differential equations in time that determine $a_0(t)$ and $a_2(t)$:

$$a_{0,t} + a_{2,t} + 2q\, a_0 + (4 + 2q)\, a_2 = 0 \tag{1.37}$$

$$a_{0,t} - (1/2)a_{2,t} - q\, a_0 - (1/2)(4 - q)\, a_2 = 0 \tag{1.38}$$

Solving these is straightforward; for the special case $q = 1$,

$$u_2(x) = \{0.95 - 0.37 \cos(2x)\} \exp[0.54t] + \{0.05 + 0.37 \cos(2x)\} \exp[-5.54t] \tag{1.39}$$

The corresponding exact solution is

$$
\begin{aligned}
u(x) \ =\ & \{0.916 - 0.404 \cos(2x) + 0.031 \cos(4x) - \cdots\} \exp[0.454t] \\
& +\{0.091 + 0.339 \cos(2x) - 0.030 \cos(4x) + \cdots\} \exp[-4.370t] + \cdots
\end{aligned}
\tag{1.40}
$$

Comparing the two solutions, we see that the low-order collocation approximation is at least qualitatively correct. It predicts that one mode will grow with time while the rest decay; the growth rate of the growing mode is about 20 % too large. The dominant Fourier coefficients are of the growing mode are fairly close — 0.95 versus 0.916, -0.37 versus - 0.404 — while the coefficients of higher degree cosines ($\cos[4x]$, $\cos[6x]$, etc.), which are completely neglected in this approximation, have amplitudes of 0.03 or less.

This example is typical of many time-dependent problems we shall solve: the pseudospectral method is applied to the *spatial* dependence to reduce the problem to a set of coupled ordinary differential equations in time. The ODE's in time will often be nonlinear, however, and it is usually easier to integrate them through finite differences in time even when a (complicated!) analytic solution is possible.

1.11 FAQ: Frequently Asked Questions

1. Are spectral methods harder to program than finite difference or finite element methods?

   Sometimes. However, our first example took just six Maple statements. Spectral methods are only a little more difficult to program than finite differences.

2. Is the high, many-decimal place accuracy of spectral methods even needed in the real world of engineering?

   Sometimes. I was called in as a consultant by KMS Fusion because they needed to model the flows around a pellet of frozen deuterium to about five decimal places. Small imperfections in the spherical shape, on the order of 1%, drastically altered nuclear fusion when the pellet was hit with high intensity laser beams. A two or three decimal place solution would not necessarily have revealed anything about the role of the bumps because the numerical errors of such crude solutions would be comparable with the size of the bumps themselves.

   Long-term hydrodynamic integrations and transition-to-turbulence are often wrecked by computational instability. Common strategies for preserving stability include (i) adding lots of dissipation and (ii) energy-conserving difference or finite element schemes. However, both strategies can greatly distort the solution. A highly accurate solution should not need strong artificial damping or explicit imposition of energy conservation. Spectral solutions are often stable even without damping or imposed energy conservation.

3. Are spectral methods useful only when high accuracy is needed?

   No, because spectral methods also are "memory-minimizing". In three dimensions, one can typically resolve a flow crudely, to within 1% accuracy or so, using only 1/8 as many degrees of freedom as needed by a second or fourth order finite difference method.

4. Are spectral methods useful for flows with shock waves or fronts?

   Yes. It's true, however, that spectral methods for shock problems do not have the sophistication of some low order finite difference, finite volume and finite element codes that have been tailored to shock flows. However, much progress has been made in adapting these ideas to spectral methods.

1.12 The Chrysalis

In numerical analysis, many computations, even in the most sophisticated models, is still performed using the same second order differences employed by Lewis Richardson in 1910, and Sir Richard Southwell and his group in the 1930's. There are some good reasons for this conservatism. When computer modelling attacks new challenges, such as shocks or the complicated geometry of ocean basins or auto frames, it is only sensible to begin by applying and refining low order methods first. The challenging task of customizing old algorithms to new vector and parallel hardware has also (sensibly) begun with simple differences and elements. Lastly, for weather forecasting and many other species of models, the physics is so complicated — photochemistry, radiative transfer, cloud physics, topographic effects, air-sea interaction, and finite resolution of observations — that purely numerical errors are a low priority.

Nevertheless, high order models displaced second order codes for operational forecasting in the 70's, and seem likely to do the same in other engineering and science fields in the twenty-first century. Even when the physics is complicated, there is no excuse for poor numerics. A rusted muffler is no excuse for failing to change the oil. Another reason is that we can, with high order algorithms, explore numerical frontiers previously unreachable. The Space Shuttle has a much greater reach than a clipper ship. Too much of numerical modelling is still in the Age of Sail.

A final reason is that low order methods are like the chrysalis of a butterfly. As shown later, inside every low order program is a high order algorithm waiting to burst free. Given a second order finite difference or finite element boundary-value solver, one can promote the code to spectral accuracy merely by appending a single subroutine to spectrally evaluate the residual, and then calling the boundary value solver repeatedly with the spectral residual as the forcing function. Similarly, the structure and logic of an initial value solver is very much the same for both low and high order methods. The central question is simply: Will one approximate the spatial derivatives badly or well?

Chapter 2

Chebyshev & Fourier Series

"Fourier's Theorem is not only one of the most beautiful results of modern analysis, but it may be said to furnish an indispensable instrument in the treatment of nearly every recondite question in modern physics."
— William Thompson & P. G. Tait (1874)

2.1 Introduction

The total error in solving differential equations is the sum of several contributions which are defined below. These errors are distinct from the spectral coefficients $\{a_n\}$, which in turn are not the same as the terms in the series, which are coefficients multiplied by a basis function. Our first major theme is that all these quantities, though distinct, have the property of decaying to zero with increasing N at the same qualitative rate, usually exponentially.

Our second theoretical keystone is Darboux's Principle. This states that the convergence of a spectral series for $u(x)$ is controlled by the singularities of $u(x)$ where "singularity" is a catch-all for any point in the complex x-plane where $u(x)$ ceases to be analytic in the sense of complex variable theory. Square roots, logarithms, simple poles, step function discontinuities and infinities or abrupt discontinuities in any of the derivatives of $u(x)$ at a point are all "singularities".

The reason that Darboux's Principle is a keystone is that it implies that two functions which have convergence-limiting singularities in the same place, of the same strength and type, will have spectral series whose coefficients a_n asymptote to the same values as $n \to \infty$. This justifies the "Method of Model Functions": We can understand a lot about the success and failure of spectral methods by first understanding the spectral series of simple, explicit model functions with various types of logarithms, poles, and jumps.

The third keystone is that from Darboux's Principle, and limited knowledge about a function, such as whether it is or is not pathological on the solution domain, we can predict rates of convergence for spectral series and spectral approximations to differential equations. Several qualitatively different rates are possible: algebraic, geometric, subgeometric, and supergeometric.

The fourth keystone is that from model functions and Darboux's Principle, we can develop some rules-of-thumb that allow us to qualitatively estimate *a priori* how many degrees of freedom N are needed to resolve a given physical phenomenon. These heuristics are useful to identify both errors and unexpected physics, and also to answer the question: Is a given calculation feasible on my machine?

We will return to each of these four key themes in the middle of the chapter, though not in the same order as above. First, though, a brief review of Fourier series.

2.2   Fourier series

The Fourier series of a general function $f(x)$ is

$$f(x) = a_0 + \sum_{n=1}^{\infty} a_n \cos(nx) + \sum_{n=1}^{\infty} b_n \sin(nx) \tag{2.1}$$

where the coefficients are

$$
\begin{aligned}
a_0 &= (1/2\pi) \int_{-\pi}^{\pi} f(x) dx \\
a_n &= (1/\pi) \int_{-\pi}^{\pi} f(x) \cos(nx) dx \\
b_n &= (1/\pi) \int_{-\pi}^{\pi} f(x) \sin(nx) dx
\end{aligned} \tag{2.2}
$$

First note: because the sines and cosines are periodic with a period of 2π, we can also compute the Fourier expansion on the interval $x \in [0, 2\pi]$. The only alteration is that the limits of integration in the coefficient integrals (2.2) are also changed from $[-\pi, \pi]$ to $[0, 2\pi]$.

Second note: the general Fourier series can also be written in the complex form

$$f(x) = \sum_{n=-\infty}^{\infty} c_n \exp(inx) \tag{2.3}$$

where the coefficients are

$$c_n = (1/2\pi) \int_{-\pi}^{\pi} f(x) \exp(-inx) dx \tag{2.4}$$

The identities

$$\cos(x) \equiv (\exp(ix) + \exp(-ix))/2; \qquad \sin(x) \equiv (\exp(ix) - \exp(-ix))/(2i), \tag{2.5}$$

show that (2.3) and (2.1) are completely equivalent, and we shall use whichever is convenient. The coefficients of the two forms are related by

$$
\begin{aligned}
c_0 &= a_0, \qquad n = 0 \\
c_n &= \begin{cases} (a_n - ib_n)/2, & n > 0 \\ (a_n + ib_n)/2, & n < 0 \end{cases}
\end{aligned}
$$

Often, it is unnecessary to use the full Fourier series. In particular, if $f(x)$ is known to have the property of being *symmetric* about $x = 0$, which means that $f(x) = f(-x)$ for all x, then all the sine coefficients are zero. The series with only the constant and the cosine terms is known as a "Fourier cosine series". (A Chebyshev series is a Fourier cosine series with a change of variable.) If $f(x) = -f(-x)$ for all x, then $f(x)$ is said to be *antisymmetric* about $x = 0$ and all the $a_n = 0$. Its Fourier series is a sine series. These special cases are extremely important in applications as discussed in the Chapter 8.

Definition 1 (PERIODICITY)
A function $f(x)$ is PERIODIC with a period of 2π if

$$f(x) = f(x + 2\pi) \tag{2.6}$$

for all x.

To illustrate these abstract concepts, we will look at four explicit examples. These will allow us to develop an important theme: The smoother the function, more rapidly its spectral coefficients converge.

EXAMPLE ONE: "Piecewise Linear" or "Sawtooth" Function

Since the basis functions of the Fourier expansion, $\{1, \cos(nx), \sin(nx)\}$, all are periodic, it would be reasonable to suppose that the Fourier series would be useful only for expanding functions that have this same property. In fact, this is only half-true. Fourier series work *best* for periodic functions, and whenever possible, we will use them only when the boundary conditions are that the solution be periodic. (Geophysical example: because the earth is round, atmospheric flows are *always* periodic in longitude). However, Fourier series will converge, albeit slowly, for quite *arbitrary* $f(x)$.

In keeping with our rather low-brow approach, we will prove this by example. Suppose we take $f(x) = x$, evaluate the integrals (2.2) and sum the series (2.1). What do we get?

Because all the basis functions are periodic, their sum must be periodic even if the function $f(x)$ in the integrals is not periodic. The result is that the Fourier series converges to the so-called "saw-tooth" function (Fig. 2.1).

Since $f(x) \equiv x$ is antisymmetric, all the a_n are 0. The sine coefficients are

$$
\begin{aligned}
b_n &= (1/\pi) \int_{-\pi}^{\pi} x \sin(nx)\,dx \\
&= (-1)^{n+1}(2/n)
\end{aligned}
\tag{2.7}
$$

Since the coefficients are decreasing as $O(1/n)$, the series does not converge with blazing speed; in fact, this is the worst known example for an $f(x)$ which is continuous. Nonetheless, Fig. 2.2 shows that adding more and more terms to the sine series does indeed generate a closer and closer approximation to a straight line.

The graph of the error shows that the discontinuity has polluted the approximation with small, spurious oscillations everywhere. At any given fixed x, however, the amplitude of these oscillations decreases as $O(1/N)$. Near the discontinuity, there is a region where (i) the error is *always* $O(1)$ and (ii) the Fourier partial sum overshoots $f(x)$ by the same amount, rising to a maximum of about 1.18 instead of 1, *independent* of N. Collectively, these facts are known as "Gibbs' Phenomenon". Fortunately, through "filtering", "sequence acceleration" and "reconstruction", it is possible to ameliorate some of these

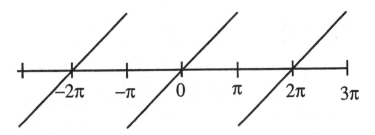

Figure 2.1: "Sawtooth" (piecewise linear) function.

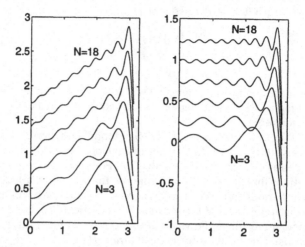

Figure 2.2: Left: partial sums of the Fourier series of the piecewise linear ("sawtooth") function (divided by π) for N=3 , 6, 9, 12, 15, 18. Right: errors. For clarity, both the partial sums and errors have been shifted with upwards with increasing N.

problems. Because shock waves in fluids are discontinuities, shocks produce Gibbs' Phenomenon, too, and demand the same remedies.

EXAMPLE TWO: "Half-Wave Rectifier" Function

This is defined on $t \in [0, 2\pi]$ by

$$f(t) \equiv \begin{cases} \sin(t), & 0 < t < \pi \\ \\ 0, & \pi < t < 2\pi \end{cases}$$

and is extended to all t by assuming that this pattern repeats with a period of 2π. [Geophysical note: this approximately describes the time dependence of thermal tides in the earth's atmosphere: the solar heating rises and falls during the day but is zero at night.]

Integration gives the Fourier coefficients as

$$a_0 = (1/\pi); \qquad a_{2n} = -2/[\pi(4n^2 - 1)] \qquad (n > 0); \qquad a_{2n+1} = 0 (n \geq 1) \qquad (2.8)$$

$$b_1 = 1/2; \qquad b_{2n} = 0 \qquad (n > 1) \qquad (2.9)$$

Fig. 2.3 shows the sum of the first four terms of the series, $f_4(x) = 0.318 + 0.5 \sin(t) - 0.212 \cos(2t) - 0.042 \cos(4t)$. The graph shows that the series is converging much faster than that for the saw-tooth function. At $t = \pi/2$, where $f(t) = 1.000$, the first four terms sum to 0.988, an error of only 1.2 %.

This series converges more rapidly than that for the "saw-tooth" because the "half-wave rectifier" function is smoother than the "saw-tooth" function. The latter is discontinuous and its coefficients decrease as $O(1/n)$ in the limit $n \to \infty$; the "half-wave rectifier" is *continuous* but its *first derivative is discontinuous*, so its coefficients decrease as $O(1/n^2)$. This is a general property: *the smoother a function is, the more rapidly its Fourier coefficients will decrease, and we can explicitly derive the appropriate power of $1/n$.*

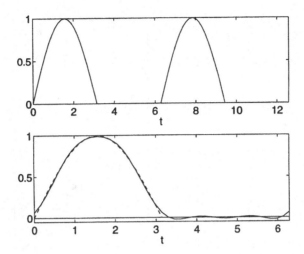

Figure 2.3: Top: graph of the "half-wave rectifier" function. Bottom: A comparison of the "half-wave rectifier" function [dashed] with the sum of the first four Fourier terms [solid]. $f_4(x) = 0.318 + 0.5\ \sin(t) - 0.212\ \cos(2\ t) - 0.042\ \cos(4\ t)$. The two curves are almost indistinguishable.

Although spectral methods (and all other algorithms!) work best when the solution is smooth and infinitely differentiable, the "half-wave rectifier" shows that this is not always possible.

EXAMPLE THREE: Infinitely Differentiable but Singular for Real x

$$f(x) \equiv \exp\{-\cos^2(x)/\sin^2(x)\} \tag{2.10}$$

This function has an essential singularity of the form $\exp(-1/x^2)$ at $x = 0$. The power series about $x = 0$ is meaningless because all the derivatives of (2.10) tend to 0 as $x \to 0$. However, the derivatives exist because their limit as $x \to 0$ is well-defined and bounded. The exponential decay of $\exp(-1/x^2)$ is sufficient to overcome the negative powers of x that appear when we differentiate so that none of the derivatives are infinite. Boyd (1982a) shows that the Fourier coefficients of (2.10) are asymptotically of the form

$$a_n \sim [\ ] \exp(-1.5n^{2/3}) \cos(2.60n^{2/3} + \pi/4) \tag{2.11}$$

where [] denotes an algebraic factor of n irrelevant for present purposes. Fast convergence, even though the power series about $x = 0$ is useless, is a clear signal that spectral expansions are more potent than Taylor series (Fig. 2.4).

This example may seem rather contrived. However, "singular-but-infinitely-differentiable" is actually the most common case for functions on an infinite or semi-infinite interval. Most functions have such bounded singularities at infinity, that is, at one or both endpoints of the expansion interval.

EXAMPLE FOUR: "Symmetric, Imbricated-Lorentzian" (SIP) Function

$$f(x) \equiv (1 - p^2)/\left\{(1 + p^2) - 2p\cos(x)\right\} \tag{2.12}$$

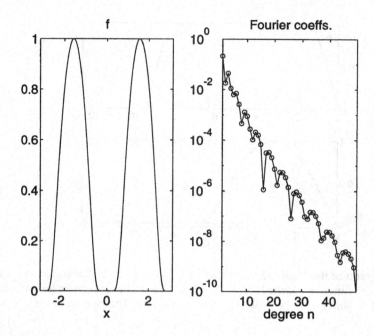

Figure 2.4: Left: graph of $f(x) \equiv \exp(-\cos^2(x) / \sin^2(x))$. Right: Fourier cosine coefficients of this function. The sine coefficients are all zero because this function is symmetric with respect to $x = 0$.

where $p < 1$ is a constant. This $f(x)$ is a *periodic* function which is *infinitely differentiable* and *continuous* in all its derivatives. Its Fourier series is

$$f(x) = 1 + 2 \sum_{n=1}^{\infty} p^n \cos(nx) \tag{2.13}$$

This example illustrates the "exponential" and "geometric" convergence which is *typical* of solutions to differential equations in the absence of shocks, corner singularities, or discontinuities.

We may describe (2.13) as a "geometrically-converging" series because at $x = 0$, this *is* a geometric series. Since $|\cos(nx)| \le 1$ for all n and x, each term in the Fourier series is bounded by the corresponding term in the geometric power series in p for all x. Because this rate of convergence is generic and typical, it is important to understand that it is qualitatively different from the rate of the convergence of series whose terms are proportional to some inverse power of n.

Note that each coefficient in (2.13) is smaller than its predecessor by a factor of p where $p < 1$. However, if the coefficients were decreasing as $O(1/n^k)$ for some finite k where $k = 1$ for the "saw-tooth" and $k = 2$ for the "half-wave rectifier", then

$$\begin{aligned}
a_{n+1}/a_n &\sim n^k/(n+1)^k \\
&\sim 1 - k/n \qquad \text{for} \quad n \gg k \\
&\sim 1 \qquad \text{[Non-exponential Convergence]}
\end{aligned} \tag{2.14}$$

Thus, even if k is a very large number, the ratio of a_{n+1}/a_n tends to 1 from below for large n. This *never* happens for a series with "exponential" convergence; the ratio of $|a_{n+1}/a_n|$ is always bounded away from one — by p in (2.13), for example.

Clearly, series can converge at qualitatively different rates.

2.3 Orders of Convergence

It is useful to have precise definitions for classifying the rate of convergence (Boyd, 1982a).

Warning: these are all *asymptotic* definitions based on the behavior of the series coefficients for *large n*. They may be highly misleading if applied for *small* or *moderate n*.

Definition 2
The ALGEBRAIC INDEX OF CONVERGENCE k is the largest number for which

$$\lim_{n \to \infty} |a_n| \, n^k < \infty, \qquad n \gg 1 \tag{2.15}$$

where the a_n are the coefficients of the series. (For a Fourier series, the limit must be finite for both the cosine coefficients a_n and the sine coefficients b_n.)
Alternative definition: if the coefficients of a series are a_n and if

$$a_n \sim O[1/n^k], \qquad n \gg 1 \tag{2.16}$$

then k is the algebraic index of convergence.

The two definitions are equivalent, but the roundabout form (2.15) also gives an unambiguous definition of the order even for exotic cases for which the asymptotic form is $a_n \sim O(\log[n]/n^k)$ or the like.

Examples: the algebraic convergence order k is $k = 1$ for the Fourier series of the "sawtooth" function and $k = 2$ for that of the "half-wave rectifier" function, whose coefficients are proportional to $1/n^2$ for $n \gg 1$. This definition provides a guideline: One should choose the spectral algorithm so as to *maximize* the *algebraic convergence order* for a given problem; the method with the largest k will always give fastest *asymptotic* convergence.

Definition 3
If the algebraic index of convergence k is unbounded – in other words, if the coefficients a_n decrease faster than $1/n^k$ for ANY finite power of k – then the series is said to have the property of "INFINITE ORDER", "EXPONENTIAL", or "SPECTRAL" convergence.
Alternative definition: If

$$a_n \sim O[\exp(-qn^r)], \qquad n \gg 1 \tag{2.17}$$

with q a constant for some r > 0, then the series has INFINITE ORDER or EXPONENTIAL convergence.

The equivalence of the second definition to the first is shown by the identity

$$\lim_{n \to \infty} n^k \exp(-qn^r) = 0, \qquad \text{all k, all r > 0} \tag{2.18}$$

(Abramowitz and Stegun, 1965, pg. 68). The reason for giving two definitions is that (2.17), which is more obvious and easier to understand, does not cover all possible cases. The terms "exponential" and "infinite order" are synonyms and may be used interchangeably.

The term "spectral accuracy" is widely used. However, we shall avoid it because algebraically-converging Chebyshev series are "spectral", too. The popularity of this term is a reminder that infinite order convergence is usual for any well-designed spectral algorithm.

Definition 4

The EXPONENTIAL INDEX OF CONVERGENCE r is given by

$$r \equiv \lim_{n \to \infty} \frac{\log |\log(|a_n|)|}{\log(n)} \tag{2.19}$$

An equivalent definition is that if s and $q > 0$ are constants and

$$a_n \sim O(s \exp[-qn^r]), \qquad n \gg 1, \tag{2.20}$$

then the EXPONENTIAL INDEX OF CONVERGENCE is the exponent r.

Example: the coefficients of the function (2.12) are $a_n = 2 \exp(-\{-\log p\}\, n)$, which is of the form of (2.20) with the constant $s = 2, q = -\log(p)$, and the exponential index of convergence $r = 1$.

Not all functions have series coefficients in the form of (2.20); the coefficients of the j-th derivative of $f(x) = (1 - p^2)/[(1 + p^2) - 2p \cos(x)]$ are $O[n^j \exp(-qn)]$ whereas the j-th integral has coefficients that are $O[\exp(-qn)/n^j]$. The rigorous definition (2.19) allows for such algebraic factors of n multiplying the exponential — by ignoring them.

The reason for ignoring algebraic factors is that n^k varies very slowly in comparison with the exponential for *large n*. (This is why (2.18) is true even for very large positive values of k). We shall adopt the convention of representing such slowly-varying algebraic factors of n by empty brackets [].

Definition 5 (Rates of Exponential Convergence) *A series whose coefficients are a_n is said to have the property of SUPERGEOMETRIC, GEOMETRIC, or SUBGEOMETRIC convergence depending upon whether*

$$\lim_{n \to \infty} \log(|a_n|)/n = \begin{cases} \infty & SUPERGEOMETRIC \\ constant & GEOMETRIC \\ 0 & SUBGEOMETRIC \end{cases} \tag{2.21}$$

Alternative definitions:

1. *If $a_n \sim O([]\exp\{-(n/j)\log(n)\})$, convergence is SUPERGEOMETRIC*

2. *If $a_n \sim O([]\exp\{-qn\})$, convergence is GEOMETRIC*

3. *If the exponential index of convergence $r < 1$, then the convergence is SUBGEOMETRIC.*

(The empty brackets [] denote factors that vary more slowly with n than the exponentials.)

The motive for these definitions is that the Fourier and Chebyshev series of so-called "entire functions" [functions without singularities anywhere in the complex plane except at ∞] have "supergeometric" convergence. For the expansion of functions with poles or branch points which are a finite distance off the expansion interval — the usual case — *geometric* convergence is *normal*. Both entire functions and functions with singularities at finite x (but off the expansion interval) have $r = 1$, so the exponential index of convergence cannot discriminate between them even though these are quite distinct classes of functions. "Subgeometric" convergence rarely occurs when solving problems on a *finite* interval, but it is normal for series on *infinite* or *semi-infinite* intervals in x.

These definitions have been expressed in terms of the sequence of coefficients a_n. However, normalized basis functions are always $O(1)$ so that the *magnitude* of the *term $a_n\, \phi_n(x)$* is that of the *coefficient a_n*.

The second comment is that the limits defined above are meaningless when, for example, every other coefficient is zero. Strictly speaking, the limits should be *supremum* limits (see glossary). It often suffices, however, to apply the ordinary limit to the non-zero coefficients.

Definition 6 (ASYMPTOTIC RATE OF GEOMETRIC CONVERGENCE) *If a series has geometric convergence, that is, if an expansion has an exponential index of convergence r = 1 so that*

$$a_n \sim [\,] \exp(-n\mu) \tag{2.22}$$

where a_n are the spectral coefficients, μ is a constant, and [ ] denotes unspecified factors that vary more slowly with n than the exponential (such as n^k for some k), then the ASYMPTOTIC RATE OF GEOMETRIC CONVERGENCE is μ. Equivalently,

$$\mu = \lim_{n \to \infty} \{-\log|a_n|/n\} \tag{2.23}$$

This definition is meaningful only for geometrically converging series; it does not apply when the algebraic index of convergence is $< \infty$ nor when the exponential index of convergence $r < 1$.

For power series, μ is simply the logarithm of the radius of convergence. Later in the chapter, we shall explain how to calculate μ for Fourier and Chebyshev series in terms of the singularities of the function being expanded.

2.4 Graphical Interpretation of Orders of Convergence

These abstract concepts become clearer with graphs. On a LOG-LINEAR graph, for example, the coefficients of a GEOMETRICALLY converging series will ASYMPTOTE to a STRAIGHT LINE as shown by the solid curve in Fig. 2.5. "Supergeometric" convergence can then be graphically defined as coefficients whose curve develops a more and more negative slope (rather than a constant slope) on a log-linear graph. Similarly, subgeometric and

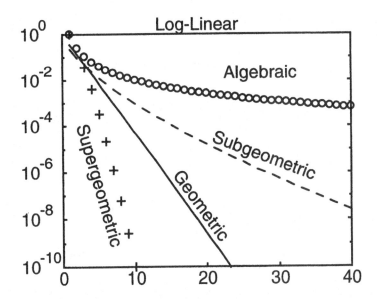

Figure 2.5: $\log|a_n|$ versus n for four rates of convergence. Circles: algebraic convergence, such as $a_n \sim 1/n^2$. Dashed: subgeometric convergence, such as $a_n \sim \exp(-1.5\, n^{2/3})$. Solid: geometric convergence, such as $\exp(-\mu\, n)$ for any positive μ. Pluses: supergeometric, such as $a_n \sim \exp(-n\,\log(n))$ or faster decay.

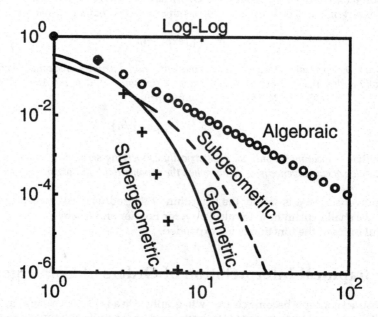

Figure 2.6: Same as previous figure except that the graph is log-log: the degree of the spectral coefficient n is now plotted on a logarithmic scale, too.

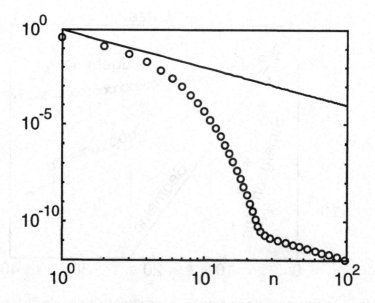

Figure 2.7: Spectral coefficients for two series. Upper curve: $a_n = 1/n^2$. Lower curve: $a_n = \exp(-n) + 10^{-8}/n^2$.

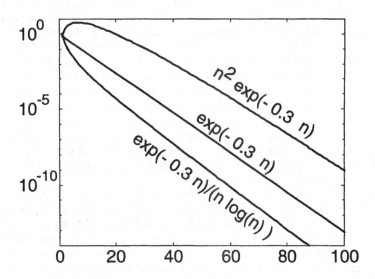

Figure 2.8: Spectral coefficients for three geometrically converging series. Although the three sets of coefficients differ through algebraic coefficients — the top curve is larger by n^2 than the middle curve, which in turn is larger by a factor of $n \log(n)$ than the bottom curve — the *exponential* dependence on n is the same for all. Consequently, all three sets of coefficients asymptote to parallel lines on this log-linear plot.

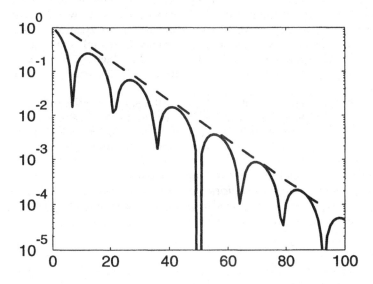

Figure 2.9: Solid: logarithm of the absolute value of the spectral coefficients of a geometrically-converging series whose coefficients *oscillate* with degree n. Dashed: the "envelope" of the spectral coefficients.

algebraic convergence rates produce curves which bend upward away from the straight line of geometric convergence: Their slopes tend to zero from below on a log-linear plot.

If we change the graph format from log-linear to log-log, then algebraic convergence with algebraic index of convergence k can be defined as coefficients whose curve asymptotes to a straight line whose slope is $- k$ (Fig. 2.6). All species of exponential convergence bend away with ever-increasing negative slopes on a plot of the logarithm of the absolute value of $\{ a_n \}$ versus the logarithm of n. Graphically, "infinite order" convergence is a synonym for "unbounded negative slope" on a log-log plot.

Four qualifiers are necessary. First, these orders of convergence are meaningfully only *asymptotically*. Fig. 2.7 compares the coefficients of two spectral series that have the same (poor) asymptotic convergence rate. However, the lower curve shows that its coefficients have fallen to $O(10^{-8})$ before entering the asymptotic regime, so its slow convergence is likely to be irrelevant in a practical sense.

Although the figure is contrived, the same thing is common in real-life. Boyd(1986c) compares the coefficients of the two-dimensional Chebyshev series for the solution of the nonlinear eigenvalue problem known as Bratu's equation with the corresponding series for the solution to Poisson's equation with constant forcing. Asymptotically, both spectral series are sixth order because of weak singularities in the four corners of the square domain. Although the Poisson series clearly exhibits this algebraic convergence on a log-log graph, the Bratu coefficients appear to fall off *exponentially* on Boyd's graph. The reason is that the nonlinearity introduces additional singularities – off the computational domain but strong – which dominate the early spectral coefficients. Only after the Bratu coefficients have fallen to less than $O(10^{-5})$ of the maximum of the solution does the slope of the Bratu series alter to match that of the Poisson equation. We cannot repeat this too often: Asymptotic concepts like "convergence order" are meaningful only for large n. For small n, who knows!

The second point is that the definitions above ignore algebraic factors of n which modulate the leading exponential; we have replaced these by empty square brackets in many of the formulas of this chapter. Fig. 2.8 compares three series which differ only through such algebraic factors. Obviously, the three series are not identical. A factor of n^2 may be anything but ignorable when $n = 100$! However, the graph shows that all three series *asymptote* to *parallel* straight lines. It is in this sense that algebraic factors of n, such as powers and logarithms of n, are asymptotically irrelevant. The exponential factor of n must dominate for sufficiently large n.

Third, what happens when the coefficients of a spectral series *oscillate* with degree n? Fig. 2.9 shows, consistent with asymptotic theories of spectral coefficients, that it is possible to tightly bound the spectral coefficients by a monotonically-decaying curve even when the individual coefficients oscillate. This motivates the following.

Definition 7 (Envelope of the Spectral Coefficients)
The ENVELOPE of the spectral coefficients is defined to be that curve, of the form of the leading asymptotic term in the logarithm of the absolute value of the spectral coefficients, which bounds the coefficients from above as tightly as possible.

Lastly, the asymptotic order may be defeated by the "Roundoff Plateau" illustrated in Fig 2.10.

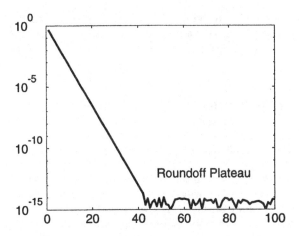

Figure 2.10: Numerically-generated spectral coefficients for a typical function. Let a_{max} denote the maximum absolute value of the spectral coefficients a_j for all j. Let ϵ denote a constant proportional to the roundoff error or "machine epsilon", typically around 10^{-16} on most computers, but somewhat larger, perhaps by a factor of 100 or more. Then when the exact coefficients fall below ϵa_{max}, spectral algorithms including interpolation will compute roundoff-corrupted coefficients that will flatten out at roughly $a_n \sim \epsilon a_{max}$ for all sufficiently large n. ("Roughly" means that coefficients in the "Roundoff Plateau" fluctuate randomly about the indicated magnitude.)

2.5   Assumption of Equal Errors

Definition 8 (Truncation Error)
The TRUNCATION ERROR $E_T(N)$ is defined to be the error made by neglecting all spectral coefficients a_n with $n > N$.

Definition 9 (Discretization Error)
The DISCRETIZATION ERROR $E_D(N)$ is the difference between the first $(N + 1)$ terms of the exact solution and the corresponding terms as computed by a spectral or pseudospectral method using $(N + 1)$ basis functions.

Definition 10 (Interpolation Error)
The INTERPOLATION ERROR $E_I(N)$ is the error made by approximating a function by an $N + 1$-term series whose coefficients are chosen to make the approximation agree with the target function exactly at each of $N + 1$ "interpolation" or "collocation" points, or it is the error in differential equation approximations that use similar principles, i. e., the "pseudospectral" method.

It is generally impossible to estimate these various errors precisely for the unknown solution to a differential equation. We have two useful alternatives. One is to look at the numerically-computed spectral coefficients, as described in Sec. 12. The other strategy, and the only one which can be applied *a priori* is defined by the following.

Definition 11 (Method of Model Functions)
Truncation error is estimated by computing the expansion of a KNOWN function which is "similar", in a technical sense to be explained later, to the solution of the target differential equation, and then truncating the known expansion after $N + 1$ terms.

It is rather disheartening to be forced to look at spectral coefficients and truncation errors for models when what we want is the *total* error in computing the solution to a differential equation. A numerical method that has a low truncation error is useless if the corresponding discretization error is horrible. However, many years of numerical experience has established the following assumption which will underlie the rest of the chapter.

Rule-of-Thumb 1 (ASSUMPTION OF EQUAL ERRORS)
The discretization and interpolation errors are the same order-of-magnitude as the truncation error. Therefore, we can roughly compare the effectiveness of various algorithms and estimate the smallest truncation N that gives a specified accuracy by inspection of the truncation error alone.

This is a strong assumption, and one that cannot be "proved" in any sense. A few counter-examples are known. However, the assumption is clearly true in a negative sense: A method that gives a poor truncation error (unless N is huge) is a terrible way to solve the differential equation. The truncation error is therefore a very safe way to determine what numerical procedures *not* to use.

Later, we will evaluate both types of error for a couple of examples and show that the assumption is true at least for these particular cases.

2.6   Darboux's Principle

Theorem 1 (DARBOUX'S PRINCIPLE: SINGULARITIES & CONVERGENCE)
For all types of spectral expansions (and for ordinary power series), both the DOMAIN *of* CONVERGENCE *in the complex plane and also the* RATE *of* CONVERGENCE *are controlled by the* LOCATION *and* STRENGTH *of the* GRAVEST SINGULARITY *in the complex plane. "Singularity" in this context denotes poles, fractional powers, logarithms and other branch points, and discontinuities of $f(z)$ or any of its derivatives.*

Each such singularity gives its own additive contribution to the coefficients a_n in the asymptotic limit $n \to \infty$. The "gravest" singularity is the one whose contribution is larger than the others in this limit; there may be two or more of equal strength.

For the special case of power series, this is "DARBOUX's THEOREM" (Darboux, 1878a, b, Dingle, 1973, Hunter, 1986).

Darboux's Principle also justifies the assertion made in Sec. 2: the smoother $f(x)$, the faster the convergence of its spectral series. Table 2.2 (end of the chapter) catalogues the relationship between the type of a singularity (logarithm, square root, etc.) and the asymptotic spectral coefficients.

Darboux's Principle implies a corollary that clarifies the Method of Model Functions.

Corollary 1 (SINGULARITY-MATCHING)
If two functions $f(z)$ and $g(z)$ have convergence-limiting singularities of the same type and strength, then their asymptotic spectral coefficients are the same in the limit $n \to \infty$. If $f(z) - g(z)$ is singular, but more weakly singular than either $f(z)$ or $g(z)$, the difference between the spectral coefficients of $f(z)$ and $g(z)$ decreases as an algebraic function of the degree n. If $f(z) - g(z)$ is not singular at the location of the common gravest singularity of $f(z)$ and $g(z)$, then the difference between spectral coefficients decreases exponentially fast with n.

This corollary is a direct consequence of Theorem 1: If we can compute asymptotic expansions for the series coefficients which depend only on the gravest singularity, then when the singularity is eliminated by taking the difference $f(z) - g(z)$, the leading term in the asymptotic approximation for the spectral coefficients is also eliminated. The spectral series for the difference must therefore converge more rapidly than the expansions for the more singular functions $f(z)$ and $g(z)$.

This corollary defines in what sense an explicit model function should resemble the unknown solution to a differential equation: The model and the unknown $u(x)$ should have the same gravest, convergence-limiting singularities.

This theorem implies that the examples below are not merely special cases. Rather, each is typical of a whole *class* of functions that have singularities of the same type at the same point. For example, $1/(x + a)$ where a is an arbitrary complex parameter is representative of all functions whose convergence is limited by a first order pole at $x = -a$.

An example is useful in explaining Darboux's Principle. The geometrically converging Fourier series (2.24) has a partial fraction expansion which explicitly shows that this function is singular only through simple poles at $x = \pm ia$ along the imaginary axis and the images of these two poles under the periodicity shift, $x \to x + 2 \pi m$ where m is an arbitrary integer:

$$\lambda(x; p) \equiv \frac{(1 - p^2)}{(1 + p^2) - 2\, p\, \cos(x)}$$

$$= 1 + 2 \sum_{n=1}^{\infty} p^n \, \cos(n\, x) \tag{2.24}$$

$$= 2\, a \sum_{m=-\infty}^{\infty} \frac{1}{(a^2 + (x - 2\pi\, m\,)^2)} \tag{2.25}$$

where $a \equiv -\log(p) \to p = \exp(-a)$.

In contrast, the elliptic function Dn has an *infinite* number of simple poles on the imag-

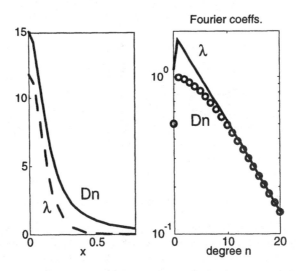

Figure 2.11: Left panel: Graphs of Dn$(x; p)$ (solid) and $\lambda(x; p)$ (dashed). Right panel: Fourier coefficients. Top (solid): coefficients of λ. Bottom (circles): coefficients of Dn.

inary axis, instead of just two.

$$Dn(x;p) \equiv (1/2) + 2 \sum_{n=1}^{\infty} [p^n/(1+p^{2n})] \cos(nx) \tag{2.26}$$

$$= B \sum_{m=-\infty}^{\infty} \text{sech}[B(x - 2\pi m)] \tag{2.27}$$

$$B(p) \equiv \pi/[2 \log(1/p)] \tag{2.28}$$

Dn has simple poles at

$$x = 2\pi m + ia(2j+1) \qquad \text{all} \quad \text{integers} \quad m, j \tag{2.29}$$

Either by matching poles and residues or by Taylor expanding the $1/(1+p^{2n})$ factor, one can show

$$Dn(x;p) = (1/2) + \sum_{j=0}^{\infty} (-1)^j \{\lambda(x; p^{2j+1}) - 1\} \tag{2.30}$$

This is a partial fraction expansion, modified so that all poles at a given distance from the real axis are combined into a single term, and it converges rather slowly. Graphed as a function of x, this elliptic function $Dn(x;p)$ resembles $\lambda(x;p)$ only in a general way (Fig. 2.11).

However, the *Fourier coefficients* of the elliptic function are more and more dominated by those of $\lambda(x;p)$ as the degree of the coefficient increases because

$$p^n/(1+p^{2n}) \approx p^n \{1 + O(p^{2n})\}, \qquad p << 1 \tag{2.31}$$

If p is only slightly smaller than 1 (a slowly-converging series), then the lowest few elliptic coefficients will be roughly half those of the "imbricated-Lorentzian" λ. As the degree of the coefficient n increases, however, the approximation (2.31) will eventually become accurate. If we truncate the series at $n = N$ and choose N sufficiently large so that we obtain an accuracy of 10^{-t}, then the relative error in the last retained coefficient, a_n, is $O(10^{-2t})$.

The reason that Eq. (2.31) is so accurate is that the difference,

$$\delta(x;p) \equiv Dn(x;p) - \lambda(x;p), \tag{2.32}$$

is a function which is no longer singular at $x = \pm i\,a$, but instead converges for all $|\Im(x)| < 3a$. It is a general truth that whenever the difference between the "model" and "target" functions has a *larger* domain of convergence than $f(x)$, the difference between the Fourier (or Chebyshev or any kind of spectral) coefficients will decrease *exponentially* fast with n.

Of course, the situation is not always so favorable. For example, if a function is approximated by

$$f(x) \approx \log(x - ia)\{1 + b_1(x - ia) + b_2(x - i\,a)^2 + \dots\} \tag{2.33}$$

in the vicinity of its convergence-limiting poles, it is easy to match the gravest branch point, $\log(x - i\,a)$. Unfortunately, the difference between $f(x)$ and the model will still be singular at $x = ia$ with the weaker branch point, $(x - i\,a)\log(x - i\,a)$. The difference $f(x) - \log(x - i\,a)$ has a Fourier series that converges more rapidly than that of $f(x)$ by a factor of n, so the error in approximating the Fourier coefficients of $f(x)$ by those of the logarithmically singular model will decrease as $O(1/n)$.

Even so, it is clear that each function is representative of a whole class of functions: The class of all functions that have singularities of that type and at that convergence-limiting location. In a weaker sense (that is, if algebraic factors of n are ignored), each function is representative of all functions that have convergence-limiting singularities at a given point, regardless of the type of singularity (pole, square root, logarithm or whatever). It follows that to understand a few representative functions is to understand spectral series in general.

2.7 Convergence Domains: Why Taylor Series Don't Work

For all the standard basis functions, the spectral series converges in a domain of known shape in the complex plane. Since the singularities of $f(x)$ control the asymptotic form of the spectral coefficients (Darboux's Principle), it follows that the size of the convergence domain and the rate of convergence at a given point within the domain are both controlled by the singularities of $f(x)$, too.

Theorem 2 (CONVERGENCE DOMAIN in COMPLEX PLANE)
Barring a few exceptional cases, a spectral series converges in the largest domain in the x-plane, of a certain basis-dependent shape, which is free of singularities. For power series, the domain shape is a disk. For Fourier and Hermite series, it is an infinite strip, symmetric around the real x-axis. For Chebyshev, Legendre, and Gegenbauer polynomials, the convergence domain is the interior of an ellipse with foci at $x = \pm 1$.

The exceptional cases are entire functions which grow very rapidly in certain parts of the complex plane as described in Chapter 17 and Boyd (2001).

PROOF: Given, for various basis sets, in classical treatises on Fourier series and orthogonal polynomials.

Appendix A graphs the convergence domain for several basis sets, but we usually have little direct interest in summing a spectral series for complex x! These convergence domains are significant only in the indirect sense that the larger the convergence region, the faster the spectral series converges for that interval of real x that we really care about. Later in the chapter, we shall show that knowledge of the complex singularities of a function $u(x)$ allows us to explicitly calculate the asymptotic Fourier or Chebyshev coefficients of the function.

A power series' disk-shaped convergence domain is a fatal weakness for many applications, such as approximating the solution to a boundary value problem. Figure 2.12 compares the domains of convergence for the power series and the Fourier series of the function $f(x) \equiv (1 - p^2) / \{(1 + p^2) - 2p\cos(x)\}$. This function is smooth, well-behaved and infinitely differentiable for all real x. Nevertheless, both expansions have finite domains of convergence because $f(x)$ has simple poles, where it blows up as $1/(x - x_0)$, along the imaginary x-axis.

Because the convergence domain for a Taylor series is a disk, the power series only converges for a finite interval along the real x-axis. If we want to solve a boundary value problem on an interval $x \in [-\pi, \pi]$ whose solution is $f(x)$, or a function similar to it, a power series will fail. In contrast, the strip of convergence for the Fourier series embraces *all* real x.

Similarly, Chebyshev and Legendre polynomials are normally applied to solve problems on the canonical interval [-1, 1]. (If the domain is a different interval $y \in [a, b]$, one can always make the trivial change of variable $x \equiv (2y - (b + a))/(b - a)$ to map the problem into $x \in [-1, 1]$.) Since the foci of an ellipse always lie *inside* it, the convergence domain for

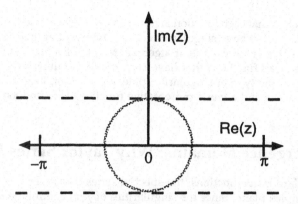

Figure 2.12: A comparison of the regions of Fourier and Taylor series convergence in the complex z-plane for the function $f(z) \equiv (1 - p^2)/\{(1 + p^2) - 2p\cos(z)\}$ for $p = 0.35$. The Fourier series converges within the strip bounded by the two horizontal dashed lines. The power series converges only within the disk bounded by the dotted circle.

these orthogonal polynomials, which is bounded by an ellipse with foci at $x = \pm 1$, always includes the whole interval [-1, 1]. Thus, singularities of $f(x)$ outside this interval (whether at real or complex locations) can *slow* the convergence of a Legendre or Chebyshev series, but can never destroy it.

It is for this reason that our series, instead of being the more familiar Taylor expansions, will be Fourier or Chebyshev or Legendre series instead. If its disk of convergence is too small to include the boundaries, the power series will give nonsense. In contrast, the success of spectral methods is *guaranteed* as long as the target interval, $x \in [-1, 1]$, is free of singularities.

2.8 Stalking the Wild Singularity or Where the Branch Points Are

It is sometimes possible to identify the type and nature of the singularities of the solution to a differential equation *a priori* without knowing the solution.

Theorem 3 (SINGULARITIES of the SOLUTION to a LINEAR ODE)
The solution to a linear ordinary differential equation is singular only where the coefficients of the differential equation or the inhomogeneous forcing are singular, and it may be analytic even at these points.

PROOF: Given in undergraduate mathematics textbooks.

An equivalent way of stating the theorem is to say that at points where all the coefficients of the differential equation and the inhomogeneous term are analytic functions, one may expand the solution as a Taylor series (Frobenius method) about that point with a non-zero radius of convergence.

Thus, the solution of

$$u_{xx} + \frac{1}{1 + x^2}u = 0 \tag{2.34}$$

on $x \in [-1, 1]$, is singular only at the poles of the coefficient of the undifferentiated term at $x = \pm i$ and at infinity. The Chebyshev and Legendre series of $u(x)$ is, independent of the boundary conditions, guaranteed to converge inside the ellipse in the complex x-plane with foci at ± 1 which intersects the locations of the poles of $1/(1 + x^2)$. By using the methods described in a later section, we can calculate the rate of geometric convergence to show that the series coefficients $|a_n|$ must decrease as $O([\ ] (0.17)^{n/2})$. The empty bracket represents an algebraic factor of n; this can be deduced by performing a local analysis around $x = \pm i$ to determine the type of the singularity of $u(x)$ [this type need not match that of the coefficients of the differential equation] and applying Table 2.2. All without actually knowing $u(x)$ itself or even specifying boundary conditions!

Unfortunately, the theorem does *not* extend to *partial* differential equations or to *nonlinear* equations even in one dimension.

EXAMPLE: First Painlevé Transcendent
The differential equation

$$u_{xx} - u^2 = x \tag{2.35}$$

$$u(0) = u_x(0) = 0 \tag{2.36}$$

has coefficients and an inhomogeneous term which have no singularities. Nonetheless, numerical integration and asymptotic analysis show that $u(x)$ has poles at $x=3.742$, 8.376, 12.426, etc.: an infinite number of poles on the real axis. (Bender and Orszag, 1978, pg. 162). These poles are actually "movable" singularities, that is, their location depends on the *initial conditions*, and not merely upon the form of the differential equation.

Movable singularities are generic properties of solutions to nonlinear differential equations. The reason can be understood by examining an ODE closely related to the Painlevé transcendent:

$$u_{xx} - U(x)u = x \tag{2.37}$$

where $U(x)$ is arbitrary. Because this is linear, Theorem 3 tells us that $u(x)$ is singular only where $U(x)$ is singular. If we chose $U(x)$ to be a solution of the Painlevé equation, however, then the linear ODE simply becomes (2.35). Thus, for a nonlinear ODE, the solution $u(x)$ itself furnishes the spatially-variable coefficients. Theorem 3 actually still applies; unfortunately, it is useless because we cannot apply it until we already know the singularities of $u(x)$, which is of course the very information we want from the theorem.

For many nonlinear equations, a problem-specific analysis will deduce some *a priori* information about the location of singularities, but no general methods are known.

On a more positive note, often the physics of a problem shows that the solution must be "nice" on the problem domain. This implies, if the interval is rescaled to $x \in [-1, 1]$ and a Chebyshev or Legendre expansion applied, that the spectral series is guaranteed to converge exponentially fast. The only catch is that the asymptotic rate of convergence μ, where the spectral coefficients are asymptotically of the form $a_n \sim [\ ] \exp(-n \mu)$, could be very large or very small or anywhere in between — without advance knowledge of the singurities of $u(x)$ *outside* the problem domain, we cannot be more specific than to assert "exponential convergence". (However, some rules-of-thumb will be offered later.)

2.8.1 Corner Singularities & Compatibility Conditions

Unfortunately, for *partial* differential equations, it is usual for the solution to even a linear, constant coefficient equation to be *weakly singular* in the *corners* of the domain, if the boundary has sharp corners.

EXAMPLE: Poisson equation on a rectangle. If

$$\nabla^2 u = -1; \qquad u = 0 \qquad \text{on the sides of the rectangle,} \tag{2.38}$$

then the solution is weakly singular in the four corners. In terms of a polar coordinate system (r, θ) centered on one of the corners, the singularity is of the form

$$u = (\text{constant})\, r^2 \log(r) \sin(2\theta) + \text{other terms} \tag{2.39}$$

(Birkhoff and Lynch, 1984). The singularity is "weak" in the sense that $u(x, y)$ and its first two derivatives are bounded; it is only the third derivative that is infinite in the corners.

Constant coefficient, constant forcing, singular solution? It seems a contradiction. However, the boundary curve of a square or any other domain with a corner cannot be represented by a smooth, infinitely differentiable curve. At a right-angled corner, for example, the boundary curve must abruptly shift from vertical to horizontal: the curve is continuous, but its slope has a jump discontinuity.

This argument suggests, correctly, that corner singularities can be eliminated by slighly rounding the corners so that both the boundary curve and the values of u upon it can be parameterized by smooth, infinitely differentiable curves. This is true physically as well as mathematically.

In solid mechanics, corners are regions of very high stress, and the corner singularities are merely a mathematical reflection of this. In a house, cracks in paint or drywall often radiate from the corners of windows and door openings. The first commercial jet aircraft, the British *Comet*, was grounded in the early 1950s after three catastrophic, no-survivor crashes. One of the surviving airframes was tested to destruction in a water tank that was repeatedly pressurized. After six weeks, the airframe failed abruptly. A crack began at a corner of one of the airliner's *square* windows and then propagated backwards until it was ten meters long!

Modern airliners all have windows with *rounded* corners. Unfortunately, in other contexts, it is often necessary to solve problems with unrounded corners.

There are two pieces of good news. First, corner singularities are often so weak as to be effectively ignorable even for high accuracy Chebyshev solutions (Boyd, 1986c, Lee, Schultz and Boyd, 1989b). Second, there are good methods for dealing with singularities including mapping and singularity subtraction as will be described in Chapter 16.

EXAMPLE: One-Dimensional Diffusion Equation

$$u_t = u_{xx}, \qquad u(0) = u(\pi) = 0 \tag{2.40}$$

$$u(x, t = 0) = Q(x) \tag{2.41}$$

Although the diffusion equation is a partial differential equation, it is one-dimensional and linear. The particular initial condition

$$Q(x) = x(\pi - x) \tag{2.42}$$

satisfies the boundary conditions and is a polynomial which can be exactly represented by the sum of the lowest three Chebyshev polynomials. Nevertheless, $u(x, t)$ is *singular*.

The reason is that the solution to the problem as posed is actually the restriction to the interval $x \in [0, \pi]$ of the diffusion equation on the infinite interval, subject to an initial condition which is spatially periodic and antisymmetric with respect to both the origin and $x = \pi$. We can create such an initial condition by either (i) expanding the initial $Q(x)$ as a sine series or (ii) defining it directly as

$$P(x) = \text{sign}(\sin(x))Q(x) \qquad \forall x \in [-\infty, \infty] \qquad (2.43)$$

where the sign function equals one when its argument is positive and is equal to minus one when its argument is negative. Fig 2.13 shows the initial condition (Eq. 2.42) and its second derivative. The latter has jump discontinuities at $x = \pm 0, \pi, 2\pi, \ldots$.

At $t = 0$, these discontinuities cause no problems for a Chebyshev expansion because the Chebyshev series is restricted to $x \in [0, \pi]$ (using Chebyshev polynomials with argument $y \equiv (2/\pi)(x - \pi/2)$). On this interval, the initial second derivative is just the constant -2. For $t > 0$ but very small, diffusion smooths the step function discontinuities in u_{xx}, replacing the jumps by very narrow boundary layers. As $t \to 0+$, the layers become infinitely thin, and thus a Chebyshev approximation for any fixed truncation N *must converge slowly* for sufficiently *small t*.

Fortunately, this pathology is often not fatal in practice because these diffusive boundary layers widen rapidly so that the evolution for later times can be easily tracked with a Chebyshev spectral method for small or moderate N. Indeed, many scientists have happily solved the diffusion equation, graphing the answer only at longish time intervals, and missed the narrow transient boundary layers with no ill effect.

EXAMPLE: One-Dimensional Wave Equation

$$u_{tt} = u_{xx}, \qquad u(0) = u(\pi) = 0 \qquad (2.44)$$

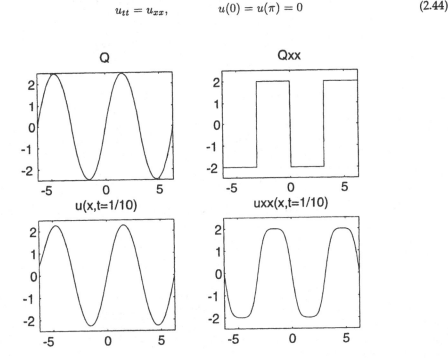

Figure 2.13: A representative solution to the diffusion equation. Upper panels: Initial condition (left) and its second derivative (right). Bottom panels: Solution at $t = 1/10$ (left) and its second derivative (right).

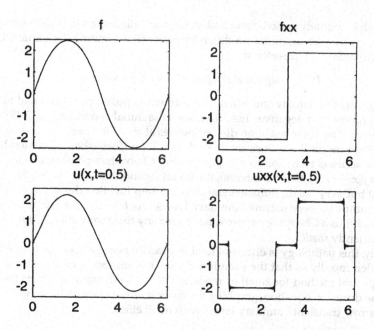

Figure 2.14: Wave equation. Upper panels: Initial condition (left) and its second derivative (right). Bottom panels: Solution at $t = 1/2$ (left) and its second derivative (right). The singularities, which are initially at $x = 0, \pi, 2\pi$, etc., propagate both left and right.

$$u(x, t = 0) = f(x), \qquad u_t(x, t = 0) = g(x) \tag{2.45}$$

The particular initial condition

$$f(x) = x(\pi - x), \qquad g(x) = 0 \tag{2.46}$$

is the same as for the diffusion example, and also yields a singular solution

Again, the reason is that the problem as posed is actually the restriction to the interval $x \in [0, \pi]$ of the same equation on the infinite interval with the initial condition:

$$P(x) = \text{sign}(\sin(x))f(x) \ \forall x \in [-\infty, \infty] \tag{2.47}$$

Fig 2.14 shows the initial condition and its second derivative. The latter has jump discontinuities at $x = \pm 0, \pi, 2\,\pi, \ldots$.

The general solution to the wave equation ($u_t(x, t = 0) = 0$) is

$$u(x, t) = (1/2)\{\, f(x - t) \,+\, f(x + t)\,\} \tag{2.48}$$

Instead of diffusing away, the jumps in the second derivative, initially at $t = m\pi$ where m is an any integer, propagate both left and right. Thus, at the later time illustrated in the lower panels of Fig. 2.14, the exact solution has a second derivative which is discontinuous at two points in the interior of $x \in [0, 1]$, even though the initial condition was super-smooth: a quadratic polynomial.

The singularities of the diffusion and wave equations are similar to those of Poisson's equation in that they, too, are located at the corners, at least initially. However, the corners are now in space-time, that is, at the corners of a domain in the $x - t$ plane.

In contrast to the singularities of an elliptic equation, the corner singularities of a hyperbolic or parabolic equation are usually unphysical. The reason is that the choice of the initial time is usually arbitrary. In weather forecasting, for example, the forecast is usually begun at midnight Greenwich mean time. Why should the singularities be located at the boundary of the atmosphere at this time and no other? Would physical singularities move if the forecast were begun an hour earlier or later? Boyd and Flyer(1999) discuss how initial conditions can be slightly modified to satisfy "compatibility" conditions and thus eliminate the space-time corner singularities.

These three examples should chill the blood of any aspiring computational scientist. Albert Einstein noted "Subtle is the Lord", which is the great man's way of saying that your computer program is always trying to kill you. Fear is a very good thing for a numerical analyst. One may wish mournfully for faster silicon, but the only absolutely fatal disease to a scientist is a deficiency of thinking.

With thinking, spectral methods do very well indeed, even for nasty problems. In the rest of the book, we explain how.

2.9 FACE: Integration-by-Parts Bound on Fourier Coefficients

The coefficients of the complex-exponential form of a Fourier series are:

$$c_n = \frac{1}{2\pi} \int_{-\pi}^{\pi} f(x) \, \exp(-in \, x) \, dx \qquad (2.49)$$

If we integrate by parts, repeatedly integrating the exponential and differentiating $f(x)$, we obtain without approximation after $(J+1)$ steps:

$$c_n = \frac{1}{2\pi} \sum_{j=0}^{J} (-1)^{j+n} \left(\frac{i}{n}\right)^{j+1} \left\{ f^{(j)}(\pi) - f^{(j)}(-\pi) \right\}$$
$$+ \frac{1}{2\pi} \left(-\frac{i}{n}\right)^{J+1} \int_{-\pi}^{\pi} f^{(J+1)}(x) \, \exp(-in \, x) \, dx \qquad (2.50)$$

where $f^{J+1}(x)$ denotes the $(J+1)$-st derivative of $f(x)$. (Note that these integrations-by-parts assume sufficient differentiability of $f(x)$.) The integral can be bounded by the length of the integration interval times the maximum of the integrand, which is the maximum of the absolute value of $|f^{J+1}|$. In the limit $n \to \infty$ for fixed J, it follows that the integral is $O(1/n^{J+1})$; ignoring it gives what Lyness (1971, 1984) has dubbed the "Fourier Asymptotic Coefficient Expansion" (FACE):

$$c_n \sim \frac{1}{2\pi} \sum_{j=0}^{J} (-1)^{j+n} \left(\frac{i}{n}\right)^{j+1} \left\{ f^{(j)}(\pi) - f^{(j)}(-\pi) \right\} + O(n^{-(J+1)}), \; n \to \infty, \text{fixed } J \quad (2.51)$$

This expansion is an asymptotic series, diverging in the limit of fixed n, $J \to \infty$.

By taking real and imaginary parts and recalling that the cosine and sine coefficients are related to the complex coefficients via (for real-valued $f(x)$)

$$a_n = 2 \, \Re(c_n); \qquad b_n = -2 \, \Im c_n \qquad \forall \, n > 0 \qquad (2.52)$$

one obtains an equivalent form of the FACE as $(n > 0)$

$$a_n \sim \frac{1}{\pi} \sum_{j=0}^{J} (-1)^{n+j} \left(\frac{f^{(2j+1)}(\pi) - f^{(2j+1)}(-\pi)}{n^{2j+2}} \right) + O(n^{-(2J+4)}), \ n \to \infty, \text{ fixed } J \quad (2.53)$$

$$b_n \sim \frac{1}{\pi} \sum_{j=0}^{J} (-1)^{n+1+j} \left(\frac{f^{(2j)}(\pi) - f^{(2j)}(-\pi)}{n^{2j+1}} \right) + O(n^{-(2J+3)}), \ n \to \infty, \text{ fixed } J \quad (2.54)$$

If a function has discontinuities in the interior of the interval $x \in [-\pi, \pi]$, then the integrals can be split into segments and integration-by-parts applied to derive similar asymptotic expansions.

Although sometimes used to accelerate the calculation of Fourier coefficients for functions whose series lack exponential convergence, the FACE is most important as the proof of the following theorem.

Theorem 4 (INTEGRATION-BY-PARTS COEFFICIENT BOUND) *If*

1.

$$f(\pi) = f(-\pi), f^{(1)}(\pi) = f^{(1)}(-\pi), ..., f^{(k-2)}(\pi) = f^{(k-2)}(-\pi) \quad (2.55)$$

2. *$f^{(k)}(x)$ is integrable*

then the coefficients of the Fourier series

$$f(x) = a_0 + \sum_{n=1}^{\infty} a_n \cos(nx) + \sum_{n=1}^{\infty} b_n \sin(nx) \quad (2.56)$$

have the upper bounds

$$|a_n| \le F/n^k; \qquad |b_n| \le F/n^k \quad (2.57)$$

for some sufficiently large constant F, which is independent of n.

An equivalent way of stating the theorem is that, if the two conditions above are satisfied, then the algebraic index of convergence is at least as large as k.

Notes: (a) $f^{(k)}$ denotes the k-th derivative of f(x) (b) The integrability of $f^{(k)}$ requires that $f(x)$, $f^{(1)}(x), \ldots, f^{(k-2)}(x)$ must be continuous.

PROOF: Under the conditions of the theorem, the first $(k - 1)$ terms in the FACE are zero. For sufficiently large n, the lowest nonzero term in the series will dominate, implying that the Fourier coefficients are asymptotically $O(n^k)$. Q. E. D.

A few remarks are in order. First, the usual way to exploit the theorem is to integrate-by-parts as many times as possible until lack of smoothness or a mismatch in boundary values forces us to stop. At that point, the number k that appears in the theorem is a lower bound on the algebraic convergence index defined earlier.

However, the index of convergence need not be an integer. For example, if $f(x)$ is a periodic function with a cube root singularity for real x, such as $\sin^{1/3}(x)$, one integration by parts shows that the Fourier coefficients decrease at least as fast as $O(1/n)$. Because the second derivative of a cube root singularity has a branch point proportional to $x^{-5/3}$, which is not integrable, we cannot integrate-by-parts a second time. However, one can show by other arguments that the integral

$$\frac{1}{n} \int_{-\pi}^{\pi} f^{(1)}(x) \exp(-inx)\, dx \qquad (2.58)$$

is actually $O(1/n^{4/3})$. (It would take us too far afield to give the detailed proof of this.) The theorem gives the largest integer which is less than or equal to the actual algebraic index of convergence. For the sawtooth function and the half-wave rectifier function, the index is an integer and then the theorem gives it precisely.

Second, for pure sine or pure cosine series, every other integration-by-parts is trivial. (Note that the FACE for the cosine coefficients proceeds only in *even* powers of n and *odd* derivatives of $f(x)$ while the sine coefficient expansion involves only *odd* powers of n but *even* derivatives of $f(x)$.) For example, if $f(x) = f(-x)$, then the function is symmetric about $x = 0$ and all its sine coefficients are zero. The even order boundary terms for the cosine coefficients vanish *independent* of whether $f^{(2k)}(\pi) = f^{(2k)}(-\pi)$; as if to reinforce the point, the symmetry condition ensures that all the even derivatives are equal at the boundaries anyway. Thus, for a cosine series, it is sufficient to examine whether $f^{(2k-1)}(\pi) = f^{(2k-1)}(-\pi)$. Similarly, for a *sine* series, boundary matching of the odd order derivatives is irrelevant and the process of integration–by–parts can only be blocked after an *odd* number of integrations.

Third, if the function is periodic and differentiable to all orders, we can integrate–by–parts an arbitrary number of times. In that case, the theorem implies that for large n, the Fourier coefficients are decreasing faster than any finite power of n. This is the property of "infinite order" or "exponential" convergence defined above.

Fourth, since a Chebyshev series in x becomes a Fourier series in y only after we make the substitution $x = \cos(y)$, a Chebyshev series – after the change of variable – is automatically periodic in y. Therefore, a CHEBYSHEV SERIES ALWAYS HAS THE PROPERTY OF INFINITE ORDER CONVERGENCE EVEN FOR FUNCTIONS THAT ARE NON-PERIODIC (in x, the Chebyshev argument) if all its derivatives are bounded on the expansion interval, $x \in [-1, 1]$.

Fifth, the constant F increases with k, the number of times we integrate-by-parts. Consequently, we cannot take the limit of $k \to \infty$ for fixed n. (Taking this limit with the false assumption of k-independent F would imply that all the coefficients are zero!) Rather, the theorem tells us how rapidly the coefficients decrease for large n as $n \to \infty$ for fixed k.

Sixth, this theorem can be strengthened to the statement that if $f(x)$ is in the Lipschitz space L^λ "on the circle", then $a_n, b_n \sim O(n^{-\lambda})$. A function is in the Lipschitz space L^λ (for $0 < \lambda < 1$) if $|f(x) - f(y)| = O(|x - y|^\lambda)$ for all x, y on the interval; Lipschitz spaces for $\lambda > 1$ are defined by taking higher order differences of f. The phrase "on the circle" means that continuity is analyzed as if the interval $x \in [-\pi, \pi]$ is bent into a circle so that $x = \pi$ is continuous with $x = -\pi$; for example, $f(-\pi) \neq f(\pi)$ implies that the function is not in any Lipschitz space with $\lambda > 0$.

EXAMPLE: Let us calculate the index of convergence of the Fourier series solution to

$$u_{xx} + Q(x)\, u = f(x); \qquad u(0) = u(\pi) = 0 \qquad (2.59)$$

Since the boundary conditions do not impose periodicity, the solution to this problem is usually *not* periodic even if the functions $Q(x)$ and $f(x)$ are periodic themselves.

Because of the lack of periodicity, we really ought to use Chebyshev polynomials for this problem. Nonetheless, many old papers have applied Fourier series to non-periodic problems; certain analytical techniques, such as the method of separation-of-variables, fail otherwise. It therefore is useful to see just how bad this choice of basis functions is.

Because the boundary conditions demand $u(0) = 0$, we must exclude the cosine terms (and the constant) from the Fourier series and use only the sine functions:

$$u(x) = \sum_{n=1}^{\infty} b_n \sin(nx) \qquad (2.60)$$

Since the sum of the sine series and the individual sine functions are both antisymmetric with respect to $x = 0$, we halve the interval of integration to $[0, \pi]$ and double the result. This is natural for this problem since $[0, \pi]$ is the interval between the boundaries. Thus,

$$b_n = (2/\pi) \int_0^{\pi} u(x) \sin(nx) dx \qquad (2.61)$$

One integration-by-parts gives

$$b_n = -(2/[n\pi]) \Big|_0^{\pi} u(x) \cos(nx) + (2/[n\pi]) \int_0^{\pi} u^{(1)}(x) \cos(nx) \qquad (2.62)$$

The boundary conditions $u(0) = u(\pi)$ guarantee that the boundary term in (2.62) is zero *independent* of $Q(x)$ and $f(x)$. Integrating–by–parts a second time gives

$$b_n = (2/[n^2\pi]) \Big|_0^{\pi} u^{(1)}(x) \sin(nx) - (2/[n^2\pi]) \int_0^{\pi} u^{(2)}(x) \sin(nx) dx \qquad (2.63)$$

Since $\sin(0) = \sin(n\pi) = 0$, it does not matter whether $u^{(1)}(0) = u^{(1)}(\pi)$; the boundary term in (2.63) is 0 anyway.

Integrating-by-parts a third time gives

$$b_n = (2/[n^3\pi]) \Big|_0^{\pi} u^{(2)}(x) \cos(nx) - (2/[n^3\pi]) \int_0^{\pi} u^{(3)}(x) \cos(nx) dx \qquad (2.64)$$

Since $\cos(n\pi) \neq \cos(0)$ for odd n, we are stuck with $k = 3$ as the algebraic index of convergence unless $u^{(2)}(0) = u^{(2)}(\pi)$. Since one can only impose one pair of boundary conditions on the solution of a second order differential equation, we would seem to be out of luck. Evaluating the differential equation at $x = 0$, however, gives

$$\begin{aligned} u^{(2)}(0) &= -Q(0)u(0) + f(0) \\ &= f(0) \end{aligned} \qquad (2.65)$$

since the boundary conditions on the differential equation require $u(0) = 0$. Similarly, $u^{(2)}(\pi) = f(\pi)$. We see that if

$$f(0) = f(\pi) = 0, \qquad (2.66)$$

then the combination of the differential equation with its associated boundary conditions forces the boundary term in (2.64) to vanish. We can integrate twice more before being defeated by the fact that $u^{(4)}(0) \neq u^{(4)}(\pi)$. Thus, whether the index of convergence is $k = 3$ or 5 depends entirely upon whether $f(x)$ satisfies condition (2.66) (as is often true in practical applications). A $k = 5$ geophysical example is Eliasen (1954), who used a Fourier sine series to solve the "barotropic instability problem" for a cosine jet.

The important point is that we deduced the index of convergence simply from the *form* of the differential equation and the boundary conditions. We did not have to specify $Q(x)$

at all, and $f(x)$ only at two points to compute k. The integration–by–parts theorem is a very powerful tool because we can apply it *before* we solve the differential equation.

This theorem is also important because it *generalizes* to all other spectral basis sets. That is to say, integration–by–parts can be used to provide similar coefficient bounds for any of the standard basis functions including Chebyshev and Legendre polynomials, Hermite functions and spherical harmonics.

2.10 Asymptotic Calculation of Fourier Coefficients

Theorem 5 (STRIP OF CONVERGENCE FOR FOURIER SERIES) *Let* $z = x + iy$ *and let* ρ *denote the absolute value of the imaginary part of the location of that singularity of* $f(z)$ *which is nearest the real z-axis. In other words, if* $z_j, j = 1, 2, \ldots$ *denotes the location of the singularities of* $f(z)$, *then*

$$\rho = \min_{j} |\Im(z_j)| . \tag{2.67}$$

Then the Fourier series converges uniformly and absolutely within the STRIP in the complex z-plane, centered on the real axis, which is defined by

$$| y | < \rho, \qquad x = arbitrary \qquad [convergence] \tag{2.68}$$

and diverges outside the strip,

$$| y | > \rho, \qquad x = arbitrary \qquad [divergence] \tag{2.69}$$

If the limit exists, then the asymptotic coefficients of the Fourier series

$$f(z) = a_0 + \sum_{n=1}^{\infty} a_n \cos(nz) + \sum_{n=1}^{\infty} b_n \sin(nz) \tag{2.70}$$

are related to ρ, *the half-width of the strip of convergence, by*

$$\limsup_{n \to \infty} \log | a_n / a_{n+1} | = \rho \tag{2.71}$$

and similarly for the sine coefficients b_n.

The asymptotic rate of convergence for real z is given by $\mu = \rho$.

If the series has a finite algebraic index of convergence, then $\rho = 0$ *and the series converges only on the real axis.*

If the function is periodic and entire, i. e., has no poles or branch points except at infinity, then usually $\rho = \infty$ *and the coefficients are* $O[\exp(-(n/k)\log(n) + O(n))]$ *for some constant k. (The exceptions with finite* ρ *are entire functions that grow very rapidly with* $| z |$ *such as exponentials-of-exponentials, as discussed in Boyd (1994b).)*

Note: In this context, "singularity" includes not only poles and branch points, but also discontinuities caused by a lack of periodicity, as for the saw-tooth function.

PROOF: We omit a rigorous proof, but offer the following argument. Let $z = x + iy$ so that $| \exp(i n z) | = \exp(- n y)$. Thus, in the complex form of the Fourier series

$$f(z) = \sum_{n=-\infty}^{\infty} c_n \exp(inz) = \sum_{n=-\infty}^{\infty} c_n \exp(inx) \exp(-ny) \tag{2.72}$$

all terms with negative n grow exponentially as $\exp(|\,n\,|y)$ in the upper half-plane while those with $n > 0$ grow in the lower half-plane. It follows that the series can converge at $|\,y\,| = \rho$ if and only if $|\,c_{\pm n}\,|$ decay at least as fast as $\exp(-\,n\,\rho)$.

On the other hand, if $f(z)$ has a pole at $y = \rho$, the series must converge to ever-larger values as $\Im(z) \to \rho$. This implies that the coefficients cannot decrease faster than $\exp(-\,n\rho)$ (*modulo* the usual algebraic factor of n). (Note that if the coefficients could be bounded by $\exp(-n(\rho - \epsilon))$ for some small positive ϵ, for example, then one could bound the Fourier series, term-by-term, by a convergent geometric series to show it was finite even at the pole.)

Thus, the series coefficients must decrease as $\exp(-\,n\,\rho)$ when $f(z)$ has a singularity at $|\,\Im(z)\,| = \mu$. Q. E. D.

The theorem shows that *"geometric" convergence is normal* for a Fourier series. It is only when the function has a discontinuity or other singularity for *real z* that the series will have sub-geometric or algebraic convergence.

2.11   Convergence Theory for Chebyshev Polynomials

Since a Chebyshev polynomial expansion is merely a Fourier cosine series in disguise, the convergence theory is very similar to that for Fourier series. Every theorem, every identity, of Chebyshev polynomials has its Fourier counterpart. Nonetheless, the mapping does produce some noteworthy consequences.

The mapping is

$$z = \cos(\theta) \tag{2.73}$$

and then

$$T_n(z) \equiv \cos(n\theta) \tag{2.74}$$

The following two series are then equivalent under the transformation:

$$f(z) = \sum_{n=0}^{\infty} a_n T_n(z) \tag{2.75}$$

$$f(\cos\theta) = \sum_{n=0}^{\infty} a_n \cos(n\theta) \tag{2.76}$$

In other words, the coefficients of $f(z)$ as a Chebyshev series are *identical* with the Fourier cosine coefficients of $f(\cos(\theta))$.

FIRST IMPLICATION: EXPONENTIAL CHEBYSHEV CONVERGENCE for NON-PERIODIC FUNCTIONS.

Even if $f(z)$ itself is not periodic in z, the function $f(\cos\theta)$ must inevitably be periodic in θ with period 2π. As we vary θ over all real θ, the periodicity of $\cos(\theta)$ implies that z (= $\cos[\theta]$) merely oscillates between -1 to 1 as shown in Fig. 2.15. Since $f(\cos\theta)$ is periodic, its Fourier series must have exponential convergence unless $f(z)$ is singular for $z \in [-1, 1]$. It does not matter if $f(z)$ is periodic in z nor does it matter if $f(z)$ has singularities for *real z* *outside* the interval [-1, 1]. The Fourier cosine series in (2.76) sees $f(\cos\theta)$ only as a periodic function. For real θ, the Fourier series sees only those variations in $f(z)$ that occur for $z \in [-1, 1]$. The exponential convergence of the Fourier series (2.76) then implies equally fast convergence of the Chebyshev series since the sums are term-by-term identical.

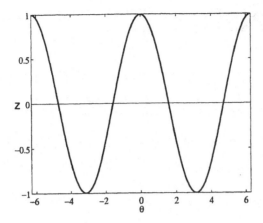

Figure 2.15: The Chebyshev polynomials, $T_n(z)$, are related to the terms of a Fourier co-sine series through the identity $T_n(\cos[\theta]) = \cos(n\,\theta)$. The graph shows the relationship between z and θ, $z = \cos(\theta)$.

SECOND IMPLICATION: the FOURIER EQUIVALENT of a CHEBYSHEV SERIES is a FOURIER *COSINE* SERIES.

This is implicit in (2.76). Since $\cos(\theta)$ is symmetric about $\theta = 0$, $f(\cos\theta)$ is forced to be symmetric in θ, too, even if $f(z)$ has no symmetry whatsoever with respect to z. Consequently, we need only the cosines.[1]

THIRD IMPLICATION:

Theorem 6 (CHEBYSHEV TRUNCATION THEOREM) *The error in approximating $f(z)$ by the sum of its first N terms is bounded by the sum of the absolute values of all the neglected coefficients. If*

$$f_N(z) \equiv \sum_{n=0}^{N} a_n T_n(z) \tag{2.77}$$

then

$$E_T(N) \equiv |f(z) - f_N(z)| \le \sum_{n=N+1}^{\infty} |a_n| \tag{2.78}$$

for all $f(z)$, all N, and all $z \in [-1, 1]$.

PROOF: The Chebyshev polynomials are bounded by one, that is, $|T_n(z)| \le 1$ for all $z \in [-1, 1]$ & for all n. This implies that the n-th term is bounded by $|a_n|$. Subtracting the truncated series from the infinite series, bounding each term in the difference, and summing the bounds gives the theorem. Q. E. D.

The mapping alters the shape of the convergence region as described by the following.

---

[1] If $f(z)$ does have symmetry, the basis can be further reduced. If $f(z) = f(-z)$, then only the even cosines, $1, \cos(2\,\theta), \cos(4\,\theta), \dots$, are needed. Similarly, only the odd cosines (or odd degree Chebyshev polynomials in z) are needed if $f(z) = -f(-z)$.

Theorem 7 (CONVERGENCE ELLIPSE: CHEBYSHEV SERIES) *Let* $z = x + iy$. *Introducing the elliptical coordinates* (μ, η) *via the equations*

$$
\begin{aligned}
x &= \cosh(\mu)\cos(\eta), & \mu &\in [0, \infty] \\
y &= -\sinh(\mu)\sin(\eta), & \eta &\in [0, 2\pi],
\end{aligned}
\tag{2.79}
$$

the surfaces of constant μ define ellipses in the complex z-plane with foci at $z = -1$ and $z = 1$. (See Fig. 2.16.) For large μ, elliptical coordinates tend to polar coordinates with μ proportional to $\log(r)$ and $\eta \to -\theta$.)

Suppse $f(z)$ has poles, branch points or other singularities at the points (in elliptical coordinates) $(\mu_j, \eta_j), j = 1, 2, \ldots$ in the complex z-plane. Define

$$
\mu_0 \equiv min_j |\mu_j|
\tag{2.80}
$$

Then the Chebyshev series

$$
f(z) = \sum_{n=0}^{\infty} a_n T_n(z)
\tag{2.81}
$$

converges within the region bounded by the ellipse $\mu = \mu_0$,

$$
\mu < \mu_0, \text{ for all } \eta, \qquad [\text{convergence}]
\tag{2.82}
$$

and diverges for all (x, y) outside this ellipse.

The coefficients are related to μ_0, the quasi-radial coordinate of the ellipse, by the equation

$$
\limsup_{n \to \infty} \log(|a_n/a_{n+1}|) = \mu_0, \qquad [\text{if the limit exists}]
\tag{2.83}
$$

Note: the "ellipse" $\mu = 0$ is the real interval [-1, 1]. If the coefficients a_n decrease algebraically with n or if the convergence is exponential but subgeometric, then the Chebyshev series converges only for $\mu = 0$, i. e., only on the real interval $z \in [-1, 1]$.

PROOF: Let the complex variable $\theta = \eta + i\mu$. Now a trigonometric identity shows that for any (μ, η),

$$
\cos(\eta + i\mu) = \cosh(\mu)\cos(\eta) - i\sinh(\mu)\sin(\eta)
\tag{2.84}
$$

(Abramowitz and Stegun,1965). Thus, the transformation $z = cos(\theta)$ relates (x, y) to (μ, η) just as defined in (2.79). The straight lines which are the boundaries of the strip of Fourier convergence in the complex θ-plane must be mapped into curves of constant μ in the z-plane.

To show that $\mu = \mu_0$ is indeed the equation of an ellipse, substitute Eq.(2.79) into

$$
x^2/a^2 + y^2/b^2 = 1 \qquad [\text{central equation of ellipse}]
\tag{2.85}
$$

where the axes of the ellipse are $a = \cosh(\mu_0)$ and $b = \sinh(\mu_0)$. Cancelling the hyperbolic functions reduces (2.85) to the identity $\cos^2(\eta) + \sin^2(\eta) = 1$. Q. E. D.

It is unusual to worry about imaginary values of z in applications, but since complex singularities will limit the convergence of the series, we need to understand that a singularity at one point on an ellipse will produce the same (asymptotic!) rate of convergence as a pole or branch point on any other part of the same ellipse. This has two important implications.

First, Fig. 2.16 shows that when μ is small, the ellipses of constant μ are long and narrow, pass very close to $z = \pm 1$ and then cross the imaginary axis at much larger distances.

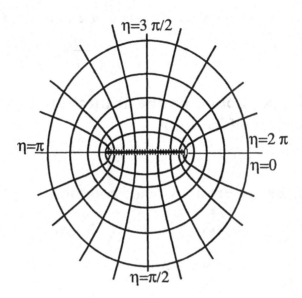

Figure 2.16: Elliptical coordinates (μ, η) in the complex z-plane; $z = x + i\,y$ via $x = \cosh(\mu)\cos(\eta)$ & $y = -\sinh(\mu)\sin(\eta)$. which implies $T_n(x + i\,y) = \cos(\,n\,[\eta + i\,\mu]) = \cosh(n\,\mu)\,\cos(n\,\eta) - i\,\sinh(n\,\mu)\,\sin(n\,\eta)$. The surfaces of constant μ are *ellipses*; the contours of constant "angle" η are *hyperbolas*. If a function $f(z)$ has a pole or other singularity anywhere on the ellipse $\mu = \mu_0$ and no singularities on ellipses of smaller μ, then the Chebyshev series of $f(z)$ *diverges* everywhere *outside* the ellipse $\mu = \mu_0$ and *converges* everywhere *inside* this ellipse.

This implies that singularities very close to $z = \pm 1$ have no worse effect on convergence than singularities on the imaginary axis that are much farther from the interval [-1, 1]. Chebyshev polynomials have a kind of inherent coordinate-stretching that makes them much better at resolving singularities or boundary layers near the *endpoints* of [-1, 1] than near the *middle*.

Second, Fig. 2.16 shows that singularities for *real z* will not destroy the property of exponential convergence for the Chebyshev series as long as they lie outside the interval [-1, 1]. Conversely, convergence on the "canonical interval", even geometric convergence, does not imply convergence for all real z, but no further than $|\,z_0\,|$ where z_0 is the location of the nearest real singularity, and maybe not even that far if there are also poles or branch points of $f(z)$ for complex z.

If we know (or can estimate) the location of the convergence-limiting singularity, then the following is very useful.

Theorem 8 (CHEBYSHEV RATE of CONVERGENCE) *The asymptotic rate of convergence of a Chebyshev series for $z \in [-1, 1]$ is equal to μ, the quasi-radial coordinate of the ellipse of convergence. This is related to the location of the convergence-limiting singularity at (x_0, y_0) in the complex plane via*

$$\mu = \mathrm{Im}\{\arccos[x_0 + iy_0]\} \tag{2.86}$$
$$= \log|\,z_0 \pm (z_0^2 - 1)^{1/2}\,| \tag{2.87}$$
$$= \log\left(\alpha + \sqrt{\alpha^2 - 1}\right) \tag{2.88}$$

where the sign in the second line is chosen to make the argument of the logarithm > 1 so that $\mu > 0$ and where

$$\alpha \equiv \frac{1}{2} \sqrt{(x_0 + 1)^2 + y_0^2} + \frac{1}{2} \sqrt{(x_0 - 1)^2 + y_0^2} \tag{2.89}$$

2.12 An Upper Bound on the Truncation Error: Last Coefficient Rule-of-Thumb

Theorem 9 (FOURIER TRUNCATION ERROR BOUND) *Let*

$$f(x) = a_0 + \sum_{n=1}^{\infty} a_n \cos(nx) + \sum_{n=1}^{\infty} b_n \sin(nx) \tag{2.90}$$

and let $f_N(x)$ be the approximation to $f(x)$ which is obtained by truncating each of the series in (2.90) after N terms:

$$f_N(x) \equiv a_0 + \sum_{n=1}^{N} a_n \cos(nx) + \sum_{n=1}^{N} b_n \sin(nx) \tag{2.91}$$

Then the truncation error defined by

$$E_T(N) \equiv \mid f(x) - f_N(x) \mid, \qquad [\text{``truncation error''}] \tag{2.92}$$

is bounded from above by the sum of the absolute values of all the neglected coefficients:

$$E_N(N) \leq \sum_{n=N+1}^{\infty} \mid a_n \mid + \mid b_n \mid \tag{2.93}$$

PROOF: Similar to the analogous Chebyshev result, Theorem 6.

Other basis functions satisfy similar theorems with slightly different numerical factors; for Hermite functions, for example, we multiply the R. H. S. of (2.93) by 0.785.

This theorem is very powerful because it tells us that the coefficients of the spectral series are also (roughly) the errors, or at least an upper bound on the errors. It is also seemingly quite useless because we normally do not know what all the neglected coefficients are, but let us look at a representative case.

EXAMPLE: For the "Symmetric, Imbricated-Lorentzian" function of the previous section, the expansion is

$$f(x) = 1 + 2 \sum_{n=1}^{\infty} p^n \cos(nx) \tag{2.94}$$

The Truncation Error Bound Theorem then implies

$$\mid f(x) - f_N(x) \mid \leq 2 \sum_{n=N+1}^{\infty} p^n = 2p^{N+1}/(1-p) \tag{2.95}$$

since the geometric series can be explicitly summed. Because the terms of (2.94) are all positive at $x = 0$, Eq. (2.95) is a *tight* upper bound.

When computing the hitherto-unknown solution of a differential equation, we do not know the exact form of the neglected coefficients. However, we have already seen that almost all Fourier and Chebyshev series for smooth, infinitely differentiable functions converge *asymptotically* like geometric series. In other words, it is the rule rather than the exception that

$$a_n \sim [\ ] p^n, \qquad n >> 1 \tag{2.96}$$

where $[\ ]$ is a slowly varying (algebraic) function of n or a constant. This implies that while we cannot get a numerical bound on the truncation error for unknown functions, we can predict that they will behave *qualitatively* like the geometrically-converging Fourier series (2.95).

The pay-off is the following assertion:

Rule-of-Thumb 2 (LAST COEFFICIENT ERROR ESTIMATE) *The truncation error is the same order-of-magnitude as the* last COEFFICIENT RETAINED *in the truncation for series with geometric convergence. Since the truncation error is a quantity we can only estimate anyway (in the absence of a known, exact solution), we can loosely speak of the last retained coefficient as being the truncation error, that is:*

$$E_T(N) \sim O(|\, a_N\,|) \tag{2.97}$$

Extension: if the series has algebraic convergence index k, i. e., if $a_n \sim O(1/n^k)$ for large n, then

$$E_T(N) \sim O(N\,|\,a_N\,|) \tag{2.98}$$

JUSTIFICATION: If we accept that a large class of Fourier and Chebyshev series converge geometrically, then (2.97) follows immediately from (2.95). Note that the last retained coefficient is $a_N = 2\,p^N$, so that (2.97) is true with the proportionality constant $p/(1-p) \sim O(1)$.

For algebraically converging series, note that

$$\sum_{n=N+1}^{\infty} \frac{1}{n^k} \sim \frac{1}{(k-1)N^{k-1}}, \qquad N >> 1 \tag{2.99}$$

(Bender and Orszag, 1978, pg. 379.) Thus, the bound in the error truncation theorem is $O(1/N^{k-1})$ while the last retained coefficient is $O(1/N^k)$, and this implies (2.98).

The ultimate test of a numerical solution is to repeat the calculation with different N and compare results. The rule-of-thumb is not intended to substitute for that. Rather, it has two uses. First, it provides a quick-and-dirty way of estimating the error from a single calculation; if the last retained coefficient is *not* small in comparison to the desired error, then we need larger N. If it is small, and the lower coefficients decrease smoothly towards a_n, the calculation is probably okay. Second, it provides order-of-magnitude guidelines for estimating the feasibility of a calculation. Without such guidelines, one would waste lots of time by attempting problems which far exceed the available computing power.

2.13 Convergence Theory for Legendre Polynomials

When the computational domain is split into a large number of subdomains with a separate spectral series on each subdomain, a popular strategy is to mimic the finite element method

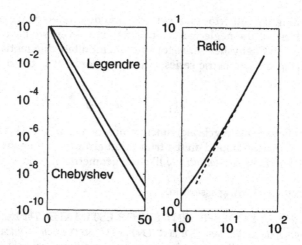

Figure 2.17: Left: absolute values of the coefficients of $f(x; p) = 1/(1 - 2\ p\ x + p^2)^{1/2}$. Legendre coefficients: upper curve. Chebyshev coefficients: bottom curve. Right: the solid curve is the ratio of the absolute value of the n-th Legendre coefficient to the n-th Chebyshev coefficient. The dashed line is the $n^{1/2}$.

with a so-called "weak" formulation. One crucial point is that the weak formulation is greatly simplified by using a basis of Legendre polynomials instead of Chebyshev.

The good news is that the convergence theory for Legendre polynomials is virtually identical with that of Chebyshev polynomials — indeed, it *is* identical if one ignores *algebraic* factors of n, as done so often above. The bad news is that for a given arbitrary function $f(x)$, the maximum pointwise error of a Legendre series (or interpolant), truncated after N terms, is worse than that of the corresponding Chebyshev series by a factor of the square root of N.

Theorem 10 (LEGENDRE POLYNOMIAL RATE of CONVERGENCE)

- *The domain of convergence for Legendre series is the interior of the largest ellipse, with foci at ± 1, which is free of singularities of the function being expanded. This domain is identical with the domain of convergence of the Chebyshev series for the same function.*

- *The asymptotic rate of convergence μ of a Legendre series is identical with that of the corresponding Chebyshev series, that is, it is the elliptical coordinate of the boundary of the domain of convergence in the complex plane.*

- *In the asymptotic limit $N \to \infty$, the maximum pointwise error of a Legendre series is larger than that of the Chebyshev series of the same order, also truncated after N terms, by $O(N^{1/2})$.*

PROOF: The first part is proved in Davis (1975). The convergence domain implicitly specifies the asymptotic rate of convergence. Fox and Parker (1968, pg. 17) prove an asymptotic relation that suggests Legendre series are worse than Chebyshev by a factor of $\sqrt{2/(\pi N)}$.

An example is instructive. The function

$$\frac{1}{\sqrt{1 - 2\,p\,x + p^2}} = \sum_{n=0}^{\infty} p^n\, P_n(x) \tag{2.100}$$

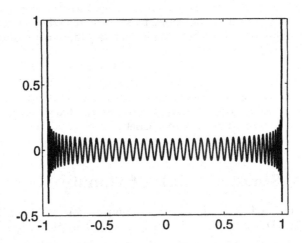

Figure 2.18: Graph of $P_{100}(x)$, the Legendre polynomial of degree one hundred. It is also a schematic illustrating the form of all Legendre polynomials: Oscillations over most of the interval whose amplitude at a fixed x is $O(1/n^{1/2})$ plus narrow boundary layers near the endpoints where the polynomial rises to ± 1.

has reciprocal-of-square root branch points in the complex plane which limit convergence. The coefficients of the Chebyshev series for this function are not known in closed form, but it is known that they are asymptotically of the form

$$a_n^{(Chebyshev)} \sim \text{constant} \, \frac{p^n}{\sqrt{n}}, \qquad n \to \infty \qquad (2.101)$$

Fig 2.17 compares the Chebyshev and Legendre coefficients for this function. The left panel shows that both sets asymptote to parallel straight lines, but the Chebyshev coefficients are smaller. (Since $P_n(1) = 1$ and $|P_n(x)| \leq 1$ for all n, just as for the Chebyshev polynomials, these differences in coefficients translate directly into differences in errors between the two basis sets.) The right panel shows that the ratio of the coefficients rapidly asymptotes to $\sqrt{(\pi/2)N}$.

Another way of stating the same thing is to inspect the Legendre polynomials themselves (Fig. 2.18). In contrast to the Chebyshev polynomials, which oscillate *uniformly* over the interval $x \in [-1, 1]$ (as obvious from the relation $T_n(\cos(\theta)) \equiv \cos(n\,\theta)$), the Legendre polynomials are nonuniform with small amplitude over most of the interval except in extremely narrow boundary layers where the polynomial rises to one or falls to minus one. The even polynomials have local maxima at both $x = 1$ and $x = 0$; the ratio of these is

$$\rho_n \equiv |P_n(0)/P_n(1)| \sim \sqrt{\frac{2}{\pi\,n}}, \qquad n \text{ even}, \qquad n \to \infty \qquad (2.102)$$

There are two reasons why this nonuniformity and the poorer error of Legendre polynomials are not particularly upsetting to practitioners of spectral elements. First, the degree of the polynomials on each element is rather small, rarely greater than $N = 8$. Second, Fig. 2.18 shows that the boundary layers where the Legendre polynomial is large are *very thin*. For *arbitrary* functions, this is still an embarrassment. However, the solutions to boundary value problems are *constrained* by the boundary conditions. For spectral elements, some constraint exists even on interior subdomain walls because the expansions in each adjoining element must have common values on the common walls. Thus, spectral element expansions in Legendre polynomials are constrained by matching and boundary conditions to be close to the exact solution in precisely those regions where the errors in a Legendre series would *otherwise* be large.

The Chebyshev-Legendre theory also applies to the larger family of Gegenbauer polynomials. All members of the Gegenbauer family converge in the same ellipse-bounded domain at the same asymptotic rate of convergence. However, for Gegenbauer polynomials $C_n^m(x)$, which are orthogonal with respect to the weight factor $w(x) = (1 - x^2)^{m-1/2}$, the maximum pointwise errors are worse than Chebyshev by a factor of $O(n^m)$ for sufficiently large m. Legendre polynomials are the special case $m = 1/2$.

2.14   Quasi-Sinusoidal Rule of Thumb

It is easy to determine the cost of an accurate computation after the fact. (One can repeat with different N to be sure the calculation *is* accurate.) In planning a computation, however, it is important to estimate *in advance* the approximate cost because this may determine whether the computation is feasible.

Rule-of-Thumb 3 (TELLER'S LAW) *A state-of-the-art calculation requires 100 hours of CPU time on the state-of-the-art computer,* independent *of the decade.*

The point of this maxim is that people push the current technology to the limit. This will be as true in 2100 as it was in the time of Charles Babbage. However, there have always been bureaucrats to put some limits on the enthusiasts (for which moderate users should be grateful). Because of these constraints, one can get in enormous trouble if a proposed 20 hour calculation in fact requires 200.

There are several strategies for estimating costs *a priori*. One is library research: looking up the costs and resolution of previously published calculations for similar problems.

A second strategy is the "Method of Model Functions". One chooses a simple analytic function which has the same behavior as is expected of the unknown solution — boundary layers, fronts, oscillations, whatever. One can then expand this as a series of basis functions by performing a matrix multiplication or FFT. If the model has the same scales of variation as the true solution, then one can obtain a good estimate of how many basis functions will be needed. The rate at which the model's $\{a_n\}$ decrease will be roughly the same as that of the coefficients of the unknown solution even though the individual numerical values will be different.

A third strategy is to use rules-of-thumb.

The "quasi-sinusoidal" rule-of-thumb is based on the observation that a wide variety of phenomena have oscillations. If we can estimate or guess the approximate wavelength — strictly speaking, we should say *local* wavelength since the scale of oscillation may vary from place to place — then we ought to be able to estimate how many Chebyshev polynomials are needed to resolve it. Gottlieb and Orszag's method of estimation is quite simple: solve the partial differential equation

$$u_t + u_x = 0 \qquad\qquad x \in [-1, 1] \qquad\qquad (2.103)$$

with initial and boundary conditions such that the exact solution is a steadily translating sine wave, and then compare different numerical methods.

Table 2.1 shows the results. The solution is

$$u(x, t) = \sin[M\,\pi(x - t - 1)] \qquad\qquad (2.104)$$

Table 2.1: RMS (L_2) errors in the solution of the one-dimensional wave equation. The exact solution is $u(x) = \sin(M \pi[x - t - 1])$. There are a total of M wavelengths on the interval $[-1, 1]$. N is the total number of grid points or Chebyshev polynomials. To achieve a 1% error, the second order method requires $N/M > 40$, the second order method needs $N/M > 15$, while the spectral method needs $N/M > 3.5$. The Chebyshev-tau algorithm was used here, but another graph in Gottlieb & Orszag (1977) shows that error is approximately the same for Galerkin's method and the pseudospectral method. The striking differences are between the spectral methods and the finite difference methods, not between different variants within the family of spectral algorithms.

SECOND ORDER			FOURTH ORDER			CHEBYSHEV		
N	M	ERROR	N	M	ERROR	N	M	ERROR
40	2	0.1	20	2	0.04	16	4	0.08
80	2	0.03	30	2	0.008	20	4	0.001
160	2	0.008	40	2	0.002	28	8	0.2
40	4	1.	40	4	0.07	32	8	0.008
80	4	0.2	80	4	0.005	42	12	0.2
160	4	0.06	160	4	0.0003	46	12	0.02

so that M = total number of wavelengths on the interval $[-1, 1]$. For all the numerical methods, N is the total number of degrees of freedom.

As might be expected, the accuracy does not depend on either N or M alone, but only on their ratio

$$\frac{N}{M} \equiv \frac{\text{number of grid points}}{\text{wavelengths}} \qquad (2.105)$$

We find

Rule-of-Thumb 4 (QUASI-SINUSOIDAL RESOLUTION REQUIREMENT)
To achieve 5% accuracy for a wave-like solution, one needs roughly

20 grid pts./wavelength	*SECOND ORDER FINITE DIFFERENCE*	*[40 pts/wave for 1 % error]*
10 grid pts./wavelength	*FOURTH ORDER FINITE DIFFERENCE*	*[15 pts/wave for 1 % error]*
3.5 polys./wavelength	*CHEBYSHEV PSEUDOSPECTRAL*	*[3.5 polys./wave for 1 % error]*

To estimate the number of wavelengths M, let $\triangle x$ be the smallest scale that one must resolve on $x \in [-1, 1]$. Then

$$M \approx \frac{\pi}{\triangle x}$$

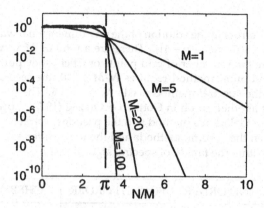

Figure 2.19: The errors in the approximation of $f(x) \equiv 2^{-1/2}(\cos(M\pi x) + \sin(M\pi x))$, graphed versus N/M, that is, the number of Chebyshev polynomials retained in the truncation divided by M, the number of wavelengths of $f(x)$ on $x \in [-1, 1]$. The dashed vertical line is the asymptotic limit $M \to \infty$, which is at $N/M = \pi$.

This is but a rule-of-thumb — an unkind wit might define "rule-of-thumb" as a synonym for "educated guess" — but it is representative of the extraordinary efficiency of pseudospectral methods versus finite difference methods for large N. The rule-of-thumb lists the needed resolution for 1% error to emphasize that the advantage of pseudospectral over finite difference methods increases very rapidly as the error decreases, even for rather moderate accuracy.

Fig. 2.19 shows the errors in the Chebyshev expansion of $\sin(M\pi x)$. There is almost no accuracy at all until $N/M > \pi$, and then the error just falls off the table.

One must be a little careful in applying the rule when M and N are small. The reason is that when we increase N by 1 or 2, we typically decrease the error (for a sine function) by an order-of-magnitude whether N is 10 or 10,000. This increase of N (for the sake of more accuracy) has a negligible effect on the ratio N/M when M is huge, but N/M must be significantly greater than 3 to obtain high accuracy when M is small. A more conservative rule-of-thumb is to use

$$N = 6 + 4(M - 1) \tag{2.106}$$

coefficients, allocating 6 polynomials to resolve the first wavelength and 4 for each additional wavelength.

2.15   Witch of Agnesi Rule–of–Thumb

Another caution is that $\sin(M\pi x)$ is an *entire* function, free of singularities, whereas most real-life solutions have poles or branch-points which determine the asymptotic rate of convergence. For this reason, we will derive a second rule-of-thumb, which we call poetically the "Witch of Angesi" Rule-of-Thumb' because it is derived by examing the analytical Chebyshev coefficients for the function known variously as the "Lorentzian" or the "Witch of Agnesi", which is

$$\frac{\epsilon^2}{(x^2 + \epsilon^2)} = \sum_{n=0}^{\infty} a_n T_n(x) \tag{2.107}$$

where the odd coefficients are zero by symmetry and where the even coefficients are asymptotically

$$|a_n| \sim 2\epsilon e^{-n\epsilon} \qquad n \gg 1 \,\,\&\,\, \epsilon \ll 1 \qquad (2.108)$$

For simplicity, assume ϵ is small. Bounding the error by the sum of all the neglected coefficients as in (2.95) shows that the error in truncating after a_N is approximately $|a_N|/\epsilon$ for $\epsilon \ll 1$. If we demand the truncation error $= 10^{-d}$ so that d is the number of decimal places of accuracy, then we obtain the following.

Rule-of-Thumb 5 (WITCH-OF-AGNESI RESOLUTION ESTIMATE)
If a function has an internal boundary layer of width $2\,\epsilon$ due to poles at $x = \pm i\,\epsilon$, then the value of N needed to reduce the truncation error for that function to 10^{-d} is roughly

$$N \approx \left(\frac{d}{\epsilon}\right) \log(10) \qquad \epsilon \ll 1 \qquad (2.109)$$

What this rule-of-thumb asserts is that for more realistic functions — with singularities — the slope of the error will asymptote to a straight line on a log/linear plot as shown in Fig. 2.20 rather than diving more steeply off the bottom of the graph, as true for the entire function $\sin(x)$.

By "width of the internal boundary layer", we mean that the spoor of the complex singularities is that the witch-of-Agnesi function rises from $1/2$ to 1 and then falls to $1/2$ again on the small interval $[-\epsilon, \epsilon]$ as shown on the inset in Fig. 2.20. Presumably, some small-scale features in hydrodynamic flows (or whatever) are similarly associated with poles or branch points for complex x close to the center of the feature.

A couple of *caveats* are in order. First, singularities a distance ϵ from the real axis are most damaging to convergence when near the *center* of the interval. As one moves to $x = \pm 1$, the pole becomes progressively less harmful until N becomes a function of $1/\sqrt{\epsilon}$ — not $1/\epsilon$ — in the immediate neighborhood of the endpoints. This special endpoint effect is described in the next section, but it significantly modifies (2.109) only when the branch point or pole is quite close to $x = \pm 1$. (Observe the equiconvergence ellipses in Fig. 2.16; the innermost curves (small ϵ) are very eccentric and approximately parallel the real axis except in the small, semi-circular caps around ± 1.)

Second *caveat*: Eq.(2.108) should be multiplied by a factor of n^k when the singularities are other than simple poles where $k = 1$ for a second order pole (stronger singularity) and $k = -1$ for a logarithm (weaker singularity, the integral of a simple pole). In this case, one can show that $N(d, \epsilon)$ should be determined from the more complex rule

$$N\epsilon + k \log(N) \approx d \log(10) \qquad (2.110)$$

Often, however, especially when d is large, the logarithm of N may be neglected. Usually the simpler rule is best because one rarely knows the precise type or location of a complex singularity of the solution to a differential equation; one merely *estimates* a scale, ϵ.

2.16   Boundary Layer Rule-of-Thumb

Viscous flows have boundary layers next to solid surfaces where the tangential velocity is reduced to zero. When the boundary layer is of width ϵ where $\epsilon \ll 1$, it is normally necessary to increase the number of grid points as $O(1/\epsilon)$ as $\epsilon \to 0$. This is extremely expensive for high Reynolds number flows.

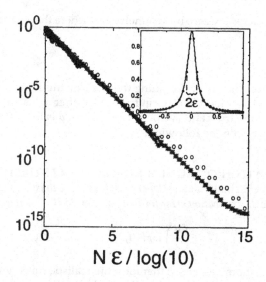

Figure 2.20: Witch-of-Agnesi Rule-of-Thumb. This predicts that if a function has a peak in the *interior* of the interval of width 2ϵ due to singularities at $\Im(x) = \epsilon$, then the error will be 10^d where $d = N\,\epsilon/log(10)$. The errors in the Chebyshev approximation of $f(x) \equiv \epsilon^2/(x^2 + \epsilon^2)$ is plotted versus $N\,\epsilon/\log(10)$, i. e., against the expected number of decimal digits in the error for ϵ varying from 1 (circles) to $1/100$. As predicted, the curves lie approximately on top of one another (with some deviation for $\epsilon = 1$. When $N\,\epsilon/\log(10) \approx 15$, the error is indeed $O(10^{-15})$. The inset graph illustrates the "Witch of Agnesi" function itself along with the meaning of the width parameter ϵ.

A much better approach is to use a mapping, that is, a change of variable of the form

$$y = f(x) \tag{2.111}$$

where y is the original, unmapped coordinate and x is the new coordinate. Unfortunately, there are limits: If the mapping varies too rapidly with x, it will itself introduce sharp gradients in the solution, and this will only aggravate the problem. However, the higher the order of the method, the more rapidly the transformation may vary near the boundary. Orszag and Israeli (1974) assert (without giving a proof) that with an "optimal mapping", this change-of-coordinates trick can reduce the error (for a given total number of grid points) by $O(\epsilon^2)$ for a second order method and $O(\epsilon^4)$ for a fourth order scheme.

One effective mapping for the most common case — boundary layers near the endpoints — was employed by Kalnay de Rivas(1972) who used

$$y = \cos(x) \tag{2.112}$$

for a "physical" coordinate $y \in [-1, 1]$. This is precisely the change-of-variable that converts a Chebyshev series in y into a Fourier series in x. It follows that her evenly spaced grid in x is the same grid that we would use with an *un-mapped* Chebyshev series in y. To put it another way, the uneven grid in y that is sometimes such a burden with Chebyshev methods, particularly for time-integration schemes, is in fact just what we need to resolve boundary layers.

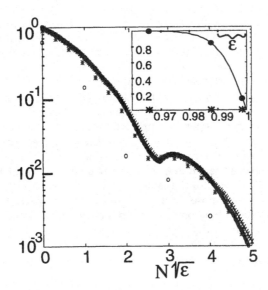

Figure 2.21: Boundary Layer Rule-of-Thumb. The errors in the approximation of $f(x) = \tanh([x-1]/\epsilon)$ are plotted versus $N_{scaled} \equiv N\,\epsilon^{1/2}$ for $\epsilon = 1$ (lower circles), $\epsilon = 1/10$ (asterisks), $\epsilon = 1/100$ (plus signs), $\epsilon = 1/1000$ (x's) and $\epsilon = 1/10000$ (upper circles). Except for $\epsilon = 1$, which is offset a little below the others, the curves for different ϵ are superimposed one atop another. Regardless of the thickness ϵ of the boundary layer, when $N_{scaled} \geq 3$, that is, when the number N of Chebyshev polynomials is greater than $3/\epsilon^{1/2}$, the approximation has at least moderate (1%) accuracy. The inset shows $f(x) = \tanh((x-1)/\epsilon)$ for $\epsilon = 1/100$, together with the grid points (circles) for $N_{scaled} = 3$, i. e., $N = 30$. As long as there are a couple of grid points in the boundary layer, the Chebyshev approximation is okay.

Rule-of-Thumb 6 (BOUNDARY LAYER RESOLUTION REQUIREMENT)

To resolve a boundary layer of width ϵ at $y = \pm 1$ (as opposed to an internal boundary layer, such as a front), one should use approximately

$$N = \frac{3}{\sqrt{\epsilon}} \; polynomials \qquad\qquad \epsilon \ll 1 \qquad\qquad (2.113)$$

to obtain a moderate accuracy of O(1 %).
Source: Orszag & Israeli (1974).

This rule-of-thumb, like all such empirical principles, must always be applied with caution. (See Fig. 2.21.) It is interesting, however, that it agrees exactly with Kalnay's (1972) rule-of-thumb for her second order finite difference method, even to the numerical factor of 3, although hers was expressed in terms of grid points rather than polynomials.

The reason that N is proportional to the square root of $1/\epsilon$, instead of $1/\epsilon$ itself, is that the Chebyshev polynomials have a "built-in" variable stretching which is "quadratic" in the sense that the interior grid points closest to the boundary are only $O(1/N^2)$ away from the endpoint [rather than $O(1/N)$ as for an evenly spaced grid]. This is easily proved by recalling that the Chebyshev grid is the map of an evenly spaced grid in x via $y = \cos x$, and then using the Taylor expansion of the cosine near $x = 0, \pi$.

The conclusion is that with Chebyshev polynomials and Fourier series, the best way to deal with boundary layers may well be to do nothing! (When Fourier series are used, the boundary condition is spatial periodicity; a periodic interval is equivalent to a ring-shaped interval without boundaries or boundary layers.) For some problems, however, the natural, quadratic boundary layer stretching which is, as it were, "built-in" to Chebyshev methods may not be optimum. Then, an additional mapping may be helpful.

For boundary value and eigenvalue problems, it follows that boundary layers are not a major problem. For time-dependent problems, alas, they are a serious complication because the narrow grid spacing near the wall implies that a very small time step is needed with an explicit time-differencing scheme. This is not a problem peculiar to spectral methods, however; Kalnay's finite difference scheme was subject to exactly the same restrictions as a Chebyshev method, and for the same reason.

Thus, boundary layers pose no special problems for spectral methods. Rather, the Chebyshev grid is unfortunate only when boundary layers are *not* present. With Chebyshev polynomials, the high density of grid points near the walls requires either a very small timestep or an implicit time integration even when there is no *physical* reason for high resolution near the endpoints.

Table 2.2: Relationships Between Singularities and Asymptotic Spectral Coefficients. Note: all rows apply to Fourier series as well as Chebyshev and Legendre polynomials except where marked by (*), which exclude trigonometric series.

Form of singularity	Type of singularity	Asymptotic form spectral coefficients
$1/(x-a)$	Simple pole	$[\ ]p^n$
$1/(x-a)^2$	Double pole	$[\ ]n\,p^n$
$\log(x-a)$	Logarithm	$[\ ]n^{-1}\,p^n$
$1/\sqrt{x-a}$	Reciprocal of square root	$[\ ]n^{-1/2}\,p^n$
$(x-a)^{1/3}$	Cube root	$[\ ]p^n/n^{4/3}$
$\exp(-q/|x|)$	Infinitely differentiable but singular at $x=0$	$[\ ]\,\exp(-p\,n^{1/2})$
x^ψ	Branch point within [-1,1]	$[\ ]\,1/n^{\psi+1}$
$f(x)=sign(x)$	Jump discontinuity	$[\ ]\,/n$
$df/dx=sign(x)$	Discontinuous first derivative (assuming f continuous)	$[\ ]\,/n^2$
$(1-x)^\psi$	Branch point at endpoint	$[\ ]\,1/n^{2\,\psi+1}$ *
$\exp(-q/|x+1|)$	Infinitely differentiable but singular at endpoint	$[\ ]\,\exp(-p\,n^{2/3})$

Chapter 3

Galerkin Method, the Mean Weighted Residual Method & Inner Products

"Six months in the lab can save you a day in the library."
   Albert Migliori, quoted by J. Maynard in *Physics Today 49*, 27 (1996)

3.1   Mean Weighted Residual Methods

One difference between algorithms is the "error distribution" principle, that is, the means of making the residual function $R(x; a_0, a_1, \dots, a_N)$ small where

$$R(x; a_0, a_1, \dots, a_N) \equiv Hu_N - f \tag{3.1}$$

where H is the differential or integral operator and $u_N(x)$ is the approximation

$$u_N \equiv \sum_{n=0}^{N} a_n \, \phi_n(x) \tag{3.2}$$

for some suitable basis functions $\phi_n(x)$.

   Finlayson (1973) has pointed out that most methods of minimizing $R(x; \{a_n\})$ can be lumped into a common framework as the "method of mean weighted residuals" (MWR). If the $\phi_n(x)$ already satisfy the boundary conditions, MWR determines the spectral coefficients a_n by imposing the $(N + 1)$ conditions

$$(w_i, R[x; a_0, a_1, \dots, a_N]) = 0, \qquad i = 0, 1, \dots, N \tag{3.3}$$

for some suitable "test functions" $w_i(x)$ where the "inner product" is defined by

$$(u, v) \equiv \int_a^b \omega(x) dx \, u(x) \, v(x) \tag{3.4}$$

61

for a given non-negative weight function $\omega(x)$ and any two functions $u(x)$ and $v(x)$.
Four choices of test functions are popular.

PSEUDOSPECTRAL alias COLLOCATION alias METHOD OF SELECTED POINTS:

$$w_i(x) \equiv \delta(x - x_i) \tag{3.5}$$

where the x_i are some suitable set of "interpolation" or"collocation" points [the two terms
are used interchangeably] and where $\delta(x)$ is the Dirac delta-function. Computationally,
one would impose the condition that

$$R(x_i; a_0, a_1, \dots, a_N) = 0, \qquad i = 1, \dots, N \tag{3.6}$$

for each collocation point.

METHOD OF MOMENTS:

$$w_i(x) \equiv x^i, \qquad i = 0, 1, \dots, N \tag{3.7}$$

The reason for the name is that $(x^i, f(x))$ is said to be, in statistics, the "i-th moment" of
$f(x)$.

This technique has been very popular in engineering. It works well if a small number of
terms (say $N = 2$ or 3) is used. Unfortunately, the method of moments is almost invariably
a disaster if N is large. The difficulty is that high powers of x become linearly dependent
due to round-off error. For this reason, the powers of x are unsuitable as basis functions.
For example, fitting a polynomial to a function will fail when N is greater than 6 if the polynomial is expressed as powers of x. The matrix of this least-squares problem, the infamous
"Hilbert matrix", is a popular test for matrix software because it maximizes the effect of
round-off! The method of moments generates matrices that are just as ill-conditioned.

The method of moments works well if (i) N is very small or (ii) calculations are performed in exact rational arithmetic using a symbolic manipulation language like Maple or
Mathematica. Otherwise, the best way to avoid disastrous accumulation of roundoff error
is reject the method of moments. Instead, use Chebyshev or Legendre polynomials as the
"test" and basis functions.

LEAST SQUARES:
If H is a *linear* operator, then

$$w_i(x) \equiv H\phi_i(x) \tag{3.8}$$

The reason for the name is that one can show that this minimizes the "norm" of the residual
where the norm is defined by

$$\|R\| \equiv \sqrt{(R, R)} \tag{3.9}$$

In other words, "least squares" determines the coefficients $a_0, a_1, \dots, a_N$ in such a way
that the inner product of $R(x; a_0, a_1, \dots, a_N)$ with itself is made as small as possible for the
class of functions defined by an $(N + 1)$-term spectral series. The mathematical statement
of this is

$$(R, R) \leq (Hv - f, Hv - f) \quad \text{for all} \;\; v(x) \;\; \text{of the form} \tag{3.10}$$

$$v(x) = \sum_{n=0}^{N} d_n \phi_n(x) \quad \text{for some} \quad \{d_n\} \tag{3.11}$$

It is possible to make (R, R) smaller only by using more terms in the series, not by using different numerical values of the $(N + 1)$ coefficients.

If the basis functions have the property of completeness — i. e., if we can approximate the exact solution $u(x)$ to arbitrarily high accuracy by using a sufficiently large number of terms in the spectral series, then the "least squares" method is guaranteed to work.

One great advantage of "least-squares" is that it always generates symmetric matrices even if the operator H lacks the property of "self-adjointness". This is helpful in three ways. First, it greatly simplifies the theoretical numerical analysis. Second, symmetric matrix problems can be solved by a variety of special tricks which are faster and use less storage than the algorithms for non-symmetric matrices. Third, symmetric matrices are more stable in time-marching schemes.

It is possible to generalize the "least squares" method to *nonlinear* problems by defining

$$w_i(x) \equiv \partial R / \partial a_i, \quad i = 0, 1, \ldots, N \tag{3.12}$$

which includes (3.8) as a special case.

Despite its many virtues, the "least squares" method is relatively uncommon. For self-adjoint problems — and this includes all of quantum physics and chemistry, for example — the Galerkin method is simpler and has the same virtues: Minimum norm of the residual and a symmetric matrix. *Nonlinear* problems are usually solved by collocation/pseudospectral methods because "least squares" gives a very unwieldy set of algebraic equations.

However, the least squares method is still the safest way of generating approximations that depend *nonlinearly* on the unknowns.

Galerkin METHOD:

$$w_i(x) \equiv \phi_i(x) \tag{3.13}$$

This choice of test function is successful because any well-chosen set of basis functions $\phi_n(x)$ will have all the properties desired of weighting functions, too, including linear independence and completeness. (We will take up the properties of basis sets in the next two sections.)

There are many workable choices of "test functions" $w_i(x)$ besides the four described here. This book, however, will concentrate on the Galerkin method and the pseudospectral/collocation algorithm.

One should also briefly note that terminology is not too well standardized. In this book, "collocation" is a synonym for "pseudospectral". However, the former term is also used for a class of finite element methods whereas "pseudospectral" is applied only when collocation is used with a basis of *global* (Fourier, Chebyshev ...) rather than *local* (tent, pyramid, splines ...) functions.

In addition, some authors such as Birkhoff and Lynch (1984) and Strang and Fix (1973) use "Rayleigh–Ritz" as a synonym for what we have called "Galerkin" and "Galerkin" as a synonym for "mean weighted residual". Finlayson's terminology is preferable since "mean weighted residual" is descriptive in a way that a surname is not.

When the test functions differ from the basis functions, the label "Petrov-Galerkin" is common.

In addition, some authors like Canuto *et al.* (1988) restrict "Galerkin" to basis sets that individually satisfy the boundary conditions while using "tau-method" when $\phi_i(x)$ is a

Chebyshev polynomial. Our reasons for avoiding this usage are partly historical and partly pragmatic. The historical reason is given in Chapter 21. The pragmatic reason is that for some problems, such as those with differential equations which are singular at the endpoints, Chebyshev polynomials do satisfy the boundary conditions. Thus, a tau-method sometimes is a Galerkin method, too. We shall therefore avoid the label "tau–method" and use "Galerkin" to refer to any method which uses basis functions, either before or after modifications, as the "test" functions.

3.2   Completeness and Boundary Conditions

Useful sets of basis functions have a number of properties. First, they must be easy to compute. Trigonometric functions and polynomials both certainly meet this criterion.

A second requirement is *completeness*. This means that the basis functions must be sufficient to represent all functions in the class we are interested in with arbitrarily high accuracy. A rigorous completeness proof is too complicated to discuss here. However, all the basis functions in this book — Fourier series, Chebyshev polynomials, Hermite functions, spherical harmonics and so on — do have the property of completeness.

When explicitly imposing boundary conditions, however, it is possible — indeed, very useful — to use a basis set which is *incomplete* for *arbitrary* functions. For instance, suppose the boundary conditions are homogeneous such as:

$$u(-1) = u(1) = 0 \tag{3.14}$$

We have two options: (i) impose the constraint on the basis functions by adding two rows to the matrix problem we solve for the a_n:

$$\sum_{n=0}^{\infty} a_n T_n(\pm 1) = 0 \tag{3.15}$$

or (ii) choose basis functions that *independently* satisfy the boundary conditions.

EXAMPLE: Chebyshev basis functions that vanish at the endpoints

The set of functions defined by

$$\phi_{2n}(x) \equiv T_{2n} - T_0; \qquad \phi_{2n+1}(x) \equiv T_{2n+1}(x) - T_1, \qquad n \geq 1 \tag{3.16}$$

have the property

$$\phi_n(\pm 1) = 0, \qquad \text{for all } n \tag{3.17}$$

as can be proved by converting the Chebyshev polynomials in (3.16) into the corresponding cosines and evaluating at $\theta = 0, \pi$, which correspond to $x = 1, -1$.

This basis set omits two degrees of freedom since T_0 and $T_1(x)$ are no longer independent, but instead appear only in combination with Chebyshev polynomials of higher degree. However, $u(x)$ is not arbitrary either; it must satisfy the two boundary conditions. Therefore, the basis functions defined by (3.16) *are complete* for the most general function $u(x)$ that *satisfies the boundary conditions* (3.14).

Warning: a common programming error is to set up a test problem whose solution does not have homogeneous boundary conditions. If the basis vanishes at the endpoints and the target function $u(x)$ does not, then the coefficients are meaningless, oscillatory, and rather large even in the N-th coefficient.

This device of taking linear combinations of Chebyshev polynomials (or something else) is extremely useful with *homogeneous* boundary conditions. We no longer have to complicate the programming by using two rows of our matrix equation for boundary conditions; now, all the rows of the matrix come from minimizing the residual $R(x; a_n)$.

Another advantage of basis recombination, noted first by W. Heinrichs, is that the recombined basis functions are better conditioned than the Chebyshev polynomials themselves. That is to say, accumulated roundoff error is much lower with basis recombination, especially when solving high order differential equations. A full explanation is deferred until a later chapter. The short explanation is that high order derivatives of Chebyshev (and Legendre) polynomials grow rapidly as the endpoints are approached. The mismatch between the large values of the derivatives near $x = \pm 1$ and the small values near the origin can lead to poorly conditioned matrices and accumulation of roundoff error. Basis recombination greatly reduces this because the basis functions are tending rapidly to zero near $x = \pm 1$, precisely where the polynomials (and their derivatives) are oscillating most wildly. (However, the ill-conditioning of an uncombined basis is usually not a problem unless $N \sim 100$ or the differential equation has third or higher derivatives.)

3.3 Properties of Basis Functions: Inner Product & Orthogonality

Definition 12 (INNER PRODUCT) *Let $f(x)$ and $g(x)$ be arbitrary functions. Then the inner product of $f(x)$ with $g(x)$ with respect to the weight function $\omega(x)$ on the interval $[a, b]$ is defined by*

$$(f, g) \equiv \int_a^b f(x)g(x)\omega(x)dx \tag{3.18}$$

Definition 13 (ORTHOGONALITY) *A set of basis functions $\phi_n(x)$ is said to be ORTHOGONAL with respect to a given inner product if*

$$(\phi_m, \phi_n) = \delta_{mn}\, \nu_n^2 \tag{3.19}$$

where δ_{mn} is the "Kronecker delta function" defined by

$$\delta_{mn} = \begin{cases} 1, & m = n \\ & \quad \quad [Kronecker\ \delta - function] \\ 0, & m \neq n \end{cases} \tag{3.20}$$

and where the constants ν_n are called "normalization constants".

The great virtue of orthogonality is that it gives a procedure for computing the spectral coefficients. Let the expansion be

$$f(x) = \sum_{n=0}^{\infty} a_n\phi_n(x) \tag{3.21}$$

The inner product of both sides with the m-th basis function is

$$(f, \phi_m) = \sum_{n=0}^{\infty} a_n(\phi_m, \phi_n) \tag{3.22}$$

However, if (and only if) the basis functions are orthogonal, all of the inner products are zero except the single term $n = m$. This proves the following.

Theorem 11 (INNER PRODUCT for SPECTRAL COEFFICIENTS) *If a function $f(x)$ is expanded as a series of orthogonal functions,*

$$f(x) \equiv \sum_{n=0}^{N} a_n \phi_n(x) \tag{3.23}$$

then for arbitrary N,

$$a_n = (\phi_n, f)/(\phi_n, \phi_n) \qquad \forall n \tag{3.24}$$

It is sometimes convenient to eliminate the "normalization constants" $\nu_n \equiv \sqrt{(\phi_n, \phi_n)}$ by rescaling the basis functions by defining

$$\overline{\phi}_N(x) \equiv \phi_n(x)/\nu_n \tag{3.25}$$

Definition 14 (ORTHONORMAL) *A basis set is said to be ORTHONORMAL if the basis functions have been rescaled so that*

$$(\phi_n, \phi_n) = 1 \tag{3.26}$$

This implies that the coefficients of an arbitrary function f(x) are simply

$$a_n = (\phi_n, f) \tag{3.27}$$

The property of orthonormality, as opposed to mere orthogonality, is much more useful for some basis functions than for others. Since the normalization factor is $1/\sqrt{\pi}$ for all Fourier and Chebyshev terms (except the constant), it is usually not worth the bother to "normalize" these basis sets. The simplest definition of Hermite functions (integer coefficients) unfortunately leads to normalization factors which grow wildly with n, so it is quite helpful to use orthonormalized Hermite functions. ("Wildly" is not an exaggeration; I have sometimes had overflow problems with unnormalized Hermite functions!) From a theoretical standpoint, it does not matter in the least whether the basis is orthonormal or not so long as one is consistent.

Orthogonality, however, is always useful. It seems so special that one might think it hard to obtain orthogonal sets of functions, but in fact it is quite easy. All the sets which we shall use are the eigenfunctions of a class of eigenvalue equations known as "Sturm-Liouville" problems. These two 19th century mathematicians proved that all problems of this class generate complete, orthogonal basis sets.

Alternatively, we can choose a set of arbitrary functions. By applying a simple recursive procedure called "Gram-Schmidt" orthogonalization, which is discussed in most linear algebra texts, one can systematically find linear combinations of the chosen functions which are mutually orthogonal. For example, the new basis functions defined by (3.16) are not orthogonal, but can be made so by the Gram-Schmidt process. The orthogonal polynomials in this book — Chebyshev, Legendre, Hermite, and so on — can be described as the result of Gram-Schmidt orthogonalization of the set of powers of x, $\{1, x, x^2, \dots\}$, on various intervals with various weight functions.

Since the properties of Sturm-Liouville eigenfunctions are not important in applications, the theoretical details will be left to Morse and Feshbach(1953) and Carrier and Pearson (1968). However, Sturm-Liouville theory is of enormous *conceptual* importance because it is a general way of obtaining basis functions that are both *complete* and *orthogonal*. Furthermore, the Sturm-Liouville methodology supplies the weight function $\omega(x)$ and the interval $[a, b]$ for each basis set as summarized in Appendix A.

One final comment on orthogonality: as noted earlier, it is important that the basis functions be as different as possible; adding terms that can be closely approximated by a sum of the other basis functions is a waste of time. Obviously, two functions that are very similar will *not* be orthogonal, but instead will have an inner product little different from the inner product of either one with itself.

Orthogonality guarantees that a set of basis functions will be as different from one another as possible; it is a sort of maximization of linear independence.

3.4 Galerkin Method

The residual function $R(x; a_0, a_1, \ldots, a_N)$ can be expanded as a series of basis functions like any other function of x,

$$R(x; a_0, a_1, \ldots, a_N) = \sum_{n=0}^{\infty} r_n(a_0, a_1, \ldots, a_N)\, \phi_n(x) \qquad (3.28)$$

where the coefficients are given by the usual inner product

$$r_n = (R, \phi_n) \qquad (3.29)$$

The Galerkin method employs the "error distribution principle" that $R(x)$ should be small in the sense that the first $(N+1)$ terms of its spectral series are 0. The Fourier and Chebyshev series of smooth functions decrease exponentially fast with n, so all the r_n for $n > N$ will presumably be very, very tiny if N is large enough. Thus, forcing the lower degree r_n to be 0 should make $R(x)$ very, very small over the whole interval. In the limit that $N \to \infty$, $R(x)$ must $\to 0$ and therefore the approximation must converge exponentially fast to the exact solution.

This strategy is identical to the Galerkin method as previously defined since the "weighted residual" conditions $(\phi_n, R) = 0$ are the same as

$$r_n = 0, \qquad n = 0, 1, 2, \ldots, N \qquad \text{[Galerkin method]} \qquad (3.30)$$

in virtue of (3.29).

Because the weighting functions are a complete set, (3.30) must force $R(x) \to 0$ as $N \to \infty$. In contrast to the "method of moments", Galerkin's method is a very robust algorithm; the "test" functions are orthogonal, and therefore are highly linearly independent.

When the basis set is not complete for arbitrary functions, but rather is restricted to basis functions that satisfy homogeneous boundary conditions, one might wonder whether it is legitimate to limit the "test functions" to these modified basis functions, such as the set defined by (3.16). The answer is yes. The reason is that, at least in the limit $N \to \infty$, the residual function $R(x)$ is converging towards the trivial function $f(x) \equiv 0$. Since this limiting function and all its derivatives vanish at the endpoints (and everywhere else!), it follows that $f(x) \equiv 0$ must satisfy any and all reasonable homogeneous boundary conditions that we may have imposed upon the basis functions. Thus, any set of basis functions that are complete for $u(x)$ will also be complete for the residual function $R(x)$.

For simplicity, we assume here and in the next section that the operator H is a linear operator. (This is not as restrictive as it might seem as explained in Appendix C.) In that case, the Galerkin conditions can be expressed in the form of a matrix equation. Let $\vec{\vec{H}}$ denote an $(N+1) \times (N+1)$ square matrix whose elements are given by (3.31), $\vec{f}$ be the column matrix whose elements are given by (3.32), and $\vec{a}$ be the column vector whose elements are the $(N+1)$ spectral coefficients:

$$H_{ij} = (\phi_i, H\phi_j) \qquad i, j = 1, 2, \dots, (N+1) \tag{3.31}$$

$$f_i = (\phi_i, f) \qquad i = 1, 2, \dots, (N+1) \tag{3.32}$$

Then the spectral coefficients a_n are the solution of the matrix equation

$$\vec{\vec{H}} \ \vec{a} = \vec{f} \qquad [\text{Galerkin \quad method}] \tag{3.33}$$

The i-th row of (3.33) is the inner product of $\phi_i(x)$ with the residual function

$$R(x; a_0, a_1, \dots, a_N) \equiv -f(x) + \sum_{j=0}^{N} a_j H\phi_j \tag{3.34}$$

If the basis functions do not individually satisfy the boundary conditions, then it is necessary to replace some of the rows of (3.33) by equations that express the boundary condtions. The endproduct is still a matrix equation like (3.33).

The final step is to solve (3.33) via standard linear algebra libraries such as LINPACK and EISPACK. The *user's* chore is simply to (i) calculate the matrix elements in (3.31) and (3.32) and (ii) evaluate $u_N(x)$ by summing the spectral series. It's that simple.

3.5   Weak & Strong Forms of Differential Equations: the Usefulness of Integration-by-Parts

The "strong" form of a differential equation is the one we all learned as undergraduates: a relationship between derivatives that applies at each point on an interval, for example

$$u_{xx} - q(x)u = -f(x) \tag{3.35}$$

In the mathematical and numerical literature, there are frequent references to the "weak" form, which is based on the Galerkin method.

For simplicity, assume that the differential equation is subject to homogeneous Dirichlet boundary conditions. (Quite general boundary conditions can be used without altering what follows.) Then the "weak" form of Eq.(3.35) is: For all "test" functions $v(x)$ in the appropriate Sobolev space, the solution of the "weak" form of the differential equation is that function $u(x)$ such

$$(v, u_{xx} - q(x)u) = -(v, f) \tag{3.36}$$

where the parentheses denote the usual integral inner product. "Appropriate" means that the test functions must satisfy the boundary conditions.

The weak form is merely Galerkin's method. In practical applications, the Sobolev space would be restricted to polynomials up to and including degree N for both the test functions and the solution.

The usefulness of the weak form arises from a trick: Integration-by-parts. If we multiply by (-1), the weak form can be rewritten without approximation as

$$(v_x, u_x) - \{v(b)u_x(b) - v(a)u_x(a)\} + (v, qu) = (v, f) \tag{3.37}$$

where $x \in [a, b]$. Because the "test" functions $v(x)$ satisfy the homogeneous boundary conditions, the endpoint terms in the braces are zero, simplifying the "weak" form to

$$(v_x, u_x) + (v, qu) = (v, f) \tag{3.38}$$

The reason that this is useful is that although the original differential equation (in its "strong" form) is second order, through integration-by-parts we have found an equivalent statement which involves only *first* derivatives. This has several benefits.

First, the "weak" form is meaningful even for problems with shock waves or other pathologies. As long as the first derivative is *integrable*, even if it has a discontinuity at some point on the interval, the "weak" form is well-defined.

Second, the weak form widens the choice of basis functions. To solve the strong form through collocation or the Galerkin method, we need basis functions that have non-zero second derivatives, so piecewise-linear functions are excluded. Nevertheless, piecewise linear or "tent" functions are very widely used in finite elements and give second order accuracy. "Tent" functions are possible only with the weak form.

The high-brow way of saying this is to assert that the appropriate Sobolev space is H_0^1. The subscript "0" means that the elements of the Sobolev space satisfy homogeneous Dirichlet boundary conditions, that is, $v(a) = v(b)$ for all test and basis functions. The superscript "1" means that the first derivatives of the functions in the space must be in the space L_2, that is, the square of v_x must be *integrable* on the interval $x \in [a, b]$.

The third advantage is important only in multiple dimensions. To solve a problem on an irregularly-shaped domain, one must make a change of coordinates to map the domain from the physical coordinates (x, y) to a square domain in the computational coordinates (r, s). One can then expand the solution in a "tensor product" basis, i. e., in basis functions of the form $T_m(r)T_n(s)$. Unfortunately, the change of coordinates alters the coefficients of the differential equation by inserting "metric factors" that depend on the mapping. The transformation of first derivatives is messy but not too bad. The transformation of second derivatives is *very* messy. (Appendix E, Sec. 9.) The transformation of third and fourth derivatives — the programmer shudders and turns away in horror.

The weak form, however, needs only *first* derivatives even though the original differential equation is of second order. Similarly, a fourth order differential equation can be represented in the weak form using no derivatives higher than second order. With the weak form, it is no longer necessary to drown under a flashflood of the metric factors of high order derivatives.

This is important not only for preserving the sanity of the programmer. Complicated metric factors also greatly increase the operation count and slow the computations.

The fourth advantage is that if the Sobolev space is restricted to the span of a finite number N of linearly independent basis functions, the Galerkin matrix for the weak form is automatically *symmetric*:

$$H_{ij} = (\phi_{x,i}, \phi_{x,j}) + (\phi_i, q(x)\phi_j) \tag{3.39}$$

Because the test functions and basis functions both enter as the first derivative, interchanging the indices i, j does not change the numerical value of the Galerkin matrix for the weak form.

The symmetry of the Galerkin matrix for the weak form has some important implications. First, it explains the relative unpopularity of the least squares method, which also generates symmetric matrices. The Galerkin weak form is also symmetric but much simpler.

Second, symmetric matrices can be solved by special methods which are much cheaper than the more general algorithms that must be used for nonsymmetric matrices. For example, a matrix equation with a symmetric matrix can be solved by a "Cholesky" decomposition (Appendix B), which requires only half the floating point operations and half the memory storage of an LU factorization.

If the matrix equation is solved by an iterative scheme, one can use the "conjugate gradient" method if the matrix is symmetric. Its unsymmetric counterpart is the "biconjugate gradient" method which is roughly twice as expensive per step in both operations and storage, and also more likely to converge very slowly or fail.

Matrix symmetry requires that the differential equation should be "self-adjoint". Fortunately, this is true of many important physical problems.

Lastly, the weak form can be connected with the calculus of variations to justify some unconventional (but cheap and efficient) schemes for patching solutions on separate subdomains together. Consequently, the spectral element method, which is the most popular scheme for applying spectral methods with domain decomposition, almost always is based on the "weak" form rather than collocation.

It turns out, however, that the integrals of the weak form are usually approximated by a Gaussian quadrature scheme which is very closely related to interpolation and the pseudospectral method as explained in the next chapter.

3.6 Galerkin Method: Case Studies

EXAMPLE ONE:

$$u_{xx} - (1/2)u = -(3/2)\cos(x) - (9/2)\cos(2x) \tag{3.40}$$

with periodic boundary conditions. This is in the form $Hu = f$ with

$$H \equiv \partial_{xx} - (1/2); \qquad f(x) = -(3/2)\cos(x) - (9/2)\cos(2x) \tag{3.41}$$

Since the boundary conditions are periodic, Fourier functions are the obvious basis set. We assert without proof that a Fourier cosine series (without the constant) is sufficient for this problem; note that only cosines appear on the R. H. S. [A full discussion is given in Chap. 8 on parity and symmetry.] The matrix equivalent of (3.41) with the basis $\cos(x), \cos(2x)$ is

$$\begin{vmatrix} (\cos[x], H\cos[x]) & (\cos[x], H\cos[2\,x]) \\ (\cos[2x], H\cos[x]) & (\cos[2x], H\cos[2\,x]) \end{vmatrix} \begin{vmatrix} a_1 \\ a_2 \end{vmatrix} = \begin{vmatrix} (\cos[x], f) \\ (\cos[2\,x], f) \end{vmatrix}$$

where

$$(g, h) \equiv \int_{-\pi}^{\pi} dx\, g(x)\, h(x) \tag{3.42}$$

for any two functions $g(x)$ and $h(x)$.

To evaluate the matrix elements, note that

$$\begin{aligned} H\cos(nx) &\equiv [\cos(nx)]_{xx} - (1/2)\cos(nx), & \text{for all } n \\ &= -\{n^2 + 1/2\}\cos(nx), & \text{for all } n \end{aligned} \tag{3.43}$$

and

$$(\cos[mx], \cos[nx]) = \pi\delta_{mn}, \qquad \text{all} \quad m, n > 0 \tag{3.44}$$

Thus, (3.6) becomes

$$
\begin{vmatrix} (-3/2) & 0 \\ 0 & (-9/2) \end{vmatrix} \begin{vmatrix} a_1 \\ a_2 \end{vmatrix} = \begin{vmatrix} (-3/2) \\ (-9/2) \end{vmatrix}
$$

which gives $\vec{a} = (1,1)^T$ [the superscript "T" denotes the "transpose" of the row vector into a column vector] and the solution

$$u_2(x) = \cos(x) + \cos(2x) \tag{3.45}$$

For this carefully contrived example, the approximation is exact.

EXAMPLE TWO:

$$u_{xx} + \{\cos(x) + \cos^2(x)\}u = \exp[-1 + \cos(x)] \tag{3.46}$$

subject to periodic boundary conditions. We assert without proof that a Fourier cosine series is sufficient to solve (3.46). [A rigorous justification can be given through the discussion of "parity" in Chapter 8]. The R. H. S. of (3.46) is also the exact solution, which is the special case $\Delta = 1$ of the expansion

$$\exp\{\Delta[-1 + \cos(x)]\} = \exp(-\Delta)\{I_0(\Delta) + 2\sum_{n=1}^{\infty} I_n(\Delta)\cos(nx)\} \tag{3.47}$$

Let

$$u_3(x) = a_0 + a_1\cos(x) + a_2\cos(2x) + a_3\cos(3x) \tag{3.48}$$

$$H \equiv \partial_{xx} + \cos(x) + \cos^2(x); \qquad f(x) = \exp[-1 + \cos(x)] \tag{3.49}$$

The Galerkin method gives the following 4 x 4 matrix problem:

$$
\begin{vmatrix} (1/2) & (1/2) & \vdots & (1/4) & (0) \\ (1) & (-1/4) & \vdots & (1/2) & (1/4) \\ \cdots & \cdots & \vdots & & \\ (1/2) & (1/2) & & (-7/2) & (1/2) \\ (0) & (1/4) & & (1/2) & (-17/2) \end{vmatrix} \begin{vmatrix} a_0 \\ a_1 \\ \text{---} \\ a_2 \\ a_3 \end{vmatrix} = \begin{vmatrix} f_0 \\ f_1 \\ \text{----} \\ f_2 \\ f_3 \end{vmatrix} \tag{3.50}
$$

As noted earlier, the acid test of the accuracy of any numerical calculation is to repeat it with different truncations. The two-term Galerkin approximation to (3.46) can be obtained by simply truncating (3.50) to the upper halves of the column vectors $\vec{a}$ and $\vec{f}$ — the parts above the dashes — and truncating the square matrix H to its upper left 2 × 2 block, which is the block partitioned from the rest of the matrix by the dotted lines.

The matrix elements in (3.50) are given symbolically by

$$H_{jk} \equiv (\cos(jx), \{-k^2 + \cos(x) + \cos^2(x)\}\cos(kx)) \tag{3.51}$$

$$f_j \equiv (\cos(jx), \exp(-1 + \cos(x))) = \begin{cases} \exp(-1)I_0(1) & [j = 0] \\ 2\exp(-1)I_j(1) & [j \neq 0] \end{cases}$$

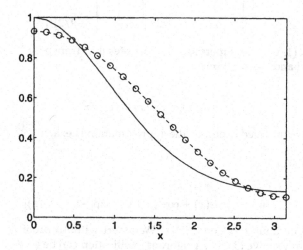

Figure 3.1: Example Two: a comparison between the exact solution (solid) and the two–term approximation (dashed with circles).

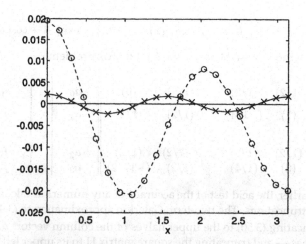

Figure 3.2: Example Two: the errors for the three-term approximation (dashed with circles) and four-term Galerkin solution (solid with x's).

Table 3.1: Comparison of exact and approximate spectral coefficients for EXAMPLE TWO for various truncations, and also the errors in these coefficients.

n	Exact	N=1 (two terms)		N=2 (three terms)		N=3 (four terms)	
	a_n	a_n	Errors	a_n	Errors	a_n	Errors
0	0.4658	0.5190	-0.0532	0.4702	-0.0044	0.4658	0.0
1	0.4158	0.4126	0.0032	0.4126	0.0032	0.4159	-0.0001
2	0.0998	–	–	0.0976	0.0022	0.0998	0.0
3	0.0163	–	–	–	–	0.0162	0.0001
4	0.0020	–	–	–	–	–	–

Table 3.2: Comparison of upper bounds on the discretization error E_D and truncation error E_T with the maximum pointwise error [error in the L_∞ norm].

	[N=1]	[N=2]	[N=3]
E_D(bound)	0.0564	0.0098	0.0002
E_T(bound)	0.1181	0.0183	0.0020
L_∞ error	0.153	0.021	0.0023

$$\text{upper bound on} \quad E_D = \sum_{n=0}^{N} |a_n^{\text{exact}} - a_n^{\text{approx.}}|$$
$$\text{upper bound on} \ E_T = \sum_{n=N+1}^{\infty} |a_n|$$
$$L_\infty \quad \text{error} = \max_{\forall x \in [-\pi, \pi]} |u(x) - u_N(x)|$$

where $j, k = 0, \dots, N$. Numerical evaluation of these integrals then gives (3.50).

The coefficients as calculated with two, three, and four terms are in Table 3.1 along with the differences between these coefficients and those of the exact solution. Table 3.2 of the table shows upper bounds on the discretization error and the truncation error that were calculated by taking the absolute value of each error (or neglected term) and summing. Fig. 3.1 compares the exact solution with the two term approximation. The errors in the higher approximations are so small that it is difficult to see the difference between the curves, so Fig. 3.2 graphs only the absolute errors for $u_2(x)$ and $u_3(x)$. The table and figures support many previous assertions.

First, it was claimed that the discretization error E_D and the truncation error E_T are of the same order-of-magnitude so that it is legitimate to use the latter as an estimate of the total error. Table 3.1 shows that this is true; the total error (sum of E_D plus E_T) is only half again as large as E_T for all three truncations.

Second, Table 3.2 shows that the maximum pointwise error is only a little smaller than the sum of the absolute values of the discretization and truncation errors. Put another way, the Fourier Truncation Theorems give fairly tight bounds on the pointwise error rather than wild overestimates.

Third, we asserted that the last retained coefficient a_N was a loose but useful estimate of the total error. Table 3.2 shows that this is true although rather conservative. For $N = 3$, for instance, the last calculated coefficient is 0.0162, but the maximum error in $u_2(x)$ is about eight times smaller.

Fourth, it was claimed that spectral methods are much more accurate than finite difference methods. Fig. 3.1 shows that even the $N = 1$ (two term) approximation is qualitatively similar to $u(x)$; Fig. 3.2 shows that using just four terms gives a maximum error of only 1 part in 500 relative to the maximum of $u(x)$. It would be quite impossible to obtain anything like this accuracy with finite difference methods. The Quasi-Sinusoidal Rule-of-Thumb suggests that one would need about 20 grid points with a second order method to achieve the same accuracy as the $N = 3$ spectral approximation.

Fifth, Fig. 3.2 shows that the error is *highly uniform* over the interval. The *error oscillates* with several changes of sign, and different peaks of error differ in size by no more than a factor of two for $N = 2$ and are almost identical in size for $N = 3$. These are very general properties of spectral approximations: Basis function methods do not favor one part of the domain over another.

EXAMPLE THREE:

The eigenvalue problem

$$(1 - x^2)\, u_{xx} - \lambda x^2 u = 0 \tag{3.52}$$

where λ is the eigenvalue and where the homogeneous boundary conditions are that $u(x)$ should be *analytic* at $x = \pm 1$ in spite of the singularities of the differential equation at those points. This problem is a special case of Laplace's Tidal Equation for zero zonal wavenumber (Leovy, 1964, & Longuet-Higgins,1968).

We will use the Chebyshev approximation

$$u_2 = a_0 T_0 + a_2 T_2(x) + a_4 T_4(x) \tag{3.53}$$

(The reason for using only the symmetric polynomials is explained in Chapter 8: All the eigenfunctions have either even or odd parity.) We do not need to impose explicit conditions on the approximation because every term in (3.53) is analytic at the endpoints. This situation — holomorphic eigensolutions at endpoints where the differential equation is singular — is the only case where explicit boundary conditions are unnecessary in a problem which is (a) nonperiodic and (b) defined on a finite interval. (The boundary conditions are discussed further below.)

The algebraic equivalent of the differential equation is

$$\begin{vmatrix} (T_0, HT_0) & (T_0, HT_2) & (T_0, HT_4) \\ (T_2, HT_0) & (T_2, HT_2) & (T_2, HT_4) \\ (T_4, HT_0) & (T_4, HT_2) & (T_4, HT_4) \end{vmatrix} \begin{Vmatrix} a_0 \\ a_2 \\ a_4 \end{Vmatrix} = \begin{vmatrix} 0 \\ 0 \\ 0 \end{vmatrix}$$

where the operator H is

$$H \equiv (1 - x^2)(d^2/dx^2) - \lambda x^2 \tag{3.54}$$

The integrals can be evaluated by making the change of variable $x = \cos(\theta)$ and then applying a table of trigonometric integrals, or by numerical integration. The result, dividing by a common factor of π, is

$$\begin{vmatrix} (-\lambda/2) & (2 - \lambda/4) & (4) \\ (-\lambda/2) & (-2 - \lambda/2) & (8 - \lambda/4) \\ (0) & (-\lambda/4) & (-12 - \lambda/2) \end{vmatrix} \begin{Vmatrix} a_0 \\ a_2 \\ a_4 \end{Vmatrix} = \begin{vmatrix} 0 \\ 0 \\ 0 \end{vmatrix}$$

Since this algebraic system of equations is homogeneous, it has a non-trivial solution if and only if the determinant of the matrix is 0, that is, if

$$\lambda\, (-\lambda^2 - 96\lambda - 768) = 0 \tag{3.55}$$

which has the three roots

$$\lambda = 0, -8.808, -87.192 \tag{3.56}$$

One difficulty which is peculiar to eigenvalue problems is to determine which eigenvalues are accurate and which are hopelessly inaccurate for a given truncation N. A full discussion is postponed to Chapter 7, but we anticpate one rule-of-thumb: For a "nice" eigenvalue problem, typically about $N/2$ eigenvalues are accurate.

If N is large, the error in the lowest mode may be only 1 part in 10^{10}, but will increase rapidly with the mode number until there is no accuracy at all for (roughly!) the largest $N/2$ modes. Repeating the calculation with different N is a good idea for any class of problem to *check* the answer; linear eigenvalue problems are the only class where one has to rerun the code simply to know what the answer *is*. (More precisely, to know how many modes one has actually succeeded in computing.)

For this problem, the lowest mode is exact but useless: $\lambda = 0$ corresponds to $u(x) =$ constant, which is a mathematically legitimate solution with no physical significance. The first non-trivial mode has the exact eigenvalue $\lambda_1 = -8.127$, so that the three-term spectral solution has an error of only 7.7%. The second root, -87.192, is an absolutely atrocious approximation to the second mode, $\lambda_3 \approx -12$. However, we can calculate λ_3 as accurately as desired by using a sufficiently large truncation.

Besides the eigenvalue rule-of-thumb, this example is useful in illustrating two additional general principles.

First, this is our first example of a *"behavioral"* (as opposed to *"numerical"*) boundary condition for a non-periodic problem. "Numerical" boundary conditions are those that we must explicitly impose on the numerical approximation to force it or its derivatives to equal certain numbers at the boundaries. The Dirichlet conditions $u(0) = 1, u(1) = -1/2$ are "numerical" conditions. For Laplace's Tidal Equation, however, the differential equation itself forces the solution to behave properly at the boundaries without the need for any intervention on our part. The only requirement is that the numerical solution must be analytical at the endpoints even though the differential equation has singular coefficients there; this behavioral boundary condition is *implicitly* satisfied by a truncated Chebyshev series because these polynomials are analytic at $x = \pm 1$.

Behavioral boundary conditions are the rule rather than the exception in *spherical* or *cylindrical* geometry, or whenever the coordinate system introduces singularities that do not reflect the physics of the problem. (Laplace's Tidal Equation is usually expressed in terms of colatitude ϕ on a sphere; we have converted it to a form suitable for Chebyshev polynomials by making the change-of-variable $x = \cos(\phi)$.) Spherical coordinates are fully discussed with numerical illustrations in Boyd (1978b) and in Chapter 18.

Another example is Bessel's equation, which arises in polar coordinates or cylindrical coordinates. The function J_n has an n-th order zero at the origin, but one can blindly assume a solution in Chebyshev polynomials in radius r, and obtain good results. An approximation table for $J_7(r)$ is given in Gottlieb and Orszag (1977).

A third class of examples is furnished by problems on an unbounded interval. "Behavioral" versus "numerical" boundary conditions for this case are discussed in Chapter 17.

Second, because this is a problem in spherical coordinates, the natural basis set would be spherical harmonics — Legendre polynomials for our special case of zero zonal wavenumber. Nonetheless, we bravely applied Chebyshev polynomials to compute the smallest eigenvalue to within less than 8% error by solving a quadratic equation.

As claimed earlier, Chebyshev polynomials work just fine even when they are not the obvious or the optimum basis set.

3.7   Separation-of-Variables & the Galerkin Method

In the pre-computer age, partial differential equations were solved almost exclusively by "separation–of–variables". While systematizing this procedure, Sturm and Liouville discovered that eigenvalue problems generate eigenfunctions which have all the properties that make them desirable as spectral basis functions. Conversely, one can look backwards and see that "separation–of–variables" is just a special case of spectral methods.

EXAMPLE: Diffusion Equation

$$u_t = u_{xx} \tag{3.57}$$

with boundary conditions of spatial periodicity with a period of 2π and the initial condition $u(x, t = 0) = Q(x)$ where $Q(x)$ is arbitrary. Since the boundary conditions are periodicity, the obvious basis is a Fourier series:

$$u(x,t) = a_0(t) + \sum_{n=1}^{\infty} a_n(t) \cos(nx) + \sum_{n=1}^{\infty} b_n(t) \sin(nx) \tag{3.58}$$

Since the basis functions are individually periodic, all rows of the Galerkin matrix come from the differential equation. The cosine rows are

$$\begin{align}
H_{mn} &= (\cos(mx), \{\partial_t - \partial_{xx}\} \cos(nx)) \tag{3.59}\\
&= (\cos(mx), \{\partial_t + n^2\} \cos(nx)) \tag{3.60}\\
&= \{\partial_t + n^2\} \delta_{mn} \pi \tag{3.61}
\end{align}$$

The sine rows are similar. It is not necessary to bother with a particular truncation because the Galerkin matrix is *diagonal* and each sine or cosine function is uncoupled from all the others. The spectral method in x reduces the problem to the uncoupled, independent ordinary differential equations

$$a_{0,t} = 0; \qquad a_{n,t} + n^2 a_n = 0; \qquad b_{n,t} + n^2 b_n = 0 \tag{3.62}$$

$$a_0(t) = a_0(0); \qquad a_n(t) = a_n(0)\,\exp(-n^2 t); \qquad b_n(t) = b_n(0)\exp(-n^2 t) \tag{3.63}$$

where the values of the coefficients at $t = 0$ are obtained by expanding the initial condition $Q(x)$ as a Fourier series.

We could alternatively solve this same problem with Chebyshev polynomials. The Chebyshev series would also generate a set of ordinary differential equations in time. The difference is that with a basis of anything other than trigonometric functions, all the differential equations would be *coupled*, just as were the pair obtained by Fourier series for the variable coefficient diffusion equation in Chapter 1. Instead of the explicit solutions of (3.63), we are left with a mess.

The heart of "separation–of–variables" is to make a clever choice of basis functions so that the differential equations in the remaining variable are *uncoupled*. For the constant coefficient diffusion equation, sines and cosines are the "clever" choice, independent of the boundary conditions.

Separation-of-variables series often converge at only an algebraic rate. However, this slow convergence frequently reflects corner singularities in space or space-time, so non-separable basis sets like Chebyshev and Legendre polynomials converge algebraically, too.

For more complicated equations, alas, it is necessary to use basis sets more complicated than trigonometric functions to "separate variables" — wierd creatures named Mathieu,

Lamé, and parabolic cylinder functions, to list but a few. We shall spare the reader these exotica. Nevertheless, since most textbooks on the physical sciences written before 1960 are little more than catalogues of successful applications of "separation-of-variables", it is important to recognize that it is but a special case of spectral methods.

3.8 Galerkin Method in Quantum Theory: Heisenberg Matrix Mechanics

In quantum mechanics, one computes the energy levels of atoms and molecules by solving the eigenvalue problem

$$H\psi = E\psi \tag{3.64}$$

where the linear operator H is the "Hamiltonian", E is the energy and also the eigenvalue, and $\psi(x)$ is the wave function. Until very recently, spectral methods were the *only* tool used on Schroedinger's equation. Indeed, the first half-century of wave mechanics is mostly a catalogue of increasingly elaborate applications of spectral methods to calculate energy levels.

Why spectral methods? The first reason is the curse of dimensionality. The wavefunction for the helium atom, which has two electrons, is a function of the coordinates of *both* electrons and thus is *six*-dimensional even though physical space has only three dimensions. An N-electron wavefunction is a native of a $3N$-dimensional configuration space.

The second reason is that one special case, the hydrogen atom, can be solved in closed, analytical form by separation-of-variables. The eigenfunctions are the product of Laguerre functions in radius with spherical harmonics in longitude and latitude; although a bit messy, the eigenfunctions can evaluated by recurrence on a four-function calculator. This is a great gift because the orbits of electrons in molecules and multi-electron atoms are not too much distorted from those of the hydrogen atom. In other words, it is possible to build up good approximations to molecular structure from a small number of hydrogenic wave functions.

The third reason is the Rayleigh-Ritz variational principle. This states that for a linear, self-adjoint eigenvalue problem, the solution for the lowest eigenvalue is that which minimizes the "Rayleigh functional"

$$RR \equiv (\psi, H\psi)/(\psi, \psi) \tag{3.65}$$

(Physicists call the mode with the lowest eigenvalue the "ground state".) If we substitute a *trial* wavefunction ψ_{trial} into $RR(\psi)$, then $E \approx RR$. The special virtue of the Rayleigh-Ritz principle is that one can prove that

$$E \le RR(\psi) \qquad \text{for all } \psi \tag{3.66}$$

In words, any calculation, even a crude one, will always give an *upper bound* on the true energy.

The Rayleigh-Ritz principle allowed the early quantum chemists to search for approximate wave functions with wild abandon. It was rather like breaking an Olympic record; each guess that broke the previous minimum for E was a triumph since the Rayleigh-Ritz principle guaranteed that this *had* to be a better approximation than any wavefunction that had gone before it. In particular, one did not need to worry about completeness, or even linearity. If the trial wavefunction depended nonlinearly on certain parameters, one could still minimize R with respect to those parameters and perhaps obtain a very good approximation.

EXAMPLE: H_2^+ MOLECULAR ION

This is a one-electron ion with two attracting protons which we shall label "A" and "B". Now the ground state for the hydrogen atom is

$$\psi_A = [\text{constant}] \quad \exp[-r_A/a_0] \tag{3.67}$$

where r_A is the distance from the electron to proton A and where a_0 is the "Bohr radius". To simplify notation, we shall ignore the normalizing constants as in (3.67). An obvious approach is to use hydrogen 1s orbitals like (3.67), but to take

$$\psi \approx \psi_A \pm \psi_B \tag{3.68}$$

where ψ_A is given by (3.67) and where ψ_B has the same functional form except that r_A is replaced by r_B, the distance from the electron to the second proton. This trial solution allows for the electron's freedom to orbit either proton or to trace figure-eights. The symmetry — both protons have equal charge — implies that ψ_A and ψ_B must come in with equal amplitude, but we must allow for the possibility of the minus sign. Graphing the Rayleigh functional as a function of the internuclear separation r_N then gives a prediction for E, and simultaneously for the distance between the two protons.

Crude as this approximation is, it correctly predicts negative energy for the plus sign in (3.68); the hydrogen ion is a stable species. The observed internuclear separation distance is such that contours of wavefunction amplitude are nearer the protons in the ion than for the hydrogen atom, so this suggests improving (3.68) by taking ψ_A to be of the same shape as (3.67), but with a re-scaled radial dependence:

$$\psi_A \equiv \exp[-c r_A/a_0] \tag{3.69}$$

and similarly for ψ_B. Minimizing R with respect to both c and r_N gives the lowest energy for $c = 1.24$.

The third improvement is to note that the pull of the other proton tends to distort the hydrogenic orbitals so that they are elongated in the direction of the other positive charge. This "polarity effect" suggests

$$\psi_A = \exp[-c r_A/a_0] + \lambda x \exp[-c r_A/a_0] \tag{3.70}$$

where the x-axis is aligned along the line between the two protons. The result of minimizing the Rayleigh functional with respect to c and λ is shown as a function of r_N in Fig. 3.3, which is taken from Coulson (1961). In Coulson's words, "the true energy differs inappreciably from (iii) [7.7]".

This is a stirring success for an approximation with just three parameters. When more terms are included, as is necessary for more complicated molecules, it becomes a great burden to use nonlinear parameters, so chemists are forced to use approximations that depend linearly on the unknowns, that is, a spectral series. The Rayleigh-Ritz method is then equivalent to the Galerkin. The "valence bond" and "molecular orbital" theories are merely different choices of spectral basis functions. The acronym often used for the latter — L. C. A. O.–M. O. theory — sums up the underlying philosophy of both: "Linear Combinations of Atomic Orbitals - Molecular Orbitals". In other words, the approximations for molecules are assembled using atomic orbitals as the basis functions.

For some types of bonds, the appropriate atomic orbitals are the "s-orbitals" of hydrogen; for others, the p-, d- or higher orbitals are appropriate. For the benzene ring, for example, which consists of six carbon atoms at the corners of a hexagon, the simplest approximation is

$$\psi \approx a_1\psi_1 + a_2\psi_2 + a_3\psi_3 + a_4\psi_4 + a_5\psi_5 + a_6\psi_6 \tag{3.71}$$

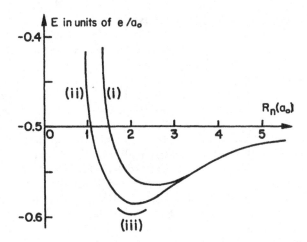

Figure 3.3: A graph of the energy for the hydrogen molecular ion H_2^+, as a function of the internuclear separation R_N in units of the Bohr radius a_0. (i) Sum of two hydrogen orbitals, $\psi = \psi_A + \psi_B$ (ii) Sum of two "screened" hydrogen orbitals which incorporate the screening parameter, c (iii) $\psi_A = \exp(-cr_A/a_0) + \lambda x \exp(-cr_A/a_0)$ and similarly for ψ_B where λ is known as the "polarity" constant. The true energy differs negligibly from the minimum of (iii). After Coulson (1961) and Dickinson(1933).

where the ψ_i's are hydrogen p-orbitals centered on each of the six carbon atoms and where the a_i's are the spectral coefficients. The separation of the carbon atoms, although it can be calculated as for the hydrogen molecular ion, is usually taken as an observational given at this (crude) order of approximation so that the problem is linear in the variational parameters. Minimizing the Rayleigh functional with respect to the a_i's gives the matrix problem

$$
\begin{vmatrix}
(\psi_1, H - E\psi_1) & (\psi_1, H - E\psi_2) & \cdots & (\psi_1, H - E\psi_6) \\
(\psi_2, H - E\psi_1) & \cdots & \cdots & \cdots \\
\vdots & & & \\
(\psi_6, H - E\psi_1) & \cdots & \cdots & (\psi_6, H - E\psi_6)
\end{vmatrix}
\begin{vmatrix}
a \\
a_2 \\
\\
a_6
\end{vmatrix}
=
\begin{vmatrix}
0 \\
0 \\
\\
0
\end{vmatrix}
\tag{3.72}
$$

Chemists, for historical reasons, call determinants of this form "secular determinants", but the matrices are the same as those created by the Galerkin method. Eq. (3.72) is a standard matrix eigenvalue equation with E as the eigenvalue.

As Fig. 3.3 shows, heuristic guessing of basis functions which are *not* standard basis functions can be very effective when coupled with the Rayleigh-Ritz principle and an appropriate measure of desperation. This strategy is a good way of generating low-order analytical approximations in other fields, too, but it can be risky. With the Rayleigh-Ritz principle, a bad guess is immediately evident: its lowest energy is larger than that of previous calculations. Victory is as clear and well-defined as breaking a record in track-and-field.

In other scientific areas, this strategy has also been tried, but the risks are much higher. In the absence of the Rayleigh-Ritz principle, guessing trial functions may make Galerkin and collocation approximations worse instead of better (Finlayson, 1973).

Fortunately, it is possible to regain security of the Rayleigh-Ritz principle. The remedy is to minimize the square of the residual as explained in Appendix G of the first edition of this book (1989). The cost is that the programming is more complicated and the execution time and storage are somewhat increased.

3.9   The Galerkin Method Today

In the next chapter, we will discuss pseudospectral/collocation algorithms, so it is a little premature to compare different methods. However, a few preliminary remarks are in order.

First, the Galerkin method is more difficult to program than pseudospectral algorithms. The reason is simply that it is easier to *evaluate* a function than it is to *integrate* it.

It is often possible to replace the numerical integration by use of trigonometric identities, but this makes it more difficult to test the program because a simple problem with a known, exact solution may not use all the identities which are needed to solve more complex equations (O'Connor, 1996). In contrast, a collocation program (or Galerkin method with numerical evaluation of the matrix elements) does not use special tricks, and therefore will either succeed or fail for both the test problem and the real one.

Second, the Galerkin method with a truncation of N is usually equal in accuracy to a pseudospectral method with a truncation of $N+1$ or $N+2$. When N is large, this difference is negligible, and pseudospectral algorithms are better. When N is small, however, that extra one or two degrees of freedom can be very important.

We have already seen that spectral methods are often accurate even for N so small that one can substitute paper-and-pencil for a computer. Thus, spectral methods are not merely a tool for *numerical* analysis, but for *theoretical* analysis, too. For theoretical purposes, the smallness of N usually implies that one needs that extra accuracy given by the Galerkin method. Furthermore, if a problem is simple enough so that one can analytically perform the linear algebra, one can usually analytically evaluate the integrals for the Galerkin matrix elements, too. Finlayson (1973) is full of such examples, drawn mostly from engineering. A full discussion with programs in the algebraic manipulation languages REDUCE and Maple is given in Chapter 20 and Boyd (1993).

Another exception arises in time-marching. Implicit methods require solving a boundary value problem at each step. However, if only some of the terms are treated implicitly so as to give a "semi-implicit" algorithm, the boundary value problem may be *linear* and *constant coefficient*. For such problems, as explained in Chapter 15, the Galerkin matrix is *banded* and Galerkin's method is much faster than the pseudospectral scheme. Galerkin's method is an important component of spherical harmonic weather forecasting and climate models for this reason.

In addition to these exceptions, quantum chemistry is an area where the Galerkin method is still widely used, and for all the right reasons. Complex molecules require so many degrees of freedom that most of the computations must be done by machine. However, one is usually forced to use a small number of basis functions *per electron*, so that is one is in effect carrying out a low order simulation even when $N = 400$! The extra accuracy of Galerkin's method is therefore very important.

Normally, one should use only the standard, simple basis sets like Fourier series and Chebyshev polynomials. However, quantum chemists are quite sensible in using LCAO ("Linear Combinations of Atomic Orbitals") as the trial functions; it would be quite impossible to obtain such good results for the H_2^+ molecular ion with just three degrees of freedom unless the choice of trial solution was guided by physical intuition. For similar reasons explained Chapter 20 on symbolic calculations, it is sometimes better to use polynomials other than Chebyshev for certain types of *analytical, low-order* calculations.

For most high-resolution numerical calculations, however, the best advice is still this: use pseudospectral methods instead of spectral, and use Fourier series and Chebyshev polynomials in preference to more exotic functions.

Chapter 4

Interpolation, Collocation & All That

"In the past several years, there has been extensive activity in both the theory and application of spectral methods. This activity has been mainly concentrated in the area of pseudospectral methods."
— Gottlieb, Hussaini, & Orszag (1984)

4.1   Introduction

The pseudospectral family of algorithms is closely related to the Galerkin method. To show this and to understand the mechanics of pseudospectral methods, it is necessary to review some classical numerical analysis: polynomial interpolation, trigonometric interpolation, and Gaussian integration.

Definition 15 (INTERPOLATION)
An INTERPOLATING approximation to a function $f(x)$ is an expression $P_{N-1}(x)$, usually an ordinary or trigonometric polynomial, whose N degrees of freedom are determined by the requirement that the INTERPOLANT agree with $f(x)$ at each of a set of N INTERPOLATION points:

$$P_{N-1}(x_i) = f(x_i) \qquad i = 1, 2, \ldots, N \qquad (4.1)$$

In the rest of this chapter, we will discuss the choice of interpolation points and methods for computing the interpolant.

A note on terminology: we shall use "collocation points" and "interpolation points" as synonyms. However, "interpolation" has the connotation that $f(x)$, the function which is being approximated by a polynomial, is already a known function. "Collocation" and "pseudospectral" are applied to interpolatory methods for solving differential equations for an unknown function $f(x)$. The label "pseudospectral" is narrower than "collocation" in that the former is rarely used except when the basis functions are global, such as Chebyshev or Fourier functions. Thus, almost any journal article with "pseudospectral" in the title is relevant to this book, but many "collocation" methods are finite element procedures.

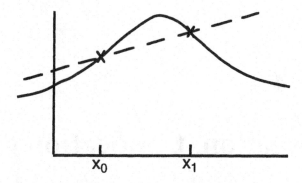

Figure 4.1: Linear interpolation. The dashed line is that linear polynomial which intersects the function being approximated (solid curve) at the two interpolation points.

4.2 Polynomial interpolation

Before hand-held calculators, tables of mathematical functions were essential survival equipment. If one needed values of the function at points which were not listed in the table, one used interpolation. The simplest variant, linear interpolation, is to draw a straight line between the two points in the table which bracket the desired x. The value of the linear function at x is then taken as the approximation to $f(x)$, i. e.

$$f(x) \approx \frac{(x - x_1)}{(x_0 - x_1)} f(x_0) + \frac{(x - x_0)}{(x_1 - x_0)} f(x_1) \qquad \text{[Linear Interpolation]} \qquad (4.2)$$

Fig. 4.1 is a graphic proof of the high school theorem: a straight line is completely determined by specifying any two points upon it; the linear interpolating polynomial is unique. A more abstract definition is that $P_1(x)$ is that unique linear polynomial which satisfies the two interpolation conditions

$$P_1(x_0) = f(x_0) \qquad ; \qquad P_1(x_1) = f(x_1) \qquad (4.3)$$

Linear interpolation is not very accurate unless the tabulated points are very, very close together, but one can extend this idea to higher order. Fig. 4.2 illustrates quadratic interpo-

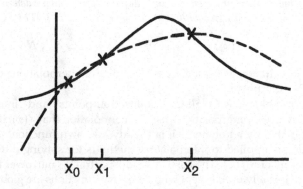

Figure 4.2: Schematic of quadratic interpolation. The dashed curve is the unique quadratic polynomial (i. e., a parabola) which intersects the curve of the function being approximated (solid) at three points.

lation. A parabola is uniquely specified by giving any three points upon it. Thus, we can alternatively approximate $f(x)$ by the quadratic polynomial $P_2(x)$ which satisfies the *three* interpolation conditions

$$P_2(x_0) = f(x_0) \quad ; \quad P_2(x_1) = f(x_1) \quad ; \quad P_2(x_2) = f(x_2) \tag{4.4}$$

$$\begin{aligned}P_2(x) \equiv{}& \frac{(x-x_1)(x-x_2)}{(x_0-x_1)(x_0-x_2)}f(x_0) + \frac{(x-x_0)(x-x_2)}{(x_1-x_0)(x_1-x_2)}f(x_1) \\ {}+{}& \frac{(x-x_0)(x-x_1)}{(x_2-x_0)(x_2-x_1)}f(x_2)\end{aligned} \tag{4.5}$$

In general, one may fit any $N+1$ points by a polynomial of N-th degree via

$$P_N(x) \equiv \sum_{i=0}^{N} f(x_i)\,C_i(x) \qquad \text{[Lagrange Interpolation Formula]} \tag{4.6}$$

where the $C_i(x)$, the "cardinal functions", are polynomials of degree N which satisfy the conditions

$$C_i(x_j) = \delta_{ij} \tag{4.7}$$

where δ_{ij} is the Kronecker δ-function. The cardinal functions are defined by

$$C_i(x) = \prod_{j=0,j\neq i}^{N} \frac{x-x_j}{x_i-x_j} \qquad \text{[Cardinal Function]} \tag{4.8}$$

The N factors of $(x-x_j)$ insure that $C_i(x)$ vanishes at all the interpolation points *except* x_i. (Note that we omit the factor $j = i$ so that $C_i(x)$ is a polynomial of degree N, not $(N+1)$.) The denominator forces $C_i(x)$ to equal 1 at the interpolation point $x = x_i$; at that point, every factor in the product is $(x_i - x_j)/(x_i - x_j) = 1$. The cardinal functions are also called the "fundamental polynomials for pointwise interpolation", the "elements of the cardinal basis", the "Lagrange basis", or the "shape functions".

The cardinal function representation is not efficient for computation, but it does give a proof-by-construction of the theorem that it is possible to fit an interpolating polynomial of any degree to any function. Although the interpolating points are often evenly spaced — surely this is the most obvious possibility — no such restriction is inherent in (4.6); the formula is still valid even if the $\{x_i\}$ are unevenly spaced or out of numerical order.

It seems plausible that if we distribute the interpolation points evenly over an interval $[a, b]$, then the error in $P_N(x)$ should $\rightarrow 0$ as $N \rightarrow \infty$ for any smooth function $f(x)$. At the turn of the century, Runge showed that this is not true.

His famous example is

$$f(x) \equiv \frac{1}{1+x^2} \qquad x \in [-5, 5] \tag{4.9}$$

Runge proved that for this function, interpolation with evenly spaced points converges only within the interval $\mid x \mid \leq 3.63$ and diverges for larger $\mid x \mid$ (Fig. 4.3). The 15-th degree polynomial does an excellent job of representing the function for $\mid x \mid \leq 3$; but as we use more and more points, the error gets worse and worse near the endpoints. The lower right panel of Fig. 4.3, which plots the logarithm (base 10) of the error for the 30-th degree polynomial, makes the same point even more strongly.

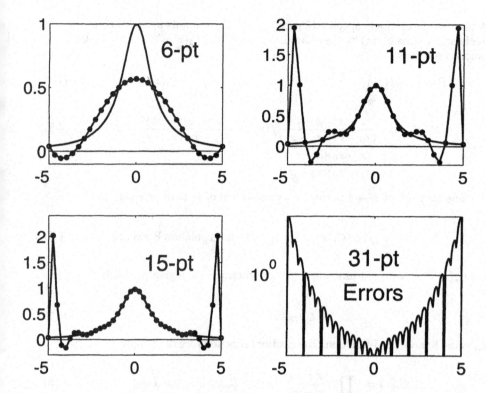

Figure 4.3: An example of the Runge phenomenon.
a. Solid curve without symbols: $f(x) \equiv 1/(1 + x^2)$, known as the "Lorentzian" or "witch of Agnesi". Disks-and-solid curve: fifth-degree polynomial interpolant on $x \in [-5, 5]$. The six evenly spaced interpolation points are the locations where the dashed and the solid curves intersect.
b. Interpolating polynomial [disks] of tenth degree.
c. The interpolating polynomial of degree *fifteen* is the dashed curve. d. Same as previous parts except that only the error, $\log_{10}(|f(x) - P_{30}|)$, is shown where $P_{30}(x)$ is the polynomial of degree *thirty* which interpolates $f(x)$ at 31 evenly spaced points on $x \in [-5, 5]$. Because the error varies so wildly, it is plotted on a logarithmic scale with limits of 10^{-3} and 10^2.

In the interior of the interval, the error is only 1 /1000. Just inside the endpoints, however, $P_{30}(x)$ reaches a maximum value of 73.1 even though the maximum of $f(x)$ is only 1! It is not possible to graph $f(x)$ and $P_{30}(x)$ on the same figure because the curve of $f(x)$ would be almost invisible; one would see only the two enormous spikes near the endpoints where the polynomial approximation goes wild.

Since $f(x) = 1/(1 + x^2)$ has simple poles at $x = \pm i$, its power series converges only for $|x| \leq 1$. These same singularities destroy the polynomial interpolation, too; the wonder is not that Lagrangian interpolation fails, but that it succeeds over a much larger interval than the Taylor series.

This in turn hints that the situation is not hopeless if we are willing to consider an *uneven* grid of interpolation points. The lower panels of Fig. 4.3 show that, as Runge proved, the *middle* of the interval is *not* the problem. The *big errors* are always near the *endpoints*. This suggests that we should space the grid points relatively far apart near the middle of the interval where we are getting high accuracy anyway and increase the density of points as we approach the endpoints.

Unfortunately, as its degree N increases, so does the tendency of a polynomial of degree N to oscillate wildly at the endpoints. Consequently, the enpoint squeezing must increase with N. The Chebyshev interpolation procedure, which does indeed make Lagrangian interpolation successful for any function which is analytic on the interval, has the grid points only $O(1/N^2)$ apart near the endpoints versus the $1/N$ spacing of our (failed!) effort to approximate Runge's example with an evenly spaced grid.

But what distribution of points is best? The answer is given by a couple of very old theorems.

Theorem 12 (CAUCHY INTERPOLATION ERROR THEOREM)

Let $f(x)$ have at least $(N + 1)$ derivatives on the interval of interest and let $P_N(x)$ be its Lagrangian interpolant of degree N. Then

$$f(x) - P_N(x) = \frac{1}{[N+1]!} f^{(N+1)}(\xi) \prod_{i=0}^{N} (x - x_i) \qquad (4.10)$$

for some ξ on the interval spanned by x and the interpolation points. The point ξ depends on the function being approximated, upon N, upon x, and upon the location of the interpolation points.

PROOF: Davis (1975).

If we want to optimize Lagrangian interpolation, there is nothing we can do about the $f^{(N+1)}(\xi)$ factor (in general) since it depends on the specific function being approximated, but the magnitude of the polynomial factor depends upon our choice of grid points. It is evident that the coefficient of x^N is 1, independent of the grid points, so the question becomes: What choice of grid points gives us a polynomial (with leading coefficient 1) which is as small as possible over the interval spanned by the grid points? By a linear change of variable, we can always rescale and shift the interval $[a, b]$ to $[-1, 1]$, but what then? Ironically, this question was answered half a century before the Runge phenomenon was discovered.

Theorem 13 (CHEBYSHEV MINIMAL AMPLITUDE THEOREM)

Of all polynomials of degree N with leading coefficient [coefficient of x^N] equal to 1, the unique polynomial which has the smallest maximum on $[-1, 1]$ is $T_N(x)/2^{N-1}$, the N-th Chebyshev polynomial divided by 2^{N-1}. In other words, all polynomials of the same degree and leading coefficient unity satisfy the inequality

$$\max_{x \in [-1,1]} |P_N(x)| \geq \max_{x \in [-1,1]} \left| \frac{T_N(x)}{2^{N-1}} \right| = \frac{1}{2^{N-1}} \qquad (4.11)$$

PROOF: Davis (1975, pg. 62).

Now any polynomial of degree N can be factored into the product of linear factors of the form of $(x - x_i)$ where x_i is one of the roots of the polynomial, so in particular

$$\frac{1}{2^N} T_{N+1}(x) \equiv \prod_{i=1}^{N+1} (x - x_i) \qquad (4.12)$$

To minimize the error in the Cauchy Remainder Theorem, the polynomial part of the remainder should be proportional to $T_{N+1}(x)$. This implies that the OPTIMAL INTERPOLATION POINTS are the ROOTS of the CHEBYSHEV POLYNOMIAL of DEGREE $(N+1)$.

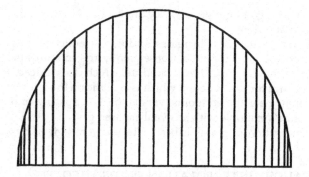

Figure 4.4: Graphical construction of the *unevenly* spaced Chebyshev interpolation grid. If a semicircle of unit radius is cut into evenly spaced segments, and then vertical lines are drawn from these "Fourier" gridpoints to the line segment [−1, 1], which is the base of the semicircle, the vertical lines will intersect the horizontal line at the Chebyshev grid points. The polar coordinates of the grid points on the semicircle are unit radius and angle $\theta = \pi(2i - 1)/(2N)$ where $i = 1, \ldots, N$. The Chebyshev grid points are $x_i = \cos(\theta_i)$.

Since the Chebyshev polynomials are just cosine functions in disguise, these roots are

$$x_i \equiv -\cos\left[\frac{(2i-1)\pi}{2(N+1)}\right] \qquad i = 1, 2, \ldots, N+1 \qquad (4.13)$$

By Taylor expansion of the cosine function, we verify the assertion made above that the grid spacing is $O(1/N^2)$ near the endpoints:

$$x_1 \approx -1 + \frac{\pi^2}{8N^2} \qquad ; \qquad x_2 \approx -1 + \frac{9\pi}{8N^2} \qquad [N \gg 1] \qquad (4.14)$$

and similarly near $x = 1$.

The grid is illustrated in Fig. 4.4.

A number of important questions still remain. First, how accurate is Chebyshev interpolation? Does it converge over as wide a region as the usual Chebyshev expansion in which we compute a polynomial approximation by *integration* instead of *interpolation*? Second, many problems (such as those in spherical coordinates or an infinite domain) require something exotic like spherical harmonics or Hermite functions. What interpolation scheme is best for those geometries? Third, it is obvious that (4.13) defines a choice of interpolation points that we can use to solve differential equations via the pseudospectral method, but what is the relationship, if any, between such an algorithm and the Galerkin method?

In the next section, we generalize the choice of a best grid to other basis sets appropriate for other geometries. In later sections, we will return to the Chebyshev and Fourier methods and show that the price we pay for interpolation instead of integration is modest indeed.

4.3   Gaussian Integration & Pseudospectral Grids

The reason that "collocation" methods are alternatively labelled "pseudospectral" is that the optimum choice of the interpolation points, such as (4.13) for Chebyshev polynomials, makes collocation methods *identical* with the Galerkin method if the inner products are evaluated by a type of numerical quadrature known as "Gaussian integration".

Numerical integration and Lagrangian interpolation are very closely related because one obvious method of integration is to fit a polynomial to the integrand $f(x)$ and then integrate $P_N(x)$. Since the interpolant can be integrated exactly, the error comes entirely from the difference between $f(x)$ and $P_N(x)$. The standard textbook formulas are of the form

$$\int_a^b f(x)\,dx \approx \sum_{i=0}^N w_i f(x_i) \tag{4.15}$$

The weight functions w_i are given by

$$w_i \equiv \int_a^b C_i(x)\,dx \qquad \text{["quadrature weight"]} \tag{4.16}$$

where the $C_i(x)$ are the cardinal functions on the set of points $\{x_i\}$ as defined by (4.8) above. As if there were not already enough jargon, the interpolation points are "abscissas" and the integrals of the cardinal functions are "quadrature weights" in the context of numerical integration.

The *order* of a numerical method is directly related to the highest polynomial for which it is exact. For example, the usual three-point and five-point centered difference formulas (Chapter 1) are exact when applied to general second-degree and fourth-degree polynomials, respectively, so the errors are $O(h^2)$ and $O(h^4)$, respectively. Similarly, a quadrature formula with $(N+1)$ points will be exact if the integrand is a polynomial of degree N.

Gauss made the observation that an evenly spaced grid is nothing sacred. In particular, if we allow the interpolation points $\{x_i\}$ as well as the weights $\{w_i\}$ to be unknowns, we have twice as many parameters to choose to maximize the accuracy of the method, and we can make it exact for polynomials of degree $(2N+1)$.

Theorem 14 (GAUSS-JACOBI INTEGRATION)

If the $(N+1)$ "interpolation points" or "abscissas" $\{x_i\}$ are chosen to be the zeros of $P_{N+1}(x)$ where $P_{N+1}(x)$ is the polynomial of degree $(N+1)$ of the set of polynomials which are orthogonal on $x \in [a, b]$ with respect to the weight function $\rho(x)$, then the quadrature formula

$$\int_a^b f(x)\,\rho(x)\,dx = \sum_{i=0}^N w_i\,f(x_i) \tag{4.17}$$

is exact for all $f(x)$ which are polynomials of at most degree $(2N+1)$.

PROOF: Davis (1975, pg. 343)

As noted earlier, we can take almost any smooth function $\rho(x)$, insert it into the integral inner product, and then apply Gram-Schmidt orthogonalization to 1, x, x^2, ... to create an infinite set of polynomials which are orthogonal with respect to the inner product

$$(f, g) = \int_a^b f(x)\,g(x)\,\rho(x)\,dx \tag{4.18}$$

The Chebyshev, Legendre, Gegenbauer, Hermite, and Laguerre polynomials merely correspond to different choices of weight functions and of the interval $[a, b]$.

The general theory of orthogonal polynomials shows that the $(N+2)$-th member of a family of polynomials is always of degree $(N+1)$ and always has exactly $(N+1)$ *real* zeros *within* the interval $[a, b]$. We might worry about the possibility that, for exotic basis sets like Hermite functions, the zeros might be complex, or otherwise outside the physical domain, but this fear is groundless. The only difficulty is that the roots are not known in closed form for general N except for Chebyshev polynomials or trigonometric interpolation. However, the roots and weights w_i for various N and various weight functions $\rho(x)$ can be found in most mathematical handbooks such as Abramowitz and Stegun (1965) and also are easily calculated by the FORTRAN routines in the appendix of Canuto *et al.*(1988). The Legendre-Lobatto roots up to degree N are known in analytical form and are given in Appendix F, Sec. F.8.

Theorem 15 (PERIODIC GAUSSIAN QUADRATURE)

The "Composite Trapezoidal Rule" and "Composite Midpoint Rule", which both use abscissas which are uniformly spaced, are Gaussian quadratures in the sense that these formulas are exact with N points for TRIGONOMETRIC polynomials of degree $2N - 2$, that is, for polynomials composed of the constant plus the first $(N - 2)$ cosines plus the first $(N - 2)$ sines.

PROOF: Implicit in the discrete orthogonality relationships between the sines and cosines given in the next section.

The Trapezoidal Rule and Midpoint Rule are described in introductory numerical analysis courses as simple but crude quadrature schemes with an accuracy of $O(1/N^2)$. This is a libel, true only when integrating a NON-PERIODIC $f(x)$. For PERIODIC functions, these numerical integration schemes are GAUSSIAN QUADRATURES with an error which decreases GEOMETRICALLY FAST with N for $f(x)$ which are analytic for real x.

What does all this have to do with interpolation and the pseudospectral method? The answer is that since quadrature formulas are obtained by analytical integration of the interpolating polynomial, it follows that the best choice of quadrature points is also the best choice of interpolation points and vice-versa. This implies the following principle that applies to Fourier series, Chebyshev, Legendre and Gegenbauer polynomials, Hermite and Laguerre functions and so on.

Rule-of-Thumb 7 (CHOICE OF PSEUDOSPECTRAL GRIDS)

The grid points should be the abscissas of a Gaussian quadrature associated with the basis set.

The vague phrase "a Gaussian" quadrature is necessary because there are actually two useful Gaussian quadratures associated with each basis. In the Fourier case, one can use either the Trapezoidal Rule or the Midpoint Rule; which are equally accurate. The Trapezoidal Rule, as used in the next section, is the default choice, but the Midpoint Rule (alias Rectangle Rule), which does not include either $x = 0$ or $x = 2\pi$ as grid points, is convenient when solving differential equations which have singularities at either of these endpoints.

Lobatto showed that one could obtain a numerical integration formula with all the good properties of Gaussian quadrature by using the zeros of the first derivative of a Chebyshev or Legendre polynomial plus the endpoints $x = \pm 1$ as the quadrature points. This "Lobatto" or "endpoints-plus-extrema" grid is now more popular than the "Gauss" or "roots" grid for solving boundary value problems because the boundary conditions instantly furnish two grid point values for the unknown $u(x)$. When the differential equation is singular at the endpoints — for example, in cylindrical coordinates, differential equations usually have singularities at $r = 0$ even when the solution is perfectly smooth and well-behaved at the origin — then the "roots" grid is preferable . There is little difference between the Trapezoidal and Midpoint Rule, or between the Gauss and Lobatto grids, in terms of accuracy or other theoretical properties.

The one mild exception to this Rule-of-Thumb is that weather forecasting models, which use a spherical harmonics basis, usually employ a single latitudinal grid for all zonal wavenumbers. The spherical harmonics for a given zonal wavenumber m are Gegenbauer polynomials whose order is $m + 1/2$. Technically, the optimum grid for a given m is a function of m. However, using different grids for different m would entail enormous amounts of interpolation from one grid to another to calculate the nonlinear terms where different wavenumbers m interact. So, a single grid is employed for all m. However, this grid is the Gauss grid for the Legendre polynomials, so interpreted broadly, the rule applies to this case, too.

4.4 Pseudospectral: Galerkin Method via Gaussian Quadrature

The integrals that define the matrix elements in the Galerkin method usually must be evaluated by numerical quadrature. What scheme is most efficient? The answer is Gaussian quadrature, using the roots of the very same orthogonal polynomials (or trigonometric functions) that are the basis functions. This choice connects the collocation and Galerkin methods.

Theorem 16 (GALERKIN's with QUADRATURE)

If the matrix elements are evaluated by $(N+1)$-point Gaussian quadrature, then for linear problems the spectral coefficients $\{a_n\}$ in

$$u_{N,G}(x) = \sum_{n=0}^{N} a_n \, \phi_n(x) \tag{4.19}$$

as calculated by the Galerkin method will be identical with those computed by the collocation method using the same number of terms in the series and the same set of interpolation points.

It is because of this close relationship between the collocation and Galerkin methods that collocation-at-the-Gaussian-quadrature-abscissas is also known as the "pseudospectral" method. (Historically, before the development of collocation algorithms, "spectral" was a synonym for Galerkin.)

PROOF: Let the pseudospectral matrix equation be $\vec{\vec{L}}\,\vec{a} = \vec{f}$. The matrix elements are

$$L_{ij} \equiv L\,\phi_j(x_i) \qquad ; \qquad f_i \equiv f(x_i) \tag{4.20}$$

where L (without vector symbol) is the operator of the differential or integral equation and $f(x)$ is the forcing function. A single row of the matrix equation is then

$$\sum_{j=0}^{N} L_{ij}\, a_j = f_i \tag{4.21}$$

or written out in full,

$$\sum_{j=0}^{N} [L\,\phi_j(x_i)]\, a_j = f(x_i) \tag{4.22}$$

Multiply each row of (4.22) by $w_i\, \phi_n(x_i)$ for some particular choice of n and then add. (For simplicity, assume the basis functions are orthonormal.) The constants w_i are the Gaussian quadrature weights. This gives

$$\sum_{j=0}^{N} H_{nj}\, a_j = g_n \tag{4.23}$$

where

$$H_{nj} \equiv \sum_{i=0}^{N} w_i \, \phi_n(x_i) \, [L \, \phi_j(x_i)] \tag{4.24}$$

But this is simply the Galerkin matrix element

$$H_{nj} \approx (\phi_n, \, L \, \phi_j) \tag{4.25}$$

if the inner product integral is evaluated using Gaussian quadrature with $(N + 1)$ points. Similarly,

$$g_n \equiv \sum_{i=0}^{N} w_i \, \phi_n(x_i) \, f(x_i) \tag{4.26}$$

$$g_n \approx (\phi_n, \, f) \tag{4.27}$$

which is the n-th row of the column vector on the R. H. S. of the Galerkin matrix equation, $\vec{\vec{H}} \, \vec{a} = \vec{g}$.

If some rows of the matrix are used to impose boundary conditions, then the proof is only slightly modified. The boundary rows are the same for either the Galerkin or pseudospectral methods. With two boundary rows, for example, we impose only $(N - 1)$ collocation conditions. We then add those collocation rows using the weights w_i appropriate to Gaussian integration with $(N - 1)$ points, and again reproduce the corresponding $(N - 1)$ rows of the Galerkin matrix. Q. E. D.

Theorem 16 implies that collocation — with the right set of points — must inherit the aura of invincibility of the Galerkin method.

Theorem 17 (ORTHOGONALITY under the DISCRETE INNER PRODUCT)
  If a set of $(N + 1)$ basis functions are orthogonal under the integral inner product

$$(\phi_i, \, \phi_j) = \delta_{ij} \tag{4.28}$$

where δ_{ij} is the Kronecker δ-function, then they are still orthogonal with respect to the discrete inner product

$$(\phi_i, \, \phi_j)_G = \delta_{ij} \tag{4.29}$$

where the discrete inner product is defined by

$$(f, \, g)_G \equiv \sum_{i=0}^{N} w_i \, f(x_i) \, g(x_i) \tag{4.30}$$

for arbitrary functions $f(x)$ and $g(x)$, that is, where the discrete inner product is the $(N+1)$-point Gaussian quadrature approximation to the integral inner product.

PROOF: The product of two polynomials, each of degree at most N, is of degree at most $2N$. Gaussian quadrature with $(N + 1)$ points, however, is exact for polynomials of degree $(2N + 1)$ or less. Therefore, Gaussian quadrature evaluates without error the integrals that express the mutual orthogonality of the first $(N + 1)$ basis functions among themselves. Q.E.D.

This second theorem implies that the reward for exact evaluation of inner product integrals, instead of approximation by Gaussian quadrature, is modest indeed. If we write

$$f(x) = \sum_{n=0}^{N} a_n \, \phi_n(x) + E_T(N; x), \tag{4.31}$$

then the property of exponential convergence implies that for large N, $E_T(N; x)$ will be very small. Precisely how small depends upon $f(x)$ and N, of course, but the virtue of exponential convergence is that it is not a great deal more difficult to obtain 15 decimal places of accuracy than it is to get 1% accuracy. (For $\ln(1 + x)$ on $x \in [0, 1]$, for example, $N = 19$ gives a relative error of at worst 1 part in 10^{15} !) Theorem 17 implies that

$$a_{n,\text{G}} \equiv (f, \, \phi_n)_\text{G} \tag{4.32}$$

must evaluate to

$$a_{n,\text{G}} = a_n + (E_T, \, \phi_n)_\text{G} \tag{4.33}$$

In words, the discrete inner product (that is, Gaussian integration) correctly and exactly integrates that part of (f, ϕ) which is due to the truncated series portion of f, and all the error in the integration comes from the truncation error in the $(N + 1)$-term approximation. If $E_T \approx 1/10^{15}$, it is obvious that the difference between $a_{n,\text{G}}$ and the exact spectral coefficient a_n must be of this same ridiculously small magnitude. (The exceptions are the coefficients whose degree n is close to the truncation limit N. The *absolute* error in approximating these tiny coefficients by quadrature is small, but the *relative* error may not be small.) However, Gaussian integration is a well-conditioned numerical procedure, so the *absolute* errors in the $\{a_n\}$ will always be small for all n. It follows that if we blindly calculate $(N + 1)$ spectral coefficients via $(N + 1)$-point Gaussian quadrature and then sum to calculate $f(x)$, we will always get very high accuracy for $f(x)$ if N is sufficiently large.

Thus, to represent a known function $f(x)$ as a spectral series, we can safely dump our table of integrals and simply multiply the grid point values of $f(x)$ by a square matrix:

$$\vec{a} \approx \vec{\vec{M}} \, \vec{f} \tag{4.34}$$

where $\vec{a}$ is the column vector of spectral coefficients and where the elements of the other two matrices are

$$M_{ij} \equiv \phi_i(x_j) \, w_j \qquad ; \qquad f_j \equiv f(x_j) \tag{4.35}$$

where the $\{x_j\}$ are the $(N + 1)$ Gaussian quadrature abscissas — the roots of $\phi_{N+1}(x)$ — and the w_j are the corresponding quadrature weights. Note that if the basis functions are not orthonormal, we should divide the i-th row of $\vec{M}$ by (ϕ_i, ϕ_i). This transformation from the grid point values $f(x_j)$ to the corresponding spectral coefficients a_j is discussed further in Chapter 5, Sec. 5, and Chapter 10, Sec. 4, as the "Matrix Multiplication Transformation" (MMT).

When solving a differential equation, we merely apply the same procedure as in (4.34) and (4.35) to approximate the residual function $R(x; a_0, a_1, \ldots, a_N)$ by interpolation. Again, if N is large enough so that the coefficients r_n in the residual series are small for $n > N$, the error in using Gaussian quadrature instead of exact integration to compute the r_n must be small.

Although we have referred to "polynomials" throughout this section, Theorems 16 and 17 not only apply to all the standard types of orthogonal polynomials but to trigonometric polynomials (that is, truncated Fourier series) as well. For most types of polynomials, it is difficult to estimate the error in (4.34) except to conclude, as we have already done, that if N is large, this error will be absurdly small. Numerical experience has confirmed that the pseudospectral method is a very accurate and robust procedure for Hermite functions, Jacobi polynomials and so on.

For Fourier series and Chebyshev polynomials, however, it is possible to analyze the error in approximating coefficient integrals by Gaussian quadrature (4.34) in a simple way. This "aliasing" analysis turns out to be essential in formulating the famous "Two-Thirds Rule" for creating "un-aliased" pseudospectral codes for nonlinear fluid mechanics problems, so it is the topic of the next section.

We began this chapter with a discussion of interpolation and then turned to inner products and numerical integration. The next theorem shows that these two themes are intimately connected.

Theorem 18 (INTERPOLATION BY QUADRATURE)
  Let $P_N(x)$ denote that polynomial of degree N which interpolates to a function $f(x)$ at the $(N+1)$ Gaussian quadrature points associated with a set of orthogonal polynomials $\{\phi_n(x)\}$:

$$P_N(x_i) = f(x_i) \qquad\qquad i = 0, 1, \ldots, N \tag{4.36}$$

$P_N(x)$ may be expanded without error as the sum of the first $(N+1)$ $\phi_N(x)$ because it is merely a polynomial of degree N. The coefficients $\{a_n\}$ of this expansion

$$P_N(x) = \sum_{n=0}^{N} a_n \phi_n(x) \tag{4.37}$$

are given WITHOUT APPROXIMATION *by the discrete inner product*

$$a_n = \frac{(f, \phi_n)_G}{(\phi_n, \phi_n)_G} \tag{4.38}$$

that is to say, are precisely the coefficients calculated by (4.34) and (4.35) above, the Matrix Multiplication Transform (MMT).

PROOF: Since the interpolating polynomial *is* a polynomial, and since Theorem 18 shows that the discrete inner product preserves the orthogonality of the first $(N+1)$ basis functions among themselves, it is obvious that applying the discrete inner product to the finite series (4.37) will exactly retrieve the coefficients a_n. What is not obvious is that we will compute the same coefficients when we use $f(x)$ itself in the inner product since $f(x)$ is not a polynomial, but a function that can be represented only by an infinite series.

However, the Gaussian integration uses *only* the $(N+1)$ values of $f(x)$ at the *interpolation points* — and these are the *same* as the values of $P_N(x)$ at those points. Thus, when we use Gaussian quadrature to approximate $f(x)$, we are really expanding the interpolating polynomial $P_N(x)$ instead. Q. E. D.

Since it is easy to sum a truncated spectral series like (4.37) by recurrence, it is far more efficient to perform Lagrangian interpolation by calculating the coefficients as in (4.34) and (4.35) than it is to use the cardinal function representation (4.6), even though the two are mathematically identical (ignoring round-off error).

One must be a bit careful to understand just what the theorem means. The coefficients computed by Gaussian integration are *not* the *exact* spectral coefficients of $f(x)$, but only good approximations. The pseudospectral coefficients are the exact expansion coefficients only of $P_N(x)$, the interpolating polynomial. For large N, however, such subtleties are academic: $P_N(x)$ is a ridiculously good approximation to $f(x)$, and therefore its coefficients are exceedingly good approximations to those of $f(x)$.

4.5   Pseudospectral Errors: Trigonometric & Chebyshev Polynomials

Theorem 19 (TRIGONOMETRIC INTERPOLATION)
Let the collocation points $\{x_k\}$ be defined by

$$x_k \equiv -\pi + \frac{2\pi k}{N} \qquad\qquad k = 1, 2, \ldots, N \qquad (4.39)$$

Let a function $f(x)$ have the exact, infinite Fourier series representation

$$f(x) \equiv \frac{1}{2}\alpha_0 + \sum_{n=1}^{\infty} \alpha_n \cos(nx) + \sum_{n=1}^{\infty} \beta_n \sin(nx) \qquad (4.40)$$

Let the trigonometric polynomial which interpolates to $f(x)$ at the N collocation points (4.39) be

$$S_N(x) = \frac{1}{2}a_0 + \sum_{n=1}^{N/2-1} a_n \cos(nx) + \sum_{n=1}^{N/2-1} b_n \sin(nx) + \frac{1}{2}a_M \cos(Mx) \qquad (4.41)$$

where $M \equiv N/2$ and where

$$S_N(x_k) = f(x_k), \qquad\qquad k = 1, 2, \ldots, N \qquad (4.42)$$

Then the coefficients of the interpolant can be computed without error by the Trapezoidal Rule

$$a_n = \frac{2}{N}\sum_{k=1}^{N} f(x_k) \cos(nx_k) \qquad \text{[Trapezoidal Rule]} \qquad (4.43a)$$

$$b_n = \frac{2}{N}\sum_{k=1}^{N} f(x_k) \sin(nx_k) \qquad \text{[Trapezoidal Rule]} \qquad (4.43b)$$

and these coefficients of the interpolant are given by infinite series of the exact Fourier coefficients:

$$a_n = \alpha_n + \sum_{j=1}^{\infty} (\alpha_{n+jN} + \alpha_{-n+jN}) \qquad n = 0, 1, \ldots, \frac{N}{2} \qquad (4.44a)$$

$$b_n = \beta_n + \sum_{j=1}^{\infty} (\beta_{n+jN} - \beta_{-n+jN}) \qquad n = 1, 2, \ldots, \frac{N}{2} - 1 \qquad (4.44b)$$

PROOF: Young and Gregory (1972).

The factor of $(1/2)$ multiplying both a_0 and α_0 is a convention. The reason for it is that $(\cos[nx], \cos[nx]) = (\sin[nx], \sin[nx]) = \pi$ for any $n > 0$, but $(1, 1) = 2\pi$. There are two ways of dealing with this factor of 2. One, which is normal in working with a Fourier series, is to simply insert a denominator of $(1/2\pi)$ in front of the integral that gives the constant in the Fourier series and a factor of $(1/\pi)$ in front of all the other coefficient integrals. The alternative, which is used in the theorem, is to employ a factor of $1/\pi$ in the definition of all coefficients — which means computing a coefficient a_0 which is a factor of two larger than the constant in the Fourier series — and then inserting the compensating factor of $(1/2)$ directly into the Fourier sum (4.40) or (4.41).

This second convention is quite popular with the pseudospectral crowd because it eliminates factors of 2 from (4.44) as well as from (4.43).

Eq. (4.43) is labelled the "Trapezoidal Rule" is because it is equivalent to applying that simple integration formula to the Fourier coefficient integrals. [1] The Trapezoidal Rule is a very crude approximation with a relative accuracy of only $O(h^2)$ for *general*, that is to say, for *non-periodic* functions. For *periodic* $f(x)$, however, the *Trapezoidal Rule* is equivalent to *Gaussian integration*. Unlike the optimum quadrature methods associated with Legendre or Hermite polynomials, it is not necessary to look up the weights and abscissas in a table. The weights (with the convention of the factor of $(1/2)$ in (4.40) and (4.41)) are $w_k \equiv 2/N$ for all k, and the interpolation points are evenly spaced as in (4.39). (Parenthetically, note that in computing Fourier transforms, the Trapezoidal Rule also gives accuracy that increases exponentially with N for sufficiently nice functions.)

Most of Theorem 19 merely repeats the trigonometric equivalent of the interpolation-at-the-Gaussian-abscissas ideas of the previous section, but (4.44) is something remarkably different. Theorems 17 and 18 imply that for *any* set of orthogonal polynomials,

$$a_n = \alpha_n + \sum_{j=N+1}^{\infty} e_{n,j}(N)\,\alpha_j \tag{4.45}$$

where the $\{\alpha_n\}$ are the exact spectral coefficients and where the $\{a_n, n = 0, 1, \dots, N\}$ are the coefficients of the interpolating polynomial. What is surprising about (4.44) is that it shows that for Fourier series (and through change-of-variable, Chebyshev polynomials), only one e_j out of each group of $(N/2)$ is different from 0.

This simple fact turns out to be profoundly important for coping with "nonlinear aliasing instability", which is a vice of both finite difference and pseudospectral hydrodynamic codes. For pseudospectral algorithms and quadratic nonlinearity, aliasing can be cured by evaluating the nonlinear products using $2N$ interpolation points [double the number of terms in the series] so that the expansion of the nonlinear term can be computed exactly. S. A. Orszag pointed out that this is wasteful for Fourier and Chebyshev methods. The special form of the Gaussian quadrature error in (4.44) makes it possible to use only about $(3/2)\,N$ points instead of $2N$. This has become known as the "3/2's Rule" or "Two-Thirds Rule" where the latter name reflects that the yield of coefficients is only $(2/3)$ the number of points used for quadrature. Unfortunately, the Two-Thirds Rule applies *only* for these two types of basis functions: Fourier and Chebyshev. We will discuss the "Two-Thirds Rule" in some detail in Chapter 11.

Another payoff of (4.44) is a proof of the following.

Theorem 20 (ERROR: FOURIER INTERPOLATION)
Let $S_N(x)$ denote the trigonometric polynomial that interpolates to a function $f(x)$ at N points. Let $f_N(x)$ denote the corresponding truncation of the exact Fourier series. Let the $\{\alpha_n\}$ denote the exact spectral coefficients of the infinite Fourier series. Then, as stated in Theorem 10 of Chapter 2 (with slightly different notation),

$$|f(x) - f_N(x)| \leq |b_{N/2}| + \sum_{n=1+N/2}^{\infty} (|a_n| + |b_n|), \tag{4.46}$$

that is to say, the error is bounded by the sum of the absolute value of all the neglected coefficients.

The new theorem is that the corresponding bound for trigonometric interpolation is, for all N

---

[1]Strictly speaking, the classical Trapezoidal Rule would add the quadrature point $x = -\pi$ and multiply $f(-\pi)$ and $f(\pi)$ by one-half the weight used for the interior points. Because of the periodicity, however, $f(-\pi) \equiv f(\pi)$, so it suffices to weight all the points the same, and use only $f(\pi)$.

and all real x,

$$|f(x) - S_N(x)| \leq 2 \left\{ |b_{N/2}| + \sum_{n=1+N/2}^{\infty} (|a_n| + |b_n|) \right\}, \tag{4.47}$$

that is to say, the error is bounded by TWICE the sum of the absolute values of all the neglected coefficients.

Comparing (4.46) and (4.47), we conclude:
the PENALTY for using INTERPOLATION instead of TRUNCATION is at WORST a

FACTOR of TWO.

PROOF:
Similar to that of Theorem 10 (Chap. 2). When we sum the absolute values of the terms in (4.44) that are the errors due to interpolation, we find that each neglected coefficient appears in (4.44) exactly once. Each neglected coefficient also appears in the difference between $f(x)$ and $f_N(x)$ exactly once. When we add the absolute values of all the errors, truncation and interpolation, we obtain (4.48).

To a finite difference modeller, the factor of two in Theorem 20 is a big deal. With a second order scheme, it would be necessary to increase the number of grid points by 40% to reduce the error by that factor.

A spectral modeller has a rather different perspective. For example, the coefficients of $\ln(1 + x)$ for $x \in [0, 1]$ are proportional to p^n/n where $p = 0.18$. This implies that each coefficient is more than five times smaller than its immediate predecessor. It follows that if the first N terms of the Chebyshev series for $\ln(1 + x)$ gives an error everywhere less than ϵ, then the Chebyshev interpolating polynomial with $(N + 1)$ terms will have an even smaller error bounded by 0.23 ϵ. When N is large, adding one or two more terms increases the cost only slightly. The penalty we accept for using the pseudospectral method instead of the Galerkin method is completely insignificant[2].

The only reason the Galerkin method has not completely disappeared for numerical computations is that it *sometimes* generates *sparse* matrices which can be inverted much more easily than the *full* matrices generated by the pseudospectral method, as discussed in the chapter on matrix-solving.

For completeness, we shall state the analogous theorems for Chebyshev polynomials.

Theorem 21 (CHEBYSHEV INTERPOLATION & its ERROR BOUND)
Let the "Chebyshev-extrema" ("Gauss-Lobatto") grid $\{x_k\}$ be given by

$$x_k = -\cos\left(\frac{k\pi}{N}\right) \qquad k = 0, 1, \dots, N \qquad [\text{"Chebyshev-extrema" grid}] \tag{4.48}$$

Let the polynomial $P_N(x)$ which interpolates to $f(x)$ at these grid points be

$$P_N(x) = \sum_{n=0}^{N}{}'' b_n T_n(x) \qquad\qquad [\text{Extrema Grid}] \tag{4.49}$$

---

[2]The exception is a *paper-and-pencil* or Maple calculation for very *small* N. A 4×4 determinant is approximately five times harder to evaluate analytically than a 3×3 determinant, so the Galerkin method is a good choice for low order analytical work.

where the (") on the summation means that the first and last terms are to be taken with a factor of (1/2). [Compare this with the constant and $\cos(Mx)$ term in the Fourier interpolant, $S_N(x)$.] The coefficients of the interpolating polynomial are given by

$$b_n = \frac{2}{N} \sum_{k=0}^{N} {}'' f(x_k) \, T_n(x_k) \qquad \text{[Extrema Grid]} \qquad (4.50)$$

Let the "Chebyshev-roots" grid be defined by

$$x_k = -\cos\left[\frac{(2k+1)\pi}{2(N+1)}\right] \qquad k = 0, 1, \ldots, N \qquad \text{["Chebyshev-roots" Grid]} \qquad (4.51)$$

and let $Q_N(x)$ denote the interpolating polynomial of degree N which interpolates to $f(x)$ on this alternative grid:

$$Q_N(x) = \sum_{n=0}^{N} {}' c_n \, T_n(x) \qquad \text{[Roots Grid]} \qquad (4.52)$$

where the (') on the sum means that the first term $[c_0 \, T_0]$ is to be divided by (1/2), then the coefficients are

$$c_n = \frac{2}{N+1} \sum_{k=0}^{N} f(x_k) \, T_n(x_k) \qquad \text{[Roots Grid]} \qquad (4.53)$$

Let $\{\alpha_n\}$ denote the exact spectral coefficients of $f(x)$, that is, let

$$f(x) = \sum_{n=0}^{\infty} {}' \alpha_n \, T_n(x) \qquad (4.54)$$

The coefficients of the interpolating polynomials are related to those of $f(x)$ via

$$b_n = \alpha_n + \sum_{j=1}^{\infty} (\alpha_{n+2jN} + \alpha_{-n+2jN}) \qquad \text{[Extrema Grid]} \qquad (4.55)$$

$$c_n = \alpha_n + \sum_{j=1}^{\infty} \left(\alpha_{n+2j(N+1)} + \alpha_{-n+2j(N+1)}\right)(-1)^j \qquad \text{[Roots Grid]} \qquad (4.56)$$

For all N and all real $x \in [-1, 1]$, the errors in either of the interpolating polynomials is bounded by TWICE *the sum of the absolute values of all the neglected coefficients:*

$$|f(x) - P_N(x)| \leq 2 \sum_{n=N+1}^{\infty} |\alpha_n| \qquad (4.57)$$

$$|f(x) - Q_N(x)| \leq 2 \sum_{n=N+1}^{\infty} |\alpha_n|, \qquad (4.58)$$

that is, the penalty for using interpolation instead of truncation is the same for Chebyshev series as for Fourier series:

a FACTOR *of* TWO.

PROOF: Fox and Parker (1968).

Canuto *et al.* (1988) show that for incompressible (or weakly divergent) fluid mechanics, it is often useful to employ *both* these Chebyshev grids simultaneously. On this pseudospectral "staggered" grid, the pressure is defined on the "roots" grid while the velocities are defined on the "extrema" grid.

Although two alternative optimum Chebyshev grids may seem one too many, both sets of points are useful in practice.

Chapter 5

Cardinal Functions

"Understanding grows only logarithmically with the number of floating point operations."
— J. P. Boyd

5.1 Introduction

Mathematics courses have given the average physics or engineering student a rather warped view of global expansion methods: The coefficients are everything, and values of $f(x)$ at various points are but the poor underclass. Indeed, some massive tomes on Fourier series never even mention the the the word "interpolation".

To understand pseudospectral methods, one must take a more broad-minded view: The series coefficients $\{a_n\}$ and the values of the function at the interpolation points $\{f(x_i)\}$ are *equivalent* and co-equal representations of the pseudospectral approximation to $f(x)$. Given the $(N+1)$ coefficients of the approximating polynomial $P_N(x)$, we can certainly evaluate the truncated series at each grid point to calculate the $\{f(x_i)\}$. It is equally true, however, that given the $\{f(x_i)\}$, we have all the information needed to calculate the $\{a_n\}$ by applying the discrete inner product, i. e. interpolation. This operation, which is simply the multiplication of a column vector by a square matrix[1] introduces no additional errors because the discrete inner product computes the expansion coefficients of $P_N(x)$ exactly.

Because of this, pseudospectral methods may use *either* the series coefficients or the grid point values as the unknowns. Employing the $\{u(x_i)\}$ is very convenient because a pseudospectral technique then becomes a grid point method — just like a finite difference procedure. The only alteration is that derivatives are computed through an N-point scheme instead of the three- or five-point formulas of conventional finite differences.

There are a couple of different ways of working directly with grid point values, and this chapter will discuss each in turn. In Chapter 9, we will then show that this freedom to jump between the $\{a_n\}$ and $\{u(x_i)\}$ representations of $u(x)$ is absolutely essential to coping with nonlinearity.

A major reason for discussing cardinal functions, however, is not computational but conceptual. Pseudospectral algorithms are simply N-th order finite difference methods in disguise.

---

[1]For Fourier and Chebyshev series, this transformation can alternatively be performed by using a Fast Fourier Transform (FFT) routine.

5.2 Whittaker Cardinal or "Sinc" Functions

Sir Edmund Whittaker (1915) showed that for an *infinite* interval, the analogues to the "fundamental polynomials of Lagrangian interpolation" are what he called the "cardinal functions". The collocation points are evenly spaced:

$$x_j \equiv j\,h \qquad\qquad j = 0, \pm 1, \pm 2, \ldots \qquad\qquad (5.1)$$

Whittaker's cardinal functions are

$$C_k(x;\, h) \equiv \frac{\sin\{\pi(x - kh)/h\}}{\pi\,(x - kh)/h} \qquad [\text{"Whittaker Cardinal Functions"}] \qquad (5.2)$$

and have the property

$$C_k(x_j;\, h) = \delta_{jk} \qquad\qquad (5.3)$$

One great simplification is that, through a linear shift and rescaling of the argument, the Whittaker cardinal functions can all be expressed in terms of a single, universal function

$$\text{sinc}(x) \equiv \frac{\sin(\pi x)}{(\pi x)} \qquad\qquad (5.4)$$

The sine function in the numerator has zeros at all integral values of x, $x = 0, \pm 1, \pm 2$, etc. However, the root at the origin is cancelled by the denominator factor of (πx). Near $x = 0$, Taylor expansion of the sine gives

$$\text{sinc}(x) \approx \frac{(\pi x) - (\pi x)^3/6 + (\pi x)^5/120 - \ldots}{\pi x} \qquad\qquad (5.5)$$

$$\approx 1 - \frac{(\pi x)^2}{6} + \ldots \qquad\qquad |x| \ll 1 \qquad\qquad (5.6)$$

Eq. (5.6) shows that sinc(0) = 1 from which (5.3) follows. The sinc function is illustrated in the upper left panel of Fig. 5.1.

The property that each cardinal function is non-zero at one and only one of the interpolation points implies that for arbitrary $f(x)$, the function defined by

$$\tilde{f}(x) \equiv \sum_{j=-\infty}^{\infty} f(x_j)\, C_j(x) \qquad\qquad (5.7)$$

interpolates $f(x)$ at every point of the grid. However, this hardly guarantees a good approximation; recall the Runge phenomenon for polynomials. In Chapter 17, Sec. 3, however, we show that, for a sufficiently small grid spacing h and for a sufficiently large truncation of the infinite sum (5.7), the sinc series is an accurate approximation if $f(x)$ decays sufficiently fast as $|x| \to \infty$.

Because of this, Sir Edmund Whittaker described sinc(x) as "a function of royal blood in the family of entire functions, whose distinguished properties separate it from its bourgeois brethren." Strong words indeed, but the sinc(x) function is not only successful, but also a good deal simpler than a Chebyshev polynomial or a spherical harmonic. In the next two sections, we shall see that Fourier and Chebyshev grids also have their "royal functions".

The lack of boundaries on an infinite interval makes the Whittaker cardinal function the simplest case.

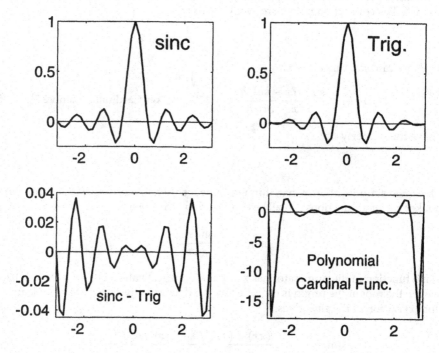

Figure 5.1: a. The Whittaker cardinal function, sinc $([x - \pi]/h)$, with $h = \pi/6$.
b. [Upper Right] Trigonometric cardinal function associated with the same interpolation point $(x = \pi)$ and grid spacing $(h = \pi/6)$ as in (a).
c. [Lower Left] Difference between the Whittaker cardinal function (graphed upper left) and the trigonometric cardinal function (graphed upper right)
d. Polynomial cardinal function associated with interpolation at $x = \pi$ and evenly spaced grid points, same as for the three other panels. Because of the large "wings" of this function, the value of $f(x)$ at $x = \pi$ has a greater effect on the interpolant in the remote region near the endpoints $(x = 0, 2\pi)$ than in the local neighborhood around $x = \pi$.

5.3   Cardinal Functions for Trigonometric Interpolation

Gottlieb, Hussaini, and Orszag (1984) show that for the collocation points

$$x_j = \frac{2\pi j}{N} \qquad ; \qquad j = 0, 1, \ldots, N - 1, \tag{5.8}$$

the trigonometric interpolant to $f(x)$ can be written as

$$f(x) \approx \sum_{j=0}^{N-1} f(x_j) C_j(x) \tag{5.9}$$

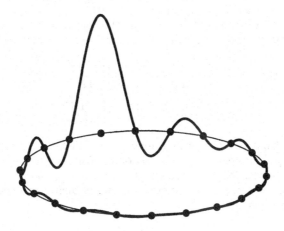

Figure 5.2: Schematic of a spatially periodic interval with nodes and a typical Fourier cardinal function.

where the cardinal functions are

$$C_j(x) \equiv \frac{1}{N} \sin\left[\frac{N(x - x_j)}{2}\right] \cot\left[\frac{(x - x_j)}{2}\right] \tag{5.10}$$

Just like the terms of the Whittaker sinc series, all the cardinal functions (5.10) are *identical* in *shape*, and *differ* only in the *location* of the peak. This shape similarity of all the trigonometric cardinal functions, regardless of whether x_j is in the middle or near the endpoints of $[0, 2\pi]$, is a reminder that for *periodic* functions, there are in effect *no endpoints*. A spatially-periodic interval can always be conceptualized as a ring, rather than a line segment (Fig. 5.2).

The same is emphatically not true for polynomial interpolation of a non-periodic function. Elementary courses in numerical analysis stress that *centered* finite difference approximations are much more accurate than *one-sided* approximations. This implies that, all other things being equal, Lagrangian interpolation will give much more accurate approximations to df/dx (and to $f(x)$ itself) at the *center* of the interval than at the *endpoints* where the approximation is completely one-sided. This is exactly what happens with an *even* grid spacing in the Runge example — good accuracy near the center of the interval and divergence in the regions of mostly one-sided approximations near the ends of the interval.

Chebyshev interpolation gives very *uniform* approximation over the whole interval [-1, 1] because it compensates for the one-sided approximation near the endpoints by increasing the density of the grid near the endpoints. We have the symbolic relationship

(near $x \approx \pm 1$) [Small h; One-Sided] $\longleftrightarrow$ [Large h; Centered] (near $x \approx 0$)

For *periodic* functions with a uniform grid, however, we have the accuracy of a fully-centered approximation even at the endpoint $x = 0$. Although the grid point values to the left of the origin ($x \in [-\pi, 0]$) are outside the interval of approximation, they are equal to the corresponding values of $f(x)$ on $x \in [\pi, 2\pi]$, which *is* part of our grid. Consequently, the cardinal functions for trigonometric interpolation on an evenly spaced grid have the same shape for each point on the grid. Because of this uniformity of shape, the accuracy of the approximation is uniform over the whole interval.

An analytical way to show the close relationship between the trigonometric cardinal functions and the sinc functions is to Taylor expand the cotangent in (5.10), which gives

$$C_j(x) = \frac{\sin[N(x-x_j)/2]}{N(x-x_j)/2} \left\{ 1 - \frac{[x-x_j]^2}{12} + O\left([x-x_j]^4\right) \right\} \tag{5.11}$$

$$= \text{sinc}\left[\frac{N(x-x_j)}{2}\right] \left\{ 1 - \frac{[x-x_j]^2}{12} - \frac{[x-x_j]^4}{720} + \dots \right\} \tag{5.12}$$

Near its peak, the trigonometric cardinal function is indistinguishable from Whittaker's cardinal function, and differences appear only on the fringes where $C_j(x)$ has decayed to a rather small amplitude. Fig. 5.1 illustrates the sinc function, the trigonometric cardinal function, and (lower left) the small difference between them. The maximum pointwise *absolute* error in the approximation

$$C_j(x) \approx \text{sinc}\left[\frac{N(x-x_j)}{2}\right] \tag{5.13}$$

is never worse than 0.05 even for $N = 6$; for larger N, the error falls as $O(1/N)$. There is clearly an intimate relationship between sinc series on $x \in [-\infty, \infty]$ and trigonometric interpolation on $x \in [0, 2\pi]$.

An analytical expression of this kinship is the identity:

$$C_0(x; h) \equiv \frac{1}{2N} \sin(Nx) \cot(x/2) \tag{5.14}$$

$$= \sum_{m=-\infty}^{\infty} \text{sinc}\left(\frac{x - 2\pi m}{h}\right)$$

In words, the trigonometric cardinal function is the result of duplicating an infinite number of copies of the sinc function, spacing them evenly over all space, and summing as shown schematically in Fig. 5.3.

For comic relief, Fig. 5.1 [lower right] shows the graph of the cardinal function for Lagrangian *polynomial* interpolation at the same grid points. The sinc function and the trigonometric cardinal function both *decay* away from their peak so that $f(x_j)$ for some particular j has a smaller and smaller influence as we move away from $x = x_j$. In contrast, the Lagrangian cardinal function is much larger near the endpoints than at $x = \pi$, the one grid point where it differs from 0. The value of $f(x)$ at that one point has a much larger influence *away* from that point than in its neighborhood. This makes no sense at all and is another manifestation of the Runge phenomenon.

The trigonometric cardinal function has the alternative complex representation

$$C_j(x) \equiv \frac{1}{N} \sum_{k=-N/2}^{N/2} \frac{1}{c_k} \exp[i\,k(x - x_j)] \tag{5.15}$$

where $c_k = 1$ except for $k = \pm N$; $c_{-N/2} = c_{N/2} = 2$. Although the appearance of (5.15) is wildly different from (5.10), the two definitions are in fact equivalent.

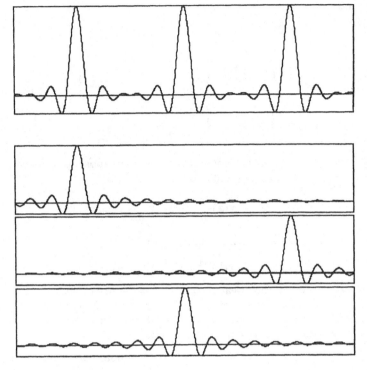

Figure 5.3: Top: the trigonometric cardinal function, shown over three periods. Bottom three panels: three copies of the Whittaker cardinal function which are added, together with similar copies spaced evenly over all space, to give the trigonometric cardinal function shown at the top.

Since the approximation (5.9) is *identical* with the trigonometric interpolant $S_N(x)$, we can use the cardinal functions *directly* to solve differential equations *without* explicitly using the truncated Fourier series form at all. A linear differential equation $Lu = f(x)$, becomes the matrix problem

$$\vec{\vec{L}}\,\vec{u} = \vec{f} \tag{5.16}$$

where the matrix elements are given by

$$u_j \equiv u(x_j) \qquad\qquad j = 0, 1, \ldots, N-1 \tag{5.17}$$

$$f_i \equiv f(x_i) \qquad\qquad i = 0, 1, \ldots, N-1 \tag{5.18}$$

$$L_{ij} \equiv [LC_j(x)]|_{x=x_i} \qquad\qquad i, j = 0, 1, \ldots, N-1 \tag{5.19}$$

Because the cardinal functions are so simple, it is easy to evaluate their derivatives. Thus, introducing a generalization of the usual Kronecker δ-function notation as in Stenger

(1981), we find

$$C_j(x_i) = \delta_{ij} \tag{5.20}$$

$$C_{j,x}(x_i) \equiv \delta_{ij}^{(1)} = \begin{cases} 0 & i = j \\ \frac{1}{2}(-1)^{i+j} \cot\left[\frac{(x_i - x_j)}{2}\right] & i \neq j \end{cases} \tag{5.21}$$

$$C_{j,xx}(x_i) \equiv \delta_{ij}^{(2)} = \begin{cases} -\frac{(2N^2+1)}{6} & i = j \\ \frac{1}{2}\frac{(-1)^{i+j+1}}{\sin^2[(x_i - x_j)/2]} & i \neq j \end{cases} \tag{5.22}$$

where δ_{ij} is the usual Kronecker δ-function and where the subscript x denotes x-differentiation. Thus, a general second order differential operator such as

$$L \equiv a_2(x)\frac{\partial}{\partial x^2} + a_1(x)\frac{\partial}{\partial x} + a_0(x) \tag{5.23}$$

is represented by a matrix $\vec{L}$ whose elements are simply

$$L_{ij} = a_2(x_i)\,\delta_{ij}^{(2)} + a_1(x_i)\,\delta_{ij}^{(1)} + a_0(x_i)\,\delta_{ij} \tag{5.24}$$

Non-periodic problems can always be transformed into a periodic problem on $x \in [0, \pi]$ by the Chebyshev-to-Fourier mapping. Since we need only the cosine terms, rather than a general Fourier basis, we can still apply (5.8) – (5.24) except that we drop all grid points such that $x_j > \pi$. The explicit boundary conditions

$$u(-1) = u(1) = 0 \qquad \text{["Dirichlet boundary conditions"]} \tag{5.25}$$

become transformed into $u(0) = u(\pi)$; we simply *omit* the cardinal functions corresponding to those two points. (In the original non-periodic coordinate, these cardinal functions are polynomials.)

Thus, in the cardinal function basis, it is just as easy to solve problems with Dirichlet boundary conditions as to solve those with boundary conditions of periodicity. Unlike the algorithm where the spectral coefficients of the equivalent Chebyshev series are the unknowns, it is not necessary to solve a matrix problem with two rows reserved to impose (5.25) nor is it necessary to form new basis functions which are linear combinations of Chebyshev polynomials. The cardinal function representation makes Dirichlet boundary conditions a snap.[2]

Appendix F lists properties of trigonometric cardinal functions both for a general Fourier series and for the special cases when the solution can be expanded in a Fourier cosine series or a Fourier sine series. The only complication with the two special cases is that the derivative of a cosine series is a sine series, so the matrix of the second derivatives of the cosines at the grid points is the product of the first derivative sine matrix with the first derivative cosine matrix. Assembling the differentiation matrices to all orders is quite trivial, however, as long as this is kept in mind.

5.4   Cardinal Functions for Orthogonal Polynomials

It is quite easy to construct cardinal functions for general orthogonal polynomials. Recall that the grid points are the zeros of $\phi_{N+1}(x)$. Thus, $\phi_{N+1}(x)$ is similar to what we need

---

[2]Neuman boundary conditions, that is to say, conditions on the *derivative* of u at the boundaries, require reserving two rows of the matrix for these conditions using the $\delta_{ij}^{(1)}$ matrix elements. The cardinal function representation, however, is still just as easy to use as the truncated Chebyshev series.

except that it vanishes at *all* the grid points, and we want it to differ from 0 at $x = x_j$. In the vicinity of the j-th root, Taylor expansion gives

$$\phi_{N+1}(x) \approx \phi_{N+1}(x_j) + \phi_{N+1,x}(x_j)(x - x_j) + O\left([x - x_j]^2\right) \qquad (5.26)$$

Since the first term on the right in (5.26) is 0, one can eliminate the root at $x = x_j$ and normalize the value of the cardinal function to 1 at $x = x_j$ by defining

$$C_j(x) \equiv \frac{\phi_{N+1}(x)}{\phi_{N+1,x}(x_j)(x - x_j)} \qquad \text{[Cardinal Function]} \qquad (5.27)$$

where subscript x denotes x-differentiation. This prescription works for Chebyshev polynomials of the first and second kinds, Legendre polynomials, Laguerre functions, Hermite functions, etc.

As noted in Chapter 4, a useful alternative grid consists of the *extrema* of a basis function together with the endpoints, the "Lobatto" grid. The cardinal functions for this are

$$C_j(x) \equiv \frac{(1 - x^2)\,\phi_{N,x}(x)}{[(1 - x_j^2)\,\phi_{N,x}(x_j)]_x (x - x_j)} \qquad j = 0, \ldots, N \qquad (5.28)$$

where we have assumed that the endpoints are at $x = \pm 1$. As in (5.26), the numerator is a function of x while, except for the linear factor of $(x - x_j)$, all the terms in the denominator

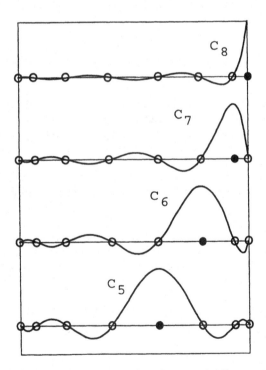

Figure 5.4: Cardinal functions for the Legendre pseudospectral method on an 8-point Lobatto grid. The grid point associated with each cardinal function is solid; the other grid points are marked by open circles. Only the cardinal functions associated with positive x are shown since those for negative x are just the reflections of those illustrated.

Legendre-Lobatto

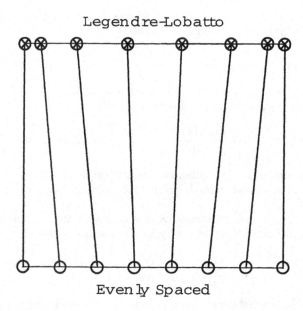

Evenly Spaced

Figure 5.5: Relationship between the evenly spaced grid and the endpoint-concentrated Legendre-Lobatto grid. The points of both grids are marked by circles. The Legendre-Lobatto grid is at the top.

are constants. It is often possible to simplify (5.28) so as to eliminate the second derivative by using the differential equation which the orthogonal polynomials satisfy.

The shape of polynomial cardinal functions is very similar to that of their trigonometric counterparts. The Chebyshev cardinal functions are in fact *identical* with the cosine cardinal functions except for the change of coordinate. This mapping causes the Chebyshev cardinal functions to oscillate very rapidly near the endpoints and slowly near the center. The cardinal functions for Gegenbauer polynomials of higher degree, such as the Legendre polynomials $P_N(x)$, are qualitatively similar except that the interpolation points are a little closer to the center of the interval, so the oscillations near the endpoints are larger than for Chebyshev functions.

Our description of polynomial cardinal functions is brief because they can be used exactly like their trigonometric counterparts. In particular, Dirichlet boundary conditions are very easy to impose *if* the "extrema-plus-endpoints" ("Gauss-Lobatto") grid is used, as in (5.28), which is why this choice of interpolation points has become very popular.

The differentiation matrices for Chebyshev and Legendre cardinal functions, analogous to $\delta_{ij}^{(k)}$ of the previous section, are given in Gottlieb, Hussaini, and Orszag (1984) and Appendix F.

Fig. 5.4 illustrates the cardinal functions for a low order Legendre discretization on the Lobatto grid, a typical choice for spectral elements and $h - p$-finite elements. The wild oscillations of power-of-x, non-trigonometric polynomials on an *evenly* spaced grid are suppressed because, as illustrated in Fig. 5.5, the pseudospectral grid is not evenly spaced, but has points concentrated near both endpoints.

5.5 Transformations and Interpolation

It is possible to apply the pseudospectral method using either the grid point values $\{u(x_i)\}$ or the series coefficients $\{a_n\}$ as the unknowns. The relationship between the two formalisms is important because efficient algorithms for nonlinear problems must jump back and forth between the grid point representation and the truncated series.

For simplicity, assume that $L u = f$ is a *linear* equation. Let the pseudospectral-generated matrix problems take the equivalent forms

$$\vec{\vec{L}}\,\vec{u} \;=\; \vec{f} \qquad \text{[grid point/cardinal function form]} \tag{5.29}$$

$$\vec{\vec{H}}\,\vec{a} \;=\; \vec{f} \qquad \text{[series coefficient form]} \tag{5.30}$$

where the matrix elements of $\vec{a}$ are the coefficients in the truncated sum

$$u(x) \approx \sum_{n=0}^{N} a_n\,\phi_n(x) \tag{5.31}$$

and where the other matrix elements are given by

$$u_i \;\equiv\; u(x_i) \tag{5.32}$$
$$f_i \;\equiv\; f(x_i) \tag{5.33}$$
$$L_{ij} \;\equiv\; (LC_j[x])|_{x=x_i} \tag{5.34}$$
$$H_{ij} \;\equiv\; (L\phi_j[x])|_{x=x_i} \tag{5.35}$$

where the $\{x_i\}$ are the interpolation points, the $\{\phi_j(x)\}$ are the spectral basis functions, and the $\{C_j(x)\}$ are the cardinal functions.

The coefficients $\{a_n\}$ and the grid point values are related through the following equation from Chapter 4, which we repeat:

$$\vec{\vec{M}}\,\vec{u} = \vec{a} \tag{5.36}$$

where the elements of $\vec{\vec{M}}$ are

$$M_{ij} \equiv \frac{\phi_i(x_j)\,w_j}{(\phi_i,\,\phi_i)} \qquad \text{[transformation matrix } u(x_j) \to a_i\text{]} \tag{5.37}$$

where the $\{w_j\}$ are the Gaussian quadrature weights and where $(\phi_i,\,\phi_i)$ denotes the inner product of the i-th basis function with itself (Appendix A).

Comparing (5.36) with (5.29) and (5.30), we see that

$$\vec{\vec{L}} \;=\; \vec{\vec{H}}\,\vec{\vec{M}} \tag{5.38}$$

$$\vec{\vec{H}} \;=\; \vec{\vec{L}}\,\vec{\vec{M}}^{-1} \tag{5.39}$$

Thus, we can use the gridpoints as unknowns even if we only have subroutines to compute the basis functions $\{\phi_j(x)\}$, and don't want the bother of writing codes to evaluate the cardinal functions. We simply set up the matrix $\vec{\vec{H}}$ and then multiply it on the right by $\vec{\vec{M}}$ to obtain $\vec{\vec{L}}$. Conversely, we can convert from a cardinal function representation to a form that uses the series coefficients as the unknowns by multiplying the matrix $\vec{\vec{L}}$ by $\vec{\vec{M}}^{-1}$. It is not necessary to invert $\vec{\vec{M}}$ via Gaussian elimination because $\vec{\vec{M}}^{-1}$ satisfies the equation

$$\vec{u} = \vec{\vec{M}}^{-1}\vec{a} \tag{5.40}$$

This shows that the inverse transformation matrix has elements given by

$$M_{ij}^{-1} \equiv \phi_j(x_i) \qquad \text{[inverse transformation } a_j \to u(x_i)] \qquad (5.41)$$

i. e. $\vec{M}^{-1}$ is simply the matrix that sums the spectral series to obtain grid point values of the solution.

Consequently, we can move freely back and forth from one representation to another merely by multiplying one square matrix by another. Such transformations-by-matrix are dubbed, in both directions, the Matrix Multiplication Transform (MMT) and are discussed further in Chapter 10, Sec. 4. Unfortunately, shift-of-representation is a relatively expensive calculation: $O(N^3)$ to calculate $\vec{L}$ from $\vec{H}$ or vice versa, and $O(N^2)$ merely to calculate $\vec{u}$ from $\vec{a}$ or the reverse.

For the special case of Fourier series and its clone, the Chebyshev expansion, the work can be reduced by using the Fast Fourier Transform (FFT) discussed in Chapter 10.

Chapter 6

Pseudospectral Methods for Boundary Value Problems

"One must watch the convergence of a numerical code as carefully as a father watching his four year old play near a busy road."
— J. P. Boyd

6.1 Introduction

The goal of this chapter is to apply what we have learned in the previous five chapters to solving boundary value problems. Time-dependent equations require additional techniques developed later. There will be some repetition, with amplification, of key ideas already presented.

It is sufficient to restrict ourselves to *linear* boundary value equations. Nonlinear problems are usually solved through an iteration, the so-called Newton-Kantorovich method, in which a linear differential equation is solved at each step (Appendices C and D).

6.2 Choice of Basis Set

Table A–1 is a flow chart for choosing the basis set. On a sphere, use spherical harmonics for latitude and longitude. If the solution is periodic, use Fourier series. If the domain is finite, but $u(x)$ is not a periodic function of x, Chebyshev polynomials are best. If the interval is unbounded, i. e., one or both endpoints are at $|x| = \infty$, then use the appropriate species of rational Chebyshev functions ($TL_n(x)$ or $TB_n(x)$) (Chapter 17).

6.3 Boundary Conditions: Behavioral & Numerical

Boundary conditions may be divided into two broad categories: behavioral and numerical. Periodicity is a behavioral condition: it demands that the solution $u(x)$ have the property that $u(x) = u(x + 2\pi)$ for any x, but this behavior does not impose any specific numerical value upon $u(x)$ or its derivatives at any particular point. Another behavioral condition is that of being bounded and infinitely differentiable at a point where the differential equation is singular. In contrast, conditions such as $u(-1) = 3$ and $u(1) + du/dx(0) = 1.7$ are numerical.

The significance of these categories is that it is almost always necessary to *explicitly* impose numerical boundary conditions. In contrast, behavioral conditions may be satisfied *implicitly* by choosing basis functions such that each have the required property or behavior.

For example, when the boundary conditions are that $u(x)$ is *periodic* in x, then the appropriate basis functions are the sines and cosines of a Fourier series. Since each basis function is itself periodic, their sum will be periodic, too. Consequently, it is unnecessary to *explicitly* impose periodicity on the numerical solution. Our choice of basis function has *implicitly* satisfied the boundary conditions.

Similarly, the two-dimensional basis functions known as "spherical harmonics" automatically have the proper behavior on the surface of a sphere. It is never necessary to impose additional explicit constraints after choosing spherical harmonics to represent the latitude/longitude dependence of functions in spherical coordinates.

When the problem is posed on an unbounded interval, the coefficients of the equation are usually singular at the infinite endpoint or endpoints. As explained Chapter 17 on infinite interval problems, this usually implies that only one of the linearly independent solutions of the differential equation is bounded and infinitely differentiable at the endpoints. By choosing a basis of analytic functions, we force the numerical approximation to have the desired behavior at infinity. On a doubly infinite interval, $x \in [-\infty, \infty]$, the rational Chebyshev functions $TB_n(x)$ are a good basis. If, for example, the exact solution decays exponentially fast as $|x| \to \infty$, the differential equation will then force the numerical solution to have the proper exponential decay even though the individual $TB_n(x)$ do not decay at all!

Behavioral boundary conditions are possible on a bounded interval, too. The differential equation

$$(1 - x^2)\,u_{xx} - x\,u_x + \lambda\,u(x) = 0 \qquad\qquad x \in [-1, 1] \qquad\qquad (6.1)$$

is singular at both endpoints. Nonetheless, $u(x)$ may be nonsingular at $x = \pm 1$ if the eigenvalue λ is equal to any one of an infinite number of discrete values. Because the Chebyshev polynomials are individually analytic at the endpoints, their sum satisfies the behavioral boundary condition of analyticity at $x = \pm 1$. Consequently, it is possible to solve this singular differential equation, with an exponential rate of convergence, merely by substituting unmodified Chebyshev polynomials into (6.1) without any explicit constraints. In fact, with $\lambda_n = n^2$, the eigenfunctions of (6.1) *are* the Chebyshev polynomials $T_n(x)$.

Numerical boundary conditions, in contrast, must be explicitly imposed. There are two strategies: (i) reducing the number of collocation conditions on the residual of the differential equation, and using rows of the pseudospectral matrix to explicitly impose the constraint or (ii) modifying the problem (if need be) so that the boundary conditions of the modified problem are *homogeneous* and then altering the basis set so that the basis functions individually satisfy these conditions. We will refer to these two strategies as "boundary-bordering" and "basis recombination", respectively, and discuss them at length in the next two sections.

In finite element theory, boundary conditions are classified as either "essential", which must be explicitly imposed, or "natural" conditions, which are automatically satisfied by the approximate solution. This dual classification is not quite the same as the division between "behavioral" and "numerical" conditions discussed here because in most instances, both "essential" and "natural" conditions are numerical and the distinction is only between whether $u(x)$ or its derivative is set equal to a particular number at the endpoints.

Nevertheless, from a programmer's standpoint, this duality is the same: "behavioral" and "natural" conditions can be ignored in the code whereas "numerical" and "essential" conditions require work. It is for this reason that Boyd (1978b, 1987a, b) borrowed the finite element terminology: for global expansion methods, "numerical" boundary conditions are most definitely "essential"!

6.4 "Boundary-Bordering" for "Numerical" Boundary Conditions

To generate a "boundary-bordered" pseudospectral matrix of dimension N for a second order ordinary differential equation boundary value problem, we demand that the residual of the differential equation should vanish at $(N-2)$ interpolation points on the interior of the interval, and then allocate the top two rows of the matrix to impose the boundary conditions. The reason for the name "boundary-bordering" is that the "border" of the matrix — in this case, the first and second rows — explicitly enforces the boundary conditions. For example, suppose the problem is

$$u_{xx} + q(x)\,u = f(x) \qquad \& \qquad u_x(-1) = \alpha \quad \text{and} \quad u(1) = \beta \qquad (6.2)$$

The Chebyshev grid is

$$x_i = \cos\left(\frac{\pi i}{N-1}\right) \qquad\qquad i = 1, 2, \ldots, (N-2) \qquad (6.3)$$

The matrix discretization of (6.2) is

$$\vec{\vec{L}}\,\vec{a} = \vec{F} \qquad (6.4)$$

where the elements of the column vector $\vec{a}$ are the Chebyshev coefficients of $u(x)$,

$$u(x) \approx \sum_{j=1}^{N} a_j\, T_{j-1}(x) \qquad (6.5)$$

and where

$$L_{i+2, j} = T_{j-1, xx}(x_i) + q(x_i)\, T_{j-1}(x_i) \qquad i = 1, \ldots, (N-2) \qquad (6.6)$$
$$j = 1, \ldots, N$$
$$F_{i+2} = f(x_i) \qquad\qquad\qquad i = 1, \ldots, (N-2) \qquad (6.7)$$

where $T_{j-1, xx}$ denotes the second x-derivative of the $(j-1)$-st Chebyshev polynomial. The top two rows of the matrix impose the boundary conditions:

$$L_{1, j} = T_{j-1, x}(-1) \qquad \& \qquad F_1 = \alpha \qquad (6.8a)$$
$$L_{2, j} = T_{j-1}(1) \qquad \& \qquad F_2 = \beta \qquad (6.8b)$$

where $j = 1, \ldots, N$ and subscript x denotes differentiation with respect to x.

Eqs. (6.6) to (6.8) easily generalize. A fourth order differential equation, for example, would require four boundary rows instead of two. A Dirichlet boundary condition on $u(-1)$ would simply replace the first derivative of the Chebyshev polynomials by the polynomials themselves in (6.8a). The non-boundary elements of the square matrix $\vec{\vec{L}}$, given by here by (6.7), are always the contribution of the j-th basis function to the residual of the differential equation at the i-th interpolation point where j is the column index and $(i + 2)$ is the row index. There are some subtleties for partial differential equations which will be discussed later in the chapter.

The "boundary-bordering" method is applicable to even the most bizarre boundary conditions. For a given N, this approach does give a slightly larger matrix (dimension N versus $(N - 2)$ for a second order ODE) in comparison to the basis recombination method, but for large N, this is compensated by the higher cost of calculating the matrix elements in the latter technique.

6.5 "Basis Recombination" and "Homogenization of Boundary Conditions"

If the problem

$$L u = f \tag{6.9}$$

has inhomogeneous boundary conditions, then it may always be replaced by an equivalent problem with homogeneous boundary conditions, so long as the boundary conditions are linear. The first step is to choose a simple function $B(x)$ which satisfies the inhomogeneous boundary conditions. One may then define a new variable $v(x)$ and new forcing function $g(x)$ via

$$u(x) \equiv v(x) + B(x) \tag{6.10}$$
$$g(x) \equiv f(x) - L B(x) \tag{6.11}$$

so that the modified problem is

$$L v = g \tag{6.12}$$

where $v(x)$ satisfies homogeneous boundary conditions. For simplicity, (6.11) and (6.12) implicitly assume that L is a linear operator, but (6.10) and this technique for "homogenizing the boundary conditions" are applicable to nonlinear differential equations, too, if the boundary conditions are linear.

The shift function $B(x)$ is arbitrary except for the constraint that it must satisfy all the inhomogeneous boundary conditions. However, the simplest choice is the best: polynomial interpolation of the lowest order that works. For example, consider a two-point ODE boundary value problem with the mixed conditions

$$u_x(-1) = \gamma \quad \& \quad \frac{du}{dx}(1) - u(1) = \delta \tag{6.13}$$

Assume

$$B(x) = \Lambda + \Omega \, x \tag{6.14}$$

and then choose the two undetermined coefficients by demanding that $B(x)$ satisfy the boundary conditions. For example, in case (ii)

$$\frac{dB}{dx}(-1) = \gamma \quad \leftrightarrow \quad \Omega = \gamma$$
$$\frac{dB}{dx}(1) - B(1) = \delta \quad \leftrightarrow \quad \Omega - (\Lambda + \Omega) = \delta \quad \rightarrow \Lambda = -\delta \tag{6.15}$$

Once $B(x)$ has been found and the problem transformed to the new unknown $v(x)$, the next step is to actually do "basis recombination": Choosing simple linear combinations of the original basis functions so that these combinations, the new basis functions, individually satisfy the homogeneous boundary conditions. For example, suppose that

$$v(-1) = v(1) = 0 \tag{6.16}$$

The identity $T_n(\cos(t)) = \cos(nt)$ for $t = 0, \pi$ implies

$$T_{2n}(\pm 1) = 1 \qquad \& \qquad T_{2n+1}(\pm 1) = \pm 1 \tag{6.17}$$

so a good choice of basis functions such that $\phi_n(\pm 1) = 0$ for all n is

$$\phi_{2n}(x) \equiv T_{2n}(x) - 1 \qquad\qquad n = 1, 2, \ldots \tag{6.18a}$$

$$\phi_{2n+1}(x) \equiv T_{2n+1}(x) - x \qquad\qquad n = 1, 2, \ldots \tag{6.18b}$$

Note the indices in (6.18): Because of the two boundary constraints, $T_0(\equiv 1)$ and $T_1(\equiv x)$ are no longer independent basis functions, but rather are determined by the coefficients of all the higher functions.

The collocation points are the same as for boundary bordering: the $(N - 2)$ interior points of an N-point Gauss-Lobatto grid. The residual is not evaluated at the endpoints because the boundary conditions are implicitly used at these points instead.

We write

$$v(x) \approx \sum_{n=2}^{N-1} b_n \, \phi_n(x) \tag{6.19}$$

If we define a column vector $\vec{b}$ of dimension $(N - 2)$ whose i-th element is b_{i-1}, then the differential equation becomes the matrix problem

$$\vec{\vec{H}} \vec{b} = \vec{G} \tag{6.20}$$

where

$$H_{ij} \equiv \phi_{j+1,xx}(x_i) + q(x_i)\,\phi_{j+1}(x_i) \qquad i, j = 1, \ldots, N - 2 \tag{6.21}$$

where the double subscript x denotes double x-differentiation and

$$G_i \equiv g(x_i) \qquad\qquad i = 1, 2, \ldots, (N - 2) \tag{6.22}$$

After solving (6.20) to compute the coefficients of the ϕ-series for $v(x)$, it is trivial to convert the sum to an ordinary Chebyshev series

$$v(x) \equiv \sum_{n=0}^{N-1} a_n \, T_n(x) \tag{6.23}$$

by noting

$$a_n = b_n \qquad\qquad n \geq 2 \tag{6.24}$$

$$a_0 = - \sum_{n=1}^{(2n) \leq (N-1)} b_{2n} \qquad \& \qquad a_1 = - \sum_{n=1}^{(2n+1) \leq (N-1)} b_{2n+1} \tag{6.25}$$

By using the identity

$$\left.\frac{d^p T_n}{dx^p}\right|_{x=\pm 1} = (\pm 1)^{n+p} \prod_{k=0}^{p-1} \frac{n^2 - k^2}{2k + 1}, \tag{6.26}$$

one can derive similar basis sets for other boundary conditions. As an illustration, we give a set of functions that individually satisfy homogeneous Neuman conditions:

$$\phi_{2n}(x) \quad \equiv \quad \begin{cases} 1 & n = 0 \\ T_{2n}(x) - \left[\dfrac{n^2}{(n+1)^2}\right] T_{2n+2}(x) & n = 1, 2, \ldots \end{cases}$$

$$\{\phi_{n,x}(\pm 1) = 0\} \tag{6.27}$$

$$\phi_{2n+1}(x) \quad \equiv \quad T_{2n+1}(x) - \left[\frac{2n+1}{2n+3}\right]^2 T_{2n+3}(x) \qquad n = 0, 1, \ldots$$

This method is superior to "boundary-bordering" for eigenvalue problems whose boundary conditions do *not* involve the eigenvalue because "basis recombination" gives a spectral matrix which contains the eigenvalue in every row, and therefore can be attacked with standard library software. "Boundary-bordering" would give two rows which do not contain the eigenvalue, and this wrecks some library eigenvalue-solvers.

"Basis recombination" also gives a smaller matrix than boundary bordering: the dimension of $\vec{H}$ is only $(N-2)$ although the final answer (6.23) is the sum of the first N Chebyshev polynomials. In practice, because the matrix elements for the modified basis functions are more complicated to compute than those involving the unmodified basis functions of the "boundary-bordering" method, this advantage is non-existent except for very small N. (In Chapter 20, we show that "basis recombination" is almost always the method of choice for *analytical*, paper-and-pencil or algebraic manipulation language calculations.)

In the general case, the great advantage of "basis recombination" is conceptual simplicity. After we have shifted to the modified basis set, we can thenceforth ignore the boundary conditions and concentrate solely on the differential equation. The form of the pseudospectral matrix is the same for numerical boundary conditions as for behavioral boundary conditions: All the rows are derived from setting the residual of the differential equation equal to zero at a given collocation point.

In many situations, the differences between basis recombination and boundary bordering have little practical significance. Karageorghis (1993b) has discussed the relationship between basis recombination and boundary bordering formulations for the pseudospectral method in rectangles.

6.6   Basis Recombination for PDEs: Transfinite Interpolation

For partial differential equations, the principles are quite similar. Suppose, for example, that on the sides of the unit square $[-1, 1] \times [-1, 1]$, the boundary conditions are that

$$u(-1, y) = f_W(y), \quad u(1, y) = f_E(y), \quad u(x, -1) = f_S(x), \quad u(x, 1) = f_N(x) \tag{6.28}$$

We assume that the boundary conditions are continuous around the wall so that $f_W(-1) = f_S(-1)$, or in other words that two boundary functions which share a corner give the same values at that corner. (This condition is necessary to avoid strong singularities in the corners; the solution may be weakly singular at the corners even when the boundary conditions are continuous.) Then the so-called "transfinite" interpolation formula gives a quadratic polynomial in x and y which satisfies the boundary conditions:

$$
\begin{aligned}
B(x, y) \quad \equiv \quad & \frac{1-x}{2} f_W(y) + \frac{x+1}{2} f_E(y) + \frac{1-y}{2} f_S(x) + \frac{y+1}{2} f_N(x) \\
& -\frac{(1-x)(1-y)}{4} u(-1,-1) - \frac{(1-x)(y+1)}{4} u(-1,1) \\
& -\frac{(x+1)(1-y)}{4} u(1,-1) - \frac{(x+1)(y+1)}{4} u(1,1)
\end{aligned}
\tag{6.29}
$$

The linear terms (first row) are the same as for two independent one-dimensional interpolation problems. The second and third rows of quadratic terms correct for "interference" between the linear terms in the first row. Along the eastern boundary $x = 1$, for example,

$$
\begin{aligned}
B(1, y) \quad = \quad & f_E(y) + \frac{1-y}{2} f_S(1) + \frac{y+1}{2} f_N(1) - \frac{1-y}{2} u(1,-1) - \frac{y+1}{2} u(1,1) \\
= \quad & f_E(y)
\end{aligned}
\tag{6.30}
$$

as desired because the assumption of continuity around the boundary implies that $f_S(1) = u(1,-1)$ and $f_N(1) = u(1,1)$.

Nakamura (1996) gives a good treatment of transfinite interpolation with additional generalizations of the concept.

6.7   The Cardinal Function Basis

A cardinal function basis can always be used in place of the Chebyshev polynomials:

$$
u(x) = \sum_{j=1}^{N} u_j \, C_j(x)
\tag{6.31}
$$

where the cardinal functions, as in Chapter 5, are defined by the requirement that $C_j(x_i) = 1$ if $i = j$ and 0 otherwise. The unknowns u_j are now the values of $u(x)$ at the interpolation points, i. e., $u_j \equiv u(x_j)$.

The cardinal basis is very convenient with Dirichlet boundary conditions. With the "extrema-plus-endpoints" grid, one may enforce the boundary conditions $v(x_1) = v(x_N) = 0$ merely by omitting the two cardinal functions corresponding to the endpoints, simplifying (6.31) for the new unknown $v(x)$ to

$$
v(x) \approx \sum_{j=2}^{N-1} v_j \, C_j(x)
\tag{6.32}
$$

If we define a column vector $\vec{v}$ of dimension $(N-2)$ whose i-th element is v_{i-1}, then the differential equation becomes the matrix problem

$$
\vec{\vec{H}} \vec{v} = \vec{\vec{G}}
\tag{6.33}
$$

where

$$
H_{ij} \equiv C_{j+1,\,xx}(x_{i+1}) + q(x_{i+1}) \, C_{j+1}(x_{i+1}) \qquad i, j = 1, \ldots, N-2
\tag{6.34}
$$

where the double subscript x denotes double x-differentiation and the elements of $\vec{G}$ are $g(x_i)$ as in (6.22). The Gauss-Lobatto grid is:

$$x_i = \cos\left[\frac{\pi(i-1)}{N-1}\right] \qquad i = 2, \ldots, N-1 \qquad (6.35)$$

What is most striking, however, is how little is changed by the shift from the polynomial basis to the cardinal function basis. In particular, the unknowns are the coefficients of a spectral series (either polynomial or cardinal function), the matrix elements of $\vec{H}$ are identical except for the replacement of one set of basis functions by another, and the the column vector on the right-hand side is unchanged.

The truth is that if the basis functions individually satisfy the boundary conditions, the abstract formalism — the formulas for the matrix elements — are *independent* of the basis set. Only the numerical values of the matrix elements change when we replace Chebyshev polynomials by Legendre polynomials, or $T_j(x)$ by the corresponding cardinal function. The matrix problem is always

$$\vec{\vec{H}}\,\vec{a} = \vec{G} \qquad (6.36)$$

where

$$G_i = g(x_i) \qquad (6.37)$$
$$H_{ij} = (L\,\phi_j)\big|_{x=x_i} \qquad (6.38)$$

where $g(x)$ is the inhomogeneous term in the differential equation, L is the operator of the differential equation, and the notation in (6.38) means the elements of $\vec{H}$ are the result of applying the operator L to the j-th basis function and then evaluating the resulting residual at the i-th grid point. The column vector $\vec{a}$ contains the unknown coefficients of the spectral series. In the cardinal function basis, these coefficients are also the grid point values of $u(x)$.

6.8   The Interpolation Grid

As explained in Chapter 4, every standard basis function has one or two optimal grids associated with it. The Fourier, Chebyshev, and rational Chebyshev grids are given by analytical formulas involving nothing more exotic than trigonometric functions (Appendix F). When a given basis function has two "good" grids, the choice between them is strictly one of convenience. For Legendre polynomials, Hermite functions, Laguerre functions and spherical harmonics, the grid points are not known in analytical closed form except for small N. (The Legendre-Lobatto grid up to nine points are given by previously unpublished analytical formulas in Appendix F.) However, numerical tables for these points are given in mathematical handbooks such as Abramowitz and Stegun (1965) and may be calculated *ab initio* using the subroutines given in Canuto *et al.* (1988), Appendix C.

6.9   Computing the Basis Functions and Their Derivatives

All the standard basis functions *without exception* (including Fourier) may be computed via three-term recurrence relations as given in Appendix A. These recurrences may be evaluated as a single DO loop. What could be simpler?

Of course, to solve differential equations we also need the *derivatives* of the functions. However, the derivative of the n-th Hermite polynomial is proportional to the $(n-1)$-st

Hermite polynomial. The Chebyshev and Legendre polynomials belong to a family of polynomials known as the "Gegenbauer polynomials", which may be labelled by a parameter α. The derivative of any member of this family is another polynomial within the family: n is reduced by one while α is increased by 1. It follows that one may always evaluate the *derivatives* of Chebyshev and Legendre polynomials to *arbitrary order* by three-term recurrence. All the needed formulas are in Appendix A.

Texts tend to discuss these recurrences as if they were the only option because for Legendre, Hermite, Laguerre, and spherical harmonics, they *are* the only option. For those basis sets which are a disguised Fourier series, i. e. Chebyshev polynomials and the rational Chebyshev functions for infinite and semi-infinite intervals, we may alternatively compute the basis functions and their derivatives using trigonometric functions.

If x is the argument of the Chebyshev polynomials and t the argument of the trigonometric functions, then

$$x = \cos(t) \longleftrightarrow t = \arccos(x) \qquad x \in [-1, 1] \quad \& \quad t \in [0, \pi] \tag{6.39}$$

$$T_n(x) = \cos(nt) \tag{6.40}$$

$$\frac{dT_n}{dx} = \left[\frac{-1}{\sin(t)}\right] \frac{d}{dt}[\cos(nt)] = n\frac{\sin(nt)}{\sin(t)} \tag{6.41}$$

$$\frac{d^2T_n}{dx^2} = \left[\frac{-1}{\sin(t)}\right] \frac{d}{dt} \left\{\left[\frac{-1}{\sin(t)}\right] \frac{d}{dt}[\cos(nt)]\right\} \tag{6.42}$$

$$= \frac{-n^2}{\sin^2(t)} \cos(nt) + \left\{\frac{n \cos(t)}{\sin^3(t)}\right\} \sin(nt) \tag{6.43}$$

and so on, repeatedly applying the elementary identity $d/dx \leftrightarrow [-1/\sin(t)]d/dt$ when x and t are connected via (6.39). The strategy for the rational Chebyshev functions is similar; such change-of-coordinate tricks are so useful that we will devote a whole chapter to the subject.

The first table of Chapter 16 on coordinate-transformation methods gives a FORTRAN routine for computing $T_n(x)$ and its first four derivatives. It contains only 11 executable statements. All the details of the change-of-coordinate are buried in this subroutine; the calling program is blissfully ignorant.

There is one minor complication: the derivative formulas have numerators and denominators that are singular as $t \to 0, \pi$, which correspond to $x \to \pm 1$. Analytically, these formulas have finite limits that can be derived by l'Hopital's rule, which is simply the Taylor expansion of both numerator and denominator about their common zero, but blind numerical application of (6.40) and (6.42) will give overflow. In practice, this is no difficulty because (6.26) is a simple analytical expression for the exact endpoint derivatives of the p-th derivative of $T_n(x)$ for arbitrary p. Consequently, we may avoid overflow merely by adding an IF statement to switch to the boundary derivative formula at the endpoints.

After using recurrences early in my career (Boyd, 1978a) and trigonometric derivative formulas in recent years, I personally find the latter are simpler. However, the choice is mostly a matter of habit and preference. The trigonometric method requires fewer loops, but demands the evaluation of transcendentals; however, the cost of evaluating the basis functions is usually a negligible fraction of total running time. The trigonometric formulas become less accurate near the boundaries because both numerator and denominator are tending to 0, especially for the higher derivatives, but the derivative (Gegenbauer) recurrences are mildly unstable, especially for the higher derivatives. In spite of these mild deficiencies, however, both approaches are usually accurate to within one or two digits of machine precision for first and second order derivatives. Schultz, Lee and Boyd(1989) show that the trigonometric form is a little more accurate than recurrence.

Both because it is my personal preference and also because this method has been largely ignored in the literature, trigonometric derivative formulas are used in the sample program at the end of the chapter.

It is most efficient to write a single subroutine that will return *arrays* containing the values of all N polynomials and their derivatives at a single point; this simultaneous computation reduces the number of subroutine calls.

To implement basis recombination, it is easy to appropriately modify the basis-computing subroutine. Spectral methods lend themselves well to "data-encapsulation": the main program never needs to know even the identity of the basis set — only that there is a subroutine BASIS that can be called to return arrays with the values of the basis functions at a given x.

In the cardinal basis, the trigonometric formulas are unnecessary. *Analytical* formulas for derivatives at the grid points for the standard Fourier, Legendre and Chebyshev cardinal functions are given in Appendix F.

6.10 Special Problems of Higher Dimensions: Indexing

To solve a boundary value problem in two dimensions, it is most efficient to use a *tensor product* basis, that is, choose basis functions that are products of one-dimensional basis functions. Thus, on the unit square $[-1, 1] \times [-1, 1]$,

$$\Phi_{mn}(x, y) \equiv \phi_m(x)\, \phi_n(y) \qquad m = 1, \dots, Nx \ \& \ n = 1, \dots, Ny \qquad (6.44)$$

As in one dimension, one may apply either "boundary-bordering" or "basis recombination" to deal with the boundary conditions. Similarly, the interpolation grid is a tensor product, the $Nx\, Ny$ points

$$\vec{x} \equiv (x_i, y_j) \qquad i = 1, \dots, Nx \ \& \ j = 1, \dots, Ny \qquad (6.45)$$

One complication is that the grid points, the basis functions, and the coefficients of the spectral series now require *two* indices. The rules of matrix algebra, however, allow only one index for a vector and two for a square matrix, not four (i, j, m, n). The simplest remedy is to write a short subroutine to perform the preprocessing step of collapsing two indices into one and vice versa.

Thus, let ix and iy denote one-dimensional indices and let I vary over the range $1, 2, \dots, Nx$. Let $x(ix)$ and $y(iy)$ denote functions that compute the points on the one-dimensional grids. Then trace the double loop

```
I = 0
for ix = 1 to Nx, for iy = 1 to Ny
     I = I + 1
     XA(I) = x(ix),    YA(I) = y(iy)
        MA(I) = ix,    NA(I) = iy
          IA(MA(I),NA(I)) = I
  end loops
```

For example, suppose $Nx = Ny = 10$. The loop will then compute four one-dimensional arrays which each contain 100 elements. Each value of the "collapsed" index I specifies a unique point on the two-dimensional grid, and also a unique, two-dimensional basis function. When we compute the 100×100 pseudospectral matrix, the 70th row of the matrix is determined by evaluating the residual of the differential equation at the 70th grid point. What are the x and y values of that grid point? XA(70) and YA(70), respectively.

Similarly, the 31-st *column* of the square matrix is the contribution of the 31-st basis function to the differential equation residual at all the points on the grid. MA(31) and NA(31) give us the subscripts (m, n) in (6.44) so that we can compute this basis function. In symbols,

$$\vec{x}_I \equiv (XA(I), YA(I)) \tag{6.46}$$
$$I = 1, \ldots, Nx \cdot Ny$$
$$\Phi_I(x, y) \equiv \phi_{MA(I)}(x)\, \phi_{NA(I)}(y) \tag{6.47}$$

Thus, the four one-dimensional arrays associate a single index I with pairs of one-dimensional indices of smaller range. The two-dimensional array IA is used to go in the opposite direction: to compute I given (m, n). This is useful in printing a table of the spectral coefficients as functions of m and n.

As discussed in Chapter 8, it is often possible to greatly reduce the work by exploiting the *symmetries* of the problem. In higher spatial dimensions, these can become rather complicated. Boyd (1986c) describes an eigenvalue problem with an eight-fold symmetry. One may reduce the size of the basis by this same factor of eight by choosing a new basis, formed from linear combinations of Chebyshev polynomials, which all possess the same eight-fold symmetry as the solution. However, the symmetry-respecting basis is no longer a simple tensor product basis.

Another motive for rejecting a tensor product basis is to apply a "circular" rather than "rectangular" truncation. It is a little inconsistent to retain basis functions such as $\phi_{10}(x)\,\phi_{10}(y)$ while omitting functions such as $\phi_{11}(x)\,\phi_0(y)$ because the latter is actually smoother than the former. If we were using complex Fourier terms as the building blocks of our basis, then $\exp(i[10x + 10y])$ has a much larger total wavenumber than a deleted term such as $\exp(i\,11\,x)$ where the total wavenumber of $\exp(imx + iny)$ is $k \equiv \sqrt{m^2 + n^2}$. For this reason, many prefer the "circular" truncation,

$$\sqrt{m^2 + n^2} \leq Nx \qquad [\text{"circular"}] \tag{6.48}$$

instead of the simpler "rectangular" truncation, which keeps all basis functions whose one-dimensional indices satisfy the inequalities

$$m \leq Nx \qquad \& \qquad n \leq Ny \qquad [\text{"rectangular"}] \tag{6.49}$$

As one might imagine, symmetry-modified basis sets and circular truncation generate a complicated relationship between the one-dimensional indices (m, n) and the "collapsed" index I. However, the double loop strategy makes it quite unnecessary to have an analytical relationship between indices. It is merely necessary to insert the proper IF statements into the preprocessing double LOOP (to skip the assignment statements if $\sqrt{ix^2 + iy^2} > Nx$, for example). The rest of the program — everything outside this preprocessing subroutine and the code that computes the basis functions — is shielded from the complexities of the modified basis set. The basis function subroutine never performs any index calculations; given I, it reads the necessary values of (m, n) from the arrays MA(I) and NA(I).

It may seem silly to discuss such a humble matter as indexing in such detail, but the philosophy of this book is ruthlessly practical. The reason that a multi-dimensional BVP solver may take weeks or months to develop is not because of the intricacies of Sobolev spaces, but because of the care and fussing that is needed to correctly use the !*#$%&[expletive deleted] matrix indices.

6.11   Special Problems of Higher Dimensions: Boundary Conditions, Singular Matrices, and Over-Determined Systems

In two dimensions, a *singular* pseudospectral matrix seems to be a common experience. Part of the problem is boundary conditions: the four corner points are each a part of two different walls. Thus, imposing the condition that $u = 0$ at each of the Nx grid points on the top and bottom walls and each of the Ny grid points on the sides does not lead to $(2\,Nx + 2\,Ny)$ independent conditions; the corners are counted twice. A similar problem exists even for non-interpolating algorithms as shown by Eq. (14.8) of Gottlieb and Orszag (1977).

Another difficulty, especially when using a symmetry-modified basis or a "circular" truncation, is that some of the basis functions may lose linear independence on the grid. (Note that circular truncation requires reducing the number of grid points to match the reduction in the number of basis functions.) For example, $T_4(x)$ can be exactly interpolated by basis functions of lower degree on a grid that contains only four points. In one dimension, it is easy to avoid such absurdities; with complicated basis sets in higher dimensions, it may be quite difficult (Haupt and Boyd, 1988). (A Galerkin method, with numerical quadrature for the integrals, will solve these difficulties because one can use a rectangularly truncated quadrature grid even if the basis set itself has a circular truncation.)

Schultz, Lee and Boyd (1989), who obtained a singular pseudospectral matrix in attempting to solve the fourth order, two-dimensional partial differential equation for Stokes' flow in a box, evaded this problem by computing a least squares answer to the over-determined linear algebra system.

Indeed, experience has shown that large pseudospectral matrices are often mildly ill-conditioned even if non-singular. Fourteen decimal place precision or better is recommended, but occasionally this is not enough. (Increasing N and also increasing order of derivatives both amplify roundoff problems.) The Householder method, alias "QR" factorization, has lower round-off error than Gaussian elimination, and thus may be a helpful option for solving high order differential equations even if the pseudospectral matrix is not over-determined. Numerical checks described in Lee, Schultz and Boyd (1989b) show that the solution to the over-determined system has all the desirable properties of spectral solutions: very rapid convergence and extremely high accuracy for moderate cost.

Thus, redundant boundary conditions, loss of linear independence on the grid, and the weak ill-conditioning of very large pseudospectral systems are all difficulties that are almost trivially removed. It is essential, however, that the reader be prepared for these problems; the alternative is despair and cursing when a Gaussian elimination subroutine bombs out with the error message: SINGULAR MATRIX.

6.12   Special Problems in Higher Dimensions: Corner Singularities

When the walls of a computational domain meet at a sharp angle — the corners of a rectangle or a triangle as opposed to the ever-smooth walls of a circle — the solution to a partial differential equation will almost invariably be singular at the corner. The classic example is Poisson's equation $\triangle u = -1$ on the unit square with the boundary conditions that $u = 0$ on all four walls. Even though the boundary conditions are continuous, the coefficients of the differential equation and the forcing are constant, and the equation is linear, the solution is weakly singular. If we define a local polar coordinate system centered on one of the corners, then the most strongly singular term is of the form

$$r^2 \log(r) \sin(2\theta) \tag{6.50}$$

This geometrical difficulty occurs only at sharp corners and has no one-dimensional counterpart. If we "round" the corners even a little, then $u(x, y)$ is smooth everywhere within the domain including the boundaries. In the absence of rounding, however, the corner singularities will *dominate* the asymptotic behavior of the Chebyshev coefficients for sufficiently large (m, n), and the convergence will be algebraic rather than exponential.

However, articles such as Boyd (1986c) and Schultz, Lee and Boyd (1989) have already shown that for relatively weak singularities like that in (6.50), the corner branch points may be *irrelevant* unless one is interested in very high accuracy (greater than five decimal places). Note that the function in (6.50) is everywhere bounded; only its higher derivatives are infinite at $r = 0$, and it is in this sense that the singularity is "weak".

For stronger singularities, such as cracks in a solid material or obtuse angles such as that in the oft-studied "L-shaped domain eigenvalue problem", one needs special methods and tricks. Changes-of-coordinate and "singularity-subtraction" are explained in Chapter 16 and "global sectorial elements" in Chapter 22. The generic recommendation is to ignore the singularities unless either (i) one has prior knowledge that $u(x, y)$ is discontinuous or has other strongly pathological behavior or (ii) poor convergence and rapid variation of the numerical solution near the corners suggests *a posteriori* that the solution is strongly singular.

The vagueness of this advice is unsatisfactory. However, it is a fact of life that shock waves and other types of discontinuities always require special treatment whether the basic algorithm is spectral, finite difference, or finite element. If the solution is sufficiently well-behaved so that it can be computed by finite differences without tricks, then it can be computed by unmodified spectral methods, too, even with corners.

6.13 Matrix methods

Because inverting the (dense) pseudospectral matrix is usually the rate-determining step, we devote a whole chapter (Chapter 15) to good strategies for this. A couple of preliminary remarks are in order.

First, if the total number of basis functions is small ($N < 200$), use Gaussian elimination. The LU factorization of a 100×100 matrix requires less than a third of a second on a personal computer with a megaflop or better execution rate, so there is no excuse for being fancy. Even Boyd (1986c), which solves a two-dimensional nonlinear problem with 200 basis functions, used Gaussian elimination on a lowly IBM AT. For over-determined or ill-conditioned systems, a QR factorization should be used instead.

For large, multi-dimensional problems, Gaussian elimination may be too expensive. In this case, the iterations and multi-grid strategies described in Chapter 15 are very important, perhaps essential. They are almost *never* justified for one-dimensional problems. No one is awarded style points in numerical analysis: Keep-It-Simple,Stupid and use direct, non-iterative Gaussian elimination or Householder's method whenever possible.

6.14 Checking

The most reliable way to verify the *correctness* of a spectral code is to evaluate the residual by an independent algorithm. The recommendation is to use finite differences with a very small grid spacing h. Observe that since the pseudospectral approximation $u_N(x)$ is a series expansion, we can evaluate it and its derivatives for *arbitrary* x, not merely at values of x on the pseudospectral interpolation grid.

Thus, to verify that $u_N(x)$ is a good approximation to the solution to $u_{xx} + q(x)u = f(x)$, evaluate

$$R_{fd}(x) \equiv \frac{[u_N(x+h) - 2 u_N(x) + u_N(x-h)]}{h^2} + q(x) u_N(x) - f(x) \tag{6.51}$$

where $h \sim O(1/1000)$. The finite difference residual should be small over the whole computational interval, and especially small [equal to the finite differencing error] at the N points of the pseudospectral grid.

There is one mild caveat: for high order differential equations, the residual may be $O(1)$ even when $u_N(x)$ is a terrifically good approximation to $u(x)$. For example, suppose a differential equation of sixth order has an exact solution which is a trigonometric cosine polynomial of degree 30. Suppose that the pseudospectral method exactly computes all of the Fourier coefficients except for a_{30}, which is in error by 1×10^{-6} because of roundoff. The maximum pointwise error of the approximation to $u(x)$ is everywhere less than or equal to 1×10^{-6} [recall that all cosines are bounded by 1]. However, the sixth derivative of $u_{30}(x)$ multiplies each coefficient of $u_{30}(x)$ by n^6. The error in $u_{30,xxxxxx}(x)$ is $30^6 \times 10^{-6} = 729$! The residual will be large compared to one even though the pseudospectral solution is terrifically accurate.

It follows that calculating the residual of a numerical approximation is a very conservative way to estimate its error. By "conservative", we mean that the error in $u_N(x)$ will usually be much smaller than the residual. If k is the order of the differential equation, then $R_{fd}(x)$ is typically $O(N^k)$ [Fourier] or $O(N^{2k})$ [Chebyshev] times the error in $u_N(x)$ (Boyd, 1990c).

To verify the adequacy of a given truncation N, i. e. to estimate the error in $u(x)$, there are two good tactics. The reliable-but-slow method is to repeat the calculation with a different value of N. The error in the run with larger N is smaller (probably *much* smaller) than the difference between the two computations. The quick-and-dirty method is to simply make a log-linear plot of $|a_n|$ versus n. The error will be roughly the order of magnitude of the highest computed Chebyshev coefficient (or the magnitude of the "envelope" of the coefficients (Chapter 2) at $n = N$, if the a_n oscillate with n as they decay).

Definition 16 (IDIOT) *Anyone who publishes a calculation without checking it against an identical computation with smaller N OR without evaluating the residual of the pseudospectral approximation via finite differences is an IDIOT.*

The author's apologies to those who are annoyed at being told the obvious. However, I have spent an amazing amount of time persuading students to avoid the sins in this definition.

Gresho *et al.*(1993) describe an embarrassment in the published literature. The flow over a backward-facing step was used as a benchmark for intercomparison of codes at a 1991 minisymposium. Except for the spectral element code, all programs gave steady, stable flow . Despite the disagreement, the spectral element team published their unsteady, fluctuating solutions anyway in the *Journal of Fluid Mechanics*. This prompted the Gresho *et al.*(1993) team to reexamine the problem with multiple algorithms at very high resolution. They found that spectral elements did indeed give unsteady flow when the number of elements was small and the degree N of the polynomials in each element was moderate. However, when the resolution was increased beyond that of the *Journal of Fluid Mechanics* paper, either by using more subdomains or very high N within each subdomain, the flow

Table 6.1: Checking Strategies

1. Geometric decrease of coefficients a_n with n. (least reliable)
2. Finite difference residual (most conservative)
3. Varying the truncation N

became steady and in good agreement with the other finite difference and finite element simulations. It is easy for even experienced and knowledgeable people to be fooled!

One must watch the convergence of a numerical code as carefully as a father watching his four year old play close to a busy road.

The quick-and-not-so reliable tactic for estimating error is to simply look at the rate of decrease of the spectral coefficients (for a non-cardinal basis). If a_N is only 10 or 100 times smaller than a_0, then the calculation is certainly accurate to no more than 1 decimal place and may be completely in error. On the other hand, if a_N is $O(10^{-8})a_0$, then the solution is probably accurate to many decimal places.

One must be cautious, however, because this is an *optimistic* means of estimating error. To solve a differential equation of order k, one may need good accuracy in approximating the k-th derivative, and this is hard to estimate from the series for $u(x)$ itself. (A rough rule-of-thumb is to multiply a_n by n^k for Fourier or Chebyshev series.) A quick inspection of the Chebyshev coefficients is reliable for rejecting calculations whose accuracy is poor because N is too small. Coefficient-inspection is suggestive but *not reliable* for certifying that a computation is highly accurate. When inspection suggests that a run *might* be accurate, repeat with different N and be sure!

6.15 Summary

I assert that a well-organized pseudospectral code to solve a differential equation is only a little harder to write than its finite difference counterpart. Table 6.2 illustrates a sample program to support this contention. For clarity, comments are appended to the end of many lines (although not allowed in FORTRAN) and also deleted some type conversion statements.

Excluding comments, the output block, and the user-supplied functions which would have to be included in a finite difference code, too, the program contains only 58 lines. A two-dimensional boundary value solver would have an extra subroutine to compute the auxiliary indexing arrays, and the basis subroutine would be slightly longer. Nevertheless, one cannot escape the conclusion that solving BVP's using spectral methods is simply not very difficult.

Real world nonlinear eigenvalue solvers, of course, may be rather complicated. However, the complexity comes from path-following algorithms (Appendices C & D), subroutines to initialize the continuation, blocks of code to compute complicated equation coefficients and so on. These blocks would have to be included in the corresponding finite difference or finite element codes as well.

It follows that approximating the derivatives in the differential equation is almost never the hard part of writing a scientific code. Given that this is so, one might as well approximate the derivatives well instead of badly. One should solve the problem to high accuracy so that one can then forget the numerics, banish truncation worries from the mind, and concentrate on the *physics* of the solution. Pseudospectral methods do just this.

Table 6.2: A sample FORTRAN program for solving a two-point boundary value problem:

$$d_2(x)\, u_{xx} + d_1(x)\, u_x + d_0(x)\, u = f(x) \qquad x \in [-1, 1]$$

$$u(-1) = \alpha \qquad u(1) = \beta$$

```
      DIMENSION XI(20),APHI(20),G(20),H(20,20),UGRAPH(101)
      COMMON/BASIS/PHI(20),PHIX(20),PHIXX(20)
      PI = 3.14159...
C     SPECIFY PARAMETERS
      ALPHA = 1.              [ u(-1) ]
      BETA = -0.5             [ u(1) ]
      N = 22                  [No. of Chebyshev polynomials]
      NBASIS = N - 2          [No. of basis functions φⱼ(x); φⱼ(±1) = 0]
C     COMPUTE THE INTERIOR COLLOCATION POINTS AND THE FORCING VECTOR G
      DO 100 I=1,NBASIS
      XI(I)= COS(PI*I/(NBASIS+1))
      X = XI(I)
      B = ALPHA*(1-X)/2.+BETA*(1+X)/2.   [Function added to v(x) so that
                                          v(x) ≡ u(x) − B(x) satisfies
                                          homogeneous boundary conditions.]
      BX = (-ALPHA + BETA)/2.            [x-derivative of B(x)]
      G(I)=F(X) - D0(X)*B - D1(X)*BX     [modified inhomogeneous term]

  100 CONTINUE


C     COMPUTE THE SQUARE MATRIX
      DO 200 I=1,NBASIS
      X = XI(I)
      CALL BASIS(X,NBASIS)    [Compute all N basis functions & derivatives.
                              Results are returned via arrays in COMMON.]
      DD0 = D0(X)
      DD1 = D1(X)             [These three lines avoid needless calls to
      DD2 = D2(X)             D0(x), etc., inside the J loop.]
      DO 200 J=1,NBASIS
      H(I,J)=DD2*PHIXX(J)+DD1*PHIX(J)+DD0*PHI(J)
  200 CONTINUE
C     CALL A LINEAR ALGEBRA ROUTINE TO SOLVE THE MATRIX EQUATION

          H * APHI = G

      CALL LINSOLV(H,G,APHI,NBASIS)

C     The array APHI now contains the NBASIS coefficients of the
C     expansion for v(x) in terms of the basis functions φⱼ(x).
C     We may convert these coefficients into those of the ordinary
C     Chebyshev series as described in Sec. 5, but this is optional.
```

```
C       Make a graph of u(x) at 101 points by summing the basis function
C       [not Chebyshev] series.

        UGRAPH(1) = ALPHA
        UGRAPH(101) = BETA
        DO 300 I = 2,100
        X = -1 + (I-1)*0.02
        CALL BASIS(X,NBASIS)
C       Recall that u(x) = v(x) + B(x) where v(x) is given by the
C       computed series and B(x) is the R. H. S. of the next line.
        UGRAPH(I) = (ALPHA*(1-X) + BETA*(1+X))/2
        DO 350 J=1,NBASIS
  350   UGRAPH(I)=UGRAPH(I) + APHI(J) * PHI(J)
  300   CONTINUE                      [At the end of this loop, UGRAPH contains
                                      the desired values of u(x) ... input them to
                                      your favorite plotter!]
        END


        SUBROUTINE BASIS(X,NBASIS)
        COMMON/BASIS/PHI(20),PHIX(20),PHIXX(20)
C       After this call, the arrays PHI, PHIX, and PHIXX contain the
C       values of the basis functions (and their first two derivatives,
C       respectively, at X.
        IF (ABS(X).LT.1) THEN   [block below is executed only on the
                                interior. This IF branches to a later
                                block to evaluate functions at x = ±1.]
        T = ACOS(X)             [T is the argument of the trig. functions]
          C = COS(T)
          S = SIN(T)
        DO 100 I=1,NBASIS
        N = I+1
        TN = COS(N*T)           [Trig. form of T_N(x)]
        TNT = - N * SIN(N*T)    [Derivative of T_N with respect to t]
        TNTT= - N*N * TN        [Second t-derivative of T_N]


C       Convert t-derivatives into x-derivatives

        TNX = - TNT / S         [x-derivative of N-th Chebyshev polynomial.]
        TNXX= TNTT/(S*S) - TNT*C/(S*S*S) [Second x-derivative]


C       Final step: convert T_Ns into the basis functions φ_Ns.
C       We subtract 1 from the even degree TN's, x from the odd degree
C       T_N and 1 from the first derivative of the odd polynomials only


        IF (MOD(N,2).EQ.0) THEN
            PHI(I) = TN - 1.
            PHIX(I) = TNX
        ELSE
            PHI(I) = TN - X
            PHIX(I)= TNX - 1.
```

```
      ENDIF
         PHIXX(I) = TNXX
100 CONTINUE

      ELSE                          [Alternative formulas when X = ±1]
      DO 200 I=1,NBASIS
      PHI(I) = 0.
      N = I+1
      IF (MOD(N,2).EQ.0) THEN
      PHIX(I)= SGN(X,1.)*N*N
      ELSE
      PHIX(I)= N*N - 1
      ENDIF
      PHIXX(I) = (SGN(X,1.))**N * N*N * (N*N-1.)/3.
200 CONTINUE
      ENDIF
      RETURN
      END
```

```
[ADDITIONAL USER-SUPPLIED ROUTINES: D0(X), D1(X), D2(X), F(X) PLUS TH
LINEAR ALGEBRA ROUTINE LINSOLV.]
```

Chapter 7

Linear Eigenvalue Problems

"It is not the process of linearization that limits insight. It is the nature of the state that we choose to linearize about."
— Erik Eady

7.1 Introduction: the No-Brain Method

The default, almost-no-thinking-required method for solving linear eigenvalue problems is summarized in Table 7.1. There are two difference from a linear boundary value problem: (i) the matrix equation is solved by a different linear algebra library routine (eigensolve instead of matrixsolve) and (ii) comparisons for two different N must be mode-by-mode. There would seem to be no need for a separate chapter on this topic.

Indeed, often eigenproblems are very simple. Unfortunately, there are a variety of complications that can arise. This chapter might be more appropriately titled: "Damage Control for Eigenproblems". No need to read it when solving easy, standard Sturm-Liouville eigenproblems, but as valuable as a lifevest and a waterproof radio after your code has been torpedoed by one of the difficulties explained below.

Some of these traps, mines and torpedoes are the following:

1. The QR/QZ algorithm, which costs $O(10N^3)$ for an $N \times N$ matrix, may be too expensive for multidimensional problems.

Table 7.1: Default Method for Eigenproblems

Step No.	Procedure
One	Apply spectral method to convert differential or integral equation to a matrix problem, just as for BVP
Two	Call a matrix eigensolver from linear algebra library (The QR/QZ algorithm is a robust blackbox that finds all eigenvalues and eigenfunctions without user input except the matrix.)
Three	Repeat for different N. Trust only those eigenvalues and eigenfunctions which are the same for both resolutions

2. Instead of a discrete, countable infinity of eigenmodes, there may be a continuous spectrum, or a few discrete modes plus a continuous spectrum.

3. It may be tedious to compare eigenvalues for different N to reject those which are numerically inaccurate.

4. Generalized eigenproblems often have a couple of very large eigenvalues which are *physically* spurious in the sense of resulting from inconsistent application of boundary conditions.

5. In hydrodynamic stability theory, weakly unstable or neutral modes may be singular or very nearly singular on the interior of the computational domain

6. The eigenparameter may occur as a polynomial in an otherwise linear problem.

7. High order differential equations seem to be quite common in eigenproblems and may have poor numerical conditioning, i. e., large roundoff errors.

In the rest of this chapter, we explain how to survive these shipwrecks. First, though, we begin with some basic definitions and comments on the "no-brain" strategy of Table 7.1.

7.2   Definitions and the Strengths and Limitations of the QR/QZ Algorithm

Definition 17 (Eigenvalue Problem) *A LINEAR EIGENVALUE problem is an equation of the form*

$$Lu = \lambda M u \qquad (7.1)$$

where L and M are linear differential or integral operators or are square matrices and λ is a number called the EIGENVALUE. The boundary conditions, if any, are homogeneous so that Eq.(7.1) has solutions only when λ is equal to a set of discrete values or is on a continuous range. When M is the identity operator so the right side of (7.1) simplifies to λu, the problem is said to be a regular eigenvalue problem; otherwise, it is a "generalized" eigenproblem.

Definition 18 (Eigenvectors/Eigenfunctions) *The solutions u to a linear eigenvalue problem are called "EIGENVECTORS" (if the operators L and M are matrices) or "EIGENFUNCTIONS" (if L and M are differential or integral operators). The eigenvectors are labelled by an index "j", the "MODE NUMBER", which is usually chosen so that the corresponding discrete eigenvalues are ordered from smallest to largest with the lowest j (usually either 0 or 1) corresponding to the smallest eigenvalue. When the eigenfunctions exist for a continuous interval in λ (not possible for matrix problems), the eigenfunctions are labelled by λ and are said to be "CONTINUOUS" eigenfunctions.*

A simple illustration of a differential eigenproblem, later dubbed "Example One", is

$$u_{xx} + \lambda u = 0, \qquad u(-1) = u(1) = 0 \qquad (7.2)$$

which has the exact eigenmodes

$$u_j(x) = \begin{cases} \cos\left(j\,\frac{\pi}{2}x\right), & j = \text{positive odd integer} \\ \sin\left(j\,\frac{\pi}{2}x\right), & j = \text{positive even integer} \end{cases} \qquad (7.3)$$

$$\lambda_j = j^2 \frac{\pi^2}{4}, \qquad j = 1, 2, 3, \ldots \qquad (7.4)$$

When L and M are matrices, there is good and widely available software for solving eigenproblems. Indeed, such commands are built-in to languages such as Matlab, Maple and Mathematica. The Matlab command to find both eigenfunctions and eigenvalues is simply: **[Eigenfunctions,Eigenvectors]=eig(L,M)**. FORTRAN subroutines to do the same may be found in the public domain EISPACK and LINPACK libraries as well in proprietary libraries like NAG and IMSL. These codes are very robust, and rarely fail except by being too expensive.

Library software for the algebraic eigenproblem usually employs an algorithm called the QZ method (for the generalized eigenproblem) or its cousin the QR scheme (for the regular eigenproblem). The good news is that (i) these algorithms require as input nothing but the square matrices that define the eigenproblem and (ii) reliably compute all the matrix eigenvalues and eigenfunctions. The bad news is that QR/QZ is *slow*: for an $N \times N$ matrix, the cost is $O(10N^3)$ operations. This is an order of magnitude slower than Gaussian elimination (LU factorization) of a dense matrix of the same size. Unlike almost all other matrix algorithms, QR/QZ schemes cannot exploit matrix sparsity because zero elements are all *filled in* as the algorithm proceeds.

Gary and Helgason (1970) pointed out that this implies: Hurrah for high order discretizations! The reason is that high order methods make it possible to resolve a given number of eigenmodes with much smaller N than with lower order methods. Replacing a 100-point second order computation by an equally accurate tenth order matrix which is only 40×40 reduces the cost of the QR algorithm by a factor of 15!

Orszag (1971b) noted that Chebyshev pseudospectral methods are the ultimate in high order, and reduce N still further from the values needed by the eighth and tenth order finite difference methods of Gary and Helgason.

It is ironic: the QR/QZ algorithm has no direct connection with spectral methods. Nevertheless, when this is the chosen matrix eigensolver, the QR algorithm cries out for a spectral discretization.

On modern workstations, almost all one-dimensional and some two-dimensional and three-dimensional eigenproblems can be solved efficiently by the pseudospectral/QZ combination. Unfortunately, when one needs to resolve multidimensional modes with lots of fine structure, or worse still, needs to compute the eigenvalues throughout a multidimensional parameter space, the QR algorithm may be unaffordable.

In later sections, we therefore describe two representative alternatives: the power and inverse power methods. Both are "local" methods in the sense that they compute only one eigenvalue at a time. In contrast, the QR/QZ method is "global" because it computes all N matrix eigenvalues without requiring any input from the user except the matrices themeselves. With local methods, it is easy to miss eigenvalues; we offer some cautionary tales below. However, the cost per point in parameter space is usually an order of magnitude less than the QR scheme.

7.3   Numerical Examples and the Eigenvalue Rule-of-Thumb

Fig. 7.1 illustrates the errors in computing eigenvalues of Example One (Eq. 7.2). The error tolerance is of course both user-dependent and problem-dependent. For the sake of discussion, we shall arbitrarily define a "good" eigenvalue as one whose absolute error is less than or equal to 0.01, as marked by the horizontal dividing line in the graph.

By this criterion, the 16-point discretization returns seven "good" eigenvalues and nine "bad" eigenvalues. Fig. 7.2 compares a "good" (top) and "bad" eigenfunction as computed

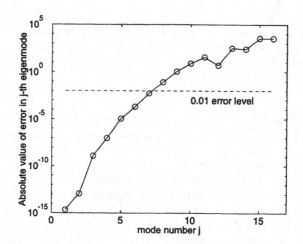

Figure 7.1: Example One: Absolute errors in the eigenvalues as given by a 16-point Cheby-shev pseudospectral discretization. The horizontal dividing line separates the "good" eigenvalues from the "bad" where the "good" eigenvalues are defined (arbitrarily) to be those whose absolute error is 0.01 or less.

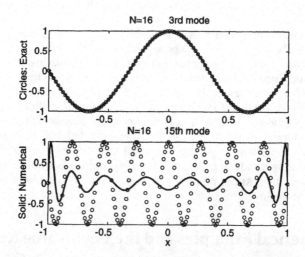

Figure 7.2: Example One: 16-point Chebyshev pseudospectral discretization. Exact (cir-cles) and numerical (solid) approximations to the eigenmodes. Upper panel: Third Mode. Lower panel: Fifteenth mode.

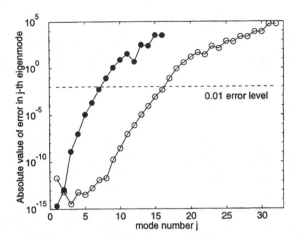

Figure 7.3: Example One: Absolute errors in the eigenvalues for 16-point (solid) and 32-point (circles) Chebyshev pseudospectral discretization.

numerically with the corresponding exact modes. The third mode, because it oscillates slowly with x, is well-resolved and the exact and numerical eigenfunctions are graphically indistinguishable. The fifteenth mode, however, is oscillating so rapidly that it cannot be resolved by the first sixteen Chebyshev polynomials. To be sure, $T_{15}(x)$ has fifteen roots, just like the exact fifteenth mode, but the Chebyshev oscillations are not uniform. Instead, $T_{15}(x)$ oscillates slowly in the center of the interval and very rapidly near to the endpoints, in contrast to the uniform oscillations of the eigenmode it is vainly trying to mimic.

To resolve more eigenmodes, merely increase N as illustrated in Fig. 7.3, which compares the errors for 16-point and 32-point calculations. The number of good eigenvalues has risen from seven to sixteen. The graph is flat for small j because of roundoff error. In multiple precision, the errors for the first three or four modes would be smaller than 10^{-15}, off the bottom of the graph!

For "nice" eigenvalue problems, this behavior is typical. To show this, consider a second example which is posed on an *infinite* interval and is solved using not the Chebyshev polynomials but rather the "rational Chebyshev" functions $TB_n(x)$, which are a good basis for an unbounded domain (Chapter 17).

EXAMPLE TWO:

$$u_{xx} + (\lambda - x^2)\, u = 0, \qquad |u| \to 0 \text{ as } x \to \infty \qquad (7.5)$$

The exact eigenfunctions are the Hermite functions,

$$u_j(x) = \exp\left(-\frac{1}{2}x^2\right) H_j(y) \qquad (7.6)$$

where H_j is a polynomial of degree j, the j-th Hermite polynomial. The eigenvalues are

$$\lambda_j = 2j + 1, \qquad j = 0, 1, 2, \dots \qquad (7.7)$$

Fig. 7.4 shows that again the lowest few modes are accurately approximated. The infinite interval problem is harder than a finite interval problem, so there are only four "good"

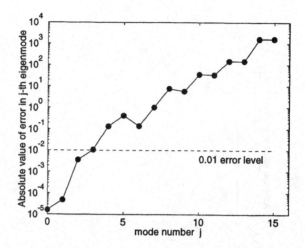

Figure 7.4: Example Two (Eigenfunctions of parabolic cylinder equation on an infinite interval): Absolute errors in the eigenvalues as given by a 16-point Rational Chebyshev (TB_n) discretization with the map parameter $L = 4$.

eigenvalues versus seven for the 16-point discretization of Example One. Fig. 7.5 shows that the second eigenmode is well-approximated, but the seventh eigenmode is poorly approximated. The most charitable comment one can make about the seventh numerical eigenfunction is that it vaguely resembles the true eigenmode by having a lot of wiggles. Fig. 7.6 confirms that increasing N also increases the number of "good" eigenvalues, in this case from four to ten.

It is important to note that for both examples, the approximation to the lowest mode is extremely accurate with errors smaller than 10^{-11} for $N = 32$ for both examples. The error *increases exponentially fast with mode number* j until finally the error is comparable in magnitude to the eigenvalue itself.

These examples suggest the following heuristic.

Rule-of-Thumb 8 (EIGENVALUE RULE-OF-THUMB)

In solving a linear eigenvalue problem by a spectral method using $(N + 1)$ terms in the truncated spectral series, the lowest $N/2$ eigenvalues are usually accurate to within a few percent while the larger $N/2$ numerical eigenvalues differ from those of the differential equation by such large amounts as to be useless.

Warning #1: the only reliable test is to repeat the calculation with different N and compare the results.

Warning #2: the number of good eigenvalues may be smaller than $(N/2)$ if the modes have boundary layers, critical levels, or other areas of very rapid change, or when the interval is unbounded.

Although this rule-of-thumb is representative of a wide class of examples, not just the two shown above, nasty surprises are possible. We attempt a crude classification of linear eigenvalue problems in the next section.

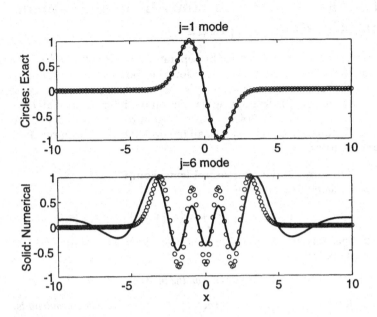

Figure 7.5: Example Two (Exact modes are Hermite functions): 16-point Rational Chebyshev pseudospectral method. Exact (circles) and numerical (solid) approximations to the eigenmodes. Upper panel: Second Mode. Lower panel: Seventh mode.

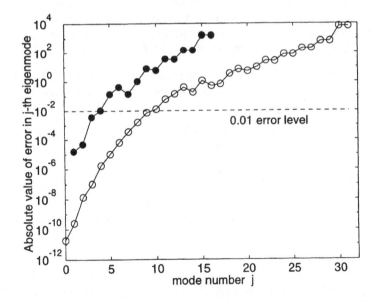

Figure 7.6: Example Two (Infinite Interval): Absolute errors in the eigenvalues for 16-point (solid disks) and 32-point (open circles) TB_n discretization.

7.4   Four Kinds of Sturm-Liouville Eigenproblems and Continuous Spectra

Solving partial differential equations by separation-of-variables generates eigenproblems of the form of Eq. 7.8 below. Sturm and Liouville showed in the mid-nineteenth century that as long as the equation coefficients $p(x)$ and $r(x)$ are positive everywhere on the interval $x \in [a, b]$ and $q(x)$ is free of singularities on the interval, then all eigenfunctions are discrete and orthogonal. Such "nice" classical eigenproblems are very common in applications, but unfortunately do not exhaust all the possibilities. This motivated the following classification scheme.

Definition 19 (Sturm-Liouville Eigenproblems: Four Kinds)
   A Sturm-Liouville eigenproblem is

$$[p(x)u_x]_x + \{q(x) + \lambda r(x)\} u = 0, \qquad x \in [a, b] \tag{7.8}$$

subject to various homogeneous boundary conditions. There are four varieties in the classification scheme of Boyd(1981a):

- *First Kind: p, q, r analytic everywhere on the interval*

- *Second Kind: p, q, r analytic everywhere on the interval except the endpoints.*

- *Third Kind: differential equation has singularities on the interior of the interval, but the singularities are only "apparent".*

- *Fourth Kind: differential equation and eigenfunctions are singular on the interior of the interval.*

SL problems of the First Kind are guaranteed to be "nice" and well-behaved. However, problems of the Second Kind, which includes most eigenproblems on an unbounded domain, may be either regular, with a discrete infinity of eigenvalues all of the same sign and orthogonal eigenfunctions, or they may have only a finite number of discrete eigenvalues plus a continuous eigenspectrum.

The differential equation satisfied by the Associated Legendre functions is a good illustration of the Second Kind. The Legendre functions are the solutions on $x \in [-1, 1]$ of

$$D^2 u + \{ \lambda(1 - x^2) + E\}u = 0 \tag{7.9}$$

where the differential operator D is

$$D \equiv (1 - x^2) \frac{d}{dx} \tag{7.10}$$

The eigenfunctions are those solutions which are regular at the endpoints except perhaps for a branchpoint:

$$u = P_n^m(x), \qquad \lambda = n(n + 1), \qquad E = -m^2 \tag{7.11}$$

When derived from problems in spherical coordinates, x is the cosine of colatitude and the periodicity of the longitudinal coordinate demands that m be an integer. Then Eq.(7.9) is a regular Sturm-Liouville problem of the second kind and λ is the eigenvalue. For each longitudinal wavenumber m, there is a countable infinity of discrete eigenvalues: n must be an integer with the further requirement $n \geq |m|$.

However, the same differential equation is also an important exactly-solvable illustration of the stationary Schroedinger equation of quantum mechanics with radically different behavior: a continuous spectrum plus a finite number of discrete eigenmodes. The crucial difference is that the roles of the parameters λ and E are interchanged in quantum mechanics: E is now the eigenvalue ("energy") and λ measures the strength of the specified potential.

The unimportant difference is that quantum problems are usually posed on an infinite interval, which requires the change of variable

$$x = \tanh(y) \qquad [\text{"Mercator coordinate"}] \qquad (7.12)$$

The Legendre equation (7.9) becomes

$$u_{yy} + \{ \lambda \operatorname{sech}^2(y) + E\}u = 0, \qquad y \in [-\infty, \infty] \qquad (7.13)$$

For all positive E, the equation has solutions which are everywhere bounded; these are the continuous spectrum or "continuum". Since the sech^2 potential decays exponentially fast to zero as $|y| \to \infty$, the continuum eigenmodes are asymptotically proportional to $\exp(iE^{1/2})$.

The discrete modes are spatially localized around the origin and have negative E. Introducing the auxiliary parameter $\nu(\lambda)$ via

$$\lambda \equiv \nu(\nu+1) \qquad (7.14)$$

the discrete eigenvalues are

$$E_j = -(\nu - j)^2, \quad j = 0, 1, ..., j_{max} \qquad (7.15)$$

where j_{max} is the largest integer smaller than ν, i. e., the number of allowed modes is one plus the integer part of ν.

The good news is that both the continuum and discrete eigenmodes can be calculated by spectral methods. The continuous spectrum requires special tricks; see Sec. 4 of Chapter 19 for an example. The discrete modes can be computed by exactly the same tricks as used for the quantum harmonic oscillator, whose countable infinity of modes are the Hermite functions. The only complication is that the number of "good" modes is now fixed, and will not increase above $j_{max} + 1$ even if we used 10,000 spectral basis functions.

Laplace's Tidal Equations are even trickier, an eigenproblem of the "Third Kind" in Boyd's terminology:

$$
\begin{aligned}
\lambda u - xv - s\zeta &= 0 \\
xu - \lambda v + D\zeta &= 0 \\
su - Dv - \epsilon\lambda(1 - x^2)\zeta &= 0
\end{aligned}
\qquad (7.16)
$$

where x is the coordinate (cosine of colatitude) and $D \equiv (1 - x^2)d/dx$ as for Legendre's equation, u and v are the horizontal fluid velocities, and ζ is the sea surface height (in oceanographic applications) or the pressure (in meteorological use). Depending on the application, any of the set of three parameters (λ, s, ϵ) may be the eigenvalue where λ is the nondimensional frequency, s is the zonal wavenumber and ϵ is "Lamb's parameter", which is proportional to the depth of the water or (in meteorology) to the vertical wavelength.

The eigenproblem is of the Third Kind because for modes of low frequency, the differential system is singular at those latitudes ("inertial latitudes" or "critical latitudes") where $\lambda = x$. However, the eigenfunctions themselves are well-behaved and analytic even at the inertial latitudes.

When the goal is to calculate the free tidal oscillations of a global ocean, s and ϵ are known and the frequency is the eigenvalue. All modes are discrete and this case is relatively uncomplicated.

For forced atmospheric tides, s and the frequency λ are known. However, the earliest calculations agreed badly with observations, even the limited data of the 1930's. All sorts of imaginative (but wrong!) theories were proposed, but the correct explanation was given independently after thirty years of confusion by Lindzen and by Kato. It had been assumed by analogy with Sturm-Liouville eigenproblems of the First Kind that all the eigenvalues ϵ were positive. Actually, because of the inertial latitudes, the diurnal tide also has an infinite number of modes with negative ϵ. One class of modes is oscillatory at low latitudes (between the inertial latitudes) and decays exponentially beyond the inertial latitudes as the poles are approached. The other class of modes is confined mostly poleward of the inertial latitudes and has little amplitude in the tropics. The singularities of the differential equation are crucial even though the eigenfunctions themselves are paragons of mathematical regularity.

A third application (O'Connor, 1995, 1996, Boyd, 1996c) is ocean oscillations in a basin bounded by meridians where now the eigenparameter is the zonal wavenumber s. Complex eigenvalues are possible, and merely describe oscillations whose amplitude decays exponentially with increasing distance from the coast. The surprise, as yet unsupported by rigorous proof, is that the eigenvalues are a mixture of a discrete and a continuous spectrum. In addition, unless s is an integer, the eigenfunctions have weak singularities at the poles. This implies that for the spectral series of an eigenmode (using any standard basis), the asymptotic rate of convergence is algebraic rather than exponential unless special tricks such as an exponential change-of-coordinate are used (Chapter 16, Sec. 5).

Lastly, Boyd (1981a, 1982a, 1985a) has studied Sturm-Liouville eigenproblems of the Fourth Kind such as

$$u_{xx} + (1/x - \lambda)u = 0, \qquad u(a) = u(b) = 0 \qquad (7.17)$$

where the interval $x \in [a, b]$ *spans* the origin. Superficially, this looks like a self-adjoint SL problem of standard form, which would be expected to have only real discrete eigenvalues. In reality, the presence of the pole in the coefficient of the undifferentiated term changes everything. The pole must be interpreted as the limit

$$1/(x - i\delta) \qquad (7.18)$$

as δ tends to zero where δ represents viscosity. For non-zero δ, there is a small viscous layer of thickness proportional to δ. Outside this layer, the eigenfunctions are effectively inviscid. In the limit $\delta \to 0$, the thickness of the viscous layer shrinks to zero and the eigenfunctions are singular as $x \log(x)$ at $x = 0$. The eigenfunctions and eigenvalues are both complex-valued; one must circle the branchpoint below the real axis to obtain the branch which is the correct limit of the viscous solution.

The good news is that Boyd (1981a, 1982a) and Gill and Sneddon(1995, 1996) developed special spectral methods to compute the singular eigenfunctions. The bad news is that Boyd (1985a) showed that the author's earlier papers had missed a root.

Clearly, Sturm-Liouville eigenproblems can be full of surprises. Hydrodynamic stability problems, which are not self-adjoint and usually have complex-valued eigenvalues and nearly-singular, complex-valued eigenfunctions, merely reiterate this theme: Eigenproblems can be nasty, and it is terribly easy to be speared by internal layers or singularities, or miss modes entirely.

7.5 Winnowing the Chaff: Criteria for Rejecting Numerically Inaccurate or Physically Non-Existent Eigenvalues

It is always a good idea to repeat each calculation twice with different N to verify that the solution is well-resolved. With eigenproblems, this is doubly important because, as expressed by the Eigenvalue Rule-of-Thumb, many of the eigenvalues of the discretization matrix are numerical nonsense unrelated to the eigenvalues of the underlying differential or integral eigenproblem. To know how many modes are "good", that is, have been computed accurately, one must compare the list of eigenvalues for two different N and accept only those which are the same (within some user-set tolerance) on both lists.

One difficulty is that if one is performing a large number of eigencalculations, comparing-by-eye can lead to eyestrain, headaches, and a firm resolve to find a new, more exciting profession like accounting or tax law. This is one motive for inventing some simple graphical and numerical methods for reliably separating "good" eigenvalues from trash.

Another is that we often do not know a priori how many discrete eigenmodes even exist, as illustrated by the Legendre equation.

Our recommended strategy is to make a plot on a logarithmic scale of the reciprocal of the difference between corresponding eigenvalues as calculated at different resolutions N, scaled by some measure of the size of the eigenvalues. The reciprocal of the difference is plotted so that the "good" eigenvalues are at the top of the graph; a semilogarithmic plot is recommended because the accuracy varies exponentially with mode number j as already seen above. There are two minor complications.

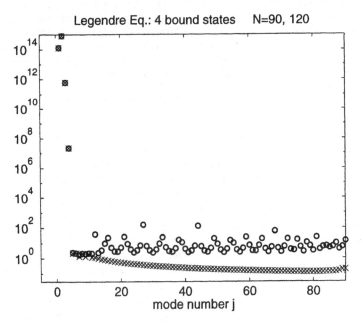

Figure 7.7: The reciprocal eigenvalue drift ratios $1/\delta_{j,nearest}$(circles) and $1/\delta_{j,ordinal}$ (x's) are plotted on a logarithmic scale versus mode number j for Legendre's equation $u_{yy} + \{\,15.75\,\mathrm{sech}^2(y) + E\}u = 0$ where E is the eigenvalue (the energy). The modes are ordered by real part of E with $j = 1$ the smallest. For this case, there are precisely four bound states which are the four numerical eigenvalues with $E < 0$.

First, the obvious scaling for the j-th eigenvalue is $|\lambda_j|$ itself. However, an eigenvalue may accidentally be very close to zero. Furthermore, for a "nice" Sturm-Liouville problem, the eigenvalues typically grow as $O(j^2)$, but the *difference* between adjacent eigenvalues grows only as fast as $O(j)$. We can solve both these difficulties by defining an "intermodal separation" via

$$\sigma_1 \equiv |\lambda_1 - \lambda_2|$$
$$\sigma_j \equiv \frac{1}{2}\left(|\lambda_j - \lambda_{j-1}| + |\lambda_{j+1} - \lambda_j|\right), \qquad j > 1 \tag{7.19}$$

and then conservatively scale the eigenvalues by σ_j. (In the case of degeneracy, that is, two or more eigenfunctions with the same eigenvalue, it is assumed that the differences in Eq. (7.19) are taken between the non-degenerate modes nearest to mode j.)

Second, in many problems, the eigenvalue ordering is invariant when the resolution is changed — at least after the eigenvalues for each run have been sorted by magnitude or real part — and it is sufficient to plot the reciprocal of what we shall dub the (scaled) "ordinal" difference

$$\delta_{j,ordinal} \equiv |\lambda_j(N_1) - \lambda_j(N_2)| / \sigma_j \tag{7.20}$$

where the arguments denote the number of degrees of freedom N in the low and high resolution computations. However, for many problems, such as Laplace's Tidal Equation (O'Connor, 1995, 1996, Boyd, 1996c), the well-resolved eigenvalues are not evenly spaced. (Tidal oscillations fall into two main classes: Rossby waves, which have low frequency, and gravity waves, which have high frequency. Thus, the gravest Rossby and gravity modes (well-resolved) are separated by an infinite number of higher Rossby modes of intermediate frequency, which are not as accurately computed.) For such more intricate relationships between eigenvalue magnitude and the spatial scale of the wave, one needs to compare the j-th low resolution mode with whatever eigenvalue of the high resolution computation agrees most closely with it. This "nearest" difference (scaled) is

$$\delta_{j,nearest} \equiv \min_{k \in [1, N_2]} |\lambda_j(N_1) - \lambda_k(N_2)| / \sigma_j \tag{7.21}$$

Fig. 7.7 is a plot of the scaled differences or "drift-with-N" for the equation solved by the Associated Legendre equation using 90 and 120 rational Chebyshev functions on the infinite interval. The exact solution has four discrete modes plus a continuous spectrum. The discrete modes of the spectral matrix cannot converge to the continuous spectrum, so that the best we can do is to resolve the four discrete modes. The graph shows that with $N = 90$, these modes are very well-resolved in the sense that these eigenvalues, scaled by the smaller of $|\lambda_j|$ or σ_j, change by less than one part in a million when the basis size is increased to $N = 120$. For this case, the "ordinal ratio" would have been sufficient because the discrete modes are the four smallest numerically computed eigenvalues at both resolutions; the "ordinal" and "nearest" ratios are so close that the circles and x's are superimposed. For the unresolved modes of the continuous spectrum (bottom of the graph), the "nearest" ratios are consistently a little larger than the "ordinal" ratios. The reason is that by coincidence, two inaccurate eigenvalues for different resolution may be rather close, producing a nearest ratio which is small even though the spectral method is generating random numbers for these modes. It is both quicker and less confusing to use the "ordinal" ratio wherever possible. As shown in Boyd(1996c), however, other eigenproblems such as Laplace's Tidal equation absolutely require the "nearest" ratio.

Fig. 7.7 is easy to interpret — four discrete modes, all the rest nonsense — because the reciprocal ratios $1/\delta_j$ are so many orders of magnitude larger for the "good" modes than for the bad. A low precision finite difference calculation would be much more ambiguous – is this eigenvalue real or just a numerical artifact that would jump around at higher resolution?

7.6   The Curse of "Spurious" Eigenvalues

Generalized eigenproblems may have some eigenvalues which are *physically* spurious rather than merely numerically underresolved. The difficulty was first characterized by Gottlieb and Orszag (1977, pg. 145), who noted "low modes are given accurately ... but there appear spurious unstable modes with large growth rates. Similar spurious unstable modes appear in finite-difference solution of the Orr-Sommerfeld equation." (pg. 145).

Much time and energy has been expended in the invention of slight modifications to standard spectral methods to solve the "spurious eigenvalue" difficulty: Gottlieb and Orszag (1977, pg. 143-146), Brenier, Roux and Bontoux(1986), Zebib(1987b), Gardner, Trogdon and Douglass(1989), McFadden, Murray and Boisvert (1990), Huang and Sloan (1994), and Dawkins, Dunbar, and Douglass (1998). However, a major theme of the chapter is that convert-to-matrix methods *always* have *lots* of nonsense eigenvalues, so what is so special about these "spurious eigenvalues"?

The short answer is that because the "spurious eigenvalues" are spurious due to misrepresented physics, rather than mere underresolution of genuine eigenmodes, there are some differences between them and the other unacceptable eigenvalues which are worth discussing. Before we can discuss these differences, however, we must first note that the bland label "spurious eigenvalues", used in many previous studies, is a semantic atrocity because it blurs the distinction between these eigenvalues and those which are in error merely because N is too small.

To correct this semantic sloppiness, we offer the following.

Definition 20 (SPURIOUS EIGENVALUES)
   PHYSICALLY SPURIOUS EIGENVALUES are numerically-computed eigenvalues which are in error because of misapplication of boundary conditions or some other misrepresentation of the physics.
   NUMERICALLY SPURIOUS EIGENVALUES are poor approximations to exact eigenvalues because the mode is oscillating too rapidly to be resolved by N degrees of freedom. A given numerically spurious eigenvalue can always be computed accurately by using sufficiently large N.

An example, first studied by Gottlieb and Orszag (1977, pg. 143-145), will help. The equations for a viscous, incompressible fluid can, in the limit of small amplitude and one-dimensional flow, be written as the system

$$\zeta_t = \nu \zeta_{xx}$$
$$\zeta = \psi_{xx} \tag{7.22}$$

where ν is the constant viscosity. The bounary conditions are

$$\psi(\pm 1) = \psi_x(\pm 1) = 0 \tag{7.23}$$

Note that these are entirely on the streamfunction ψ with no boundary conditions on the vorticity ζ. This system can be reduced to the single equation:

$$\psi_{txx} = \nu \, \psi_{xxxx} \tag{7.24}$$

The Galerkin method, called the "tau method" in Gottlieb and Orszag, is to write

$$\psi(t) \approx \sum_{j=0}^{N} a_n(t)T_n(x) \tag{7.25}$$

and then impose the $N - 3$ conditions that the residual should be orthogonal, in the usual Chebyshev inner product, to $T_k(x)$ for $k = 0, 1, \ldots, N - 4$. These conditions are supplemented with the four boundary conditions (7.23).

Unfortunately, the resulting system of ODEs in time is a disaster wrapped in a catastrophe: all time-marching schemes for the system are unstable. Note that both (7.24) and its Chebyshev discretization can be solved exactly by expansion in the solutions (exact or Chebyshev-discretized, respectively) of the eigenvalue problem

$$\nu u_{xxxx} = \lambda u_{xx} \tag{7.26}$$

where each mode depends upon time as $\exp(\lambda t)$ where λ is the eigenvalue. The exact eigenvalues of the original PDE are $\lambda = -\nu\mu$ where either $\mu = n\pi$ or μ is any nonzero root of $\tan\mu = \mu$; the eigenvalues are negative real so that the exact eigenmodes all decay monotonically with time. The Chebyshev discretized problem, however, has two positive real eigenvalues whose magnitude increases as roughly $O(N^4)$. Thus, the Chebyshev-discretized system of ODEs in time has two eigenmodes which blow up very fast, and no time-marching scheme can forestall disaster.

These two physically-spurious eigenvalues arise because (7.24) is an "explicitly-implicit" problem, that is, in order to apply standard software for solving the system of ordinary differential equations in time, we must rewrite the equation as

$$\psi_t = \frac{1}{\partial_{xx}^2}\psi_{xxxx} \tag{7.27}$$

The problem is that we impose *four* boundary conditions on the streamfunction ψ, consistent with the fact that the differential equation is fourth order, but the differential operator that must be inverted to compute the right-hand side of (7.27) is only of *second* order. In general, a second order differential equation with *four* boundary conditions is insoluble. A similar inconsistency arises when converting (7.26) from a generalized to a standard eigenproblem:

$$\nu \frac{1}{\partial_{xx}^2}u_{xxxx} = \lambda u \tag{7.28}$$

This suggests – as McFadden, Murray and Boisvert (1990) have confirmed through many numerical experiments – that these physically spurious eigenvalues only arise for *generalized* eigenproblems, and are absent for standard eigenproblems which do not involve the inversion of a differential operator. Furthermore, this difficulty happens only for *non-periodic* problems. In a periodic problem, one cannot have inconsistencies in the number of boundary conditions because the boundary condition of periodicity is the same for differential operators of all orders. It is only generalized eigenproblems in non-periodic geometry where physically spurious eigenvalues may arise.

Dawkins, Dunbar and Douglass (1998) have rigorously proved the existence of the large positive eigenvalues for the eigenproblem (7.26). They show that the overspecification of boundary conditions technically creates two L_2 eigenmodes of the differential eigenproblem with infinite eigenvalues, which the discretization approximates as positive eigenvalues of magnitude $O(N^4)$.

Several remedies have been proposed, but all seem to be closely related, so we shall describe only the variant due to Huang and Sloan (1994). Their spectral basis is composed of Chebyshev-Lobatto cardinal functions on an N-point grid $x_k, k = 1, \ldots, N$. Dirichlet homogeneous boundary conditions are imposed by omitting $C_1(x)$ and $C_N(x)$ from the basis. The twist is that for (7.26), a non-standard basis is used to represent the fourth derivatives:

$$h_j(x) \equiv \frac{(1 - x^2)}{(1 - x_j^2)} C_j(x) \tag{7.29}$$

where the $C_j(x)$ are the standard cardinal functions. Note that the factor of $(1-x^2)$ enforces the homogeneous Neuman condition so that a sum of h_j from $j = 2, \ldots, N - 1$ must have a double zero at both endpoints. The coefficients of both basis sets are the same: the values of $u(x)$ at the interior grid points. However, the modified functions h_k are used only to represent the fourth derivative while the standard basis functions with "lazy" imposition of boundary conditions (just two) are used for the second derivatives. Thus, the discrete generalized eigenvalue problem $\vec{A}\vec{u} = \lambda \vec{B}\vec{u}$ for Eq.(7.26) has the matrix elements

$$A_{ij} = h_{j,xxxx}(x_i), \qquad B_{ij} = C_{j,xx}(x_i) \tag{7.30}$$

where $\vec{u}$ is the column vector of grid point values $\{u(x_2), \ldots, u(x_{N-1})\}$.

This works, but the question still remains: With so many numerically underresolved eigenvalues to battle, do a couple of physically spurious eigenvalues matter? We shall offer three positive answers, each followed by a rejoinder.

The first argument is that these physically spurious eigenvalues are very important because they cause unconditional instability of time-marching schemes. However, this is really an issue to be fought out in a discussion of time-marching. It is not a "spurious eigenvalues" problem so much as it is a "screwing up the bounary conditions" problem. Further, Gottlieb and Orszag note that "this version of the tau method [with physically spurious eigenvalues] may be suitable for eigenvalue problems even though it is unconditionally unstable for the initial-value problem".

Spurious modes are often generated when a *system* of equations is reduced to a single equation, as often happens in setting up eigenproblems as illustrated by Eq.(7.26). The most famous example of "spurious" modes is the generation of unreal pressure modes when the equations of incompressible fluid flow are reduced to a single equation for the pressure. Although there are no physical boundary conditions for the pressure, the Poisson-like equation for it requires boundary conditions – and evaluating the Navier-Stokes equations at solid walls actually gives more boundary conditions than are needed.

The remedy is now well-understood: discretize the *system*, impose boundary conditions on the *system*, and then *reduce* the system to a single equation *after discretization*. Another important tactic is to use polynomials of different degrees where necessary. Again, the most well-known illustration is the practice of approximating the pressure in incompressible flow by a polynomial whose degree is two less than that of the velocity. Similarly, Huang and Sloan's strategy approximates the second derivative using basis functions whose degree is two less than that of the modified cardinal functions h_j which are used for the fourth derivative.

So, physically spurious eigenvalues do have some significance because they imply temporal instability. However, this is a well-understood phenomena now, and besides it is a tale that belongs in another chapter. When the goal is only eigenvalues, temporal instability does not matter.

The second argument in favor of modified eigenvalue-solving is that if the target is unstable modes, then physically spurious modes of large positive real part might be confused with genuine instabilities, and so waste a lot of time. However, as explained in the previous section, this shouldn't be a difficulty if one plots the change in eigenvalues with resolution. The physically-spurious eigenvalues seem to scale as $O(N^4)$ for most problems and thus are among the very easiest "bad" eigenvalues to detect through sensitivity to N.

The third argument for purging physically spurious eigenvalues is that misrepresentation of boundary conditions worsens the condition number of the matrices. This is perfectly true, and is particularly annoying since eigenproblems – again because of the reduction of a system of equations to one differential equation – often involve high order derivatives. Since the derivatives of Chebyshev polynomials oscillate more and more wildly in narrow boundary layers as the order of the derivative increases, Chebyshev differentiation matrices become more and more ill-conditioned for problems with high order derivatives, with or without physically spurious eigenvalues. We shall describe how to ameliorate this ill-conditioning in the next section.

7.7   Reducing the Condition Number

One unpleasant difficulty with a Chebyshev or Legendre basis is that the derivatives of the polynomials oscillate near the endpoints with increasing amplitude and decreasing wavelength as the degree and order of differentiation increase. To be precise,

$$\left| \frac{d^p T_N}{dx^p}(\pm 1) \right| \sim N^{2p} \prod_{k=1}^{p} \frac{1}{2k+1} \left\{ 1 + O(1/N^2) \right\} \tag{7.31}$$

In turn, this implies that the condition number of the Chebyshev and Legendre matrices for a differential equation are $O(N^{2p})$. Thus, for a sixth order equation, the matrix is blowing up as $O(N^{12})$, and the linear algebra subroutine is likely to crash spectacularly even for N little larger than 10.

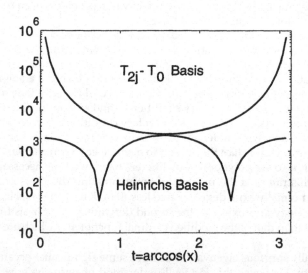

Figure 7.8: Absolute values of the second derivative of the highest basis function versus the trigonometric argument $t = \arccos(x)$ for two different basis sets ($N = 50$ collocation points). Upper curve: Each basis function is the difference of a Chebyshev polynomial with either 1 or x. Its maximum is 6.9E5. Bottom curve: Heinrichs' basis: $\phi_{50}(x) = (1 - x^2) T_{49}(x)$. The maximum is only 2.4E3.

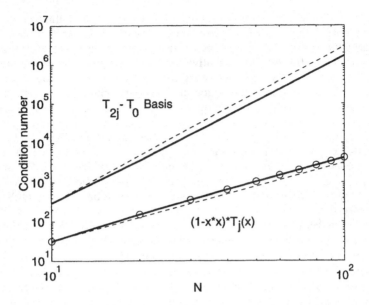

Figure 7.9: Solid: condition numbers for the matrices which discretize the second derivative operator in two different basis sets, plotted versus N, the size of the matrix. Top: the "difference" basis in which each function is $T_j(x)$ minus either 1 or x, depending on parity. Bottom with circles: Heinrichs' basis: $\phi_j(x) = (1 - x^2) T_j(x)$. The condition number is defined to be the ratio of the largest to the smallest singular value of the matrix. The dotted lines have slopes of N^4 (top) and N^2 (bottom), illustrating that the rate of growth of the condition number closely follows these respective power laws.

Fortunately, there are a couple of remedies. Wilhelm Heinrichs (1989a, 1991b, 1991c) observed that if we apply basis recombination to create basis functions that satisfy homogeneous boundary conditions, it is possible to choose particular sets that greatly reduce the condition number. For example, suppose the target is to solve a second order differential equation with the homogeneous Dirichlet boundary conditions $u(\pm1) = 0$. The simplest choice of basis functions, and one that works very well except for $N \gg 100$, is the "difference" basis

$$\phi_{2j}(x) \equiv T_{2j}(x) - T_0, \qquad \phi_{2j-1}(x) \equiv T_{2j-1}(x) - T_1 \qquad (7.32)$$

but the second derivative of these is equal to that of a Chebyshev polynomial and thus this basis gives poor condition number.

Heinrichs proposed instead

$$\phi_j \equiv (1 - x^2)T_j, \qquad j = 0, 1, \dots \qquad (7.33)$$

The endpoint and near-the-endpoint values of the second derivatives of these functions, which we shall dub the "Heinrichs" basis, are much smaller than those of the Chebyshev polynomials or the difference basis: $O(N^2)$ basis versus $O(N^4)$. His reasoning is that the second derivative of his basis functions is

$$\phi_{j,xx} = (1 - x^2)T_{j,xx} - 4xT_{j,x} - 2T_j \qquad (7.34)$$

Now $T_{j,xx}$ is $O(N^4)$ near the endpoints, but this is precisely where the factor $(1 - x^2)$ tends to zero. The contribution of the first derivative is not cancelled near the boundary, but since the first derivative is only $O(N^2)$, the second derivative of the Heinrichs basis is no larger.

Fig. 7.8 compares the second derivatives of two basis functions of the same degree in the two different basis sets. It is necessary to plot the absolute values on a logarithmic scale, and to plot them versus the trigonometric coordinate t (= $\arccos(x)$) because the oscillations grow so rapidly in magnitude and wavelength near the endpoints that the obvious plot of the second derivatives on a linear scale versus x shows little but some wiggly, indecipherable spikes near $x = \pm 1$. The lower plot shows that the $(1 - x^2)$ factor in the Heinrichs basis turns the magnitude back towards zero so that the maximum amplitude is roughly the *square root* of the maximum for the "difference" basis. Fig. 7.9 shows that this translates into a great difference in the condition number of the Chebyshev discretization matrices for the second derivative: $O(N^2)$ for the Heinrichs basis versus $O(N^4)$ (ouch!) for the "difference" basis.

There are a couple of little white lies hidden in these graphs. First, the Heinrichs' basis makes the condition number of the matrix which discretizes the *undifferentiated* basis *worse*. The result is that when we convert $u_{xx} + \lambda u = 0$ into the standard matrix eigenproblem $\vec{G}\vec{u} = \lambda \vec{u}$ with $\vec{G} = \vec{B}^{-1}\vec{A}$ where $\vec{A}$ is the discretization of the second derivative and $\vec{B}$ that of the undifferentiated term, it turns out the condition number of $\vec{G}$ is almost *identical* for the two basis sets. The other white lie is that for a second order problem, ill-conditioning is a problem only when $N > 100$, which is to say it is not a problem.

For fourth order and higher order differential equations, however, ill-conditioning is more of a problem, and the Heinrichs' basis comes increasingly into its own. The idea generalizes in an obvious way to higher order equations. For example, for a fourth order problem with the boundary conditions $u(\pm 1) = u_x(\pm 1) = 0$, the Heinrichs basis is

$$\phi_j(x) \equiv (1 - x^2)^2 T_j(x) \tag{7.35}$$

where the quartic multiplier enforces a double zero at both endpoints and thus lowers the condition number of the fourth derivative from $O(N^8)$ to $O(N^4)$.

Huang and Sloan (1994) propose a related strategy which is more suitable for a cardinal function basis. For a fourth order problem, their cardinal functions are

$$C_k(x) = \frac{(1 - x^2)^2}{(1 - x_k^2)^2} \frac{\pi(x)}{\pi_x(x_k)(x - x_k)} \tag{7.36}$$

where $\pi(x)$ is the product of monomial factors constructed from the *interior* points of the usual Chebyshev-Lobatto grid:

$$\pi(x) \equiv \prod_{i=2}^{N-1} (x - x_i) \tag{7.37}$$

After the usual spectral discretization using these basis functions to give the generalized matrix eigenproblem $\vec{A}\vec{u} = \lambda \vec{B}\vec{u}$, Huang and Sloan multiply both matrices by the diagonal matrices whose elements are

$$D_{jj} = (1 - x_j^2)^2, \qquad j = 2, \ldots, N - 1 \tag{7.38}$$

Like Heinrichs' improvement for the non-cardinal basis, Huang and Sloan's method reduces the condition number to its square root. For a differential equation of order p, the condition number is reduced from $O(N^{2p})$ to $O(N^p)$. Thus, it is possible to solve differential equations of quite high order with large N.

Although we have chosen to discuss condition number in the midst of a chapter on eigenproblems because condition number is measured as a ratio of eigenvalues, the Heinrichs and Huang and Sloan basis sets can be equally well employed for high order BOUNDARY VALUE problems.

7.8 Alternatives for Costly Problems, I: The Power Method

When the QR/QZ algorithm is too expensive, one needs iteration schemes which compute only a single eigenvalue at a time. The simplest is the "power method", which can be applied both to the matrix discretization of an eigenproblem and also to the original time-dependent equations from which the eigenproblem was derived. A couple of examples will make the idea clearer.

First, suppose the goal is to calculate hydrodynamic instabilities. The flow is split into a user-specified "basic state" plus a perturbation, which is assumed to be initially small compared to the basic state but is otherwise unrestricted. The flow is unstable if the perturbation amplifies in time.

A useful illustration is the barotropic vorticity equation in a "beta-plane" approximation on a rotating planet:

$$\psi_{xxt} + \psi_{yyt} + \psi_x(\psi_{xxy} + \psi_{yyy}) - \psi_y(\psi_{xxx} + \psi_{xyy}) + \beta\psi_x = 0 \tag{7.39}$$

where $\psi(x, y, t)$ is the streamfunction for two-dimensional flow and β is the latitudinal derivative of the Coriolis parameter. For simplicity, we shall consider only a basic state which is independent of x, that is, denoting the basic state by upper case letters,

$$\Psi(y) = -\int^y U(z)dz \tag{7.40}$$

where $U(y)$ is the "mean flow", a jet moving parallel to the x-axis.

The time-dependent variant of the power method is to integrate the equations of motion from an *arbitrary initial condition* until the fastest-growing mode completely dominates the solution, at which point the solution is the desired eigenmode and its growth rate and spatial translation rate are the imaginary and real parts of the complex-valued phase speed c, which is the eigenvalue.

There are two variants. The first is to *linearize* the equations of motion about the basic state:

$$\psi = \psi' + \int^y U(z)dz, \qquad \psi' \sim O(\epsilon), \qquad \epsilon << 1 \tag{7.41}$$

Since the perturbation is assumed to be very small, we can always expand the equations of motion in a power series in the amplitude ϵ of the perturbation; the linearized equation of motion is simply the $O(\epsilon)$ term in this expansion. The linearized barotropic vorticity equation, for example, is

$$\psi'_{xxt} + (\beta - U_{yy})\psi'_x + U(y)(\psi'_{xxx} + \psi'_{xyy}) = 0 \tag{7.42}$$

where the prime on ψ denotes the perturbative part of the streamfunction.

In the linearized equation, feedback from the perturbation to the basic state, as happens for the fully nonlinear equations of motion, is suppressed. This is a detriment because it restricts the model to the early stages of the instability when the perturbation is very small. However, it is also a virtue. The linearized stability problem is completely specified by the specification of the basic state.

The second variant of the time-dependent power method is to solve the full, nonlinear equations of motion by starting from a perturbation of amplitude ϵ and integrating until the amplitude is $O(a)$ where $a >> \epsilon$, but a is still small compared to the basic state. The result will be the same as that of solving the linearized problem (on the same time interval) to within an absolute error of $O(a^2)$ [relative error $O(a)$] because this is the leading order of the terms neglected in the linearization. One virtue of this method is that it is unnecessary to create a second computer program, with slightly different (i. e., linearized) equations of motion, to explore linearized stability. Another is that one can easily extend the integration further in time to explore how nonlinear feedback alters the growth and structure of the amplifying disturbances. The disadvantage is that if there are unstable modes of similar growth rates, one may be forced to stop the integration (to keep $a << 1$) before the fastest growing mode has become large compared to all other modes.

In either variant, the growth rate and amplitude are estimated by fitting the approximation

$$\psi(x, y, t_2) - \psi(x, y, t_1) \quad \approx \quad \Xi(y) \exp(ikx) \{-ikct_2) - \exp(-ikct_1)\}$$
$$\text{+complex conjugate} \tag{7.43}$$

where t_1 and t_2 are two large but otherwise arbitrary times and $\Xi(y)$ gives the complex-valued latitudinal structure of the instability while c is the complex phase speed and k the x-wavenumber, which must also be determined from the perturbation for large time.

Because the fitting is a little tricky, and it sometimes requires a huge number of time steps before the fastest-growing mode dominates its slower-growing brethren, it is often more convenient to convert the time-dependent equations into a matrix eigenproblem. The justification for this step is that as long as the linearized equations have coefficients which are independent of time, all discretizations of the spatial dependence will collapse the linearized PDE into a system of ordinary differential equations in time of the form $d\vec{u}/dt = \vec{A}\vec{u}$ where $\vec{A}$ is a square matrix whose elements are independent of time. The usual theory of constant coefficient ODEs, found in most undergraduate differential equation texts, then asserts that the general solution is a superposition of the eigenmodes of $\vec{A}$, each oscillating as an exponential of time with a complex frequency which is the corresponding eigenvalue of $\vec{A}$, assuming these are all distinct. The matrix formulation of linearized stability theory is simply the computation of these eigenvectors and eigenvalues of $\vec{A}$.

When the coefficients of the linearized PDE do not vary with all the spatial coefficients, then one can separate variables still further. For the barotropic vorticity equation, one can expand the solution in a Fourier series in x; if linearized with respect to a mean current which varies only with y, the x-wavenumbers are uncoupled. A large two-dimensional eigenvalue problem therefore collapses into a set of differential eigenproblems in y only, one for each wavenumber in x. Denoting the wavenumber by k, the eigenproblem is

$$\psi_{yy} + \left\{ \frac{\beta - U_{yy}}{U - c} - k^2 \right\} \psi = 0 \tag{7.44}$$

(In a minor abuse of notation that is almost ubiquitous in the literature, $\psi(y)$ is used for the eigenfunction even though the same symbol has already been employed for the total time-dependent streamfunction.)

The power method can be applied directly to the matrix form of an eigenvalue problem by first discretizing the differential operator and then multiplying through by the appropriate inverse matrix to convert the problem into a standard (as opposed to a generalized) matrix eigenproblem:

$$\vec{A}u = \lambda \vec{u} \tag{7.45}$$

Note that if we assume that $\vec{A}$ has as complete set of eigenvectors $\vec{e}_j$ and expand a vector u_0

$$u_0 = \sum_{j=1} a_j \vec{e}_j \tag{7.46}$$

then the result of multiplying u_0 by the square matrix $\vec{A}$ will be a vector whose eigencoefficients will be the same except that a_j is multiplied by λ_j, the eigenvalue associated with $\vec{e}_j$:

$$\vec{A}u_0 = \sum_{j=1} \lambda_j a_j \vec{e}_j \tag{7.47}$$

If we iterate by repeatedly multiplying by powers of $\vec{A}$, it follows that as the iteration number k increases, the eigencoefficients will be multiplied by λ_j^k and therefore the eigenfunction associated with the eigenvalue of largest magnitude will more and more dominate the k-th power of $\vec{A}$. As it does so, the ratio $\|\vec{A}\vec{u}\|/\|\vec{u}\|$ will tend more and more towards the magnitude of the largest eigenvalue where $\| \|$ denotes any reasonable matrix/vector norm.

This suggests the following algorithm for computing the eigenfunction and eigenvalue for the mode of largest (absolute value) of eigenvalue. First, choose an arbitrary starting vector u_0 and then divide it by its norm so as to rescale it to unit norm. (The algorithm will fail if u_0 is orthogonal to the target eigenmode, but if u_0 is a column of numbers from a random number generator, the probability of such failure is negligibly small.) Then, repeat the following loop over the iteration number k until the eigenfunction and eigenvalue have converged to within the user-chosen tolerance:

$$\begin{aligned} \vec{u}_{k+1} &= \vec{A}\vec{u}_k, \qquad k = 0, 1, \dots \\ \lambda_{k+1} &= \|u_{k+1}\| \\ \vec{u}_{k+1} &= \vec{u}_{k+1} / \|\vec{u}_{k+1}\| \end{aligned} \tag{7.48}$$

This is simply the matrix form of the power method: iterate until the fastest-growing mode dominates. The third line of the loop is meant in the sense of a computational assignment statement rather than mathematical equality — u^{k+1} is replaced by the same vector divided by its norm to rescale it to unit norm. This renormalization avoids overflow problems. When λ is complex-valued, the second line must be replaced by the average of the element-by-element division of the new eigenvector by the old:

$$\lambda_{k+1} = (1/N) \sum_{j=1}^{N} u_{k+1,j}/u_{k,j} \tag{7.49}$$

The matrix power method can be applied to any matrix, whether connected with stability problems or differential eigenproblems or not. It is very fast and cheap per iteration because the most costly step is merely a matrix-vector multiply, which is a highly vectorizable operation. However, the power method has the disadvantage that it can find only the mode of largest absolute value of eigenvalue, which may not be the only mode of interest. In stability problems, it is quite common to have two unstable modes whose growth rates switch dominance at some point in parameter space. In such transition regions where two modes have the same or almost the same growth rates, the power method will fail.

In the next section, we therefore describe another algorithm which fixes many of the defects of the power method, albeit at a much higher cost per iteration. First, though, we

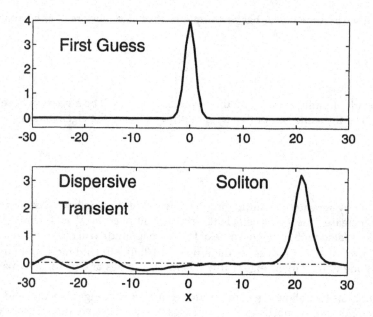

Figure 7.10: Illustration of the power method applied to the KdV equation. The initialization ("first guess") is $u = 4\exp(-x^2/2)$. At $t = 20$, this has split into a dispersing transient (which moves left) and a single soliton (which moves rightward).

must note that the power method is not limited only to instability problems or only to linear eigenvalue problems.

For example, the power method is a good way to compute solitary waves of the Korteweg-deVries (KdV) equation:

$$u_t + uu_x + u_{xxx} = 0 \tag{7.50}$$

The solitary waves are the solutions to the *nonlinear* eigenvalue problem

$$(u - c)u_X + u_{XXX} = 0 \tag{7.51}$$

where $X \equiv x - ct$ is the coordinate in a frame of reference which is travelling with the wave and c is the phase speed. By the power method, the ODE nonlinear eigenproblem can be bypassed and solitons computed directly from the time-dependent equation: Integrating forward in time from an arbitrary initial condition until the dominant eigenmode has separated itself from the transients.

In stability problems, the separation is by separation-by-amplitude: the dominant mode is fastest-growing and therefore eventually becomes larger than the other modes everywhere in space. For the KdV equation, the separation is *spatial*: the dispersing transients travel leftward (because the group velocity for small amplitude waves of all wavenumbers is *negative*) whereas the velocities of all solitary waves are *positive*. Because the amplitude of a solitary wave decreases exponentially fast with distance from the center of the soliton, the overlap between a given solitary wave and the rest of the solution decreases *exponentially* with time. Thus, even though there is no instability and the eigenvalue problem is nonlinear, the power method converges geometrically for the KdV equation.

7.9 Alternatives for Costly Problems, II: Inverse Power Method

The matrix variant of the power method has the disadvantage that it can find only the mode of largest eigenvalue. The inverse power method, also sometimes called simply "inverse iteration", lacks this limitation. The algorithm is to apply the ordinary power method to the matrix

$$\vec{M} \equiv (\vec{A} - \Lambda\vec{I})^{-1} \tag{7.52}$$

where $\vec{I}$ is the identity matrix and Λ is a user-specified constant. The eigenvalues of $\vec{M}$ are

$$\mu_j = \frac{1}{\lambda_j - \Lambda} \tag{7.53}$$

where the λ_j are the eigenvalues of the original matrix $\vec{A}$, so the largest eigenvalue of the shifted-and-inverted matrix $\vec{M}$ is the reciprocal of whichever eigenvalue of A is closest to the shift Λ. It follows that the inverse power method will converge geometrically to *any* eigenvalue of the matrix $\vec{A}$, provided we have a sufficiently good first guess (and set Λ equal to it).

As usual, the inverse is not explicitly computed. Instead, one iterates

$$(\vec{A} - \Lambda_k\vec{I})\vec{u}_{k+1} = \vec{u}_k, \qquad k = 0, 1, \ldots$$

$$\mu_{k+1} = (1/N) \sum_{j=1}^{N} u_{k+1,j}/u_{k,j} \tag{7.54}$$

$$\Lambda_{k+1} = 1/\mu_{k+1} + \Lambda_k \tag{7.55}$$

$$\vec{u}_{k+1} = \vec{u}_{k+1}/ \|\vec{u}_{k+1}\|$$

The disadvantage of the inverse power method is that the cheap matrix-vector multiply is replaced by solving a matrix equation, which is $O(N)$ more expensive (by direct methods). However, as explained in Chapter 15, one can often solve the matrix equation implicitly (and cheaply) through a preconditioned Richardson iteration instead.

The inverse power method is very powerful because it can be applied to compute any eigenmode. Indeed, the routines in many software libraries compute the eigenfunctions by the inverse power method after the eigenvalues have been accurately computed by a different algorithm.

The power and inverse power algorithms are but two members of a wide family of iterative eigenvalue solvers. The Arnoldi method, which does not require storing the full matrix $\vec{A}$, is particularly useful for very large problems. Although the algorithm is too complicated to describe here, Navarra(1987) used Arnold's method to solve geophysical problems with as many as 13,000 spectral coefficients.

7.10 Mapping the Parameter Space: Combining Global and Local Methods

In stability calculations, one often wants to compute growth rates as a function of two or more parameters as shown schematically in Fig. 7.11. The so-called "neutral curve", which is the boundary between stability and instability in parameter space, is usually an important goal. Unfortunately, for problems such as the barotropic vorticity equation, the

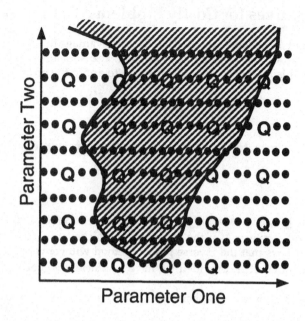

Figure 7.11: Schematic of a stability calculation in a two-dimensional parameter space. The unstable region is shaded; it is bounded by the "neutral curve". To map the parameter space, it is efficient to use the QR/QZ algorithm to solve the matrix eigenproblem on a *coarse* grid (large Q's) and fill in the gaps with an inexpensive local iteration, such as the power or inverse power method, applied to the pseudospectral matrix (solid disks). This dual strategy is much cheaper than using the QR/QZ algorithm everywhere (which has an $O(10N^3)$ cost versus $O(N^2)$ for either the power method or the inverse power method (with finite difference preconditioned iteration for the latter).) At the same time, it is much safer than using a one-mode-at-a-time eigensolver everywhere because this may easily miss important eigenmodes.

neutral curve is precisely where the linearized equation is singular on the interior of the domain similar to a Sturm-Liouville eigenproblem of the Fourth Kind.

In the next section, we describe a good strategy for computing the eigenvalues even when the differential equation is singular. However, there is a more serious problem: to map a two-dimensional parameter space on a relatively coarse 32×32 grid requires solving more than a thousand cases. If the QR/QZ algorithm is used for each, this may be prohibitively expensive, especially in multiple space dimensions, even though a good workstation never sleeps.

However, it is easy to miss easy modes with a local iterative method. Many an arithmurgist has traced one branch of unstable modes with loving care, only to find, years later, that another undiscovered mode had faster growth rates in at least part of the parameter space. A safer strategy is to apply the expensive global algorithm (QR/QZ) on a *coarse* grid and then "connect-the-dots" by using an iterative scheme as illustrated in Fig. 7.11.

7.11   Detouring into the Complex Plane: Eigenproblems with Singularities on the Interior of the Computational Domain

Hydrodynamic stability and wave problems often have solutions with branch points on or near the computational domain. These singularities, usually called "critical latitudes" or "critical points", create severe numerical difficulties. However, a good remedy (Boyd, 1985a) is to make a transformation from the original variable y such that the problem is solved on a *curve* in the *complex* y-plane rather than on the original interval on the real y-axis. With the proper choice of map parameter, one can loop the curve of integration away from the singularity so that it does not degrade numerical accuracy.

To illustrate, consider

$$u_{yy} + \left[\frac{1}{y} - \lambda\right] u = 0 \qquad \text{with } u(a) = u(b) = 0; \quad a < 0 \ \& \ b > 0 \qquad (7.56)$$

where λ is the eigenvalue. If a and b were of the same sign so that the singularity was not on the interior of the interval, $y \in [a, b]$, then (7.56) would be a normal, self-adjoint Sturm-Liouville problem. As it is, not only the differential equation but also the solution are *singular* on the interior of the interval.

After we have made a simple linear stretching to shift the interval from $[a, b]$ to $[-1, 1]$, an effective transformation for (7.56) is

$$y = x + i \triangle (x^2 - 1) \qquad (7.57)$$

where $\triangle$ is a (possibly complex) mapping parameter. We solve the problem using a Chebyshev series in $x \in [-1, 1]$ in the standard way. Because of the change of variable, however, the real interval in x is an arc in the complex y-plane which detours away from the singularity. Since $u(y)$ has a branch point at $y = 0$, the choice of looping the contour above or below the real y-axis is an implicit choice of branch cut. The correct choice can be made only by a careful physical analysis; for geophysical problems, Boyd (1982c) explains that the proper choice is to go *below* the real axis by choosing $\triangle > 0$ in (7.57); this implicitly forces the branch cut to lie in the upper half of the y-plane.

Fig. 7.12a shows the contour of integration (dashed) while Table 7.2 illustrates the results for (7.56). The basis functions are sums of Chebyshev polynomials which vanish at the endpoints so that each $\phi_n(x)$ individually satisfies the boundary conditions at $x = \pm 1$. Despite the singularity, only 6 basis functions — a 6×6 matrix eigenproblem — is sufficient to yield both the real and imaginary parts of the lowest eigenvalue to within an error of less than 1.4%.

The rightmost column also is a personal embarrassment. Boyd (1981b) solved this same problem using an artificial viscosity combined with an iterative finite difference method — and missed two eigenvalues with very small imaginary parts (Modes 3 and 7 in the table). The mapped Chebyshev procedure does not require an artificial viscosity or any first guesses; the QR algorithm will find all the eigenvalues of the matrix eigenvalue problem automatically.

It is also a method with a weakness in that the Chebyshev series for $u(y)$ converges most rapidly for real x — but this is an arc of complex y. How is one to interpret an imaginary latitude? The series converges more and more slowly as we move away from the arc that corresponds to real x and it must diverge at the singularity at $y = 0$. Therefore, the detour into the complex plane is directly useful *only* for computing the *eigenvalues*.

Once we have λ, of course, we can use a variety of methods to compute the corresponding eigenfunction. Power series expansions about the branch point at $y = 0$ in combination

Table 7.2: Eigenvalues of a Singular Sturm-Liouville Problem: The First 7 Eigenvalues of $u_{yy} + \{1/y - \lambda\}u = 0$ Subject to $u(\pm 6) = 0$ as Computed Using the Complex Parabolic Mapping $y = x + i\Delta(x^2 - 1)$ with $\Delta = 1/2$.

Note: N is the number of Chebyshev polynomials retained in the truncation. The errors for $N < 40$ are the differences from the results for $N = 40$ for the modes shown. Modes that are wildly in error or violate the theorem that the imaginary part of the eigenvalue is always positive have been omitted; thus only one eigenvalue is listed for $N = 6$ although the Chebyshev-discretized matrix eigenproblem had five other eigenvalues.

n	$\lambda(N=6)$	Error $(N=6)$	$\lambda(N=20)$	Error $(N=20)$	$\lambda(N=40)$
1	0.1268	1.4 %	0.125054	Negligible	0.125054
	+ i 0.2852	0.09 %	+i 0.284986	Negligible	+ i 0.284986
2			-0.296730	Negligible	-0.296730
			+ i 0.520320	Negligible	+ i 0.520320
3			-0.638802	Negligible	-0.638802
			+i 0.000262	Negligible	+ 0.000262
4			-1.21749	0.0008 %	-1.21750
			+ i 0.563350	0.0004 %	+ i 0.563349
5			-1.60228	0.024 %	-1.60190
			+ i 0.007844	0.24 %	+ i 0.007825
6					-2.72598
					+ i 0.490012
7					-3.10114
					+ i 0.070176

with shooting for larger y should work quite well. Since λ is known, there is no need for iteration; the Runge-Kutta method, initialized near $y = 0$ via the power series, will automatically give $u(a) = u(b) = 0$. Still, it is unpleasant to be forced to compute the eigenfunctions in a second, separate step.

For hydrodynamic instabilities which are *strongly* unstable, the critical points are at complex y. As shown through a barotropic instability calculation in Boyd (1985b), the detour into the complex y-plane may be unnecessary near the points of maximum growth. However, most stability studies map the "neutral curve", which is defined to be the boundary of the unstable region in parameter space. On the neutral curve, the critical points are on the real axis just as for (7.56). Thus, the mapping trick, despite its inability to compute the eigenfunctions, is very useful for stability calculations.

In more complicated cases, multiple critical points may require maps that loop both above and below the real y-axis as shown in Fig. 7.12b. In addition, when the interval is unbounded, it may be necessary to combine the complex detour with the stretching map that transforms an infinite interval into a finite interval so that we can apply Chebyshev polynomials as shown in Fig. 7.12d. Boyd (1985b) gives a thorough discussion of these variants.

Gill and Sneddon(1995, 1996) have given some useful extensions of Boyd's paper. Their first article gives an analytic formula for optimizing the quadratic map: If there is only one critical latitude and its location is $y = y_c$, then the mapping $y = x + i\Delta(1 - x^2)$ is optimized by

$$y_c = -i\left\{y_c \pm \sqrt{y_c^2 - 1}\right\}/2 \tag{7.58}$$

When more than one critical latitude is sufficiently close to the domain to be troublesome, more complicated transformations are useful. Gill and Sneddon show that the cubic mapping

$$y = x - (\alpha + i)(\beta_0 + \beta_1 x)(x^2 - 1) \tag{7.59}$$

is free of cusps and loops, that is, self-intersections, for all real values of the three parameters α, β_0, β_1.

When the singularity is *very* close to an endpoint, Gill and Sneddon (1996a) show that the quadratic map is still effective: Even with $y_c = 0.99$, the pseudospectral error is proportional to $O(1.44^{-N})$. This can be improved still further by iterating the quadratic map using an analytical formula for optimizing the composite transformation. However, each iteration of the composition raises the order of the poles of the transformed solution, so the composite map is useful only for large N and very high accuracy. In multiple precision, they give good illustrations of "cross-over": the composite mapping, which is always superior to the standard quadratic mapping in the asymptotic limit $N \to \infty$, is worse for small N.

Gill and Sneddon (unpublished preprint) extends the analysis to semi-infinite intervals, weaving together ideas from Boyd(1987b) as well as Boyd(1985a). The good news is that they obtained quite useful analytical formulas for optimizing two different families of mappings. The theoretical estimates of the total error were not very accurate for reasonable N, but the convergence rates were gratifyingly high for their test problems.

Singular problems, and other difficult eigenproblems like the Orr-Sommerfeld equation, require much higher N than our first example. (Although not singular for real y, the Orr-Sommerfeld eigenfunctions have thin internal layers and boundary layers.) Nevertheless, with $N \geq 40$ and a complex-plane mapping when needed, even nasty eigenproblems can be solved with high accuracy.

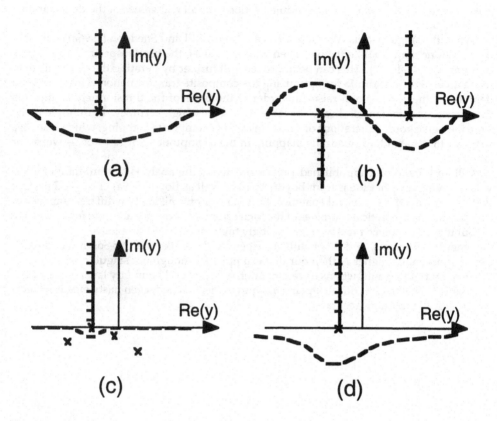

Figure 7.12: Four representative mappings for coping with singularities on or near the interior of the integration interval. For each case, the real and imaginary y-axes are shown as solid lines, the transformed path of integration as a dashed line, the branch cut proceeding away from the singularity as a cross-hatched line, and the singularities are x's.

(a) The solution has a branch point on the real axis. (The original and transformed integration paths intersect at the boundaries, $y = a, b$.)

(b) Two singularities on $y \in [a, b]$ with branch cuts that must be taken in opposite directions (typical for a symmetric jet).

(c) Identical with (a) except for the presence of additional singularities near (but not on) the interval which force the transformed path to hug the real axis.

(d) Identical with (a) except that the integration interval is *infinite*. The mappings for unbounded intervals (Chapter 17) may be freely combined with detours in the complex y-plane.

7.12   Common Errors: Credulity, Negligence, and Coding

"Credulity" is a shorthand for believing that a calculation is accurate when it in fact is nonsense. It is easier to be fooled than one might imagine.

In graduate school, I solved an eigenproblem with Chebyshev polynomials and beamed happily at the smooth graphs of the eigenmodes. To be sure, the eigenvalues did seem to jump around a bit when the resolution was varied, but the graphs of even the higher modes were so smooth – I walked around in a fool's paradise for weeks. Actually, I was solving a Sturm-Liouville problem of the Fourth Kind with a nasty singularity right on the expansion interval. Enlightment came when I graphed the *second derivative* of the eigenmodes, and found it resembled the fingerpainting of a crazed pre-schooler. I recovered from this fiasco by applying a change-of-coordinate as described earlier. Plotting the reciprocal of differences in eigenvalues for two different numerical resolutions should protect one from a similar blunder. Well, *most* of the time.

Boyd (1996b) solved the so-called "equatorial beta-plane" version of the tidal equations, and obtained two modes whose eigenvalues, equal in magnitude but opposite in sign, changed almost negligibly with N. One was the so-called "Kelvin" wave, which is known to have the simplest structure of all tidal modes and therefore should indeed be resolved most accurately by any numerical solution. The other, however, was a sort of "anti-Kelvin" wave that has no counterpart in reality.

The proof lies in the graph: Fig. 7.13 shows the zonal velocity for the two modes. The wild oscillations of the lower mode, which become only wilder with increasing N, show that it is a numerical artifact. Yet its eigenvalue is unchanged through the first twelve nonzero digits when N is increased from 36 to 50!

Obviously, the prudent arithmurgist will compare the eigenmodes, and not merely the eigenvalues, for different N, at least occasionally. As pilots say, even when flying on instruments, it is a good idea to occasionally look out the window.

"Negligence" is a shorthand for missing important modes. The QR/QZ algorithm computes all the eigenvalues of a given matrix, so with this method for the algebraic eigenvalue problem, failure is possible for a given mode only by choosing N too small. When QR is

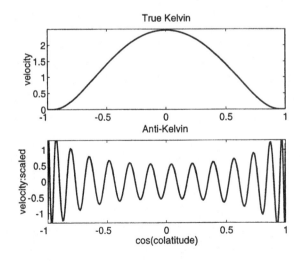

Figure 7.13: Example Two (Infinite Interval): Absolute errors in the eigenvalues for 16-point (solid disks) and 32-point (open circles) TB_n discretization.

replaced by an iterative, find-one-mode-at-a-time algorithm, however, it is very easy to overlook eigenmodes even if N is large enough to resolve them.

Boyd (1985a) records an embarrassing example. By combining the QR algorithm with a Chebyshev pseudospectral method and a detour into the complex plane, I found that my earlier calculations (Boyd, 1981a) had missed the third and seventh eigenmodes. Oops! The earlier work had combined a local iterative method with continuation in the parameter. However, these modes have very small imaginary part, and the older numerical method, which did not use a complex-plane mapping, was sufficiently inaccurate to miss them.

It really is a good idea to use the QR/QZ method whenever possible. If one must iterate, one must worry over the workstation like a parent watching over a toddler asleep with a high fever.

Coding errors have also sent eigenvalue computations into the Black Swamp of Published Errors. Of course, the misplaced "if" statement or the ever-dangerous sign error can imperil any kind of calculation. Galerkin calculations, however, seem to be especially prone to slips.

A good program-checking strategy is to apply the code to a simple problem with a known solution. Unfortunately for Galerkin algorithms, solvable test problems usually have *diagonal* Galerkin discretization matrices, and therefore do not test the off-diagonal elements at all.

In ye olden days when computers were young and full of vacuum tubes, it was common to derive the Galerkin matrix elements by hand calculations using recurrence relations. Fox & Parker (1968) and Fox, Hayes and Mayer (1973) are good exemplars of this old style; these recurrences are used in Orszag's (1971b) classic article and furnish a couple of pages of appendix in Gottlieb and Orszag (1977). It is so beautiful and elegant, revealing whatever sparsity is present in the Galerkin matrix, that it is still widely used.

Unfortunately, these recurrences are risky. Each change of problem, even from the target differential equation to a test equation, requires redoing much of the algebra. Consequently, a successful test is no guarantee that the algebra – much of it different – is correct for the target problem, too.

O'Connor (1995) applied recurrences to Laplace's Tidal Equation. This seemed particularly appropriate because the exact Galerkin matrix is sparse. However, the published article contains one spurious mode and slightly incorrect eigenvalues for all the others as corrected in O'Connor (1996). I know of other examples where the published work eventually fell under suspicion, but cannot name names because the authors never found their errors.

Galerkin-by-recurrence-relation has been used successfully, but one should be very cautious. Exact quadrature is no excuse for taking risks.

Table 7.3: A Selected Bibliography of Spectral Calculations for Eigenproblems

References	Comments
Longuet-Higgins(1968)	Laplace's Tidal Equation; spherical harmonics; special continued fraction algorithm for tridiagonal Galerkin matrix
Birkhoff&Fix(1970)	Fourier and Hermite function Galerkin methods
Orszag (1971b)	Chebyshev; 4th order Orr-Sommerfeld stability eqn. Noted usefulness of spectral combined with QR/QZ
Fox&Hayes &Mayers(1973)	Double eigenvalue problem: 2d order ODE with 3 boundary conditions and 2 eigenparameters
Boyd (1978a)	Chebyshev domain truncation for quantum quartic oscillator Chebyshev curve-fitting for analytical approximations to eigenvalue
Boyd (1978c)	more remarks on Chebyshev/QR connection
Boyd (1978b)	Chebyshev, Fourier & spherical harmonic bases on sphere
Banerjee (1978) Banerjee et al. (1978)	Eigenvalue problem: quantum anharmonic operator Hermite func. $\psi_n(\alpha y)$ with variable α
Boyd (1981a)	Eigenproblem with interior singularity
Boyd (1982b, c)	Atmospheric waves in shear flow
Liu&Ortiz(1982)	Singular perturbations; tau method
Boyd (1983d)	Analytic solutions for continuous spectrum (fluids)
Lund&Riley(1984)	sinc basis with mapping for radial Schroedinger equation
Liu&Ortiz&Pun(1984)	Steklov PDE eigenproblem
Boyd (1985a)	Change-of-coordinate to detour around around interior singularity into the complex plane
Brenier&Roux &Bontoux(1986)	Comparison of Chebyshev tau and Galerkin for convection
Liu&Ortiz(1986)	PDE eigenproblems; tau method
Liu&Ortiz(1987a)	complex plane; Orr-Sommerfeld equation
Liu&Ortiz(1987b)	powers of eigenparameter in ODE
Boyd (1987a)	TB_n basis: $x \in [-\infty, \infty]$
Boyd (1987b)	TL_n basis: $x \in [0, \infty]$, Charney stability problem
Eggert&Jarratt &Lund(1987)	sinc basis;finite and semi-infinite intervals, too, through map singularities at end of finite interval
Zebib (1987b)	Removal of spurious eigenvalues
Navarra(1987)	Very large meteorological problem (up to 13,000 unknowns) via Arnoldi's algorithm
Lin & Pierrehumbert (1988)	Tensor product of $TL_n(z) \otimes TB_m(y)$ for two-dimensional baroclinic instability
Gardner&Trogdon &Douglas(1989)	Modified tau scheme to remove "spurious" eigenvalues
Boyd (1990d)	Chebyshev computation of quantum scattering (continuous spectrum)
Malik(1990)	Hypersonic boundary layer stability; spectral multidomain
Jarratt&Lund&Bowers(1990)	sinc basis; endpoint singularities, finite interval
McFadden&Murray &Boisvert(1990)	Elimination of spurious eigenvalues, tau method

Table 7.3: Bibliography of Spectral Calculations for Eigenproblems[continued]

References	Comments
Boyd (1992a)	Arctan/tan mapping for periodic eigenproblems with internal fronts
Falques & Iranzo (1992)	TL_n and Laguerre, edge waves in shear
Su&Khomami (1992)	Two-layer non-Newtonian fluids
Mayer&Powell(1992)	Instabilities of vortex trailing behind aircraft One-dimensional in r in cylindrical coordinates Integration along arc in complex plane for near-neutral modes
Khorrami&Malik (1993)	Spatial eigenvalues in hydrodynamic instability
Chen (1993)	$TL_n(x)$ basis for semi-infinite interval; nonparallel flow
Boyd (1993)	Symbolic solutions in Maple & REDUCE
Huang&Sloan(1994b)	Pseudospectral method; preconditioning
Boyd (1996b)	Legendre, quantum and tidal equations; traps and snares in eigencalculations
Gill& Sneddon (1995,1996)	Complex-plane maps (revisited) for eigenfunctions singular on or near interior of (real) computational domain
Dawkins&Dunbar &Douglass (1998)	Prove tau method, for $u_{xxxx} = \lambda u_{xx}$, always has 2 spurious eigenvalues larger than N^4
O'Connor(1995,1996)	Laplace's Tidal Equation in meridian-bounded basin
Sneddon (1996)	Complex-plane mappings for a semi-infinite interval
Straughan&Walker(1996)	Porous convection; compound matrix & Chebyshev tau
Dongarra&Straughan &Walker(1996)	Chebyshev tau/QZ for hydrodynamic stability Comparisons: 4th order vs. lower order systems
Boomkamp&Boersma &Miesen&Beijnon(1997)	Pseudospectral/QZ algorithm for eigenvalue problem; stability of two-phase flow; 3 subdomains

Chapter 8

Symmetry & Parity

"That hexagonal and quincuncial symmetry ... that doth neatly declare how nature Geometrizeth and observeth order in all things"
— Sir Thomas Brown in *The Garden of Cyrus* (1658)

8.1 Introduction

If the solution to a differential equation possesses some kind of symmetry, then one can compute it using a *reduced* basis set that omits all basis functions or combinations of basis functions that lack this symmetry. The branch of mathematics known as "group theory" is a systematic tool for looking for such symmetries. Group theory is a mandatory graduate school topic for solid-state physicists, physical chemists, and inorganic chemists, but it is also important in fluid mechanics and many other fields of engineering.

However, in fluid mechanics, the formal machinery of group theory is usually not worth the bother. Most of the observed symmetries can be found by inspection or simple tests.

8.2 Parity

The simplest symmetry is known as "parity".

Definition 21 (PARITY) *A function* $f(x)$ *is said to be SYMMETRIC about the origin or to possess "EVEN PARITY" if for all* x,

$$f(x) = f(-x) \tag{8.1}$$

A function is said to be ANTISYMMETRIC with respect to the origin or to possess "ODD PARITY" if for all x,

$$f(x) = -f(-x) \tag{8.2}$$

A function which possesses one or the other of these properties is said to be of DEFINITE PARITY. The word "PARITY" is used as a catch-all to describe either of these symmetries.

Note that trivial function $f(x) \equiv 0$ *is of BOTH EVEN & ODD parity — the only function with this property.*

Parity is important because most of the standard basis sets — the sines and cosines of a Fourier series, and also Chebyshev, Legendre and Hermite polynomials — have definite parity. (The exceptions are the basis sets for the semi-infinite interval, i. e., the Laguerre functions and the rational Chebyshev functions $TL_n(y)$.) If we can determine in advance that the solution of a differential equation has definite parity, we can HALVE the basis set by using only basis functions of the SAME PARITY.

The terms of a Fourier series not only possess definite parity with respect to the origin, but also with respect to $x = \pi/2$. Consequently, it is sometimes possible to reduce the basis set by a factor of *four* for Fourier series if the differential equation has solutions that also have this property of double parity. The most studied example is Mathieu's equation, whose eigenfunctions fall into the same four classes that the sines and cosines do.

Theorem 22 (PARITY OF BASIS FUNCTIONS)

(i) *All cosines,* $\{1, \cos(nx)\}$, *are SYMMETRIC about the origin.*

All sines $\{\sin(nx)\}$ *are ANTISYMMETRIC about* $x = 0$.

(ii) *The EVEN degree cosines* $\{1, \cos(2x), \cos(4x), \dots\}$ *and the ODD sines* $\{\sin(x),$ $\sin(3x), \sin(5x), \dots\}$ *are SYMMETRIC about* $x = \pm\pi/2$.

The cosines of ODD degree $\{\cos(x), \cos(3x), \cos(5x), \dots\}$ *and the sines of EVEN degree* $\{\sin(2x), \sin(4x), \dots\}$ *are ANTISYMMETRIC about* $x = \pm\pi/2$.

(iii) *All orthogonal polynomials of EVEN degree (except Laguerre) are SYMMETRIC:* $\{T_{2n}(x), P_{2n}(x), C_{2n}^{(m)}(x), \text{ and } H_{2n}(x)\}$.

All orthogonal polynomials of ODD degree (except $L_n(y)$) *are ANTISYMMETRIC.*

(iv) *The rational Chebyshev functions on the infinite interval,* $TB_n(y)$, *have the same symmetry properties as most orthogonal polynomials (EVEN parity for EVEN subscript, ODD parity for ODD degree), but the rational Chebyshev functions on the semi-infinite interval,* $y \in [0, \infty]$, $TL_n(y)$, *have no parity.*

The double parity of the sines and cosines is summarized in Table 8.1 and illustrated in Fig. 8.1.

PROOF: (i) and (ii) are obvious. (iii) is a consequence of the fact that all the polynomials of even degree (except Laguerre) are sums only of *even* powers of x, and all those of odd degree are sums of $\{x, x^3, x^5, \dots\}$. [This can be rigorously proved by induction by using the three-term recurrences of Appendix A.] Eq. (iv) may be proved along the same lines as (iii); each rational function $TB_n(y)$ has a denominator which is symmetric about the origin and a numerator which is a polynomial of only even or only odd powers of y. Theorem 22 then follows from the theorem below.

Table 8.1: Symmetry classes for trigonometric basis functions.
"Even" parity with respect to $x = \pi/2$ means that the functions are symmetric with respect to that point, that is,

$$f(-x + \pi) = f(x) \qquad \longleftrightarrow \qquad \text{Even parity about} \quad x = \frac{\pi}{2}$$

or equivalently, $f(\pi/2 - y) = f(\pi/2 + y)$ for the shifted variable $y = x - \pi/2$. Similarly, antisymmetry with respect to $\pi/2$ implies that

$$f(-x + \pi) = -f(x) \qquad \longleftrightarrow \qquad \text{Odd parity about} \quad x = \frac{\pi}{2}$$

Trig. Functions	Parity with respect to $x = 0$	Parity with respect to $x = \dfrac{\pi}{2}$
$\cos([2n]x)$	Even	Even
$\cos([2n+1]x)$	Even	Odd
$\sin([2n]x)$	Odd	Odd
$\sin([2n+1]x)$	Odd	Even

These four symmetry classes are illustrated in Fig. 8.1. The dotted lines on each graph denote the symmetry planes at the origin and at $x = \pi/2$.

Theorem 23 (PARITY OF THE POWERS OF X)

(i) *All EVEN powers of x are SYMMETRIC about the origin:*

$$\{1, \, x^2, \, x^4, \, x^6, \, \dots\} \quad \text{are of EVEN PARITY}$$

(ii) *All ODD powers of x are ANTISYMMETRIC:*

$$\{x, \, x^3, \, x^5, \, x^7, \, \dots\} \quad \text{are of ODD PARITY}$$

PROOF: Replace x^n by $(-x)^n$ and see what happens.

Although trivial to prove, this theorem justifies both Theorem 22 and the following.

Theorem 24 (POWER SERIES WITH DEFINITE PARITY)

A function of EVEN parity, i. e. $f(x) = f(-x)$ for all x, has a power series expansion containing only EVEN powers of x. A function of ODD parity, that is, one such that $f(x) = -f(-x)$, can be expanded in a power series that contains only ODD powers of x.

PROOF: Odd powers of x change sign under the replacement $x \to (-x)$, so it is impossible for a function to be symmetric about the origin unless its power series contains only even powers of x. (Note that the powers of x are all linearly independent; there is no way

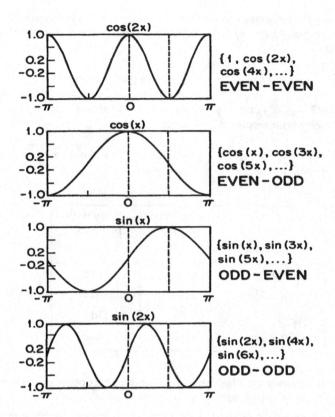

Figure 8.1: Schematic of the four symmetry classes of the terms of a general Fourier series along with the simplest member of each class. All Fourier functions can be classified according to their symmetry with respect to (i) x=0 and (ii) $x = \pi/2$. These symmetry points are marked by the dashed vertical lines.

that possible cancellations among various odd powers can keep them from violating the symmetry condition for *all* x.). Similarly, even powers of x would wreck the condition of antisymmetry except perhaps at a few individual points.

Theorem 25 (DIFFERENTIATION AND PARITY)
Differentiating a function $f(x)$ which is of definite parity *REVERSES the parity if the differentiation is performed an ODD number of times and leaves the parity UNCHANGED if $f(x)$ is differentiated an EVEN number of times.*

PROOF: A function of even parity has a power series that contains only even powers of x by Theorem 24. Differentiating x^{2n} gives $(2n)\,x^{2n-1}$ which is an *odd* power of x for any n. Similarly, differentiating x raised to an odd number gives x raised to an even power. Differentiating *twice*, however, merely restores the original parity.

These four theorems are elementary, but useful. One way of determining the parity of a function with respect to $x = 0$ is to calculate its power series expansion about that point. One can show rigorously that a function $f(x)$ must retain its parity, if any, outside the radius of convergence of the power series.

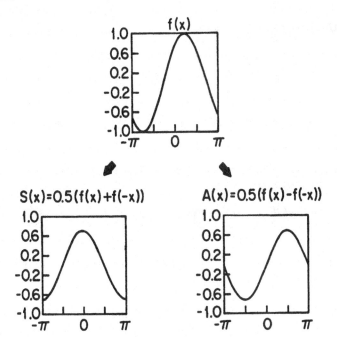

Figure 8.2: Schematic illustrating the decomposition of an *arbitrary* function into its symmetric [S(x)] and antisymmetric [A(x)] parts.

It is usually possible to determine if the solution to a differential equation has parity *without* first solving it by inspecting the coefficients of the equation and the boundary conditions. We need a few preliminary results first.

Theorem 26 (PARITY DECOMPOSITION)

An arbitrary function $f(x)$ can be decomposed into the sum of two functions of definite parity,

$$f(x) = S(x) + A(x) \tag{8.3}$$

where $S(x)$ is symmetric about the origin [$S(x) = S(-x)$ for all x] and $A(x)$ is antisymmetric [$A(x) = -A(-x)$]:

$$S(x) \equiv \frac{f(x) + f(-x)}{2} \qquad [Symmetric] \tag{8.4}$$

$$A(x) \equiv \frac{f(x) - f(-x)}{2} \qquad [Antisymmetric] \tag{8.5}$$

PROOF: It is trivial to verify that the sum of $S(x)$ and $A(x)$ correctly adds up to $f(x)$ since the terms in $f(-x)$ cancel. It is just as simple to verify that $S(x)$ is symmetric since the subsitution $x \rightarrow (-x)$ gives us back the same function except that $f(x)$ and $f(-x)$ have swapped places with respect to the plus sign.

Fig. 8.2 gives a graphical proof of the theorem for the unsymmetric function $f(x) = (\cos(x) + \sin(x))/\sqrt{2}$.

Theorem 27 (SYMMETRY PROPERTIES of an ODE)
Consider an ordinary differential equation in the form

$$\sum_{n=N}^{0} a_n(x) \frac{d^n u}{dx^n} = f(x) \tag{8.6}$$

Then $u(x)$ is a symmetric function if and only if the boundary conditions are compatible with symmetry and if also either

(i) *every even coefficient in (8.6) is even and the coefficient of every odd derivative has odd parity and also $f(x)$ is symmetric or*

(ii) *every even coefficient in (8.6) is odd and the coefficient of every odd derivative has even parity and also $f(x)$ is antisymmetric.*

Similarly, $u(x)$ is an antisymmetric function if and only if the boundary conditions change sign under the replacement of x by $(-x)$ and also if either

(iii) *every coefficient of an even derivative in (8.6) is odd and the coefficient of every odd derivative has even parity and also $f(x)$ is symmetric or if*

(iv) *every even coefficient in (8.6) is even and the coefficient of every odd derivative has odd parity and also $f(x)$ is antisymmetric.*

PROOF: We will illustrate the argument with a second order equation, but the theorem is true for general N. The ODE is

$$a_2(x)\, u_{xx} + a_1(x)\, u_x + a_0(x)\, u = f(x) \tag{8.7}$$

Split all the functions into their symmetric and antisymmetric parts as

$$u(x) = S(x) + A(x); \qquad\qquad f(x) = s(x) + a(x) \tag{8.8}$$

$$a_2(x) = S_2 + A_2 \;\; ; \;\; a_1(x) = S_1 + A_1 \;\; ; \;\; a_0(x) = S_0 + A_0 \tag{8.9}$$

Define $\tilde{u}(x) \equiv u(-x)$, which need not have any definite parity; this solves

$$a_2(-x)\, \tilde{u}_{xx} - a_1(-x)\, \tilde{u}_x + a_0(-x)\, \tilde{u} = f(-x) \tag{8.10}$$

If we add (8.10) to (8.7) and divide by 2, similar to the way the symmetric function $S(x)$ is created from an arbitrary $f(x)$ in Theorem 26, then we obtain an equation in which all the terms are symmetric. Similarly, subtracting (8.10) from (8.7) gives an equation in which all terms are antisymmetric. These two coupled equations are

$$\begin{aligned} S_2\, S_{xx} + A_2\, A_{xx} + S_1\, A_x + A_1\, S_x + S_0\, S + A_0\, A &= s(x) \quad \text{[Symm. Eqn.]} \\ S_2\, A_{xx} + A_2\, S_{xx} + S_1\, S_x + A_1\, A_x + S_0\, A + A_0\, S &= a(x) \quad \text{[Antisymm.]} \end{aligned} \tag{8.11}$$

We can solve this coupled pair of equations in no more operations than for the original single differential equation (8.6) — but no fewer unless the coefficients satisfy the symmetry conditions of the theorem. If we assume that $u(x)$ is purely symmetric so that $A(x) \equiv 0$, then (8.11) implies that $S(x)$ must simultaneously satisfy *two* equations. This is an impossible task for a *single* function unless one of the pair of equations degenerates into $0 = 0$.

With $A(x) \equiv 0$, (8.11) is

$$S_2(x)\, S_{xx} + A_1(x)\, S_x + S_0(x)\, S = s(x) \tag{8.12}$$

$$A_2(x)\, S_{xx} + S_1(x)\, S_x + A_0(x)\, S = a(x) \tag{8.13}$$

There are only two possibilities. One is that $\{S_2, A_1, S_0\}$ are non-zero while $\{A_2, S_1, A_0, a(x)\}$ are all 0. In words, this means that the coefficients of the second derivative and the undifferentiated term are symmetric about $x = 0$ while the coefficient of the first derivative is antisymmetric; the forcing function $f(x)$ must also be symmetric or 0. This is (i) of the theorem. The alternative is that all the coefficients and the forcing function in (8.12) are zero, and this is (ii) of the theorem. The proof for antisymmetric $u(x)$ is similar. Q. E. D.

Thus, we can predict in advance whether or not a linear differential equation with a particular set of boundary or initial conditions will have solutions with definite parity. The same line of reasoning can be extended to nonlinear ordinary differential equations and to partial differential equations, too.

8.3 Modifying the Grid to Exploit Parity

When the basis set is halved because the solution has definite parity, the grid must be modified, too, as illustrated in Fig. 8.3

For example, if we use a "half-basis" of cosines only, applying collocation conditions on $x \in [-\pi, \pi]$ is disastrous. The reason is that because all the included basis functions have definite parity and so, by assumption, does the residual, the collocation conditions at $x = -jh$ where h is the grid spacing are *identical* to the collocation conditions at $x = jh$ (except for a sign change if the residual is antisymmetric). It follows that the pseudospectral matrix will be *singular* because it has only $N/2$ linearly independent rows; each row has the same elements (or the negative of the elements) of one of the other rows in the square matrix.

This disaster can be avoided by restricting the collocation points to *half* of the original interval as shown by middle panel in Fig. 8.3. Similarly, if $u(x)$ has double parity, one must restrict the collocation points to $x \in [0, \pi/2]$, one quarter of the spatial period, in order to avoid redundant collocation conditions.

8.4 Other Discrete Symmetries

Parity is by far the most useful of symmetries because it is the simplest. More exotic examples are possible, however.

In the language of group theory, symmetry with respect to the origin means that $f(x)$ is "invariant under the actions of the group C_2 in the complex plane". This is a highbrow way of saying that we can rotate a symmetric function $f(x)$ through 180 degrees in the complex x-plane, which is equivalent to replacing x by $-x$, without changing anything. Functions may also be invariant under C_n, the group of rotations through any multiple of 360 degrees/n. For example, if a function is invariant under C_4, then its power series expansion is of the form

$$f(x) = \sum_{n=0}^{\infty} a_{4n} x^{4n} \qquad \text{[invariant under } C_4 \text{]} \tag{8.14}$$

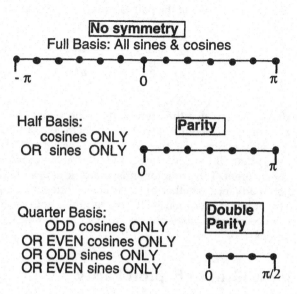

Figure 8.3: Grids for Fourier series.

Although this invariance is not an automatic property of basis functions, it is straightforward to take linear combinations of Chebyshev polynomials (or whatever) which are invariant under C_4.

A real world example of such a function is the third derivative of the similarity solution for a laminar boundary layer at separation (Boyd, 1997a). Functions which have C_3 symmetry in the complex x-plane or are the products of such functions with a power of x include the Airy functions Ai(x) and Bi(x) and the general, unseparated similarity solution for boundary layer flow.

Rotational symmetry is sometimes found in connection with parity and/or reflection symmetry. Fig. 8.4 shows a familiar example: a snowflake, which is not only invariant under the "rotation" or "cyclic" group C_6 but also possesses the property that each of its six points is symmetric about its midpoint. In the language of group theory, snowflakes belong to the "dihedral" group D_6. The significance of "dihedral" as opposed to "cyclic" symmetry is the we can reduce the basis set still further as shown below:

$$C_n \quad : \quad \text{basis set is } \{1, \cos(nx), \sin(nx), \cos(2nx), \sin(2nx), \dots\}$$

$$D_n \quad : \quad \text{basis set is } \{1, \cos(nx), \cos(2nx), \dots\}$$

where x is an angular coordinate. Unfortunately, dihedral symmetry is rare: Travelling waves will disrupt this symmetry unless we shift to a coordinate system that is travelling with the waves.

Two- and three-dimensional symmetries also exist, and some are not simple combinations of one-dimensional symmetries. A famous example is the Taylor-Green vortex, which has been widely studied as a model for a three-dimensional turbulent cascade. This is the

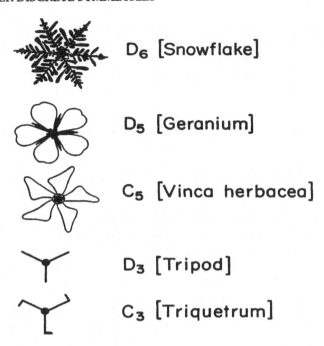

D_6 [Snowflake]

D_5 [Geranium]

C_5 [Vinca herbacea]

D_3 [Tripod]

C_3 [Triquetrum]

Figure 8.4: Illustration of the discrete rotation group C_n [no mid-sector symmetry] and the dihedral group D_n [symmetric about some axis in each sector] for various n. (After Weyl, 1952.)

flow that develops from the initial condition:

$$
\begin{aligned}
u &= \sin(x)\cos(y)\cos(z) \\
v &= -\cos(x)\sin(y)\cos(z) \\
w &= \sqrt{\frac{2}{3}}\cos(x)\cos(y)\sin(z)
\end{aligned}
\tag{8.15}
$$

with boundary conditions of periodicity of 2π in all three coordinates and constant viscosity. One can show that the symmetries inherent in the initial condition — u is antisymmetric about $x = 0$ and symmetric about $y = 0$ and $z = 0$ — are preserved for all time, even though the equations of motion are nonlinear. Consequently, the solution for all times can be expanded as

$$
\begin{aligned}
u &= \sum_{m=0}^{\infty}\sum_{n=0}^{\infty}\sum_{p=0}^{\infty} u_{mnp}(t)\sin(mx)\cos(ny)\cos(pz) \\
v &= \sum_{m=0}^{\infty}\sum_{n=0}^{\infty}\sum_{p=0}^{\infty} v_{mnp}(t)\cos(mx)\sin(ny)\cos(pz) \\
w &= \sum_{m=0}^{\infty}\sum_{n=0}^{\infty}\sum_{p=0}^{\infty} w_{mnp}(t)\cos(mx)\cos(ny)\sin(pz)
\end{aligned}
\tag{8.16}
$$

The symmetries inherent (8.16) allow a reduction of a factor of *eight* in the total number of basis functions since we need only half of the possible one-dimensional basis functions in each of x, y, and z as factors for our three-dimensional functions. This alone is a huge savings, but Appendix I of Brachet et $al.$ (1983) shows that the flow is also symmetric with respect to rotations through 180 degrees about any of the three axes: (i) $x = y = \pi/2$ (ii) $x = z = \pi/2$ and (iii) $y = z = \pi/2$. In mathematical terms, this implies that 7 out of every 8 coefficients in (8.16) are zero: m, n, and p must differ by an $even$ integer (including 0). The flow is also invariant under a rotation by 90 degrees about the vertical axis $x = y = \pi/2$, but the computer code did not exploit this final symmetry.

The total number of degrees of freedom is reduced by a factor of 64. The result is a three-dimensional code which, at 64^3 unknowns per field, is equivalent in accuracy to a $general$ spectral code with 256 grid points in each of x, y, and z. Even 64^3 taxed the memory of the best supercomputer of 1983, the Cray-1, so that the coefficients and the corresponding grid point representations would not both fit in core memory at the same time, forcing an intricate data shuttle to disk storage.

Brachet et $al.$ (1983) note: "The present work has been possible because of the development of new algorithms that take advantage of the symmetries of the Taylor-Green flow". The "new algorithm", given in their Appendix III, is an efficient FFT for real Fourier series whose non-zero coefficients are all odd degree such as $\{1, \cos(x), \cos(3x), \cos(5x), \ldots\}$.[1]

Boyd (1986c) exploits a two-dimensional symmetry that has no counterpart in one dimension. In addition to parity in both x and y, the solution of the nonlinear PDE known as "Bratu's equation" is invariant under the C_4 rotation group, which is equivalent to the replacement of x by y and y by x. His article shows how to exploit the combined double parity/C_4 symmetries to reduce the basis set by a factor of eight, which allowed him to solve the nonlinear two-dimensional BVP to five decimal places on a 1985-vintage personal computer.

Similarly, L. F. Richardson, who made great contributions to numerical analysis, the theory of war and conflict, meteorology, and fractals, solved a two-dimensional partial differential in 1908 $without$ the aid of a digital computer. He attacked Laplace's equation in the unit square subject to the boundary condition $V(x, y = \pm 1) = 1$ on the top and bottom boundaries while $V(x = \pm 1, y) = 0$ on the sides. As shown in Fig. 8.5, the square can be divided into eight triangles by horizontal, vertical, and diagonal lines. The vertical and horizontal lines are lines of symmetry: the solution is symmetric with respect to both $x = 0$ and $y = 0$. The diagonals are lines of antisymmetry, but there is a subtlety: it is $V(x, y) - 1/2$, not $V(x, y)$ itself, which is antisymmetric with respect to the diagonals:

$$V(x + \delta, x) - 1/2 = -\{V(x, x + \delta) - 1/2\} \qquad (8.17)$$

where δ is arbitrary and similarly for the other diagonal (Fig. 8.6).

Marcus(1984a,b, 1990) employs a "reflect-and-shift" symmetry for the velocities that halves the number of Fourier modes (and the cost!).

Weyl(1952) and Hargittai and Hargittai(1994) are good popular treatments of symmetry in nature and art.

---

[1]Kida (1985) set a new "record" for symmetry by identifying a flow in which storage requirements are only 1/192 of what they would be for a general Fourier series with the same resolution.

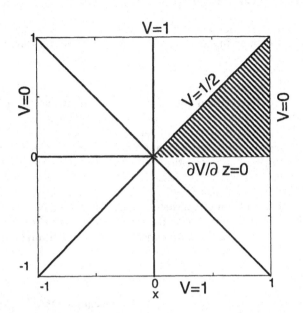

Figure 8.5: Richardson's Laplace equation problem. By exploiting symmetry; it is sufficient to solve the problem only in the shaded triangle with the indicated boundary conditions. Symmetry then gives the solution in the other seven triangles.

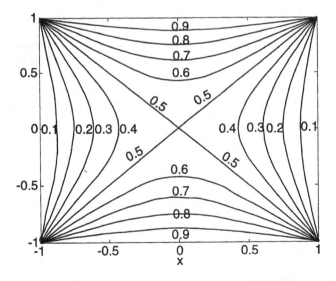

Figure 8.6: Solution to Richardson's Laplace problem.

8.5   Axisymmetric, Hemispheric & Apple-Slicing Models

If a physical problem does not have *exact* symmetries like the Taylor-Green problem, it may still be possible to obtain big savings by making the *approximation* that various symmetries exist. Although unacceptable in operational forecasting, approximate symmetries are widely used in meteorological research.

The crudest approximation is axisymmetry: The flow is independent of longitude. In spectral terms, the three-dimensional atmospheric flow can always be represented without error as a Fourier series in longitude with Fourier coefficients that are functions of latitude, height, and time. The assumption of axisymmetry is equivalent to truncating the Fourier series at $N = 0$, i. e. keeping only the constant term. The longitude-independent approximation is terrible for day-to-day forecasting, but gives useful insights into climate.

Another simplification is the hemispheric model: A grid that spans only the northern hemisphere is justified by the fiction that the flow has equatorial parity. "Symmetry about the equator" has a special meaning to a meteorologist: From the Navier-Stokes equations, one can show that it is not possible for all three flow variables to be simultaneously symmetric about the equator. For example, the (linearized) x-momentum equation is

$$u_t - f\,v = -\frac{1}{\rho}\,p_x \tag{8.18}$$

Since the Coriolis parameter is antisymmetric about the equator, it follows that v must have parity *opposite* to that of u and p. Hemispheric models using spherical harmonics therefore use symmetric basis functions for u, vertical velocity w, and p, but employ antisymmetric harmonics to expand the north-south velocity v and the potential vorticity q.

A third approach is "apple-slicing": Assuming that the spectrum is dominated by one longitudinal wavenumber, and keeping only that wavenumber and its harmonics in the basis set. The theory of "baroclinic instability" shows that when the global weather patterns are expanded in a Fourier series, the coefficients do not monotonically decrease with zonal wavenumber k but rather show a large peak in the range of $k \approx 6 - 8$. A crude-but-quick spectral approach is to keep just zonal wavenumbers from the set $\{0, 8, 16, 24, \ldots\}$.

The reason for the name "apple-slicing" is that this sort of truncation is equivalent to assuming that the earth can be cut into eight wedges, bounded by meridians, with identical flow on each wedge. Consequently, solving for the weather on a single wedge gives the flow on all the other wedges. The "apple-slicing" approximation is equivalent to assuming that the weather is invariant under rotations through $360/m$ degrees for some integer m. The spectral series in longitude is then

$$u = \sum_{j=0} u_{mj}(\phi,\, z,\, t)\, e^{imj\lambda} \tag{8.19}$$

Fig. 8.7 illustrates this idea.

Simmons and Hoskins (1975) show that their "apple-sliced" model with $m = 8$, which they used for preliminary experiments, needed only 0.084 seconds on the CDC7600 per time step versus the 0.67 seconds/time step of their full model, which kept all spherical harmonics up to a zonal wavenumber of 21.

Apple-slicing is a very good way of testing new paramerization schemes, vertical resolution and so on. Half a dozen "apple-sliced" experiments have the same cost as a single full-basis simulation. The disadvantages are obvious: it only applies to special problems where the spectrum is dominated by *intermediate* wavenumber, and it is usually a rather crude approximation.

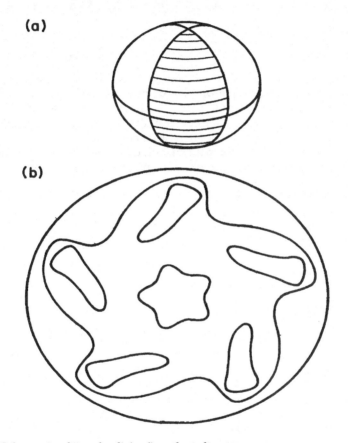

Figure 8.7: Schematic of "apple-slicing" on the sphere.
(a) Side view; the dynamics in the shaded sector [which is the computational domain] is assumed to repeat periodically in all the other sectors of the sphere.
(b) A polar projection with the north or south pole at the center of the diagram, illustrating C_5 symmetry in a contour plot of a variable such as pressure or zonal velocity.

Chapter 9

Explicit Time-Integration Methods

" ... In my experience, many people who do computing are reluctant to look at numbers. At Stanford, the general level of our students has been pretty high, but ... their main weakness is in their inability to look at outputs and extract the meaningful information ... In fact, they are generally less efficient than the assistants I used to have ... in the ACE days [1948], in spite of having far superior mathematical qualifications. Most of those assistants had experience with desk computers and had learned to "look at numbers"."

"I certainly do not want to suggest that the way to acquire this habit is to serve an apprenticeship on desk computers, but we have yet to learn how to instill the relevant knowledge."

— James Wilkinson, interviewed in *BYTE*, Feb. '85

9.1 Introduction

An "implicit" time-marching scheme is one which requires solving a boundary value problem at every time step. "Explicit" schemes give the solution at the next time level in terms of an explicit formula: u^{n+1} = stuff where "stuff" is terms that can be evaluated by using the solution at previous time levels. Implicit schemes are always more costly per time step than their explicit counterparts, but implicit schemes allow a long time step, which often makes them more efficient. Semi-implicit schemes, which treat some terms in the PDE explicitly so as to simplify the boundary value problem, have become the norm in operational weather forecasting and climate prediction.

Unfortunately, there are some schemes which defy such simple classifications. Semi-Lagrangian algorithms, which have very good front-resolving and stability properties, are explicit, but one has to solve an auxiliary problem, as in implicit schemes. We will describe such algorithms in Chapter 14. The Regularized Long Wave equation (water waves) and the quasi-geostrophic equation (meteorology, oceanography, and plasma physics where it is the "Hasegawa-Mima" equation) have a differential operator acting on the time derivative, and thus rather confusingly require the solution of a boundary value problem at every time step even when the time-marching algorithm is explicit. (See Sec. 6 below.)

In this chapter, we focus on explicit time-marching schemes. We shall assume that the spatial dependence has already been discretized by the pseudospectral or Galerkin algorithm so that the remaining problem is to integrate a system of ordinary differential equations in time of the form

$$u_t = F(u, x, t), \tag{9.1}$$

where u and F are vectors, either of grid point values or spectral coefficients.

Explicit time-marching methods are inevitably limited by an unphysical, purely computational instability first identified by Courant, Friedrichs and Lewy in 1928.

Definition 22 (Courant-Friedrichs-Lewy (CFL) Instability)

When the timestep τ for an EXPLICIT time-marching method exceeds a limit τ_{max} that depends on the time-marching algorithm, the physics of the underlying partial differential equation, and the spatial resolution, the error grows exponentially with time. This exponential error growth is the CFL Instability. All explicit methods have a finite τ_{max}, but many implicit methods (discussed in later chapters) are stable for arbitrarily large timesteps.

To obtain a precise value for τ_{max}, one must analyze each individual combination of spatial and temporal discretizations for each specific problem. However, it is possible to offer an estimate.

Rule-of-Thumb 9 (CFL STABILITY LIMIT: PHYSICS)

The maximum timestep is the same order of magnitude as the time scale for advection or diffusion or wave propagation or whatever across the SMALLEST distance h between two grid points. Thus, for wave propagation and for diffusion, the limits are

$$\tau_{max} = d\,\frac{h}{c_{max}} \quad \text{[Waves]} \qquad \tau_{max} = d'\nu\,h^2 \quad \text{[Diffusion]} \tag{9.2}$$

where d, d' are $O(1)$ constants, c_{max} is the speed of the FASTEST-MOVING waves and ν is the diffusion or viscosity coefficient, and h is the SMALLEST distance between two adjacent grid point,

$$h \equiv min_j |x_j - x_{j+1}| \tag{9.3}$$

The two most popular time-integration methods (with spectral algorithms) are the fourth-order Runge-Kutta ("RK4") scheme, which is given in all elementary numerical analysis texts, and the third-order Adams-Bashforth method, which is

$$u^{n+1} = u^n + \tau \left\{ \frac{23}{12} F(u^n, x, t^n) - \frac{4}{3} F(u^{n-1}, x, t^{n-1}) + \frac{5}{12} F(u^{n-2}, x, t^{n-2}) \right\} \tag{9.4}$$

where τ is the time step and where $t^n = n\tau$.

The Runge-Kutta algorithm is rather expensive because the fourth order version requires evaluating the vector F four times per time step. Since F must be computed by calculating spatial derivatives and nonlinear products, usually with extensive use of Fast Fourier Transforms, the evaluation of F is the most costly part of any time-marching calculation. However, Runge-Kutta schemes have considerable virtues, too. First, RK4 is fourth order, that is, has an error decreasing as $O(\tau^4)$: one does not wish to sacrifice the high accuracy with which spatial derivatives are computed in a spectral method by combining it with a lousy time-marching routine.[1] Second, it is stable with a rather long time step compared to other explicit methods for both wave propagation and diffusion. Third, Runge-Kutta methods are "self-starting" and do not require a different time-marching scheme to

---

[1] There is a tendency for Runge-Kutta schemes to have relatively large errors near the boundaries when numerical (as opposed to behavior) boundary conditions are imposed. This is not a serious problem and never arises with a Fourier basis; some remedies are known (Fornberg, 1996).

compute the first couple of time steps, as is true of Adams-Bashforth. Fourth, one may freely vary the length of the time step as the flow evolves because the algorithm is self-starting. Fifth, for only a little extra expense, one can compute Runge-Kutta methods of adjacent orders *simultaneously* to obtain error estimates that can be used to dynamically and adaptively vary the time step. Most ODE libraries offer a routine in which RK4 and RK5 are combined ("RK45") to vary the time step on the fly to keep the error within a user-specified tolerance. In contrast to most other algorithms, it is not necessary for the user to specify a time step; the adaptive RK code will automatically choose one so that the calculation is stable.

There are a couple of *caveats*. If one *does* know the time step in advance, then it is much faster to use RK4 with that τ and no adaptation. Also, although the library adaptive codes are quite reliable and robust, the RK4/RK5 code occasionally blows up anyway if the time scale shortens abruptly. Even so, RK4 and adaptive RK4/RK5 are good time-marching methods when the cost of evaluating $F(u, x, t)$ is not excessive.

The third order Adams-Bashforth scheme requires an initialization phase in which u^1 and u^2 are computed from the initial condition u^0 by some other procedure, such as RK4 or a first or second order scheme with several short time steps. One must also specify the time step τ since adaptively varying the time step, to allow the code to choose its own time step, requires restarting the march with a different time step. However, AB3 is stable, robust and less costly per time step than RK4. Consequently, it, too, is widely used. Durran (1991) praises AB3 as the most efficient of the many algorithms he compares in his very readable review of dissipation and phase errors in time-marching schemes.

Many alternatives have also been tried. In ye olde days, leapfrog was very popular because of its simplicity:

$$u^{n+1} = u^{n-1} + 2\tau F(u^n, x, t^n) \qquad \text{[Leapfrog]} \qquad (9.5)$$

It has the disadvantage that the solution at odd time steps tends to drift farther and farther from the solution for even time steps, so it is common to stop the integration every twenty time steps or so and reinitialize with the first order forward Euler method, or else average u^n and u^{n-1} together. The method is only second order even without the averaging or Euler steps. It is also unstable for diffusion for all τ, so it was common in old forecasting models to "lag" the diffusion by evaluating these terms at time level $(n-1)$, effectively time-marching diffusion by a first order, forward Euler scheme instead of leapfrog. With Chebyshev polynomials, leapfrog is also unstable even for the simple wave equation

$$u_t + u_x = 0, \qquad (9.6)$$

which may be taken as a prototype for *advection* (Gottlieb, Hussaini and Orszag, 1984). With Fourier series, leapfrog is okay, but its popularity has fallen like leaves in the first winter snow.

The second order Adams-Bashforth scheme

$$u^{n+1} = u^n + \tau \left\{ \frac{3}{2} F(u^n, x, t^n) - \frac{1}{2} F(u^{n-1}, x, t^{n-1}) \right\} \qquad (9.7)$$

has also been used with spectral methods, but has fallen from grace because it has a very weak computational instability for all τ. The growth rate is so slow that AB2 is quite satisfactory for short-time integrations, but for long runs, AB2 is slow poison.

Canuto *et al.* (1988) and Durran (1991) give good reviews of other explicit schemes.

Often, however, the major worry is adequate *spatial* resolution, especially in computational fluid dynamics. A second or even first order time marching scheme may then be adequate and the time step τ will be chosen to be the largest step that is *stable* rather than the largest that achieves a certain time-marching error.

In any event, the big worry is not choosing a time integration scheme, but rather finding a cheap way to evaluate spatial derivatives. In the next section, we discuss the problem and its cure.

9.2 Differential Equations with Variable Coefficients

Trouble appears even for a simple problem like

$$u_t + c(x)\, u_x = 0 \tag{9.8}$$

with periodic boundary conditions and the initial condition

$$u(x,\, t = 0) = q(x) \tag{9.9}$$

If $c(x) = c$, a constant, then the solution of (9.8) is the trivial one

$$u = q(x - c\, t) \tag{9.10}$$

(This illustrates a general principle: *constant coefficient* differential equations are best solved by special algorithms. We shall illustrate this point for Chebyshev solutions of non-periodic problems when we discuss matrix-solving methods in Chapter 15, Sec. 10.)

When $c(x)$ is variable, the obvious collocation algorithm is very expensive. With a cardinal function basis $\{C_j(x)\}$, leapfrog is

$$u_j^{n+1} = u_j^{n-1} - 2\,\tau\, c(x_j) \left\{ \sum_{k=0}^{N-1} u_k^n\, C_{k,x}(x_j) \right\} \qquad j = 0,\, \ldots,\, N - 1 \tag{9.11}$$

where the second subscript denotes differentiation with respect to x. This is identical with a finite difference scheme except that the derivative is evaluated using N points instead of just three. This N-point formula is much more accurate than three-point differences, but also much more expensive. Since the sum in (9.11) has N terms and we must evaluate such a sum at each of N grid points at each and every time step, the leapfrog/cardinal method has an operation count which is $O(N^2/\text{time step})$. Ouch! An explicit-in-time/finite-difference-in-space code has a cost of only $O(N)/\text{time step}$.

Galerkin's method is no improvement unless $c(x)$ is of special form. If $c(x)$ is a trigonometric polynomial (or an ordinary polynomial if solving a non-periodic problem using Chebyshev polynomials), then Galerkin's method yields a sparse matrix, i. e. the sum analogous to (9.11) has only a small, N-independent number of terms. To show this, let

$$u(x) = \sum_{m=0}^{N-1} a_m(t)\, \phi_m(x) \tag{9.12}$$

If the basis functions $\phi_n(x)$ are orthogonal, then the spectral form of (9.8) is

$$a_m^{n+1} = a_m^{n-1} - 2\,\tau \left\{ \sum_{k=0}^{N-1} a_k^n(t)\, (\phi_m,\, c(x)\, \phi_{k,x}) \right\} \tag{9.13}$$

If $c(x)$ is a polynomial, then the matrix

$$M_{mk} \equiv (\phi_m,\, c(x)\, \phi_{k,x}) \tag{9.14}$$

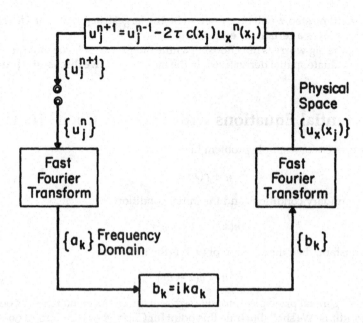

Figure 9.1: Schematic for the leapfrog/Fourier pseudospectral method. The arrows show how the algorithm processes grid point values of $u(x)$ at time level n [lower double circle, left side] to generate $\{u(x_j)\}$ at time level $(n+1)$ [upper double circle]. The algorithm proceeds in a counterclockwise direction. (Inspired by a diagram of Vichnevetsky & Bowles (1982).)

will be sparse and banded. (An explanation for the banded structure will be given in later chapters.) Unless $c(x)$ has such a special form, however, the matrix $\vec{\vec{M}}$ is a full matrix, and leapfrog/Galerkin method will also cost $O(N^2)$ per time step.

This is poor in comparison to the $O(N)$ operation count for a spatial finite difference algorithm. We seem to be trapped between Scylla and Charybdis[2], but the pseudospectral and Galerkin's method are costly for *different* reasons. The cardinal function representation (9.11) is not bothered in the least by the fact that $c(x)$ is a function of x instead of constant; the value of $c(x)$ is merely *multiplied* into the value of u_x at the j-th grid point. What wrecks the cardinal function method is that evaluating the *derivative* requires a sum over the values of u at *all* the grid points.

In contrast, Galerkin's method is untroubled by differentiation because the derivative of a trigonometric function is always a trigonometric function. Consequently, if $c(x)$ were constant, the matrix M defined by (9.14) would have only a *single non-zero* element in each row: the Fourier-Galerkin representation of differentiation (to any order) is a *diagonal* matrix. What ruins the Galerkin method is the need to *multiply* the derivative by a complicated function and then reexpand the product as a series of basis functions.

The remedy is a *hybrid* algorithm — part cardinal function, part Galerkin's. This combination is usually dubbed the "pseudospectral" method (even though it contains Galerkin ideas) because this hybrid algorithm is the only practical way to solve time-dependent problems with non-constant coefficients and large N.

The basic idea is to modify (9.11), the cardinal function representation, in one crucial

---

[2]In Greek mythology, two monsters that guarded either side of the [narrow!] Straits of Messina between Sicily and Italy.

respect. Instead of evaluating $\partial u/\partial x$ by summing up all the derivatives of the cardinal functions, use

$$u_x = \sum_k i\, k\, a_k\, e^{ikx} \tag{9.15}$$

The coefficients of the derivative can be evaluated from those of $u(x)$ in only N operations. We can then evaluate the derivative series (9.15) at each of the N grid points to obtain $u(x_j)$. We may now extrapolate in time with a final step that needs only $O(N)$ operations:

$$u_j^{n+1} = u_j^{n-1} - 2\tau\, c(x_j)\, u_x^n(x_j) \qquad j = 0, \ldots, N-1 \tag{9.16}$$

This is more like it!

For Chebyshev algorithms, differentiation is also an $O(N)$ operation in spectral space. Let

$$\frac{d^q u}{dx^q} = \sum_k^N a_k^{(q)}\, T_k(x) \tag{9.17}$$

so that the superscript "q" denotes the coefficients of the q-th derivative. These may be computed from the Chebyshev coefficients of the $(q-1)$-st derivative by the recurrence relation (in descending order)

$$a_N^{(q)} = a_{N-1}^{(q)} = 0 \tag{9.18}$$

$$a_{k-1}^{(q)} = \frac{1}{c_{k-1}}\left\{2\,k\,a_k^{(q-1)} + a_{k+1}^{(q)}\right\}, \qquad k = N-1, N-2, N-3, \ldots, 1$$

where $c_k = 2$ if $k = 0$ and $c_k = 1$ for $k > 0$.

The one flaw is that there are still two expensive procedures that must be executed at each step: (i) the calculation of the coefficients $\{a_k\}$ from the grid point values $\{u(x_j)\}$ (interpolation) and (ii) the evaluation of the derivative from the series (9.15) at each of the N grid points (summation). If we execute these two steps the easy way — by matrix multiplication — both cost $O(N^2)$ operations, and we are worse off than before.

What makes the hybrid pseudospectral method efficient is that we can replace the costly, rate-determing matrix multiplications by the Fast Fourier Transform (FFT) algorithm. This requires only $O(5N \log_2 N)$ real operations to compute either grid point values from coefficients or vice versa. The bad news is that the FFT is applicable only to Fourier series and to its offspring-by-change-of-variable, Chebyshev polynomials; this is a major reason for the primacy of Fourier and Chebyshev expansions over all their competitors.

The FFT has been the foundation for the modern rise of spectral methods and is therefore the theme of the next chapter.

Fig. 9.1 is a schematic of the pseudospectral algorithm. The figure shows clearly that we jump back and forth between the grid point representations of $u(x)$ [to multiply by $c(x)$] and the spectral coefficients [to differentiate $u(x)$].

9.3 The Shamrock Principle

A millenium and a half ago, a Welsh bishop named Patrick was chiefly responsible for converting the Irish people from paganism to Christianity. According to tradition, the doctrine of the Trinity was an especially hard one to understand: One God, yet three Persons: Father, Son and Holy Spirit.

St. Patrick devised an ingenious audiovisual aid: a little plant called the shamrock, rather like a clover but with squarish leaves, which grows wild all over the country. Plucking a shamrock, he said, "Look! It is one plant, but it has three leaves. Three can be One."

Pseudospectral time-marching require a similar conceptual awakening:

Principle 1 (SHAMROCK PRINCIPLE) *In the limit $N \to \infty$, the set of spectral coefficients $\{a_n\}$ and the set of gridpoint values $\{u(x_j)\}$ are fully equivalent to each other and to the function $u(x)$ from whence they come. These three different, seemingly very different representations, are really just a single function.*

Fig. 9.2 shows a shamrock with each leaf appropriately labeled. The grid point values and the spectral coefficients of a function $u(x)$ and the function itself are all equivalent to one another. We must freely use whichever form is most convenient.

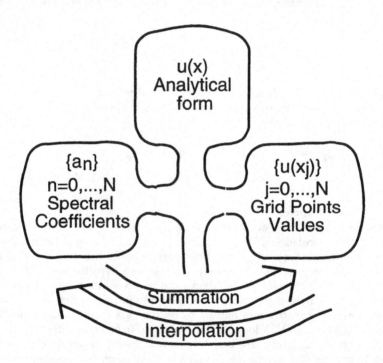

Figure 9.2: A visualization of the Shamrock Principle, or the trinity of spectral methods. The first N Chebyshev coefficients, $\{a_n\}$, and the set of N values on the interpolating grid, $\{u(x_j)\}$, are *equivalent* to the analytical definition of $u(x)$, represented by the top leaf, in the sense that either of these two sets allows one to compute $u(x)$ to arbitrarily high precision in the limit $N \to \infty$.

Via interpolation, one may pass from the "grid point representation", $\{u(x_j)\}$, to the "spectral" representation, $\{a_n\}$. Summation of the Chebyshev series transforms $\{a_n\} \to \{u(x_j)\}$.

9.4   Linear and Nonlinear

Reviews of spectral methods in hydrodynamics (e. g., Orszag and Israeli, 1974) often justify the pseudospectral algorithm by showing the inefficiency of Galerkin's method for *nonlin-*

Table 9.1: MATLAB code for Right-Hand Side of the System of ODEs in Time

```
function F=KdVRHS(t,u);      global k kcub
a=ifft(u);                   % Compute Fourier coefficients a_n from grid point values u(x_j)
ax=k .* a;       axxx = kcub .* a; % Compute coefficients of 1st and 3rd derivative
ux=real(fft(ax));    uxxx=real(fft(axxx) ); % Reverse FFT to get grid point values
                             % of first and third derivatives
F= - u .* ux - uxxx;         % RHS of KdV ODE system. Nonlinear term evaluated by
                             %pointwise multiplication of u by u_x

% In a preprocessing step, either in the main program or in a subroutine called once,
% one must execute the following to initialize the vectors k and kcub with the product
% of i with the wavenumber k and with the (negative) of its cube, respectively
for j=1:(n/2), k(j)=-i*(j-1); end;      for j=(n/2+1):n, k(j)=-i*(j-1 - n); end
for j=1:n, kcub(j)=k(j)*k(j)*k(j), end
```

ear equations. This is a little misleading because the real problem is not the nonlinearity per se, but rather the fact that *coefficients* of the differential equation *vary with x*.

Consequently, Sec. 2 deliberately discussed a *linear* equation. The methodology is exactly the same for a nonlinear problem.

9.5 The Time-Dependent KdV Equation: An Example

To illustrate explicit time-marching with spectral discretization in space, examine the Korteweg-deVries (KdV) equation:

$$u_t + u\,u_x + u_{xxx} = 0 \tag{9.19}$$

subject to the boundary condition of spatial periodicity with period 2π:

$$u(x + 2\pi) = u(x) \tag{9.20}$$

for all x. If the grid point values of u are used for time-advancement, as in a finite difference code, the system of ordinary differential equations in time is

$$\frac{u(x_j, t)}{dt} = F_j(t) \equiv -u(x_j, t)\,u_x(x_j, t) - u_{xxx}(x_j, t) \tag{9.21}$$

A complete code must create the initial conditions, declare parameters, loop forward in time and call the Runge-Kutta subroutine. However, all these steps are necessary with finite difference or finite element discretizations of x, too. The only part of the overall code that needs any knowledge of spectral methods is the subroutine that computes the right-hand side of the system of ODEs in time, that is, the vector F. *A finite difference code can be upgraded to spectral accuracy merely by replacing the subroutine which calculates F.*

Table 9.1 shows a Fourier pseudospectral subroutine (in the language MATLAB) for evaluating this vector of grid point values for the KdV equation. It calls "fft" and "ifft", which are built-in MATLAB Fast Fourier Transform routines. It employs MATLAB's operation for elementwise multiplication of one vector by another vector, which is denoted by ".*". A FORTRAN 77 subroutine would be a little longer because the elementwise multiplications would have to be replaced by DO loops and so on. Nevertheless, the brevity of the subroutine is startling.

The first line computes the Fourier coefficients, a vector a, by taking a Fourier Transform. The second line computes the coefficients of the first and third derivatives by multiplying those of $u(x)$ by (ik) and $(-ik^3)$, respectively. (It is assumed that in a preprocessing step, the vectors "k" and "$kcub$" have been initialized with the appropriate wavenumbers). The third step is to take inverse Fourier Transforms to convert these coefficients for the derivatives into the corresponding grid point values. The final step is to add the grid point values together to form the vector F. Note that the nonlinear term is evaluated by point-by-point multiplication of the grid point values of u with those of u_x.

Tables 9.2 and 9.3 are short bibliographies of the simplest class of time-marching problems: PDEs with periodic boundary conditions, solved by Fourier series.

Table 9.2: Time-Marching with a One-Dimensional Fourier Basis

References	Comments
Abe&Inoue (1980)	Korteweg-deVries (KdV) nonlinear wave equation
Chan&Kerkhoven (1985)	Explicit & semi-implicit schemes for KdV eqn.
Flå (1992)	Energy-conserving scheme for Derivative-NLS equation
Fornberg(1975)	Hyperbolic equations
Fornberg&Whitham(1978)	Nonlinear waves.: KdV, MKdV, Benjamin-Ono & others
Fornberg (1987)	Elastic waves; comparisons with finite differences
Frutos $et\ al.$ (1990)	"Good" Boussinesq; Hamiltonian (symplectic) time-marching
Frutos&Sanz-Serna (1992)	4-th order time integration for KdV & other wave eqns.
García-Archilla(1996)	'Equal Width' equation
Guo&Manoranjan (1985)	Regularized Long Wave (RLW) equation
Herbst&Ablowitz (1992)	Sine-Gordon eqn.; numerical instabilities; integrable-to-chaos transition because of numerical errors
Herbst&Ablowitz (1993)	Symplectic time-marching, numerical chaos, exponentially small splitting of separatrices
If&Berg&Christiansen & Skovgaard(1987)	Split-step spectral for Nonlinear Schrödinger Eq. with absorbing (damping) boundary conditions
Li&Qin (1988)	Symplectic time-marching
Ma&Guo (1986)	Korteweg-deVries equation; uses "restrain operator"
Mulholland&Sloan(1991)	Filtering and its effects on solitons for RLW, KdV, etc.
Mulholland&Sloan(1992)	Implicit & semi-implicit with preconditioning for wave equations
Nouri&Sloan(1989)	Comparisons between pseudospectral algorithms for KdV
Qin&Zhang (1990)	Multi-stage symplectic schemes for Hamiltonian systems
Salupere&Maugin &Engelbrecht&Kalda(1996)	KdV; many-soliton flows
Sanders&Katopodes &Boyd (1998)	KdV; RLW, Boussinesq eqs.
Sanugi&Evans (1988)	Nonlinear advection equation
Sloan (1991)	Regularized Long Wave (RLW) eqn.
Wang (1991)	Hamiltonian systems and conservation laws
Weideman&James (1992)	Benjamin-Ono equation
Zheng&Zhang&Guo(1989)	SLRW equation

Table 9.3: A Selected Bibliography of Time-Marching with Multi-Dimensional Fourier Bases

References	Comments
Chen *et al.*(1993)	Three-dimensional turbulence: 512^3 resolution
Fox&Orszag(1973)	Two-dimensional turbulence
Fornberg (1977)	Two-dimensional turbulence in doubly periodic box
Ghosh&Hossain &Matthaeus(1993)	2D MHD turbulence (non-quadratic nonlinearity)
Haidvogel(1977, 1983)	Review of quasi-geostrophic models
Hald(1981)	Navier-Stokes equations
Hua&Haidvogel(1986)	Quasi-geostrophic turbulence; details of algorithms
Hua(1987)	Review of quasi-geostrophic ocean models
Kosloff&Reshef &Loewenthal(1984)	Elastic waves (seismology)
Ma&Guo(1987a)	Two-dimensional vorticity equation
Tan&Boyd(1997) Boyd&Tan(1998)	Two-dimensional generalization of quasi-geostrophic eq. Solitary vortices, topographic deformations
Rashid&Cao &Guo(1993, 1994a,b)	Low Mach number compressible fluid flow
Vallis(1985)	Doubly-periodic quasi-geostrophic flow

9.6 Implicitly-Implicit: the Regularized Long Wave equation and the Quasi-Geostrophic equation

Some partial differential equations have time derivatives multiplied by a differential operator L, i. e.,

$$L \, u_t = G(u, x, t) \tag{9.22}$$

To restate this in the canonical form

$$u_t = F(u, x, t), \tag{9.23}$$

we must solve a boundary value problem for F:

$$L \, F = G \tag{9.24}$$

For such an equation, it is as laborious to apply an explicit method as an implicit method because one must solve a boundary value problem at every time step. Hence, we shall dub such equations "implicitly-implicit" because the computational labor of an implicit time-marching algorithm is inevitable.

Two examples are the Regularized Long Wave (RLW) equation of water wave theory,

$$\{1 - \partial_{xx}\}u_t = -u_x - uu_x \tag{9.25}$$

and the quasi-geostrophic equation of meteorology and physical oceanography,

$$\{\partial_{xx} + \partial_{yy}\}\psi_t = \psi_y \left(\psi_{xxx} + \psi_{xyy}\right) - \psi_x \left(\psi_{xxy} + \psi_{yyy}\right) \tag{9.26}$$

When the boundary conditions are spatial periodicity, the inversion of the operator L is trivial for both examples.

For the RLW equation, for example, the first step is to evaluate

$$G(x_j, t) = - (1 + u(x_j)) \, u_x(x_j) \tag{9.27}$$

The second step is compute the Fourier coefficients g_n by a Fast Fourier Transform (FFT). In a Galerkin representation, the boundary value problem Eq.(9.24) is

$$(1 + n^2)f_n = g_n \qquad \rightarrow f_n = \frac{g_n}{1 + n^2} \qquad (9.28)$$

where the $\{f_n\}$ are the spectral coefficients of $F(u, x, t)$ and where we have the used the fact that the second derivative of $\exp(inx)$ is $-n^2 \exp(inx)$. An inverse FFT then gives the grid point values of F.

Fourth order Runge-Kutta requires four evaluations of F per time step. One might think that this need for four boundary value solutions per time step would make "implicitly-implicit" equations rather expensive to solve. The irony is that both our examples were originally proposed as *cheaper* alternatives to other PDEs that could be solved by fully explicit methods!

For the RLW equation, the resolution of the apparent paradox is that the differential operator acting on u_t drastically slows the phase speed of short waves, allowing a *much longer time step* than for KdV without computational instability. The quasi-geostrophic equation was invented for similar reasons and used for the first computer weather forecasts in 1950. It continued in use until 1965 when the first CDC6600s made it possible to forecast with the "primitive equations".

However, modern semi-implicit algorithms allow as long a time step for the primitive equations as for the quasi-geostrophic equation. The boundary value problem of the semi-implicit code is just as cheap to solve as that for quasi-geostrophy. The difference is that the primitive equations model allows more accurate forecasts because it does not filter out low frequency gravity waves and Kelvin waves, as quasi-geostrophy does.

Similarly, the KdV equation has been exorcised of its former frightfulness by a combination of fast workstations, which can solve one-dimensional problems quickly even with a tiny time step, and semi-implicit time-marching algorithms, which allow it to be integrated as quickly as its understudy, the RLW equation.

One parenthetical note of caution: when the boundary conditions are not periodic and the operator which is inverted is of *lower order* than the rest of the equation, some spatial discretizations of implicitly-implicit equations can generate modes with infinite eigenvalues due to imposing the full set of boundary conditions on the matrix that discretizes the lower order operator. For such dangerous "L-lower-order-than-G" problems, there are simple remedies which are described in the section on the "Spurious Eigenvalue" problem in Chapter 7.

"Implicitly-implicit" equations, that is, PDEs which require the solution of a boundary value problem even for explicit time-marching, are now falling from favor. Implicit and semi-implicit algorithms, which will be discussed at length in later chapters, can do the same even better. Smart algorithms have replaced clever approximation.

Chapter 10

Partial Summation, the Fast Fourier and Matrix Multiplication Transforms and Their Cousins

"The derivative calculation [by Matrix Multiplication Transform] is the single most computationally intensive step [in my spectral element code]; as a result; a five-fold reduction in Navier-Stokes solution time is obtained when standard vectorized FORTRAN code is substituted with library matrix-matrix product routines on the Cray X-MP. Similar performance improvements are obtained by employing hand coded matrix-matrix products on the iPSC/VX and iPSC/860 hypercubes."

— Paul F. Fischer in an unpublished report (1991)

10.1 Introduction

Spectral time-marching algorithms are prohibitively expensive for large N unless the program jumps back and forth between the cardinal function basis and the Chebyshev polynomial basis at every timestep. The transforms between one representation and the other are usually the *rate-determining* parts of a spectral calculation. It is therefore vital to perform these transforms efficiently.

The Fast Fourier Transform (FFT) is the most efficient algorithm for one-dimensional transformations. However, the biggest savings are realized by performing multidimensional transforms by "partial summation". Instead of summing a series, one point at a time, by a single DO or FOR/NEXT loop running over the entire two-dimensional basis, the transform is factored into a nested sequence of one-dimensional transforms. Partial summation and the FFT are closely related because both are based on a "divide and conquer" strategy of breaking a sum into smaller parts, evaluating each separately, and then recombining.

Both these wonder-transforms impose restrictions. The FFT is applicable only to Fourier series and Chebyshev polynomials. Partial summation requires that both the basis set and the grids be *tensor products*.

We shall also describe alternatives to the FFT. The Matrix Multiplication Transform (MMT) is simply Gaussian quadrature (to interpolate) and direct summation (to evaluate). It turns out that the MMT is suprisingly competitive with the FFT for small to moderate N. By exploiting parity, the cost of the MMT can be halved. This is very important for the non-FFT basis sets, which is all basis sets except Fourier series and Chebyshev polynomials, and the images of Fourier and Chebyshev under a change-of-coordinate.

Fast algorithms for non-FFT basis sets have been extensively studied for two different reasons. The first is to develop an efficient transform for large N. For example, as the resolution of weather forecasting models increases — the 2000 vintage has $N = 512$ — the cost of the MMT transform for Associated Legendre functions becomes an increasingly large fraction of the total cost of the forecast.

The second reason is to interpolate and sum on a "non-canonical" grid. Even with Fourier and Chebyshev series, the FFT is useless if the grid points are spaced irregularly. Unfortunately, semi-Lagrangian time-marching schemes, which have wonderful front-resolving and stability properties, require just such an "off-grid" interpolation at every time step.

We shall therefore describe non-FFT fast transforms at some length.

10.2   Partial Summation in Two or More Dimensions

Orszag (1980) saved a factor of 10,000 (!!) in the running time of his turbulence code CENTICUBE ($128 \times 128 \times 128$ degrees of freedom) merely by evaluating the multidimensional spectral transforms through partial summation. We will illustrate the idea through a two-dimensional example.

It is important to note that partial summation applies only when both the basis functions and grid are tensor products of one-dimensional functions and grids, respectively, as will be assumed throughout the rest of this section. Non-tensor product bases of *low* polynomial order are widely (and successfully) employed for finite elements on triangular subdomains. However, the cost of summations and interpolations rises so rapidly that triangular spectral elements are often implemented by mapping the physical triangle to a computational square and applying a tensor product basis on the square. There is only one reason for the mapping: to gain the enormous cost reductions possible, when N is large, by partial summation.

To evaluate the sum

$$f(x, y) = \sum_{m=0}^{M-1} \sum_{n=0}^{N-1} a_{mn} \, \phi_m(x) \, \psi_n(y) \tag{10.1}$$

at an *arbitrary* point as a double DO loop, a total of MN multiplications and MN additions are needed even if the values of the basis functions have been previously computed and stored[1].

Since there are MN points on the collocation grid, we would seem to require a total of $O(M^2 N^2)$ operations to perform a two-dimensional transform from series coefficients to grid point values. Thus, if M and N are the same order-of-magnitude, the operation count for each such transform increases as the fourth power of the number of degrees in x — and we have to do this at least twice per time step. A finite difference method, in contrast, requires only $O(MN)$ operations per time step. Good grief! With $M = N = 32$, the pseudospectral method would seem to be 1,000 times more expensive than a finite difference method with the same number of grid points.

---

[1] When the basis functions are orthogonal polynomials or trigonometric functions, we can exploit three-term recurrence relations to evaluate the double sum in $O(MN)$ operations without prior computation of the basis functions — but this still gives a cost of $O(MN)$ per point.

Table 10.1: Partial Summation in Two Dimensions

Goal: To evaluate $f(x,y) = \sum_{m=1}^{M} \sum_{n=1}^{N} a_{mn}\phi_m(x)\psi_n(y)$ at each point of the tensor product grid (x_i, y_j) given the spectral coefficients a_{mn}.

Cost $\sim O(MN^2 + M^2N)$ **Operations,**
$O(3MN + M^2 + N^2)$ **Storage**

%Comment: F(i,j)=$f(x_i, y_j)$, A(m,n)= a_{mn}, PHI(i,m)=$\phi_m(x_i)$,
% PSI(j,n)=$\psi_n(y_j)$, B(j,m)=$\alpha_m^{(j)}$
for m=1 to Mmax
    for j=1 to Nmax
    B(j,m)=0
    For n=1 to Nmax
    B(j,m)=B(j,m) + PSI(j,n)*A(m,n)
End triple loop in n,j,m
for j=1 to Nmax
    for i=1 to Mmax
    F(i,j)=0
    for m=1 to Mmax
    F(i,j)=F(i,j) + PHI(i,m)*B(j,m)
End triple loop in m,i,j

Note: The reverse grid-to-spectral transform may be evaluated via partial summation in the same way.

Suppose, however, that we rearrange (10.1) as

$$f(x,\, y) = \sum_{m=0}^{M-1} \phi_m(x) \left\{ \sum_{n=0}^{N-1} a_{mn}\,\psi_n(y) \right\} \tag{10.2}$$

Let the points of the collocation grid be given by the tensor product (x_i, y_j) where $i = 0, \ldots, M-1$ and $j = 0, \ldots, N-1$. A key to the partial summation method is to evaluate the function on the full two-dimensional grid by first pursuing the more limited goal of evaluating $f(x,y)$ along a single line parallel to the x-axis which passes through the M grid points for which $y = y_j$. Define the "line functions" via

$$f_j(x) \equiv f(x,\, y_j) \qquad [\text{"Line Functions"}] \tag{10.3}$$

It follows from (10.2) that

$$f_j(x) = \sum_{m=0}^{M-1} \phi_m(x)\,\alpha_m^{(j)} \tag{10.4}$$

where the constants $\alpha_m^{(j)}$ are simply the value of the braces $\{\}$ in (10.2) at $y = y_j$:

$$\alpha_m^{(j)} \equiv \sum_{n=0}^{N-1} a_{mn}\,\psi_n(y_j) \qquad\qquad \begin{aligned} m &= 0, \ldots, M-1 \\ j &= 0, \ldots, N-1 \end{aligned} \tag{10.5}$$

There are MN coefficients $\alpha_m^{(j)}$, and each is a sum over N terms as in (10.5), so the expense of computing the spectral coefficients of the auxiliary functions $f_j(x)$ is $O(MN^2)$.

Each auxiliary function describes how $f(x, y)$ varies with respect to x on a particular grid *line*, so we can evaluate $f(x, y)$ everywhere on the grid by evaluating each of the N auxiliary functions at each of M grid points in x. However, the auxiliary functions are *one-dimensional*, so each can be evaluated at a single point in only $O(M)$ operations. The cost for the sums in (10.4) over the whole grid is therefore $O(M^2 N)$.

Conclusion: by defining auxiliary "line" functions that represent the original function for a particular y-value on the grid, we have reduced the cost from

$$O(M^2 N^2)[\text{direct sum}] \quad \rightarrow\rightarrow\rightarrow \quad O(M\,N^2) + O(M^2 N)[\text{partial sums}], \qquad (10.6)$$

a savings of a factor of $O(N)$. Pseudocode for the transform is given in Table 10.1.

In three dimensions, the reward is even greater. Let

$$f(x, y, z) = \sum_{l=0}^{L-1} \sum_{m=0}^{M-1} \sum_{n=0}^{N-1} a_{lmn}\, \phi_l(x)\, \phi_m(y)\, \phi_n(z) \qquad (10.7)$$

Direct summation costs $O(L^2 M^2 N^2)$. The partial summations must now be performed in two stages. First, define the two-dimensional auxiliary functions

$$
\begin{aligned}
f_k(x, y) &\equiv f(x, y, z_k) \quad k = 0, \ldots, N-1 \quad [\text{"Plane functions"}] \\
&= \sum_{l=0}^{L-1} \sum_{m=0}^{M-1} \alpha_{lm}^{(k)}\, \phi_l(x)\, \phi_m(y)
\end{aligned}
\qquad (10.8)
$$

where

$$\alpha_{lm}^{(k)} = \sum_{n=0}^{N-1} a_{lmn}\, \phi_n(z_k) \qquad
\begin{aligned}
l &= 0, \ldots, L-1 \\
m &= 0, \ldots, M-1; \\
k &= 0, \ldots, N-1
\end{aligned}
\qquad (10.9)$$

Since there are N auxiliary "plane" functions, each with LM coefficients which must be evaluated by summing over N grid points in z, the cost of forming the $f_k(x, y)$ is $O(L M N^2)$. To evaluate these two-dimensional auxiliary functions, use (10.3) – (10.5), that is, define

$$
\begin{aligned}
f_{jk}(x) &\equiv f(x, y_j, z_k) \quad j = 0, \ldots, M-1 \; ; \; k = 0, \ldots, N-1 \\
&= \sum_{l=0}^{L-1} \beta_l^{(jk)}\, \phi_l(x) \qquad [\text{"Line functions"}]
\end{aligned}
\qquad (10.10)
$$

where

$$\beta_l^{(jk)} = \sum_{m=0}^{M-1} \alpha_{lm}^{(k)}\, \phi_m(y_j) \qquad
\begin{aligned}
l &= 0, \ldots, L-1; \\
m &= 0, \ldots, M-1; \\
n &= 0, \ldots, N-1
\end{aligned}
\qquad (10.11)$$

Since there are MN one-dimensional "line" functions, and each has L coefficients which must be evaluated by summing over M terms as in (10.10), the expense of setting up the $f_{jk}(x)$ is $O(L M^2 N)$. The cost of evaluating these MN "line" functions at each of the L points on each grid line requires summing over L terms, so the cost of performing the sums in (10.9) is $O(L^2 M N)$.

Conclusion: in three dimensions,

$$O(L^2 M^2 N^2) \quad \rightarrow\rightarrow\rightarrow\rightarrow\rightarrow \quad O(L M N^2) + O(L M^2 N) + O(L^2 M N) \qquad (10.12)$$

When $L = M = N$, partial summation reduces the cost by $O(N^2/3)$.

We have deliberately left the basis set unspecified because this argument is *general*, and independent of the basis set. Even if the Fast Fourier Transform is inapplicable, we can still apply partial summation.

Unfortunately, there is still a "transform" penalty: if we have N degrees of freedom in each dimension, the cost of the spectral-to-grid transform will be a factor of N greater than the total number of grid points, N^d, where d is the dimension. A pseudospectral time-marching scheme, alas, will require at least one spectral-to-grid transform and one grid-to-spectral transform per time step. In contrast, an explicit finite difference method has a per-time-step cost which is directly proportional to the number of grid points.

For small to moderate N, the smaller N of the pseudospectral method (to achieve a given error) may outweigh the higher cost per-degree-of-freedom. Unfortunately, the ratio $N_{\text{pseudospectral}}/N_{\text{finitedifference}}$ for a given accuracy is *independent* of N [typically 1/2 to 1/5 as explained in Chapter 2, Sec. 14] whereas the cost per time step grows linearly with N. However, when the solution has sufficient small-scale variability, large N is needed even for moderate accuracy. Because of the factor-of-N "transform penalty", the pseudospectral method must invariably lose the competition for efficiency at a moderate error tolerance when the required N is sufficiently large.

However, there is an escape: if the basis functions are trigonometric functions, Chebyshev polynomials, or rational Chebyshev functions, then the "transform penalty" can be lowered still further by performing the one-dimensional transforms via the Fast Fourier Transform. This reduces the "transform penalty" for FFT-able functions to $\log_2 N$. Since the logarithm grows very slowly with N, Chebyshev and Fourier pseudospectral methods easily defeat their finite difference brethren in the war of efficiency.

To avoid drowning in formulas, we have only discussed the spectral-to-grid transform in this section. However, exactly the same trick can be applied in the opposite direction to calculate the coefficients from $f(x_i, y_j)$.

10.3 The Fast Fourier Transform: Theory

The FFT is restricted to Fourier and Chebyshev functions, but it is extremely powerful since it removes almost all of the "transform penalty" discussed in the previous section. The standard FFT has a cost of (assuming N is a power of 2)

$$\text{N-pt. complex FFT} \sim 5N \log_2[N] \text{ total real operations} \qquad (10.13)$$

where "total operations" includes both multiplications and additions. To sum the same one-dimensional series by direct summation requires the multiplication of a column vector (of coefficients or grid point values) by a square matrix. The cost of this Matrix Multiplication Transform (MMT) is

$$\text{N-pt. complex matrix transform} \sim 8N^2 \text{ real operations} \qquad (10.14)$$

Costs will later be analyzed more carefully, but it is obvious that at least for large N, the FFT is much faster than the MMT. In addition, the FFT both vectorizes well (for Cray supercomputers) and parallelizes well (for massively parallel machines like the IBM SP2). Therefore, its cost-savings will only become larger as the relentless march to larger and larger N continues.

All software libraries offer a variety of canned FFT routines, so it is never necessary to write an FFT code from scratch. For example, *Numerical Recipes* (Press *et al.*, 1986) gives a complete listing of an FFT code in FORTRAN, C, or Pascal, depending on the edition. Both one-dimensional and two-dimensional FFTs are built-in commands in MATLAB. And so it goes. Nevertheless, it is useful to understand the theoretical basis of the FFT, and to appreciate its similarities to multi-dimensional partial summation

Define the discrete transform as

$$X_j = \sum_{k=1}^{N} x_k \exp\left[-i\,k\left(\frac{2\pi j}{N}\right)\right] \qquad j = 1, \ldots, N \qquad (10.15)$$

and the inverse by

$$x_k = \frac{1}{N}\sum_{j=1}^{N} X_j \exp\left[i\,k\left(\frac{2\pi j}{N}\right)\right] \qquad k = 1, \ldots, N \qquad (10.16)$$

(The notation and treatment follow unpublished notes by P. N. Swarztrauber; see Swarztrauber(1986).) What is striking about (10.15) is that it is *not* in the form of a standard complex Fourier series for which the sum variable would run over both the positive and negative integers as in

$$u_j = \sum_{m=-N/2}^{N/2-1} a_m \exp\left[-i\,m\left(\frac{2\pi j}{N}\right)\right] \qquad j = N/2, \ldots, N/2 - 1 \qquad (10.17)$$

One reason for this odd convention is that if we introduce the parameter

$$w \equiv \exp\left(-i\left[\frac{2\pi}{N}\right]\right), \qquad (10.18)$$

we can rewrite (10.15) as a power series in w:

$$X_j = \sum_{k=1}^{N} x_k\, w^{jk} \qquad j = 1, \ldots, N \qquad (10.19)$$

Now the key to our success with multi-dimensional sums was that the three-dimensional basis functions could be factored into the product of three one-dimensional functions. The key to the FFT is that powers of w can be factored, too.

For simplicity, suppose that N is even. We can then split $\{x_k\}$ into two sequences of length $N/2$ which consist of the even and odd parts, respectively:

$$y_k \equiv x_{2k} \qquad ; \qquad z_k \equiv x_{2k-1} \qquad k = 1, \ldots, N/2 \qquad (10.20)$$

These have their own transforms which are

$$Y_j \equiv \sum_{k=1}^{N/2} y_k\, w^{2jk} \qquad ; \qquad Z_j \equiv \sum_{k=1}^{N/2} z_k\, w^{2jk} \qquad j = 1, \ldots, N/2 \qquad (10.21)$$

If we could somehow deduce the transform of the original sequence from $\{Y_j\}$ and $\{Z_j\}$, we would save a factor of 2; when evaluated by matrix multiplication, the cost of a transform of length N is N^2 operations, so the two half-transforms (10.21) cost $N^2/4 + N^2/4 = N^2/2$.

We can retrieve the transform of the original sequence X_j by segregating the even and odd terms in (10.19) and then applying the definitions of the half transform (10.21):

$$X_j = \sum_{k=1}^{N/2}\left[x_{2k}\, w^{j(2k)} + x_{2k-1}\, w^{j(2k-1)}\right] \qquad (10.22)$$

$$= Y_j + w^{-j}Z_j \qquad j = 1, \ldots, N/2 \qquad (10.23)$$

Table 10.2: A Selected Bibliography on the Fast Fourier Transform

References	Comments
Carse&Urquhart (1914)	[Historical] Early but efficient FFTs ("Runge grouping"); printed computing forms used in place of digital computers
Deville&Labrosse(1982)	FORTRAN listings and a little theory for Chebyshev FFT in one, two, and three space dimensions
Brachet *et al.*(1983)	Efficient odd cosine transform
Temperton (1983a)	Explains why inclusion of factors of 4 and 6 in factorization of N reduces cost by 15% for $N = 64$ also shows simultaneous computation of many transforms is highly vectorizable and parallelizable
Temperton (1983b)	Fast real-valued transforms
Temperton (1985, 1992)	General prime-number factorization of N
Press *et al.*(1986) (*Numerical Recipes*)	Program listing for FFT code: C, FORTRAN or Pascal
Swarztrauber (1986)	Efficient cosine (Chebyshev) and sine transforms
Canuto *et al.*(1988)	Appendix B with detailed theory and FORTRAN programs for cosine, Chebyshev transforms
Swarztrauber (1987)	Parallel FFT
van Loan (1992)	Highly-regarded monograph on FFT
Swarztrauber (1993)	Vector spherical harmonics; also cosine & sine transforms
Pelz (1993)	Parallel FFTs for real data
Briggs&Henson(1995)	Excellent, readable book on the FFT and its applications
Fornberg(1996)	Careful discussion of Chebyshev transform, etc.

There is one minor snag: Y_j and Z_j are only defined for $j \leq N/2$. However,

$$w^{-j-N/2} = -w^{-j},$$

which implies

$$X_{j+N/2} = Y_j - w^{-j} Z_j \qquad\qquad j = 1, \ldots, N/2 \qquad\qquad (10.24)$$

Together, (10.22) and (10.24) gives X_j for $j = 1, \ldots, N$ from Y_j and Z_j which are defined only for $j \leq N/2$.

Thus, we can indeed save a factor of two (for N even) by splitting the transform into two parts and computing the partial sums of each. The trick is the same as for the two-dimensional sums of the previous section if the "y" coordinate is imagined as having only two degrees of freedom.

If $N = 2^M$, then we can repeat this factorizing trick M times (until the Z_j and Y_j are trivial transforms of length 1), and save *roughly* a factor of 2 at each step. (A more careful count reveals the $\log_2(N)$ factor in (10.13).)

We can modify this factorizing trick by separating the original sequence into three parts of length $N/3$, or five parts of length $N/5$, or seven parts of length $N/7$, etc., if N is divisible by 3, 5, or 7, respectively. The most general library subroutines will factor an arbitrary N into primes and then subdivide the sums appropriately. The factoring of N requires additional overhead, and the splitting of the transform saves less and less time as the prime factors become larger. Consequently, most hydrodynamics codes choose N to be a power of 2, or a power of 2 multiplied by a single factor of 3.

The articles and books in Table 10.2 give additional theory and advice.

10.4 Matrix Multiplication Transform (MMT) and Matrix Multiplication Transform with Parity (PMMT)

In the grid-point-to-spectral-coefficients direction, the Matrix Multiplication Transform (MMT) is simply Gaussian quadrature. For example, a four-point transform would be

$$
\begin{vmatrix} a_0 \\ a_1 \\ a_2 \\ a_3 \end{vmatrix} = \begin{vmatrix} w_1\phi_0(x_1) & w_2\phi_0(x_2) & w_3\phi_0(x_3) & w_4\phi_0(x_4) \\ w_1\phi_1(x_1) & w_2\phi_1(x_2) & w_3\phi_1(x_3) & w_4\phi_1(x_4) \\ w_1\phi_2(x_1) & w_2\phi_2(x_2) & w_3\phi_2(x_3) & w_4\phi_2(x_4) \\ w_1\phi_3(x_1) & w_2\phi_3(x_2) & w_3\phi_3(x_3) & w_4\phi_3(x_4) \end{vmatrix} \begin{vmatrix} f_1 \\ f_2 \\ f_3 \\ f_4 \end{vmatrix}
$$

where the w_j are the Gaussian quadrature weights multiplied by normalization factors and the $\phi_j(x)$ are the basis functions, which could be Fourier, Chebyshev, Hermite, Associated Legendre, etc. (The normalization factors are chosen so the square matrix above is the inverse of the square matrix below, i. e., such that $a_j = 1$, all other coefficients zero when $f(x) = \phi_j(x)$; numerical values are given in the Appendices.)

The backward, spectral-to-grid-point MMT is merely summation of the interpolant, such as

$$
\begin{vmatrix} f_1 \\ f_2 \\ f_3 \\ f_4 \end{vmatrix} = \begin{vmatrix} \phi_0(x_1) & \phi_1(x_1) & \phi_2(x_1) & \phi_3(x_1) \\ \phi_0(x_2) & \phi_1(x_2) & \phi_2(x_2) & \phi_3(x_2) \\ \phi_0(x_3) & \phi_1(x_3) & \phi_2(x_3) & \phi_3(x_3) \\ \phi_0(x_4) & \phi_1(x_4) & \phi_2(x_4) & \phi_3(x_4) \end{vmatrix} \begin{vmatrix} a_1 \\ a_2 \\ a_3 \\ a_4 \end{vmatrix}
$$

The cost of an MMT is that of a single matrix-vector multiply: $8N^2$ real operations for complex-valued matrices and vector and $2N^2$ operations for real-valued matrices and vectors. However, these cost estimates are a bit misleading. Vector supercomputers and RISC workstations have the hardware to perform matrix-vector multiplies very efficiently. Furthermore, because the matrix-vector multiply is a key building block of many algorithms, assembly-langauge routines to perform it are available on many machines, which roughly halves the execution time.

The cost of the MMT can be halved for any *any* basis set whose elements have *definite parity* to give the faster PMMT algorithm.

Theorem 28 (PARITY MATRIX MULTIPLICATION (PMMT)) *If all the elements of a basis set have definite parity, that is, are all either symmetric or antisymmetric with respect to $x = 0$, then one may halve the cost of a matrix multiplication transform from grid space to spectral coefficients or vice versa by using the following algorithm.*

Suppose that the even degree basis functions are symmetric while the odd degree are antisymmetric with respect to $x = 0$. Sum the series

$$
f_S(x_i) = \sum_{n=0}^{N/2} a_{2n}\,\phi_{2n}(x_i) \qquad \& \qquad f_A(x_i) = \sum_{n=0}^{N/2} a_{2n+1}\,\phi_{2n+1}(x_i) \tag{10.25}
$$

at half of the grid points: all those for positive x, for example. The sum of both halves of the series at all points of the grid is then given by

$$
f(x_i) = \begin{cases} f_S(x_i) - f_A(x_i) & i = 1, \dots, N/2 \\ f_S(x_i) + f_A(x_i) & i = (N/2) + 1, \dots, N \end{cases} \tag{10.26}
$$

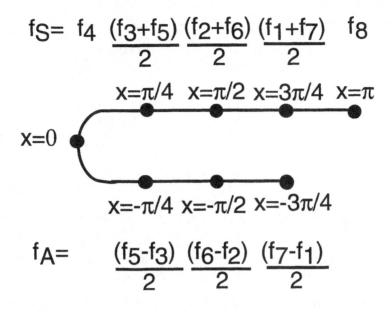

Figure 10.1: Schematic of how the grid point values of a general function $f(x)$ can be decomposed into grid point values of its symmetric part $f_S(x)$ and its antisymmetric part f_A. In the case illustrated, the eight-point interpolant of f by a weighted sum of $\{1, \cos(x), \cos(2x), \cos(3x), \cos(4x), \sin(x), \sin(2x), \sin(3x)\}$ is replaced by the five point interpolant of $f_S(x)$ by $\{1, \cos(x), \cos(2x), \cos(3x), \cos(4x)\}$ and the three point interpolant of $f_A(x)$ by $\{\sin(x), \sin(2x), \sin(3x)\}$.

The cost is only $O(N^2/2)$ multiplications plus an equal number of additions (ignoring terms linear in N). This is half the cost of the standard N-point transform because we have split the problem into two transforms of length $N/2$.

This "parity transform" is applicable to Legendre polynomials, Hermite functions, rational Chebyshev function $(TB_n(x))$ on the infinite interval and all other basis sets whose elements are all either even or odd with respect to $x = 0$. It is not applicable to Laguerre functions or the rational Chebyshev functions on the semi-infinite interval, $TL_n(x)$.

PROOF: As noted in Chapter 2, an arbitrary function $f(x)$ may always be split into its symmetric and antisymmetric parts, $f_S(x)$ and $f_A(x)$, respectively. $f_S(x)$ is the sum of the even degree basis functions only, so it may be computed by summing only $(N/2)$ terms, and similarly for the antisymmetric component.

Fig. 10.1 illustrates the PMMT in the opposite direction, grid point values $\{f_j\}$ to spectral coefficients. The $N = 4$ general MMTs given above simplify to

$$\begin{vmatrix} a_0 \\ a_2 \end{vmatrix} = \begin{vmatrix} w_3\phi_0(x_3) & w_4\phi_0(x_4) \\ w_3\phi_2(x_3) & w_4\phi_2(x_4) \end{vmatrix} \begin{vmatrix} f_2 + f_3 \\ f_1 + f_4 \end{vmatrix}$$

and

$$\begin{vmatrix} a_1 \\ a_3 \end{vmatrix} = \begin{vmatrix} w_3\phi_1(x_3) & w_4\phi_1(x_4) \\ w_3\phi_3(x_3) & w_4\phi_3(x_4) \end{vmatrix} \begin{vmatrix} f_3 - f_2 \\ f_4 - f_1 \end{vmatrix}$$

for the forward transform while the reverse transform is

$$\begin{vmatrix} (f_3 + f_2)/2 \\ (f_4 + f_1)/2 \end{vmatrix} = \begin{vmatrix} \phi_0(x_3) & \phi_2(x_3) \\ \phi_0(x_4) & \phi_2(x_4) \end{vmatrix} \begin{vmatrix} a_0 \\ a_2 \end{vmatrix}$$

$$\begin{vmatrix} (f_3 - f_2)/2 \\ (f_4 - f_1)/2 \end{vmatrix} = \begin{vmatrix} \phi_1(x_3) & \phi_3(x_3) \\ \phi_1(x_4) & \phi_3(x_4) \end{vmatrix} \begin{vmatrix} a_1 \\ a_3 \end{vmatrix}$$

plus

$$\begin{vmatrix} f_1 \\ f_2 \\ f_3 \\ f_4 \end{vmatrix} = \begin{vmatrix} (f_1 + f_4)/2 - (f_4 - f_1)/2 \\ (f_2 + f_3)/2 - (f_3 - f_2)/2 \\ (f_3 + f_2)/2 + (f_3 - f_2)/2 \\ (f_4 + f_1)/2 + (f_4 - f_1)/2 \end{vmatrix}$$

Because the terms of a general Fourier series satisfy a *double* parity symmetry, we can apply the above reasoning *twice* to split a 24-point transform (for example) into four 6×6 matrix multiplies: one for the even cosines, one for the odd cosines, one for even sines, and one for the odd sines.

Theorem 28 appeared on pg. 255 of the 1989 edition of this book, but Solomonoff (1992) published a note on the same idea and factor-of-two savings. It is just as well that he did, because later references to the PMMT are sparse and generally refer to Solomonoff's paper. However, parity-in-transforms has a long history.

Runge (1903, 1905, 1913) used this double-parity reduction as the first two steps of his transform, which is a variant of the FFT. Sir Edmund Whittaker had special computing forms printed for "Runge grouping" for his computing lab at Edinburgh (Carse and Urquhart, 1914); this was computer programming before there were digital, electronic computers!

Runge's ideas were forgotten and the FFT was reinvented several times before the rediscovery by Cooley and Tukey (1965).

No doubt the parity transform will be independently rediscovered many times. Halving both execution time and storage is a useful savings.

10.5 Costs of the Fast Fourier Transform: FFT versus PMMT

Table 10.3 lists the cost of various algorithms for one-dimensional transforms. All the FFT variants have a cost which is $O(N \log_2(N))$ while all the Matrix Multiplication Transforms have costs which are $O(N^2)$. Except for the complex FFT, the cost of the FFT is always $O((5/2)N \log_2(N))$ where N is the number of points which are actually transformed. For an odd cosine transform which computes $\cos(x), \cos(3x), ..., \cos([2N - 1x])$, i. e., N odd cosines, only N points on $x \in [0, \pi/2]$ are needed. The periodic grid spans $x \in [-\pi, \pi]$ with a total of $4N$ points, but it is N, not $4N$, that appears in the cost estimate.

The reason that the table has so many entries is that there are several factors which can reduce the total cost by a factor of two:

- real-valued data

- symmetry with respect to the origin

  metry with respect to $\pi/2$

    ting parity in matrix multiplication transforms

Table 10.3: FFT costs

Note: all costs are expressed in terms of *real* operations for an N-point transform where N is the number of basis functions of the proper symmetry which are transformed; complex multiplications and additions have been converted to their equivalent costs in real operations. FFT is the Fast Fourier Transform, MMT is the Matrix Multiplication Transform and PMMT is the MMT split into two smaller matrix-vector multiplies, one for the symmetric part of the function or series and the other for the antisymmetric.

FFT of complex data	$5N \log_2(N)$
MMT of complex data	$8N^2$
PMMT of complex data	$4N^2$
FFT of real-valued data	$(5/2)N \log_2(N)$
MMT of real-valued data	$2N^2$
PMMT of real-valued data	N^2
FFT cosine,sine or Chebyshev transform	$(5/2)N \log_2(N)$
MMT cosine,sine or Chebyshev transform	$2N^2$
PMMT cosine,sine or Chebyshev transform	N^2
FFT odd cosine, odd sine or odd Chebyshev trans.	$(5/2)N \log_2(N)$
MMT odd cosine, odd sine or odd Chebyshev trans.	$2N^2$

To apply the full, complex-valued FFT on $4N$ points to compute the first N odd cosines, for example, costs eight times more than an efficient odd cosine transform. One factor of two is for real data, another for parity with respect to the origin (a cosine series), and the third for parity with respect to $x = \pi/2$ (*odd* cosine series).

However, these formal operation counts must be interpreted with caution. First, the FFT has a lot of nested operations which create overhead that is not apparent in the count of multiplications and additions. The MMT is just a matrix-vector multiply, which is a simpler algorithm. Second, both the FFT and the matrix-vector multiply are such fundamental algorithms that assembly-language, highly optimized codes to perform them are available for many machines. These are typically faster than a FORTRAN subroutine by a factor of two. Third, the relative efficiency of the algorithms is hardware-dependent.

The net result is that for small and moderate N, the MMT is much more competitive than operation counts indicate (Fig. 10.2), at least for *single* transforms. The graph also illustrates the strong hardware and software dependence of the relative costs of the algorithm. On the personal computer, the FFT matches the efficiency of MMT at about $N = 48$, but on the Cray, not until $N = 512$, a difference of a factor of 10! Fornberg's figure contains another curve, not reproduced here, that shows the relative efficiency and execution times are strongly dependent on software, too. The FFT in the Cray library is much faster than its FORTRAN equivalent and lowers the point where the FFT and MMT curves meet from $N = 512$ to $N = 128$.

Clive Temperton (private communication, 1998) has noted that in large models, one can usually arrange to compute many one-dimensional FFTs *simultaneously*. For example, in a multi-level weather forecasting model, the transforms of velocities, temperature and so on at each level are independent (i. e., logically in parallel). One can then program so that "the innermost loops of the code step 'across' the transform, so that all the messy indexing is banished to the outermost loops ... [where the cost is amortized over many one-dimensional transforms]. I bet the break-even point between FFT and PMMT is lower than $N = 8$, perhaps even $N = 2$!" The algorithmic details are given in Temperton(1983a) where it is shown that the multi-transform strategy is very efficient for both vector (Cray-YMP) and parallel machines. Indexing and if statements, Because they are logical and integer operations, are omitted from the usual floating point operation counts, but are nevertheless

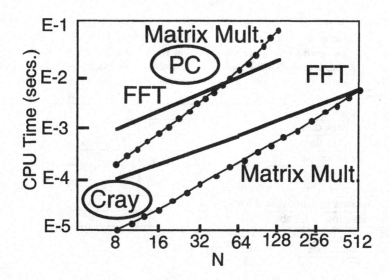

Figure 10.2: A comparison of the execution times of the Chebyshev (or cosine) transform by FFT versus Matrix Multiplication Transform (MMT). (Lower is better!) In each pair, the execution time of the FFT is the solid curve while the MMT is marked with disks. The top pair of graphs were obtain on an IBM personal computer with an Intel 486DX33 central processor while the lower pair was run on a Cray X-MP. Data from Fig. F-3.1 of Fornberg (1996). Note that the FFT curves are for a *single* one-dimensional transform. When many transforms are computed simultaneously (logically in parallel, even if not necessarily implemented on parallel hardware), the indexing can be done once and amortized over many transforms, greatly reducing the cost per transform.

a major expense of one-at-a-time FFTs for small to moderate N.

Thus, simultaneous transforms may be much faster than a single transform, but only if coded carefully — off-the-shelf library software is often not optimized to perform many transforms in a single subroutine call using Temperton's inner loop/outer loop strategy. (Exceptions include a routine in the Cray Scientific Subroutine Library and Roland Sweet's VFFTPACK, which is available on the Web.) This reiterates the point made earlier: the relative efficiencies of the MMT and FFT are highly hardware and software dependent. The cost of the FFT can be greatly reduced by machine coding, simultaneous computation of many transforms, exploitation of symmetries (such as foreknowledge that the grid point values are real-valued) and so on.

Spectral element models, with typically $N \sim 8$, use a Legendre basis and the MMT without tears: for such small N, the FFT would probably be slower than the MMT so a Chebyshev basis has no advantage over Legendre. Orszag (1980) estimated that using the FFT instead of MMT only reduced the cost of his 128^3 turbulence model by a factor of two.[2] Still, a factor of two is a significant savings for a supercomputer code. The virtues of the Fast Fourier Transform will continue to improve as the relentless march to larger and larger values of N continues.

---

[2]Secret confession: I have never used the FFT for a pseudospectral program in BASIC, only for large codes that needed FORTRAN or MATLAB, and then only sometimes.

10.6 Fast Multipole Methods, the Orszag Transform, and Other Generalized FFTs

Fast Multipole Methods (FMM) were originally developed for problems in stellar dynamics, such as the motion of N stars in a model of an evolving star cluster. The naive approach is to compute the total force on the j-th star by adding up the forces exerted upon it by each of the other stars individually. The cost is $O(N^2)$ operations per time step, which is very expensive when $N \sim 100,000$, as would be appropriate for a globular cluster.

The idea behind FMM is illustrated by the schematic (Fig. 10.3) of two interacting star clusters. The thousands of stars in the right cluster can be replaced by a series expansion which converges geometrically. Thus, to calculate the forces on the stars in the left cluster due to stars in the right cluster, we only need to sum a few terms of the multipole series instead of thousands of individual interactions. If the force of the left cluster on the right is calculated similarly, then the cost is reduced by a factor of two, assuming that interactions between stars in the same cluster are still computed individually. By careful partitioning of the cluster and the use of many overlapping multipole expansions in a nested hierarchy, one can reduce the cost to $O(N \log_2(N))$ — an enormous saving.

Boyd (1992c) pointed out that the FMM can also be applied to cardinal function series. This means that (i) FMM can be used as a fast transform for Hermite, Laguerre, spherical

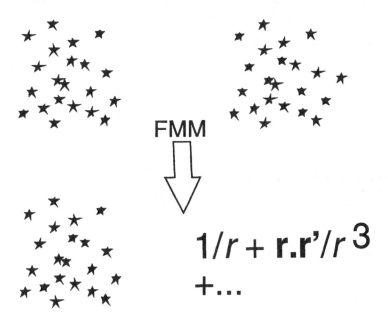

Figure 10.3: Calculating the gravitational forces exerted by two star clusters (top half) is rather expensive because every star interacts with every other. The multipole expansion replaces many individual stars by a few terms of a series, greatly reducing the cost of computing the pull of the right cluster on the stars in the left. $1/r$ is simply the distance from a star in the left cluster (different for each star) to the center of mass of the right cluster. The lowest term is Newton's approximation: all the mass pulls on external objects as if the whole cluster were shrunk to a point at its center of mass.

Table 10.4: A Selected Bibliography of Generalized FFTs

Note: "FD" is an abbreviation for finite difference.

References	Comments
Taylor,Hirsh, Nadworny(1981,1984)	Comparison of FFT with MMT & conjugate gradient to estimate penalty for nonstandard (non-FFT) grid
Orszag (1986)	Fast FFT for almost any function with 3-term recurrence
Greengard&Strain (1991)	Fast Gauss transform (useful with Green's functions,etc.)
Alpert & Rokhlin (1991)	Fast Legendre-to-Chebyshev transform
Süli&Ware (1991,1992)	Nonequispaced fast transform for semi-Lagrangian time-marching
Boyd (1992c)	Showed FFM could be applied to (i) non-Chebyshev basis (ii) non-equispaced grid for Fourier and Chebyshev
Boyd (1992d)	off-grid interpolation through two-step algorithm (i) FFT-to-fine-grid (ii) Lagrange interpolation on fine grid
Moore&Healy&Rockmore(1993)	$O(N\log_2(N))$ exact Fourier transform for nonuniform grid
Dutt&Rokhlin (1993,1995)	FMM nonequispaced FFT
Jakob-Chien & Alpert (1997), Jakob (1993)	FMM for spherical harmonics
Anderson & Dahleh(1996)	Taylor series about nearest equispaced point for off-grid interpolation. Also do reverse off-grid interpolation (from irregular grid to Fourier coefficients) by iteration
Driscoll&Healy&Rockmore(1997)	$O(N\log_2(N))$ transform for any orthogonal polynomials

harmonics and other functions for which the FFT is inapplicable and (ii) for Fourier series and Chebyshev polynomials, the FMM provides a fast method for off-grid interpolation.

To illustrate the first application, let the grid points x_j be the roots of $\phi_N(x)$ and let $C_j(x)$ denote the corresponding cardinal function. Then

$$
\begin{aligned}
f(x) &= \sum_{j=1}^{N} f_j C_j(x_i) \\
&= \phi_N(x) \sum_{j=1}^{N} \frac{f_j/\phi_{N,x}(x_j)}{x - x_j}
\end{aligned}
\tag{10.27}
$$

by virtue of the general construction of cardinal functions given in Chapter 5. However, the lower sum in (10.27) is of the same form as the electrostatic or gravitational potential of a sum of point charges or masses with $(f_j/\phi_{N,x}(x_j))$ playing the role of the charge of the j-th object.

It is trivial to evaluate the cardinal function series to compute $f(x)$ on the canonical grid because the j-th cardinal function is zero at all grid points except $x = x_j$, but the evaluation of the cardinal series for the first derivative costs $O(N^2)$ operations. Differentiation of the cardinal series gives

$$
\begin{aligned}
\frac{df}{dx}(x_i) &= \phi_{N,x}(x_i) \sum_{j=1}^{N} \frac{f_j/\phi_{N,x}(x_j)}{x_i - x_j} + \phi_N(x) \sum_{j=1}^{N} \frac{f_j/\phi_{N,x}(x_j)}{-(x_i - x_j)^2} \\
&= \phi_{N,x}(x_i) \sum_{j=1,i\neq j}^{N} \frac{f_j/\phi_{N,x}(x_j)}{x_i - x_j}
\end{aligned}
\tag{10.28}
$$

since the second sum is zero except for contributing a singular term which cancels the $i = j$ term in the first sum. (To avoid unpleasant cancellation of singularities, one could have noted that since the i-th cardinal function has a maximum at $x = x_i$, its derivative

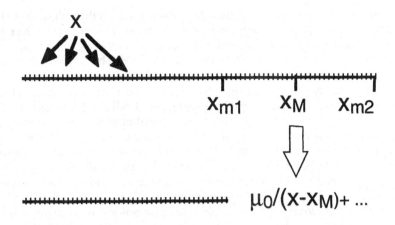

Figure 10.4: The grid point contributions on a typical interval, $x \in [x_{m1}, x_{m2}]$, can be replaced by one or a few terms of a multipole expansion when the target point x is sufficiently far away, such as any of the grid points (short vertical lines) pointed to by the black arrows.

is zero at that point and one can therefore omit the i-th term in the sum in (10.27) before performing the differentiation.) Thus, differentiation is a sum of multipole form, too. The common factor of $\phi_{N,x}(x_i)$ is computed and stored during the initialization of the program.

The multipole idea, defining the "charges" $q_j \equiv f_j/\phi_{N,x}(x_j)$, is to replace a portion of the sum where $j \in [m_1, m_2]$ by a geometric series in inverse powers of $(x - x_M)$ where $M = (m_1 + m_2)/2$. It is implicitly assumed that this particular series will be used only for $x = x_i$ which are far from the interval $[x_{m_1}, x_{m_2}]$ (Fig. 10.4). For different ranges of x_i, one must use different multipole series, which makes the bookeeping a bit tricky. Nevertheless, the rewards are great when N is very large. Without approximation,

$$\sum_{j=m_1}^{m_2} \frac{q_j}{(x - x_j)} = \frac{1}{x - x_M} \sum_{j=m_1}^{m_2} q_j \left(1 + \frac{x_M - x_j}{x - x_M} \right)^{-1}$$

$$= \sum_{k=0}^{\infty} \mu_k \left(\frac{1}{x - x_M} \right)^{k+1} \tag{10.29}$$

where the "multipole moments" are

$$\mu_k = \sum_{j=m_1}^{m_2} q_j \left(x_M - x_j \right)^k \tag{10.30}$$

In practice, the infinite series of multipoles (which is exact) must be truncated to just a handful of multipoles, which requires some careful thought about error tolerances. One drawback of FMM is that the cost increases roughly linearly with the number of correct digits.

A number of variations on this "divide-and-conquer" strategy have been devised. Alpert and Rohklin (1991) have developed a fast (i. e., $O(N \log_2(N))$) algorithm for converting a Chebyshev series to a Legendre series or vice versa. The heart of the algorithm is that the elements of the transformation matrix vary smoothly with row and column indices, which allows one to replace blocks of the usual matrix-vector multiply by expanding the matrix elements as low order polynomials in the indices and then reversing the order of summation. Although their scheme does not directly compute derivatives, one can evaluate derivatives using the regular FFT on the Chebyshev series, so their algorithm allows fast solutions to differential equations using a Legendre basis. Unfortunately, their algorithm does not generalize to Associated Legendre functions because, except for the special case of Legendre polynomials, the Associated Legendre-to-Chebyshev matrix is not sufficiently smooth.

However, a variety of other schemes have been invented to transform spherical harmonics including Dilts (1985), Brown (1985), and Jakob (1993) and Jakob-Chien & Alpert(1997) (Table 10.4). Orszag (1986) devised an ingenious algorithm which can be applied to almost any series of functions which satisfy three-term recurrence relations. This includes all orthogonal polynomials plus some types of Bessel function series. Driscoll, Healy and Rockmore (1997) have developed a faster transform that for any set of orthogonal polynomials.

Unfortunately, the FMM/Fast Generalized Transform revolution has been long on ingenuity and short on applications to spectral methods. The reason is that although these algorithms have costs which are $O(N \log_2(N))$, the proportionality constant, which is $(5/2)$ for a Chebyshev FFT, is very large for these algorithms. When N is 10,000, as in stellar simulations, these algorithms generate vast savings. For the $N = 213$ resolution of 1996 spherical harmonics forecasting models, the savings relative to the Matrix Multiplication Transform are so small — and perhaps nonexistent — that the FMM and its brethren have not been used. As N rises in hydrodynamics models, however, such algorithms will become increasingly attractive as alternatives to the MMT for basis sets like spherical harmonics for which no FFT is known.

10.7   Transforms on an Irregular, Nonequispaced Grid: "Off-grid" Interpolation

In Semi-Lagrangian time-marching schemes (Staniforth and Côté, 1991, and Ritchie, 1987, 1988, 1991), one must interpolate velocities and other fields to the departure points of trajectories, which are spaced irregularly with the respect to the fixed, evenly spaced Fourier grid (or the uneven Chebyshev grid). Because the spectral series must be summed at N unevely spaced, irregularly distributed points, the Fast Fourier Transform cannot be used even with a Fourier or Chebyshev basis. Worse still, this "off-grid" interpolation must be performed every time step.

The naive strategy is to simply sum the spectral series term-by-term, but this costs $O(N^2)$ operations versus the $O(N \log_2(N))$ cost of an FFT. However, there is an added cost: because the grid points shift from time step to time step, the sines and cosines or Chebyshev polynomials cannot be precomputed during program initialization, but must be computed afresh at each step. This means that the cost of the MMT is $O(N^2)$ with a large proportionality constant instead of the usual factor of two. Since all other steps in a Chebyshev or Fourier semi-Lagrangian code can normally be performed by FFT and other fast procedures, the "off-grid" interpolation is the rate-determining step, the most expensive step, in the entire algorithm.

Fortunately, five strategies for fast off-grid interpolation are known. The first is to apply the FMM; first published by Boyd (1992c), the FMM/off-grid procedures have been systematically developed by Dutt and Rohklin (1993, 1994). The second strategy is a Chebyshev polynomial expansion procedure invented by Süli and Ware (1991, 1992). Anderson and Dahleh (1996) devised a similar algorithm with Taylor series instead of Chebyshev series; they are unique by also providing an efficient algorithm for the reverse off-grid transform (from grid point values at a set of irregularly spaced points to Fourier coefficients.) Moore, Healy and Rockmore(1993) invented a fast $O(N \log_2(N))$ algorithm for discrete Fourier transforms on nonuniform grids. Ware (1998) is a good review with some original comparisons between different algorithms.

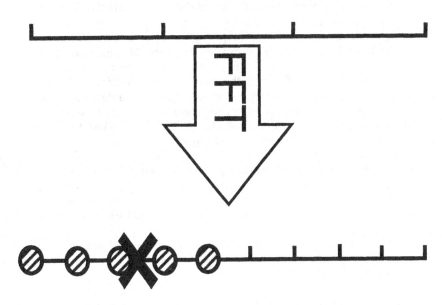

Figure 10.5: Boyd's off-grid interpolation scheme, illustrated through a Fourier basis. The first step is a standard FFT (arrow) from an evenly spaced coarse grid with N points to a finer grid (bottom). The second step is low order polynomial interpolation to obtain an approximation at the off-grid point (marked by the large "X"), using a small number of grid points centered on the target (heavy circles).

The fourth strategy is a two-part procedure: (i) interpolate to a fine grid, typically with $3N$ points, from the N-point canonical grid and (ii) apply low order polynomial interpolation or Euler summation of the cardinal series on the fine grid. It is clear that polynomial interpolation, using the $2M + 1$ points nearest the target point, is relatively inexpensive with a cost of $O(M^2 N)$ operations at most. (We say "at most" because for large M, the usual Lagrangian formulas can be replaced by more sophisticated algorithms with some saving.) However, if polynomial interpolation is performed on the *original* grid, much accuracy is lost: the precision of a hundred-point approximation to derivatives is replaced by perhaps a seven-point approximation to velocities at the departure points.

On the *fine* grid, however, the shortest waves in the function being interpolated have a wavelength equal to $6h_{fine}$, assuming $h_{fine} = h/3$. Boyd shows that each increase in M by one will reduce the error in interpolating the $6h_{fine}$ waves by a factor of four; for other wavelengths, the ratio is $1/\sin^2(\pi\, h/\text{Wavelength})$, which is even larger. By choosing the proper combination of M and the ratio of h_{fine}/h, one can recover spectral accuracy at a cost which is proportional to $O(N \log_2(N))$. The cost of the interpolation to the fine grid of $3N$ points is three times the cost of an FFT on N points, but this still gives a relatively small proportionality constant. The programming is also much easier than the Fast Multipole Method; Lagrangian interpolation and the FFT are the key building blocks, and both are FORTRAN library software or built-in commands in MATLAB.

10.8 Fast Fourier Transform: Practical Matters

The first point is: It is never necessary to write an FFT routine yourself. Not only are FFTs available in most software libraries, such as the NAG and IMSL libraries, and as a built-in command in MATLAB, but also many books give FORTRAN or C codes (Canuto *et al.*, 1988, Appendix B, Fornberg, 1996, Appendix F, Press *et al.*, 1986). All three books give special FFTs for (i) real data (ii) cosine transforms and (iii) sine transforms.

The second point is that unfortunately there is no universally accepted convention for the FFT. Some software defines it as the complex conjugate of what other routines compute; others begin the summation with zero instead of one, and thus multiply the transform by $\exp(2\pi i/N)$, etc. It is very important to carefully read the documentation (or comment statements) in your chosen library software to understand what it is computing.

For example, most FFTs sum over wavenumbers from $k = 1$ to N (as in our Eq. (10.19) above) instead of $k = -N/2$ to $k = N/2 - 1$, which is the usual convention in solving differential equations. The sum over positive wavenumbers is an "aliased" form in the sense that $k = (3/4)N$ and $k = -N/4$ have identical grid point values on a grid of only N points, and similarly for all wavenumbers $k > N/2$. To correctly compute derivatives, one must multiply the coefficient of a wavenumber $k > N/2$ by its "alias" whose wavenumber is on $k \in [-N/2, N/2]$.

There are many formulas for extracting an optimal cosine transform from a general, complex FFT if a cosine transform is not available. Details are given in the books by Canuto *et al.*, Fornberg, and Briggs and Henson (1995).

10.9 Summary

- For large N and a basis of Fourier functions or Chebyshev polynomials or rational Chebyshev functions, the most efficient way to pass from grid points to spectral coefficients or vice versa is through the Fast Fourier Transform (FFT)

- If the solution has symmetries, such as periodic data which can be represented by a cosine series instead of a full Fourier series, these symmetries can be exploited by the FFT.

- The cost of the FFT is approximately $((5/2) \log_2(N)$ for all transforms of real-valued data including cosine, sine, odd cosine, odd sine and Chebyshev transforms if N is interpreted as the appropriate fraction of points on the periodicity interval which are actually transformed

- The FFT, usually in several forms including separate, optimized subroutines for the sine and cosine transform, is available in most software libraries and is also given in *Numerical Recipes* (Press *et al.*, 1986, 1992) and in Appendix B of Canuto *et al.* and Appendix F of Fornberg (1996).

- Both the grid-point-to-spectral coefficient transform (interpolation) and the spectral-to-grid transform (summation of spectral series) can be performed by a square matrix/vector multiply at a cost of $2N^2$ operations for real-valued data; this is the Matrix Multiplication Transform (MMT).

- For small to moderate N, the Matrix Multiplication Transformation is more efficient than the FFT for Chebyshev and Fourier calculations.

- The "cross-over" point where the FFT becomes faster than MMT is *highly* dependent on both hardware and software and may range from $N = 8$ to $N = 512$.

- When many one-dimensional transforms are computed simultaneously, the cost of the FFT can be greatly reduced by banishing the indexing to the outermost loops.

- The FFT is restricted to the Fourier basis, Chebyshev polynomials, and basis functions obtained from these by a change of coordinates.

- The MMT can be applied to any orthogonal polynomial or Fourier basis including Hermite functions, Laguerre functions and others for which the FFT is inapplicable.

- For basis sets whose elements have definite parity, the MMT can be split into two transforms of size $N/2$ to obtain the Parity Matrix Multiplication Transformation (PMMT), which costs only half as much as the MMT.

- Generalized FFTs, many based on the Fast Multipole Method (FMM), have been developed for most non-FFT basis sets.

- The generalized FFTs, although faster than MMT or PMMT in the asymptotic limit $N \to \infty$, have such large proportionality constants p in cost estimates of the form $O(pN \log_2(N))$ that these algorithms have only rarely been used in pseudospectral calculations.

- Fast algorithms for interpolating to a set of N points which are spaced irregularly with respect to the standard pseudospectral grid have been developed; these have a cost of $O(N \log_2(N))$ like the FFT, but with a proportionality constant three or four times larger than for the FFT. (Needed for semi-Lagrangian time-marching algorithms, among others.)

- Definitions of the FFT vary; read your documentation *carefully* and adjust the output of the library software to the transform you want to compute.

Chapter 11

Aliasing, Spectral Blocking, & Blow-Up

"Blowup of an aliased, non-energy-conserving model is God's way of protecting you from believing a bad simulation."
— J. P. Boyd

"If a problem ... for which the solution develops a singularity in finite time ... is simulated with a fixed resolution, then a dealiasing procedure is advisable. Moreover, if a fully turbulent flow is simulated with marginal resolution, then dealiasing may also be useful."
– Canuto *et al.*(1988, pg. 123)

11.1   Introduction

On a grid of N points, Fourier components $\exp(ikx)$ with $|k| > N/2$ appear as low wavenumbers; the high wavenumber is said to be "aliased" to the low. Phillips (1959) attributed the blowup of his model of the earth's general circulation (1956) to an instability specifically caused by aliasing. Because of his work, many modern codes employ a "dealiasing" procedure or special numerical algorithms that conserve the discretized energy so that a runaway to supersonic winds is impossible.

There are precursors to blow-up. One is the appearance of "$2h$" waves in physical space, that is, waves whose wavelength is twice the grid spacing h. This is accompanied by a tendency for the Fourier coefficients to curl up near $k = \pi/h$, the "aliasing limit"; this growth with k, instead of the expected decay, is called "spectral blocking".

In this chapter, we review all these matters. Although "aliasing instability" and dealiasing algorithms are well-known, there is still a major controversy as to whether heroic measures to control aliasing and/or enforce strict energy conservation are helpful or merely misleading.

11.2   Aliasing and Equality-on-the-Grid

Although our real topic is spatial aliasing, the most familiar examples of aliasing are in frequency. In a TV western, watch the six-spoked wheels of the stagecoach as it begins to

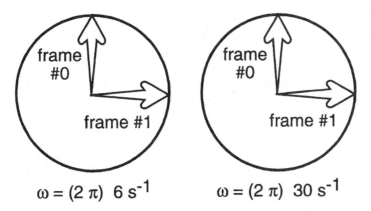

$$\omega = (2\,\pi)\ 6\ s^{-1} \qquad\qquad \omega = (2\,\pi)\ 30\ s^{-1}$$

Figure 11.1: Frequency aliasing. Both panels show the motion of the hands of two clocks as recorded by a conventional movie camera, which takes a frame every 1/24 of a second. The left clock hand is turning at a frequency of 6 revolutions/second, so if the hand is vertical in frame 0, it will be a quarter-turn clockwise in frame 1. The right clock hand is rotating at 30 revolutions/second — five times faster. However, on film, the two stopwatches appear *identical*. The hand of the right watch has turned through one-and-a-quarter revolutions in the 1/24 second interval between frames, so it, too, is pointing to "three o'clock" in frame 1.
The frequency of the high-speed stopwatch has been "aliased" to the lower frequency, 6 turns/second, because a person who watched the film, even in slow motion, would see the clock hand turn through only a quarter of a turn [versus the actual five quarters of a turn] between frames.

move. If you look carefully, you may see the spokes turning in the *wrong* direction for a few moments, then come to rest, and finally begin to rotate in the opposite sense. The same phenomenon is seen when the propeller of an airplane is started from rest.

The reason for this highly unphysical behavior is that a movie camera cannot *continuously* photograph a scene; instead, the shutter snaps 24 times per second. This discrete sampling creates errors that are mathematically identical to those caused by interpolating a continuous function on a discrete grid of points. Frequencies whose absolute values are higher than $|\omega| = 2\pi(12)s^{-1}$ cannot be properly represented with such a limited sampling. Instead, the eye — and mathematics — interprets the high frequency as some lower frequency on the range

$$\omega \in (2\pi)[-12,\ 12]\ s^{-1} \tag{11.1}$$

Fig. 11.1 illustrates the problem. The two figures appear identical on film — the clock hand is pointing to 12 o'clock in frame #0 and to 3 o'clock in frame #1. The frequency in the left frame is only $\omega = (2\pi)6/s$ while it is five times higher in the right panel. The discrete sampling is unable to distinguish between rotation through 90 degrees per frame and rotation through a full turn plus 90 degrees. To the eye, the higher frequency appears as the lower frequency $\omega_A = \omega - (2\pi)24/s$.

One can show that all frequencies outside the range given by (11.1) will be "aliased" to a frequency within this range. The upper limit in (11.1) is a motion which rotates through 180 degrees between frames. It might seem as though only frequencies sufficiently high to rotate through a *full* turn between frames would be aliased. However, if the clock hand turns through, say, 190 degrees per frame, it is impossible to distinguish this motion from one which rotates through - 170 degrees per frame. Indeed, it is these motions which rotate through more than a half turn but less than a full turn per frame that produce the bizarre effect of propellers and wheels (apparently) rotating in the wrong direction.

Our real interest is pseudospectral methods rather than cinematic trivia, but the geometric argument is useful because it indicates that high frequencies are not aliased to arbitrary frequencies, or to some mix of frequencies. Instead, the aliasing is always through a *positive* or *negative multiple* of one full turn/frame. This is obvious from inspecting the clock diagram because the eye will always perceive the clock hand as having rotated through less than half a turn per frame, either clockwise or counterclockwise, regardless of the true speed of rotation. If the number of rotations (through 360 degrees) per frame is r, then the viewer will interpret the frequency as

$$(r - m) \text{ turns/frame} \tag{11.2}$$

where m is that integer (positive, negative, or zero) such that

$$|r - m| < \frac{1}{2} \tag{11.3}$$

Exactly the same applies to spatial aliasing.

Definition 23 (ALIASING) *If the interval $x \in [-\pi, \pi]$ is discretized with uniform grid spacing h, then the wavenumbers in the trigonometric interpolant lie on the range $k \in [-K, K]$ where $K \equiv \pi/h$ and is the so-called "aliasing limit" wavenumber. The elements of the Fourier basis are $\exp[i\,k\,x]$ with k an integer. Higher wavenumbers k such that $|k| > K$ will appear in the numerics as if they were the wavenumbers*

$$k_A = k \pm 2\,m\,K \qquad\qquad m = integer \tag{11.4}$$

where

$$k_A \in [-K, K] \tag{11.5}$$

The wavenumbers outside the range $[-K, K]$ are said to be "aliased" to wavenumbers within this range and k_A is the "alias" of k.

Eq. (11.4) is merely a restatement of the error law (4.44) for trigonometric interpolation (Theorem 19) in Chapter 4. There, we stated without proof that if we used N points in trigonometric interpolation so that $\cos([N/2]x)$ was the most rapidly oscillating function in the basis set, then the computed sine and cosine coefficients, $\{b_n\}$ and $\{a_n\}$ were contaminated by those neglected coefficients of the exact series, $\{\alpha_n\}$ and $\{\beta_n\}$, as

$$a_n = \sum_{m=-\infty}^{\infty} \alpha_{|n+mN|} \qquad \begin{array}{l}\text{[Aliasing Law for Cosines in}\\ \text{Trigonometric Interpolation]}\end{array} \tag{11.6}$$

(on the endpoint/trapezoidal rule grid) and similarly for the sine coefficients. Because $K = N/2$, (11.6) and (11.4) are equivalent. We have made good the lack of an analytical proof of the theorem in Chapter 4 by here giving a geometric proof (through clock-faces).

Aliasing may also be stated directly in terms of sines and cosines. Two trigonometric identities imply

$$\cos([j + mN]\,x) = \cos(jx)\,\cos(mNx) - \sin(jx)\,\sin(mNx) \tag{11.7a}$$

$$\sin([j + mN]\,x) = \sin(jx)\,\cos(mNx) + \cos(jx)\,\sin(mNx) \tag{11.7b}$$

for arbitrary j and m. When j and m are integers and x is a point *on the grid*, $x_i = 2\pi i/N, i = 0, 1, \ldots, (N-1)$,

$$\cos(mNx_k) = \cos\left(mN\frac{2\pi k}{N}\right) = \cos(2\pi k\, m) = 1 \quad \text{all integral } k, m \qquad (11.8a)$$

$$\sin(mNx_k) = 0 \qquad\qquad\qquad\qquad\qquad \text{all integral } k, m \qquad (11.8b)$$

In consequence,

$$\cos(jx) \stackrel{G}{=} \cos([j+mN]x) \stackrel{G}{=} \cos([-j+mN]x) \qquad j = 0, 1, \ldots \qquad (11.9a)$$

$$\sin(jx) \stackrel{G}{=} \sin([j+mN]x) \stackrel{G}{=} -\sin([-j+mN]x) \qquad j = 1, \ldots \qquad (11.9b)$$

where the "G" above the equals sign denotes that the equality is true *only* on the N-pt. evenly spaced *grid*. Similar relationships exist for the interior (rectangle rule) grid, too.

This special notion of equality — linearly independent functions that are point-by-point equal on the grid — is the essence of aliasing. Canuto *et al.* (1988, pg. 40) offer a graphic illustration of (11.9).

Recall that the error in solving differential equations has two components: a truncation error $E_T(N)$ and a discretization error $E_D(N)$. The truncation error is that of approximating an infinite series as a sum of only N terms. However, there is a second source of error $E_D(N)$ because the N spectral coefficients that we explicitly compute differ from those of the exact solution.

Aliasing is the reason for these differences in coefficients. "Discretization error" and "aliasing error" are really synonyms.

Aliasing can cause numerical instability in the time integration of *nonlinear* equations. For example, a typical quadratically nonlinear term is

$$u\, u_x = \left(\sum_{p=-K}^{K} a_p e^{ipx}\right)\left(\sum_{q=-K}^{K} i\, q\, a_q e^{iqx}\right) \qquad (11.10)$$

$$= \sum_{k=-2K}^{2K} b_k e^{ikx} \qquad (11.11)$$

where the b_k are given by a sum over products of the a_k. The nonlinear interaction has generated high zonal wavenumbers which will be aliased into wavenumbers on the range $k \in [-K, K]$, creating a wholly unphysical cascade of energy from high wavenumbers to low.

When there is a numerical instability, the earlier assertion that the truncation and discretization errors are the same order of magnitude is no longer correct; E_D can be arbitrarily large compared to E_T. (Note that the truncation error, which is just the sum of all the neglected terms of the *exact* solution, is by definition independent of all time-stepping errors.) The statement $E_D \sim O(E_T)$ is true only for *stable* numerical solutions.

11.3 "2 h-Waves" and Spectral Blocking

The onset of aliasing instability can often be detected merely by visual inspection because the waves at the limit of the truncation — $k = K$ — are preferentially excited.

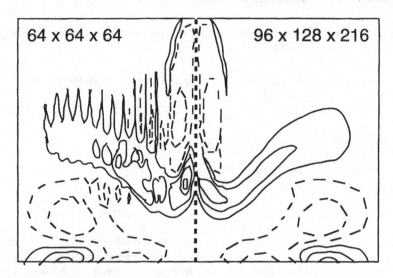

Figure 11.2: Streamwise vorticity contours at $t = 22.5$ in a transition-to-turbulence model. The low resolution results are in the left half of the figure; the high resolution at right. The plot should be symmetric about the dashed line down the middle, so the differences between left and right halves are due almost entirely to numerical errors in the low resolution solution (left). Redrawn from Fig. 6 of Zang, Krist and Hussaini (1989).

Definition 24 (TWO-H WAVES) *Oscillations on a numerical grid which change sign between adjacent grid points — and therefore have a wavelength equal to twice the grid spacing h — are called "$2h$-waves". They are the shortest waves permitted by the numerical grid.*

Assertion: a solution with noticeable $2h$-waves has almost certainly been damaged by aliasing or CFL instability.

Fig. 11.2 is an illustration of $2h$-waves in a three-dimensional simulation of transition-to-turbulence. As the flow evolves, it develops finer and finer structures and consequently requires higher and higher resolution. When solved at fixed resolution, $2h$-waves appear as seen in the left half of the figure. These are suppressed, without any of the special tricks described below, merely by using higher resolution (right half of the figure), which is why these special tricks are controversial.

The $2h$-waves are a precursor to breakdown. Although the 64^3 model still captures some features at the indicated time, the corresponding plot at the later time, $t = 27.5$ (Zang, Krist, and Hussaini, 1989), is much noisier and accuracy continues to deteriorate with increasing time.

The time-dependent increase in $2h$-waves must be matched by a growth in the corresponding Fourier coefficients, namely those near the aliasing limit $K = \pi/h$. This is shown schematically in Fig. 11.3; an actual spectrum from the same transition-to-turbulence model is illustrated at different times in Fig. 11.4. The latter graph shows clearly that spectral blocking is a "secular" phenomenon, that is, it gets worse with time. A numerical simulation which is smooth in its early stages, with monotonically decreasing Fourier or Chebyshev coefficients, may be very noisy, with lots of $2h$-waves and a pronounced upward curl in the coefficient spectrum, for later times.

This curl-up in the absolute values of the Fourier or Chebyshev spectrum near the highest resolvable wavenumber (or near $T_N(x)$) has acquired the following name.

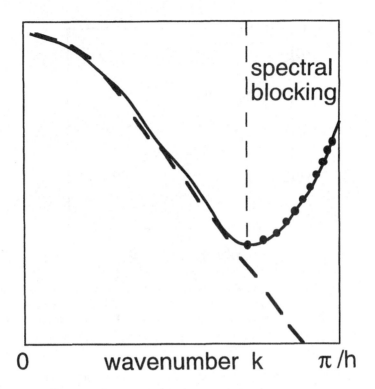

Figure 11.3: Schematic of "spectral blocking". Dashed line: logarithm of the absolute values of Fourier coefficients ("spectrum") at $t = 0$. Solid: spectrum at a later time. The dashed vertical dividing line is the boundary in wavenumber between the decreasing part of the spectrum and the unphysical region where the amplitude increases with k due to numerical noise and aliasing. The corrupted coefficients are marked with disks.

Definition 25 (SPECTRAL BLOCKING) *If the spectral coefficients, when graphed on the usual logarithm-of-the-absolue value graph, rise with increasing wavenumber or degree near the highest wavenumber or degree included in the truncation, then this is said to be "spectral blocking".*

The name was coined by fluid dynamicists. In turbulence, nonlinear interactions cascade energy from smaller to larger k. Very high wavenumbers, with $|k| > O(k_{diss})$ for some dissipation-scale wavenumber k_{diss}, will be damped by viscosity, and the coefficients will fall exponentially fast in the dissipation range, $k \in [k_{diss}, \infty]$. Unfortunately, k_{diss} for a high Reynolds flow can be so large that even a supercomputer is forced to use a truncation $K \ll k_{diss}$. Aliasing then blocks the nonlinear cascade and injects energy back into smaller wavenumbers. This spurious reverse cascade affects all wavenumbers, but is especially pronounced near the truncation limit $k = K$ because these wavenumbers have little amplitude except for the erroneous result of aliasing. The numerical truncation has "blocked" the cascade, and the blocked energy piles up near the truncation limit.

In some ways, the term is misleading because the tendency to accumulate numerical noise near $k = \pi/h$ is generic, and not merely a property of turbulence. Nevertheless, this term has become widely accepted. It is more specific than "high wavenumber noise accumulation", which is what spectral blocking really is.

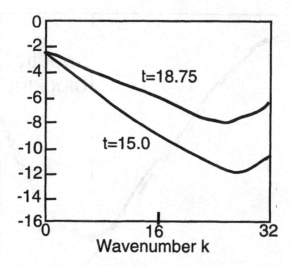

Figure 11.4: Spectral blocking in a three-dimensional transition-to-turbulence model. As time increases, more and more resolution is needed; by $t = 27.5$, the vorticity contour plot (for this $64 \times 64 \times 64$ run) is a sea of noise. Redrawn from Fig. 4 ("z resolution spectra") of Zang, Krist and Hussaini (1989).

A wave equation with constant coefficients will have a truncation error E_T because the exact solution is, except for special cases, an infinite series whereas the numerical approximation includes only a finite number of terms. However, the Fourier pseudospectral discretization of a linear, constant coefficient wave equation does *not* display spectral blocking because waves of different wavenumbers are represented exactly as Fourier coefficients. If the time step is sufficiently short, the Fourier algorithm will advect these components without amplitude or phase error so that the high wavenumbers in the truncation will not grow, but rather will remain the same in amplitude. The lack of amplitude and phase errors in propagation and advection is one of the great virtues of spectral methods.

If the equation has spatially-varying coefficients, either due to nonlinearity or to spatial variations in the wave environment such as a change in the depth of the water, then even a spectral method will have aliasing error. The product of an x-dependent coefficient with $u(x)$ has a Fourier expansion with components outside the truncation limit which are aliased to smaller wavenumbers.

Although some spectral blocking is almost inevitable in a long time integration of a nonlinear system (unless the dissipation is large), there is also a really stupid way to generate spectral blocking: Use too large a time step. For an explicit time-marching scheme for the diffusion equation, for example, the high wavenumbers become unstable before the low wavenumbers as the time step is increased. It is thus possible to accurately and stably integrate the low wavenumbers while the wavenumbers near the truncation limit are spuriously amplifying because the time step is too long. Since these high wavenumber components have exponentially small amplitudes at $t = 0$ (if the initial condition is smooth), the amplitudes of the high wavenumber components will remain exponentially small for a finite time interval. However, the instability will cause the graph of the amplitudes of the spectral coefficients (on a log-linear plot) to curl up more and more with time, i. e., exhibit spectral blocking. This result of violating the Courant-Friedrichs-Lewy (CFL) timestep may be dubbed "CFL Blocking" or alternatively, "Really Stupid Blocking".

Fig. 11.5 is an illustration. There is nothing special about the partial differential equation, boundary conditions, or initial condition; spectral blocking can occur for any linear equation when the phase speed or diffusion rate of a mode increases with wavenumber or degree so that the CFL criterion for $a_j(t)$ becomes increasingly restrictive as j increases.

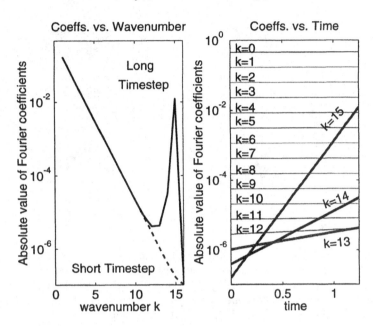

Figure 11.5: Spectral blocking in a LINEAR differential equation due to violation of the CFL timestep limit for LARGE wavenumbers (although the small wavenumbers are STABLE). The low wavenumbers $k = 0, 1, 2, \ldots , 11$ for basis functions $\exp(ikx)$ are stable (as are the corresponding negative wavenumbers, not shown), but the three highest wavenumbers grow exponentially with time for the long timestep used (solid curve). The dashed line illustrates the amplitudes of the exact solution, which are independent of time. The solid curve was generated by the adaptive fourth/five order Runge-Kutta subroutine **ode45** in Matlab 5.2 with 12 subroutine calls on the time interval $t \in [0, 1.25]$. Because the high wavenumber Fourier coefficients are initially very small, inaccuracies in integrating them had only a negligible impact on the accuracy of the initial step and so the adaptive routine happily picked a time step which was unstable for the highest three wavenumbers. When the calculation was repeated with much more frequent calls to the subroutine **ode45**, the adaptive routine was forced to use a short timestep, and all wavenumbers were accurately integrated. The PDE is the linear Benjamin-Davis-Ono equation, $u_t = -\mathcal{H}(u_{xx})$ where $\mathcal{H}$ is the linear operator known as the Hilbert transform; in wavenumber space, $\mathcal{H}\{\exp(ikx)\} = i\text{sign}(k)\exp(ikx)$. Arbitrarily, the spatial period was chosen to be 2π and the initial condition was $u(x, 0) = 2\tanh(1)/(1 - \text{sech}(1)\cos(x))$.

The CFL criterion can sometimes be exceeded in the middle of an integration even though the computation is stable at $t = 0$. The reason is that the advective time-stepping limit depends on the maximum current, and this can increase during the time evolution of a flow. Fig. 11.6 shows contour plots of four different time levels in the evolution of a dipolar vortex. At $t = 10$, the flow is still perfectly smooth, but then $2h$-waves develop in the zone of intense shear between the two vortices and rapidly amplify until the calculation becomes nonsense. The culprit is not lack of spatial resolution or a blocked turbulent cascade,

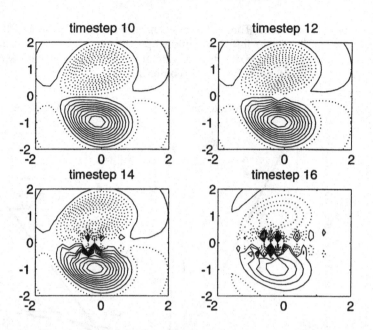

Figure 11.6: Streamlines of the evolution of two contra-rotating vortices. (Negative contours are dashed). At the tenth time step (upper left), the flow is very smooth, but $2h$-waves appear at the twelfth step and amplify rapidly. Although spectral blocking is occurring in the sense that the high wavenumbers are rapidly amplifying after the tenth step, the difficulty is completely cured by merely halving the time step.

however: the calculation can be extended indefinitely merely by halving the timestep.

Elsaesser(1966a) describes a similar example in numerical weather prediction. His Fig. 12 shows a forecast two time steps before overflow; like our Fig. 11.6 [timestep 14, lower left panel], Elsaesser's flow is smooth everywhere except for one small region where $2h$-waves have appeared, and his difficulty was cured by shortening the time step from four hours to three hours.

If the timestep is sufficiently short, however, spectral blocking can still arise, but only due to spatially-varying coefficients in the differential equation. Nonlinear terms are merely a special case of spatially-varying coefficients.

11.4   Aliasing Instability: History and Remedies

Three-dimensional hydrodynamic codes known as "general circulation models" (GCMs) have become a major tool in understanding future climate, such as the effects of an increase in greenhouse gases in the atmosphere. The ur-GCM, the ancestral Adam from whence all later models came, was Phillips (1956). Starting from rest, and imposing no forcing except a mean north-south temperature gradient (warm at the equator, cold at the poles), the model spontaneously developed mean jets and travelling vortices rather closely resembling those of the real atmosphere. Hurrah!

Unfortunately, after a few more days of model time, the winds became supersonic. The GCM had blown up!

Phillips tried to stabilize his model by greatly decreasing both the time step and the spatial grid size, but at roughly the same time (about 20 days) the model blew up anyway. Although his model employed finite differences, he reasoned spectrally. He noticed that the first warning of the impending disaster was the appearance of $2h$-waves. In Fourier space, this means that energy is building up near the aliasing limit,

$$K \equiv \pi h \qquad \leftrightarrow \qquad \text{wavelength} = 2h \qquad (11.12)$$

Now the hydrodynamic equations are quadratically nonlinear. The interaction of a wave with $|k| > K/2$ with another wave of similar magnitude must generate waves with $k > K$. However, these cannot be resolved on a grid with spacing h, but are instead "aliased" to waves of lower wavenumber. Phillips conjectured that this spurious, high-to-low wavenumber transfer of energy was the source of his instability.

He tested his hypothesis (Philips, 1959) by repeating his earlier experiment with a twist. Every few steps, he calculated the Fourier transform of the grid point values of each field and then set the upper half of the resolved spectrum to zero. That is to say, he applied the all-or-nothing filter

$$a_k \rightarrow [filtered] \begin{cases} a_k & |k| < K/2 \\ 0 & |k| > K/2 \end{cases} \qquad (11.13)$$

It worked! With this filtering, it was possible to integrate his model for arbitrarily long periods of time.

In a note less than a full page long, Orszag (1971a) pointed out that Phillips' filtering was wasteful: if only the upper third of the spectrum were filtered, aliasing would still be eliminated. (To put it more precisely, aliasing would still occur, but only into wavenumbers purged by the filtering. Thus, all the *surviving* wavenumbers are alias-free.)

Akio Arakawa (1966) proposed another solution. Aliasing caused the discretized energy to grow without bound when it should be conserved. Arakawa therefore developed a modified finite difference formula for the Jacobian operator (advection) which exactly conserved the discretized energy. This also prevented aliasing instability.

Yet another remedy is to use special finite difference or spectral algorithms that have been modified to better cope with fronts or shock waves. (The "synoptic scale" vortices in Phillips' GCM should form narrow fronts as they mature; it was during this "frontogenesis" stage that Phillips' model went bad.)

All three strategies — filtering, energy-conservation, and shock-resolution — have their uses. However, if a flow is being well-resolved by an exponentially-convergent spectral model, shouldn't the energy be conserved to a high approximation anyway? Are these alias-fixers really necessary? Moreover, it turns out that these remedies offer seductive possibilities for self-deception. We will return to these difficult questions after first describing a couple of these procedures in more detail in the next two sections.

11.5 Dealiasing and the Orszag Two-Thirds Rule

Phillips suggested a dealiasing procedure: apply a spatial filter so as to suppress all waves with wavelengths between $2h$ and $4h$. For spectral methods, we filter by simply deleting all the offending wavenumbers just before each transform to grid point values.

Orszag (1971a) showed that purging half the spectrum is wasteful. If we filter all waves such that $|k| > (2/3)K$, then the quadratic interaction of two wavenumbers p, q that survive the filter will alias only to wavenumbers that are purged. The reason is that the aliasing shift in k must always be by a *multiple* of $2K$ (Fig. 11.7). The conclusion is that it is only necessary to filter waves with wavelengths between $2h$ and $3h$ to suppress aliasing.

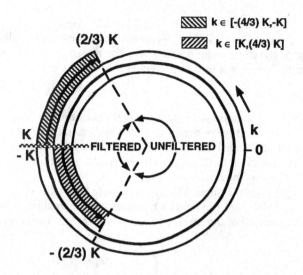

Figure 11.7: The "Aliasing Wheel". The wavenumber k for Fourier basis functions $\exp(i\,k\,x)$ is depicted as an angle in a polar coordinate system, scaled so that the basis set $k \in [-K, K]$ spans the full circle. Orszag's "Two-Thirds Rule" filter has no effect on waves in the range $k \in [-(2/3)K, (2/3)K]$, but all Fourier coefficients with $|k| > (2/3)K$ are set equal to zero at the end of each time step.

Quadratically nonlinear terms have Fourier components that span the range $k \in [-(4/3)K, (4/3)K]$, which is represented by the double-headed arrow that completes one-and-one-third complete turns of the circle. As shown by the diagram, wavenumbers on $k \in [K, (4/3)K]$ are aliased to $k \in [-K, -(2/3)K]$ (lower shaded band). Similarly, wavenumbers on $k \in [-(4/3)K, -K]$ are aliased to $k \in [(2/3)K, K]$ (upper shaded area). The filtering removes all effects of aliasing because the misrepresented wavenumbers — the shaded bands — lie within the "death zone" of the filter. The dotted horizontal ray is the branch cut: all $k > K$ or $< -K$ will be aliased.

Rule-of-Thumb 10 (TWO-THIRDS RULE) *To obtain an alias-free computation on a grid of N points for a* quadratically nonlinear *equation, filter the high wavenumbers so as to retain only* $(2/3)N$ *unfiltered wavenumbers.*

This precept is alternatively (and confusingly) known as the "Three-Halves Rule" because to obtain N unfiltered wavenumbers, one must compute nonlinear products in physical space on a grid of $(3/2)N$ points.

This procedure is also called "padding" because one must add $(N/2)$ zeros to the spectral coefficients before making the coefficient-to-grid Fast Fourier Transform.

Whatever the name, the filtering or padding is always performed just before taking coefficient-to-grid transforms so that the shortest unfiltered waves on the grid have wavelength $3h$.

A total of $(3/2)N$ basis functions are used during intermediate stages of the calculation, but only the lowest N wavenumbers are retained in the final answer at each time level.

Two alternative dealiasing methods have also been tried. Patterson & Orszag (1971) evaluated grid point sums with a phase shift. However, this trick has fallen from favor because it is always more expensive than the Two-Thirds Rule (Canuto *et al.* 1988, pg. 85).

Rogallo (1977, 1981, Rogallo and Moin, 1984), who is better known for inventing the wing used in the first modern hang-glider, showed that for a two-step time scheme, it is possible to apply the phase shift method at negligible extra cost to reduce aliasing errors to $O([\Delta t]^2)$ of those of the standard pseudospectral method. This is very appealing, even though a small residue of aliasing remains, but for a decade, it was never mentioned except in Rogallo's own work until praised by Canuto *et al* (1988).

Other strategies which control aliasing indirectly are described in the next section.

11.6 Energy-Conserving Schemes and Skew-Symmetric Advection: Interpolation with Constraints

Arakawa (1966) proposed a very different strategy: modifying the numerical algorithm so that it exactly conserves the discretized energy. (By "energy", we mean the total energy of the model.) He reasoned that blow-up implies a spectacular (and wholly spurious) increase in energy. If the discretized energy is constant, however, then supersonic winds are obviously impossible.

Similarly, for Hamiltonian systems, one can use special time-marching schemes ("symplectic" algorithms) which preserve the Hamiltonian structure even in discrete form. These have been very useful in megayear-long integrations of solar system orbital dynamics problems (Sanz-Serna and Calvo, 1994).

As noted by Canuto et al.(1988), it is often straightforward to obtain spectral schemes which are energy-conserving. Indeed, Galerkin's method is automatically energy-conserving for some equations, such as for the barotropic vorticity equation with a triangular truncation of the spherical harmonic basis (Haltiner and Williams, 1980). However, for most problems, pseudospectral schemes conserve the discrete energy only after (small) modifications (Canuto *et al.*, 1988, Secs. 4.5 and 4.6).

Zang (1991a) has shown that the most important such modification is to use the skew-symmetric form of advection for the Navier-Stokes equations. A skew-symmetric matrix or differential operator is one whose eigenvalues are all pure imaginary. Skew-symmetry is a property of advection operators and also the operators of wave equations: the imaginary eigenvalues (paired with a first order time derivative) imply that operator *advects* or *propagates* without causing growth or decay. It is desirable to preserve skew-symmetry in numerical approximation because pure advection and propagation is precisely what *doesn't* happen when a model is blowing up.

The usual forms of advection are

$$\vec{u} \cdot \nabla \vec{u}, \qquad \text{[Standard form]}$$
$$\vec{\omega} \times \vec{u} + \nabla \left((1/2)\,\vec{u} \cdot \vec{u}\right), \qquad \text{[Rotation form]} \qquad (11.14)$$

However, Zang (1991a) shows that skew-symmetry is preserved only by the more complicated form

$$\frac{1}{2}\,\vec{u} \cdot \nabla \vec{u} + \frac{1}{2}\nabla\left(\vec{u}\,\vec{u}\right), \qquad \text{[Skew-Symmetric Advection]} \qquad (11.15)$$

where $\nabla\left(\vec{u}\,\vec{u}\right)$ is a vector whose components are $\nabla \cdot \left(\vec{u}\,u_j\right)$ where u_j is one of the three vector components of $\vec{u}$. The property of skew-symmetry also depends upon boundary conditions, but we refer the reader to his paper for details.

Blaisdell, Spyropoulos and Qin(1996), through a mixture of theoretical analysis and experiments for Fourier pseudospectral codes for turbulence and Burgers' equation, offer additional support for the skew-symmetric form. They show that this form cancels much of the differentiation error of either the conservative form or the usual non-conservative advection form for the largest aliased components.

More recently, a wide variety of algorithms have been developed to avoid spurious oscillations near fronts and shock waves, such as Van Leer's MUSCL scheme, Roe's algorithm, and the Piecewise-Parabolic Method (Carpenter *et al.*, 1990). These do not attempt to control the energy, but rather use analytical theories of shock structure to suppress the growth of $2h$-waves.

Some of these schemes have the disadvantage of being more costly than conventional schemes. Arakawa's finite difference Jacobian requires twice as many operations as approximating the advection operation using standard differences. The Two-Thirds Rule requires dumping (1/3) of the numerical degrees of freedom in *each* dimension, doing the work of a grid which is say, 96^3 to obtain only a 64^3 set of nonzero spectral coefficients.

Is the additional cost worth it? The answer, to borrow the title of a science fiction novel by the Strugasky brothers, is Definitely Maybe.

Optimist's Viewpoint: Energy-conserving and front-resolving schemes are interpolation-with-physics-constraints. Instead of merely using the grid point values of a function, as if it were completely arbitrary, these schemes exploit the fact that the function is not arbitrary, but rather is constrained by physics. Building constraints into interpolation schemes is perfectly legitimate and should lead to an improved approximation.

Pessimist's Viewpoint: The Arakawa scheme guarantees that the numerical solution will stay *bounded*, but it does not guarantee that it is *accurate*. In Phillips (1956), for example, the calculation blew up only at the stage where the baroclinic cells were forming fronts. Since his coarse grid could not have resolved the fronts anyway, it can be argued that the blow-up of his model really did him a favor by aborting the computation as soon as it became inaccurate.

Similarly, dealiasing is a dumb idea: if the nonlinear interactions are so strong that a lot of energy is being aliased, one probably needs to double the grid spacing.

Likewise, the PPM method and its cousins guarantee smoothness but not accuracy. Who is right?

11.7   Energy-Conserving and Shock-Resolving Schemes: Discussion

The workshop published by Straka *et al.*(1993) computed a two-dimensional density current using a wide variety of schemes. The comparison supports both the optimistic and pessimistic viewpoints.

Fig. 11.8 compares a standard, second order finite difference method at various resolutions. The most important point is that one needs a grid spacing $h = 100m$ or less to get a decent solution; the plot for $h = 200m$ is filled with noise while the numerical solution for $h = 400m$ is completely useless.

Fig. 11.9 compares the high accuracy solution (the $h = 25m$ solution from the previous figure, which Straka *et al.* use as a benchmark) with standard finite differences and the PPM scheme at $h = 400$. The optimist is encouraged because the PPM solution is still stable, smooth, and resembles the accurate solution even though the grid spacing is very coarse. The PPM's incorporation of wave physics has made it far more successful at this resolution than the finite difference scheme, which is junk. The spectral model was unstable at this resolution.

However, the pessimist can find solace, too, because the smoothness and stability of the PPM solution is deceiving: it is not a very good numerical solution. Two smaller vortices and several other features in the right half of the disturbance have been simply combined into a single smooth vortex by PPM. The truth is that $h = 400m$ is simply too coarse to resolve the smaller features.

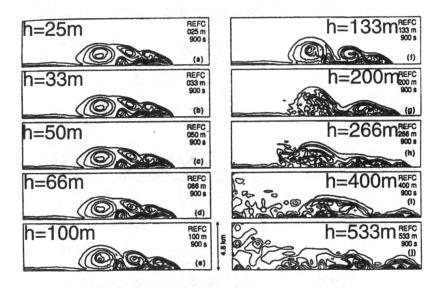

Figure 11.8: Comparison of second order finite difference solutions at different grid spacings h. The algorithm has no particular front-resolving or monotonicity properties. From Straka *et al.*(1993) with permission of the authors and the American Meteorological Society.

At $h = 200m$, the spectral solution is noisy, but much more accurate than any of the others in computing the total energy and enstrophy, even though it is not exactly energy-conserving. PPM is smoother, but only because this algorithm has strong computational dissipation, losing almost half the enstrophy at the time shown in the figures versus only about 13% for the spectral code. At $h = 100m$, the spectral and PPM schemes are both accurate, but the spectral method is considerably better than PPM at resolving the smaller scale features and preserving energy and enstrophy (Fig. 11.10).

The moral of this and similar comparisons is the following

Rule-of-Thumb 11 (DEALIASING/ENERGY-CONSERVING)

1. *For well-resolved flows, dealiasing and energy-conserving and front-resolving algorithms are unnecessary. A spectral method will conserve energy to a very high accuracy and will faithfully resolve a shock zone or frontal zone with only two or three grid points within the zone.*

2. *For marginally-resolved flows, dealiasing and other remedies may prevent blow-up and yield acceptable results when blind application of spectral or other methods yields only a sea of noise or worse. However, some fine structure in the solution may be missed, and the special algorithms can do nothing about this.*

3. *For poorly-resolved flows, nothing helps, really, except using more grid points. Energy-conserving and front-tracking schemes may generate smooth solutions, but these will only vaguely resemble the true solution.*

Special algorithms and tricks are "Plan B", fallbacks to implement when the straightforward algorithm isn't working well and the computation already fills all the available computer memory. Filtering and skew-symmetric advection are often useful; conserving energy is helpful; dealiasing is a last resort because of its high cost in two or three dimensions. Front-resolving schemes are quite useful for shocks.

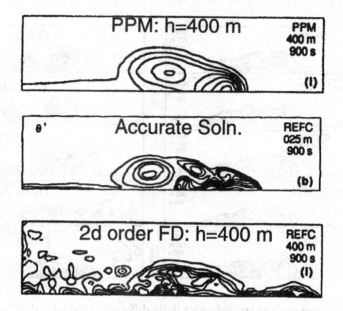

Figure 11.9: Comparison of the piecewise-parabolic method (PPM) at $h = 400$ [top] with the standard second order finite difference model [bottom] at the same resolution and with the highly accurate reference solution [middle]. Although the PPM solution is smooth and has captured some of the features of the true solution, it has only two vortices instead of three. Redrawn from Straka *et al.*(1993).

A turbulence calculation almost by definition is poorly resolved, so dealiasing is not particularly rare for modelling such flows. Even in turbulence, however, dealiasing is often unnecessary; even in turbulence, dealiasing is not a cure for a flow that is badly underresolved.

Aliasing error is likely to be important only when, as in Phillips' pioneering model, the calculation is becoming highly inaccurate. Aliasing blow-up has saved many modellers from believing lousy simulations. There have been many energy-conserving calculations published — it would be tactless to name names! — which have smooth solutions that are complete nonsense.

11.8 Aliasing Instability: Theory

A number of papers have tried to identify specific mechanics for blow-up such as Briggs, Newell and Sarie (1981). Aoyagi (1995) is a recent article with many references. These studies have successfully identified particular modes of instability for particular algorithms. However, the studies have not been successful in developing any sort of general theory. It may be that model blow-up, like any other train wreck, creates casualties of many sorts and not any specific, aliasing-related injury.

Phillips (1959), who explained the blow-up of his model in terms of aliasing and the resulting spurious, wrong-way transfer of energy, believed that he had diagnosed a specific illness with a specific cure. He did not think that aliasing instability was merely a lack of resolution because he repeated his GCM calculation with a much shorter time step and spatial grid spacing and the model still blew up at about the same time!

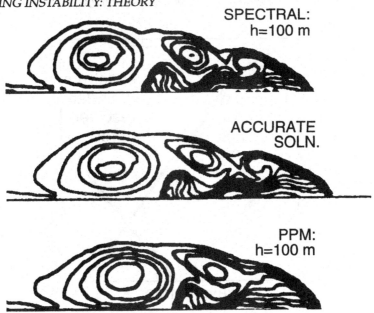

Figure 11.10: Comparison of spectral and piecewise-parabolic method (PPM) at $h = 100m$ with each other and the reference solution. From Strata *et al.*(1993).

However, his experiments do not completely resolve the issue. When a shock wave forms, the solution is smooth for a finite time interval and then develops a jump discontinuity which — in the absence of sufficient viscosity — the numerical method cannot resolve. The transition from exponential accuracy to blow-up may be rapid, occurring on a time scale short compared to the total time of integration. Even with an abrupt onset, blow-up may be simply the result of underresolution of small scale features. [1] As we saw earlier, spectral blocking can also be the result of too long a time step (although Phillips' tests excluded this in his model.)

Spectral blocking would seem to support Phillips' belief that aliasing instability is not merely underresoution. If the flow is evolving narrower and narrower fronts, one would expect that the Fourier spectrum would flatten out, rather than develop a curl at high wavenumber. Actually, in simple one-dimensional PDEs like the Regularized Long Wave equation and Burgers' equation, this "flatten-to-blow-up" scenario is observed. On a log-linear plot, the coefficients asymptote to a straight line for all t, implying that the convergence rate is always geometric, but the slope of the asymptotic line becomes flatter and flatter. When the plotted coefficients asymptote to a horizontal line, i. e., the spectrum resembles "white noise", disaster is quick and spectacular.

In Phillips' simulation, and also in the Zang-Krist-Hussaini transition-to-turbulence experiment illustrated earlier, however, spectral blocking develops while the large scale features are still well-resolved and the amplitude of the "blocked" wavenumbers is small compared to that of low wavenumbers. Doesn't this imply a numerical instability rather than underresolution as the cause of blocking? Not necessarily.

Theories of two-dimensional and quasi-geostrophic turbulence predict that (i) the flow

---

[1]Cloot, Herbst and Weideman, 1990, give a good discussion of a one-dimensional wave equation that develops unbounded solutions in finite time. Their adaptive Fourier pseudospectral method increases N close to the time of the singularity. This is a good strategy for fighting blow-up when the model physics is taking a rapid turn for the nasty.

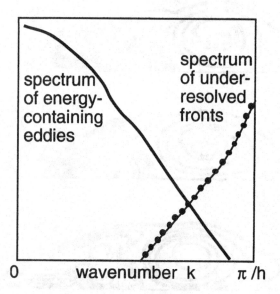

Figure 11.11: Solid: spectrum of energy-containing eddies ("synoptic eddies" in Phillips' simulation). Circles: Fourier coefficients of the unresolved fronts.

will spontaneously develop fronts too thin to be numerically resolved (if the viscosity is very weak) and (ii) the energy spectrum will fall off as $O(k^{-3})$ for wavenumbers smaller than k_{diss}, the beginning of the dissipation range, where k is the total wavenumber. For the Fourier coefficients (as opposed to the energy, whose spectrum is proportional to the absolute value *squared* of the Fourier coefficients), this implies $|a_k| \sim O(1/k^{3/2})$. The crucial point is that the Fourier coefficients of a jump discontinuity (like a front or the "sawtooth" function) decrease more slowly as $O(1/k)$. The implication is that the *fronts* have rather *small energy*, or otherwise the energy spectrum would decay as $O(1/k^2)$.

Thus, it is possible for the underresolved small features to contain little energy compared to large, well-resolved features. Thus, spectral blocking in Phillips' model could be described (conjecturily) by Fig. 11.11.

On the other hand, Brown and Minion(1995), Minion and Brown(1997) and the studies reviewed by Aoyagi (1995) have identified some specific computational instabilities. These seem to occur only for marginally-resolved or poorly-resolved flow, thank goodness. When sideband resonance is the villain, the bland statement "too little resolution" seems insufficient. Perhaps with a little tinkering of the algorithm, it might be stabilized against sideband resonace (or whatever) so that the integration could be extended without a shift to a finer spatial grid.

The theory of aliasing instability and blow-up is on the whole in a pretty sorry state. It matters because many important flows can only be marginally resolved, forcing the use of energy-conserving algorithms, dealiasing and so on. We could fight the instability much better if we only understood what we were fighting!

11.9   Summary

Spectral blocking, $2h$-waves and blow-up are as much a problem today as in the time of Phillips' first GCM experiment. The bad news is that despite Phillips' own demonstration

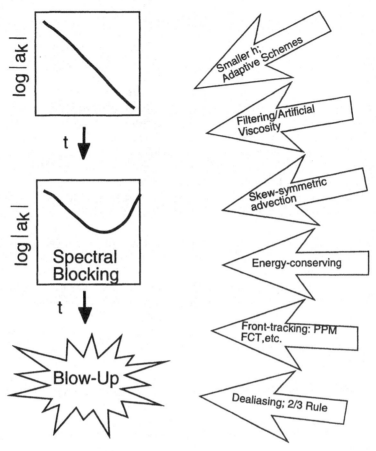

Figure 11.12: Schematic of the time evolution of a model from a smooth initial condition (top left) to a flow that has lost some accuracy to spectral blocking (middle left) before blowing up (bottom left) and six types of remedies for spectral blocking and blow-up (right arrows).

of the wrong-way energy cascade created by aliasing, the causes of blow-up are still not well understood. The good news is that we have a quiver full of arrows to shoot at the problem (Fig. 11.12).

The first arrow is to increase the resolution. Phillips found that this didn't help very much, buying only a few hours of additional time. However, this may only reflect the fact that frontogenesis happens very rapidly once the eddies have grown to a reasonable size. It is dangerous to always blame blow-up on a numerical villain, a computational Bogeyman named Aliasing Instability. It may just be that the flow is *physically* generating features whose width is $1/1000$ of the size of the computational domain, and a model with $N = 100$ fails simply because it can't resolve it.

Adaptive schemes are very useful in this context because often a flow can be resolved with rather small N over much of its lifetime before the front develops. With a variable resolution, that is, a fine grid around the front only, the narrow features may be resolved with only moderate N. Cloot, Herbst, and Weideman (1990), Bayliss, Gottlieb, Matkowsky and Minkoff (1989), Bayliss and Turkel(1992), Bayliss, Garbey and Matkowsky (1995) and

Augenbaum (1989, 1990) are illustrations.

The second arrow is filtering or the use of an artificial, computational viscosity (Chapter 18, Sec. 20).

The third arrow is skew-symmetric advection. This forces the numerical discretization of advection to provide pure translation without adding a little bit of spurious numerical growth or decay. It seems to be helpful with marginally-resolved flows.

The fourth arrow is an energy-conserving algorithm. Canuto *et al.* (1988, Sec. 4.6) give an example of a magnetohydrodynamic flow where this is helpful. Energy-conserving schemes are likely to be useful for climate modelling, too, where the flow is integrated over a very long period of time. Symplectic integration schemes for Hamiltonian systems conserve not only energy but the Hamiltonian structure.

The fifth arrow is to use a front-resolving scheme like PPM or MUSCL or Flux-Corrected Transport (FCT) or the like. Spectral generalizations of most of these algorithms are now available.

The sixth arrow is the Two-Thirds Rule. This is really just a special case of filtering, but we list it separately because it is often discussed in the literature as something separate with a different motive from conventional filters: to dealias a quadratically nonlinear computation. However, Orszag's procedure discards 1/3 of the numerical degrees of freedom in one dimension, 5/9 in two dimensions and 19/27 in three dimensions. One pays a very heavy price just to eliminate aliasing, so this is probably a last resort.

Note that the Two-Thirds Rule is an "all-or-nothing" filter: the coefficients are unmodified or completely discarded. To resolve fronts and other narrow features, filters which vary *smoothly* and *continuously* with wavenumber have become quite popular and successful. Experience and some theory suggests that a smooth filter is better than a step function (in wavenumber) like the Two-Thirds Rule. Since the higher wavenumbers are clobbered by the filter anyway, one is likely to gain most of the benefits of dealiasing.

There are two important things to remember before applying these remedies. The first is that for a well-resolved flow, none of them are needed! A spectral code will do a very good job of preserving the energy over a fairly long time simply because of its inherent exponential accuracy. It is only when the flow develops fronts, shocks and other pathologies that one ever needs to think about tinkering with the algorithm to enforce conservation of the discrete energy. Otherwise, one is incurring extra cost, both your time and the computer's, for no particular gain.

The second point is that when these remedies succeed, they succeed dangerously. That is to say, these interpolation-with-physics-constraints force the computed flow to be smooth and the energy to stay constant even when the numerical errors are large. If the eddies have a characteristic scale of 1000 km, it is clear that a very coarse grid with h=5000 km cannot possibly furnish an accurate solution. However, with an Arakwa-type energy-conserving scheme, the numerical solution won't blow up. Indeed, the dreadfully-underresolved solution may look quite smooth, even though it's nonsense.

For his thesis, a colleague ported the energy-conserving Arakawa-Mintz model from the atmosphere to the ocean. Unfortunately, because ocean eddies have a diameter of O(100 km),which is very small compared to the width of an ocean basin, it was not possible to explicitly resolve the eddies in 2 B. C. [Before Cray[2]]. However, it was possible to make a short run at higher resolution. The horror! The horror! With smaller h, everything changed!

Nevertheless, the production run was published. Although it certainly had major errors, my friend was unable to convince his advisor of this because the Arakawa scheme (and viscosity) made the computed solutions *look smooth* even though the flows were actually rubbish.

---

[2]The Cray-1, the first supercomputer, was introduced in 1976.

In contrast, a code which is resolving a flow ought to approximately conserve energy for a long time, even if this property is not explicitly built into the discretization. If the code isn't resolving the flow, well, perhaps it *ought* to blow up.

Smoothness and energy conservation are like the shining glitter of gold. Perhaps it really is gold, or perhaps it is only iron pyrites, "fool's gold", which gleams as brightly but is worth almost nothing. The number-cruncher must be as suspicious of glitter as the mineralogist.

Chapter 12

Implicit Time-Marching, the Slow Manifold and Nonlinear Galerkin Methods

"In the terminology which you graciously ascribe to me, we might say that the atmosphere is a musical instrument on which one can play many tunes. High notes are sound waves, low notes are long inertial [Rossby] waves, and nature is a musician more of the Beethoven than of Chopin type. He much prefers the low notes and only occasionally plays arpeggios in the treble and then only with a light hand. The oceans and the continents are the elephants in Saint-Saens' animal suite, marching in a slow cumbrous rhythm, one step every day or so. Of course there are overtones: sound waves, billow clouds (gravity waves), inertial oscillations, etc., but these are unimportant and are heard only at N. Y. U. and M. I. T."

–Jule Charney (1917-1982) (from a 1947 letter to Philip Thompson)

12.1   Introduction

When the partial differential equation is *linear* and the "method of lines" is applied to discretize the spatial dependence, the system of ordinary differential equations in time is linear, too, and can be written

$$\frac{d\vec{u}}{dt} = \vec{\Lambda}\vec{u} + \vec{f}(t) \tag{12.1}$$

where $\vec{\Lambda}$ is a large square matrix of dimension N where N is the number of elements in the column vector $\vec{u}$ and where $\vec{f}(t)$ is the discretization of the inhomogeneous terms in the differential equation. (We assume that the boundary conditions have been dealt with by basis recombination or whatever; the vector $\vec{f}$ may contain terms proportional to inhomogeneous boundary conditions in some or all of its elements.)

The "configuration space" of the model is just the N-dimensional space of the vector $\vec{u}$, the spectral coefficients or grid point values. The instantaneous state of the model is a point in configuration space; its evolution through time is a one-dimensional curve in configuration space, parameterized by the single coordinate t.

Eq.(12.1) is completely general and applies whether the spatial discretization is spectral, finite difference or finite element and also whether the unknowns are grid point values or Chebyshev polynomial coefficients. In general, the elements of the matrix $\vec{\Lambda}$ are functions of time. In most of this chapter, however, we shall make the further assumption that $\vec{\Lambda}$ is independent of time.

In a world which is highly nonlinear and time-dependent, these assumptions would seem drastic. Indeed, such approximations cannot capture the mysteries of solitary waves, chaos, aliasing instability and a wide variety of other phenomenon. Nevertheless, such a simplified model as (12.1) can teach us much about time-marching algorithms. The reason is that accurate time-stepping requires a time step which is short compared to the time scales on which the coefficients of the differential equation and u itself are changing. This implies that the matrix $\vec{\Lambda}$ will not change much between then and time $t = t_0 + \tau$ where τ is the timestep; otherwise, we should shorten the timestep.

The conceptual advantage of the linear-with-time-independent coefficient approximation is that the linear, constant coefficient ODE system (12.1) has the general solution

$$\vec{u} = \exp(\vec{\Lambda}t)\vec{u}_0) + \exp(\vec{\Lambda}t) \int^t \exp(-\vec{\Lambda}s)\,\vec{f}(s)\,ds \tag{12.2}$$

where $\vec{u}_0 \equiv \vec{u}(t = 0)$. The matrix exponential is most easily interpreted by expanding $\vec{u}$ in terms of the eigenvectors of the square matrix $\vec{\Lambda}$. If

$$\vec{\Lambda}\vec{e}_j = \lambda_j \vec{e}_j \tag{12.3}$$

and

$$\vec{u} = \sum_{j=1}^{N} b_j\,\vec{e}_j \tag{12.4}$$

then in terms of the eigenvector basis, $\vec{\Lambda}$ becomes a *diagonal* matrix, with its eigenvalues λ_j as the diagonal elements, and the ODE system collapses to the *uncoupled* set of ordinary differential equations

$$\frac{db_j}{dt} = \lambda_j\,b_j + g_j(t) \tag{12.5}$$

where $g_j(t)$ is the j-th coefficient of the eigenexpansion of the forcing function $\vec{f}(t)$.

It follows that we can learn almost everything about time-marching schemes – to the extent that neglecting nonlinearity and other time-dependence of the coefficients of the PDE is legitimate – by understanding how different algorithms integrate

$$\frac{du}{dt} = \lambda u + f(t) \tag{12.6}$$

where all the unknowns are now scalars, not vectors. For a simple wave equation like

$$u_t + cu_x = 0, \tag{12.7}$$

the usual Fourier basis ($\exp(ikx)$) shows that the equation for each wavenumber is uncoupled with

$$\lambda = -ikc, \qquad b_k(t) = b_k(0) \exp(-ikct) \tag{12.8}$$

Other constant coefficient wave equations similarly generate *pure imaginary* eigenvalues. Advection does, too; note that the one-dimensional advection equation, also known as the inviscid Burgers' equation, is identical with 12.7 except that $c \to u$.

In contrast, the diffusion equation

$$u_t = u_{xx} \tag{12.9}$$

generates *negative real* eigenvalues; in the Fourier basis, which is the eigenexpansion for the diffusion equation,

$$\lambda = -k^2; b_k = b_k(0) \exp(-k^2 t) \tag{12.10}$$

In general, numerical algorithms fail to propagate and/or diffuse at the exact rate, creating computational dispersion and/or diffusion that is superimposed on whatever is in the exact solution. One major theme of this chapter is that the choice of time-stepping algorithms and time step are both strongly constrained by the question: How much error is tolerable for a given component of the solution?

Before we turn to time-marching errors, however, it is useful to first look briefly at errors due to spatial discretization.

12.2 Dispersion and Amplitude Errors Due to Spatial Discretization

It is easy to discuss the computational dispersion and dissipation of a Fourier basis: there isn't any! Advection is corrupted only by aliasing errors; if the Two-Thirds Rule is applied, then advection is exact for all the unfiltered wavenumbers.

If no filtering is used, then advection is not exact even in a Fourier basis because of aliasing errors; high wavenumbers near the truncation limit will not be correctly advected by high wavenumber components of the velocity. However, high wavenumbers will be advected exactly by low wavenumbers including the zero wavenumber component, which is the mean flow. Even in an aliased calculation, the total error in advection is small, which is one reason why Fourier methods have become so popular for modelling turbulence in a periodic box.

Unfortunately, the same is not so for finite difference methods. One simplifying principle, much exploited by J. von Neumann, is that $\exp(ikx)$ is an eigenfunction of a finite difference approximation just as it is for differential operators. Thus, backwards and centered difference approximations to the first derivative,

$$\frac{du}{dx} \approx \frac{u(x) - u(x - h)}{h} \qquad \text{[BACKWARDS]}$$

$$\approx \frac{u(x + h) - u(x - h)}{2h} \qquad \text{[CENTERED]} \tag{12.11}$$

give the approximations, when $u = \exp(ikx)$,

$$\frac{d\exp(ikx)}{dx} \approx \frac{(1 - \exp(-ikh))}{h} \exp(ikx) \qquad \text{[BACKWARDS]}$$

$$\approx \frac{i \sin(kh)}{h} \exp(ikx) \qquad \text{[CENTERED]} \tag{12.12}$$

The corresponding approximations to the true eigenvalue of the first derivative operator can be written as

$$\lambda_k \approx ik + \frac{h}{2}k^2 + O(h^2 k^3) \qquad \text{[BACKWARDS]}$$

$$\approx ik - i\frac{h^2}{6}k^3 + O(h^4 k^5) \qquad \text{[CENTERED]} \qquad (12.13)$$

The leading terms in (12.13) correctly reproduce the exact eigenvalue of the first derivative operator, which is ik. However, the backwards difference approximation has a second term, the dominant error, which is proportional to the eigenvalue of the second derivative operator, $-k^2$. Indeed, it is explained in elementary numerical analysis texts that the error for the one-sided and centered differences are proportional to the second and third derivatives of the function being differentiated, respectively. It follows that one can give two interpretations to the finite difference approximations to the wave equation $u_t + cu_x = 0$. The first is that the replacing the space derivative by two-point backwards and centered differences approximates that equation with errors of $O(h)$ and $O(h^2)$, respectively. The unconventional interpretation (equally correct) is that the differences formulas give *second* and *third* order approximations to the *modified* wave equations:

$$u_t + cu_x = c\frac{h}{2}u_{xx} \qquad \text{[BACKWARDS]} \qquad (12.14)$$

$$u_t + cu_x = -c\frac{h^2}{6}u_{xxx} \qquad \text{[CENTERED]} \qquad (12.15)$$

Whether one prefers to conceptualize using the "modified" equations (12.14), 12.15) or the power series for the eigenvalues (12.13), the point is that the backwards difference errs by providing an artificial computational diffusion. Diffusion is scale-selective, damping the high wavenumbers much faster than low wavenumbers. The backwards difference (which is the "upwind" difference for $c > 0$) is popular in hydrodynamics because the inherent diffusivity reduces spectral blocking, that is, the build-up of noise at high wavenumbers near the truncation limit. In the presence of marginally resolved fronts, the upwind difference approximation acts as a smoother. In contrast, the forward ("downwind") difference acts an "antidiffusion", generating spectral blocking and blow-up. (Note that the waves near the truncation limit are most strongly amplified by an "antidiffusion".)

The centered difference is not dissipative to lowest order. However, the exact solution to the original wave equation is nondispersive: all wavenumbers propagate with phase speed c so that the initial disturbance translates without change in shape. The centered difference introduces a spurious computational dispersion. A crest in the initial condition will disperse into many small ripples over time because of the $O(h^2)$ dispersion. When there is a frontal zone, the lack of damping and the dispersion cause lots of small oscillations to develop. Consequently, upwind differences are more popular than centered differences when fronts are anticipated even though the centered difference is technically of higher order.

With a Fourier basis, as already noted, we don't have to worry about either computational dissipation or dispersion. Unfortunately, with the Chebyshev basis only the lowest $(2/\pi)N$ eigenvalues of an N-term Chebyshev approximation are good approximations to the corresponding eigenvalues of the derivative operator. However, the error in the lower eigenvalues is exponentially small in N, so the computational dissipation and dispersion of a Chebyshev algorithm is very small compared to a finite difference algorithm.

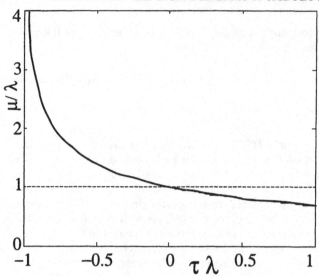

Figure 12.1: Numerical/exact eigenvalue ratio, μ/λ, for the forward Euler solution to $du/dt = \lambda u$, plotted as a function of the scaled time step $\tau\lambda$. The horizontal dividing line at $\mu/\lambda = 1$ is the exact decay rate.

12.3 Time-Marching Errors and the CFL Stability Limit for Explicit Schemes

Even when there is no error due to spatial discretization, as true with a Fourier basis, there will be still be computational dispersion and/or dissipation because of time-marching errors. The ODE which results from switching to a basis of the eigenvectors of the spatial discretization operators is (12.6), which we repeat for clarity:

$$\frac{du}{dt} = \lambda u + f(t) \tag{12.16}$$

The simplest explicit scheme is the "forward Euler" (FE) method:

$$u^{n+1} = (1 + \tau\lambda)\,u^n + \tau\,f(t^n) \tag{12.17}$$

where the superscripts denote the time level (not powers) and τ is the time step.

If λ is negative, as for diffusion, then for sufficiently small τ, the FE solution damps out, too. If f is a constant, then the finite difference solution will damp out to the steady state

$$u^\infty = -f/\lambda \tag{12.18}$$

which is *exact*.

The bad news is that the FE solution does not damp out at the correct rate. The exact solution decays as $\exp(\lambda t)$ whereas the finite difference solution decays towards the steady state as $\exp(\mu t)$ where

$$\mu = \lambda\frac{\log(1+\tau\,\lambda)}{\tau\,\lambda} \approx \lambda\left\{1 + \frac{\tau\,\lambda}{2} + \dots\right\} \tag{12.19}$$

Fig. 12.1 shows that the numerical solution damps out too rapidly; when $\tau\lambda = -1$, the numerical damping rate is infinitely large! When $\lambda > 0$, the finite difference solution grows (as appropriate for "anti-diffusion"), but grows too slowly: at only 70 % of the true growth rate when $\tau\lambda = 1$.

When $\tau\lambda < -1$ (not shown on the graph), something worse happens: the numerical solution grows with time even though the exact solution decays. This sort of computational instability was first discovered by Courant, Friedrichs and Lewy (CFL) in 1928. In this case, the "CFL Stability Criterion" is that $\tau < |1/\lambda|$ for computational stability.

Stability is usually presented as an "all-or-nothing" concept, but this is misleading. First, nothing bad happens when $\tau > |1/\lambda|$ and $\lambda > 0$: the numerical solution blows up, but the exact solution blows up, too. The only difficulty with $\tau > |1/\lambda|$ is that the numerical solution is growing at only two-thirds or less of the true rate of growth, which is usually an unacceptably large error.

Second, even when $\lambda < 0$, Fig. 12.1 shows that the error becomes large even before the CFL criterion is violated. Blow-up seems to the modeller a disaster that appears discontinuously when the CFL criterion is exceeded, but really the instability is just the outlands of a larger region where the time-stepping is inaccurate.

There is one good defense for marching with a time step which is close to the stability limit: When the only goal is to approximate the steady-state, large errors in the decay rate are irrelevant. As noted earlier, the Forward Euler method always gives the steady state *exactly*. Consequently, it is perfectly sensible to solve the Poisson equation

$$u_{xx} = f(x) \tag{12.20}$$

by applying the FE scheme to the diffusion equation

$$u_t = u_{xx} - f(x) \tag{12.21}$$

with a time step which is only a little below the stability limit. The steady-state limit of the diffusion equation is the solution to the Poisson equation.

If, however, we wish to follow the decay of the *transients* to the steady state, then the errors in the decay rate matter, and one should use a time step which is much shorter than the stability limit.

The diffusion equation with a steady forcing has two time scales: the short time scale on which transients decay towards the steady solution and the infinitely long scale of the steady solution and forcing. Inaccurate resolution of the "fast" time scale [decay] is okay provided that one is only interested in the "slow" time scale. This line of reasoning leads to the concept of the "slow manifold" which will dominate the second half of this chapter.

Other explicit time-marching schemes have stability limits which are similar to that of the Forward Euler method:

$$\tau < q/|\lambda| \tag{12.22}$$

where the constant q depends on the algorithm, but is always $O(1)$. (For the Fourth-Order Runge-Kutta (RK4) method, for example, q varies with $\arg(\lambda)$, but is no smaller than about 2.8.) Unfortunately, these time-stepping constraints can be rather severe.

For the diffusion equation, for example, the eigenvalues in a Fourier basis $\{\exp(ikx)\}$ are $\lambda_k = -k^2$. If we truncate to $k \in [-N/2, N/2]$, then it follows that the FE will be stable for all wavenumbers only if

$$\tau < 4/N^2 \tag{12.23}$$

Thus, doubling N requires four times as many time steps to reach a given time. It follows that applying an explicit time-marching scheme to the diffusion equation is not a very fast way to solve Poisson equations: we need $O(N^2)$ time steps to get a good approximation to the steady state. Much faster iterations will be described in Chapter 15.

Chebyshev polynomials have an even more deplorable characteristic: their largest numerical eigenvalues tend to be $O(N^{2j})$ where j is the order of the highest derivative. Thus, an explicit time-marching scheme for the diffusion equation with Chebyshev spatial discretization is stable only when τ is $O(1/N^4)$ or smaller. Ouch! For this reason, implict and semi-implicit time-marching algorithms, which allow long time steps, are especially important in the spectral world.

12.4 Implicit Time-Marching Algorithms

Implicit time-marching algorithms for $du/dt = F(u, t)$ are formulas that require us to solve a boundary value problem (BVP) at every time step to advance the solution to the next time level. The reward for this costly BVP solution is that implicit methods are stable for much longer time steps than explicit schemes.

There is a very wide range of implicit algorithms. Lie (1993) has successfully used implicit Runge-Kutta schemes with spectral methods, for example, but implicit RK time-stepping is rather expensive. The most popular implicit schemes include the Backward-Differentiation Formulas (BDF) of of various orders and the "Crank-Nicholson" (CN) method. The first order BDF scheme is more commonly known as the "Backwards Euler" method and also as the Adams-Moulton scheme of order one. The Crank-Nicholson scheme is also known as the "Trapezoidal" method or the second order Adams-Moulton algorithm.

For $du/dt = F(u, t)$ where both F and u may be vectors, these algorithms are

$$\frac{u^{n+1} - u^n}{\tau} = F(u^{n+1}, t^{n+1}) \qquad \text{[Backwards Euler ("BE")]} \qquad (12.24)$$

$$\frac{(3/2)u^{n+1} - 2u^n + (1/2)u^{n-1}}{\tau} = F(u^{n+1}, t^{n+1}) \qquad \text{[BDF2]} \qquad (12.25)$$

$$\frac{(11/6)u^{n+1} - 3u^n + (3/2)u^{n-1} - (1/3)u^{n-2}}{\tau} = F(u^{n+1}, t^{n+1}) \qquad \text{[BDF3]} \qquad (12.26)$$

$$\frac{\frac{25}{12}u^{n+1} - 4u^n + 3u^{n-1} - \frac{4}{3}u^{n-2} + \frac{1}{4}u^{n-3}}{\tau} = F(u^{n+1}, t^{n+1}) \ \text{[BDF4]} \qquad (12.27)$$

$$\frac{u^{n+1} - u^n}{\tau} = \frac{F(u^{n+1}, t^{n+1}) + F(u^n, t^{n+1})}{2} \qquad \text{[Crank-Nicholson (CN)]} \qquad (12.28)$$

For the special case of the linear, scalar-valued ODE, $du/dt = \lambda u + f(t)$, it is easy to solve the implicit formulas:

$$u^{n+1} = \frac{u^n}{1 - \tau\lambda} + \tau \frac{f(t^{n+1})}{1 - \tau\lambda} \qquad \text{[BE]} \qquad (12.29)$$

$$u^{n+1} = \frac{1 + \tau\lambda/2}{1 - \tau\lambda/2}u^n + \frac{\tau}{2}\frac{f(t^{n+1}) + f(t^n)}{1 - \tau\lambda/2} \qquad \text{[CN]} \qquad (12.30)$$

If λ is negative and real, then the solutions to the difference equations decay geometrically to the steady state as $\exp(\mu t)$ where

$$
\frac{\mu_{BE}}{\lambda} = -\frac{\log(1 + \tau\lambda)}{\tau\lambda}
$$

$$
\frac{\mu_{CN}}{\lambda} = -\frac{\log\left(\frac{1+\tau\lambda/2}{1-\tau\lambda/2}\right)}{\tau\lambda} \tag{12.31}
$$

Since $1 + \tau\lambda$ must always be smaller than unity (since $\lambda < 0$), it follows that the logarithms in (12.31) are always negative so that the ratios of μ/λ are always positive. In other words, the numerical solutions decay, as does the true solution, for all $\tau \in [0, \infty]$. The *conditional* stability of the Forward Euler scheme (where the condition is $\tau < 1/|\lambda|$) has been replaced, for these two implicit schemes, by unconditional stability.

It is left to the reader to demonstrate – if not already convinced by earlier numerical analysis texts or courses – that the unconditional stability extends to λ anywhere in the left half-plane including the imaginary axes, which is called the property of being "A-stable".

Higher order schemes are not so well-behaved; $BDF3$ and $BDF4$, for example, are unstable for small imaginary λ. However, these algorithms are still sometimes used because weak dissipation ($\Re(\lambda) < 0$) is sufficient to stabilize them. Fornberg(1996, Appendix G) gives stability diagrams for a wide variety of schemes.

12.5   Semi-Implicit Methods

For nonlinear equations, an implicit algorithm has a high cost-per-timestep because one must solve a *nonlinear* boundary value problem at every time step. For the Navier-Stokes equations and for numerical weather forecasting and climate modelling, it is therefore common to use an algorithm in the following class.

Definition 26 (SEMI-IMPLICIT) *If some terms in a differential equation are approximated by an explicit scheme while others are approximated implicitly, then the resulting time-marching algorithm is said to be SEMI-IMPLICIT.*

For example, for the nonlinear PDE

$$
u_t = F(u, x, t) + L(u, x, t) \tag{12.32}
$$

where F and L denote the nonlinear and linear parts of the differential equation, a popular semi-implicit scheme is ("[AB3CN]")

$$
u^{n+1} = u^n + \tau\left\{\frac{23}{12}F\left(u^n, t^n\right) - \frac{4}{3}F\left(u^{n-1}, t^{n-1}\right) + \frac{5}{12}F\left(u^{n-2}, t^{n-2}\right)\right\}
$$

$$
+ \frac{\tau}{2}\left\{L(u^{n+1}, t^{n+1}) + L(u^n, t^n)\right\} \tag{12.33}
$$

In general, the implicit and explicit schemes may be of the same or different order.

It may seem illogical to treat some terms explicitly while other terms implicitly, and even more bizarre to use schemes of different orders, as in the AB3CN's mix of third order Adams-Bashforth with the Crank-Nicholson scheme, which is only second order. However, there are major advantages.

First, because the nonlinear term is treated explicitly, it is only necessary to solve a *linear* boundary value problem at each time step. Second, the viscous terms, which involve second derivatives, impose much stiffer time step requirements ($\tau \sim O(1/N^4)$ for a Chebyshev basis) than the advective terms, which involve only first derivatives and impose a timestep limit proportional to $1/N^2$. In other words, the semi-implicit algorithm stabilizes the most dangerous terms. Third, in weather forecasting and other fluid dynamics, advection is crucial. It is important to use a high order time-marching scheme with a short timestep to accurately compute frontogenesis, turbulent cascades, advection of storm systems and so on. There is little advantage to treating the nonlinear terms implicitly because a timestep longer than the explicit advective stability limit would be too inaccurate to be acceptable.

Semi-implicit algorithms have therefore become almost universal in weather forecasting and climate models.

12.6   Speed-Reduction Rule-of-Thumb

Rule-of-Thumb 12 (IMPLICIT SCHEME FORCES PHYSICS SLOWDOWN)
Implicit schemes obtain their stability by slowing down the time evolution of the numerical solution so that explicit stability criteria are satisfied. In other words, if the implicit algorithm approximates the true eigenvalue λ by μ where the homogeneous solution evolves as $\exp(\lambda t)$, then

$$\tau|\mu(\tau)| < O(1) \tag{12.34}$$

where τ is the time step. (Recall that the stability condition for the Forward Euler method is $\tau|\lambda| < 1$ and for RK4 is $\tau\lambda < 2.8$.)

The Rule-of-Thumb has not been demonstrated for all possible implicit algorithms, but no counterexamples have been published. The Rule-of-Thumb is also not quite a theorem because it ignores the distinction between logarithmic growth and a constant.

For example, it is easy to show from (12.31) that

$$\tau|\mu_{BE}| = |\log(1 + \tau\lambda)|$$
$$\tau|\mu_{CN}| = \left|\log\left(\frac{1 + \tau\lambda/2}{1 - \tau\lambda/2}\right)\right| \tag{12.35}$$

Thus, the slowed-down numerical eigenvalue μ does not quite satisfy a stability requirement of the form

$$\tau|\mu| < q \tag{12.36}$$

for a constant q for all τ but only for q which is allowed to grow logarithmically with the time step. However, when $\tau|\lambda| = 1000$, that is, when the time step is 1000 times larger than would be stable for the Forward Euler method, $q < 8$, that is, the homogeneous solution to $du/dt = \lambda u + f(t)$ in the implicit difference schemes are evolving roughly 8/1000 times as fast as for the exact solution.

Obviously, only a lunatic would use so long a time step if tracking evolution on the time scale of $1/\lambda$ is physically significant. This can be formalized as the following.

Rule-of-Thumb 13 (EXPLICIT-DEMANDING)
Suppose that the stability requirement for an explicit time-marching scheme for a time-dependent PDE is, for some constant q,

$$\tau|\lambda_{max}| < q \tag{12.37}$$

where, after discretization using N degrees of freedom for the spatial derivatives, $\lambda_{max}(N)$ is the eigenvalue of the discretization matrix which is largest in magnitude. If it is important to follow wave propagation, damping or other time behavior on the time scale of $\exp(i\lambda_{max}\ t)$, then one should use an EXPLICIT *scheme with a time step which is* SMALL *compared to*

$$\tau_{limit} \equiv \frac{q}{|\lambda_{max}|}\ . \tag{12.38}$$

The first point in justifying this Rule-of-Thumb is the observation that if $\tau \sim \tau_{limit}$, that is, if the time step is near the stability limit, then any explicit or implicit algorithm will have large errors in following decay or propagation on a time scale of $O(1/|\lambda_{max}|)$. (It may be highly accurate for *slower* components of the solution, but the Rule only applies when these slow components are not enough, and one needs to accurately resolve the fastest component, too.) It follows that we must use a time step which is *small* compared to the stability limit to accurately track the dynamics of the fast component that imposes the stability limit.

Given that the time step is so small, we may use either an implicit or explicit method. Since implicit methods are usually much more expensive than explicit schemes for the same time step, however, the second point is that it is usually cheaper to use an explicit scheme.

There are some exceptions. First, implicit methods require solving a Boundary Value Problem (BVP) at every time step, but if the basis is Fourier or spherical harmonics and the operator of the BVP is constant coefficient or a Laplace operator, solving the BVP may be trivial. In that case, the Crank-Nicholson scheme, which is second order with a smaller proportionality constant than most explicit schemes of the same order, is quite attractive.

Second, a Chebyshev basis usually generates some eigenvalues which are poor approximations to those of the corresponding differential operator. For the Laplacian, for example, some of the Chebyshev eigenvalues are $O(N^4)$, implying a time step which is no larger than q/N^4 for an explicit scheme, while the first N true eigenvalues are bounded by N^2. In this case, stability is limited by garbage modes, that is, by large spurious eigenvalues whose magnitude is unrelated to anything physical. It is sensible to use an implicit method with a time step which is $O(1/N^2)$ because then all the modes whose eigenvalues are good approximations to those of the differential operator will be integrated accurately in time; the modes which are advanced with poor time accuracy are nonsense anyway.

Even so, it is clear that implicit methods must be used with discretion. In meteorology, where implicit methods have been wisely employed for many years, this has led to the concept of the "slow manifold".

12.7 Slow Manifold: Meteorology

Weather forecasting is done by solving the "primitive equations", which filter sound waves so that only two important classes of waves remain. Rossby waves (and also nonlinear advection, which has roughly the same time scale) are responsible for the travelling high-pressure and low-pressure centers that are the large-scale weather. Gravity waves are generated by small-scale phenomena like thunderstorms and have frequencies an order of magnitude higher than Rossby waves or the advective time scale.

Observations supply initial conditions for the numerical code. Rossby waves are well-resolved by the radiosonde and satellite network, but gravity waves are not. However, it has been known for many years that the amplitude of gravity waves, when averaged over the globe, is very small compared to Rossby waves. If we imagine a configuration space of $3N$ dimensions where each coordinate is the amplitude of a wave mode in a numerical model, then the instantaneous state of the atmosphere should lie close to the N-dimensional subspace of pure Rossby wave motion with all the gravitational modes equal to zero.

It follows that a good global forecasting model does not need to accurately track movements in the whole phase space, but only flow on the "slow manifold". Implicit methods are natural tools to integrate on the slow manifold. Their stability allows a long time step. On the slow manifold – but *only* on the slow manifold – the long time step does not compromise the accuracy of the numerical solution.

Gravity waves limit the timestep for explicit algorithms to less than ten minutes. However, with an implicit or semi-implicit algorithm, one can make good five-day forecasts with a timestep of an hour. Because of the errors introduced by subgridscale turbulence, cumulus convection, photochemistry and so on, the total forecast error cannot be significantly reduced by using a shorter timestep.

Because the advective limit is roughly an hour and a timestep longer than an hour would damage accuracy anyway, it is usual in forecasting to treat the nonlinear terms explicitly. The linear terms associated with gravity waves are treated implicitly, but in a spherical harmonics basis, the (linear) BVP is easy to solve as explained in Chapter 18. The semi-implicit algorithm costs only a little more per timestep than an explicit scheme, but permits a time step six times longer.

Because the goal is to forecast only slow motion, implicit methods make sense for global meteorology.

"Global" means a model whose goal is to forecast weather over a large area such as the continental United States. In small areas (such as the rain shadow of a thunderhead), local "fast" dynamics such as thunderstorms may be very important. (In densely-populated areas, local Doppler radar and high-density instrument arrays can provide enough "fast" data to make mesoscale modelling feasible *locally*, though not globally.) It follows that so-called "mesoscale" models, which attempt to forecast severe weather for a few hours ahead, must resolve the fast motion, not just the slow manifold, and therefore usually employ explicit time-marching.

The concept of a "slow manifold" and the usefulness of implicit time-stepping methods are *situational*.

12.8   Slow Manifold: General Definition and Physical Examples

Many scientific problems have multiple time scales. A major branch of perturbation theory, the "method of multiple scales", exploits these different time scales. Implicit time-marching schemes are a numerical strategy for doing the same thing.

As explained in the previous section, however, implicit methods are only sensible when the goal is to track motion on the "slow manifold".

Definition 27 (SLOW MANIFOLD)
In the phase space of a numerical model which allows motion on both a fast time scale and a slow time scale, the SLOW MANIFOLD is the submanifold where the time evolution is on the longer time scale only.

EXAMPLE ONE: "TRIVIAL" Slow Manifold: STEADY-STATE

If a partial differential equation (and its discretization) have a steady state, then this is the ultimate slow manifold. If we wish to integrate in time to allow the flow to find its own way to the steady state, we may use inaccurate time-marching schemes because only the final steady state matters.

EXAMPLE TWO: FORCED LINEAR OSCILLATOR

$$du/dt + iu = f(\epsilon t), \qquad \epsilon \ll 1 \tag{12.39}$$

The homogeneous solution of the oscillator, $\exp(-it)$, is the "fast" motion with a time scale of unity; the $O(1/\epsilon)$ time scale of the forcing is the "slow" scale. The "slow manifold" is that particular solution which varies only on the slow time scale; the general solution to the ODE contains fast oscillations also.

Although the forced, linear oscillator seems too simple to be useful, it contains hidden depth. First, the more general problem $du/dt + i\omega u = f(\epsilon\omega t)$ can be converted into the form of (12.39) merely by rescaling the time through $t \to \omega t$. Second, as noted in the first section, any linear partial differential equation with time-independent coefficients can be converted into the form of (12.39) by shifting to a basis of eigenfunctions of the matrix of the discretized spatial derivatives.

Third, even a *nonlinear* problem can be cast in the form of (12.39) if the forcing $f(\epsilon t)$ is reinterpreted as a symbol for the nonlinear terms that couple different eigenmodes of the linear part of the partial differential equation together. This point of view is popular in weather forecasting because the nonlinear terms are an order of magnitude smaller than the linear terms (for gravity modes); thus, global weather can be conceptualized as a sea of independently propagating wave modes with weak wave-wave coupling.

Actually, the nonlinear interactions of the Rossby waves among themselves are rather strong, but to a first approximation the gravity waves are linear oscillators forced by the nonlinear Rossby-Rossby interactions. These depend very weakly on the gravity waves themselves, so that these Rossby-Rossby couplings may be regarded as external forcing for the gravity waves.

These ideas are clarified by the following.

EXAMPLE THREE: LK QUINTET

This model is described by Lorenz and Krishnamurthy (1987). It is obtained by truncating an atmospheric model to the most ruthless extreme that allows nontrivial interactions: three Rossby modes and two gravity waves. If we denote the (large) Rossby amplitudes by uppercase letters (U, V, W) and the (small) gravity mode amplitudes by (x, z) – note that these are Fourier coefficients and not velocities or coordinates – then the LK Quintet is

$$
\begin{aligned}
U_t &= -VW + \beta V z - aU \\
V_t &= UW - \beta U z - aV + F \\
W_t &= -UV - aW \\
x_t &= -z - ax \\
z_t &= bUV + x - az
\end{aligned}
\tag{12.40}
$$

where subscript t denotes time differentiation and where (a, F, b, β) are constant parameters. We shall mostly concentrate on the "inviscid" quintet, which is the special case that

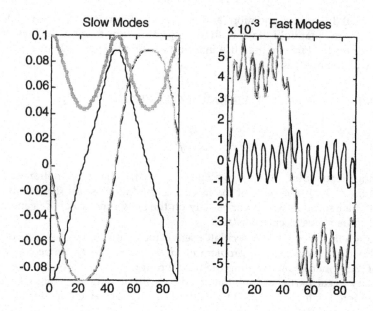

Figure 12.2: Example of solution to the "one-way coupled" inviscid LK Quintet ($b = 1, \beta = 0$ for the initial conditions $[U, V, W, x, z] = [1/10, 0, -9/100, 0, 0]$. Left panel: amplitudes of the three Rossby waves. (i) U: Thickest curve (never changes sign). V: Medium thickness ("tent-shaped"). W: Thinnest curve. Right panel: amplitudes of the two gravity waves with $x(t)$ as the thicker, larger amplitude curve.

the damping coefficient a and the forcing F are both zero:

$$
\begin{aligned}
U_t &= -VW + \beta Vz \\
V_t &= UW - \beta Uz \\
W_t &= -UV \\
x_t &= -z \\
z_t &= bUV + x
\end{aligned}
\tag{12.41}
$$

It is often convenient to set $\beta = 0$ to obtain the "one-way coupled" model because an exact analytical solution has been given by Boyd(1994c).

A representative solution is shown in Fig. 12.2.

EXAMPLE FOUR: KdV Solitary Wave

The Korteweg-deVries equation is

$$
u_t + u\, u_x + u_{xxx} = 0
\tag{12.42}
$$

The general solution with spatially periodic boundary conditions consists of solitary waves and quasi-sinusoidal travelling waves. The simplest solitary wave or "soliton" is an exact nonlinear solution which steadily translates at a constant phase speed c:

$$
u_{sol}(x, t) = 12\epsilon^2 \mathrm{sech}(\epsilon\, [x - ct] + \phi); \qquad c = 4\,\epsilon^2
\tag{12.43}
$$

where ϵ and ϕ are constants. Strictly speaking, the soliton is defined only for a spatially unbounded interval, but its periodic generalization, called a cnoidal wave, is well approximated by (12.43) when $\epsilon \sim O(1)$ or larger.

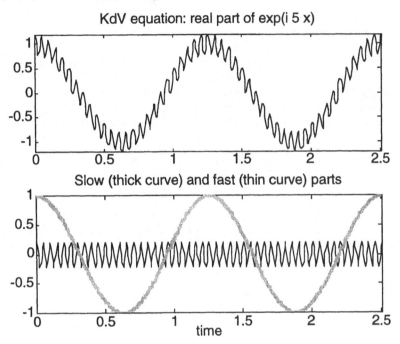

Figure 12.3: The real part of the Fourier coefficient of $\exp(i5x)$ as a function of time for a Korteweg-deVries (KdV) solution which is the sum of a solitary wave of unit phase speed and a small amplitude sine wave of wave number $k = 5$. The amplitude has been scaled to unity for the soliton contribution to this Fourier coefficient.

If the initial conditions are slightly perturbed from a soliton, then the flow will evolve as a solitary wave plus small amplitude sine waves which satisfy the linear KdV equation:

$$u_{sine}(x, t) = \alpha \cos(k[x - k^2 t]) \tag{12.44}$$

More general solutions with multiple solitons and large amplitude non-soliton waves are also possible, but for our purposes, it will suffice to consider a small amplitude solitary wave perturbed by very small sine waves. The interaction between the soliton and the rest of the flow, although not zero, can be neglected to a first approximation.

Fig. 12.3 illustrates the time evolution of a typical Fourier coefficient. The slow component, shown separately in the lower panel, is the solitary wave. The fast component, which controls the maximum time step for an explicit method, is the perturbing sine wave. The crucial point is that because the frequency of a sine wave of wavenumber k grows rapidly as k^3, these small waves are extremely fast-moving.

If the goal is to resolve the fast sine waves, then one might as well use an explicit method. If K is the maximum wavenumber, then one must use $\tau \sim O(1/K^3)$ to accurately track its propagation.

If the goal is only to understand the evolution of the solitary waves — historically true of most KdV numerical studies — then one can use an implicit or semi-implicit scheme with a much longer time step. For example, the fourth order Runge-Kutta method, which has a rather large stability limit, is stable only when

$$\tau < 2.8/K^3 \tag{12.45}$$

However, $a_K(t)$ varies due to the solitary wave as $\cos(Kct)$, which has a temporal period of $2\pi/(K\,c)$. Thus, the largest timestep which is stable for RK4 is $2.2K^2/c$ timesteps per temporal period. For $c = 1$ and $K = 16$, this is 570 time steps per period, and the number grows quadratically with K. Implicit algorithms allow one to accurately model the solitary wave with a much longer time step than $1/570$ of a temporal period.

There is actually a close relationship between this example and the forced linear oscillator because the Fourier-Galerkin ODEs in time for the KdV equation are

$$\frac{da_k}{dt} - ik^3 a_k = -i \sum_{m=-\infty}^{\infty} a_{k-m}\, a_m \tag{12.46}$$

This is identical in form to our second example (12.39) with $\omega = -k^3$ and with the nonlinear terms in the infinite series serving as the forcing for the k-th Fourier coefficient. Of course, one must be careful not to push the analogy too far because (12.46) is a coupled system, not merely a single isolated ODE. However, for large k, the nonlinear self-interaction of a_k can be neglected and the forced linear oscillator is a good first approximation.

12.9   Numerically-Induced Slow Manifolds: The Stiffness of Chebyshev Differentiation Matrices

As noted in the CFL Stability Limit Rule-of-Thumb in Chapter 9, the maximum timestep is the same order of magnitude as the time scale for advection or diffusion or wave propagation or whatever across the *smallest* distance h between two grid points. For wave propagation and for diffusion, the limits are

$$\tau_{max} = d\,\frac{h}{c_{max}} \quad \text{[Waves]} \qquad \tau_{max} = d'\nu\, h^2 \quad \text{[Diffusion]} \tag{12.47}$$

where d, d' are $O(1)$ constants and where c_{max} is the speed of the fastest-moving waves and ν is the diffusion or viscosity coefficient.

For Chebyshev and Legendre methods, these stability limits are very bad news because these use highly nonuniform grids with the smallest h — right next to the endpoints — being proportional to $1/N^2$. In contrast, an evenly spaced grid has $h = 2/N$ everywhere. It follows that explicit time-marching schemes are subject to very short time step limits when applied to ODEs derived from Chebyshev or Legendre spatial discretizations. For this reason, implicit time-marching schemes are very widely used with Chebyshev and Legendre spectral methods. The longest chapter in the book (Chap. 15) will focus on efficiently solving the boundary value problems, such as those from implicit algorithms.

An even more pressing concern is: Are implicit schemes sufficiently accurate to resolve everything that needs to be resolved when used with Chebyshev spatial discretizations? An example is useful: the one-dimensional diffusion equation with homogeneous Dirichlet boundary conditions is

$$u_t = u_{xx} + f(x); \qquad u(-1) = u(1) = 0; \qquad u(x,0) = Q(x) \tag{12.48}$$

where for simplicity $f(x)$ is independent of time.

We can exploit symmetry by splitting $u = S(x,t) + A(x,t)$ where S and A are u's symmetric and antisymmetric parts with respect to $x = 0$. We can define symmetric cardinal functions as a basis for S via

$$\phi_j(x) \equiv C_j(x) + C_{-j}(x) \tag{12.49}$$

where the $C_j(x; 2N)$ are the usual Chebyshev cardinal functions on the $(2N + 2)$-point Lobatto grid, numbered so that the cardinal functions which are one at the endpoints are $C_{\pm(N+1)}$.

The equation for the symmetric part is then reduced by the Chebyshev cardinal function spatial discretization to

$$\vec{S}_t = \vec{D}\,\vec{S} + \vec{f} \tag{12.50}$$

where the elements of the N-dimensional column vector $\vec{S}(t)$ are the time-dependent values of $S(x,t)$ at each of the *positive* collocation points, excluding the endpoint $x = 1$, which is omitted so that $S(1,t) = 0$ for all t to satisfy the boundary conditions, and $\vec{f}$ is similarly the grid point values of $f(x)$ for $x \geq 0$. The elements of the square matrix $\vec{D}$, which is the discrete representation of the second derivative operator, are

$$D_{ij} \equiv \phi_{j,xx}(x_i), \qquad i,j = 1,2,\ldots,N \tag{12.51}$$

where the boundary conditions have been implicitly included by omitting the cardinal function and grid point associated with $x = 1$.

If we compute the eigenvectors and eigenvectors of $\vec{D}$, we can switch to an eigenvector basis. In terms of this, the second derivative matrix is diagonal and (12.50) reduces to the uncoupled set of equations

$$\frac{db_j}{dt} = -\lambda_j b_j + g_j \tag{12.52}$$

which is identical to (12.5) except that we have changed the sign of the eigenvalue so that λ_j is positive for diffusion; the $\{b_j(t)\}$ and $\{g_j\}$ are the spectral coefficients in the eigenvector basis. The solution is

$$\vec{S}(t) = \vec{f} + \sum_{j=1}^{N} b_j(0)\,\exp(-\lambda_j t)\,\vec{e}_j \tag{12.53}$$

Table 12.1: Eigenvalues of the Second Derivative with Homogeneous Dirichlet Boundary Conditions Including Both Symmetric and Antisymmetric Modes

Method	Eigenvalues $\lambda_j, j = 1, 2, ..., N$	$\max(\lambda)$
Exact	$j^2\pi^2 / 4$	$N^2\pi^2 / 4$
2d order Finite Difference	$(N+1)^2 \sin^2\left(\frac{j\pi}{2(N+1)}\right)$	N^2
Chebyshev: Lobatto grid	accurate for $j < N\,(2/\pi)$	$0.048\,N^4$
Chebyshev: Gauss grid	accurate for $j < N\,(2/\pi)$	$0.303\,N^4$
Legendre: Gauss grid	accurate for $j < N\,(2/\pi)$	$0.102\,N^4$

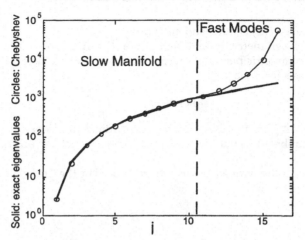

Figure 12.4: The exact eigenvalues of the second derivative operator with homogeneous Dirichlet boundary conditions on $x \in [-1, 1]$ are shown as the heavy solid line; the corresponding eigenvalues of the Chebyshev collocation matrix are the circles-with-thin-line. Roughly $(2/\pi) \approx 0.64$ of the Chebyshev eigenvalues — roughly 10 of the 16 eigenmodes illustrated — are good approximations. Eigenvalues 11 to 16 are way off, and there is no point in accurately tracking the time evolution of these modes of the Chebyshev pseudospectral matrix because they have no counterpart in the solution of the diffusion equation. The slow manifold is composed of the modes left of the vertical dividing line whose Chebyshev eigenvalues do approximate modes of the diffusion equation. The largest numerical eigenvalue is about 56,000 whereas the largest *accurately approximated* eigenvalue is only about 890, a ratio of roughly 64, and one that increases quadratically with N. Only *symmetric* eigenmodes are shown.

The exact solution to the diffusion equation is of the same form except that the eigenvectors are trigonometric functions and the eigenvalues are different as given by the first line of Table 12.1. The good news of the table is: the first $(2/\pi)N$ eigenvalues are well-approximated by any of the three spectral methods listed. This implies that we can follow the time evolution of as many exact eigenmodes as we wish by choosing N sufficiently large.

The bad news is: the bad eigenvalues are really bad because they grow as $O(N^4)$ even though the true eigenvalues grow only as $O(N^2)$. In the parlance of ODE theory, the Chebyshev-discretized diffusion equation is very "stiff", that is, it has an enormous range in the magnitude of its eigenvalues. The stability limit (set by the largest eigenvalue) is very small compared to the time scale of the mode of smallest eigenvalue.

The slow manifold is the span of the eigenmodes which are *accurately* approximated by the discretization, which turns out to be the first $(2/\pi)N$ modes for this problem. The fast-and-we-don't-give-a-darn manifold is the subspace in configuration space which is spanned by all the eigenmodes that are are *poorly* resolved by N Chebyshev polynomials (Fig. 12.4).

However, the physics is also relevant: if only the steady-state solution matters, than the slow manifold is the steady-state, and the time evolution of *all* modes can be distorted by a long time step without error. The physics and numerics must be considered jointly in identifying the slow manifold.

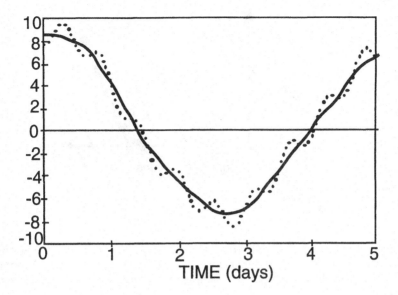

Figure 12.5: The time evolution of the north-south wind at 50.4^0 N. on the Greenwich meridian for the barotropic forecast of Richardson (1922). The dotted curve is the forecast made from the raw data without initialization. The solid curve is not the result of a running time average, but rather is the forecast from the initial conditions obtained by adjusting the raw data onto the slow manifold. [Redrawn from Fig. 8c of Lynch(1992) with permission of the author and the American Meteorological Society.]

12.10 Initialization

Meteorological jargon for the statement that "the large scale dynamics is near the slow manifold" is to say that the flow is "in quasi-geostrophic balance". Unfortunately, raw observational data is corrupted by random instrumentation and sampling errors which, precisely because they are random, do not respect quasi-geostrophic balance. Initializing a forecast model with raw data generates a forecast with unrealistically large gravity wave oscillations superimposed on the slowly-evolving large-scale dynamics (Daley, 1991, Lynch, 1992, Fig. 12.5). The remedy is to massage the initial conditions by using an algorithm in the following category.

Definition 28 (SLOW MANIFOLD INITIALIZATION)
A numerical algorithm for adjusting unprocessed initial values from a point in configuration space off the slow manifold to the nearest point on the slow manifold is called a SLOW MANIFOLD INITIALIZATION or simply (especially in meteorology) an INITIALIZATION procedure.

Fig. 12.6 schematically illustrates initialization onto the slow manifold.

Meteorological simulations of large-scale weather systems can be subdivided into forecasting and climate modelling. For the latter, the model is run for many months of time until it "spins up" to statistical equilibrium. The time-averaged temperature and winds then define climate of the model. The choice of initial conditions is irrelevant because the flow will forget them after many months of integration; indeed, independence of the initial conditions is an essential part of the very definition of climate.

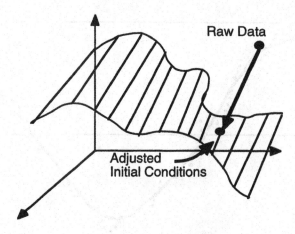

Figure 12.6: Schematic of an initialization onto the slow manifold, which is shown as the cross-hatched surface in a three-dimensional configuration space. (A real configuration space might have millions of dimensions where each coordinate is a spectral coefficient or grid point value.)

A slow manifold initialization is *unnecessary* for a *climate* model. The inherent dissipation of the model gradually purges the fast transients, like white blood cells eating invading germs. Similarly, the backwards Euler scheme, which damps all frequencies but especially high frequencies, will pull the flow onto the slow manifold for almost any physical system, given long enough to work. The BE algorithm, viscosity, and other physical or computational damping mechanisms are a kind of "Poor Man's Slow Manifold Initialization".

When the flow must be slow from the very beginning, however, stronger measures are needed. Meteorologists have tried a variety of schemes including:

1. Modifying the partial differential equations so that the approximate equations support only "slow" motions (quasi-geostrophic and various "balanced" systems, reviewed in Daley, 1991).

2. "Dynamic initialization", which is integrating a few steps forward in time, followed by an equal number back to $t = 0$, using special algorithms designed to damp high frequencies, rather than accurately integrate in time (Daley, 1991, Chap. 11).

3. Time-frequency filtering via Laplace transforms (Lynch and Huang, 1992).

4. Method of multiple scales ["Normal mode initialization"] (Machenhauer, 1977, Baer, 1977, Baer and Tribbia, 1977).

Without assessing the relative merits of these schemes for operational forecasting, we shall describe the method of multiple scales, or "normal mode initialization" as meteorologists call it, because it provides the theoretical underpinning for the Nonlinear Galerkin methods.

12.11   The Method of Multiple Scales: The Baer-Tribbia Series

When a partial differential equation with both fast and slow modes is discretized through Galerkin's method using the normal modes as the spectral basis, the result is a system of coupled ordinary differential equations in time that can be written

$$\vec{S}_t + i\,\vec{\omega}_S \cdot * \vec{S} = \vec{f}_S(\vec{S}, \vec{F})$$
$$\vec{F}_t + i\,\vec{\omega}_F \cdot * \vec{F} = \vec{f}_F(\vec{S}, \vec{F})$$

after partitioning the solution $\vec{u}$ into its slow and fast modes, $\vec{S}$ and $\vec{F}$, respectively. (In meteorology, $\vec{S}$ is the vector of amplitudes of the Rossby modes while $\vec{F}$ is the amplitude of the gravity waves.) The symbol $.*$ is used to denote elementwise multiplication of two vectors, that is, $\vec{\omega}_S \cdot * \vec{S}$ is a vector whose j-th element is the product of the frequency of the j-th mode, i. e., the j-th element of $\vec{\omega}_S$, with the j-th element of $\vec{S}$. The obvious way to integrate on the slow manifold is to simply ignore the fast modes entirely and solve the reduced system

$$\vec{S}_t + i\,\vec{\omega}_S \cdot * \vec{S} = \vec{f}_S(\vec{S}, \vec{0}) \tag{12.54}$$

Machenhauer (1977) pointed out that this is a rather crude approximation. The non-linear interaction of the slow modes amongst themselves will create a forcing for the fast modes. If the time scale for the slow modes is $O(\epsilon)$ compared to the fast frequencies, then the second line of Eq. (12.54) implies that

$$\vec{F}_j \approx -i\frac{(\vec{f}_F(\vec{S}, \vec{0}))_j}{\omega_j} \tag{12.55}$$

to within a relative error of $O(\epsilon)$. This forced motion in the fast modes varies slowly with time because it is forced only by the slow modes. Machenhauer showed that initialization schemes for weather forecasting were significantly improved by computing the initial values of the gravity modes from Eq.(12.55). If the gravity modes are initialized to zero, then the forecast must include (thrice-accursed!) high frequency gravity waves to cancel (at $t = 0$) the slow particular solution given by Eq.(12.55), and the forecast will resemble the wiggly dotted curve in Fig. 12.5.

Baer(1977) and Baer and Tribbia (1977) observed that Machenhauer's approximation was really just the lowest order of a singular perturbation expansion in ϵ, derivable by the so-called "method of multiple scales". The underlying idea can be illustrated by computing the perturbation series for a linear, uncoupled ODE:

$$\frac{du}{dt} = \lambda u + f(\epsilon\,t) \tag{12.56}$$

which is just Eq.(12.5) again except that we make one crucial assumption: that the forcing varies *slowly* with time as measured by the small parameter ϵ which appears inside its argument. (Implicitly, $\lambda \sim O(1)$, but this is not crucial; the effective perturbation parameter will turn out to be the ratio ϵ/λ which is the ratio of the "fast" and "slow" times scales for the problem.)

It is convenient to introduce the slow time variable

$$T \equiv \epsilon\,t \tag{12.57}$$

The ODE becomes

$$\epsilon \frac{du}{dT} = \lambda u + f(T) \tag{12.58}$$

It is then clear that lowest nontrivial approximation is the same as Machenhauer's: neglecting the time derivative so that there is an approximate balance between the two terms on the right-hand-side. This approximation, however, is sensible only for motion on the slow manifold or within $O(\epsilon)$ of the slow manifold; the homogeneous solution is the fast-evolving function $\exp(\lambda t)$, which is poorly approximated by neglecting the time derivative. Defining

$$u(T; \epsilon) = \sum_{j=0}^{N} u^j \, \epsilon^j \tag{12.59}$$

we obtain

$$
\begin{aligned}
u^0(T) &= -f(T)/\lambda \\
u^1(T) &= \frac{du^0/dT}{\lambda} \\
&= -\frac{df/dT}{\lambda^2} \\
u^2(T) &= \frac{du^1/dT}{\lambda} \\
&= -\frac{d^2 f/dT^2}{\lambda^3}
\end{aligned}
$$

and in general

$$u_{slow}(T; \epsilon) \sim -\frac{1}{\lambda} \sum_{j=0}^{N} \frac{d^j f}{dT^j} \left(\frac{\epsilon}{\lambda}\right)^j \tag{12.60}$$

This infinite series is noteworthy for a couple of reasons. First, the asymptotic equality sign is used because this is an asymptotic series that diverges except for special cases. However, when $\epsilon/\lambda \sim O(1/10)$ as in meteorology, one can use six to ten terms of the series to obtain an extremely accurate approximation, and the fact that the error would grow with additional terms has no practical relevance.

The second is that the asymptotic series does not contain an arbitrary constant of integration. The *general* solution to this first order ODE is

$$u(t) = A \, \exp(\lambda/\, t) + u_{slow}(\epsilon \, t; \epsilon) \tag{12.61}$$

where A is an arbitrary constant determined by the initial condition. If we pick any old $u(0)$, then $u(t)$ will be a mixture of fast and slow motion. Initialization onto the slow manifold means setting $u(0) = u_{slow}(0; \epsilon)$ so that $A = 0$.

The method is almost the same for the coupled system of fast and slow modes, Eq.(12.54), except for two differences. First, only the time derivative of the "fast" components is neglected so that the "Machenhauer" approximation is

$$
\begin{aligned}
\vec{S}_t + i\, \vec{\omega}_S \, . * \, \vec{S} &= \vec{f}_S \\
i\, \vec{\omega}_F \, . * \, \vec{F} &= \vec{f}_F
\end{aligned}
\tag{12.62}
$$

The system of ODEs has been replaced by a so-called "Differential-Algebraic" (DAE) set of equations.

The second difference is that f_S and f_F are not pre-specified, known functions of time but rather depend nonlinearly on the unknowns. Since the fast component is typically small compared to the slow modes – if it wasn't one would hardly be integrating near the slow manifold — one can build the dependence on $\vec{F}$ into the perturbation theory. However, the coupling between slow modes is often so strongly nonlinear than it cannot be approximated. The DAE must usually be solved by numerical integration of the ODE part plus algebraic relations which become increasingly complicated as one goes to higher order.

Indeed, the difficulties of separating various orders in the perturbation theory escalate so rapidly with order that Tribbia(1984) suggested replacing perturbation by iteration, which is easier numerically. With this simplification, the Baer-Tribbia method and its cousins are in widespread operational use in numerical weather prediction to adjust raw observational data onto the slow manifold. (See the reviews in Daley, 1991.) Errico (1984, 1989a) has shown that the Machenhauer approximation [lowest order multiple scales] is a good description of the relative magnitudes of terms in general circulation models, which include such complications as latent heat release, chemistry and radiative transfer, provided that the label of "fast" mode is restricted to frequencies of no more than a few hours.

12.12 Nonlinear Galerkin Methods

Although the Baer-Tribbia series and related algorithms have been used almost exclusively for initialization, the split into fast and slow modes applies for all time and is the basis for the following class of algorithms.

Definition 29 (NONLINEAR GALERKIN) *A numerical algorithm for discretizing a partial differential equation is called a "Nonlinear Galerkin" scheme if the system of ODEs in time that results from a conventional spatial discretization is modified to a Differential-Algebraic Equation (DAE) system by replacing the ODEs for the time derivatives of the fast modes by the method of multiple scales (in time).*

Nonlinear Galerkin methods can be classified by the order m of the multiple scales approximation which is used for the algebraic relations. NG(0) is the complete omission of the fast modes from the spectral basis. NG(1) is the lowest non-trivial approximation, known in meteorgy as the "Machenhauer approximation". For the single linear forced ODE, $du/dt = \lambda u + f(t)$, the first three approximations are

$$
\begin{aligned}
u &\equiv 0 && NG(0) \\
u &\sim -f/\lambda && NG(1) \\
u &\sim -f/\lambda - df/dt/\lambda^2 && NG(2)
\end{aligned}
\tag{12.63}
$$

Kasahara (1977, 1978) experimented with forecasting models in which the usual spherical harmonic basis was replaced by a basis of the Hough functions, which are the normal modes of the linear terms in the model. This was a great improvement over the old quasi-geostrophic models, which filtered gravity modes, because the Hough basis could include some low frequency gravitational modes. However, it was still only an $NG(0)$ approximation: high frequency Hough modes were simply discarded.

Daley (1980) carried out the first experiments with NG(1) for a multi-level weather prediction model. He found that the algorithm compared favorably with stanard semi-implicit algorithms in both efficiency and accuracy. Unfortunately, the Nonlinear Galerkin method was no better than semi-implicit algorithms either. Haugen and Machenhauer

Table 12.2: A Selected Bibliography of Nonlinear Galerkin Methods

Note: KS is an abbreviation for "Kuramoto-Sivashinsky" , FD for "finite difference", "CN" for the Crank-Nicholson scheme and "BDF" for Backward Differentiation time-marching schemes. Most articles with non-spectral space discretizations are omitted.

Reference	Area	Comments
Kasahara (1977, 1978)	Meteorology	NG(0)
Daley (1980)	Meteorology	NG(1)
Daley (1981)	Meteorology	Review: normal mode initialization
Foias *et al.*(1988)	Reaction-diffusion & KS	Independent rediscovery; call it "Euler-Galerkin" method
Marion & Témam(1989)	Dissipative systems	Independent rediscovery of NG
Témam (1989)	Navier-Stokes	Recursive calculation high order NG series
Jauberteau, Rosier & Témam(1990a,b)	Fluid Dynamics	
Jolly, Kevrekidis & Titi(1990, 1991)	KS Eq.	Computation of inertial manifolds; Fourier
Dubois, Jauberteau & Témam (1990)	Fluids	Fourier, 2D, 3D number of fast modes adaptively varied
Foias *et al.*(1991)	KS, forced Burgers	Blow-up
Témam (1991b)	Numerical analysis	Stability analysis
Témam (1991a)	Fluid Mechanics	Review
Promislow & Témam(1991)	Ginzburg-Landau Eq.	numerical experiments; higher order *NG*
Témam (1992)		Review
Devulder & Marion(1992)	Numerical Analysis	
Margolin & Jones(1992)	Burgers	NG with spatial FD
Pascal & Basdevant(1992)	2D turbulence	Improvements & efficiency tests
Gottlieb & Témam(1993)	Burgers	Fourier pseudospectral
Foias *et al.*(1994)	KS	Blow-up
García-Archilla & de Frutos(1995)	Numerical analysis	Comparisons of NG with other time-marching schemes
García-Archilla(1995)	Numerical analysis	Comparisons of NG with other schemes
Boyd (1994e)	Numerical Analysis	Theory for inferiority of NG to CN & BDF
Vermer&van Loon(1994)	chemical kinetics	NG(1) popular as PSSA: "Pseudo-steady-state approx."
Debussche&Dubois &Témam(1995)	Homogeneous turbulence	
Jolly&Xiong(1995)	2D turbulence	
Jones&Margolin &Titi(1995)		Comparison with Galerkin; forced Burgers & KS equations
Dubois&Jauberteau&Teémam(1998)		Comprehensive review

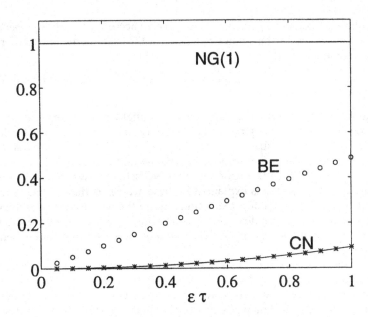

Figure 12.7: Relative errors in approximating the time derivative of $\exp(i\epsilon t)$ as a function of ϵ and the time step τ. The relative error in the Nonlinear Galerkin method, which simply discards the time derivative, is one independent of the time step (horizontal line). The errors in the Backwards Euler (BE) and Crank-Nicholson (CN) implicit methods are functions of the *product* of ϵ with τ and decrease linearly (circles) and quadratically (asterisks) as $\epsilon\tau \to 0$. However, even for $\epsilon\tau = 1$ — only 2π grid points in time per temporal period — the errors are still much smaller than $NG(1)$.

(1993) also tried an NG scheme in combination with a semi-Lagrangian (SL) treatment of advection. For long time steps (sixty minutes), the NG scheme was inaccurate unless two Machenhauer-Tribbia iterations were used. This made the NG/SL combination a little less efficient than the semi-implicit SL algorithm which has now become very popular for both global and regional models.

Nonlinear Galerkin methods were independently rediscovered by Roger Témam and his collaborators (Marion and Témam, 1989, Jauberteau, Rosier, and Témam, 1990) and also by Foias *et al.*(1988). (Parenthetically, note that similar approximations under the name of the "quasi-static" approximation or "pseudo-steady-state approximation" have also been independently developed in chemical kinetics where the ratio of time scales often exceeds a million (Vermer and van Loon, 1994)!). Table 12.2 describes a large number of applications.

The practical usefulness of Nonlinear Galerkin time-marching is controversial. Daley, Haugen and Machenhauer (1993), Boyd (1994e), García-Archilla(1995) and García-Archilla and de Frutos (1995) are negative Pascal and Basdevant (1992) conclude that NG is useful in reducing costs for turbulence modelling; most of the other authors in Table 12.2 are also optimistic about the merits of the Nonlinear Galerkin method.

12.13   Weaknesses of the Nonlinear Galerkin Method

In comparing the Nonlinear Galerkin method with implicit and semi-implicit schemes, there are two key themes.

1. All temporal finite difference approximations have an accuracy which is *proportional to* $NG(1)$ multiplied by a constant times τ^r where r is the order of the finite difference method and τ is the timestep.

2. The computational complexity and expense of Nonlinear Galerkin methods, $NG(m)$, grows very rapidly with the order m.

Demonstrating the second assertion would take us too far afield, but it is really only common sense: NG schemes are derived by perturbation theory, and the complexity of almost all perturbation schemes escalates like an avalanche. Témam (1989) gives recursive formulas for calculating high order NG formulas; Promislow and Témam(1991) is unusual in performing numerical experiments with high order NG schemes, that is, $NG(m)$ with $m > 1$. The special difficulty which is explained in these articles is that all NG schemes higher than Machenhauer's [$NG(1)$] approximation require the *time derivative* of the nonlinear terms which force the fast modes. There is no simple way to compute the time derivative of order $(m - 1)$ as needed by $NG(m)$, although complicated and expensive iterations can be implemented.

The first point is more fundamental. The sole error of Backwards Euler, Crank-Nicholson and other time-marching schemes is that they approximate the time derivatives. Since $NG(1)$ approximates the time derivative by completely ignoring it, it follows that the error in $NG(1)$ is proportional to the accuracy of this assertion that the time derivative is small. However, the relative smallness of the time derivative is an advantage for implicit time-marching schemes, too, because their error in approximating the derivative is multiplied by the smallness of the term which is being approximated.

For example, consider the simple problem

$$u_t + iu = \exp(i\epsilon t) \tag{12.64}$$

for which the exact slow manifold (Boyd, 1994e, H.-O. Kreiss, 1991) is

$$u_{slow}(t; \epsilon) \equiv - i \, \frac{1}{1 + \epsilon} \exp(i \, \epsilon \, t) \tag{12.65}$$

The exact derivative of the slow solution is $i \, \epsilon \, u_{slow}$. The relative errors in approximating this, in other words, $[u_t(exact) - u_t(approximate)]/u_t(exact)$, are graphed as Fig. 12.7. Note that because the complex exponential is an eigenfunction of the difference operators, the slow manifold for all the approximation schemes (including $NG(m)$) is proportional to $\exp(i\epsilon t)$. This factor cancels so that the relative error is independent of time.

The most remarkable thing about the graph is the fact that even when the temporal resolution is poor (no more than 6 points per period of the slow manifold), the humble BE method has only half the error of $NG(1)$ whereas the Crank-Nicholson scheme has only 1/12 the error of $NG(1)$. This seems little less than amazing because the BE and CN methods are general time-stepping schemes that in theory can be applied to any time-dependent problem. In contrast, the Nonlinear Galerkin method is specialized to track the slow manifold only; it has no meaning for time-dependent problems with but a single time scale. Why is the specialized method so bad compared to the general methods?

The answer is the obvious one that an approximation, even a bad one, is better than no approximation at all. $NG(1)$ is a *non-approximation* of the *time derivative*. In contrast, implicit schemes try to accurately approximately u_t as is also done by *higher order* $(m > 1)$ NG methods.

Of course, with a really long time step, the finite differencing schemes would give $O(1)$ errors, but this is insane. In practical applications, it is necessary to integrate the fast and slow modes as a coupled system. A time step which is long enough to make Crank-Nicholson poorer than $NG(1)$ for the fast modes will also do a terrible job of advancing

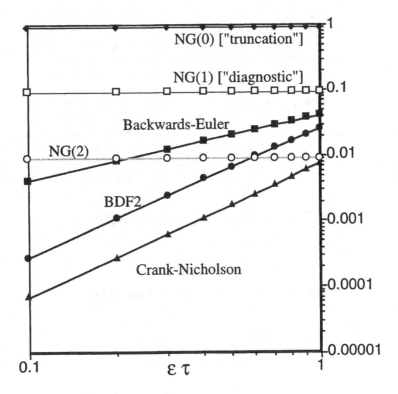

Figure 12.8: Maximum pointwise errors in approximating the slow manifold of $du/dt + i\omega u = \exp(i \, \epsilon \, t)$, plotted versus the product of ϵ with τ. The ratio of ω/ϵ is fixed at 1/10; the errors for the Nonlinear Galerkin schemes, which are independent of the time step and therefore appear as horizontal lines, are proportional to $(\omega/\epsilon)^m$ for NG(m).

the slow modes. Thus, it is always necessary to choose $\epsilon\tau < 1$ in practice, and the implicit schemes always triumph over $NG(1)$.

It also true that if we compare implicit schemes with $NG(m)$ in the limit $m \to \infty$ for *fixed* time step and ratio of time scales ϵ, the NG scheme will be superior. This is illustrated by Fig. 12.8, which compares NG schemes of various orders with implicit methods for a particular value of ϵ. However, even though $\epsilon = 1/10$, the Crank-Nicholson scheme is better than $NG(2)$ for all $\tau < 1/\epsilon$, that is, whenever there are more than six points per time period of the slow manifold. Furthermore, because of the complexity of high order NG methods, most Nonlinear Galerkin studies beginning with Daley (1980) have used $NG(1)$ only.

García-Archilla and de Frutos (1995) have compared time-marching algorithms for some simple nonlinear equations. Advection does not change the conclusion: "The results show that for these problems, the non-linear Galerkin method is not competitive with either pure spectral Galerkin or pseudospectral discretizations." Haugen and Machenhauer (1993) found that $NG(1)$ was not competitive with the semi-implicit method for a time step of an hour when advection was handled by a semi-Lagrangian scheme.

All this negative evidence raises the question: Why has so much work been done with Nonlinear Galerkin methods? Part of the answer is that many of the articles in Table 12.2 report considerable cost-savings over conventional algorithms. (However, most studies were restricted to explicit schemes versus NG schemes only.) A second answer is that numerical analysts are as opinionated as soccer fans.

A third answer is that NG schemes are self-initializing. The Crank-Nicholson scheme is a good method to track the slow manifold, yes, but it is powerless to follow the slow manifold unless some external agency adjusts the initial conditions first.

A fourth answer is that the Nonlinear Galerkin scheme is a very thoughtful algorithm. In a problem with multiple time scales that evolves only on the slow time scale, the high frequency modes are forced only by the slow modes. This, to lowest order, is what really happens. The insight underlying NG schemes has led to interesting connections between them and multigrid, hierarchical finite element bases, and a wide variety of other algorithms as outlined in the many papers of Témam and his collaborators. Perhaps in time, at least in special cases, the inefficiencies so evident in Figs. 12.7 and 12.8 will be remedied.

12.14   Following the Slow Manifold with Implicit Time-Marching: Forced Linear Oscillator

To examine implicit algorithms experimentally, it is useful to consider same model as in the previous section:

$$u_t + iu = \exp(i\epsilon t) \tag{12.66}$$

which, writing $u(t) = x(t) + i\, y(t)$, is equivalent to the pair of real-valued equations

$$x_t - y = \cos(\epsilon t) \quad \& \quad y_t + x = \sin(\epsilon t) \tag{12.67}$$

The specialization to unit frequency is no restriction because if the frequency is some different number ω, we can define a new time variable s by $\omega t = s$, and then the homogeneous solutions of Eq.(12.64) will oscillate with unit frequency in s. Note that this requires redefining ϵ: the effective perturbation parameter for general ω is ϵ/ω, that is, the ratio of the frequency of the forcing to the frequency of the homogeneous solution.

The specialization to a forcing function that is harmonic in time is a little restrictive, but only a little, because very general forcing functions can be constructed by using Fourier series or integrals to superpose components of the form of $\exp(i\epsilon t)$ for various ϵ. However, the exponential forcing is very convenient because it allows us to explicitly compute the slow manifold not only for the differential equation but also for its approximation by NG and various implicit time-marching schemes.

Fig. 12.9 shows the exact solution, the slow manifold, and the Crank-Nicholson approximation for an initial condition which excites both fast and slow motion. The timestep $\tau = 1/\epsilon$, which is 2π points per slow temporal period.

The fast component oscillates through ten periods on an interval where the Crank-Nicholson scheme has only six points, so it is absurd to suppose the time-marching method can accurately track the fast motion. Still, it fails in an interesting way. The CN scheme does oscillate around the slow manifold, just like the exact solution. Its cardinal sin is that, as asserted earlier in Rule-of-Thumb 6, it tremendously slows down the oscillation, approximating an oscillation of period 2π by an oscillation of period 2τ.

As asserted earlier, without a separate slow manifold initialization procedure, the Crank-Nicholson scheme is useless for integrating motion that is supposed to be purely slow motion. In contrast, the Backwards Euler scheme, which damps high frequencies, will converge to the slow manifold eventually.

When the initial conditions are on the slow manifold, the Crank-Nicholson scheme is very good even with a coarse grid as illustrated in Fig. 12.10. The maximum error with roughly six points per period is only 0.012.

12.15 Three Parts to Multiple Time Scale Algorithms: Identification, Initialization, and Time-Marching on the Slow Manifold

Computing slow solutions for a dynamical system that also allows fast motions requires three key steps:

1. Identification of the Slow Manifold

2. Initialization onto the Slow Manifold

3. Slow Manifold-Following Time-Stepping Algorithm.

Identification of a slow manifold is both a problem in physics or engineering and a problem in numerical analysis. The same system may or may not have a slow manifold, depending on the situation and the goals of the modeller. The atmosphere, for example, has a slow manifold for climate modelling and large-scale weather forecasting. Gravity waves, convective plumes and small-scale vortices, which can be neglected in climate studies, are essential to mesoscale meteorology (that is, small time-and-space scale studies). Therefore, mesometeorology lacks a slow manifold. A forced diffusion problem has a slow manifold — the steady state — if one is only interested in the long-time solution, but it has no slow manifold if one wants to track the high frequency transients, too.

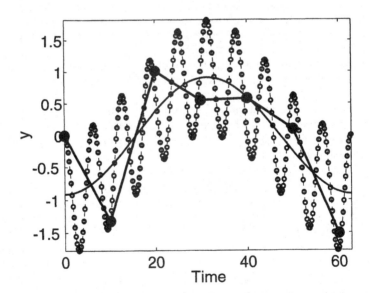

Figure 12.9: Imaginary part $y(t)$ for the forced linear oscillator with $\epsilon = 1/10$ and the initial conditions $x(0) = 0, y(0) = 0$, which excite a mixture of fast and slow motion. Solid, unlabelled curve: slow manifold. x's: exact solution. Solid disks: the Crank-Nicholson values.

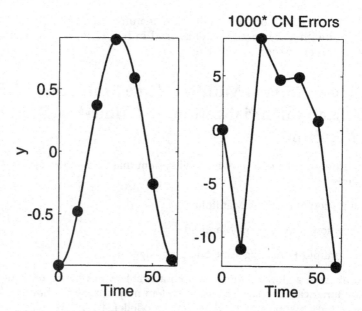

Figure 12.10: Left panel: Imaginary part $y(t)$ for the forced linear oscillator with $\epsilon = 1/10$ and the initial conditions $x(0) = 0, y(0) = -1/(1 + \epsilon)$, which excites only slow motion. Solid, unlabelled curve: slow manifold (also the exact solution). Solid disks: Crank-Nicholson values. Right panel: errors in the Crank-Nicholson approximation, multiplied by 1000.

Numerical errors can also make fast-evolving components of the discretized model very uninteresting. Even if there is no physical slow manifold, there may be a numerics-induced slow manifold where the slow components are those modes which can be accurately resolved by the Chebyshev discretization.

No kindergardener would be dumb enough to pursue a project without clear goals; even at five, my son Ian is an enthusiastic planner. The availability of fast workstations sometimes has a bad effect on inquiring minds of supposedly much superior maturity. A common disease of adult numerical analysts is: Compute First, Think Later. But it is very hard to get good results without first asking: What do I really want from the model? And what will the model allow me? It is from such metaphysical thinking that the slow manifold can be identified.

Initialization onto the slow manifold is irrelevant in climate modelling and many other problems where the dumb and mindless forces of dissipation are sufficient to move the problem onto the slow manifold soon enough for the modeller. For weather forecasting and lots of other problems, however, damping works too slowly; one must ensure that the initial data will lie on the slow manifold only by using some sort of initialization method.

The third step is to choose a numerical algorithm which is good at ignoring the fast dynamics and accurately evolving the flow along the slow manifold. The standard implicit time-marching schemes usually meet this standard; explicit schemes don't. However, implicit methods can sometimes fail because of problems of "consistency", boundary-splitting errors, and a few subtleties discussed in later chapters.

Slow manifold concepts have always been implicit in implicit time-marching schemes. Curiously, however, these ideas have been glossed over in numerical analysis texts. In meteorology, for example, gravity wave-filtering approximations have been used since the late 1940s. However, the concept of the slow manifold was not formally defined until Leith(1980). And yet the notion of filtering the gravity waves to leave only the slow flow, as embodied in Charney's 1947 quasi-geostrophic equations of motion, is as goofy as a Bugs Bunny cartoon unless the large-scale flow in the atmospheric does lie on or very near a subspace of purely slow motion in a phase space that allows both fast and slow dynamics.

In regard to time-marching algorithms, it is better to be explicit about the reason for implicit. Otherwise, implicit algorithms become a kind of vague intuition, as unreliable as a child's belief that hiding under a blanket, because it keeps out the light, will also keep out the thunder and lightning. Implicit algorithms are neither voodoo nor a lucky charm, but an efficient way to integrate along the slow manifold.

Chapter 13

"Fractional Steps" Time Integration: Splitting and Its Cousins

"Divide et impera" —"Divide and conquer"

–Louis XI

"Bender's Law: Different but reasonable schemes for the same problem require about the same work."

– Carl M. Bender

13.1 Introduction

Implicit time-marching algorithms are usually quite expensive per time step because of the need to solve a boundary value problem at each time step. The exceptions are:

1. One (spatial) dimension.

2. Semi-implicit methods with a Fourier or spherical harmonics basis which require solving only constant coefficient boundary value problems; these are trivial in a Galerkin formalism.

3. Problems which are periodic in all but one spatial dimension.

For spherical harmonic climate models, for example, it is only necessary to solve one-dimensional boundary value problems (in height), one BVP for the vertical dependence of each spherical harmonic coefficient. But what is one to do for the vast number of problems which require solving multi-dimensional BVPs by means of Chebyshev polynomials every timestep?

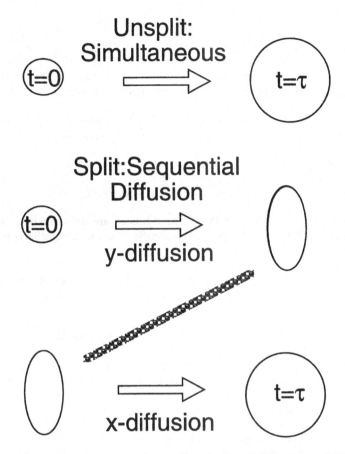

Figure 13.1: Splitting by dimension: a two-dimension process such as diffusion, which causes a drop of food coloring to widen in all directions simultaneously (top), is treated as two one-dimensional processes, happening one-after-the-other instead of simultaneously (bottom).

One strategy is to use sophisticated matrix methods. For for constant coefficient and separable BVPs, there are special fast spectral algorithms. For non-separable BVPs, preconditioned matrix iterations can be very effective.

The alternative is to attack and simplify *before* the spatial discretization is applied: The "splitting" or "fractional step" family of algorithms.

Definition 30 (SPLITTING/FRACTIONAL STEPS) *An implicit time-marching algorithm which replaces* simultaneous *processes by* sequential *steps to increase efficiency is called a SPLITTING or FRACTIONAL STEPS method. The split may be by dimensions: three-dimensional diffusion, for example, may be split into three one-dimensional substeps of diffusion in the x-direction only followed by diffusion in y and then z. The splitting may also be by physics: advection on one fractional step, pressure adjustment (waves) on another, and diffusion/viscosity on a third.*

The key idea of splitting, simultaneous-into-sequential, can be illustrated by the two-dimensional diffusion equation, $u_t = u_{xx} + u_{yy}$. If the matrices which are the spatial discretizations of the x and y derivatives are denoted by $\vec{\Lambda}_1$ and $\vec{\Lambda}_2$, then the exact solution is

$$\vec{u}(\tau) = \exp([\vec{\Lambda}_1 + \vec{\Lambda}_2] \, \tau) \, \vec{u}(0) \tag{13.1}$$

Suppose, however, that we solve the pair of equations that results from assuming that the solution diffuses in the x-direction first and then in the y-direction (Fig. 13.1):

$$\begin{aligned} v_t &= v_{xx}; & v(t=0) &= u(0) \\ w_t &= w_{yy}; & w(t=0) &= v(\tau) \end{aligned} \tag{13.2}$$

The discretized solution is

$$w(\tau) = \exp(\vec{\Lambda}_1 \, \tau) \, \exp(\vec{\Lambda}_2 \, \tau) \, u(\vec{0}) \tag{13.3}$$

In general, it is not automatic that these two solutions are the same because for matrices (or linear operators), the usual exponential identity must be generalized to, for arbitrary matrices $\vec{A}, \vec{B}$,

$$\exp(\vec{A} + \vec{B}) = \exp(\vec{A}) \, \exp(\vec{B}) \, \exp(-[\vec{A}, \vec{B}]/2) \qquad \text{Weyl formula} \tag{13.4}$$

where

$$[\vec{A}, \vec{B}] \equiv \vec{A}\,\vec{B} - \vec{B}\,\vec{A} \qquad \text{Commutator of } \vec{A}, \vec{B} \tag{13.5}$$

However, if both $\vec{A}$ and $\vec{B}$ are proportional to the time step τ, then the commutator must be $O(\tau^2)$. It follows that in the limit $\tau \to 0$, the splitting system (13.2) provides at least a first-order-in-time approximation to the diffusion equation, regardless of how the time and space derivatives are discretized. Furthermore, we can replace the matrices for diffusion by any other pair of physical processes and the argument still applies.

The great advantage of splitting for the diffusion problem is that it requires only *one-dimensional* boundary value problems. Even though we solve many such problems, one BVP in x for each interior grid point in y, for example, this is still much cheaper than solving a single two-dimensional BVP.

For the hydrodynamic equations, the advantage of splitting-by-process is that advection can be treated by a different algorithm than pressure adjustment, which in turn can be different from diffusion. We need not even use the same spatial discretization; we can use finite differences for diffusion and spectral methods for other parts (as in Lambiotte *et al.*, 1982, for example).

Because of these virtues, the splitting method, also known as the method of fractional steps, was extensively developed in the Former Soviet Union as described in books by Yanenko (1971) and Marchuk (1974). Strang (1968) independently invented this family of algorithms, which are often called "Strang splitting" in the U. S. where these ideas caught on more slowly. Although especially popular with spectral methods, splitting can be used with finite difference or finite element spatial discretizations just as well, and indeed was developed by Yanenko and Marchuk only for non-spectral methods.

Splitting sounds almost too wonderful to be true. Not quite, but there are snags. The biggest is boundary conditions. Viscosity furnishes the highest derivatives for the Navier-Stokes equations, so we cannot impose the usual rigid boundary conditions on the non-viscous fractional steps. This can and does lead to growth of error at the boundaries. (This problem does not arise in Fourier or spherical harmonics applications where there are no boundaries.) However, at the price of sacrificing the simplicity of straightforward splitting, some good remedies have been developed. Splitting is still widely used.

13.2 The Method of Fractional Steps for the Diffusion Equation

The great vice of the Crank-Nicholson method is that it is cheap only in one-dimension. In two or three dimensions, even with a finite difference method, we must invert a large matrix at each time step. With direct matrix-solving methods (LU factorization, also known as Gaussian elimination), the cost per time step in two dimensions is $O(N^4)$ and $O(N^6)$ operations for finite difference and spectral methods, respectively, where N is the number of grid points in either x or y. Ouch! These costs can be greatly reduced by clever iterative matrix-solvers as described later, but are still formidable.

The three-dimensional diffusion equation is

$$u_t = u_{xx} + u_{yy} + u_{zz} + g(x, y, z) \tag{13.6}$$

where $g(x, y, z)$ is a known forcing function.

The method of fractional steps avoids inverting full matrices by advancing the solution in one coordinate at a time. First, write

$$\Lambda = \Lambda_1 + \Lambda_2 + \Lambda_3 \tag{13.7}$$

where $\Lambda_1 = \partial_{xx}$, $\Lambda_2 = \partial_{yy}$, and $\Lambda_3 = \partial_{zz}$. Then the algorithm is

$$\frac{u^{n+1/4} - u^n}{\tau} = \frac{1}{2}\Lambda_1 \left[u^{n+1/4} + u^n \right] \tag{13.8}$$

$$\frac{u^{n+1/2} - u^{n+1/4}}{\tau} = \frac{1}{2}\Lambda_2 \left[u^{n+1/2} + u^{n+1/4} \right] \tag{13.9}$$

$$\frac{u^{n+3/4} - u^{n+1/2}}{\tau} = \frac{1}{2}\Lambda_3 \left[u^{n+3/4} + u^{n+1/2} \right] \tag{13.10}$$

$$\frac{u^{n+1} - u^{n+3/4}}{\tau} = g \tag{13.11}$$

Each fractional step is the Crank-Nicholson method for a *one-dimensional* diffusion equation. Consequently, if Λ_1 is generated by applying the usual three-point difference formula to ∂_{xx}, it is only necessary to invert tridiagonal matrices at each time step. As shown in Appendix B, a tridiagonal matrix may be inverted in $O(N)$ operations. The operation count for the fractional steps method is directly proportional to the total number of grid points even in two or three dimensions; it lacks the $O(N^2)$ penalty of the unsplit Crank-Nicholson method (versus an explicit scheme).

The pseudospectral method, alas, generates non-sparse matrices — but it is far better to invert $3N^2$ matrices each of dimension N than a single matrix of dimension N^3 since the expense of inverting a matrix scales as the cube of its dimension. Later, we shall discuss cost-reducing iterations. An alternative direct method, faster but restricted to constant coefficient differential operators, is to use Galerkin's method. For this special case of constant coefficients, the Galerkin matrix is banded and cheap to invert as explained in the Chapter 15.

It is obvious that (13.8)–(13.11) is *cheap*; it is not so obvious that it is accurate. However, eliminating the intermediate steps shows that the method of fractional steps is equivalent to the scheme

$$u^{n+1} = M(-\tau)^{-1}M(\tau)\, u^n + \tau\, M(-\tau)^{-1}g \tag{13.12}$$

in whole time steps where the operator M is given by

$$M(\tau) \equiv 1 + \frac{\tau}{2}\Lambda + \frac{\tau^2}{4}\{\Lambda_1\Lambda_2 + \Lambda_1\Lambda_3 + \Lambda_2\Lambda_3\} + \frac{\tau^3}{8}\Lambda_1\Lambda_2\Lambda_3 \qquad (13.13)$$

Eq. (13.12) is identical with the corresponding formula for the Crank-Nicholson scheme except for the terms in τ^2 and τ^3 in the definition of $M(\tau)$. Because of cancellations, the overall error in both the splitting method and Crank-Nicholson is $O(\tau^3)$ for $u(x, y, z, t)$.

At first glance, the entire premise of the method of fractional steps seems ridiculous. Physically, diffusion is *not* a sequential process. However, if we consider only a small time interval of length τ, then it is a good *mathematical* approximation to split the time interval into three, and pretend that there is only diffusion in x on the first fractional step, only diffusion in y on the second, and only diffusion in z on the third. (Note that there are no factors of $(1/3)$ in (13.8)–(13.10): we allow each second derivative to diffuse over the whole time interval even with the splitting-up.) Our fiction introduces extra error; the accuracy for a given time step τ is poorer than that of the Crank-Nicholson method. However, it is the same order-of-magnitude — $O(\tau^3)$. The cost-per-time-step of splitting is orders-of-magnitude smaller than for the Crank-Nicholson scheme.

13.3   Pitfalls in Splitting, I: Boundary Conditions

Fractional step methods introduce some new difficulties peculiar to themselves. Variables at fractional time levels are not solutions of the full set of original equations. What boundary conditions should we impose on these intermediates?

For behavioral boundary conditions, this difficulty disappears because behavioral conditions are never explicitly imposed. Instead, one merely expands all the unknowns, including intermediates at fractional time levels, in Fourier series (for spatial periodicity) or rational Chebyshev functions (on an infinite interval).

Numerical boundary conditions require careful analysis, however. If the domain is rectangular and the boundary values are independent of time, then the treatment of boundary conditions is straightforward. For simplicity, consider the two-dimensional heat equation.

With splitting, one must solve a two-point boundary value problem in x at each value of y on the interior of the grid (Fig. 13.2). The values of $u(0, y)$ and $u(1, y)$ give us just what we need, however. Similarly, the known values of u at the top and bottom of the rectangle give the boundary conditions for the second splitting step: solving a set of two-point boundary value problems in y, one for each value of x. The matrix representation of the x-derivatives is a large matrix, but it decouples into independent subproblems as illustrated in Fig. 13.3.

When the boundary conditions *vary* with *time*, we have troubles. The intermediates in splitting — $u^{n+1/2}$, for example — do not correspond to the physical solution at $t = (n + 1/2)\tau$, but rather are merely intermediate stages in advancing the solution by a whole time step.

When the domain is *irregular*, however, the proper boundary conditions require real art. Spectral elements, which subdivide a complicated domain into many subdomains only slightly deformed from rectangles, greatly alleviates this problem. Therefore, we refer the curious reader to Yanenko (1971), who explains how to apply splitting with finite differences to a problem with curvy boundaries.

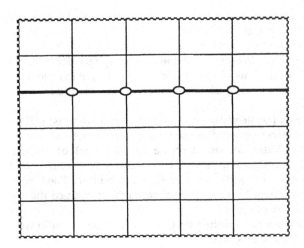

Figure 13.2: In a split scheme, it is necessary to solve one-dimensional boundary value problems on vertical and horizontal lines. One such (horizontal) line is marked by the four circles which are the unknowns. Instead of a single 20x20 matrix, splitting requires (on this 6x7 grid) the solution of five 4x4 problems and four 5x5 problems.

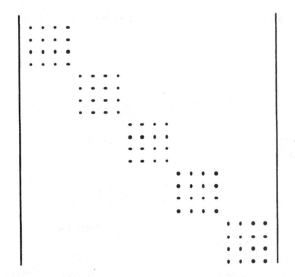

Figure 13.3: Matrix of the second x-derivative: Dots denote nonzero matrix elements. There is one block for each interior grid point in y. Each block is the matrix of a one-dimensional boundary value problem in x.

13.4   Pitfalls in Splitting, II: Consistency

The consistency problem is that *stability* does not necessarily guarantee *accuracy*. General truth: absolutely stable methods gain their stability from *being inaccurate* for the *rapidly-varying* components of the solution. For example, the Crank-Nicholson method achieves stability by artificially slowing down the extremely fast decay of the high wavenumber terms in the solution. This is acceptable if the only goal is the asymptotic, steady-state solution, but we must always ask the engineering question: do we need to correctly model the transient, too?

Even when only the steady state solution matters, we must still be careful because it is possible that the inaccuracies of the absolutely stable method will poison the asymptotic solution as well. Mathematicians describe this as a "lack of consistency". Like stability, consistency with the original differential equations may be either conditional — we get the correct answer *only* if τ is sufficiently small — or absolute. The Crank-Nicholoson scheme is "absolutely consistent" because it will correctly reproduce the asymptotic solution as $t \to \infty$ even if τ is very large.

The three-dimensional splitting method, alas, is only *conditionally consistent* with the heat equation. To show this, write

$$u = T(x, y, z, t) + v(x, y, z) \tag{13.14}$$

where $T(x, y, z, t)$ is the "transient" part of the solution and $v(x, y, z)$ is the asymptotic solution. It follows that

$$\lim_{t \to 0} T(x, y, z, t) \equiv 0 \tag{13.15}$$

$$\nabla^2 v + g = 0 \tag{13.16}$$

Similarly, the steady solution of the splitting equations will satisfy (from (13.12))

$$[M(-\tau) - M(\tau)] v = \tau g \tag{13.17}$$

Recalling that

$$M(\tau) \equiv 1 + \frac{\tau}{2}\Lambda + \frac{\tau^2}{4}\{\Lambda_1\Lambda_2 + \Lambda_1\Lambda_3 + \Lambda_2\Lambda_3\} + \frac{\tau^3}{8}\Lambda_1\Lambda_2\Lambda_3 \tag{13.18}$$

we see that only the odd powers of τ survive the subtraction in (13.17) to give (after dividing out a common factor of τ)

$$\left\{ -\Lambda + \frac{\tau^2}{4}\Lambda_1\Lambda_2\Lambda_3 \right\} v = g \tag{13.19}$$

Oops! When τ is sufficiently small, we can neglect the second term in the {} in (13.19) to obtain $-\Lambda v = g$, which is just what we want. (Recall that Λ is the Laplacian operator.) However, the "consistency" of the splitting-up scheme with the heat equation is not absolute because for sufficiently large τ, the $\Lambda_1 \Lambda_2 \Lambda_3$ term cannot be neglected, and we have in fact calculated the steady-state solution to the *wrong* differential equation:

$$-\nabla^2 v + \frac{\tau^2}{4} v_{xxyyzz} = g \tag{13.20}$$

The highest Fourier coefficients will be contaminated by the extra term even if $\tau \sim O(h^2)$.

To show this, denote the Fourier coefficients of $\exp(ikx + imy + inz)$ for g and v by g_{kmn} and v_{kmn}. Direct substitution into (13.20) implies

$$v_{kmn} = \frac{g_{kmn}}{k^2 + m^2 + n^2 - (\tau^2/4)(k^2\, m^2\, n^2)} \tag{13.21}$$

For waves near the aliasing limit, and assuming an equal grid spacing h in all directions, $k_{max} = m_{max} = n_{max} = \pi/h$ and

$$v_{kmn} = \frac{h^2}{\pi^2} \frac{g_{kmn}}{3 - (\tau^2/h^4)(\pi^4/4)} \qquad k = k_{max}, m = m_{max}, n = n_{max} \tag{13.22}$$

which confirms that the wavenumbers near the aliasing limit are seriously corrupted even for $\tau = h^2$, a timestep sufficient for the stability of explicit time-marching.

Fortunately, it is fairly straightforward to modify splitting to obtain a similar fractional step method that *is* absolutely consistent (Yanenko, 1971). Note, however, that the offending term is the product of *three* operators; the splitting-up scheme is both absolutely stable and absolutely consistent in *two* dimensions.

The spectral and pseudospectral schemes we will discuss later will always be consistent. Nonetheless, it is a complication: every fractional step algorithm must be checked for *both* stability and consistency.

13.5 Operator Theory of Time-Stepping

Now that we have seen splitting-up in a specific example, it is time to look at an abstract formalism for general integration methods. For simplicity, consider the system of ordinary differential equations in time given by

$$\vec{u}_t = \vec{\Lambda}\,\vec{u} \tag{13.23}$$

where $\vec{u}$ is a column vector and $\vec{\Lambda}$ is a square matrix whose elements are independent of time. As noted in the previous chapter, any method of discretizing a parabolic or hyperbolic PDE that is first order in time will give a system like (13.23) so long as the equation is *linear* and the coefficients are *independent of time*.

The system (13.23) has the exact solution (assuming that $\vec{\Lambda}$ has a full set of eigenvectors)

$$\vec{u} \equiv \exp\left(\vec{\Lambda}\,t\right)\vec{u}_0 \tag{13.24}$$

where $\vec{u}_0$ is the column vector containing the initial conditions. We can interpret the matrix exponential (and other functions-of-a-matrix) by expanding in a series of the eigenvectors of $\vec{\Lambda}$:

$$\vec{\Lambda}\,\vec{e}_j = \lambda_j\,\vec{e}_j \tag{13.25}$$

(Note that we omit the minus sign included earlier in the chapter.)

$$\vec{u} = \sum_{j=1}^{N} u_j\,\vec{e}_j \qquad ; \qquad \vec{u}_0 \equiv \sum_{j=1}^{N} u_{0,j}\,\vec{e}_j \tag{13.26}$$

Then

$$\exp\left(\vec{\Lambda}\,t\right)\vec{u}_0 = \sum_{j=1}^{N}\exp\left(\lambda_j t\right)u_{0,j}\,\vec{e}_j \tag{13.27}$$

From this viewpoint, the problem of devising good time-stepping schemes is equivalent to finding good ways of approximating the exponential of a matrix on the *small* interval

$$t \in [0, \tau] \tag{13.28}$$

The Euler method is the one-term Taylor series approximation:

$$\exp\left(\vec{\Lambda}\tau\right) \approx 1 + \tau\vec{\Lambda} + O\left(\tau^2\frac{\vec{\Lambda}^2}{2}\right) \qquad \text{[Euler]} \tag{13.29}$$

We could build higher approximations simply by taking more terms in the Taylor series. However, if we can put upper and lower bounds on the eigenvalues of the matrix $\vec{\Lambda}$, the truncated *Chebyshev* series of the exponential is better. This has rarely been done when the transient is interesting because one can usually obtain better time-stepping accuracy for the same amount of work simply by applying (13.29) on a smaller interval. However, when the prime goal is the asymptotic solution, the Chebyshev-of-the-exponential has been widely used under the name of the Chebyshev iteration (Chap. 15).

The Crank-Nicholson (and higher order implicit methods) are examples of a third strategy: so-called "Padé approximants". These are *rational* functions which are chosen so that their power series expansion matches that of the exponential function through however many terms are needed to determine all the coefficients in the Padé approximant. For example, the [1, 1] approximant [linear polynomial divided by a linear polynomial] gives the Crank-Nicholson scheme:

$$\exp\left(\vec{\Lambda}\tau\right) \approx \left\{\vec{I} - (\tau/2)\vec{\Lambda}\right\}^{-1}\left\{\vec{I} + (\tau/2)\vec{\Lambda}\right\} \qquad \text{[Crank-Nicholson]} \tag{13.30}$$

where $\vec{I}$ is the identity matrix. Taylor expanding gives

$$\left\{\vec{I} - (\tau/2)\vec{\Lambda}\right\}^{-1}\left\{\vec{I} + (\tau/2)\vec{\Lambda}\right\} \approx \vec{I} + \tau\vec{\Lambda} + \frac{\tau^2}{2}\vec{\Lambda}^2 + O(\tau^3\vec{\Lambda}^3) \tag{13.31}$$

which shows that the expansion in τ agrees with that of the exponential through the first three terms, thus demonstrating both that the Crank-Nicholson method is indeed the "[1, 1] Padé approximant" of the exponential and also that it is more accurate than Euler's scheme.

We can obviously generalize both the Euler and Crank-Nicholson methods by taking more terms. For example, the "fourth order" Crank-Nicholson scheme is

$$\exp\left(\vec{\Lambda}\tau\right) \approx \left\{\vec{I} - \frac{\tau}{2}\vec{\Lambda} + \frac{\tau^2}{12}\vec{\Lambda}^2\right\}^{-1}\left\{\vec{I} + \frac{\tau}{2}\vec{\Lambda} + \frac{\tau^2}{12}\vec{\Lambda}^2\right\} + O(\tau^5\vec{\Lambda}^5) \tag{13.32}$$

(Calahan, 1967, and Blue and Gummel, 1970).

The obvious advantage of Padé approximants is that they remain bounded as $\tau \to \infty$; Eq. (13.32) generates a time-stepping algorithm which, like the Crank-Nicholson scheme, is always stable as long as the eigenvalues of $\vec{\Lambda}$ are all negative. There is a second, less obvious virtue: Padé approximants are almost always more accurate — sometimes very much more so — than the Taylor series from which they are generated (Baker, 1975, Bender and Orszag, 1978).

In more than one dimension, splitting enormously reduces costs since the denominator in the rational approximations requires inverting a matrix which approximates a two- or three-dimensional differential operator. Let us write

$$\vec{\Lambda} = \vec{\Lambda}_1 + \vec{\Lambda}_2 \tag{13.33}$$

where $\vec{\Lambda}_1$ and $\vec{\Lambda}_2$ are the matrix representations of one-dimensional operators that are easily inverted.

The Alternating-Direction Implicit scheme (ADI) is the fractional steps algorithm

$$\frac{\vec{u}^{n+1/2} - \vec{u}^n}{\tau} = \frac{1}{2}\left[\vec{\Lambda}_1 \vec{u}^{n+1/2} + \vec{\Lambda}_2 \vec{u}^n\right] \tag{13.34}$$

$$\frac{\vec{u}^{n+1} - \vec{u}^{n+1/2}}{\tau} = \frac{1}{2}\left[\vec{\Lambda}_1 \vec{u}^{n+1/2} + \vec{\Lambda}_2 \vec{u}^{n+1}\right] \tag{13.35}$$

The reason for the name "Alternating Direction" is that we treat one direction implicitly on the first half-step while treating the y-derivatives explicitly and then interchange the roles of the two coordinates on the second half-step. If we introduce the auxiliary operators

$$\vec{B}_i(\tau) \equiv \vec{I} + \frac{\tau}{2}\vec{\Lambda}_i \qquad\qquad i = 1, 2 \tag{13.36}$$

where I is the identity matrix, then we can write the ADI scheme in whole steps as

$$\vec{u}^{n+1} = \vec{B}_2(-\tau)^{-1}\,\vec{B}_1(\tau)\,\vec{B}_1(-\tau)^{-1}\vec{B}_2(\tau)\,\vec{u}^n \tag{13.37}$$

Although this approximation to the exponential is the product of four factors, what matters is that it still agrees with the Taylor expansion of the exponential. By using the well-known series $1/(1+x) = 1 - x + x^2 + \ldots$, which is legitimate even when x is a matrix, we find that (13.37) is accurate to $O(\tau^2)$ just like the Crank-Nicholson scheme.

If the operators $\vec{\Lambda}_1$ and $\vec{\Lambda}_2$ commute, then we can swap $\vec{B}_1(\tau)$ and $\vec{B}_2(\tau)$ to obtain the same scheme in a different form:

$$\vec{u}^{n+1} = \vec{B}_2(-\tau)^{-1}\,\vec{B}_2(\tau)\vec{B}_1(-\tau)^{-1}\,\vec{B}_1(\tau)\,\vec{u}^n \qquad \text{[Splitting]} \tag{13.38}$$

This is the "splitting-up" form: $\vec{u}^{n+1/2} = \vec{B}_1(-\tau)^{-1}\vec{B}_1(\tau)\,\vec{u}^n$ and we have

$$\frac{\vec{u}^{n+1/2} - \vec{u}^n}{\tau} = \frac{1}{2}\vec{\Lambda}_1\left(\vec{u}^{n+1/2} + \vec{u}^n\right) \tag{13.39}$$

$$\text{[Splitting]}$$

$$\frac{\vec{u}^{n+1} - \vec{u}^{n+1/2}}{\tau} = \frac{1}{2}\vec{\Lambda}_2\left(\vec{u}^{n+1} + \vec{u}^{n+1/2}\right) \tag{13.40}$$

When the operators $\vec{\Lambda}_1$ and $\vec{\Lambda}_2$ do not commute, the splitting and ADI schemes are *not* identical and splitting is only first-order accurate in time.

13.6 Higher Order Splitting When the Split Operators Do not Commute: Flip-Flop and Richard Extrapolation

There are two ways to recover second order accuracy when the operators do not commute. The first is to reverse the sequence of fractional steps on adjacent whole steps ("flip-flop"). In other words, we solve the BVP in x first and then that in y whenever n is even and then solve the boundary value problem in y first on every odd time step. The method is then second order even with noncommuting Λ_1 and Λ_2, but with an effective time step of 2τ (rather than τ).

The other alternative is Richardson extrapolation (not be confused with Richardson's iteration of Chapter 15). This is a very general strategy for improving the accuracy of low order methods. The basic idea is that if the error in a given algorithm is $O(h^r)$ for some

r and sufficiently small h, then two computations with *different* h can be combined with appropriate weights so that the leading error terms cancel to give an answer of higher order.

Let $u_{low}(2\tau)$ and $u_{high}(2\tau)$ denote the results of integrating from $t = 0$ to $t = 2\tau$ with a single time step of $\Delta t = 2\tau$ and with two steps of length $\Delta t = \tau$, respectively. For a first order method, the errors are

$$u_{low}(2\tau) = u_{exact} + 4p\tau^2 + O(\tau^3) \tag{13.41}$$

$$u_{high}(2\tau) = u_{exact} + 2p\tau^2 + O(\tau^3) \tag{13.42}$$

The numerical value of p is unknown, but the existence of an error which is quadratic in τ is simply a definition of what it means for a scheme to "first order in time".[1] It follows that

$$u_{Richardson}(2\tau) \equiv 2u_{high}(2\tau) - u_{low}(2\tau) = u_{exact} + O(\tau^3) \tag{13.43}$$

$u_{Richardson}$ is second order because the weighted sum cancels the leading error terms in the first order approximations, u_{high} and u_{low}. This extrapolation can be performed on every sub-interval of length 2τ to refine the solution to second order accuracy as we march in time.

Consequently, when the operators of different sub-problems do not commute, we can always retrieve second order accuracy at the expense of additional programming. Often, Richardson extrapolation is omitted because hydrodynamic calculations are usually limited in accuracy by *spatial* rather than temporal resolution. At the other extreme, Strain (1995) has used repeated Richardson extrapolation of the modified Backwards Euler method to create an implicit time-marching code in which the order is easily and adaptively varied.

13.7   Splitting and Fluid Mechanics

Our example is the shallow water wave equations:

$$\frac{Du}{Dt} - fv + \phi_x = \mu \triangle u \tag{13.44}$$

$$\frac{Dv}{Dt} + fu + \phi_y = \mu \triangle v \tag{13.45}$$

$$\epsilon \frac{D\phi}{Dt} + u_x + v_y = 0 \tag{13.46}$$

where f is the Coriolis parameter and D/Dt denotes the total derivative:

$$\frac{D}{Dt} \equiv \frac{\partial}{\partial t} + u\frac{\partial}{\partial x} + v\frac{\partial}{\partial y} \tag{13.47}$$

ϵ is a nondimensional combination of depth, planetary radius, rotation rate and the gravitational constant, "Lamb's parameter".

The physical splitting is (Fig. 13.4)

$$\frac{Du^{n+1/3}}{dt} = 0 \tag{13.48a}$$

$$\frac{Dv^{n+1/3}}{dt} = 0 \qquad \text{[Advective Step]} \tag{13.48b}$$

$$\frac{D\phi^{n+1/3}}{dt} = 0 \tag{13.48c}$$

---

[1]Over many time steps, the errors accumulate so that the total error on a fixed interval in t is a linear function of τ even though the error on a single step is quadratic in τ.

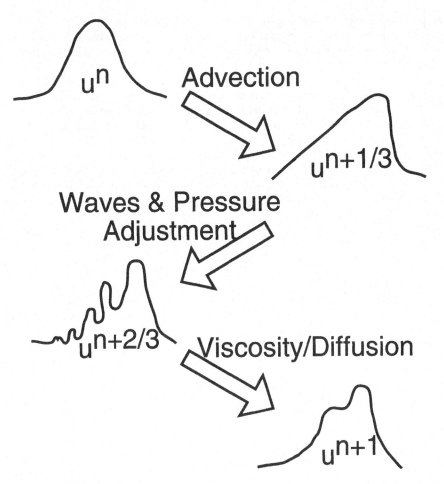

Figure 13.4: Splitting by process: Advection, adjustment of fields (linear wave dynamics and pressure), and viscosity and diffusion are split into three separate, one-after-the-other processes.

$$u_t^{n+2/3} - f\,v^{n+2/3} = -\phi_x^{n+2/3} \qquad\qquad\qquad\qquad (13.49a)$$

$$v_t^{n+2/3} + f\,u^{n+2/3} = -\phi_y^{n+2/3} \qquad \text{[Pressure Step alias} \qquad (13.49b)$$

$$\epsilon\,\phi_t^{n+2/3} + u_x^{n+2/3} + v_y^{n+2/3} = 0 \qquad \text{“Adjustment of Fields”]} \qquad (13.49c)$$

$$u_t^{n+1} = \mu\,\triangle u^{n+1} \qquad\qquad\qquad\qquad\qquad\qquad (13.50a)$$

$$v_t^{n+1} = \mu\,\triangle v^{n+1} \qquad \text{[Diffusion Step]} \qquad\qquad (13.50b)$$

$$\phi^{n+1} = \phi^{n+2/3} \qquad\qquad\qquad\qquad\qquad\qquad (13.50c)$$

In each case, the solution at the previous fractional step provides the initial condition for the equations that model the next physical process.

If we wish, we can apply an additional splitting to decompose the diffusion step into a sequence of one-dimensional diffusion problems. However, it is usually a bad idea to attempt dimensional splitting for the "adjustment of fields" step because the Coriolis force and wave dynamics strongly couple the coordinates. Splitting advection into separate x, y and z advection is unnecessary because this subproblem is normally integrated by an explicit method.

As noted earlier, the great advantage of splitting is that we can solve each fractional step using a different numerical method. The advective step is almost always advanced explicitly and the adjustment of fields is integrated implicitly. In engineering problems, diffusion is usually marched in time by an implicit algorithm, but Marchuk (1974) barely discusses diffusion because μ in his numerical weather prediction model is so small that he can use an explicit method.

It is not even necessary to be consistent by using spectral or pseudospectral algorithms for all sub-problems. Lambiotte et $al.$ (1982) compute only the advection and pressure steps pseudospectrally. Since μ is small for their high Reynolds flow, they employ $finite$ $differences$ to integrate the diffusive step.

Chapter 14

Semi-Lagrangian Advection & Other Integrating Factor Methods

"Go with the flow"

– New Age proverb

"... there is a mistaken belief that semi-Lagrangian schemes are only good for smooth flows."

– Staniforth & Côté(1991), pg. 2211

14.1 Introduction: Concept of an Integrating Factor

An "integrating factor" is a transformation of the unknowns in a differential equation. The most familiar example yields the solution of the general linear first order ODE, which is

$$u_x - q(x)\, u = f(x) \tag{14.1}$$

Let $u(x)$ be replaced by the new unknown $v(x)$ where

$$u(x) \equiv I(x)\, v(x) \tag{14.2}$$

where the integrating factor is

$$I(x) \equiv \exp\left(\int^x q(y)\, dy \right) \tag{14.3}$$

Substitution gives the transformed problem

$$I\, v_x + q(x)\, I\, v - q(x)\, I\, v = f(x) \tag{14.4}$$

using $I_x = q(x)I$. Dividing by $I(x)$ gives

$$v_x = \exp\left(- \int^x q(y)\, dy \right) f(x) \tag{14.5}$$

Thus, the integrating factor has reduced the differential equation to a quadrature, i. e., $v(x)$ is the indefinite integral of the right-hand side of (14.5):

$$v = A + \int_0^x \exp\left(-\int^z q(y)\, dy\right) f(z)\, dz \tag{14.6}$$

where $A = u(0)$.

Integrating factors come in all sorts of flavors, like ice cream, but we shall concentrate on two particular classes. The first is when a time-dependent PDE can be written in the form

$$u_t = L\, u + M(u) \tag{14.7}$$

where L is a *linear* differential operator with spatial derivatives only and where $M(u)$ is an operator, possibly nonlinear, that contains everything else in the equation including the forcing terms. The new unknown defined by

$$u \equiv \exp(Lt)\, v \tag{14.8}$$

solves the simplified equation

$$v_t = \exp(-Lt)\, M(v) \tag{14.9}$$

Just as for the first order ODE, the integrating factor has purged the transformed equation of the differential operator L except through the exponential. This is sometimes a good idea — and sometimes not.

The second class of integrating factors is to remove translation/advection by a change-of-coordinates to a moving coordinate system. Formally, this change-of-coordinate can be expressed in terms of operators, too. However, it is more convenient to perform the change-of-coordinate directly.

The simplest case is to shift into a frame of reference which is moving at a constant speed with respect to the original frame. For example,

$$u_t + c u_x = f(x,t), \qquad u(0,t) = g(t), \qquad u(x,0) = Q(x) \tag{14.10}$$

can be transformed by the shift

$$\xi \equiv x - c t \quad \leftrightarrow \quad x = \xi + c t \quad \& \quad u(x,t) \equiv v(\xi,t) \tag{14.11}$$

into

$$v_t = f(\xi + ct, t), \qquad v(-ct,t) = g(t), \qquad v(\xi,0) = Q(\xi) \tag{14.12}$$

For the special case $f = 0$, this is clearly an improvement on the original solution because then $v_t = 0$ which implies that v is a function of ξ only, that is, u depends on (x,t) only as the combination $x - ct$.

Semi-Lagrangian time-marching schemes remove nonlinear advection through a similar but more complicated transformation. Such algorithms have so many good properties that they have become standard on two important geophysical models: the NCAR Community Climate Model (CCM) and the operational forecasting code of the European Centre for Medium Range Forecasting, which provides five-day global predications to an 18-nation consortium.

However, integrating factor schemes can also be a trap. The crucial distinction is that integrating factors are useful only when the integrating factor removes an important part of the dynamics. We will explain this through a couple of bad examples in the next section. (This section can be skipped without loss of continuity.)

In the remainder of the chapter, we describe semi-Lagrangian schemes. We begin with second order schemes, which are the easiest to describe, and then explain a number of elaborations.

14.2 Misuse of Integrating Factor Methods

Our first counterexample is the ODE which was used earlier to explain the concept of the slow manifold:

$$u_t + \lambda u = f(\epsilon t) \tag{14.13}$$

Assume $\lambda/\epsilon \gg 1$, i. e., that the ratio of the "fast" time scale to the slow time scale is large, and also that the real part of λ is positive so that the homogeneous solution decays exponentially with time. We shall further assume that the goal is to compute the slow manifold of this equation, that is, the part which varies only on the same slow $O(1/\epsilon)$ time scale as the forcing f.

Unfortunately, the time step for an explicit scheme is restricted by the fast time scale of the homogeneous solution $\exp(-\lambda t)$. Implicit methods and Nonlinear Galerkin methods both remove this restriction, but have their own disadvantages. Is an integrating factor a viable alternative?

Define a new unknown by

$$u = \exp(-\lambda t)v(t) \tag{14.14}$$

The ODE becomes

$$v_t = \exp(\lambda t)f(\epsilon t) \tag{14.15}$$

Hurrah! The term λu has been removed and with it the usual CFL time-step limitation. To show this, let us analyze the Forward Euler method as applied to (14.15).

This scheme with time step τ and with superscript n denoting the time level is

$$\frac{v^{n+1} - v^n}{\tau} = \exp(\lambda n \tau)f(\epsilon n \tau) \quad \leftrightarrow \quad v^{n+1} = v^n + \tau \exp(\lambda n \tau)f(\epsilon n \tau) \tag{14.16}$$

From an arbitrary starting value v^0 and for the special case $f(\epsilon t) = F$, a constant,

$$\begin{aligned} v^n &= v^0 + \tau F \{1 + \exp(\lambda \tau) + \exp(2 \lambda \tau) + \ldots + \exp([n-1] \lambda \tau)\} \\ &= v^0 + \tau F \frac{1 - \exp(n \lambda \tau)}{1 - \exp(\lambda \tau)} \end{aligned} \tag{14.17}$$

Because of the integrating factor transformation, v^n is stable for arbitrarily large time steps τ. To be sure, the solution is blowing up exponentially, but since the transformed forcing is growing exponentially, a similar growth in v is okay.

The same solution in terms of the original unknown u is, using $v^0 \equiv u^0$,

$$u^n = \exp(-n \lambda \tau)u^0 + \tau F \frac{\exp(-n \lambda \tau) - 1}{1 - \exp(\lambda \tau)} \tag{14.18}$$

In the limit that $\tau \rightarrow 0$ and $n \rightarrow \infty$ such that $t = n \tau$ grows large, the Forward Euler/integrating factor solution simplifies to

$$\begin{aligned} u^n &\approx \tau F \frac{-1}{1 - (1 + \lambda \tau + O(\lambda^2 \tau^2))} \\ &\approx \frac{F}{\lambda} \{1 + O(\lambda \tau)\} \end{aligned} \tag{14.19}$$

where F/λ is the slow manifold for this problem. Thus, for sufficiently small time step, the scheme does successfully recover the slow part of the solution as $t \to \infty$.

However, the whole point of the integrating factor is to use a *long* time step without the bother of implicit or Nonlinear Galerkin methods. Since the stability limit of the Forward Euler scheme without integrating factor is $\tau_{max} = 1/\lambda$, it is interesting to see how well the integrating factor variant does for this τ. Unfortunately, substituting $\tau = \tau_{max}$ into (14.18) gives

$$u^n \sim \frac{F}{\lambda} \frac{1}{-1 + \exp(1)} = 0.58 \frac{F}{\lambda}, \qquad t \to \infty, \ \tau = 1/\lambda \tag{14.20}$$

The good news is that the large time solution is proportional to the true slow manifold, F/λ; the bad news is that the amplitude of the computational solution is barely half that of the true solution even when τ is no larger than the largest time step that would be stable in the absence of the integrating factor. When $\tau >> 1/\lambda$, which is of course what we really want if the integrating factor scheme is to compete with implicit schemes,

$$u^n \sim \frac{F}{\lambda} \lambda \tau \exp(-\lambda \tau), \qquad t \to \infty, \ \lambda \tau >> 1 \tag{14.21}$$

which is exponentially small compared to the correct answer.

Thus, the integrating factor method is a dismal failure. If the time step $\tau = r \, \tau_{max}$ where $\tau_{max} = 1/\lambda$ is the stability limit for the Forward Euler method (without integrating factor), then the Forward Euler/integrating factor algorithm gives a slow manifold which is too small by a factor of $r \exp(-r)$. In contrast, implicit methods like Crank-Nicholson and Backwards-Euler and also Nonlinear Galerkin schemes are accurate for the slow part of the solution even for large r.

A second bad example is the Fourier pseudospectral solution to the KdV equation

$$u_t + u u_x + u_{xxx} = 0 \tag{14.22}$$

when the goal is to model the formation and dynamics of solitary waves. For a soliton of phase speed c, the coefficient of $\exp(ikx)$ varies with time as $\exp(-ikct)$ which grows only linearly with k. However, the equation also admits small amplitude waves whose phase speeds grows as k^3, so the stability limit for an explicit time-marching scheme is inversely proportional to k_{max}^3. The KdV equation is very "stiff" in the sense that one must use a very short time step for stability reasons compared to the slow time scale of the dynamics (i. e., the solitons) that one is actually interested in.

The remedy — well, sort of a remedy — is to use the integrating factor $\exp(-\partial_{xxx} t)$ to remove the third derivative from the transformed equation:

$$v_t + v \, \exp(-\partial_{xxx} t) \, v_x = 0 \tag{14.23}$$

The stability limit of the modified equation is then controlled by nonlinear term, which involves only a first derivative and demands a time step inversely proportional to k_{max} itself rather than to its cube.

The exponential-of-derivatives operator looks forbidding, but it can be interpreted in the usual way as the exponential of the matrix that discretizes the third spatial derivative, which in turn can be interpreted as multiplication of the k-th eigenvector of the derivative matrix by the exponential of $(-t)$ times the k-th eigenvalue. In a Fourier basis, this is trivial because the basis functions, $\exp(ikx)$, are the eigenvalues of the derivative operator. It follows that for any function $v(x, t)$ with Fourier coefficients $v_k(t)$,

$$\exp(-\partial_{xxx} t) \, v(x, t) = \sum_{k=-\infty}^{\infty} \exp(i \, k^3 t) \, v_k \, \exp(ikx) \tag{14.24}$$

To put it another way, the integrating factor method for the KdV equation is equivalent to transforming the unknowns of the ODE system which comes from the usual Galerkin or pseudospectral discretization. In place of the Fourier coefficients $a_k(t)$ of $u(x,t)$, use the modified Fourier coefficients (of $v(x,t)$) given by

$$v_k(t) \equiv \exp(-ik^3t)\, a_k(t) \tag{14.25}$$

Unfortunately, this is really a terrible idea when the prime target is solitary waves. The simplest counterexample is the special case that the exact solution is a solitary wave of phase speed c. The exact Fourier coefficients — before the integrating factor — are

$$a_k = \text{constant } \exp(-ikct) \tag{14.26}$$

and thus vary rather slowly in the sense that the frequency of the highest wavenumber in the truncation grows only linearly with k_{max}. In contrast, the time dependence of the transformed coefficients is

$$v_k(t) = \text{constant } \exp(-ik^3t - ikct) \tag{14.27}$$

which varies *rapidly* in the sense that the highest frequency after transformation grows as k_{max}^3. To accurately compute $v_{kmax}(t)$, a time step proportional to k_{max}^3 is necessary — in other words, a time step as short as the explicit stability limit for the original, untransformed KdV equation.

In discussing implicit methods, we have stressed repeatedly that their usefulness is SITUATION-DEPENDENT. Implicit methods are good for computing KdV solitons because they misrepresent only the high frequency transient while accurately computing the slow-moving solitary waves. Integrating factor methods, in contrast, are bad for solitons because the integrating factor extracts a high frequency motion that is not actually present in the low frequency motion (i. e., the soliton) that we are computing. Thus, the usefulness of integrating factor methods is also situation-dependent, but often inversely to implicit methods.

Suppose our goal is to compute KdV solutions which are not solitons but rather a soup of weakly interacting waves such that the Fourier components are approximately

$$a_k = A_k(\epsilon t)\, \exp(ik^3t) \tag{14.28}$$

where the A_k are amplitudes that vary slowly under weak wave-wave interactions. Now because the integrating factor method successfully removes the high frequency oscillations, the transformed coefficients are

$$v_k(t) = A_k(\epsilon t) \tag{14.29}$$

These vary slowly with time, allowing a long time step. In contrast, implicit and Nonlinear Galerkin methods as well as standard explicit schemes all require a very short time step to be accurate for the high frequency motion. Weakly coupled wave systems ("resonant triads" and their generalization) have been widely analyzed by theorists (Craik, 1985).

Thus, it is quite wrong to conclude that integrating factor methods are counter-productive in numerical applications. Sometimes they are very valuable. The difference between the two KdV cases is that in the first (soliton) case, the integrating factor extracts a high frequency dependence that was not actually present in the slow solutions (bad!) whereas in the second (weakly interacting sine waves) case, the integrating factor extracts high frequency motion that is an essential part of the desired solution (good!).

14.3   Semi-Lagrangian Advective Schemes: Introduction

A "semi-Lagrangian" (SL) time-marching algorithm is an integrating-factor method in which the integrating factor is an advection operator that shifts into a coordinate system that travels with the fluid. In so doing, it borrows ideas from both the method of characteristics and the Lagrangian coordinate formulation of fluid mechanics. The "total" or "particle-following" derivative of any quantity "q", for simplicity given in two space dimensions, is

$$\frac{Dq}{Dt} \equiv \frac{\partial q}{\partial t} + u\,\frac{\partial q}{\partial x} + v\,\frac{\partial q}{\partial y} \tag{14.30}$$

In the moving coordinate system, the total derivative becomes the ordinary partial time derivative, and the advective terms disappear, buried in the change-of-coordinate.

The idea of tracing flow by following the motion of individual blobs or particles of fluid is at least two centuries old. Particle-following coordinates are called "Lagrangian" coordinates in fluid mechanics. However, exclusive use of a Lagrangian coordinate system is usually a disaster-wrapped-in-a-catastrophe for numerical schemes because the particle trajectories become chaotic and wildly mixed in a short period of time where "chaos" is meant literally in the sense of dynamical systems theory even for many laminar, non-turbulent flows.[1] Semi-Lagrangian (SL) algorithms avoid this problem (and earn the modifier "semi") by reinitializing the Lagrangian coordinate system after each time step. The usual choice is that the tracked particles are those which *arrive* at the points of a regular Fourier or Chebyshev spatial grid at time t^{n+1}. Since the departure points for these trajectories at $t = t^n$ are generally not on the regular grid, we follow a different set of N particles on each time interval of length τ where τ is the time step. In contrast, in a pure Lagrangian coordinate system, one follows the evolution of the same N particles forever.

There are two advantages to an SL method:

---

[1]Trajectories of fluid blobs are often chaotic even when the advecting velocity field is laminar.

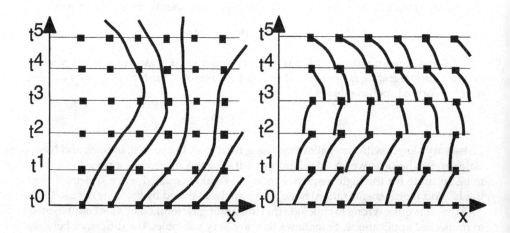

Figure 14.1: Left panel: Fluid mechanics in Lagrangian coordinates tracks the same set of particles throughout all time. Right panel: Semi-Lagrangian methods. A new set of particles to track is chosen at the beginning of each time step.

1. If the scheme is "multiply upstream", a term defined later, SL will be both stable and accurate for time steps far greater than usual explicit limit. To be precise, the advective time step limit disappears, and the maximum time step is then controlled entirely by non-advective processes like viscosity and wave propagation.

2. SL methods are very good at resolving fronts and shock waves even in the absence of artificial viscosity.

Because of these virtues, SL time-marching is now employed in the NCAR Community Climate Model and the European Centre for Medium-Range Forecasting weather prediction model. Neither is a toy model, as too often used to test algorithms; instead, both are global, three-dimensional spherical harmonic models with lots of physics.

Simple semi-Lagrangian algorithms have at least five disadvantages:

1. Difficult to parallelize (for "multiply upstream" schemes).

2. Not strictly conservative of mass, energy, etc.

3. Inherent computational dissipation (McCalpin, 1988)

4. Instability and inaccuracy for flows over mountains.

5. More expensive per time step than most competing schemes because of the need for

   - "off-grid" interpolation and
   - fixed point iteration.

None of these drawbacks, however, has proved fatal. In particular, conservative, shape-preserving and low dissipation schemes have been recently developed, and the mountain flow or "orographic" problems have been solved, too.

Semi-Lagrangian method are designed to be especially good at advection. However, the algorithms can be applied to the full Navier-Stokes equations without difficulty and can be combined with other ideas. For example, in meteorology it is common to combine semi-Lagrangian treatment of advection with a semi-implicit treatment of the viscous and Coriolis force terms ("SL/SI" schemes).

Semi-Lagrangian time-stepping can be combined with any strategy for discretization of the spatial coordinates.

14.4 A Brief Digression on the One-Dimensional and Two-Dimensional Advection Equations and the Method of Characteristics

The Navier-Stokes equations reduce to the so-called "advection equations" if we neglect all processes except advection. Restricting ourselves to two-dimensions for simplicity, the pure advection equations are

$$\frac{Du}{Dt} = 0 \quad \& \quad \frac{Dv}{Dt} = 0 \tag{14.31}$$

or written in full, where (u, v) are the x and y velocities,

$$\frac{\partial u}{\partial t} + u \frac{\partial u}{\partial x} + v \frac{\partial u}{\partial y} = 0$$

$$\frac{\partial v}{\partial t} + u \frac{\partial v}{\partial x} + v \frac{\partial v}{\partial y} = 0 \tag{14.32}$$

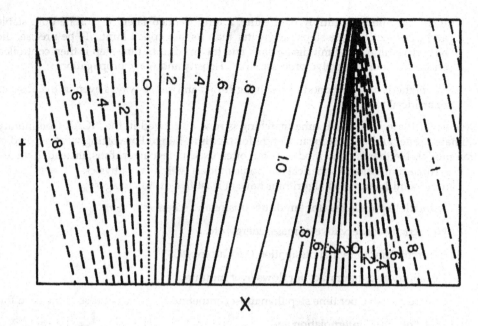

Figure 14.2: Contour plot in the space-time plane of the solution $u(x,t)$ to the one-dimensional advection equation, $u_t + uu_x = 0$. The initial condition is sinusoidal; negative contours are dashed. Where contours intersect, the solution becomes multi-valued and $u_x \to \infty$. The flow is said to have "formed a front" or experienced "wave breaking".

The reason that this drastically simplified system is interesting is that it can be solved exactly by the "method of characteristics". Semi-Lagrangian time-marching schemes, even with all the complications of pressure and viscosity, etc., merely mimic the method of characteristics.

The key observation is that the advection equations state that the velocities are conserved following the motion. In other words, if we track the fate of a particular blob of fluid, perhaps marked with a dab of red dye, the velocity of that blob will remain constant forever. Consequently, the trajectory of that blob and of all blobs is a straight line whose slope in each spatial dimension is proportional to its velocity with respect to that coordinate. (Fig. 14.2 is a one-dimensional illustration.) If the blob originates at $x = \xi$ and $y = \eta$ at $t = 0$, then its trajectory is

$$x = \xi + Q(\xi,\eta)\,t, \qquad \& \qquad y = \eta + S(\xi,\eta)\,t \tag{14.33}$$

where the initial conditions are

$$u(x,y,t=0) = Q(x,y), \qquad \& \qquad v(x,y,t=0) = S(x,y) \tag{14.34}$$

The exact solution for the velocities is

$$u(x,y,t) = Q(\xi[x,y,t],\eta[x,y,t]), \qquad \& \qquad v(x,y,t) = S(\xi[x,y,t],\eta[x,y,t]) \tag{14.35}$$

In other words, the initial conditions furnish the exact solution for all time if the arguments of the functions (Q,S) are replaced by the "characteristic coordinates" or simply "characteristics", (ξ,η).

Unfortunately, Eq.(14.33) gives explicit solutions only for (x, y) as a functions of the characteristics. To evaluate the solutions at a particular point (x, y, t), we need it the other way around. Thus, it is necessary to *solve* (14.33) to determine the departure point of a given blob where the "departure point" or "foot of the trajectory" is the location (ξ, η) of the blob at $t = 0$. However, the equations that must be solved for the characteristics are *algebraic* equations, which are much simpler than differential equations.

Semi-Lagrangian time-marching schemes also require computing the departure point of a blob by solving a system of algebraic equations. However, this is fairly easy because the blob moves only a short distance during a time interval of length τ where τ is the time step. The change of coordinates (14.33) does not trivialize the dynamics of the full Navier-Stokes equations, as it does for the pure advection equations, but it does remove advection just as thoroughly.

14.5 The Simplest SL Scheme: Three-Level, Second Order Semi-Implicit Method

A semi-Lagrangian scheme by itself is not very useful in meteorology because it only removes the advective time step limit, which is much less severe than the restriction imposed by the propagation of gravity waves. When combined with a semi-implicit treatment of the non-advective terms, however, SL becomes very powerful as first demonstrated by Robert (1981, 1982).

Let the problem be

$$\frac{Du}{Dt} + G(u, t) = F(u, t) \tag{14.36}$$

where D/Dt denotes the total, particle-following derivative including the advective terms, G is the sum of all terms which are to be treated *implicitly* while F represents all the terms which can be treated explicitly. The basic three-level, second order SL/SI algorithm is

$$\frac{u^+ - u^-}{2\,\tau} + \frac{1}{2}\,(G^+ + G^-) = F^0 \tag{14.37}$$

where the plus and minus signs denote the quantities evaluated at $t = t^{n\pm 1}$ and superscript 0 denotes a quantity evaluated at $t = t^n$. In form, this looks like a standard Crank-Nicholson/explicit method except that (i) the advective terms are missing (ii) the CN-like step is over a time interval of 2τ instead of τ where τ is the time step.

The magic is in the coordinate transform. For simplicity, assume one space dimension. Let $\alpha(x)$ denote the displacement, over the time interval $t \in [t^n, t^{n+1}]$ of a particle that arrives at point x at $t = t^{n+1}$. Then for any vector appearing in (14.37),

$$\begin{aligned}
u^+ &\equiv u(x, t^n + \tau) \\
u^0 &\equiv u(x - \alpha, t^n) \\
u^- &\equiv u(x - 2\,\alpha, t^n - \tau)
\end{aligned} \tag{14.38}$$

When the spatial dependence is discretized, u, x, and $\alpha(x)$ all turn into vectors. The displacement is defined to be the solution of the algebraic equation

$$\alpha_j = \tau\, u(x_j - \alpha_j, t^n) \tag{14.39}$$

where j is the integer index of the grid points.

The novelty of Eq.(14.39) is that when we move back one time level, we must simultaneously shift our spatial position by the displacement α_j. The predicted value $u(x_j, t^{n+1})$ is combined not with u at the same spatial location at the two previous time levels, but rather with u for that same fluid particle at earlier time levels as shown schematically in Fig. 14.3. Advection is invisible in (14.37) because it is implicit in the coordinate transform, that is, in the spatial shifts represented by α in (14.39).

This advection-hiding has two major advantages. First, since there is no advection in the time extrapolation formula (14.37), there is no advective stability limit either. (The explicitly treated term $F(u, t)$ may impose its own maximum time step, but in meteorological applications, this is huge compared to the advective limit.) Second, the SL scheme is much better at resolving steep fronts and shock waves than conventional schemes. The reason is that the frontogenetic process, advection, is computed *exactly* by the method of characteristics. In contrast, the treatment of an advective term like $u\, u_x$ in a conventional, non-SL method depends on how accurately u_x is approximated by spatial finite differences or a pseudospectral formula — poorly in the neighborhood of a front.

SL schemes, alas, require two additional steps which are unnecessary in other algorithms. First, one must evaluate u and other variables at $x = x_j - \alpha_j$, which generally is not on the canonical spectral grid. This problem of "off-grid" interpolation is sufficiently important to warrant an entire section below. However, efficient spectral off-grid interpolation methods are now available.

Second, one must solve the algebraic equation (14.39) at each grid point to compute the corresponding displacement. The good news is that because a fluid blob only moves a short distance over the short time interval of a single time step, a simple fixed point iteration is usually sufficient. That is, denoting the iteration level by superscript m:

$$\alpha_j^{m+1} = \tau\, u(x_j - \alpha_j^m, t^n) \tag{14.40}$$

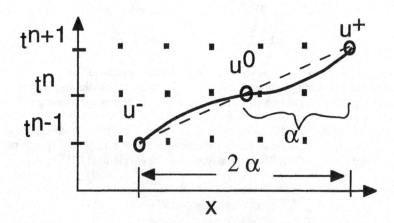

Figure 14.3: Schematic of a particle trajectory (solid curve) which arrives at one of the points of the regular, evenly spaced computational grid at $t = t^{n+1}$. The semi-Lagrangian scheme approximates this trajectory by the dashed line, using the velocity at the midpoint of the trajectory to estimate the displacement α during one time step. The SL algorithm connects the values of u at three points along a trajectory (circles) rather than at a fixed value of x, as in a conventional scheme.

At the beginning of the time integration, one can take $\alpha_j^0 = 0$; at later time steps, one can initialize the iteration with $\alpha_j(t^{n-1})$, that is, the displacement at the previous time level. Experience and theory both suggest that one never needs to take more than *two fixed point iterations* except possibly at the start of the time integration where one might need three. Each iteration increases the accuracy of the displacement by one more power of τ. The iteration is guaranteed to converge so long as

$$\tau \mid u_x \mid < 1 \qquad (14.41)$$

This condition seems to be almost always met in practice though one might worry a bit in the neighborhood of fronts where the slope u_x is very large (Pudykiewicz, Benoit and Staniforth, 1985).

The three-level SL/SI scheme generalizes easily to higher dimensions. In two dimensions, let $\alpha(x,y)$ and $\beta(x,y)$ denote the displacements in the x and y directions, respectively. The three-level SL/SI method is

$$\begin{aligned}
u_{ij}^+ &\equiv u(x_i, y_j, t^n + \tau) \\
u_{ij}^0 &\equiv u(x_i - \alpha_{ij}, y_j - \beta_{ij}, t^n) \\
u_{ij}^- &\equiv u(x_i - 2\,\alpha_{ij}, y_j - 2\beta_{ij}, t^n - \tau)
\end{aligned} \qquad (14.42)$$

and similarly for the y-momentum equation. The displacements are the solutions to

$$\alpha_{ij} = \tau\, u(x_{ij} - \alpha_{ij}, y_{ij} - \beta_{ij}, y_{ij}, t^n) \quad \& \quad \beta_{ij} = \tau\, v(x_{ij} - \alpha_{ij}, y_{ij} - \beta_{ij}, y_{ij}, t^n) \qquad (14.43)$$

where v is the y velocity.

14.6 Multiply-Upstream: Stability with a Very Long Timestep

We begin with three useful definitions.

Definition 31 (FOOT of a TRAJECTORY/DEPARTURE POINT)
The spatial location of the beginning of the trajectory of a fluid particle at some time t, usually t^{n-1} for a three-level SL scheme or t^n for a two-level method, is called the "FOOT of the TRAJECTORY" or equivalently the "DEPARTURE POINT" of the trajectory. The goal of the fixed point iteration of the SL method is to compute the displacement and through it the "foot".

Definition 32 (MULTIPLY-UPSTREAM)
A semi-Lagrangian method is "MULTIPLY-UPSTREAM" if the set of grid points used to interpolate the velocities in the computation of the "foot" of a trajectory is allowed to vary, relative to the arrival point of the trajectory, so as to center on the foot.

Definition 33 (COURANT NUMBER)
The Courant number is a nondimensional parameter which is the ratio of the time step to the smallest distance between two points on the spatial grid, multiplied by the maximum velocity:

$$Co \equiv U_{max} \frac{\tau}{\min(h)} \qquad (14.44)$$

where U_{max} is the maximum fluid velocity. Strictly speaking, this is an advective *Courant number; one can measure the maximum time step for wave propagation problems by defining a "wave" Courant number with the maximum advecting velocity replaced by the maximum phase speed c.*

In either case, explicit methods are stable only when the Courant number is $O(1)$ or less. In the rest of this chapter, we shall use only the advective Courant number.

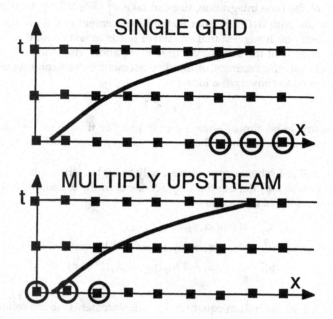

Figure 14.4: Schematic comparing the off-grid interpolation schemes of multiply upstream and single grid semi-Lagrangian methods for a large Courant number, i. e., when the particle travels through several grid points during the time interval $t \in [t^n, t^{n+1}]$. The circles show the points whose grid point values at time level t^{n-1} are used to interpolate velocities and other quantities at the foot of the trajectory. The trajectory is the heavy curve.

Fig. 14.4 illustrates this definition. One major reason for the slow acceptance of semi-Lagrangian schemes, which were used as early as Wiin-Nielsen (1959), Krishnamurti(1962) and Sawyer (1963), is that the early algorithms interpolated the departure points of the trajectory through x_j by using (u_{j-1}, u_j, u_{j+1}). That is to say, the "single grid" SL methods used the same set of points centered on the arrival point x_j irregardless of whether the foot of the trajectory was near or far, left or right, of the arrival point.

The price for this inflexibility is that when the Courant number is greater than one, single grid schemes use *extrapolation* rather than interpolation to estimate velocities, etc., at the foot of the trajectory — and are *unstable*. Consequently, the old-style semi-Lagrangian algorithms had time step restrictions which are just as strict as for conventional explicit methods.

In contrast, multiply-upstream schemes are stable and accurate even for very large Courant numbers. Indeed, as we shall see show in the next section, something very wierd and wonderful happens: multiply-upstream methods become more accurate as the time step *increases*!

Spectral and pseudospectral schemes are *always multiply-upstream*.[2] Thus, spectral SL algorithms can be used at high Courant number.

---

[2]This assertion is true if the interpolation to compute the foot is done spectrally; some weather forecasting models inconsistently use low order polynomial interpolation to compute the foot, even though the rest of the computation is spectral; such mixed schemes are not automatically multiply-stream.

14.7 Numerical Illustrations: Higher Accuracy for Longer Time Step and Superconvergence

Fig. 14.5 compares the standard Fourier pseudospectral method with the semi-Lagrangian method for the one-dimensional advection equation. This solution "breaks" at $t = 1$ and has a jump discontinuity for all larger times. The time step $\tau = 1/2$ for the semi-Lagrangian scheme — much longer than is stable for the standard, explicit Fourier method — is such that the numerical integration jumps from a very smooth flow (a sine wave!) to a front in just two time steps! Nevertheless, the semi-Lagrangian method gives a much smoother approximation to the front than the standard Eulerian coordinates algorithm, which blindly approximates all derivatives by the usual Fourier pseudospectral formulas in complete ignorance of both the method of characteristics and the hyperbolic character of the differential equation.

Fig. 14.6 shows the departure points and errors for the semi-Lagrangian calculation, same case as the previous figure but at a later time when the errors have presumably worsened. Even so, the error is quite small everywhere except for a narrow neighborhood of the front. However, large relative error in a thin zone around the front is a weakness of almost all front-resolving algorithms.

The bottom panels in Fig. 14.6 shows the same solution using ten times as many time steps. This should reduce the error, since the method is second order, by a factor of 100. Instead, the error is considerably worse. Why does a *larger* time step give *smaller* error?

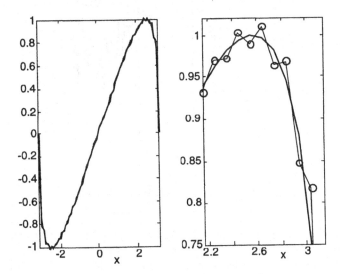

Figure 14.5: Comparison of semi-Lagrangian/semi-implicit Fourier method (thick solid curve) with standard Fourier pseudospectra method (thin line with circles) for the one-dimensional advection equation [inviscid Burgers' equation] , $u_t + uu_x = 0$ for 64 grid points, evolving from $u(x, 0) = \sin(x)$ and illustrated at $t = 1$. The right graph is just a "zoom" or blow-up of the left plot. The semi-Lagrangian time step was $\tau = 1/2$; the time step for the standard method was determined by an adaptive 4th/5th order Runge-Kutta subroutine and is not known except that it was much shorter than for SL scheme. The solution "breaks" at $t = 1$, that is, would develop an infinite slope at that time were it not for the viscosity.

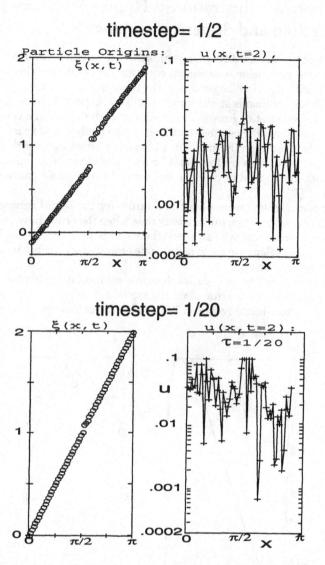

Figure 14.6: Semi-Lagrangian Fourier calculation of a moving shock wave in Burgers' equation (in the limit of vanishing viscosity) with 64 points on $x \in [0, 2\pi]$. $u(x, 0) = \sin(x)$. The solution breaks, i. e., first develops a near-discontinuity, at $t = 1$; the solution is shown at $t = 2$. Left panels: the origins of the points which lie on the pseudospectral grid at $t = 2$, i. e., the graph is a plot of $\xi(x, t) \equiv x_j - \alpha_j$ versus x_j. Right panels: error at $t = 2$. (A spike of $O(1)$ error at one point in both graphs has been truncated.) In the upper pair of plots, the time step is $\tau = 1/2$. Since the grid spacing is about $1/10$, the Courant number is roughly 5. Bottom panels: same as top except that the time step has been reduced by a factor of 10. The main changes are (i) the gap in particle origins is much smaller and (ii) the error has increased. Note that in the upper panels (where $Co \approx 5$), the error is smaller than $1/100$ almost everywhere; with the short time step, the error is larger than $1/100$ almost everywhere, and many points have errors of $1/10$ or $1/20$, even far from the shock.

The crucial data is in the left panel of each graph. When the time step is large, there is a gap in the departure points in the sense that all trajectories originate at a considerable distance from the front. This is important because the pseudospectral interpolation, which for these figures was performed without smoothing or special tricks, becomes increasingly inaccurate as the front is approached because of the oscillations of Gibbs' phenomenon. When the time step is small, some particles originate very close to the front, so the interpolation gives poor approximations to the velocities that advect the particles towards the front from either side. The result is large errors.

SL schemes that use various low order polynomial interpolations suffer the same curse; close proximity to a jump discontinuity is an ill wind for all approximation schemes. Numerical experiments such as Kuo and Williams (1990) have confirmed that SL methods have their smallest errors as a function of time step for rather large time steps — a couple of hours in meteorological applications.

Fig. 14.7 illustrates another odd property of SL schemes. If one plots only the grid point values of a pseudospectral SL scheme, one finds, for moderate τ, a good approximation to the exact solution. However, the Fourier series can be summed to evaluate $u(x)$ at arbitrary points not on the canonical grid. If we do this for many intermediate points, we obtain the very oscillatory graph shown in the right panel. The pseudospectral solution is much better on the grid than off.

Finite element methods often exhibit a similar "superconvergence" at the collocation

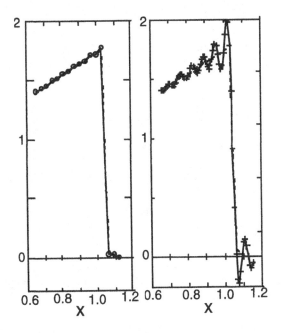

Figure 14.7: Same case as the two previous figures, but a "zoom" which shows only part of the x-interval around the shock. In both panels, the dashed line is the exact solution. On the left, the solid line-with-circles shows the semi-Lagrangian Fourier solution at the points of the collocation grid. The solid-plus-crosses on the right is the same solution except that the Fourier interpolant has been summed at many points between those of the collocation grid. The semi-Lagrangian solution exhibits "overconvergence", that is, the solution is much more accurate AT the collocation points than BETWEEN them.

points. The existence of this phenomenon has been rigorously proved for some finite element algorithms, but not as yet for spectral.

14.8   Two-Level SL/SI Algorithms

The three-level SL/SI is simple and robust, but it has drawbacks. The first is that it requires storing the solution at three different levels. The second is that the implicit terms are treated with an effective time step of 2τ whereas a standard Crank-Nicholson scheme would average the unknowns at time levels t^{n+1} and t^n with a time step of τ and an error four times smaller.

Fortunately, it is possible to replace Robert's scheme with a two-level method that has a much smaller proportionality constant. The key idea is to apply the three-level formula with u^- interpreted as the solution at time level t^n rather than t^{n-1}. The intermediate velocity at $t = (n + 1/2)\tau$ is obtained by extrapolation from previously known values of the velocity.

The displacements are calculated from

$$\alpha_j = (\tau/2)\, u_{ext}(x_j - \alpha_j, t^{n+1/2}) \tag{14.45}$$

where α is defined to be the displacement over a time interval of $\tau/2$, subscript "ext" denotes a quantity which must be extrapolated and j is the index of (spatial) grid points. The actual time-marching step for $Du/Dt + G(u,t) = F(u,t)$ is identical in form to the three-level scheme:

$$u^+ + (\tau/2)G^+ = u^- - (\tau/2)\, G^- + (\tau/2)\, F_{ext}^0 \tag{14.46}$$

As before, $u^+ = u(x, t^n + \tau)$, but

$$
\begin{aligned}
F_{ext}^0 &\equiv F(x - \alpha, t^{n+1/2}) \\
u^- &\equiv u(x - 2\,\alpha, t^n)
\end{aligned}
\tag{14.47}
$$

The most widely used extrapolation formula is

$$u_{ext}(x, t^{n+1/2}) = \frac{15}{8}u(x, t^n) - \frac{10}{8}u(x, t^{n-1}) + \frac{3}{8}\, u(x, t^{n-2}) \tag{14.48}$$

which Temperton and Staniforth (1987) found preferable to the two-level extrapolation

$$u_{ext}(x, t^{n+1/2}) = (3/2)u(x, t^n) - (1/2)u(x, t^{n-1}) \tag{14.49}$$

It is possible to extrapolate along trajectories (that is, shifting x positions by the displacement α as one moves back in time), but Temperton and Staniforth (1987) found it was more accurate (and cheaper) to extrapolate along grid points.

Extrapolation has a bad effect on stability, as noted above. However, the two-level method is stable if the extrapolation is done carefully. With complicated physics — mountains, complicated temperature dependence and so on — it is sometimes necessary to add extra damping, such as replacing Crank-Nicholson by the "theta" scheme (which is a linear combination of Crank-Nicholson with Backwards-Euler that weakly damps high frequency). It is also a good idea to transfer as many terms as possible from the extrapolated, explicit quantity F_{ext}^0 to the implicitly-treated term G. These empirical fixes are problem-specific and model-specific, so we must defer the details to the papers catalogued in the bibliographic table below.

The two-level scheme is in theory twice as cheap as the three-level scheme in the sense that one can achieve the same level of time accuracy with twice as long a time step in the two-level model as in the older algorithm. However, because the three-level scheme is more stable and does not need time-extrapolation, it remains competitive with two-level methods.

14.9 Noninterpolating SL Methods and Computational Diffusion

The error in interpolating from the evenly spaced grid to the "feet" of trajectories is a source of computational diffusion. McCalpin (1988) has given a thorough analysis for polynomial interpolation schemes of various orders, but it would take us too far afield to reproduce his analysis here. Instead, we shall offer only a heuristic argument.

Consider the simplest case: the pure advection equation in one dimension. The formula for computing the new unknown in the three-level SL scheme is then

$$u(x, t^{n+1}) = u(x - 2\alpha, t^{n-1}) \tag{14.50}$$

Interpolation errors will add a small random component to the displacement α. Sometimes the fluid blob will be advected a little too far, sometimes not far enough, because interpolation errors generate inexact particle velocities and displacements. To the SL scheme, the foot of a trajectory is not known exactly, but only to within a line segment (or circle or sphere) which is the size of the average interpolation error. The advected fields are blurred by this randomness, and blurring is equivalent to a computational damping.

McCalpin's analysis shows linear interpolation generates ordinary, second derivative diffusion; cubic interpolation gives a biharmonic (fourth derivative) damping and so on. For ocean models where the scale of the energy-containing eddies is an order of magnitude smaller than in the atmosphere, and the affordable number of grid points-per-eddy-diameter correspondingly smaller, too, McCalpin argues that, arithmurgically speaking, this computational diffusion would turn the seas into great viscous pools of honey.

Fig. 14.8 illustrates his point. The semi-Lagrangian solution is damped much more than that of the standard pseudospectral method even though the explicit, physical viscosity is the same.

Ritchie (1986) was thus motivated to invent a "non-interpolating" SL algorithm. The basic idea is very simple. At large Courant number, split the advecting velocity into two parts:

$$u_j = u_{j,SL} + u_{j,E} \tag{14.51}$$

where $u_{j,SL}$ is chosen so that the "foot" of the trajectory is exactly *on* a point of the regular grid when the particle is advected by $u_{j,SL}$ only. Of course, the actual foot of the trajectory almost never coincides with a point of the regular grid, but this is not a serious limitation since the SL algorithm is really just a transformation to a moving coordinate. There is no law of nature or arithmurgy that says that the velocity of the coordinate change must exactly match that of the fluid.

However, Ritchie's modified transformation eliminates only part of the advection, so the remainder must be treated by the standard pseudospectral method, just as if we used an Eulerian (fixed grid/trajectory-ignoring) coordinate system throughout. It is important to choose $u_{j,SL}$ to advect to the grid point *nearest* the actual foot of the trajectory so that $u_{j,E}$ is as small as possible. The residual Courant number is then always less than one so

Viscosity=h/2

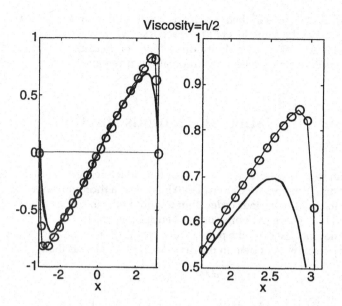

Figure 14.8: Comparison of the semi-Lagrangian/semi-implicit Fourier method (thick lower curve) with the standard Fourier pseudospectra method (upper curve with circles) for Burgers' equation, $u_t + uu_{xx} = \nu\, u_{xx}$ where $\nu = h/2$ and $h = \pi/32$ is the grid spacing. The SL time step was $\tau = 1/2$; that for the standard method was determined by an adaptive 4th/5th order Runge-Kutta subroutine. The solution from $u(x,0) = \sin(x)$ "breaks" at $t = 1$, that is, would develop an infinite slope at that time were it not for the viscosity. The solutions are shown at $t = 2$. In spite of the long time step, which tends to slow the diffusion of the νu_{xx} term (which is treated implicitly), the SL/SI algorithm is much more strongly damped than the standard pseudospectral method, which treats the viscosity honestly by using a short time step. The SL/SI damping comes primarily from interpolation errors.

that the integration is stable with very long time steps where the residual Courant number is

$$CoE \equiv max(|u_E|)\frac{\tau}{h_{min}} \qquad (14.52)$$

where h_{min} is the smallest distance between any two points on the regular pseudospectral grid.

Ritchie's later papers (1987, 1988, 1991) show that the "non-interpolating" SL method works just fine with both spectral and finite difference spatial discretizations, even in three-dimensional models. However, his idea has only partially replaced older SL algorithms.

There are two reasons. First, treating some of the advection by a non-SL scheme brings back some of the disadvantages and poor front resolution of conventional methods. Second, for atmospheric forecasting models, the resolution is sufficiently high and the length of the integration is sufficiently short that damping is not a problem. However, climate models, which are run at lower spatial resolution for long time integrations, may benefit from non-interpolating SL.

For example, his comparisons with non-SL models showed that at a resolution of "T106", then the standard for forecasting, both interpolating and non-interpolating SL models lost only about 1% of the initial energy. This was actually better than the standard non-SL spherical harmonics model of the European Centre for Medium-Range Weather Forecasting, which lost about 3% of its energy due to an (unphysical) biharmonic damping which is included to suppress small scale noise and give better front resolution. The SL models, in spite of their inherent implicit computational dissipation, were actually less damping than the non-SL algorithm because an explicit artificial viscosity was unnecessary for the SL schemes.

At lower resolution, as in a climate model or an ocean model, Robert's SL algorithm is more disspative because the interpolation errors worsen as the grid coarsens. At T63, Ritchie found that his non-interpolating model lost only 1% energy, but the interpolating SL lost 5%.

Smolarkiewicz and Rasch (1990) have generalized non-interpolating SL schemes. They show that their generalization can be applied to many existing, non-SL algorithms to relax time step limitations while retaining good properties of the methods such as that of being "positive-definite monotone".

The conclusion is that non-interpolating SL may be very useful for ocean and climate modelling, or whenever the resolution is sufficiently poor that interpolation errors create too much damping. However, the errors are a function of the order of the interpolation as well as the grid spacing. If one uses high order interpolation, such as fast spectral off-grid interpolation, or even quintic polynomial interpolation, instead of the cubic interpolation which is now the standard in atmospheric models, the computational diffusion is greatly reduced. Thus, it is only in the context of specific problems with specific grid size and a particular interpolation order that one can intelligently answer the question: Are interpolating or non-interpolating schemes better?

14.10 Off-Grid Interpolation: Shape-Preserving, Conservation and All That

14.10.1 Off-Grid Interpolation: Generalities

Off-grid interpolation is rather expensive, if done in the obvious way. For spectral methods, off-grid interpolation cannot be done by the Fast Fourier Transform because the departure points are spaced irregularly between the points of the canonical grid. Summing an N-term spectral series at each of N points costs $O(N^2)$ operations.

There are two subproblems which require off-grid interpolation: (i) calculating the displacements and (ii) calculating properties at departure points. The interpolation of midtrajectory velocities can be done by a lower order method than interpolating velocities, temperatures, ozone concentration and so on at the departure points. The reason is that the displacement is the velocity multiplied by the time step, and so errors in the displacement are smaller than those in u itself.

It is much more important to interpolate accurately when calculating the properties of a fluid blob at its departure point. The reason is that errors in this step are not multiplied by the time step, but rather are errors in the answer itself. An error of a tenth of a degree in interpolating the temperature at the departure point will be carried through the entire remainder of the computation.

The third comment is that SL schemes do not automatically conserve either mass or energy, nor preserve the positive definiteness of densities like mass and water vapor which are always non-negative. It is possible to modify SL procedures to recover many of these properties, but the "off-grid" interpolation is where the "conservation" and "positive definiteness" wars must be fought.

14.10.2   Spectral Off-grid

Spectrally-accurate off-grid interpolation algorithms include the following:

1. Fast Multipole Method (FMM) (Boyd, 1992c, Dutt & Rohklin, 1993, 1994)

2. Partial Chebyshev expansion method (Süli & Ware, 1991, 1992, Ware, 1991,1994)

3. Fine-grid Lagrange interpolation algorithm (Boyd, 1992)

As explained at greater length in Chapter 10, all three methods are asymptotically faster than direct summation. (See the review by Ware, 1998). However, the proportionality factors, which vary with the desired error tolerance, are always much greater than for the Fast Fourier Transform.

The Süli-Ware algorithm is the only one that has been explicitly applied in semi-Lagrangian calculations. It performed well in both óne and two space dimensions.

Boyd's fine-grid polynomial interpolation algorithm is probably both the fastest and the simplest to explain. The key idea is that low order polynomial interpolation can be very accurate if performed on a sufficiently fine grid. So, his first step is to apply the FFT to transform the data to a finer grid than used for the rest of the calculations. One can then use ordinary Lagrange interpolation or splines on the fine grid with little loss of accuracy compared to spectral off-grid interpolation on the original coarser grid.

His algorithm has not been *explicitly* used in semi-Lagrangian models to date. However, the ECMWF forecasting model and the NCAR Community Climate Model both employ his idea *implicitly*. For reasons of cost, and because fast off-grid interpolations were not yet known at the time, the programmers used various forms of low-order interpolation and hoped for the best. (Careful testing showed that this inconsistent use of low order off-grid interpolation in a high order model does not significantly degrade the overall forecast.) However, these models employ dealiasing so that inner products are evaluated on a grid which is much *finer* than the actual number of spectral degrees of freedom. Thus, these models *implicitly* use Boyd's strategy.

The inconsistency between spectral calculation of derivatives and non-spectral off-grid interpolation is in any event not as illogical as it seems. The reason is that to minimize dispersion, it is important to have accurate computation of derivatives in the linear terms which are not treated by semi-Lagrangian advection. The off-grid interpolation does not have to approximate derivatives, but only the velocities themselves, which is easier.

There are two open questions about spectral off-grid interpolation. First, is it cost-effective in a real climate or forecasting model, or is low order polynomial interpolation sufficient? The second is: can spectral semi-Lagrangian schemes be improved, when narrow fronts form, by skillful use of filters or sum-acceleration methods?

We discuss filters briefly in Chapter 18, Sec. 20, but it is useful to at least hint at the possibilities. In Fig. 14.9, there is a large spike of error in the immediate neighborhood of the front in both panels; it is very difficult to eliminate this because even a slight error $(O(h))$ in the location of the front will inevitably generate $O(1)$ errors, although the shape of the computed flow may look very similar to the true flow. Away from this small neighborhood, however, Fig. 14.9 shows that filtering can dramatically reduce errors in a Fourier semi-Lagrangian method. There is only a little experience with filtered spectral semi-Lagrangian schemes, and filters themselves are a hot research topic. The graph suggests that further work on filtered semi-Lagrangian algorithms would be rewarding.

14.10.3   Low-order Polynomial Interpolation

Finite difference and low order finite element models commonly use polynomial interpolation to approximate quantities off-grid. Cubic splines (Purnell, 1976) have been tried, but

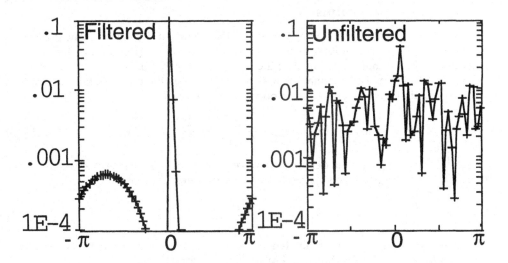

Figure 14.9: Errors in the solution to the one-dimensional advection equation by Fourier three-level semi-Lagrangian computation with $N = 64$ grid points and a time step $\tau = 1/2$. The initial condition is $u(x, 0) = 1 - \sin(x)$. Left: Vandeven filter of order $p = 6$. Right: unfiltered numerical solution.

the most common choice for interpolating particle properties (i. e., u^- in the SL schemes above) is cubic Lagrange interpolation. (Leslie and Purser, 1991, used quintic interpolation.) There is less consistency in calculating displacements because as noted in the previous subsection, the forecast error is much less sensitive to errors in the velocity in the fixed-point iteration for α than to errors in u^-, F^0, etc. Some therefore use linear interpolation of $u(x - \alpha, t^n)$ in calculating displacements. McDonald (1986) suggested a hybrid strategy: linear interpolation on the first fixed-point iteration followed by cubic interpolation on the second (and usually last) iteration.

Moisture has a very patchy distribution in the atmosphere; the air above the Mediterranean Sea is very humid because the warm water evaporates like crazy, but the Mediterranean Sea is bordered by the Sahara desert, which is not humid at all. Global spectral models, with a resolution of only O(100 km), exhibit Gibbs' phenomenon in the vicinity of such moisture fronts. The resulting regions of negative water vapor are decidely unphysical and have a bad effect on model climate. SL schemes do much better. (Rasch and Williamson,1990b,1991 and Williamson, 1990).

The moisture transport problem has motivated studies which replace ordinary interpolation by a procedure in which the interpolant is adjusted by side constraints, such as smoothness and positive-definiteness. Such "quasi-monotone" or "shape-preserving" SL schemes have been investigated by Williamson and Rasch(1989), Rasch and Williamson(1990a) and Bermejo and Staniforth (1992).

Adjusted or constrained interpolation can also be used to enforce conservation of total mass. Priestley (1993) and Gravel and Staniforth (1994) have generalized SL to be quasi-conservative.

14.10.4 McGregor's Taylor Series Scheme: an Alternative for Computing Departure Points

The only disadvantage of polynomial interpolation is that if done accurately, it is rather expensive, especially in multiple space dimensions. McGregor (1993) suggested an alternative procedure, which is to calculate the trajectory by a Taylor series in time about the departure point. The catch is that because the trajectory is the goal, the time derivatives must be *total* time derivatives. Thus, in one space dimension,

$$\alpha = -\tau\, u + \frac{\tau^2}{2}\,(u_t + u u_x) - \frac{\tau^3}{6}\left(\frac{\partial}{\partial t} + u\frac{\partial}{\partial x}\right)(u_t + u u_x) + \dots \qquad (14.53)$$

Calculating the local time derivatives above is very hard. However, McGregor argues that these can be neglected compared to the advective terms so that

$$\alpha \approx -\tau u + \frac{\tau^2}{2}\,u u_x - \frac{\tau^3}{6}u\frac{\partial}{\partial x}\,(u u_x) + \dots \qquad (14.54)$$

where the velocities are to be evaluated at the spatial location of the arrival point at the midpoint in time (t^n for the three-level scheme, $t^{n+1/2}$ for a two-level method).

His method is an order of magnitude cheaper than bicubic interpolation and accurate to second order. Adding additional terms does not increase the formal time accuracy beyond second order, but McGregor shows that empirically, including partial third order terms does improve accuracy considerably for finite τ.

14.11 Higher Order SL Methods

Purser and Leslie (1994) have developed an efficient third-order SL scheme. Earlier, Maday, Patera and Rønquist (1990) and Süli and Ware (1991, 1992) and Ware (1991, 1994) have developed a family of algorithms which can be extended to arbitrary order. Building on Holly and Preismann (1977) and Yang and Hsu (1991), Yang, Belleudy and Temperville (1991) developed a fifth order accurate SL scheme.

The bad news is that these high order schemes are all rather complicated. The assumption that a particle is advected at constant speed by the velocity at the midpoint of the trajectory is sufficient to give second order accuracy in time. At higher order, the displacement on $t \in [t^{n-1}, t^n]$ is no longer the same as $t \in [t^n, t^{n+1}]$, and a simple fixed point iteration no longer gives direct access to the trajectory.

The Maday-Patera-Rønquist and Süli-Ware algorithms require solving auxiliary time-marching problems, k problems for a $(k+1)$-th order method. Maday *et al.* show that one can "sub-cycle", integrating these auxiliary advection problems with a short time step while applying the implicit time-marching formulas (for non-advective terms) only over a long time interval.

Even so, these algorithms are uncompetitive with the second order SL schemes for meteorological applications. We shall therefore not describe them in detail. These are likely to become more important as resolutions improve so that time-truncation errors begin to catch up with space truncation errors. The latter are currently much worse for forecasting and climate models than the temporal differencing errors.

14.12 History and Relationships to Other Methods

Table 14.1: A Selected Bibliography of Semi-Lagrangian Methods

SL is an abbreviation for "semi-Lagrangian"; SI for "semi-implicit".

Reference	Comments
Fjørtoft(1952, 1955)	First Lagrangian algorithm (meteorology)
Wiin-Nielsen (1959)	Earliest SL, at least in meteorology
Sawyer (1963)	coined "semi-Lagrangian"
Holly&Preissman (1977)	SL variant using Hermite interpolation (hydrology)
Haltiner&Williams(1980)	Good review of (singly-upstream) SL
Robert (1981, 1982)	Earliest pairing of SL with SI
Douglas&Russell(1982)	Invention of an SL scheme
Pironneau (1982)	Invention of SL as "transport-diffusion" algorithm
Bates&McDonald (1982)	First multiply-upstream SL
Bates (1984)	More efficient multiply-upsteam SL
Morton (1985)	Invention of SL as "Lagrange-Galerkin" scheme
Robert&Yee&Ritchie(1985)	SL with SI
Pudykiewicz *et al.*(1985)	Convergence of fixed point iteration
Temperton & Staniforth(1987)	Independent invention of two-level SL
McDonald & Bates (1987)	2d order accuracy through extrapolated velocities
Ritchie (1986)	Invention of "non-interpolating" SL
Ritchie (1987,1988,1991)	Non-interpolating SL applied to a sequence of increasingly sophisticated spectral models
McCalpin (1988)	Analysis of dissipation inherent in SL interpolation
McDonald&Bates(1989)	Showed SL with time-splitting was inaccurate
Bates&Semazzi& Higgins&Barros(1990)	Finite difference SL shallow water model Accuracy improved by avoiding time-splitting
Côté&Gravel&Staniforth(1990)	Finite-element SL with "pseudostaggering"
Williamson&Rasch(1989) Rasch&Williamson(1990a)	SL with shape-preserving (positive definite) advection
Rasch&Williamson(1990b,1991) Williamson(1990)	moisture transport in climate & weather models
Purser&Leslie(1988,1991,1994) Leslie&Purser(1991)	Not spectral but improvements in SL latest article is 3rd order in time
Yang&Belleudy &Temperville(1991)	Fifth order SL
Yang&Hsu(1991)	Improved Holly-Preismmann & Yang*et al.* schemes
Smolarkiewicz & Rasch(1990)	Generalization of non-interpolating SL; monotone
Maday&Patera&Rønquist(1990)	SL of arbitrarily high time-order
Süli&Ware(1991,1992) Ware (1991,1994) Baker&Süli&Ware(1992)	Fourier and Chebyshev SL/SI, proofs and examples independent invention of arbitrary time-order SL & fast off-grid interpolation scheme
Staniforth& Côté(1991)	Highly readable review
Gravel&Staniforth&Côté(1993)	Stability analysis

Table 14.1: Bibliography of Semi-Lagrangian Methods (continued)

Reference	Comments
Bermejo&Staniforth(1992)	Quasi-monotone SL
Priestley(1993)	Quasi-conservative SL
Gravel&Staniforth(1994)	Quasi-conservative SL for shallow-water system
McGregor(1993)	Inexpensive $O(\tau^2)$ formula for departure pts.
Pellerin&Laprise &Zawadzki(1995)	Experiments with simple advection/ condensation system
Côté&Gravel &Staniforth(1995)	Extra time level scheme to defeat orographic resonance
Ritchie&Tanguay(1996)	Comparison of spatially averaged Eulerian & Lagrangian treatments of mountains
Böttcher (1996)	Modified exponential splines for interpolation
Makar&Karpik(1996)	Periodic B-spline interpolation on sphere for SL; comparisons with spectral methods

The Lagrangian (particle-tracking) formalism is a couple of centuries old. Fjørtoft (1952, 1955) showed that particle-tracking could be used for weather forecasting. However, Welander (1955) showed through a vivid and widely- reproduced graph that blobs of fluid were rapidly stretched and deformed into a very convoluted shapes. Today, we would say that fluids can be chaotically mixed, in the sense of "chaos" theory, and a blob of fluid initially bounded by a small square is bounded, for later times, by a "fractal" curve of ever-increasing length.

Wiin-Nielsen (1959) therefore introduced a "semi-Lagrangian" scheme: just as in modern schemes, each set of particles is tracked only over a single small time interval of length τ. A new set of particles is chosen for later time steps: those which will arrive at the points of an evenly spaced grid. Later meteorological papers by Krishnamurthi (1962) and Sawyer (1963) and others established that SL algorithms are useful, but did not identify overwhelming advantages.

The method become much more popular when Robert (1981, 1982) showed how to combine semi-Lagrangian advection with semi-implicit treatment of the other terms. This launched an explosion of activity in meteorology which has continued to the present.

Bates and McDonald (1982) made another important advance by introducing "multiply-upstream" CFL methods. These allowed the use of a time step which far exceeded the advective limits of conventional explicit schemes.

Temperton and Staniforth (1987) and McDonald and Bates (1987) independently devised two-level schemes of second order time accuracy. (Robert's method required three time levels while the earlier two-level algorithms of Bates and McDonald were only first order.) These allow a doubling of the time step with no loss of accuracy. However, two-level schemes are somewhat less stable than Robert's method, so several later articles have described slight modifications, such as a "theta" treatment of the implicit schemes, a damping of divergence, and so on to render such algorithms more stable and robust.

Ritchie (1986) developed the "non-interpolating" version of SL, and then applied it to a hierarchy of increasingly complicated models in later papers. Smolarkiewicz and Rasch (1990) have generalized this family of methods. Interpolating schemes are better for well-resolved flows, but non-interpolating SL may be very valuable in integrating climate and ocean models at coarse resolution without excessive computational damping.

Williamson and Rasch(1989), Rasch Williamson(1990a), and Bermejo and Staniforth

(1992) have studied "shape-preserving" or "quasi-monotone" algorithms. The key idea is to replace ordinary interpolation by more exotic near-interpolation schemes that are constrained to preserve essential properties of the flow such as smoothness and, for densities, non-negativity.

Priestley (1993) and Gravel and Staniforth (1994) have generalized SL to be quasi-conservative of mass. The slight adjustments to conserve the total fluid mass are made preferentially near the fronts where the Gibbs' oscillations, which cause most of the mass loss or gain, are concentrated.

Bates and McDonald (1982), Bates (1984), Bates *et al.*(1990), McDonald (1984, 1986, 1987) and McDonald and Bates (1987, 1989) have developed finite difference SL models. Bates *et al.*(1990) show that the "split-explicit" and other time-splitting schemes are not nearly as accurate an unsplit semi-implicit algorithm for the long time steps used with SL codes. Environment Canada has developed finite element SL models for forecasting (Côté, Gravel and Staniforth, 1990).

One major difficulty is that flow over mountains can trigger spurious, unphysical instability in SL/SI schemes. After much struggle, ways to fix up SL to cope with topographic instability have been found as explained in Côté, Gravel and Staniforth (1995).

SL schemes have also been independently invented under other names. "Particle-in-cell" schemes and their close connection with SL have been discussed by Bermejo (1990), who proves that a common particle-in-cell method is just SL with cubic spline off-grid interpolation.

The family of methods known variously as the "characteristic Galerkin" or "Lagrange-Galerkin" or "Eulerian-Lagrangian" was independently invented outside of meteorology by Douglas and Russell(1982), Morton(1985) and Pironneau(1982). It has been greatly developed in the finite element world.

14.13 Summary

Semi-Lagrangian methods have been a big hit in meteorology, and under other names, have had some success in engineering fluid mechanics, too. Most of the thorny problems have been solved but some issues remain.

First, can the fixers which stabilize SL against topographic instabilities be simplified and improved? Second, can high order off-grid interpolation improve, both in speed and in conservation and shape-preserving properties, so that consistent spectral SL algorithms become feasible for large models like weather forecasting codes? Third, can better and simpler high-order-in-time SL methods be devised? Going beyond second order in time seems to exact a heavy price.

Chapter 15

Iterative & Direct Methods for Solving Matrix Equations

"Practice in numerical methods is the only way of learning it"
— H. Jeffreys and B. Jeffreys

"A good scientist is one who knows enough to steal from the best people"
— J. R. Holton

15.1 Introduction

For all but the simplest problems — those with coefficients which are polynomials — Galerkin's method and the pseudospectral method lead to full matrices, unlike the sparse systems of equations generated by finite difference and finite element algorithms. This is particularly unfortunate since implicit and semi-implicit marching procedures require solving a boundary value problem at each time step. Since the expense of Gaussian elimination grows as $O(N^3)$ and N is $O(1000)$ even in two dimensions with modest resolution, spectral methods are prohibitively expensive for most problems without highly efficient iterative methods for solving the ensuing matrix equations.

For special problems, it is possible to replace iterations with direct methods. The Haidvogel-Zang algorithm for *separable* partial differential equations is the theme of one section. When the equation is *constant coefficient*, one may sometimes obtain a *sparse* matrix by applying Galerkin's method as explained in another.

In most of the chapter, we will implicitly assume that the matrix is *positive definite*, that is, all its eigenvalues are of the same sign. The reason is that most simple iterations fail when this condition is not met. Fortunately, both preconditioning and multigrid usually eliminate the divergence caused by a few eigenvalues of the wrong sign. We explain why and also describe a difficult case where *none* of our iterations can be expected to work. Fortunately, such circumstances are very rare, and there are remedies even for problems with an infinite number of both positive and negative eigenvalues.

Lastly, we describe how these iterations can be directly extended to nonlinear problems. Two key themes that dominate earlier sections of the chapter, preconditioning and iteration-as-marching-in-pseudotime, are important for nonlinear iterations, too.

15.2 Iteration-as-Diffusion, Stationary One-Step Iterations & the Richardson/Euler Iteration

If the operator L is positive definite, i. e. if all its eigenvalues are *positive*, then one way to solve the boundary value problem

$$Lu = f \tag{15.1}$$

is to integrate the time-dependent equation

$$u_t = -Lu + f \tag{15.2}$$

with the same boundary conditions and an arbitrary initial condition. Eq. (15.2) is a generalized diffusion equation. Some iterative methods are based directly on (15.2) and others are not, but the analogy of iteration-is-diffusion is very useful because all iterative methods "diffuse" the error until the solution settles down into the steady-state equilibrium described by (15.1).

Because L is positive definite, all eigenmodes of L must decay with time. If L has one or more negative eigenvalues, or a complex eigenvalue with a negative real part, then the corresponding mode will *amplify* — in words, (15.2) is unstable. Consequently, differential equations with eigenvalues of both signs require special treatment, and we will assume L is positive definite in this section and the next.

Eq. (15.2) seems to have made the problem more complex rather than simpler, but for the generalized diffusion equation, an explicit scheme costs only $O(N)$ operations per time step. In contrast, solving the discretized boundary value problem through Gaussian elimination costs $O(N^3)$ operations where N is the number of basis functions or grid points.

The simplest iterative method for (15.1) is to solve (15.2) by the first order Euler method,

$$\vec{u}^{n+1} = (\vec{\mathbf{I}} - \tau \vec{\Lambda}) \vec{u}^n + \tau \vec{f} \qquad \text{[Richardson/Euler/Jacobi]} \tag{15.3}$$

where $\vec{\Lambda}$ is the matrix which is the pseudospectral discretization of the differential operator L. This was first suggested by L. F. Richardson (1910). The parameter τ is the time step; since we are only interested in the steady state solution, the best choice for τ is only a little smaller than the time-stepping stability limit.

The Jacobi iteration, also described in many textbooks, is identical with Richardson's if the diagonal matrix elements of $\vec{\Lambda}$ are rescaled to unity. Thus, the Jacobi and Richardson iterations are conceptually identical. In this rest of the chapter, we omit the rescaling and label this iteration with the name of Richardson.

Richardson's iteration is the prototype for a whole family of methods known as "stationary one-step" iterations, which have the general form

$$\vec{u}^{n+1} = \vec{\mathbf{G}} \vec{u}^n + \vec{k} \qquad \text{[Stationary, One-Step]} \tag{15.4}$$

where $\vec{\mathbf{G}}$ is a square matrix and $\vec{k}$ is a column vector. The iteration is quite useless unless it reduces to an identity when u^n is the exact solution of $\vec{\Lambda} \vec{u} = \vec{f}$, so we have the constraint

$$\vec{\mathbf{G}} = \vec{\mathbf{I}} - \vec{\mathbf{R}}^{-1} \vec{\Lambda} \qquad ; \qquad \vec{k} = \vec{\mathbf{R}}^{-1} \vec{f} \tag{15.5}$$

for some square matrix $\vec{\mathbf{R}}$, but this still allows great diversity.

Until the early 70's, the Gauss-Seidel iteration and its refinement, Successive Over-Relaxation (SOR), were the stars. Unfortunately, they update matrix elements by using a mixture of both old values and new ones already calculated on the same time level. Neither a vector processor nor the indirect pseudospectral algorithm (Sec. 4) can cope with this mix, so we shall not discuss the Gauss-Seidel and SOR methods further.

In order to estimate the optimum τ and also to provide the framework for the generalizations of Richardson's method discussed later, we now turn to a bit of theory.

Matrix functions can always be evaluated, at least in principle, by using an expansion in terms of the eigenfunctions of the matrix. If $\vec{\phi}_n$ denotes the eigenvectors of the square matrix $\vec{\vec{G}}$ with λ_n as the corresponding eigenvalues, then successive iterates of a stationary one-step method will be related by

$$\vec{u}^{n+1} = \sum_{j=1}^{N} \lambda_j \, u_j^n \, \vec{\phi}_j + \vec{k} \tag{15.6}$$

where the u_j^n are the coefficients of the expansion of $\vec{u}^n$ in the eigenvectors of $\vec{\vec{G}}$. Define the error at the n-th iteration as

$$\vec{e}^{(n)} \equiv \vec{u} - \vec{u}^n \qquad \text{[Error at } n\text{-th step]} \tag{15.7}$$

Substituting this into (15.4) and recalling $\vec{\vec{\Lambda}} \, \vec{u} = \vec{f}$ gives

$$\vec{e}^{(n+1)} \equiv \vec{\vec{G}} \, \vec{e}^{(n)} \tag{15.8}$$

Let ρ denote the magnitude of the largest eigenvalue of the iteration matrix $\vec{\vec{G}}$, which is called the "spectral radius" of the matrix and let $\phi_{biggest}$ denote the corresponding eigenvector. For sufficiently large iteration number, the faster-decaying components of the error will be negligible and

$$|\vec{e}^{(n)}| \sim \rho |\vec{e}^{(n-1)}| \qquad \text{as } n \to \infty \tag{15.9}$$

with $\vec{e}^n \sim c\phi_{biggest}$ for some constant c. Thus, the rate of convergence is always *geometric*.

Optimizing an iteration is therefore equivalent to *minimizing* the *spectral radius* of the iteration matrix. Let λ_{min} and λ_{max} denote the smallest and largest eigenvalues of $\vec{\vec{\Lambda}}$, which for simplicity we assume are real. Then for Richardson's iteration, the corresponding eigenvalues γ_{min} and γ_{max} of the iteration matrix $\vec{\vec{G}}$ are

$$\vec{\vec{G}} \equiv \vec{\vec{I}} - \tau \vec{\vec{\Lambda}} \quad \Longrightarrow \quad \gamma_{min} = 1 - \tau \lambda_{max} \quad \& \quad \gamma_{max} = 1 - \tau \lambda_{min} \tag{15.10}$$

Fig. 15.1 shows that the spectral radius is minimized when $|\gamma_{min}| = \gamma_{max}$, which gives

$$\tau = \frac{2}{\lambda_{min} + \lambda_{max}} \qquad \text{[Optimum } \tau\text{; Richardson's]} \tag{15.11}$$

$$\rho(\vec{\vec{G}}) = \frac{\lambda_{max} - \lambda_{min}}{\lambda_{max} + \lambda_{min}} \qquad \text{[Smallest Spectral Radius]} \tag{15.12}$$

$$\approx 1 - 2\frac{\lambda_{min}}{\lambda_{max}} \qquad \lambda_{min} \ll \lambda_{max}$$

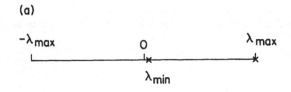

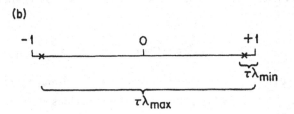

Figure 15.1: The relationship between the eigenvalues of the matrix Λ and those of the Richardson's iteration matrix, $G \equiv I - \tau\Lambda$ where I is the identity matrix and τ is the "time step". (a) The eigenvalues of Λ lie on the positive interval $\lambda \in [\lambda_{min}, \lambda_{max}]$. (b) The eigenvalues of G lie on $\lambda \in [(1 - \tau\lambda_{max}), (1 - \tau\lambda_{min})]$

One difficulty is that we normally do not know either λ_{min} or λ_{max}. However, we shall see later that it is not too hard to estimate them, especially when using a preconditioner.

The other, far more serious problem is that typically,

$$\lambda_{max} \sim O(N^2)\,\lambda_{min} \qquad\qquad \text{[2d order equations]} \qquad\qquad (15.13a)$$
$$\lambda_{max} \sim O(N^k)\,\lambda_{min} \qquad\qquad \text{[k-th order equations]} \qquad\qquad (15.13b)$$

independent of the number of spatial dimensions. This implies that for the commonest case of a 2d order equation, $\log(\rho) \sim O(1/N^2)$, which in turn means that $O(N^2)$ iterations are needed to obtain even moderate accuracy. For a fourth order equation, the number of iterations increases to $O(N^4)$.

We will show in a later section that with heavy use of the FFT, it is possible to evaluate each Richardson iteration in $O(N^2 \log N)$ operations in two dimensions, so the total cost for a second order BVP is $O(N^4 \log N)$ operations (with a large proportionality constant).

Fortunately, it is possible to tremendously improve the efficiency of the Richardson iteration through three modifications: (i) Chebyshev acceleration (ii) preconditioning and (iii) multigrid. The first was discussed in the first edition of this book; the second and third are described here. All three can be used either individually or in combination to accelerate the rate of convergence of iteration schemes.

15.3 Preconditioning: Finite Difference

The reason that Richardson's iteration converges with such lamentable slowness is that

$$\lambda_{max}/\lambda_{min} \gg 1 \qquad\qquad (15.14)$$

preconditioning is a strategy for replacing $\vec{\vec{\Lambda}}$ in the iteration by a new matrix

$$\vec{\vec{A}} \equiv \vec{\vec{H}}^{-1} \vec{\vec{\Lambda}} \tag{15.15}$$

where the "preconditioning matrix" $\vec{\vec{H}}$ is chosen to meet two criteria:

1. $\vec{\vec{H}}$ must be *easily* invertible

2. The ratio of the largest eigenvalue of $\vec{\vec{A}}$ to the smallest must be much smaller than the ratio for $\vec{\vec{\Lambda}}$:

$$\alpha_{\max}/\alpha_{\min} \ll \lambda_{\max}/\lambda_{\min} \tag{15.16}$$

From the viewpoint of the second criterion alone, the ultimate preconditioning matrix is $\vec{\vec{\Lambda}}$ itself since $\vec{\vec{A}} = \vec{\vec{I}}$, the identity matrix, whose eigenvalues are all one. Such a choice is obviously insane because if we could easily compute $\vec{\vec{\Lambda}}^{-1}$, we would not need the iteration in the first place! Still, it is suggestive: What we want is a matrix that is much simpler than $\vec{\vec{\Lambda}}$, but in some sense approximates it.

The preconditioning strategy of Orszag (1980) is to take $\vec{\vec{H}}$ to be the matrix of the usual second order *finite difference* approximation to the differential operator $\vec{L}$ for which $\vec{\vec{\Lambda}}$ is the pseudospectral approximation[1]. It is obvious that inverting the "finite difference preconditioning" is cheap. For a second order differential equation in one dimension, for example, $\vec{\vec{H}}$ is an ordinary tridiagonal matrix which can be inverted in $O(8N)$ operations. Thus, the rate-determining step is not computing the inverse of $\vec{\vec{H}}$, but rather multiplying $\vec{u}^{(n)}$ by the full matrix $\vec{\vec{\Lambda}}$, which is $O(N^2)$ without special tricks and and $O(8N \log_2 N)$ even with FFT methods. The harder question is: Do the eigenvalues of $\vec{\vec{H}}$ approximate those of $\vec{\vec{\Lambda}}$ sufficiently well so that (15.16) is satisfied?

For the *smallest* eigenvalue, the answer is always and obviously: Yes. The reason is that $\lambda_{\min}$ corresponds to the lowest mode of the original differential equation. Unless N is very small, both the finite difference and pseudospectral approximations will do a resoundingly good job of resolving this mode, and therefore

$$\alpha_{\min} = 1 \qquad \text{[Any finite difference preconditioning]} \tag{15.17}$$

The situation, alas, is quite different for the higher modes because they will not be resolved well by either the spectral or finite difference methods. Our only hope is that the erroneous large eigenvalues of the spectral and finite difference matrices will be of the *same magnitude*. It turns out that this is sometimes true and sometimes false, so it is necessary to specialize to particular cases.

EXAMPLE: Consider the ordinary differential equation

$$a_2(x)\, u_{xx} + a_1(x)\, u_x + a_0(x)\, u = f(x) \tag{15.18}$$

with the periodic boundary conditions $u(x + 2\pi) = u(x)$.

---

[1] This was suggested by D'yakonov (1961), who used a similar preconditioning to solve a non-separable PDE through an iteration that required solving a *separable* PDE at each step.

The unknowns are then $u(x_j)$ where $x_j = 2\pi j/N$, $j = 0, 1, \ldots, N-1$. Let $\vec{u}$ denote the column vector whose elements are the grid point values of $u(x)$. The finite difference approximation is

$$(\vec{\vec{H}}\vec{u})_j = a_2(x_j)\frac{u_{j+1} - 2u_j + u_{j-1}}{(\Delta x)^2} + a_1(x_j)\frac{u_{j+1} - u_{j-1}}{2\Delta x} + a_0(x_j)u_j \qquad (15.19)$$

where $\Delta x \equiv 2\pi/N$.

The eigenfunctions $\vec{u}$ and eigenvalues α_k of the matrix $\vec{\vec{A}} = \vec{\vec{H}}^{-1}\vec{\vec{\Lambda}}$ satisfy the equation

$$\vec{\vec{\Lambda}}\vec{u} = \alpha\vec{\vec{H}}\vec{u} \qquad (15.20)$$

Now if $u(x)$ is a smooth function — a low eigenmode of the differential operator $\vec{\vec{L}}$ — then both the finite difference and pseudospectral approximations will do a good job of approximating it, and $\alpha \approx 1$ as already noted.

If N is sufficiently large, then the N-th eigenmode will oscillate very rapidly in comparison to $a_2(x)$. This justifies the WKB approximation (Bender & Orszag, 1978). This shows that the eigenmodes of large λ must *locally* resemble those of a constant coefficient equation, which are $\exp[i\,k\,x]$ for both the spectral and finite difference approximations. Furthermore, if $u(x)$ is highly oscillatory — a large $\vec{\vec{\Lambda}}$ mode — then

$$u_{xx} \gg u_x \gg u \qquad (15.21)$$

which implies that the *first derivative* and *undifferentiated* terms in the differential equation are *irrelevant* to the eigenvalues α_k for large k *except* when the coefficient of the second derivative is *extremely small*: ($O[1/N]$ in comparison to the first derivative, or $O(1/N^2)$ relative to the undifferentiated term.)

Substituting

$$u_j = \exp(i\,k\,x_j) \qquad (15.22)$$

into

$$a_2(-k^2)u_j = \alpha_k a_2 \frac{u_{j+1} - 2u_j + u_{j-1}}{(\Delta x)^2} \qquad (15.23)$$

gives

$$\alpha_k = \frac{k^2[\Delta x]^2}{4\sin^2[(1/2)k\,\Delta x]} \qquad (15.24)$$

Noting that $k \in [0, \pi/\Delta x]$, the corresponding range of α is then

$$1 \le \alpha \le \frac{\pi^2}{4} \approx 2.5 \qquad \text{[Second derivative]} \qquad (15.25)$$

for all modes, *independent* of the *number* of grid points N.

Success! The optimum choice of τ for Richardson's method is

$$\tau = \frac{4}{7} \qquad (15.26)$$

The error decreases by a factor of $7/3 = 2.333$ for each Richardson's iteration. Since Chebyshev series are but a Fourier cosine series in disguise, (15.25) and (15.26) also apply when (15.18) is solved via Chebyshev polynomials.

For fourth order derivatives, one can derive

$$1 \le \alpha \le 6 \quad \left[= \frac{\pi^4}{16} \right] \tag{15.27}$$

which is good but not great. Orszag shows that if we write the problem

$$\nabla^4 u = f \tag{15.28}$$

as the equivalent *system*

$$\left| \begin{array}{cc} \nabla^2 & -1 \\ 0 & \nabla^2 \end{array} \right| \left| \begin{array}{c} u \\ v \end{array} \right| = \left| \begin{array}{c} 0 \\ f \end{array} \right| \tag{15.29}$$

the pre-conditioned matrix equivalent to the operator K, defined by the square matrix in (15.29), has eigenvalues that satisfy

$$1 \le \alpha \le 2.5 \tag{15.30}$$

When the highest derivative is *odd*, however, the preconditioned eigenvalues can become arbitrarily large. (The preconditioned eigenvalues will also become enormous when the highest derivative is even but has an extremely small coefficient, such as high Reynolds number flow.) The operator $\partial/\partial x$ has the eigenvalues $\lambda_k = i\,k$ for periodic boundary conditions while the corresponding centered finite difference approximation has the eigenvalues ($i \sin[k\triangle x]/\triangle x$) so that

$$\alpha_k \sim O \left[\frac{k\triangle x}{\sin(k\triangle x)} \right] \tag{15.31}$$

Unfortunately, this is unbounded. If we filter the high wavenumber components and thus impose the restriction

$$|k\,\triangle x| < 2\pi/3 \qquad \text{[2/3's rule]} \tag{15.32}$$

versus the maximum resolvable wavenumber $k_{\max} = \pi/\triangle x$, then

$$1 \le \alpha \le 2.4 \qquad \text{[First derivative; 2/3's Rule Filter]} \tag{15.33}$$

which is almost identical to that for the (unfiltered) second derivative. The remedy is familiar because the filter is the same as the Orszag Two-Thirds Rule (Chapter 11, Sec. 5) for dealiasing a quadratically nonlinear flow: Filter out the highest one-third of the modes by setting the large k components equal to zero immediately after taking any Fast Fourier Transform from grid point values to spectral coefficients.

Canuto *et al.* (1988, pg. 142–143) describe another alternative: a staggered grid. First order derivatives (both finite difference and pseudospectral) are evaluated on a set of points which is *shifted* from the grid on which $u(x)$ itself is evaluated. The new grid points lie halfway between the points of the original grid. All the necessary manipulations may be performed by the FFT. The pre-conditioned eigenvalues are confined to the very narrow range [1, 1.5708]. The staggered grid is often useful, independent of preconditioning considerations, to cope with the "pressure problem" in wall-bounded viscous flows.

Funaro(1994, 1996) and Funaro and Russo(1993) have described an interesting variation for convection-diffusion problems where the second derivative coefficients are very small compared to the first derivatives. By using a second collocation grid with an "upwind" shift, one can obtain a good preconditioner.

In one dimension, finite difference (and finite element) preconditioning is cheap. However, in two or more dimensions, the LU factorization of $\vec{\vec{H}}$ is expensive. Fortunately, it is often possible to use a different preconditioning which is much simpler than an "honest" finite difference matrix and yet still approximates $\vec{\vec{\Lambda}}$ in some sense. This strategy — trading a few extra iterations for a much lower cost per iteration — is the theme of Sec. 15.5.

Another difficulty is that some eigenvalues of $\vec{\vec{A}}$ may be complex-valued — systems of differential equations in complicated, high Reynolds number flows, for example. This motivates the *staggered* pseudospectral grid of Hussaini & Zang (1984). One can almost always find a filter or a grid which will give a narrow range for the span of the eigenvalues of the pre-conditioned matrix $\vec{\vec{A}}$.

One technical complication for non-Fourier applications is that the finite difference problem must be solved using the same *unevenly* spaced grid as employed by the pseudospectral method. Fornberg (1988, 1996) describes a simple recursion for computing the finite difference weights to all orders for an uneven grid. Alternatively, one can compute and differentiate the Lagrange polynomial in Maple or Mathematica.

For example, if the grid points are $x_j, j = 1, 2, \ldots$, define

$$h_j \equiv x_{j+1} - x_j, \qquad j = 1, 2, \ldots \tag{15.34}$$

The centered (unstaggered) finite difference approximations are

$$\frac{du}{dx}(x_j) = \left\{ -\frac{h_j}{h_{j-1}(h_{j-1} + h_j)} \right\} u_{j-1} + \left\{ \frac{1}{h_{j-1}} - \frac{1}{h_j} \right\} u_j + \left\{ \frac{h_{j-1}}{h_j(h_{j-1} + h_j)} \right\} u_{j+1} \tag{15.35}$$

$$\frac{d^2u}{dx^2}(x_j) = \left\{ \frac{2}{h_{j-1}(h_{j-1} + h_j)} \right\} u_{j-1} - \left\{ \frac{2}{h_{j-1} h_j} \right\} u_j + \left\{ \frac{2}{h_j(h_{j-1} + h_j)} \right\} u_{j+1} \tag{15.36}$$

Canuto *et al.* (1988) amplify on some of the points made here. They also briefly discuss finite element preconditioning.

15.4 Computing the Iterates: FFT versus Matrix Multiplication

The rate-determining step in the Richardson iteration is the matrix product $\vec{\vec{\Lambda}}\vec{u}$ where

$$\vec{\vec{\Lambda}}_{ij} \equiv a_2(x_i)\, C_{j,xx}(x_i) + a_1(x_i)\, C_{j,x}(x_i) + a_0(x_i)\, C_j(x_i) \tag{15.37}$$

for a second order ODE where the $C_j(x)$ are the appropriate cardinal functions such that $C_j(x_i) = \delta_{ij}$. There are two options for evaluating the column vector which is the product of $\vec{\vec{\Lambda}}$ with $\vec{u}$. The first is a matrix-vector multiply at a cost of about $2N^2$ operations per iteration. (We neglect the set-up cost of computing the elements of $\vec{\vec{\Lambda}}$ because this is done only once.)

The alternative uses the Fast Fourier Transform. Note that

$$\vec{\vec{\Lambda}}\vec{u}\Big)_i \equiv a_2\, u_{xx} + a_1\, u_x + a_0\, u\big|_{x=x_i} \tag{15.38}$$

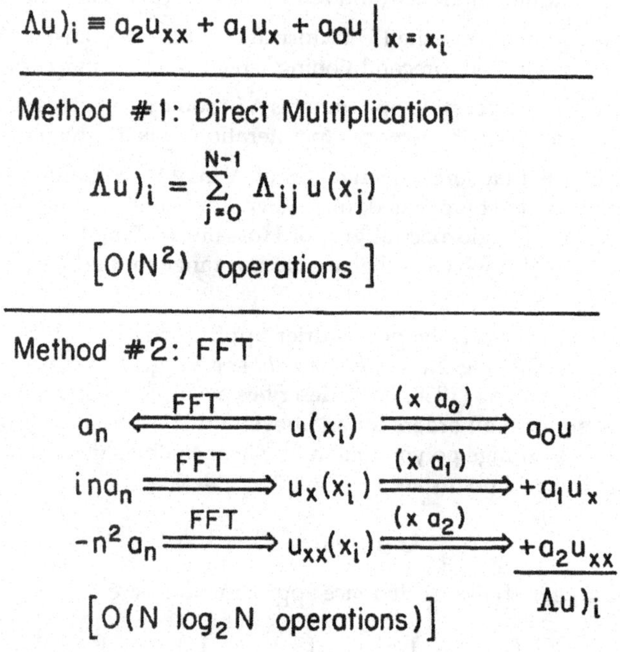

$$\Lambda u)_i \equiv a_2 u_{xx} + a_1 u_x + a_0 u \big|_{x = x_i}$$

Method #1: Direct Multiplication

$$\Lambda u)_i = \sum_{j=0}^{N-1} \Lambda_{ij} u(x_j)$$

$$\left[O(N^2) \quad \text{operations} \right]$$

Method #2: FFT

$$a_n \xleftarrow{\quad \text{FFT} \quad} u(x_i) \xrightarrow{\quad (\times\ a_0) \quad} a_0 u$$

$$i n a_n \xrightarrow{\quad \text{FFT} \quad} u_x(x_i) \xrightarrow{\quad (\times\ a_1) \quad} + a_1 u_x$$

$$-n^2 a_n \xrightarrow{\quad \text{FFT} \quad} u_{xx}(x_i) \xrightarrow{\quad (\times\ a_2) \quad} + a_2 u_{xx}$$

$$\left[O(N \log_2 N \ \text{operations}) \right] \qquad \overline{\phantom{+a_2 u_{xx}}}$$
$$\Lambda u)_i$$

Figure 15.2: Two methods for evaluating the matrix product $\vec{\Lambda}\,\vec{u}$

The FFT of $\vec{u}$ gives the series coefficients $\{a_n\}$. By differentiating the individual terms in the spectral series, we obtain (in $O(N)$ operations) the series coefficients for the first and second derivatives.

The differentiation is trivial for a Fourier basis. For a Chebyshev series, one employs a recurrence. If $a_n^{(q)}$ denotes the Chebyshev coefficients of the q-th derivative of $u(x)$, then

$$c_{n-1} a_{n-1}^{(q)} - a_{n+1}^{(q)} = 2 n a_n^{(q-1)} \qquad\qquad n \geq 1 \qquad\qquad\qquad (15.39)$$

where $c_n = 2$ if $n = 0$ and $c_n = 1$ for $n > 0$. Given the N coefficients $\{a_n^{(q-1)}\}$, we initialize by setting $a_N^{(q)} = a_{N-1}^{(q)} = 0$ and then apply (15.39) in descending order to compute all the coefficients of the derivative.

Two FFT's give $u_x(x_i)$ and $u_{xx}(x_i)$. Three multiplications and additions per grid point complete the calculation of the vector $(\vec{\Lambda}\vec{u})$. The rate-determining step is the cost of the three FFT's, which is only $O(7.5N \log_2 N)$. The two options are shown schematically in Fig. 15.2.

As explained in Chapter 10, for low to moderate resolution, matrix multiplication is as good or better than the FFT. For large N, the FFT is best.

With the FFT and an efficient preconditioning, we can solve a boundary value problem in any number of dimensions at an expense which is O(total # of grid points) multiplied by $(\log_2 N)$ where N is the number of grid intervals in one coordinate.

15.5 Alternative Pre-Conditioners for Partial Differential Equations

In two or more dimensions, Orszag's (1980) finite difference preconditioning, here dubbed $\vec{\vec{H}}_{FD}$, is much more expensive than in one coordinate. A sparse matrix direct method[2] such as block tridiagonal elimination costs $O(N^4)$ operations and $O(N^3)$ storage even on a two-dimensional $N \times N$ grid.

Zang, Wong, and Hussaini (1984) therefore experimented with alternatives. Their $\vec{\vec{H}}_{LU}$ and $\vec{\vec{H}}_{RS}$, are *approximate LU* factorizations of $\vec{\vec{H}}_{FD}$. As in an exact factorization, $\vec{\vec{L}}$ is lower triangular and $\vec{\vec{U}}$ is upper triangular with diagonal elements all equal to unity.

In their first preconditioning, $\vec{\vec{L}}_{LU}$ is the lower triangular part of $\vec{\vec{H}}_{FD}$. (They do not *factor* $\vec{\vec{H}}_{FD}$, merely identify its diagonal and lower triangular elements as the corresponding elements of $\vec{\vec{L}}_{LU}$.) The elements of $\vec{\vec{U}}_{LU}$ are chosen so that the product $\vec{\vec{L}}_{LU}\vec{\vec{U}}_{LU} \equiv \vec{\vec{H}}_{LU}$ has its two superdiagonals [elements with column numbers greater than the row numbers] agree with those of $\vec{\vec{H}}_{FD}$. Their second approximate factorization is identical except that the diagonal elements of $\vec{\vec{L}}$ are altered so that the row sums of $\vec{\vec{H}}_{RS}$ equal the row sums of $\vec{\vec{H}}_{FD}$.

Both approximate factorizations can be computed recursively. For example, let the finite difference approximation in two dimensions be of the form

$$B_{jl} u_{j,l-1} + D_{jl} u_{j-1,l} + E_{jl} u_{jl} + F_{jl} u_{j+1,l} + H_{jl} u_{j,l+1} = f_{jl} \qquad (15.40)$$

The three non-zero elements of the lower triangular matrix $\vec{\vec{L}}$ are

$$b_{jl} = B_{jl} \qquad ; \qquad d_{jl} = D_{jl} \qquad (15.41)$$

$$e_{jl} = E_{jl} - b_{jl} h_{j,l-1} - d_{jl} f_{j-1,l} - \alpha \left\{ b_{jl} f_{j,l-1} + d_{jl} h_{j-1,l} \right\} \qquad (15.42)$$

where $\alpha = 0$ for the first factorization, $\vec{\vec{H}}_{LU}$, and $\alpha = 1$ for the second factorization, $\vec{\vec{H}}_{RS}$. The non-zero elements of the upper triangular matrix (aside from the 1's on the diagonal) are

$$f_{jl} = F_{jl}/e_{jl} \qquad ; \qquad h_{jl} = H_{jl}/e_{jl} \qquad (15.43)$$

Note that we must march through the grid diagonally from top left to bottom right so that we calculate the elements of $\vec{\vec{U}}$ before we need them for the computation of the diagonal elements of $\vec{\vec{L}}$.

What is striking about (15.41)–(15.43) is that there is no fill-in: the approximate factorization still gives only five non-trivial coefficients per grid point for $\vec{\vec{L}}$ and $\vec{\vec{U}}$ together. This is in marked contrast to the exact LU decomposition of $\vec{\vec{H}}_{FD}$, which can be computed via the *block* tridiagonal algorithm: the $N \times N$ blocks of the factored $\vec{\vec{L}}$ & $\vec{\vec{U}}$ matrices are *dense*, even though the blocks of $\vec{\vec{H}}_{FD}$ are sparse, so that there are roughly $5N$ non-trivial matrix

---

[2]Unless the PDE is separable, in which case we should use the special spectral methods for this class of equations described in Sec. 15.11.

elements per grid point. Because the $\vec{\vec{L}}$ and $\vec{\vec{U}}$ matrices of the *approximate* factorizations are so sparse, the necessary backsolves can be performed in $O(N^2)$ operations — that is to say, the cost of inverting $\vec{\vec{H}}_{LU}$ or $\vec{\vec{H}}_{RS}$ is directly proportional to the total number of grid points. This is the best that we could hope for, and implies that inverting $\vec{\vec{H}}$ will not be the rate-determining part of the iteration since evaluating $(\vec{\vec{\Lambda}}\vec{u})$ is always more expensive.

The remaining issue is: How much does incomplete factorization increase the number of iterations? This is inversely proportional to the "condition number", $\kappa \equiv \alpha_{max}/\alpha_{min}$. As shown above, $\kappa \leq 2.4$ for a second order equation for finite difference preconditioning. Tables 15.1 and 15.2, show that κ is much larger for both approximations than for $\vec{\vec{H}}_{FD}$, and worse, κ also increases with N.

Table 15.1: Extreme eigenvalues for the two-dimensional Chebyshev discretization of $\nabla^2 u = f$. (From Zang *et al.*, 1984)

N	H_{FD}^{-1}		H_{LU}^{-1}		H_{RS}^{-1}	
	α_{min}	α_{max}	α_{min}	α_{max}	α_{min}	α_{max}
4	1.000	1.757	0.929	1.717	1.037	1.781
8	1.000	2.131	0.582	2.273	1.061	2.877
16	1.000	2.305	0.224	2.603	1.043	4.241
24	1.000	2.361	0.111	2.737	1.031	5.379

For calculations on a *single* grid, $\vec{\vec{H}}_{RS}$ is clearly superior to $\vec{\vec{H}}_{LU}$. For $N = 24$, the asymptotic convergence rate for $\vec{\vec{H}}_{RS}$ is about 67% of that for the finite difference preconditioning. Because it is much cheaper per iteration, approximate factorization can drastically reduce the cost for boundary value problems in two or more spatial dimensions.

Canuto *et al.* (1988) observe that these incomplete-*LU* decompositions are poor on vec-

Table 15.2: Condition Number κ for Preconditioned Chebyshev Operator in Two Dimensions $[\kappa \equiv \alpha_{max}/\alpha_{min}]$

N	Single-Grid		Multigrid	
	$H_{LU}^{-1} L$	$H_{RS}^{-1} L$	$H_{LU}^{-1} L$	$H_{RS}^{-1} L$
4	1.85	1.72	—	—
8	3.91	2.71	1.79	2.07
16	11.62	4.07	2.12	2.92
24	24.66	5.22	2.26	3.79

tor and parallel machines because they require a lot of recursive [sequential] calculations. They suggest perhaps ADI-type factorizations might be better, but so far, little has been done.

Deville and Mund(1984, 1985, 1990, 1991, 1992) and Deville, Mund and Van Kemenade (1994) and others have shown that finite elements are also a very effective preconditioner for spectral methods. One may use finite elements on triangular elements as the preconditioner even though the spectral approximation is defined on a quadrilateral or a set of quadrilateral subdomains. The condition number is even smaller than with finite difference preconditioning. Furthermore, efficient parallelization of finite elements is a well-researched and well-understood task.

We shall not discuss finite element preconditioning in detail because (i) the principles are exactly the same as for finite difference preconditioning and (ii) a detailed treatment would be too great a digression into finite elements. This lack of space, however, should not be interpreted as implying that finite elements are in any way inferior to finite differences for preconditioning.

15.6   Raising the Order Through Preconditioning

In the previous two sections, finite difference and finite element methods have been humble servants of the spectral method. One can also interpret the same preconditioned algorithm from a perspective in which the spectral method is subservient.

Suppose one has a code to solve a boundary value problem using a second order method. How can one obtain high accuracy without increasing the number of grid points to a ridiculous extreme? Or even compute a crude answer when the low order scheme needs more degrees of freedom than will fit in the available memory?

The answer is to write one additional subroutine to evaluate the residual of the boundary value problem using a high order method, such as a pseudospectral algorithm. If we set up an iteration, repeatedly calling the low order BVP-solver with the output of the high order residual as the inhomogeneous term in the boundary value problem, then the iteration will converge to an answer of *high order* accuracy even though all but the residual-evaluator uses only low order schemes. The overall procedure is identical with the preconditioned spectral iteration described in the two previous sections.

In chess, the lowly pawn can be promoted to a queen, the most powerful piece, after reaching the last row of the chessboard. In numerical analysis, preconditioned iterations provide a way to promote a low order method to spectral accuracy. This promotion does not require rewriting the low order code or negotiating a thicket of enemy chess pieces. It merely needs a single subroutine that will take the grid point values of the unknown as input, and return the residual as the output. The spectral algorithms can be completely isolated and encapsulated in this single subroutine or module.

From this perspective, EVERY LOW ORDER BVP-SOLVER CONTAINS A SPECTRAL SOLVER WITHIN IT.

15.7   Multigrid: An Overview

The reason that an un-conditioned Richardson' iteration is so expensive is that the total number of iterations,

$$N_{\text{iter}} = T/\tau \tag{15.44}$$

is large where τ is the time step and where T is the total integration time, that is, the time required for the solution of the diffusion equation to relax to the steady state. The difficulty is that different components of the flow drive the numerator and denominator of (15.44) in opposite directions. To see this, consider the forced diffusion equation

$$u_t = u_{xx} + f \tag{15.45}$$

with periodic boundary conditions, $u(x + 2\pi) = u(x)$.

The time step τ must be very small because the explicit, one-step Euler's method is unstable unless

$$\tau < \frac{2}{k_{max}^2} \tag{15.46}$$

where k_{max} is the maximum x wavenumber. This constraint on τ is due to the *high* wavenumbers in the solution.

The time interval T is large because the *low* wavenumbers decay very slowly:

$$u = a_1(0) e^{-t} \cos(x) + b_1(0) e^{-t} \sin(x) + \dots \tag{15.47}$$

so $T > 5$ is necessary for even two decimal places of accuracy. The total number of iterations, T/τ, is enormous both because the numerator is large and because the denominator is small.

Multigrid exploits the fact that these two problems — large T, small τ — come from different parts of the wavenumber spectrum. Since the high wavenumber components decay very rapidly, the error in the large k Fourier coefficients has disappeared in only a fraction of the total integration interval. The multigrid principle is simple: throw them away, and continue the integration with a *smaller* basis set and a *larger* time step.

The simplest choice is to halve the maximum wavenumber and quadruple the time step at each stage because every grid point of the reduced grid is also a part of the original fine grid. So that each of the *high wavenumber* Fourier components will decay rapidly, we need a time-step no larger than one-half the explicit stability limit. If we somewhat arbitrarily choose to reduce the error in all Fourier components between $k_{max}/2$ and k_{max} by $\exp(-10)$, then we must march 40 time steps with $\tau = \tau_{min}$, and we will repeat this pattern of taking 40 time steps at each stage. (Fig. 15.3.)

Let K denote the multigrid iteration level. Then

$$k_{max} = 2^K \qquad \text{[largest wavenumber at level } K\text{]} \tag{15.48}$$

Table 15.3: A Selected Bibliography of Spectral Multigrid

References	Comments
Zang&Wong&Hussaini(1982,1984)	Incomplete LU factorizations as smoothers
Brandt&Fulton&Taylor(1985)	Periodic problems with meteorological applications
Streett&Zang&Hussaini(1985)	Transonic potential flow (aerodynamics)
Schaffer&Stenger(1986)	sinc (Whittaker cardinal) basis
Phillips&Zang&Hussaini(1986)	Preconditioners
Zang&Hussaini(1986)	Time-dependent Navier-Stokes
Erlebacher&Zang&Hussaini(1988)	turbulence modeling
Nosenchuck&Krist&Zang(1988)	Parallel machine ("Navier-Stokes Computer")
Heinrichs(1988a)	Line relaxation
Heinrichs(1988b)	Mixed finite difference and Fourier methods
deVries&Zandbergen(1989)	Biharmonic equation
Heinrichs(1992a)	Stokes flow in streamfunction formulation
Heinrichs(1993c)	Navier-Stokes equation
Heinrichs(1993d)	reformulated Stokes equations

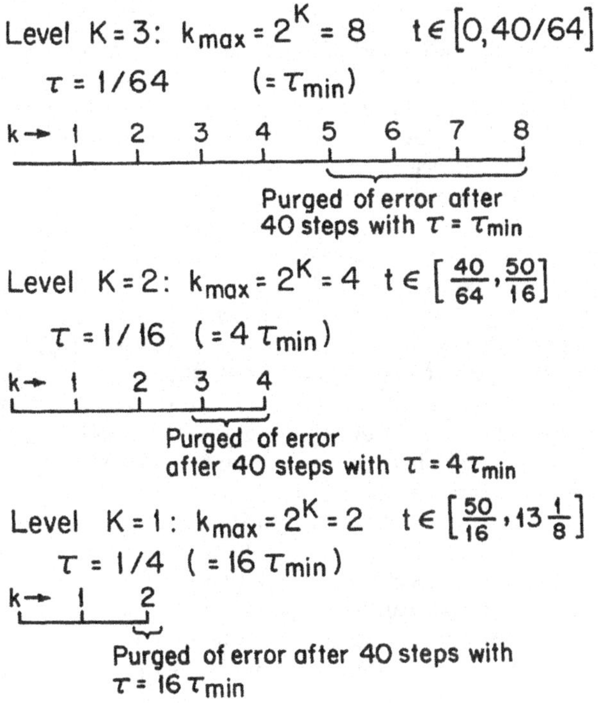

Level $K = 3$: $k_{max} = 2^K = 8$ $t \in [0, 40/64]$

$\tau = 1/64$ $(= \tau_{min})$

$k \rightarrow$ 1 2 3 4 5 6 7 8

Purged of error after
40 steps with $\tau = \tau_{min}$

Level $K = 2$: $k_{max} = 2^K = 4$ $t \in \left[\frac{40}{64}, \frac{50}{16}\right]$

$\tau = 1/16$ $(= 4\,\tau_{min})$

$k \rightarrow$ 1 2 3 4

Purged of error
after 40 steps with $\tau = 4\,\tau_{min}$

Level $K = 1$: $k_{max} = 2^K = 2$ $t \in \left[\frac{50}{16}, 13\frac{1}{8}\right]$

$\tau = 1/4$ $(= 16\,\tau_{min})$

$k \rightarrow$ 1 2

Purged of error after 40 steps with
$\tau = 16\,\tau_{min}$

Figure 15.3: Schematic of multigrid iteration for the one-dimensional diffusion equation with periodic boundary conditions. After a few iterations with shortest time step, τ_{min}, the coefficients of the upper half of the Fourier series have already converged to their exact values. We therefore store coefficients 5, 6, 7, and 8 in the column vector which holds the final answer, halve the basis set, increase the time step by four, and resume iterating. Once wavenumbers 3 and 4 have relaxed to their asymptotic values, we have the basis set and increase the time step again, and so on.

is the largest wavenumber we keep on the K-th grid (and similarly for the y and z wavenumbers),

$$\tau(K) = \frac{1}{2^{2K}} \qquad \text{[time step at level } K\text{]} \qquad (15.49)$$

A single grid using $\tau = \tau_{min}$ requires $64 \times 40 = 2560$ time steps, and each step would cost $O(N_f)$ operations where N_f is the total number of grid points on the finest grid. With multigrid, we need only 160 time steps [level $K = 0$ does not fit on the diagram], and the work is

$$N_{total} \sim O\left(40\, N_f \left\{1 + \frac{1}{2} + \frac{1}{4} + \frac{1}{8}\right\}\right) \qquad \text{operations} \qquad (15.50)$$

since each level uses a grid with only half as many points as its predecessor. The sum $\{1 + 1/2 + 1/4 + 1/8 + \dots\}$ is the geometric series which converges to 2. The total cost of removing the error from *all* wavenumbers is at worst a factor of two greater than the cost of relaxing to the correct solution for just the upper half of the wavenumber spectrum. Even with k_{max} as small as 8, the savings is a factor of 32. In principle, the multigrid iteration is cheaper than an unpreconditioned Richardson's iteration on a single grid by a factor of $O(N^2)$.

In more realistic problems, variable coefficients in the differential equation couple the high and low wavenumbers together. This makes it necessary to add a reverse cascade in K: after the solution has been computed on the coarsest grid, one must interpolate to a finer grid, apply a few more iterations and repeat. Multigrid subroutines have rather complicated control structures so that the program can iterate a sufficiently large number of times on each level to ensure a successful transition to the next. It is usually unnecessary to take 40 steps on each level when both a forward & backward cascade is used; three iterations per level is more typical, but several fine-to-coarse-to-fine transitions are usual.

The *pre-conditioned* Richardson's iteration shares with multigrid the property that the NUMBER OF ITERATIONS is INDEPENDENT OF THE GRID SIZE. Consequently, it is probably quite senseless to apply multigrid to a one-dimensional problem. In two dimensions, the high cost of the finite difference preconditioning has inspired the approximate factorizations discussed in Sec. 15.5. Their condition number does increase with N, so multigrid becomes increasingly superior as the number of grid points increases.

Note that multigrid is not a *competitor* to Richardson's iteration or preconditioning. Rather, multigrid is a strategy for *accelerating* Richardson's iteration (with or without preconditioning) or other iterations at the expense of additional programming and storage.

For further information, one can consult Zang, Wong, and Hussaini (1982, 1984), Brandt, Fulton, and Taylor (1985), and Canuto *et al.* (1988). A very readable introduction to multigrid, although it discusses finite differences only, is the tutorial by W. Briggs (1987).

15.8   The Minimum Residual Richardson's (MRR) Method

This improvement of the conventional Richardson's method, attributed to Y. S. Wong in Zang, Wong, and Hussaini (1982), has the great virtue that it is a parameter-free algorithm that does not require guessing the minimum and maximum eigenvalues of the iteration matrix. Unlike the Chebyshev acceleration of Richardson's method, the MMR algorithm does not perform poorly when the matrix has complex eigenvalues. Indeed, the only requirement for convergence of the Minimum Residual Richardson's method is that the eigenvalues of the (preconditioned) matrix lie strictly in the right half of the complex plane (Eisenstat, Elman, and Schultz, 1983).

The basic step is identical with that of the preconditioned Richardson's iteration for $\vec{\vec{A}} \vec{u} = \vec{g}$:

$$\vec{u}^{n+1} = \left(\vec{\vec{I}} - \tau \vec{\vec{A}}\right) \vec{u}^n + \tau \vec{g} \tag{15.51}$$

Like the Chebyshev-accelerated Richardson's method, the MMR algorithm varies τ to optimize the rate of convergence. However, it does not set τ to a fixed value that requires knowledge of the eigenvalues of $\vec{\vec{A}}$. Instead, the MRR varies τ on the fly, making a new choice at each iteration, so as to minimize the *residual* of the next iterate.

The residual $\vec{r}^n$ is defined by

$$\vec{r}^n \equiv \vec{f} - \vec{\vec{\Lambda}} \vec{u} \tag{15.52}$$

where $\vec{\vec{\Lambda}}$ is the discretization of the differential operator L and where $\vec{f}$ is the array of grid point values of $f(x)$ in the boundary value problem

$$L u = f \tag{15.53}$$

The matrices of the MRR are related to $\vec{\vec{\Lambda}}$ and $\vec{f}$ through the preconditioning matrix $\vec{\vec{H}}$ via

$$\vec{\vec{A}} \equiv \vec{\vec{H}}^{-1} \vec{\vec{\Lambda}} \tag{15.54a}$$

$$\vec{g} \equiv \vec{\vec{H}}^{-1} \vec{f} \tag{15.54b}$$

The goal is to choose τ^n so as to minimize the norm of the residual, i. e., create the smallest possible value of

$$\left(\vec{r}^{n+1}, \vec{r}^{n+1} \right) \tag{15.55}$$

To compute the optimum τ, it is helpful to introduce the auxiliary vector

$$\vec{z}^n \equiv \vec{\vec{H}}^{-1} \vec{r}^n = \vec{\vec{H}}^{-1} \vec{f} - \vec{\vec{H}}^{-1} \vec{\vec{\Lambda}} \vec{u}^n \tag{15.56}$$

It is easy to show (try it!) that

$$\vec{r}^{n+1} = \vec{r}^n - \tau \vec{\vec{\Lambda}} \vec{z}^n \tag{15.57}$$

The solution to this minimization problem turns out to be the same as that of a series expansion, so we briefly digress for the following:

Theorem 29 (Mean-Square Minimization with a Truncated Series) *Suppose the goal is to approximate a function $f(x)$ with a truncated series so as to minimize*

$$(R_n, R_n) \tag{15.58}$$

where the residual (error) is

$$R_N(x) \equiv f(x) - \sum_{n=0}^{N} a_n \phi_n \tag{15.59}$$

The choice that minimizes the mean-square error is

$$a_n = (f, \phi_n)/(\phi_n, \phi_n) \tag{15.60}$$

INDEPENDENT *of N so long as the basis functions are orthogonal.*

Proof: Substitute the definition of R_n into (15.58) and then differentiate with respect to each of the $(N + 1)$ coefficients in turn and set the result equal to 0. (Recall that the condition for a function of $N + 1$ variables to be a minimum is that $\partial(R_n, R_n)/\partial a_0 = \partial(R_n, R_n)/\partial a_1 = \ldots = \partial(R_n, R_n)/\partial a_n = 0$.) After the differentiations have been performed, (15.60) is obvious. QED

Eq. (15.60) is identical with the formula for the coefficients of an infinite series of orthogonal functions, (3.24). Although we are not performing a spectral expansion but rather solving a matrix algebra problem via the MRR, comparison of (15.57) with (15.59) shows that the *minimization* problem of choosing the "time step" τ^n is identical in form to that of computing a one-term, mean-square approximation to a function. Here, $\vec{r}^n$ plays the role of $f(x)$ and $\vec{\vec{\Lambda}} \vec{z}^n$ is the "basis function", impersonating $\phi_0(x)$. The choice of τ that minimizes the residual is therefore

$$\tau^n = (\vec{r}^n, \vec{\vec{\Lambda}} \vec{z}^n)/(\vec{\vec{\Lambda}} \vec{z}^n, \vec{\vec{\Lambda}} \vec{z}^n) \tag{15.61}$$

Definition 34 (Minimum Residual Richardson's (MRR) Algorithm)

To solve the matrix problem

$$\vec{\vec{\Lambda}}\,\vec{u} = \vec{f}, \tag{15.62}$$

using the preconditioning matrix $\vec{\vec{H}}$, given a first guess u^0):

Initialization:

$$r^0 \;\equiv\; \vec{f} - \vec{\vec{\Lambda}}\,u^0 \tag{15.63a}$$
$$z^0 \;\equiv\; H^{-1} r^0 \tag{15.63b}$$

and then repeat the following steps until convergence:

Iteration:

$$\tau^n \;=\; (r^n, \vec{\vec{\Lambda}}\,z^n)/(\vec{\vec{\Lambda}}\,z^n, \vec{\vec{\Lambda}}\,z^n) \tag{15.64a}$$
$$u^{n+1} \;=\; u^n + \tau^n z^n \tag{15.64b}$$
$$r^{n+1} \;=\; r^n - \tau^n \vec{\vec{\Lambda}}\,z^n \tag{15.64c}$$
$$z^{n+1} \;=\; \vec{\vec{H}}^{\,-1} r^{n+1} \tag{15.64d}$$

The iteration converges as long as the eigenvalues of $\vec{\vec{H}}^{\,-1}\vec{\vec{\Lambda}}$ lie strictly in the right half of the complex plane.

Because the only additional quantity needed at each step is the scalar τ^n, which is computed by two inexpensive inner products, the cost per iteration of the MRR is usually no more than a few percent greater than that the unmodified Richardson's method.

Fig. 15.4, from Canuto and Quarteroni (1985), compares the accuracy of the unmodified Richardson's method with the MRR method. The graphs are clear testimony to the power of minimizing the residual: four iterations of Richardson's method reduce the error by no more than a single MRR iteration.

Both cases are difficult in the sense that the first derivative terms are very large so that Orszag's estimates for the condition number are too optimistic. (Recall that, following Orszag, we neglected all derivatives except the highest in estimating the condition number.) Instead, Canuto and Quarteroni computed the exact condition numbers and used them to optimize the Richardson's method. In real-life where the eigenvalues are not known, one would try Orszag's optimal timestep and the Richardson's iteration would blow up. Trial-and-error would eventually give a stable τ, but probably not one which is optimal, so the graphs in Fig. 15.4 make the unmodified Richardson's algorithm look better than it would in practice.

In contrast, the MRR method is parameter-free, so the rapid convergence seen in Fig. 15.4 would always be achieved for these problems. This is particularly important for Case 2, which is a two-dimensional boundary value problem where the preconditioning matrix $\vec{\vec{H}}$ is an incomplete factorization of the finite difference matrix, and therefore has a rather large condition number.

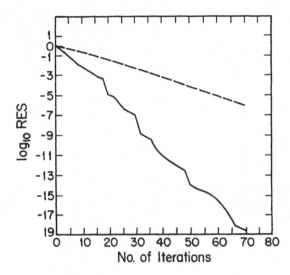

Figure 15.4: Convergence of Richardson's iteration (dashed) versus the Minimum Residual Richardson's method (solid) for two problems. The y-axis is the logarithm of $\sqrt{(\vec{r}, \vec{r})/(\vec{f}, \vec{f})}$ where the residual $\vec{r} \equiv \vec{f} - \vec{\vec{\Lambda}}\vec{u}$ as in the text and (,) denotes the usual matrix inner product. Taken from Canuto & Quarteroni (1985). The differential equation is $-[(1 + 10x)\,u_x]_x = f(x)$. It was solved with 127 grid points on $[-1, 1]$ and $f(x)$ chosen so the exact solution is $u(x) = \sin(\pi x)$.

15.9 The Delves-Freeman "Asymptotically Diagonal" Preconditioned Iteration

The monograph by Delves and Freeman (1981) and the short review of Delves (1976) analyze an alternative preconditioning which is based on two key ideas. The first idea is to precondition using a low order approximation to the *Galerkin* matrix. (In contrast, the finite difference preconditioner is a low order *grid point* approximation.)

The second key idea is to exploit the matrix property of being "asymptotically diagonal". Galerkin approximations are usually not "diagonally dominant" in the sense that is so valuable in iterations for finite difference methods (Birkhoff and Lynch, 1984). However, in many cases, the diagonal element of the j-th row of the Galerkin matrix does become more and more important relative to the other elements in the same row and column as $j \to \infty$. In this sense, the Galerkin matrix is "asymptotically diagonal".

An example will clarify this concept. Consider the Fourier cosine-Galerkin solution of

$$u_{xxxx} + q(x)\,u = f(x) \tag{15.65}$$

where $u(x)$, $q(x)$, and $f(x)$ are all smooth, periodic functions symmetric about the origin. (The symmetry is assumed only for simplicity.) The Galerkin discretization is

$$\vec{\vec{H}}\,\vec{a} = \vec{f} \tag{15.66}$$

where $\vec{a}$ is the column vector containing the Fourier cosine coefficients, a_i, and

$$H_{ij} = (\cos[ix],\ [\partial_{xxxx} + q(x)]\,\cos[jx]) \tag{15.67}$$

$$f_i = (\cos[ix],\ f) \tag{15.68}$$

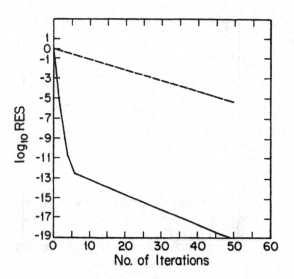

Figure 15.5: Same as Fig. 15.4 except that the differential equation is

$$-[(1 + 10\,x^2 y^2)u_x]_x - [(1 + 10\,x^2 y^2)u_y]_y = f(x,\,y)$$

where $f(x,\,y)$ is chosen so that the exact solution is $u(x,\,y) = \sin(\pi x)\sin(\pi y)$.

where

$$(a,\,b) = \int_0^\pi a(x)\,b(x)\,dx \tag{15.69}$$

The crucial point is that if we define

$$q_{ij} \equiv (\cos[ix],\,q(x)\,\cos[jx]) \tag{15.70}$$

then

$$H_{ij} = j^4\,\delta_{ij} + q_{ij} \tag{15.71}$$

where δ_{ij} is the usual Kronecker delta (=1 if $i = j$ and 0 otherwise). The fourth derivative contributes only to the diagonal elements and its contribution increases rapidly with the row or column number. In contrast, the matrix elements of $q(x)$ are $O(1)$ at most[3]. It follows that the rows and columns of the Galerkin matrix $\vec{\vec{H}}$ will be increasingly dominated by the huge diagonal elements as $i,\,j \to \infty$.

Table 15.4 shows the Galerkin matrix for a particular (arbitrary) choice of $q(x)$. To make the diagonal dominance clearer, it is helpful to define a rescaled matrix $\vec{\vec{H}}$ via

$$h_{ij} = \frac{H_{ij}}{\sqrt{H_{ii}\,H_{jj}}} \tag{15.72}$$

---

[3]Note that $|q_{ij}| \le \pi \max|q(x)|$ since the integral defining these matrix elements is bounded by the maximum of its integrand, multiplied by the length of the integration interval. This estimate is conservative: The q_{ij} for any smooth $q(x)$ decrease rapidly as either i, j, or both together increase.

Table 15.4: The upper left 8×8 block of the Galerkin matrix $\vec{\vec{H}}$. This particular case is the Fourier cosine solution of $u_{xxxx} + [1 + 10\cosh(4\cos(x) + 0.4)]\,u = \exp(-\cos(x))$.

$i \setminus j$	0	1	2	3	4	5	6	7
0	123.2	56.7	98.2	19.4	21.7	2.9	2.4	0.2
1	56.7	193.6	53.8	84.7	15.8	17.0	2.2	1.8
2	98.2	53.8	154.5	42.2	71.1	13.9	15.4	2.1
3	19.4	84.7	42.2	205.9	40.3	69.5	13.7	15.3
4	21.7	15.8	71.1	40.3	379.3	40.1	69.4	13.7
5	2.9	17.0	13.9	69.5	40.1	748.2	40.1	69.4
6	2.4	2.2	15.4	13.7	69.4	40.1	1419.2	40.1
7	0.2	1.8	2.1	15.3	13.7	69.4	40.1	2524.2

Table 15.5: The upper left 12×12 block of the rescaled Galerkin matrix $\vec{\vec{h}}$, multiplied by 1000.

$i \setminus j$	0	1	2	3	4	5	6	7	8	9	10	11
0	1000	367	712	122	100	10	6	0	0	0	0	0
1	367	1000	311	424	58	45	4	3	0	0	0	0
2	712	311	1000	236	294	41	33	3	2	0	0	0
3	122	424	236	1000	144	177	25	21	2	1	0	0
4	100	58	294	144	1000	75	95	14	12	1	1	0
5	10	45	41	177	75	1000	39	51	8	7	1	1
6	6	4	33	25	95	39	1000	21	28	4	4	0
7	0	3	3	21	14	51	21	1000	12	17	3	3
8	0	0	2	2	12	8	28	12	1000	8	11	2
9	0	0	0	1	1	7	4	17	8	1000	5	7
10	0	0	0	0	1	1	4	3	11	5	1000	3
11	0	0	0	0	0	1	0	3	2	7	3	1000

which makes the diagonal elements of $\overset{\leftrightarrow}{h}$ equal to unity while preserving the symmetry of the matrix. Table 15.5 shows $\overset{\leftrightarrow}{h}$.

Unfortunately, $\overset{\leftrightarrow}{H}$ is only "asymptotically diagonal" because when the row and column indices are small, q_{ij} is the same order of magnitude as the diagonal elements. Delves and Freeman (1981) showed that one can exploit "asymptotic diagonality" by partitioning the Galerkin matrix $\overset{\leftrightarrow}{H}$ into

$$\overset{\leftrightarrow}{H} = \overset{\leftrightarrow}{H}_D(M) + \overset{\leftrightarrow}{H}_Q(M) \tag{15.73}$$

where $\overset{\leftrightarrow}{H}_D(M)$ is the upper left-hand $M \times M$ block of $\overset{\leftrightarrow}{H}$ plus all the diagonal elements and $\overset{\leftrightarrow}{H}_Q$ contains all the off-diagonal elements outside the upper left $M \times M$ block. When $M = 3$ and $N = 6$, for example,

$$\overset{\leftrightarrow}{H}_D = \begin{vmatrix} H_{11} & H_{12} & H_{13} & 0 & 0 & 0 \\ H_{21} & H_{22} & H_{23} & 0 & 0 & 0 \\ H_{31} & H_{32} & H_{33} & 0 & 0 & 0 \\ 0 & 0 & 0 & H_{44} & 0 & 0 \\ 0 & 0 & 0 & 0 & H_{55} & 0 \\ 0 & 0 & 0 & 0 & 0 & H_{66} \end{vmatrix} \tag{15.74}$$

$$\overset{\leftrightarrow}{H}_Q = \begin{vmatrix} 0 & 0 & 0 & H_{14} & H_{15} & H_{16} \\ 0 & 0 & 0 & H_{24} & H_{25} & H_{26} \\ 0 & 0 & 0 & H_{34} & H_{35} & H_{36} \\ H_{41} & H_{42} & H_{43} & 0 & H_{45} & H_{46} \\ H_{51} & H_{52} & H_{53} & H_{54} & 0 & H_{56} \\ H_{61} & H_{62} & H_{63} & H_{64} & H_{65} & 0 \end{vmatrix} \tag{15.75}$$

The Delves-Freeman iteration is then

$$\overset{\leftrightarrow}{H}_D \, \vec{a}^{n+1} = \vec{f} - \overset{\leftrightarrow}{H}_Q \, \vec{a}^n \tag{15.76}$$

or equivalently

$$\vec{a}^{n+1} = \vec{a}^n + \overset{\leftrightarrow}{H}_D^{-1} \, \vec{r}^n \quad \& \quad \vec{r}^n = \vec{f} - \overset{\leftrightarrow}{H} \, \vec{a}^n \tag{15.77}$$

which is identical in form to the preconditioned Richardson's iteration. (Only the matrices are different.)

One crucial observation is that because $\vec{r}^n$ is the residual of the differential equation, we do not need to explicitly multiply $\vec{a}^n$ by the full Galerkin matrix $\overset{\leftrightarrow}{H}$ to compute it. Instead — just as when using the grid point values as the unknowns — the Fast Fourier Transform can be applied to evaluate $\vec{r}^n$ indirectly. The cost of computing the residual is the same as for the finite difference preconditioning.

To invert $\overset{\leftrightarrow}{H}_D(M)$, write $\overset{\leftrightarrow}{H}_D(M)$ in partitioned block form as

$$\overset{\leftrightarrow}{H}_D = \begin{vmatrix} \overset{\leftrightarrow}{\Omega} & \overset{\leftrightarrow}{0} \\ \overset{\leftrightarrow}{0} & \overset{\leftrightarrow}{\Upsilon} \end{vmatrix} \tag{15.78}$$

where $\overset{\leftrightarrow}{\Omega}$ is an $M \times M$ matrix, $\overset{\leftrightarrow}{\Upsilon}$ is a diagonal matrix whose elements are $\Upsilon_{ij} = H_{jj}\delta_{ij}$, and the 0's in (15.78) represent blocks of zeros. Then

$$\overset{\leftrightarrow}{H}_D^{-1} = \begin{vmatrix} \overset{\leftrightarrow}{\Omega}^{-1} & \overset{\leftrightarrow}{0} \\ \overset{\leftrightarrow}{0} & \overset{\leftrightarrow}{\Upsilon}^{-1} \end{vmatrix} \tag{15.79}$$

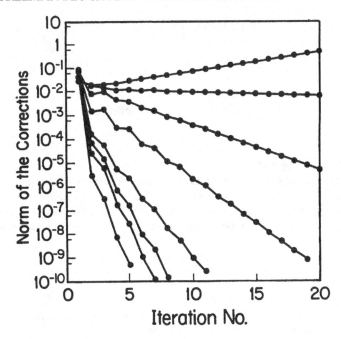

Figure 15.6: Norm of the corrections to the coefficients for the Delves-Freeman iteration for $M = 1, 2, \ldots, 8$ where M is the size of the upper left block of H_D. The top curve (iteration diverges) is $M = 1$ (H_D is diagonal). The rate of convergence increases monotonically with M. The ODE is $u_{xxxx} + [1 + 10\cosh(4\cos(x) + 4)]\,u = \exp(-\cos(x))$ [same example as Tables 15.4 & 15.5]. Fourier cosine basis with $N = 32$.

where $\vec{\vec{T}}^{-1}$ is the diagonal matrix whose elements are simply the reciprocals of the diagonal elements of $\vec{\vec{H}}$. Thus, the work in (15.79) is the cost of factoring an $M \times M$ dense matrix, which is $O([2/3]M^3)$, plus $(N - M)$ divisions.

Thus, if $M \ll N$, the cost per iteration is dominated by the $O(N\log_2 N)$ cost of computing the residual $\vec{r}^m$ via the Fast Fourier Transform — the same as for finite difference preconditioning.

Fig. 15.6 shows that when $M = 1$ (topmost graph), the iteration diverges. This is strong proof that the Galerkin matrix is not "diagonally dominant"! For $M > 2$, however, the iteration converges geometrically just like all the other variants of Richardson's method.

The rate of convergence increases rapidly with M, especially when M is small, so it is a good idea to use a moderate value for M rather than the smallest M which gives convergence.

The concept generalizes to partial differential equations and other basis sets, too. For example, $\cos(kx)\cos(my)$ is an eigenfunction of the Laplace operator, so the Fourier-Galerkin representation of

$$\nabla^2 u + q(x, y)\, u = f(x, y) \tag{15.80}$$

is "asymptotically diagonal". The only complication is that one should use the "speedometer" ordering of unknowns so that the columns and rows are numbered such that the smallest i, j correspond to the smallest values of $\sqrt{i^2 + j^2}$.

The Delves-Freeman idea works well with certain non-Fourier basis sets. For example,

Hermite functions $\psi_n(y)$ satisfy the eigenrelation

$$\frac{d^2}{dy^2}\psi_n(y) = \left[y^2 - (2n+1)\right]\psi_n(y) \tag{15.81}$$

The second derivative operator is not diagonal, but its n-dependent part contributes only to the diagonal. This implies that the Hermite-Galerkin matrix for

$$u_{yy} + q(y)\,u = f(y) \tag{15.82}$$

is asymptotically diagonal.

Unfortunately, there are complications for the most popular application — Chebyshev polynomials — because the Chebyshev-Galerkin representation of derivatives is *not* diagonal. With sufficient ingenuity, it is still possible to apply the same general idea or an extension. For example, in the next section, we show that the Chebyshev solution of

$$u_{xx} + q\,u = f(x) \tag{15.83}$$

may be converted into the solution of a bordered tridiagonal matrix (for constant q). When q varies with x, the Galerkin matrix is dense, but "asymptotically tridiagonal". One can generalize $\vec{\vec{H}}_D$ by replacing the diagonal matrix by a tridiagonal matrix; this block-plus-tridiagonal matrix is still cheap to invert (Appendix B).

In multiple dimensions with a Chebyshev basis, the natural extension of the Delves-Freeman method is to iterate using the Galerkin representation of a separable problem. As explained below in Sec. 15.11, the Galerkin representation of a separable elliptic problem has, with a little ingenuity, a simple form that allows very cheap iteration. The resulting iteration for nonseparable PDEs is preconditioned in spectral space. However, unlike Delves and Freeman's simple conception, the low degree basis functions are not treated differently from the rest of the basis.

In multiple dimensions with a Fourier or spherical harmonics basis, a simple block-plus-diagonal iteration of Delves and Freeman type may be very effective if the unknowns are ordered so that the low degree basis functions are clustered in the first few rows and columns.

The Delves-Freeman algorithm can also be extended to nonlinear problems through the Nonlinear Richardson's Iteration strategy which is described in Sec. 15.14. Boyd(1997d) is a successful illustration with a Fourier basis in one space dimension.

The moral of the story is that one can precondition either on the pseudospectral grid or in spectral space. The residual may be evaluated by the Fast Fourier Transform equally cheaply in either case. The decision between Orszag's finite difference iteration and the Delves-Freeman Galerkin iteration should be based on which of these choices leads to a preconditioning matrix, $\vec{\vec{H}}_D$, which is more easily inverted for a given problem.

15.10 Recursions & Formal Integration: Constant Coefficient ODEs

Clenshaw (1957) observed that for simple constant coefficient differential equations, Chebyshev non-interpolating methods generate *sparse* matrices. It is simplest to illustrate the idea by example. Consider

$$u_{xx} - q\,u = f(x) \tag{15.84}$$

where q is a constant. The pseudospectral discretization of Eq.(15.84) is *dense*. In contrast, Galerkin's method (using Chebyshev polynomials as "test" functions, not boundary-vanishing basis functions) gives a nearly triangular matrix. The reason is that

$$T_{k,xx} = \sum_{m=0}^{k-2} \frac{1}{c_m} k \left(k^2 - m^2 \right) T_m(x) \tag{15.85}$$

where $c_m = 2$ if $m = 0$ and $c_m = 1$ otherwise and where the sum is restricted to polynomials of the same parity as $T_k(x)$. Consequently, $(T_m, T_{k,xx}) = 0$ for all $m > (k-2)$. Unfortunately, the cost of backsolving is still $O(N^2)$. The problem is that the *derivative* of a Chebyshev polynomial involves *all* Chebyshev polynomials of the same parity and lower degree.

Clenshaw observed that the formula for the *integral* of a Chebyshev polynomial involves just *two* polynomials while the double integral involves only *three*:

$$\int T_n(x)\, dx = \frac{1}{2} \left\{ \frac{T_{n+1}(x)}{n+1} - \frac{T_{n-1}(x)}{n-1} \right\} \qquad n \geq 2 \tag{15.86}$$

This formula is actually equivalent to (15.39), the recurrence for computing the coefficients of df/dx from those of $f(x)$. Clenshaw himself exploited (15.39) to manipulate the Galerkin equations into tridiagonal form, but there are two other ways of obtaining the same result.

One is to apply the Mean-Weighted Residual method using the *second derivatives* of the Chebyshev polynomials as the "test" functions. These derivatives are the Gegenbauer polynomials of second order (as explained in Appendix A), so they are orthogonal, implying that the contributions of the second derivative to the Galerkin's matrix appear only on the diagonal, and they are complete, implying that these Gegenbauer polynomials are legitimate "test" functions. Via recurrence relations between Gegenbauer polynomials of different orders, $T_m(x)$ may be written as a linear combination of the three Gegenbauer polynomials of degrees $(m-2, m, m+2)$ so that the Galerkin matrix has only three non-zero elements in each row.

The third justification — completely equivalent to the first two — is to formally integrate the equation twice to obtain

$$u - q \iint u = \iint f(t) + A + B\, x \tag{15.87}$$

where A and B are arbitary constants determined by the boundary conditions. If we then demand that (15.87) should be orthogonal to $T_2(x), \ldots, T_n(x)$, we obtain

$$a_n + q \left\{ \left[\frac{c_{n-2}}{4n(n-1)} \right] a_{n-2} - \left[\frac{\sigma_n}{2(n^2-1)} \right] a_n - \left[\frac{\sigma_{n+2}}{4n(n+1)} \right] a_{n+2} \right\} \tag{15.88}$$

$$= \left[\frac{c_{n-2}}{4n(n-1)} \right] f_{n-2} - \left[\frac{\sigma_n}{2(n-1)} \right] f_n - \left[\frac{\sigma_{n+2}}{4n(n+1)} \right] f_{n+2}$$

where $\sigma_n = 1$ for $n < N - 1$ and 0 for all larger N and where the f_n are the Chebyshev coefficients of the forcing function, $f(x)$.

The resulting matrix contains two full rows to impose the boundary conditions, so it is only quasi-tridiagonal. However, the methods for "bordered" matrices (Appendix B) compute the coefficients a_n in "roughly the number of operations required to solve pentadiagonal systems of equations", to quote Gottlieb & Orszag (1977, pg. 120). Canuto *et al.* (1988) give a good discussion of generalizations to other boundary conditions and so on.

Clenshaw and others applied his recursive techniques to many different equations. If the coefficients are polynomials, the result will again be a banded matrix. If q includes a term in x^2, the matrix is quasi-pentadiagonal. If q has a term in x^4, the matrix is quasi-heptadiagonal, and so on.

The most complete archive of early work is Chapter 5 of Fox & Parker (1968). The authors point out that Clenshaw's strategy not only creates sparsity, but also may improve accuracy. The reason is that the Chebyshev series of a derivative always converges more slowly than that of $u(x)$ itself. After integration, accuracy is no longer limited by that of the slowly convergent series for the second derivative, but only by that of $u(x)$ itself.

Of course, as stressed many times above, factors of N^2 are irrelevant for exponentially convergent approximations when N is large. The extra accuracy and sparse matrices produced by the double integration are most valuable for paper-and-pencil calculations, or when N is small. In consequence, these integrate-the-ODE methods have gone rather out of fashion.

Zebib (1984) is a return to this idea: he assumes a Chebyshev series for the highest derivative (rather than $u(x)$ itself) and obtains formulas for the contributions of lower derivatives by applying (15.85). Although this procedure is complicated — especially for fourth order equations — it both improves accuracy and eliminates very large, unphysical complex eigenvalues from the Chebyshev discretization of the Orr-Sommerfeld equation (Zebib, 1987b).

Coutsias, Hagstrom and Torres(1996) have extended Clenshaw's method to all the standard orthogonal polynomials including Laguerre and Hermite polynomials. They also provide useful identities, recursions, and estimates of condition number. Clenshaw's method gives banded matrices when the coefficients of the ODE are *rational* functions as well as when the coefficients are polynomials. Coutsias *et al.* show that by changing the coordinate using a rational function, it is possible to resolve narrow boundary layers without sacrificing a banded matrix representation, as illustrated with numerical examples.

Thus, the integrate-the-ODE methods have a small but useful role in spectral methods even today.

15.11   Direct Methods for Separable PDE's

The earliest work on spectral methods for *separable* PDE's is Haidvogel and Zang (1979). They offer two algorithms for exploiting the separability, one iterative and one direct. We shall discuss only the latter.

The key to the Haidvogel-Zang algorithm is a procedure which can also be applied to one-dimensional problems, so we will begin with the ODE case. When

$$[p(x)\,u_x]_x - q\,u = f(x) \tag{15.89}$$

is discretized, it generates the matrix problem

$$\left(\vec{\vec{\Lambda}}_x - q\,\vec{\vec{I}}\right)\vec{a} = \vec{f} \tag{15.90}$$

where $\vec{\vec{I}}$ is the identity matrix. We could absorb the constant q into $\vec{\vec{\Lambda}}_x$, of course, but have split it off for future convenience.

If the collocation matrix $\vec{\vec{\Lambda}}_x$ is diagonalizable, we can write

$$\vec{\vec{\Lambda}}_x = \vec{\vec{E}}_x\,\tilde{\Lambda}_x\,\vec{\vec{E}}_x^{-1} \tag{15.91}$$

where $\vec{\vec{E}}_x$ is the matrix whose columns are the n eigenvectors of $\vec{\vec{\Lambda}}_x$ and where $\tilde{\Lambda}_x$ is the *diagonal* matrix whose elements are the corresponding eigenvalues. Eq. (15.90) becomes

$$\vec{\vec{E}}_x \left(\tilde{\Lambda}_x - q \vec{\vec{I}} \right) \vec{\vec{E}}_x^{-1} \vec{a} = \vec{f} \tag{15.92}$$

Multiplying through by $\vec{\vec{E}}_x^{-1}$ gives

$$\left(\tilde{\Lambda}_x - q \vec{\vec{I}} \right) \vec{\vec{E}}_x^{-1} \vec{a} = \vec{\vec{E}}_x^{-1} \vec{f} \tag{15.93}$$

Define

$$\vec{b} \equiv \vec{\vec{E}}_x^{-1} \vec{a} \qquad ; \qquad \vec{g} \equiv \vec{\vec{E}}_x^{-1} \vec{f} \tag{15.94}$$

Eq. (15.93) becomes the trivial problem

$$(\lambda_i - q) \, b_i = g_i \qquad\qquad i = 0, \dots, N+1 \tag{15.95}$$

After solving (15.95), we obtain $\vec{a}$ from $\vec{b}$ by multiplying the latter by $\vec{\vec{E}}$.

Once the eigenvectors have been computed (in $O[N^3]$ operations), each solution for new values of q and/or $\vec{f}$ costs only $O(6N^2)$ operations, that is, three multiplications of a column vector by a square matrix. In the semi-implicit time-stepping algorithm of Orszag and Patera (1982), this tactic is applied in the radial direction. It minimizes storage over the more obvious technique of Gaussian elimination because the radial boundary value problems are identical for different polar and axial wavenumbers except for a different value of the constant q. Thus, they only need to store two matrices, $\vec{\vec{E}}_x$ and its inverse, whereas with Gaussian elimination it would be necessary to store a matrix of the same size for each pair of polar and axial wavenumbers (several hundred in all!).

This trick is also a direct method of attacking two-dimensional separable problems because the diagonalization still works even if q is a *differential operator*, not a constant, so long as the operator involves only the other coordinate, y. If this condition is met, then $\vec{\vec{E}}_x$ and $\tilde{\Lambda}_x$ will be independent of y. For example, if

$$q \equiv (s(y) \, \partial_y) \, \partial_y + t(y), \tag{15.96}$$

then if we discretize only the x-derivatives, (15.95) becomes a set of $(N+1)$ *uncoupled* ordinary differential equations in y:

$$[s(y) \, b_{i,y}]_y + [t(y) - \lambda_i] \, b_i(y) = -g_i(y) \tag{15.97}$$

This is precisely what is generated by the analytical method of separation of variables, applied to the original PDE. The condition that the discretized x-derivative matrix can be diagonalized independent of y, which demands that the operator q involves only y derivatives and functions of y, is just the condition that the original problem is separable. In fact, the Haidvogel-Zang direct method *is* separation-of-variables.

An obvious question is: Why do we need this post-collocation variant of separation of variables? The answer is that unless the boundary conditions are spatial periodicity, the separation-of-variables eigenfunction series will have a slow, algebraic rate of convergence. To obtain high accuracy, we need Chebyshev polynomials in x.[4]

---

[4]In fairness, *sometimes* Chebyshev series converge as slowly as eigenseries for time-dependent problems when the initial conditions fail to satisfy sufficient "compatibility conditions" at the boundaries (Boyd and Flyer, 1998).

One way of looking at the Haidvogel-Zang procedure is that it is a method of solving the set of $(N + 1)$ coupled ordinary differential equations for the a_i by making a "smart" change-of-basis [in the space of $(N + 1)$-dimensional column vectors] to a new set of coefficients, the b_i, such that the differential equations uncouple.

Solving N ordinary differential equations is still moderately expensive, but for Poisson's equation (or any which is constant coefficient in y), Haidvogel and Zang note that Clenshaw's method (previous section!) applies at a cost of N operations per ODE in y. Thus, after an $O(N^3)$ preprocessing step, we can solve a separable PDE for different $f(x, y)$ at a cost of $O(N^2)$ per solution. The efficiency of the Haidvogel and Zang procedure can be extended to equations with y-dependent coefficients by using the preconditioned Richardson's iteration as described in earlier sections. (Note that variable x-coefficients are allowed in either case.)

The preprocessing step, which is a one-dimensional eigenproblem, costs $O(N^3)$ operations whereas fast direct finite difference methods require only $O(N^2 \log_2 N)$ operations where the tensor product grid has a total of N^2 points. However, on a 1997 personal computer, the cost of solving an $N = 100$ one-dimensional eigenproblem was only half a second! The preprocessing step was expensive in 1979, but not today.

If the separable elliptic problem is part of a semi-implicit time-marching algorithm, then the cost of the $O(N^3)$ preprocessing step for the spectral method is amortized over thousands of time steps. The actual cost per time step is merely that for a single backsolve, which is $O(N^2)$ — directly proportional to the number of grid points — for both spectral and finite difference methods.

Haldewang, Labrosse, Abboudi and Deville (1984) compared three algorithms for the Helmholtz equation. The Haidvogel-Zang strategy is to diagonalize in two dimensions and solve tridiagonal systems in the other, and proved the fastest. However, Haldewang *et al.* found moderately severe roundoff problems with this method, losing five decimal digits for N as small as 64.

Full diagonalization (in all three dimensions) was slightly slower but easier to program. This algorithm became very popular in France and perhaps deserves a wider range of applications.[5]

The third algorithm was a finite difference-preconditioned iteration. This is cheap for separable problems because the preconditioning can be done using the special "fast direct methods" for separable problems, which cost only $O(N^3 \log_2(N))$ operations on an N^3 grid in three dimensions.[6] The preconditioned iteration was an order of magnitude slower than the Haidvogel-Zang procedure at all N, but the fast direct finite difference solver can be used as a preconditioner for nonseparable problems. The iteration loses little accuracy to round-off error even at very large N.

Liffman(1996) extended the Haidvogel-Zang eigenanalysis to incorporate very general (Robin) boundary conditions. This has been widely used by the Nice group under the name of "matrix diagonalization" since Ehrenstein and Peyret (1989).

Patera (1986) combined the Haidvogel-Zang algorithm with static condensation to create a spectral element Poisson-solver with an $O(N^{5/2})$ operation count, but this algorithm has largely been replaced by multigrid.

Siyyam and Syam(1997) offer a Chebyshev-tau alternative to the Haidvogel-Zang technique. Their experiments show that their method is faster, at least for large N, but their algorithm has not been extended to PDEs with spatially-varying coefficients.

---

[5]Because there are often several algorithms of roughly equal effectiveness for a given problem, numerical analysis is prone to national and individual fashions that are determined more by accidents of education and exposure than by rigorous mathematical justification.

[6]The FISHPACK library, written by John Adams and available through the National Center for Atmospheric Research, is an excellent collection of fast direct methods.

Ierley(1997) has developed a similar fast algorithm for the Laplace operator in an arbitrary number of space dimensions. One key ingredient is to use Jacobi polynomials rather than Chebyshev polynomials.

J. Shen (1994a,1995b) and Lopez and Shen(1998) developed Legendre-Galerkin schemes which cost $O(N^3)$ operations for second and fourth order constant coefficient elliptic equations. With the basis $\phi_j(x) = L_{j+2}(x) - L_j(x)$, the weak form of the second derivative is a *diagonal* matrix. He developed similar Chebyshev algorithms in Shen(1995a). Both his Chebyshev and Legendre forms have condition numbers of $O(N^4)$ for fourth order problems, versus $O(N^8)$ for some other spectral algorithms, so that it is possible to solve the biharmonic equation with $O(500)$ unknowns in each coordinate to high accuracy. His Chebyshev algorithm costs $O(N^4)$ for fourth order problems, but Bjørstad and Tjøstheim (1997) developed an $O(N^3)$ modification, and made a number of further improvements in both Legendre and Chebyshev algorithms for fourth order problems.

Braverman, Israeli, Averbuch and Vozovoi (1998) developed a Poisson solver in three dimensions which employs a Fourier basis even though the boundary conditions are non-periodic. A clever regularization method restores an exponential rate of convergence, which would normally be lost by non-periodic application of a Fourier series.

15.12 Fast Iterations for Almost Separable PDEs

D'yakonov(1961) and Concus and Golub (1973) noted that non-separable elliptic equations can be solved very efficiently by using an iteration in which a separable differential equation is solved at each iteration. (Obviously, the coefficients of the separable BVP are chosen to approximate those of the nonseparable problem as closely as possible.)

Several groups have developed efficient spectral algorithms for nonseparable elliptic equations using a D'yakonov/Concus-Golub iteration. Guillard and Desideri(1990) combined the Minimum Residual Richardson's iteration with two different spectral preconditioners that both solved separable problems. The more efficient preconditioner required computing eigenvalues and eigenvectors in a preprocessing step as in the Haidvogel and Zang procedure. Strain(1994) solved non-separable periodic problems with a Fourier basis where the solution of the separable differential equation is almost trivial. Zhao and Yedlin(1994) showed that the efficiency of solve-a-separable-PDE iterations is as good as multigrid, at least sometimes. Dimitropoulus and Beris (1997) found that the combination of a Concus-Golub iteration with a fast spectral Helmholtz solver did not always converge for problems with strongly varying coefficients. However, when this was embedded as an inner iteration with the biconjugate gradient method, BiCGstab(m), as the outer iteration, the result was a very robust and efficient algorithm. Typically only two Concus-Golub iterations were needed between each outer iteration; the biconjugate gradient method never needed more than nine iterations.

The title of this section, "almost separable", is a little misleading. Although these iterations converge most rapidly if the PDE is only a slight perturbation of a separable problem, the examples of Zhao and Yedlin, Strain, and Dimitropoulus and Beris show that often the D'yakonov/Concus-Golub algorithm converges very rapidly even when the BVP only vaguely resembles a separable problem. This is especially true when the iteration is combined with other convergence-accelerating strategies such as the biconjugate gradient method used as an outer iteration by Dimitropoulus and Beris.

15.13   Positive Definite and Indefinite Matrices

As noted in Sec. 2, it is very important that the pseudospectral matrix be positive definite because most iterations are analogous to time-integration of the diffusion equation

$$\vec{u}_t = -\vec{\vec{\Lambda}}\,\vec{u} + \vec{f} \tag{15.98}$$

where $\vec{\vec{\Lambda}}$ is the square matrix which is the discretization of the differential operator. If $\vec{\vec{\Lambda}}$ has only positive eigenvalues, then the solution of (15.98) will decay monotonically to the desired steady-state which solves $\vec{\vec{\Lambda}}\vec{u} = \vec{f}$. However, if $\vec{\vec{\Lambda}}$ is indefinite, that is, has eigenvalues of both signs, then the negative eigenvalue modes will grow exponentially with time and the iteration will diverge.

This would seem to be a *very* serious restriction because even so simple a problem as the one-dimensional Helmholtz equation

$$-u_{xx} - q\,u = f(x) \qquad\qquad u(0) = u(\pi) = 0 \tag{15.99}$$

with periodic boundary conditions has the eigenvalues

$$\lambda_n = n^2 - q \qquad\qquad n = 1, 2, \ldots \tag{15.100}$$

The lowest eigenvalue is negative for $q > 1$, two eigenvalues are negative for $q > 4$, three for $q > 9$, and so on. Richardson's iteration would diverge.

Happily, both preconditioning and multigrid *usually* solve the problem of a *finite* number of *small* eigenvalues of the wrong sign. The reason that preconditioning is effective is that the offending modes are the ones with the simplest spatial structure. If $q = 8$, for example, only the modes $\sin(x)$ and $\sin(2x)$ are unstable. However, these modes are precisely the ones which will be accurately resolved by both the finite difference and pseudospectral approximations. Consequently, $\sin(x)$ and $\sin(2x)$ (or more precisely, the column vectors containing the grid point values of these functions) are both eigenvectors of the pre-conditioned matrix $\vec{\vec{H}}^{-1}\vec{\vec{\Lambda}}$ with eigenvalues approximately equal to 1.

There is a snare: when q is very close to n^2, one eigenvalue of $\vec{\vec{\Lambda}}$ may be very, very close to zero — smaller in magnitude than the small error in the finite difference approximation of this same mode. Canuto *et al.* (1988, pg. 143) show that the pre-conditioned eigenvalue is

$$\lambda_n^{(p)} = \frac{n^2 - q}{n^2 \sin^2(n\,\Delta x/2)/(n\,\Delta x/2)^2 - q} \tag{15.101}$$

For *fixed* q & n, $\lambda_n(p) \to 1$ in the limit $\Delta x \to 0$. However, for fixed Δx, there is always a range of q sufficiently close to n^2 so that $\lambda_n^{(p)}$ is negative and Richardson's iteration will then diverge for any time step.

Since the grid spacing Δx always is small, one has to be rather unlucky for a pre-conditioned iteration to diverge, but clearly, it can happen. It is encouraging, however, that preconditioning will remove the problem of matrix indefiniteness *most* of the time.

Multigrid is intrinsically just an efficient way of performing Richardson's iteration. It will therefore diverge due to matrix indefiniteness whenever Richardson's would, too. However, a common practice in multigrid — even when the matrix is positive definite — is to apply Gaussian elimination instead of iteration on the coarsest grid. The obvious reason is that when the coarse grid contains only nine points, Gaussian elimination is ridiculously cheap. The subtle, empirical reason is that using direct methods on the coarsest grid seems to give additional stability and improved convergence rate to the whole iteration.

When the matrix is indefinite but has only one or two negative eigenvalues, this practice of Gaussian elimination on the coarsest grid is a life-saver. Although the coarse grid cannot resolve the high modes of the operator, it can resolve — at least crudely — the lowest one or two modes which (otherwise) would make the iteration diverge.

A little care is needed. If there are several unstable modes, one needs a coarsest grid which has at least a moderate number of points.

There is, alas, a small class of problems for which neither multigrid nor preconditioning will help: those with an *infinite* number of eigenvalues of negative sign. An example is Laplace's Tidal equation for waves in a semi-infinite atmosphere (Chapman and Lindzen, 1970). The tidal equation is very messy because it includes complicated trigonometric terms due to the earth's sphericity. However,

$$u_{yy} + (\omega^2 - y^2) u_{zz} = f(y, z) \tag{15.102}$$

$$u(\pm L, z) = 0 \quad \& \quad u(y, 0) = 0 \quad \& \quad u(y, z) \text{ bounded as } z \to \infty \tag{15.103}$$

displays the same qualitative behavior.

Separation-of-variables gives the latitudinal eigenequation

$$u_{yy} + \lambda (\omega^2 - y^2) u = 0, \tag{15.104}$$

the parabolic cylinder equation. If the boundary $L > \omega$, then (15.104) has an infinite number of solutions with positive eigenvalues. The eigenmodes oscillate on $y \in [-\omega, \omega]$ and decay exponentially for larger y; they may be accurately approximated by Hermite functions (as is also true of the corresponding tidal modes). The vertical structure of these modes is oscillatory.

There is also an infinite number of modes with negative eigenvalues. These oscillate on the intervals $y \in [-L, -\omega]$ & $y \in [\omega, L]$ and are exponentially small at $y = 0$. These modes decay exponentially as $z \to \infty$.

Eq. (15.102) is rather peculiar: an equation of mixed elliptic-hyperbolic type. The boundary conditions are also a little peculiar, too, in that either oscillatory or exponential behavior is allowed as $z \to \infty$. Nevertheless, the linear wave modes of a rotating, semi-infinite atmosphere are described by a mixed elliptic-hyperbolic equation similar to (15.102), and the modes really do fall into two infinite classes: one with $\lambda > 0$ and the other with $\lambda < 0$. No iterative method has ever been successfully applied to these problems. Lindzen (1970), Schoeberl & Geller (1977) and Forbes & Garrett (1976) have all obtained good results by combining finite differences with banded Gaussian elimination.

Navarra(1987) combined finite differences in z with spherical harmonics in latitude. By applying Arnoldi's matrix-solving iteration, Navarra solved mixed-type boundary value problems with as many as 13,000 unknowns.

In principle, indefiniteness may be cured by squaring the matrix, i. e.

$$\vec{\Lambda}^T \vec{\Lambda} u = \vec{\Lambda}^T f \tag{15.105}$$

where superscript "T" denotes the matrix transpose. The disadvantage of matrix-squaring is that if the condition number of $\vec{\Lambda}$ is $O(N^2)$, then the condition number of the positive definite matrix $\vec{\Lambda}^T \vec{\Lambda}$ is $O(N^4)$.

However, we have already seen that preconditioning is strongly recommended even for the unsquared matrix. For $\vec{\Lambda}$ and $\vec{\Lambda}^T \vec{\Lambda}$, finite difference preconditioning will give a condition number for the squared matrix which is $O(1)$, independent of N. This tactic of squaring the matrix should work even for the difficult case of the Laplace Tidal equation, provided that none of its eigenvalues are too close to zero[7].

What if $\vec{\vec{\Lambda}}$ has *complex* eigenvalues? First, the transpose in (15.105) should be interpreted as the Hermitian adjoint. Second, the *Chebyshev* acceleration of Richardson's iteration (first edition of this book) should be avoided unless the imaginary parts are small. The Richardson's iteration, with or without preconditioning, will still converge rapidly as long as the real parts of the eigenvalues are all positive, and better still, so will its Minimum Residual (MRR) variant. The imaginary parts of the λ_j simply mean that the decay will be oscillatory instead of monotonic for some or all of the eigenmodes.

Alas, even squaring the matrix won't help when the difficulty, as for the Helmholtz equation (15.99), is an eigenvalue which is approximately zero. (Squaring the matrix makes the corresponding eigenvalue even smaller!) Still, the conclusion is that iterative methods can be applied — with a little care — to the vast majority of real-life problems including those with eigenvalues of both signs.

15.14   Nonlinear Iterations & the Preconditioned Newton Flow

The usual strategy to solve a nonlinear boundary value or eigenvalue problem is, after discretization, to apply Newton's iteration. When the linear matrix problems of each Newton iteration are themselves solved by iteration, it is often desirable to avoid an iteration-within-an-iteration by combining Newton's iteration and the preconditioned Richardson's iteration into a single step. To do so, it is useful to employ a conceptual device that we used to motivate Richardson's method: interpreting an iteration as the temporal discretization of a time-dependent problem[8].

Newton's iteration for the system of N equations

$$\vec{r}(\vec{u}) = 0 \tag{15.106}$$

is

$$\vec{\vec{J}}(\vec{u}^{n+1} - \vec{u}^n) = \vec{r}(\vec{u}^n) \tag{15.107}$$

where $\vec{\vec{J}}$ is the usual Jacobian matrix whose elements are

$$J_{ij} = \left. \frac{\partial r_i}{\partial u_j} \right|_{\vec{u}=\vec{u}^n} \tag{15.108}$$

This is identical with the result of integrating the system of differential equations

$$\frac{d\vec{u}}{dT} = -\left(\frac{1}{\vec{\vec{J}}}\right)\vec{r} \qquad \text{["Newton flow" equation]} \tag{15.109}$$

by the Euler forward method with unit step in the pseudotime, T.

If we expand $\vec{r}(\vec{u})$ in an N-dimensional Taylor series centered on a root $\vec{u}_0$, then (15.109) becomes locally

$$\frac{d\vec{u}}{dT} = -\frac{1}{\vec{\vec{J}}}\left[\vec{r}(\vec{u}_0) + \vec{\vec{J}}(\vec{u} - \vec{u}_0) + O\left((\vec{u} - \vec{u}_0)^2\right)\right] \tag{15.110}$$

$$= -(\vec{u} - \vec{u}_0) + O\left((\vec{u} - \vec{u}_0)^2\right) \tag{15.111}$$

---

[7]To minimize the condition number, one should solve (15.105) as the system $\vec{\vec{\Lambda}} y = f$, $\vec{\vec{\Lambda}}^T x = y$.

[8]Chu(1988) is a good review of other examples of "Continuous realizations of Iterative Processes", to quote his title.

In the language of the theory of differential equations, (15.111) is a linearization of the nonlinear system about a critical point. The linearization shows that every root of $\vec{r}(\vec{u}) = 0$ is a *local attractor*. The exact "Newton flow" converges geometrically on the root since one can prove without approximation that

$$\vec{r}(\vec{u}[T]) = e^{-T}\vec{r}(\vec{u}[0]) \tag{15.112}$$

Newton's iteration actually converges much faster than (15.112); the number of correct digits roughly doubles with each iteration. The reason for the faster convergence is that if we ignore the quadratic terms in (15.111), Euler's method with unit time step reduces $(\vec{u} - \vec{u}_0)$ to 0 instead of to $36\%(= \exp(-1))$ of its initial value. Thus, poor accuracy in time integration brings us to the root faster than solving the Newton flow accurately in pseudotime.

Unfortunately, Newton's method requires inverting the Jacobian matrix at each time step. When $\vec{r}(\vec{u}) = 0$ is the pseudospectral discretization of a differential equation, the Jacobian matrix is dense and costly to invert.

Fortunately, if the Newton flow is *modified* by replacing the inverse-Jacobian matrix by a matrix which is nonsingular at the roots of $\vec{r}(\vec{u})$, but is otherwise *arbitrary*, the roots of $\vec{r}(\vec{u})$ are still critical points of the differential equation.

Not all approximations of the Jacobian matrix will succeed because the modifications may turn some roots from attractors into repellors. Suppose, however, the inverse-Jacobian is replaced by the inverse of a *preconditioning* matrix:

$$\frac{d\vec{u}}{dT} = -\frac{1}{\overrightarrow{\overrightarrow{H}}}\vec{r}(\vec{u}) \qquad \text{["Pre-conditioned Newton flow"]} \tag{15.113}$$

Linearization near a root gives

$$\frac{d\vec{u}}{dT} = -\overrightarrow{\overrightarrow{H}}^{-1}\overrightarrow{\overrightarrow{J}}(\vec{u} - \vec{u}_0) + O\left((\vec{u} - \vec{u}_0)^2\right) \tag{15.114}$$

Orszag (1980) showed that if $\overrightarrow{\overrightarrow{H}}$ is the finite difference approximation to $\overrightarrow{\overrightarrow{J}}$, then all the eigenvalues of $(1/\overrightarrow{\overrightarrow{H}})\overrightarrow{\overrightarrow{J}}$ will lie on the interval $\alpha \in [1, 2.47]$ (as already explained in Sec. 4). If the error $(\vec{u} - \vec{u}_0)$ is decomposed into the eigenvectors of $(1/\overrightarrow{\overrightarrow{H}})\overrightarrow{\overrightarrow{J}}$, then the error in different components must decay at rates ranging between $\exp(-T)$ and $\exp(-2.47T)$.

Because of these varying rates of decay, it is not possible to find a magic time step that will keep the quadratic, digit-doubling convergence of Newton's iteration. However, the Euler forward scheme does give very rapid convergence: with a time step of $4/7$, one can guarantee that each component of the error will decrease by at least a factor of 2.33 at each iteration — just as fast as for a *linear* problem. Since the finite difference matrix can be inverted in $O(1/N^2)$ of the cost for inverting the pseudospectral Jacobian where N is the number of basis functions in each dimension, the preconditioned Newton flow is much cheaper than the classical Newton's iteration.

The iterative form of (15.114) is

$$\overrightarrow{\overrightarrow{H}}(\vec{u}^{n+1} - \vec{u}^n) = -\tau\,\vec{r}(\vec{u}^n) \qquad \text{[Nonlinear Richardson Iteration]} \tag{15.115}$$

where $\overrightarrow{\overrightarrow{H}}$ is *any* acceptable preconditioning matrix for the Jacobian matrix and τ is the pseudotime step. It is appropriate to dub (15.115) the "Nonlinear Richardson Iteration" because it reduces to the standard Richardson iteration whenever $\vec{r}$ is a linear function of $\vec{u}$.

This "preconditioned Newton flow" derivation of (15.115) is taken from Boyd (1989a), but the same algorithm was independently invented through different reasoning in the technical report by Streett and Zang (1984)[9]. Eq. (15.115) has probably been re-invented half a dozen times.

Simple though it is, the Nonlinear Richardson's Iteration (NRI) is very useful. It is silly and expensive to apply Richardson's iteration to solve the linearized differential equations of a strict Newton method; one is then wasting many iterations to converge on an intermediate Newton iterate which is of no physical interest. With (15.115), every iteration reduces the *nonlinear* residual and brings us closer to the root.

Numerical examples are given in Boyd (1989a).

15.15   Summary & Proverbs

MORAL PRINCIPLE #1: Never solve a matrix using anything but Gaussian elimination unless you really have to.

This may seem strange advice to conclude a chapter mostly on iterative methods. Nevertheless, it is more than a little important to remember that there are no style points in engineering. Iterative methods, like mules, are useful, reliable, balky, and often infuriating. When N is small enough so that Gaussian elimination is practical, don't fool around. Be direct!

Note further that for constant coefficient ordinary differential equations and for separable partial differential equations, Galerkin/recurrence relation methods yield *sparse* matrices, which are cheap to solve using Guassian elimination.

MORAL PRINCIPLE #2: Never apply an iteration without first preconditioning it.

Well, *almost* never. However, un-preconditioned iterations converge *very* slowly and are much less reliable because a single eigenvalue of the wrong sign will destroy them.

MORAL PRINCIPLE #3: Never use multigrid as a purely iterative method; apply Gaussian elimination on the coarsest grid.

Besides being the difference between convergence and divergence if the matrix is indefinite, elimination on the coarsest grid seems to improve the rate of convergence even for positive definite matrices.

Good, reliable iterative methods for pseudospectral matrix problems are now available, and they have enormously extended the range of spectral algorithms. Iterations make it feasible to apply Chebyshev methods to difficult multi-dimensional boundary value problems; iterations make it possible to use semi-implicit time-stepping methods to computed wall-bounded flows. Nevertheless, this is still a frontier of intense research — in part because iterative methods are so very *important* to spectral algorithms — and the jury is still out on the relative merits of many competing strategies. This chapter is not the last word on iterations, but only a summary of the beginning.

----

[9]See also Canuto *et al.* (1988, pg. 390).

Chapter 16

The Many Uses of Coordinate Transformations

"There are nine and sixty ways of constructing tribal lays"
— R. Kipling, *In the Neolithic Age*

16.1   Introduction

A change of coordinates is a popular tool in elementary calculus and physics. Such transformations have an equally important role in spectral methods. In the next section, the Chebyshev polynomial-to-cosine change-of-variable greatly simplifies computer programs for solving differential equations.

However, there are many other uses of mapping. In this chapter, we concentrate on one-dimensional transformations, whose mechanics is the theme of Sec. 3. In Sec. 4, we show how infinite and semi-infinite intervals can be mapped to [-1, 1] so that Chebyshev polynomials can then be applied. When the flow has regions of very rapid change — near-singularities, internal boundary layers, and so on — maps that give high resolution where the gradients are large can tremendously improve efficiency. We discuss a number of specific cases in Secs. 5-7. In Sec. 8, we discuss two-dimensional mappings and singularity subtraction for coping with "corner" singularities. Finally, in the last part of the chapter, we give a very brief description of the new frontier of adaptive-grid pseudospectral methods.

16.2   Programming Chebyshev Methods

The Chebyshev polynomials can be computed via the 3-term recurrence relation

$$
\begin{aligned}
T_0 &\equiv 1 &&; & T_1 &\equiv x \\
T_{n+1} &= 2\,x\,T_n - T_{n-1} && & n &= 1, \ldots\,.
\end{aligned}
\tag{16.1}
$$

This is numerically stable, and the cost is $O(N)$ to evaluate all the polynomials through T_N at a given x. Since the derivatives of the Chebyshev polynomials are Gegenbauer polynomials, it follows that we can evaluate these derivatives via three-term recurrences, too

(Appendix A). Indeed, through a similar three-term recurrence derived from (16.1), (Fox & Parker, 1968 and Boyd, 1978a), we can sum a Chebyshev series in $O(N)$ operations without evaluating the individual T_n.

As simple and effective as these recurrences are, however, it is often easier to exploit the transformation

$$x = \cos(t) \tag{16.2}$$

which converts the Chebyshev series into a Fourier cosine series:

$$T_n(x) \equiv \cos(nt) \tag{16.3}$$

By using the tables given in Appendix E, we can easily evaluate the derivatives of the Chebyshev polynomials directly in terms of derivatives of the cosine and sine. For example

$$\frac{dT_n(x)}{dx} = \frac{n \sin(nt)}{\sin(t)} \tag{16.4}$$

This is just the trigonometric representation of $n U_{n-1}(x)$ where $U_n(x)$ is the Chebyshev polynomial of the second kind defined in Appendix A. Similarly

$$\frac{d^2 T_n(x)}{dx^2} = \frac{1}{\sin^3(t)} \left\{ \sin(t) \left[-n^2 \cos(nt) \right] - \cos(t) \left[-n \sin(nt) \right] \right\} \tag{16.5}$$

Table 16.1 is a short FORTRAN subroutine that computes the n-th Chebyshev polynomial and its first four derivatives. Its simplicity is remarkable.

One modest defect of (16.4) and (16.5) is that the denominators are singular at the endpoints. However, this is irrelevant to the Chebyshev "roots" grid because all grid points are on the *interior* of the interval. The problem also disappears for the "extrema" (Gauss-Lobatto) grid if the boundary conditions are Dirichlet, that is, do not involve derivatives.

When boundary derivatives are needed, one must modify the subroutine in Table 16.1 by adding an IF statement to switch to an analytical formula for the endpoint values of the Chebyshev derivatives of all orders:

$$\text{(i)} \quad \frac{d}{dx} T_n(x) = n U_{n-1}(x) \tag{16.6}$$

$$\text{(ii)} \quad \left. \frac{d^p T_n}{dx^p} \right|_{x=\pm 1} = (\pm 1)^{n+p} \prod_{k=0}^{p-1} \frac{n^2 - k^2}{2k+1} \tag{16.7}$$

Instead of using formulas like (16.3) to (16.5) to compute in the original coordinate x, one may alternatively convert the original problem into an equivalent differential equations on $t \in [0, \pi]$ and then solve the result using a Fourier cosine series. For example,

$$u_{xx} - q u = f(x) \qquad u(-1) = u(1) = 0 \qquad x \in [-1, 1] \tag{16.8}$$

becomes

$$\sin(t) u_{tt} - \cos(t) u_t - q \sin^3(t) u = \sin^3(t) f(\cos[t]) \tag{16.9}$$

$$u(0) = u(\pi) = 0 \qquad t \in [0, \pi]$$

Which of these two forms is simpler is a matter of taste. My personal preference is to solve the problem in x, burying the trigonometric formulas in the subroutines that evaluate derivatives.

Table 16.1: A sample subroutine to compute the Chebyshev polynomials and their derivatives via the derivatives of the cosine functions.

```
       SUBROUTINE BASIS(N,X,TN,TNX,TNXX,TNXXX,TNXXXX)
C      INPUT: X    (on interval [-1, 1] )
C      OUTPUT:  TN, TNX, TNXX, TNXXX, TNXXXX are the N-th Chebyshev
C               polynomials and their first four derivatives.
C      T is the trigonometric argument (on interval [0, pi]).
       T = ACOS(X)
       TN = COS(FLOAT(N)*T)
C          Derivatives of cos(Nt) with respect to t:
           PT = - FLOAT(N)  * SIN(FLOAT(N)*T)
           PTT = - FLOAT(N)*FLOAT(N)  * TN
           PTTT = - FLOAT(N)*FLOAT(N)  * PT
           PTTTT = - FLOAT(N)*FLOAT(N)  * PTT
           C = COS(T)
           S = SIN(T)
C      Conversion of t-derivatives into x-derivatives
       TNX    = - PT / S
       TNXX   = ( S*PTT - C * PT) / S**3
       TNXXX  = (-S*S*PTTT + 3.*C*S*PTT - (3.*C*C + S*S) * PT )/S**5
       TNXXXX = (S*S*S*PTTTT-6.*C*S*S*PTTT+(15.*C*C*S+4.*S*S*S) * PTT
      1          - (9.*C*S*S+15.*C*C*C) * PT ) / S**7
       RETURN
       END
```

16.3 The General Theory of One-Dimensional Coordinate Transformations

Let y denote the original coordinate and x the new variable where the two are related by

$$y = f(x) \tag{16.10}$$

Elementary calculus shows

$$\frac{d}{dy} = \frac{1}{f'(x)}\frac{d}{dx} \tag{16.11}$$

where $f' \equiv df/dx$. By iterating this formula, we obtain the formulas for higher derivatives listed in Table E–1 in Appendix E. The remainder of that appendix is a collection of derivative tables, generated via the computer language REDUCE, for several particular mappings.

When a Fourier cosine series is mapped so as to create a new basis set, the orthogonality relationship is preserved:

$$\int_0^\pi \cos(mx)\cos(nx)\,dx \equiv \int_{f(0)}^{f(\pi)} \frac{\phi_m(y)\,\phi_n(y)}{f'\,(f^{-1}[y])}\,dy; \tag{16.12}$$

The Gaussian quadrature formula for the new interval with the first-derivative-of-the-inverse-of-f weight is simply the image of the rectangle rule or trapezoidal rule under the mapping: the abscissas are unevenly spaced because of the change of coordinates, but the quadrature weights are all equal.

To have the machinery for transforming derivatives is one thing; to know which mappings are useful for which problems is another. In the remainder of this chapter, we offer a variety of illustrations of the many uses of mappings.

16.4   Infinite and Semi-Infinite Intervals

When the computational interval is unbounded, a variety of options are available, and only some of them require a change-of-coordinate as discussed in much greater detail in the next chapter. For example, on the infinite interval $y \in [-\infty, \infty]$, Hermite functions or sinc (Whittaker cardinal) functions are good basis sets. On the semi-infinite domain, $y \in [0, \infty]$, the Laguerre functions give exponential convergence for functions that decay exponentially as $|y| \to \infty$.

A second option is what we shall dub "domain truncation": solving the problem on a large but *finite* interval, $y \in [-L, L]$ (infinite domain) or $y \in [0, L]$ (semi-infinite) using Chebyshev polynomials with argument (y/L) or $(L/2)[1+y]$), respectively. The rationale is that if the solution decays exponentially as $|y| \to \infty$, then we make only an exponentially-small-in-L error by truncating the interval. This method, like the three basis sets mentioned above, gives exponentially rapid convergence as the number of terms in the Chebyshev series is increased.

However, there is a complication. If we fix L and then let $N \to \infty$, the error in approximating $u(x)$ on $x \in [-L, L]$ — call this the "series" error, E_S — decreases geometrically fast with N. Unfortunately, this is not the only source of error. The "domain truncation error", $E_{DT}(L)$, arises because $u(\pm L)$ has some finite, non-zero value, but we imposed $u(\pm L) = 0$ on the numerical solution. It follows that to drive the *total* error ($= E_S + E_{DT}$) to zero, we must *simultaneously* increase *both* L and N. This in turn implies that the rate of decrease of the total error with N is "subgeometric". A full discussion is in Boyd (1982a).

The third option, and the only one that is strictly relevant to the theme of this chapter, is to use a mapping that will transform the unbounded interval into [-1, 1] so that we can apply Chebyshev polynomials without the artificiality of truncating the computational interval to a finite size. A wide variety of mappings are possible; an early paper by Grosch and Orszag (1977) compared

$$y = -L \log(1 - x) \qquad \text{"Logarithmic Map"} \qquad y \in [0, \infty] \qquad (16.13)$$

$$y = L(1 + x)/(1 - x) \qquad \text{"Algebraic Map"} \qquad x \in [-1, 1] \qquad (16.14)$$

The logarithmic map stretches the semi-infinite interval so strongly that for a function which decays exponentially with $|y|$, the change-of-coordinate creates a branch point at $x = 1$ — and a Chebyshev series that converges rather slowly. Grosch and Orszag (1977) and Boyd (1982a) offer compelling theoretical arguments that logarithmic maps will always be inferior to algebraic maps in the *asymptotic* limit $N \to \infty$. Paradoxically, however, some workers like Spalart (1984) and Boyd (unpublished) report good results with an exponential map. The resolution of the paradox is (apparently) that exponentially-mapped Chebyshev series approach their asymptotic behavior rather slowly so that for $N \sim O(50)$ or less, they may be just as good as algebraic maps.

Nevertheless, we shall discuss only algebraic maps here; more general maps are discussed in the next chapter. The logarithmic change-of-coordinate is like a time bomb — it may not blow up *your* calculation, but theory shows that it is risky.

Like domain truncation and Hermite and Laguerre functions, algebraic mapping will give exponential-but-subgeometric convergence if the function has no singularities on the expansion interval except at infinity and decays exponentially fast as $y \to \infty$.

The algebraic mapping (16.14) converts the Chebyshev polynomials in x into *rational* functions in y, dubbed the $TL_n(y)$, defined by

$$TL_n(y) \equiv T_n\left(\frac{y-L}{y+L}\right) \tag{16.15}$$

It may seem a little pretentious to introduce a new symbol to represent these "rational Chebyshev functions on the semi-infinite interval" when they are merely Chebyshev polynomials in disguise. However, that the Chebyshev polynomials themselves are merely cosine functions in masquerade. We can alternatively define the "rational Chebyshev functions" via

$$TL_n\left(L\cot^2[t/2]\right) \equiv \cos(nt) \tag{16.16}$$

It follows that just as in Sec. 2, we can evaluate the derivatives of the $TL_n(y)$ in terms of those of $\cos(nt)$ by using the transformation formulas given in Table E–6. The pseudospectral grid points are simply the images under the map $y_i = L\cot^2(t_i/2)$ of the evenly spaced roots in t. (The Gauss-Lobatto or "extrema" grid is inconvenient because the mapping is singular at the endpoints.) Thus, it is quite trivial to write programs to solve problems on an unbounded interval.

We must not forget, however, that even though we may prefer to write our program to solve the trigonometric version of the problem via a Fourier cosine series, the end result is a series for $u(y)$ as a sum of orthogonal rational functions, the $TL_n(y)$. The mapping has interesting and important consequences discussed in Chap. 17. The key both to writing the program and to understanding the subtleties of infinite and semi-infinite domains is the change of coordinates.

16.5   One-Dimensional Maps to Resolve Endpoint & Corner Singularities

The function

$$g(X) \equiv \sqrt{1 - X^2} \tag{16.17}$$

is bounded everywhere on $X \in [-1, 1]$, but because of the branch points, its Chebyshev series converges with an algebraic convergence index of two:

$$\sqrt{1 - X^2} = \frac{2}{\pi}\left\{1 - \sum_{n=1}^{\infty}\frac{2}{4n^2 - 1}T_{2n}(X)\right\} \tag{16.18}$$

Throughout this section, we will assume that the singularities at the endpoints are "weak" in the sense that the function is bounded at the endpoints.

Stenger (1981) pointed out that a mapping of $X \in [-1, 1]$ to $y \in [-\infty, \infty]$ would heal such "weak" endpoint singularities if and only if dX/dy decays *exponentially* fast as $|y| \to \infty$. The grid spacing in the original coordinate, δX, is related to the grid spacing in y by

$$\delta X \approx \frac{dX}{dy}\delta y. \tag{16.19}$$

The exponential decay of dX/dy implies a nearest neighbor separation near $X = \pm 1$ which decreases *exponentially* with N, the number of grid points.

For example, under the mapping

$$X = \tanh(y) \tag{16.20}$$

$$g(X) = \sqrt{1 - \tanh^2[y]} \tag{16.21}$$

$$= \operatorname{sech}(y) \tag{16.22}$$

using the identity $\operatorname{sech}^2(y) = 1 - \tanh^2(y)$. In startling contrast to its preimage, $\operatorname{sech}(y)$ is a remarkably well-behaved function. Stenger (1981) shows that the sinc expansion of $\operatorname{sech}(y)$ on $y \in [-\infty, \infty]$ has subgeometric but exponential convergence with exponential index of convergence $r = 1/2$. Boyd (1986a) shows that a Hermite series also has $r = 1/2$, but with even faster convergence. A rational Chebyshev basis, $TB_j(y)$, also gives subgeometric convergence.

The key is the tanh mapping (16.20); the choice of infinite interval basis is secondary. Table V of Boyd (1987a) shows that the sum of the first sixteen symmetric $TB_n(\operatorname{arctanh}[X])$ gives an error of no more than 1 part in 300,000 for the approximation of $\sqrt{1 - X^2}$. By contrast, the same table shows that the error in the unmapped Chebyshev series (16.18) to the same order is $1/50$. The derivatives of $\operatorname{sech}(y)$ are all bounded (in fact, 0!) at the endpoints whereas all derivatives of $\sqrt{1 - X^2}$, even the first, are unbounded.

The solutions to partial differential equations on domains with corners may have weak singularities; the classic example is

$$\nabla^2 u = -1 \tag{16.23}$$

on the square $[-1, 1] \times [-1, 1]$ with $u \equiv 0$ on all four walls. The solution has branch points of the form

$$u = r^2 \log(r) \sin(2\theta) + \text{less singular terms} \tag{16.24}$$

near all four corners where r and θ are the radial and angular coordinates in a local polar coordinate system centered on a particular corner. Stenger (1979) obtained five decimal place accuracy for (16.23) using the mapping (16.20) (for both coordinates) plus a 33×33 grid with sinc functions. Thus, although a corner singularity would seem to cry out for a two-dimensional mapping (as discussed later in this chapter), a simple tensor product grid with the one-dimensional tanh-map applied independently to both x and y is usually sufficient.

One *caveat* is in order: when the singularity is weak, the branch point may be irrelevant unless one needs many decimal places of accuracy. For Poisson's equation, for example, the corner branch points still permit an algebraic index of convergence of six — much weaker than the square root, which gives an index of two. The result is that Boyd (1986c) found that it was possible to match Stenger's accuracy with only 136 basis functions of the proper symmetry and *no mapping*.

Boyd (1988d) has elaborated this point by showing that there is a "cross-over" in N: for smaller N, the mapping induces variable coefficients in the differential equation, which slow convergence, but for $N > N_{\text{cross-over}}$, the mapping reduces error. It is important to apply asymptotic concepts asymptotically!

Nevertheless, it is gratifying that exponential mapping gives an "infinite order" method not only for approximating functions with weak endpoint singularities, but also for solving differential equations — even though these solutions have unbounded derivatives in the original coordinate. If the singularity is weak, try it first without a mapping. If this fails, or if the singularity is strong, i. e. produces an algebraic index of convergence of three or less, apply a change-of-coordinate such that the original coordinate varies exponentially fast with the new, computational coordinate. Lund and Riley (1984), Lund (1986), and Boyd (1987a, 1988c) give further examples.

16.6 Two-Dimensional Maps & Singularity-Subtraction for Corner Branch Points

A classic test for corner singularity-resolving algorithms is to compute the eigenvalues of the so-called "L-shaped domain membrane" problem. The domain is the interior of the region which is composed of three adjacent squares arranged in the shape of the letter *L*. The eigenfunctions are strongly singular — all derivatives are unbounded — at the corner where all three rectangles touch to form a 270-degree angle as measured from boundary to boundary on the interior of the domain.

Mason (1967) showed that one could obtain superb accuracy from the Chebyshev pseudospectral method after first applying the two-dimensional mapping, invented by Reid and Walsh (1965),

$$u = \text{Re}\left\{z^{2/3}\right\} \qquad \& \qquad v = \text{Im}\left\{z^{2/3}\right\} \qquad (16.25)$$

where u and v are new coordinates, x and y are the original coordinates, and $z \equiv x + iy$. This conformal mapping heals the singularity by mapping the 270-degree angle into a 180-degree angle, i. e. a straight line. The solution is everywhere analytic on the boundaries as a function of the new coordinate. The solution $f(x, y)$ has a $r^{2/3}$ branch point where r is a local polar coordinate centered on the offending corner. However, $f(u, v)$ is not pathological because $f(u[x, y], v[x, y])$ has the (2/3)-power singularity built into the mapping (16.25).

We refer to (16.25) as a "two-dimensional" mapping because it is not the result of independent, "tensor product" transformations of x and y, but rather strongly couples both coordinates.

The mapping introduces several complications. One is that the original, L-shaped domain is mapped into the interior of a polygon with curved sides in the $u - v$ plane. Mason solves the mapped differential equation on the unit square, $[-1, 1] \times [-1, 1]$, which contains the polygon. This allows him to use a basis whose elements are products of Chebyshev polynomials in u with Chebyshev polynomials in v. Since parts of the square are mapped into unphysical points *outside* the L-shaped domain by the inverse of (16.25), we have no *a priori* way of excluding possible poles or branch points which would wreck the convergence of the double Chebyshev series. Mason is able to prove, however, that the solution of the mapped equation is analytic everywhere within the unit u-v rectangle. For a nonlinear problem, alas, a similar proof would be difficult or impossible.

The second problem is that the boundary condition $f = 0$ must be applied on the curved sides of the polygon. Mason is able to impose the boundary conditions by writing the unknown $f(u, v)$ as

$$f(u, v) = \Phi(u, v) g(u, v) \qquad (16.26)$$

where $\Phi(u, v) = 0$ on all sides of the polygon. He then expands $g(u, v)$ as a double Chebyshev series with no additional boundary constraints imposed upon it. He was able to successfully calculate the lowest eigenvalue to five decimal places by computing the corresponding eigenvalue of an 81×81 matrix.

Mason's work is an excellent *tour de force*, but the use of a *single* mapping for the whole domain is obviously complicated and demands both analytical and numerical ingenuity. For this reason, Mason's work has had few direct imitations. However, the Liverpool school of "global element" enthusiasts has had great success in applying two-dimensional mappings to *parts* of the domain. Separate Chebyshev series are then applied in each of several subdomains (McKerrell, 1988, and Kermode, McKerrell, and Delves, 1985). This work will be discussed in more detail in the chapter on domain decomposition methods.

An alternative to mapping is to augment the basis with singular functions chosen to match the behavior of the corner singularities. Lee, Schultz, and Boyd (1988a) have applied this to the flow in a rectangular cavity; Boyd (1988d) gives one-dimensional illustrations. Much earlier, this technique has been combined with finite element methods as reviewed by Strang and Fix (1973). Although a rigorous theory is lacking, the method of singular basis functions seems to work well. The only *caveat* is that one must use only a small number of singular basis functions. The reason is that very weak branch points such as $x^k \log(x)$ where k is *large* can be approximated to many decimal places of accuracy by the N Chebyshev polynomials in the basis. Thus, the corresponding singular basis functions are indistinguishable from a sum of Chebyshev polynomials on a finite precision machine, leading to poorly conditioned matrices. The moral is that one should use singular basis functions only for mimicing those singularities which are so strong that the non-singular (Chebyshev) part of the basis set cannot cope with them.

Both two-dimensional mappings and singular basis functions require precise knowledge about the nature of the corner branch points. The exponential mapping is a brute force technique which simply concentrates lots and lots of grid points near the boundaries. It is thus the preferred method when the branch points are (i) too strong to be ignored and (ii) the analytical form of the singularities is not known.

16.7   Periodic Problems with Concentrated Amplitude and the Arctan/Tan Mapping

Many physical problems have solutions which are concentrated in a small portion of the computational interval. The "cnoidal" waves of the Korteweg-deVries equation are a good example. When the amplitude is small, this solution is indistinguishable from the cosine function. As the amplitude increases, however, the crest of the wave becomes increasingly tall and narrow. In the limit of infinite amplitude, the cnoidal wave has the shape of a delta function.

A change of coordinate which clusters the points of the pseudospectral grid in and near the peak is a good strategy for such strongly "concentrated" solutions. Boyd (1978c) describes a cubic-plus-linear stretching in conjunction with Chebyshev polynomials. Boyd (1987c) describes similar changes-of-coordinate for use when the solution is periodic. The latter article is worth discussing in greater detail because it illustrates how to choose a good mapping.

For functions that are periodic in y with a period of π, Boyd (1987c) suggests

$$y = \arctan(L\tan(x)) \qquad\qquad x,\, y \in [0,\, \pi] \qquad\qquad (16.27)$$

Here, y is the original coordinate, x is the new numerical variable, and L is a constant ("map parameter") which can be chosen to give as little or as much stretching as desired. This transformation has several advantages. First, the mapping function can be expressed in terms of elementary transcendentals, that is, functions built-in to most compilers. Second,

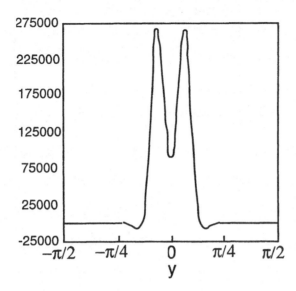

Figure 16.1: The "bicnoidal" wave of the fifth-degree Korteweg-deVries equation (Boyd, 1986b). Although the wave is spatially periodic, it is so sharply peaked about the center of the interval and decays so rapidly (exponentially) towards the endpoints that it is wasteful to use a Fourier spectral method with an evenly spaced grid.

the map has the equally trivial inverse

$$x = \arctan\left(\frac{1}{L}\tan(y)\right) \tag{16.28}$$

This explicit inverse is valuable in evaluating the numerical solution at arbitrary values of the original coordinate y. Third, the derivative of $y(x)$ is a trigonometric polynomial in x. This, together with (16.28), implies that differential equations whose coefficients are trigonometric polynomials in y will be transformed into differential equations whose coefficients are still trigonometric polynomials in the numerical coordinate x — the map is "polynomial coefficient preserving". Fourth, the mapping is *periodic* in the sense that

$$y(x) = x + P(x; L) \tag{16.29}$$

where $P(x; L)$ is a periodic function of x whose Fourier series is given explicitly in Boyd (1987c). Fifth, and perhaps most important, the mapping (16.27) is *smooth* so that it does not create additional boundary layers or regions of rapid variation.

Figs. 16.1 and 16.2 illustrate the effectiveness of the mapping. Fig. 16.1 shows the graph of the so-called "bicnoidal wave" of the Fifth-Degree Korteweg-deVries equation. Since the function is graphically indistinguishable from 0 over half of the interval, it is obviously wasteful to use an evenly spaced grid. Fig. 16.2 shows the logarithm of the relative error (relative to the maximum of the wave) as a function of the number of terms in the Fourier pseudospectral approximation for $L = 1$ (no change-of-coordinates) and $L = 0.5$. The mapping allows one to achieve a given accuracy with roughly half the number of degrees of freedom — and 1/8 the work — as without the mapping.

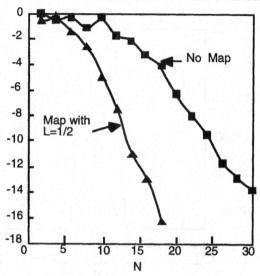

Figure 16.2: The logarithm of the relative error as a function of the number N of Fourier cosine functions without the use of a mapping and with the application of the arctan / tan map with $L = 1/2$. The change-of-coordinate allows one to achieve the same accuracy with roughly half the number of basis functions.

16.8   Adaptive Methods

Adaptive algorithms, which automatically compress and stretch the grid so as to resolve boundary layers, shocks, and other regions of rapid change, have become a "hot topic" in every area of numerical analysis. Table 16.2 catalogues spectral examples.

As a prototype, consider Burger's equation with periodic boundary conditions:

$$u_t + u\,u_y + \nu\,u_{yy} = 0 \qquad (16.30)$$

Because of the damping, the solutions are always single-valued and never break, but for small ν, $u(x, t)$ may develop a moving layer of very steep gradient, a quasi-shock wave. An obvious strategy is apply a transformation so that an evenly spaced grid in the computational coordinate x is mapped into a grid which has many grid points clustered around the shock. Because the front is propagating and the equation is nonlinear, it is almost impossible to choose a mapping non-adaptively. From scale analysis, we might be able to estimate the *width* of the frontal zone, but we cannot predict its exact location without first solving the problem.

We will discuss two different strategies for adaptation. The first is to choose arclength s along the solution curve $u(x)$ as the new coordinate (White, 1982). This is always defined even if we set $\nu = 0$; u may become a multi-valued function of x, but it is always a single-valued function of arclength.

The second strategy is to choose a particular functional form for the mapping and restrict adaptation to the more modest goal of choosing the parameters. For example, the arctan/tan mapping is

$$y = y_f + \arctan[L \tan(x)] \qquad (16.31)$$

Table 16.2: A Selected Bibliography of Adaptive Pseudospectral Methods With and Without Mappings

References	Comments
Bayliss&Matkowsky(1987)	Arctan/tan mapping; combustion fronts
Guillard&Peyret(1988)	Flame fronts and Burgers' equations
Bayliss&Gottlieb &Matkowsky&Minkoff(1989)	Arctan/tan map reaction/diffusion problems
Augenbaum (1989)	Adaption, based on Sobolev norm error indicators for both Fourier and Chebyshev algorithms
Augenbaum(1990)	Multidomain adaptation; acoustic waves in discontinuous media
Bayliss&Kuske&Matkowsky(1990)	Parameters of arctan/tan map are allowed to vary with the other spatial coordinate
Bayliss et al.(1994)	thermo-viscoplastics
Bayliss&Matkowsky(1992)	Arctan/tan map for nonlinear cellular flames
Mavriplis(1994)	Adaptive meshes for spectral elements
Gauthier et al.(1996)	Moving interface with domain decomposition
Carcione(1996)	Adaptive mapping for elastic wave equation
Renaud&Gauthier(1997)	Both map and subdomain walls are moved adaptively; compressible vortex roll-up
Mulholland,Huang&Sloan(1998)	Adaptive finite difference scheme defines map for pseudospectral solution for problems with frontal zones
Black(1998)	spectral elements; map of seminfinite domain into finite interval

Table 16.3: A Few Mapping Functions

Note: y is the physical (unmapped) coordinate; x is the computational coordinate.

References	Change-of-Coordinate Functions
Bayliss&Matkowsky(1987) Bayliss&Matkowsky(1992)	$y = \frac{\beta-1}{\beta+1} + \frac{1}{\lambda} \arctan\{\alpha x - \gamma\}$ $\beta = \frac{\arctan(\alpha(1+\gamma))}{\arctan(\alpha(1-\gamma))}$ $\lambda = [(\beta+1)/2]\arctan(\alpha(1-\gamma))$ 4 parameters: α [width] γ [location], β, λ
Bayliss&Gottlieb & Matkowsky&Minkoff(1989)	$1 + \frac{4}{\pi}\arctan\left\{\alpha\tan\left(\frac{\pi}{4}\left[\frac{\beta-x}{\beta x-1}-1\right]\right)\right\}$ 2 parameters: α [width] β [location] $\alpha > 0, -1 < \beta < 1$
Guillard&Peyret(1988)	Flame fronts and Burgers' equations
Augenbaum (1989, 1990)	$y = (2/\pi)\arctan\left\{\alpha\left(\frac{\pi}{2}X\right)\right\}$ $X \equiv (\gamma+x)/(\gamma+1)$ 2 parameters: α[width] γ[location]

where y_f and L are adaptively varied with time. The width parameter $L(t)$ allows the zone of high resolution to shrink or expand as needed. The shift parameter $y_f(t)$ allows the center of the high resolution zone to track the front.

A simple tactic for determining y_f and L is to demand that they minimize a "smoothness" function such as

$$I(y_f, L) \approx \int_0^{\pi} |u_{xx}|^2 + |u_x|^2 \, dx \tag{16.32}$$

It is easy to solve such minimization problems via Newton's method, but a first guess is needed. In a time-integration, we can use the known structure of the initial solution to generate a first guess. At later times, we use the optimum parameters for time level j as the first guess for minimizing the functional (16.32) at time level $(j+1)$. We then apply (16.31) to generate the grid that will be used to march to time level $(j+2)$.

The extra costs are obvious: We must integrate over the interval (to generate the functional) and then interpolate (with spectral accuracy!) from the old grid to the new. One harsh reality which has discouraged pseudospectral applications is that the Fast Fourier Transform *cannot* be applied to *interpolate* from the *old* grid to the slightly stretched-and-shifted *new* grid. We must use the slower "off-grid" interpolation methods described in Chapter 10, Sec. 7. (A useful Rule-of-Thumb is that these schemes, whether Boyd's or multipole-based, cost roughly four times as much as an FFT of the same N.)

There are some good coping strategies. For example, one might shift into a coordinate system moving at the approximate phase velocity of the front. One would then need to update the mapping only every ten time steps, instead of every step, because the front will not move very far relative to the center of the high-resolution zone in that short length of time. Even so, pseudospectral adaptation is a rather costly business.

The philosophy is the same, however: to minimize the residual with respect to width-and-shift parameters so as to optimize the pseudospectral solution. Adaptive grids are very promising, but because of the extra overhead, it is important that the minimization be carried out only occasionally as we march in time.

16.9   Chebyshev Basis with an Almost-Equispaced Grid: Kosloff/Tal-Ezer Arcsine Map

Kosloff and Tal-Ezer(1993) and Tal-Ezer (1994) have reduced the severe time-step requirements of standard Chebyshev methods by using a mapping which distributes the grid points more evenly over the interval $y \in [-1, 1]$. Their choice is

$$y = \frac{\arcsin([1 - \beta]x)}{\arcsin(1 - \beta)} \tag{16.33}$$

where y is the original "physical" coordinate and x is the computational coordinate. This is equivalent to replacing Chebyshev polynomials in y as the expansion functions by the new, non-polynomial basis

$$\phi_j(y) \equiv T_j \left(\frac{\sin(py)}{1 - \beta} \right) \tag{16.34}$$

where $p \equiv \arcsin(1 - \beta)$. When $\beta = 1$, this basis reduces to the usual Chebyshev polynomials. When $\beta << 1$, the collocation points, defined as the roots or extrema of the Chebyshev polynomials in the computational coordinate x, are much more uniformly spaced than those of a standard Chebyshev grid.

Define the difference between the position of one grid point and its nearest neighbor:

$$\delta_j \equiv y_{j+1} - y_j \tag{16.35}$$

The following theorem was proved by Kosloff and Tal-Ezer by substituting the appropriate expression for β into the transformation for y (16.33) and taking the limit $N \to \infty$.

Theorem 30 (KOSLOFF/TAL-EZER UNIFORMITY-of-GRID) *If*

$$\beta = C/N^2 + O(1/N^3) \tag{16.36}$$

where the number of grid points is $N + 1$, then

$$\min \delta_j \sim \frac{\pi}{\sqrt{\pi^2 + 2C} + \sqrt{2C}} \frac{2}{N} + O\left(\frac{1}{N^2}\right) \tag{16.37}$$

where C is a constant.

Table 16.4: Theory and Applications of Kosloff/Tal-Ezer Mapping

References	Comments
Kosloff&Tal-Ezer (1993)	Introduction and numerical experiments
Tal-Ezer(1994)	Theory; optimization of map parameter
Carcione(1994a)	Compares standard Chebyshev grid with Kosloff/Tal-Ezer grid
Renaut&Frohlich(1996)	2D wave equations, one-way wave equation at boundary
Carcione(1996)	wave problems
Godon(1997b)	Chebyshev-Fourier polar coordinate model, stellar accretion disk
Renaut(1997)	Wave equations with absorbing boundaries
Don&Solomonoff(1997)	Accuracy enhancement and timestep improvement, especially for higher derivatives
Renaut&Su(1997)	3rd order PDE; mapping was not as efficient as standard grid for $N < 16$
Don&Gottlieb(1998)	Shock waves, reactive flow
Mead&Renaut(1999)	Analysis of Runge-Kutta time-integration
Hesthaven, Dinesen & Lynov(1999)	Diffractive optical elements; chose $\beta = 1 - \cos(1/2)$ to double timestep versus standard grid
Abril-Raymundo & García-Archilla(2000)	Theory and experiment for convergence of the mapping

In words, the theorem states if β decreases sufficiently fast with N, then the ratio of the largest to the smallest inter-grid distance can be bounded by a constant independent of N. Though we shall not reproduce their graphs, Kosloff and Tal-Ezer calculate the eigenvalues of the differentiation matrices and show that the great reduction in the non-uniformity of the grid is indeed accompanied by an equally great reduction in the stiffness of the matrix. This implies that their Chebyshev-with-a-map scheme requires an $O(1/N)$ time step (for a PDE which is first order in space), no worse than a finite difference scheme on an equispaced grid except for an N-independent proportionality factor.

Mead and Renaut (1999) have argued that the best choice is

$$\beta_{optimum} = 1 - \cos(1/N) \leftrightarrow \beta \approx \frac{1}{2\,N^2} \tag{16.38}$$

This in turn implies that

$$\min \delta_j \approx 0.731\,\frac{2}{N} + O\left(\frac{1}{N^2}\right) \tag{16.39}$$

which is only slightly smaller than the minimum of $2/N$ for a uniformly spaced grid. The tremendous increase in timestep which is thus allowed has created many happy users of this change-of-coordinates as catalogued in Table 16.4.

The flaw of the Kosloff/Tal-Ezer transformation is that if β decrease with N, then the branch points of the mapping move closer and closer to the expansion interval as N increases. The map-induced branch points near the expansion interval deaccelerate the convergence of the spectral series. It is easiest to see this by making the transformation $t = \arccos(x)$ so the Chebyshev polynomials in x become a cosine series in t. The mapping becomes

$$y = \frac{\arcsin([1 - \beta]\cos(t))}{\arcsin(1 - \beta)} \tag{16.40}$$

The reason the trigonometric coordinate is convenient is that the convergence-slowing effects of a singularity are directly proportional to the imaginary part of t_s where t_s is the location of the singularity.

Theorem 31 (BRANCH POINTS of KOSLOFF/TAL-EZER MAP) *In terms of the trigonometric coordinate t, the Kosloff/Tal-Ezer mapping has square root branch points at*

$$t_s = m\pi \pm i\,arccosh(1 - \beta) \tag{16.41}$$

where m is an arbitrary integer. For small β, the all-important imaginary part of the location of the branch points is given approximately by

$$|\Im(t_s)| \approx \sqrt{2}\,\sqrt{\beta} + \sqrt{2}\frac{5}{2}\beta^{3/2} + O(\beta^{5/2}) \tag{16.42}$$

In particular, when β is chosen to vary inversely to the square of N, the number of Chebyshev grid points,

$$|\Im(t_s)| \approx \sqrt{2C}\,\frac{1}{N} + O(N^{-3/2}), \qquad N >> 1, \beta \sim \frac{C}{N^2} \tag{16.43}$$

Note that $\mu \equiv \Im(t_s)$ is the "asymptotic rate of geometric convergence"; the exponential factor in the usual $N \rightarrow \infty$ approximation to the spectral coefficients (Chapter 2) is $\exp(-\mu N)$, which in this case is therefore

$$\exp(-\Im(t_s)\,N) \sim \exp(-\sqrt{2C}) + O(1/N) \tag{16.44}$$

which does NOT DECREASE WITH *N.*

PROOF: The mapping is singular where the argument of the arcsin in the numerator is one. Using the identity $cos(t_r + it_{im}) = cos(t_r)\cosh(t_{im}) - i\cos(t_r)\cosh(t_{im})$ and separating real and imaginary parts, this is equivalent to the pair of equations

$$1 = (1 - \beta)\cos(t_r)\cosh(t_{im}) \qquad \& \qquad 0 = sin(t_r)\,sinh(t_{im}) \tag{16.45}$$

The second equation requires $t_r = 0, \pi$ or any other integral multiple of π. It follows that without approximation $t_{im} = arccosh(1-\beta)$ as asserted in Eq. 16.41. Expanding the inverse hyperbolic cosine for small β then gives the remaining two parts of the theorem. Q. E. D.

In the large-degree asymptotic approximation to the coefficients of the Fourier series for a function whose singularity nearest the real t-axis is at $t = t_s$, the exponential-in-degree factor is $\exp(-N|\Im(t_s)|)$ for the N-th coefficient. If the imaginary part of the mapping singularities is inversely proportional to N, then the exponential factor does not decrease at all and the Kosloff/Tal-Ezer mapping has DESTROYED the SPECTRAL ACCURACY of the Chebyshev algorithm.

To demonstrate this, Fig. 16.3 is a plot of the N-th Chebyshev or Fourier coefficient for the linear polynomial $f(y) = y$, which is simply $T_1(y)$ in the absence of the mapping. The map parameter has been varied with N according to the optimum choice of Mead and Renaut, $\beta \approx 1/(2N^2)$.

When β decreases more slowly with N than $1/N^2$, the transformed Chebyshev series still converges geometrically. Unfortunately, the timestep then increases more rapidly with N than when β is $O(1/N^2)$.

In spite of this, the Kosloff/Tal-Ezer mapping has many satisfied customers as catalogued in the table. However, the claim, made in the title of their paper, that the Kosloff/Tal-Ezer method gives an $O(1/N)$ time step limit is true only if one is willing to let spectral accuracy go to the devil.

For this reason, Hesthaven, Dinesen and Lynov(1999), who are unwilling to sacrifice spectral accuracy, choose

$$\beta = 1 - \cos(1/2) \tag{16.46}$$

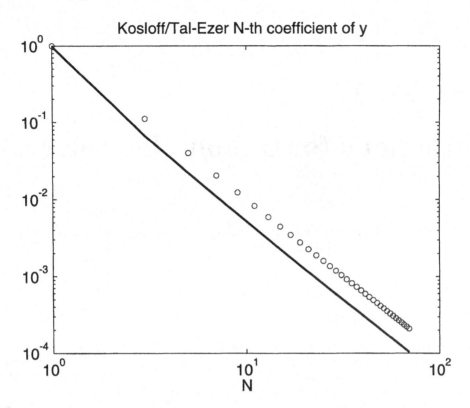

Figure 16.3: The N-the Chebyshev coefficient versus N when the map parameter is varied with N according to $\beta = 1 - \cos(1/N)$. (Note that the plotted numbers are not the coefficients of a single spectral expansion because the map parameter is different for each N.) Only odd N are plotted because the even N coefficients are all zero by symmetry. The dotted curve is a graph of $1/N^2$, which is asymptotically parallel to the solid line. Numerically, the coefficients of $f(y) = y$ asymptote to $0.488/N^2$: the usual exponential convergence of a Chebyshev series has been destroyed by the change-of-coordinate.

This "accuracy-conservative" choice nevertheless allows twice the time step of an unmapped Chebyshev algorithm with no loss of accuracy or stability.

Thus, the controversy is only about the best choice of β. There is no doubt that the Kosloff/Tal-Ezer mapping is useful and allows a longer time step.

Chapter 17

Methods for Unbounded Intervals

"The infinite! No other question has ever moved so profoundly the spirit of man."
— David Hilbert (1921)

17.1 Introduction

In this chapter, we discuss a variety of special tools for dealing with problems in infinite or semi-infinite domains. When the problem is posed on $y \in [-\infty, \infty]$, the alternatives include: (i) sinc functions (ii) Hermite functions and (iii) algebraically mapped Chebyshev polynomials (iv) exponentially mapped Chebyshev polynomials or (v) solving the problem on a large but finite interval, a technique we shall call "domain truncation". On the semi-infinite interval $y \in [0, \infty]$, we can use (i) Laguerre functions or (ii) Chebyshev polynomials with various mappings or (iii) domain truncation. In the rest of the chapter, we will describe a number of particular choices of basis set.

Rule-of-Thumb 14 (PENALTIES of UNBOUNDED INTERVAL)
The unboundedness of a computational domain extracts two penalties:

1. *The rate of convergence is usually* SUBGEOMETRIC *rather than geometric.*

2. *In addition to the truncation N, one must always choose some kind of scaling parameter L. This second numerical parameter may be the size of a truncated computational domain, a map parameter for a change-of-coordinate that transforms a finite interval to an infinite domain, or a scaling constant for the argument of the basis functions.*

The third complication of infinite intervals is that there are many good choices of spectral basis. Consequently, we shall treat in turn sinc, Hermite, Laguerre, rational Chebyshev, domain truncation, and Fourier series with a mapping. Which should one choose? The good news is: It doesn't much matter. All these basis sets work well. There are differences, but only modest differences, in efficiency, ease of programming and the relationship between the basis sets and the solutions of particular classes of physical problems. Thus, all of these basis sets are used widely.

The many options for unbounded domains fall into three broad categories:

1. Domain truncation (approximation of $y \in [-\infty, \infty]$ by $[-L, L]$ with $L >> 1$

338

2. Basis functions intrinsic to an infinite interval (sinc, Hermite) or semi-infinite (La-guerre)

3. Change-of-coordinate ("mapping") of the unbounded interval to a finite one, fol-lowed by application of Chebyshev polynomials or a Fourier series

These three strategies are not completely distinct. One can always combine mapping of the coordinate with domain truncation or any basis for an unbounded interval. Furthermore, mapping is equivalent to creating new basis functions whose natural home is the infinite or semi-infinite interval; these are simply the images of Chebyshev or Fourier functions under the change-of-coordinate. Nevertheless, it is useful to keep these key strategies in mind as these are mixed and matched to create good algorithms.

17.2 Domain Truncation

Definition 35 (DOMAIN TRUNCATION) *If a solution $u(y)$ decays rapidly in the direction or directions for which the computational interval is unbounded (or asymptotes to a simple form), then the exact solution can be calculated by solving the differential equation on a large but finite interval. This strategy for unbounded domains is "domain truncation".*

17.2.1 Domain Truncation for Rapidly-decaying Functions

Domain truncation is the "no-brain" strategy for solving problems on an unbounded in-terval because it requires no modifications of strategies for a finite interval. If the solution $u(y)$ decays exponentially with $|y|$ for sufficiently large $|y|$, then the error in approximating the infinite interval by the finite interval $y \in [-L, L]$ will decrease exponentially with the domain size L. We can then apply a standard Chebyshev, Legendre or Fourier basis on the finite domain. The joys of applying existing software for a finite domain to an unbounded computational interval should not be underestimated; programming time is much more expensive than machine time.

To estimate the error in domain truncation, the following suffices:

Definition 36 (DOMAIN TRUNCATION ERROR) *The "domain truncation error" is*

$$(i) \qquad E_{DT}(L) \equiv |u(L)| \qquad \text{if } |u(y)| \leq |u(L)| \ \forall |y| \geq L \qquad (17.1)$$

or otherwise

$$(ii) \qquad E_{DT}(L) \equiv \max_{|y| \geq L} (|u(y)|) \qquad (17.2)$$

We are forced to use the phrase "estimate" because our definition of the domain truncation error is merely the maximum value of the exact solution outside the limits of the truncated interval. The error in the approximate solution to a differential equation is another matter, and may be much larger than $E_{DT}(L)$. However, the Assumption of Equal Errors asserts that $\max |u(y) - u_N(y)| \sim O(E_{DT}(L))$ for most real-world problems, even though one can contrive examples for which this is not true. In the rest of the chapter, we shall use $E_{DT}(L)$ as our sole measure of the numerical error induced by truncating the computational inter-val.

There are a couple of subtleties even in this most obvious and unsubtle method. The first is that if L is fixed while the number of basis functions N tends to ∞, the error will not converge to zero, but rather to $E_{DT}(L)$. It follows that to improve accuracy, *both* the size of the domain L and the number of basis functions N must increase *simultaneously*.

This increase in L moves the poles and branch points of $u(y)$ closer to the real y-axis after $[-L, L]$ has been rescaled to $[-1, 1]$ for the application of a Chebyshev polynomial basis (or whatever). This movement-of-the-singularities reduces the rate of convergence of the Chebyshev series. Thus, when N is doubled, the logarithm of the total error is not doubled, but is increased by a smaller amount because of the increase in L with N. As N and L jointly increase, the error decreases to zero at a subgeometric rate in N. (If L could be fixed, then the error would decrease geometrically in N as for a finite interval, but this, alas, is only a dream when the problem is posed on an unbounded domain.)

The second subtlety is that it is more efficient to use an ordinary Fourier series instead of Chebyshev polynomials, even though the former is prone to the wild oscillations of Gibbs' Phenomenon at $y = \pm L$. The reason is that the amplitude of the Gibbs' overshoot is $0.09|u(L) - u(-L)|/N$ (Boyd, 1988e), that is to say, the size of the wiggles is proportional to but smaller than the magnitude of $u(y)$ at the ends of the interval, i. e., the domain truncation error $E_{DT}(L)$. It follows that the effects of Gibbs' Phenomenon are always smaller (by $O(1/N)$) than the domain truncation error, and thus negligible. The Fourier basis is better because for a given N, it has a higher density of points (by a factor of $\pi/2$) at the center of the interval than the Chebyshev collocation grid, which wastefully clusters points near $y = \pm L$ where $u(y)$ is very small.

Domain truncation is a good because of its simplicity. However, it can improved further by combining it with a mapping as shown by Weideman and Cloot (1990), and explained later in this chapter. Furthermore, domain truncation is more sensitive to the choice of the domain size L than the rational Chebyshev functions are to the choice of their map parameter L (Boyd, 1982a, 1987a, 1987b).

17.2.2 Domain Truncation for Slowly-Decaying Functions

If $u(y)$ decays as an inverse power of $|y|$, instead of exponentially, it is often still possible to obtain high accuracy from domain truncation by altering the boundary conditions to match the analytic, asymptotic form of the slowly-decaying solution. For example, Corral and Jiménez(1995) developed an analytic irrotational correction to apply domain domain to a flow which is unbounded in one coordinate. Rennich and Lele(1997) have applied a three-dimensional Fourier series to a vortical flow which is unbounded in two coordinates, extending Corral and Jiménez's method to an additional dimension. The analytical boundary condition makes it possible to use a computational domain of modest size even though the decay of the irrotational part of the vorticity is rather slow, which would otherwise demand a huge domain.

Domain truncation-with-matching is not the only option for slowly-decaying functions. Algorithms that combine infinite interval basis sets with special asymptotic-mimicking basis functions are the themes of Sec. 13 in this chapter (illustrated by the J_0 Bessel function) and Sec. 4 in Chapter 19 (wave scattering).

17.2.3 Domain Truncation for Time-Dependent Wave Propagation: Sponge Layers

When waves evolve far from boundaries, it is common to idealize the geometry as an infinite, boundary-free domain. Standard infinite interval basis sets have troubles because the waves do not decay for large $|y|$; instead, some waves simply propagate out of the region of interest. Domain truncation is a popular alternative. However, the usual Dirichlet boundary conditions are equivalent to rigid boundaries, which would reflect the waves back towards $y = 0$ like a mirror. The physics for large $|y|$ must therefore be modified to *absorb* the waves.

For simple wave equations (or for wave equations which asymptote to linear, constant coefficient differential equations for large $|y|$), this modification can be accomplished merely by altering the boundary conditions. For example, if the waves have the asymptotic spatial structure $\exp(\pm iky)$ for some wavenumber k where the plus sign describes outward-propagating waves for positive y, then the right boundary condition can be modified to

$$\frac{\partial u}{\partial y} - iku = 0 \qquad \text{at } y = L \qquad (17.3)$$

and similarly at the left boundary. This Robin condition filters the inward-propagating wave, thus allowing waves to pass through the boundary from small y to $y > L$ without reflection.

Unfortunately, this simple strategy fails if the far field waves are a mixture of wavenumbers or vary non-sinusoidally. It is possible to generalize this "propagate-through-the-boundary" idea; there has been much recent work on so-called "one-way wave equations", but it would take us too far afield to describe it here.

A second option is to add an artificial damping which is negligible for small y (where the real physics happens) but is allowed to increase exponentially fast for large y. The ends of the computational domain then soak up the waves like a sponge soaking up water; this is called the "sponge layer" method. Any old boundary conditions at $y = \pm L$ will do because the waves that reflect from the computational boundaries will never make it back to vicinity of the origin. If, Berg, Christiansen and Skovgaard(1987) and Zampieri and Tagliani(1997) are spectral examples.

The Fourier basis, used by If *et al.*, is superior to a Legendre or Chebyshev basis because the high resolution of the orthogonal polynomials near $y = \pm L$ is completely wasted when these regions are "sponges".

There is one peril in "sponge layers": If the artificial viscosity increases too rapidly, accuracy may be awful. If the damping coefficient is discontinuous, then the wave equation no longer has analytic coefficients and the spectral series will converge only at an algebraic rate. However, even if the viscosity is an analytic function, the sponge layer may fail to be a good absorber if the damping increases too rapidly with $|y|$. The reason is that if the damping varies more rapidly than the wave, the wave will be partially reflected — any rapid variation in the wave medium or wave equation coefficients, whether dissipative or not, will cause strong partial reflection. The physically-interesting region around $y = 0$ will then be corrupted by waves that have reflected from the sponge layers.

Fortunately, it turns out that if ϵ is defined to be the ratio of the length scale of the wave[1] to the length scale of the dissipation, then the reflection coefficient is exponentially small in $1/\epsilon$ (Boyd, 1998b). It follows that if both the length scale of the artificial viscosity and the domain size are systematically increased with the number of spectral coefficients N, then the amplitude of reflected waves at $y = 0$ will decrease exponentially fast with N. Thus, if implemented carefully, the sponge layer method is a spectrally-accurate strategy.

17.3   Whittaker Cardinal or "Sinc" Functions

Sir Edmund Whittaker (1915) showed that functions which (i) are analytic for all finite real y and (ii) which decay exponentially as $|y| \to \infty$ along the real axis could be approximated as sums of what are now called "Whittaker cardinal" or "sinc" functions. The convergence is *exponential* but *subgeometric* with an exponential convergence index usually equal to 0.5.

---

[1]The wave scale is $1/k$, i. e., the local wavelength divided by 2π.

(That is, the error typically decreases proportional to $\exp(-qN^{1/2})$ for some positive constant q.)

$$f(y) \approx \sum_{j=-N/2}^{N/2} f(y_j)\, C_j(y) \tag{17.4}$$

where

$$C_j(y;\, h) \equiv \operatorname{sinc}[(y - jh)/h] \tag{17.5}$$

$$\operatorname{sinc}(y) \equiv \frac{\sin(\pi y)}{\pi y} \tag{17.6}$$

and where the points of the associated collocation grid are

$$y_j \equiv j\, h \tag{17.7}$$

The coefficients of the expansion are simply the values of $f(y)$ on the grid because the basis functions are also cardinal functions with the property

$$C_i(y_j;\, h) = \delta_{ij} \tag{17.8}$$

What is most striking about (17.4) is that we must choose *two* independent parameters to determine the approximation: (i) the truncation N and (ii) the scaling factor h. On a finite interval, only *one* parameter — the truncation N — would suffice. This complication is not special to sinc series: all spectral methods on an infinite or semi-infinite interval require us to specify a scaling parameter of some sort in addition to N.

Spectral methods on an infinite or semi-infinite interval have other subtleties, too. A truncated Chebyshev series is exact on a finite interval only when $f(x)$ is a polynomial; what makes it useful is that one can approximate any function by a polynomial of sufficiently high degree. The analogous statement for sinc expansions is that the Whittaker cardinal series is exact for so-called "band-limited" functions; sinc series are effective for general functions that decay exponentially as $|y| \to \infty$ because one can approximate arbitrary functions in this class with arbitrary accuracy by a band-limited function of sufficiently large bandwidth.

"Band-limited" functions are defined in terms of Fourier integrals, which are yet another way of representing a function on an infinite interval:

$$f(y) = \frac{1}{\sqrt{2\pi}} \int_{-\infty}^{\infty} F(\omega)\, e^{-i\omega y}\, d\omega \qquad\qquad \text{[Fourier integral]} \tag{17.9}$$

The Fourier transform $F(\omega)$ is analogous to the function $a(n)$ that gives the coefficients of an ordinary Fourier or Chebyshev series except that $a(n)$ is meaningful only for integer n whereas $F(\omega)$ is a continuous function of ω.

In the same way that series coefficients $a(n)$ are computed by an integral which is the product of $f(t)$ with the basis functions,

$$F(\omega) = \frac{1}{\sqrt{2\pi}} \int_{-\infty}^{\infty} f(y)\, e^{i\omega y}\, dy \qquad\qquad \text{[Inverse Transform]} \tag{17.10}$$

An exponentially convergent numerical scheme to compute either the forward or inverse transform is to replace the infinite limits of integration by $\pm W$ for some sufficiently large W and then apply the rectangle rule or trapezoidal rule. We do not offer a separate chapter or even a section on Fourier integrals because this algorithm is equivalent to a Fourier *series* pseudospectral method with a very large period.

Communications engineers have introduced a special name for a truncated Fourier integral because electronic devices have maximum and minimum frequencies that physically truncate the ω-integration.

Definition 37 (BAND-LIMITED FUNCTIONS)

A function $f_W(y)$ is said to be BAND-LIMITED with BANDWIDTH W if it can be represented as the TRUNCATED Fourier integral

$$f_W(y) = \frac{1}{\sqrt{2\pi}} \int_{-W}^{W} F(\omega)e^{-i\omega y}\, d\omega \qquad (17.11)$$

In the world of Fourier integrals, band-limited functions are the analogues of polynomials. For any degree N, polynomials are "entire functions", that is to say, have no singularities for any finite x in the complex x-plane. Even so, a polynomial may be an extremely accurate approximation to a function that has poles or branch points. The resolution of this apparent contradiction is that the polynomial is a good representation of $f(x)$ only on an interval which is free from singularities; the infinities must be elsewhere. In a similar way, "band-limited" functions are always "entire functions", too. This does not alter the fact that any function $f(y)$, even if it has singularities for complex y, can be approximated for real y to within any desired absolute error by a band-limited function of sufficiently large bandwidth.

Eq. (17.4) and (17.8) show that the sinc series interpolates to $f(y)$ at every point of the grid. However, this hardly guarantees a good approximation; recall the Runge phenomenon for polynomials. For band-limited functions, success is assured by the following.

Theorem 32 (SHANNON-WHITTAKER SAMPLING THEOREM)

If $f_W(y)$ is a BAND-LIMITED function of bandwidth W, then the Whittaker cardinal series is an EXACT representation of the function if $h \le \pi/W$ where h is the grid spacing:

$$f_W(y) = \sum_{j=-\infty}^{\infty} f\left(\frac{\pi j}{W}\right) \operatorname{sinc}\left(\frac{Wy}{\pi} - j\right) \qquad (17.12)$$

PROOF: Stenger (1981) or Dym and McKean (1972).

As noted above, band-limited functions are rather special. However, under rather mild conditions on $f(y)$, one can prove that $F(\omega)$ will decrease exponentially with $|\omega|$, so the error in writing $f(y) \approx f_W(y)$ decreases exponentially with W, too. Thus, (17.12) really implies that we can use Whittaker cardinal functions to approximate any function $f(y)$ that has a well-behaved Fourier transform.

Definition 38 (SINC EXPANSION OF A FUNCTION) *An approximation to a function $f(y)$ of the form*

$$f(y) = \sum_{j=-N/2}^{N/2} f(hj) \operatorname{sinc}\left(\frac{y - jh}{h}\right) + E_W(h) + E_{DT}(Nh) \qquad (17.13)$$

is called the SINC EXPANSION of $f(y)$ where h is the grid spacing.

The error $E_W(h)$ is called the "grid-spacing" or "bandwidth" error. It arises because the grid points are spaced a finite distance h apart. Equivalently, since the sinc expansion is exact even with a non-zero h if $f(y)$ is band-limited with a sufficiently large W, we may say that this error is caused by (implicitly) approximating $f(y)$ by a band-limited function $f_W(y)$ with $W = \pi/h$.

The error $E_{DT}(Nh)$ is the "grid-span" or "domain truncation" error. It arises because we must truncate the infinite series to limits of $\pm N/2$ for some finite N. This is equivalent to interpolating $f(y)$ on a grid of $(N + 1)$ points that span only a portion of the interval $[-\infty, \infty]$. This error is $O(f[hN/2])$.

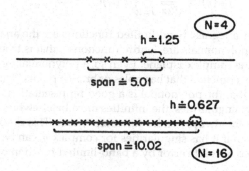

Figure 17.1: The ×'s mark the evenly spaced interpolation points used with sinc functions to solve problems on $y \in [-\infty, \infty]$. It can be shown that for most situations, best results are obtained by choosing a grid spacing h proportional to $\sqrt{N}$ where N is the number of grid points. This implies that interval spanned by the grid points — the "grid span" — imcreases simultaneously as the neighbor-to-neighbor distance decreases.

Fig. 17.1 is a schematic of the optimum relationship between h and N. If we keep h fixed, and merely use a larger and larger number of terms in (17.12), our approximation will converge not to $f(y)$ but to $f_W(y)$, and we are stuck with the fixed error $|f(y) - f_W(y)|$ even with million grid points. Conversely, if we only decrease h and truncate the sum in (17.12) by omitting all grid points such that $|y_j| > L$ for fixed L, we would be left with an $O(f[L])$ error even if the grid spacing is ridiculously small. Because of the need to simultaneously increase L and decrease h, the error is $O(\exp[-p\sqrt{N}])$ for a typical function like $f(y) = \text{sech}(y)$: Subgeometric convergence with exponential index of convergence = 1/2.

There are several alternatives for representing functions on infinite or semi-infinite intervals, but they do no better. Sub-geometric convergence is a way of life when the interval is unbounded. The grid of Gaussian quadrature points for the Hermite functions, for example, automatically becomes both wider (in total span) and more closely spaced as N increases as shown in Fig. 17.2. There is always this need to serve two masters — increasing the total width of the grid, and decreasing the spacing between neighboring points — for any method on an unbounded interval. The cardinal function series differs from orthogonal polynomial methods only in making this need explicit whereas it is hidden in the automatic behavior of the interpolation points for the Hermite pseudospectral algorithm.

For more information about sinc numerical algorithms, consult the review article by Stenger (1981) and the books by Lund and Bowers (1992) and Stenger (1993) or the articles in Table 17.1.

Table 17.1: A Selected Bibliography of Sinc (Whittaker Cardinal) Applications

References	Comments
Whittaker(1915)	A few words from our founder: theory only
Stenger (1979)	Boundary value problems
Lundin (1980)	Nonlinear PDE boundary value problem
Stenger (1981)	Review article; theory plus numerical applications
Saffman&Szeto(1981)	Vortex stability
Lund&Riley(1984)	Sinc with mapping to solve radial Schroedinger equation
Elliott&Stenger(1984)	Numerical solution of singular integral equations
Robinson&Saffman (1984)	Nonlinear eigenvalue problem in 2 space dimensions
Lund (1986)	Symmetrization of sinc-Galerkin matrix
Boyd (1986b)	Shows that exponential mapping of Stenger to infinite interval is key to solving problems with endpoint singularities on $[-1, 1]$; sinc basis is effective, but other infinite interval bases work just as well
Schaffer&Stenger (1986)	Multigrid iteration with sinc
Bowers & Lund(1987)	Singular Poisson equations
Eggert&Jarratt&Lund(1987)	Finite and semi-infinite intervals, too, through mapping; singularities at end of finite interval
Lewis&Lund&Bowers(1987)	Sinc in both x and t for parabolic PDE
McArthur&Bowers&Lund(1987a)	Second order hyperbolic PDEs
McArthur&Bowers&Lund(1987a)	Numerical methods for PDEs
Bialecki(1989, 1991)	Boundary value problems
Lund&Bowers&McArthur (1989)	Elliptic PDEs
Smith&Bowers&Lund(1989)	Fourth order problems in control theory
Jarratt&Lund&Bowers(1990)	Sinc basis; endpoint singularities , finite interval
Lund&Vogel (1990)	Inverse problem for parabolic PDE
Boyd (1991f)	Sinc solution of boundary value problems with a banded matrix obtained by Euler sequence acceleration
Smith&Bogar&Bowers&Lund(1991)	Fourth order differential equations
Lund&Bowers&Carlson(1991)	Boundary feedback stabilization (control theory)
Smith&Bowers(1991)	Inverse problem for Euler-Bernoulli beams
Smith&Bowers&Lund(1992)	Inverse problem for Euler-Bernoulli beams
Boyd (1992c)	Fast Multipole Method provides a fast sinc transform
Boyd (1992d)	Fast off-grid sinc interpolation
Lund & Bowers (1992)	Monograph
Stenger (1993)	Monograph
Yin (1994)	Poisson equation through tensor-product sinc basis
Koures (1996)	Coulomb Schroedinger equation

Appendix F gives formulas for the derivatives of the Whittaker cardinal functions at the grid points. The first derivative matrix is skew-symmetric while the second derivative is symmetric (Stenger, 1979). These are desirable properties. Skew-symmetry implies that all eigenvalues of the first derivative matrix are pure imaginary; thus the sinc discretization of the one-dimensional advection equation, $u_t + c u_x = 0$, is nondissipative. Similarly, the symmetry of the second derivative matrix implies that all eigenvalues are real; the sinc discretization of the diffusion equation gives a system of ODEs in time whose eigenmodes all decay monotonically in time.

The sinc expansion is the easiest of all pseudospectral methods to program and to understand. One weakness, shared with the Hermite functions, is that the Whittaker cardinal series is useful only when $f(y)$ decays *exponentially* as $|y| \to \infty$. In contrast, the $TB_n(y)$ will give rapid, exponential convergence for functions which decay *algebraically* with $|y|$ or asymptote to a constant, provided certain conditions are met. Thus, the mapped Chebyshev polynomials have a wider range of applicability than the sinc expansion.

A more serious weakness of sinc series is that — unlike the mapped Chebyshev functions — they cannot be summed via the Fast Fourier Transform. This is especially unfortunate for time-dependent problems and for iterative solution of multi-dimensional boundary value problems.

A third limitation is that sinc functions do not give the exact solution to any classical eigenvalue problem (to my knowledge). In contrast, the Hermite functions are the idealized or asymptotic solutions to many physical problems. Thus, the Whittaker cardinal functions, despite their simplicity, have not eliminated the competing basis sets discussed in the next two sections.

The "Big Sky School" of Stenger, Lund, and their students have applied the sinc expansion to problems on a finite interval with endpoint singularities — always in combination with a map that exponentially stretches the bounded domain to an infinite interval. As explained in Chap. 16, Sec. 5, this is a good strategy for coping with endpoint branch points. Eggert, Jarratt, and Lund (1987) also compute some solutions *without* endpoint singularities, but this is *not* recommended. Chebyshev series are much more efficient when $u(x)$ is analytic at the boundaries.

In summary, it is best to regard the Whittaker cardinal functions as a good basis for $[-\infty, \infty]$ only. Although it has some important defects in comparison to Hermite and rational Chebyshev functions, the sinc series is the simplest and it converges just as fast as the alternatives.

17.4   Hermite functions

The members of this basis set are defined by

$$\psi_n(y) \equiv e^{-0.5 y^2} H_n(y) \tag{17.14}$$

where $H_n(y)$ is the Hermite polynomial of degree n. Like the familiar Chebyshev polynomials, the n-th Hermite polynomial is of degree n. The members of the set satisfy a three-term recurrence relation. The derivative of $H_n(y)$ is simply $(2 n H_{n-1})$, so it is easy to use Hermite functions to solve differential equations.

Unlike most other sets of orthogonal functions, the normalization constants for the Hermite functions increase exponentially with n, so it is difficult to work with unnormalized functions without encountering overflow (unless N is very small). In contrast, normalized Hermite functions satisfy the bound

$$|\bar{\psi}_n(y)| \leq 0.816 \qquad \text{all } n, \text{ all real } y \tag{17.15}$$

Table 17.2: Hermite Function Theory & Applications

References	Comments
Hille(1939, 1940a,b, 1961)	Convergence theorems
Englemann&Feix & Minardi&Oxenius(1963)	1D Vlasov eq. (plasma physics); Fourier in x Hermite in velocity coordinate u
Grant&Feix(1967)	Fourier-Hermite for 1D Vlasov equation Fokker-Planck collision term to improve Hermite convergence
Armstrong(1967)	1D Vlasov equation; Fourier-Hermite Galerkin Hermite truncation reduced with increasing time to mitigate slow convergence of Hermite part of tensor basis
Birkhoff&Fix(1970)	Eigenproblems
Joyce&Knorr &Meier(1971)	1D Vlasov eq.; Fourier-Hermite basis compared with double Fourier & method-of-moments (!)
Anderson (1973)	Time-dependent model of the tropical ocean
Moore & Philander (1977)	Equatorial waves and jets in the ocean (review)
Shoucri&Gagné(1977)	2D Vlasov; tensor Hermite basis in velocity finite differences in space; plasma physics
Bain (1978)	Convergence theorems
Banerjee (1978) Banerjee et al. (1978)	Eigenvalue problem: quantum anharmonic operator $\psi_n(\alpha y)$ with variable α
Boyd (1980b)	Rate of convergence theorems
Boyd (1984a)	Asymptotic theory for Hermite coefficients through steepest descent and the calculus of residues
Tribbia (1984)	Hough-Hermite Galerkin method
Boyd & Moore (1986)	Euler sequence acceleration of algebraically-converging Hermite series with applications to oceanography
Marshall & Boyd (1987)	Equatorial ocean waves
Smith(1988)	Hermite-Galerkin model of tropical ocean
Funaro & Kavian (1991)	Time-dependent diffusion problems
Weideman (1992)	Hermite differentiation matrices are well-conditioned
Holvorcem(1992) Holvorcem&Vianna(1992) Vianna&Holvorcem(1992)	Summation methods for Hermite series including Green's functions, which depend on two variables
Boyd (1992c)	Fast Multipole Methods provide fast Hermite transform
Boyd (1993)	Hermite methods in symbolic manipulation languages
Tang (1993)	Gaussian functions; role of the scale factor
Holloway(1996a,b)	Vlasov-Maxwell equations (plasma physics) Introduces "asymmetric" Hermite basis, which is very efficient for this application
Tse&Chasnov(1997)	Hermite for vertical coordinate Convection in unstable layer surrounded by unbounded stable layers; Fourier in x, y
Schumer&Holloway(1998)	1D Vlasov equation; Hermite-Fourier basis asymmetric Hermite is unstable; standard Hermite is good with use of anti-filamentation filtering

where the overbar denotes that the $\psi_n(y)$ have been normalized. As for Chebyshev and Fourier series, this bound (17.15) implies that the error in truncating a Hermite series is bounded by the sum of the absolute values of all the neglected terms.

The optimum pseudospectral points are the roots of the $(N+1)$-st Hermite function. The points are not *exactly* evenly spaced — the outermost points are a little farther apart than the separation between points near $y = 0$ — but the asymptotic approximations to the Hermite functions show that the Hermite grid does not differ very much from the sinc grid as shown in Fig. 17.2. In the previous section, we found that it was necessary to simultaneously (i) decrease the spacing between adjacent grid points and (ii) increase the span between the outermost points on the grid if the sinc series was to give increasing accuracy as N increased. Exactly the same thing happens with the Hermite grid: the separation between grid points decreases as $1/\sqrt{N}$ while the span of the grid increases as $\sqrt{N}$. It is not necessary to choose h and the span of the grid explicitly; using the roots of the Hermite function as the grid points automatically insures the correct behavior.

Fig. 17.3 illustrates the first six (normalized) Hermite functions.

However, it is still necessary to pick a scaling factor α since we can can always use $\psi_n(\alpha y)$ as the basis set for any finite α. This freedom does not exist for a finite interval, but after a change of scale in y, an infinite domain is still an infinite domain. Banerjee (1978) and Banerjee *et al.*(1978), Boyd (1984a), and Tang(1993) have discussed the choice of scale factor. The theory (Boyd, 1984a) involves heavy use of steepest descent to asymptotically approximate the Hermite coefficient integrals as $n \to \infty$. One general conclusion is important since it is also true of mapped Chebyshev methods: The optimum α is a function of N. A good choice of scaling parameter for $N = 5$ is a mediocre choice for $N = 50$; at least for the cases studied, α should *increase* with N.

The rates-of-convergence theory for Hermite functions is now well-developed for a *fixed* (N-independent) scale factor α. It is worth summarizing since it indicates the subtleties introduced by the unboundedness of the expansion interval. We must begin with a definition because the rate of convergence depends not only on the *location* of *singularities*, as true of all expansions, but also upon the *rate* at which the function $f(y)$ *decays* as $|y|$ tends to infinity.

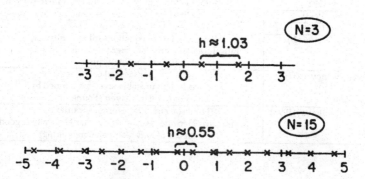

Figure 17.2: The pseudospectral grid for Hermite functions. For all N, the grid points are the roots of the $(N+1)$-st Hermite polynomial. The grid spacing varies slightly, but is *almost uniform* over the interval. The grid closely resembles the sinc grid: when N is quadrupled, the span of the grid points is roughly doubled and the grid spacing is halved.

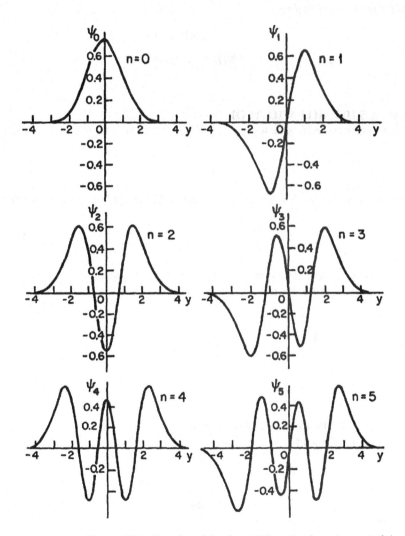

Figure 17.3: Graphs of the first 6 Hermite functions, $\psi_n(y)$.

Definition 39 (ORDER OF REAL AXIS DECAY)
    The ORDER of REAL AXIS DECAY k is the least upper bound of j for which

$$f(y) = O\left(e^{-p|y|^j}\right) \tag{17.16}$$

for some constant p as $|y| \to \infty$ along the real axis. The function is said to be "SUB-GAUSSIAN" or "SUPER-GAUSSIAN" respectively if

$$k < 2 \qquad\qquad [\text{"SUB-GAUSSIAN"}] \tag{17.17}$$

$$k > 2 \qquad\qquad [\text{"SUPER-GAUSSIAN"}] \tag{17.18}$$

Theorem 33 (HILLE'S HERMITE WIDTH-OF-CONVERGENCE)
(i) The domain of convergence of a Hermite series is the infinite STRIP about the real y-axis bounded by

$$|Im(y)| = w \tag{17.19}$$

(ii) For an ENTIRE function, that is, $f(y)$ which has no singularities except at infinity, the width is

$$w = \begin{cases} \infty & \text{if } k > 1 \\ 0 & \text{if } k < 1 \end{cases} \tag{17.20}$$

(iii) For SINGULAR functions, let τ denote the absolute value of the imaginary part of the location of that singularity which is closest to the real y-axis, and let p denote the constant in the definition of real axis decay. Then

$$w = \begin{cases} \tau & \text{if } k > 1 \\ \text{smaller of } [p, \tau] & \text{if } k = 1 \\ 0 & \text{if } k < 1 \end{cases} \tag{17.21}$$

Theorem 34 (HERMITE RATE-OF-CONVERGENCE)
(i) If the strip of convergence has a finite width w, then

$$a_n \sim O\left(e^{-w\sqrt{2n+1}}\right), \tag{17.22}$$

that is, subgeometric convergence with exponential convergence index $r = 0.5$.
(ii) For entire functions, the exponential convergence index r is

$$r = k/(2[k-1]) \qquad (\text{SUPER-GAUSSIAN, that is, } k > 2) \tag{17.23}$$

$$r = k/2 \qquad\qquad (\text{SUB-GAUSSIAN, that is, } k < 2) \tag{17.24}$$

(iii) Geometric convergence is possible only for entire functions and only for these two cases:

(a) $f(y)$ has $k = 2$ $\qquad$ [Gaussian decay; p need not equal 1/2] $\qquad$ (17.25)
(b) $f(y)$ is Super-Gaussian ($k > 2$) and the scaling factor α is varied $\qquad$ (17.26)
     with the truncation N as described in Boyd (1984a)

(iv) If the function f(y) decays ALGEBRAICALLY *with y so that*

$$f(y) \sim O\left(1/|y|^{\delta}\right) \qquad \text{as} \quad |y| \to \infty \qquad (17.27)$$

and if the function has p continuous derivatives on $[-\infty, \infty]$ *(with the (p+1)-st derivative contin-uous except at a finite number of points), then*

$$a_n \sim O\left(\frac{1}{n^{q/2}}\right) \qquad (17.28)$$

where

$$q = \text{smaller of } [p + 3/2,\ \delta - 5/6] \qquad (17.29)$$

Eq. (17.28) only gives a lower bound on the convergence rate; it is known that the Hermite coefficients for f(y) $\equiv$ 1 converge as $O(n^{-1/4})$ even though (17.28) can guarantee only that the coefficients will diverge no faster than $n^{5/12}$.

Part (i) was proved by Hille (1939, 1940a,b), (ii) and (iii) by Boyd (1984a), and (iv) by Bain (1978).

Whew! These theorems are rather complicated, but they do illustrate a couple of im-portant themes that seem to apply to other types of expansions, too. First, the rate of convergence depends on how rapidly or how slowly $f(y)$ decays as $|y| \to \infty$. When $k > 1$, that is to say, when the function is decaying faster than an exponential whose argument is linear in y, the singularities will determine the width of the strip of convergence, just as for a Chebyshev series on a finite interval. When $k < 1$, however, the slowness of the real axis decay has a more devasting effect on convergence than do the poles and branch points; when $k = 1$, either the rate of decay or the location of the nearest pole may determine the width of the strip of convergence.

Second, for entire functions the convergence is fastest when $f(y)$ is decaying at the same rate as the $\exp(-0.5y^2)$ factor built into the basis functions: Supergeometric convergence is possible for this case. The series converges more and more slowly, however, as the rate of decay varies more and more from that of $\exp(-0.5y^2)$, regardless of whether the variation is to decay more rapidly (as a "Super-Gaussian") or more slowly ("Sub-Gaussian").

Third, functions which decay algebraically with y have Hermite series whose coeffi-cients decrease algebraically with n. (Note that (17.28) gives only a lower bound on the algebraic index of convergence, so further work is needed). It is striking that whereas the continuity of one additional derivative would increase the algebraic index of convergence by 1 for a Chebyshev series, it increases the algebraic index by only (1/2) for a Hermite series — in other words, we gain only a factor of $1/\sqrt{N}$ for each additional continuous derivative for Hermite. This is another penalty of the unboundedness of the interval. The spacing between adjacent collocation points decreases only as $1/\sqrt{N}$, so (17.28) is not too surprising.

Weideman(1992) shows that the spectral radius of the Hermite differentiation matrices is $O(\sqrt{N})$ for the first derivative and $O(N)$ for the second derivative. This is equivalent to the statement that the maximum timestep for an explicit time-marching algorithm for a PDE which is second order in space is $\tau_{max} \sim O(1/N)$. The reason is that the Hermite collocation points, which are almost uniformly spaced as for a Fourier or finite difference scheme, have an average grid spacing which increases only as $N^{1/2}$ instead of N because the span of the grid points is also increasing as $O(N^{1/2})$. If we solved the same problem using Fourier domain truncation with the domain size L increasing as $O(N^{1/2})$, then the Fourier grid would be almost indistinguishable from the Hermite grid and the maximum timestep would also decrease as slowly as $O(1/N)$.

Thus, for the same N and for grid points spanning the same interval $[-L, L]$, Hermite algorithms, the sinc method, Fourier domain truncation, and finite differences with domain truncation all have similar timestep limits.

Weideman(1992) has proved that Hermite differentiation matrices have additional properties that are numerically desirable. Both the Galerkin and collocation first derivative matrices have purely imaginary eigenvalues while the second derivative matrices have only real, negative eigenvalues. All these matrices are well-conditioned. In addition, the Galerkin matrices are banded and are either skew-symmetric (first derivative) or symmetric (second derivative) for the orthonormal basis.

Although we can evaluate the Hermite functions and their derivatives through a three-term recurrence — hardly a major programming challenge — the grid points must be looked up in a mathematical handbook. Consequently, Hermite functions are not as easy to program as sinc or rational Chebyshev functions.

The latter have the major advantage that they may be evaluated via the Fast Fourier Transform. Orszag(1986), Boyd(1992c) and Dutt and Rokhlin(1993) have devised fast transforms applicable to Hermite functions. However, these are much more expensive than a standard FFT (at least a factor of four for $N \sim O(100)$), and none of these algorithms has yet been applied to any purpose except self-demonstration. The inapplicability of the standard FFT is another black mark against Hermite functions.

One virtue of Hermite functions is that these are (i) the *exact* eigenfunctions of the quantum mechanical harmonic oscillator and (ii) are the *asymptotic* eigenfunctions for Mathieu's equation, the prolate spheroidal wave equation, the associated Legendre equation, and Laplace's Tidal equation. The reason for this special role is that the Hermite functions solve the eigenvalue problem

$$u_{yy} + \left[\lambda - y^2\right] u = 0 \qquad\qquad u(-\infty) = u(\infty) = 0 \qquad\qquad (17.30)$$

where λ is the eigenvalue. If we replace (17.30) by the more general equation

$$u_{yy} + \left[\lambda - y^2 + p(y)\right] u = 0, \qquad\qquad (17.31)$$

the exponential decay $u(y)$, forced by the $-y^2$ term, makes $p(y)$ irrelevant, at least if this perturbation is small near $y = 0$. It follows that (17.30) is "generic" in the sense that its solutions will approximate those of (17.31) for small but otherwise arbitrary $p(y)$.

This close connection with the physics makes Hermite functions a natural choice of basis functions for many fields of science and engineering. One modern area of great interest is equatorial oceanography and meteorology. One or two Hermite functions give very accurate approximations to equatorially-trapped waves when linearized about a resting state. The same functions have been very valuable in solving equatorial jet and wave reflection problems as reviewed in Moore and Philander(1977).

The Vlasov-Maxwell equation of plasma physics has been another important application. The velocity u is not an unknown, as in conventional hydrodynamics, but is an independent *coordinate*. This is important because the usual fluid advection term is multiplication by u in the Vlasov equation, and the Galerkin representation is a banded matrix, eliminating a slow Hermite grid-to-spectral transform. The u-derivatives also generate a sparse matrix representation because the derivative of the j-th Hermite function is a weighted sum of the $(j - 1)$-st and $(j + 1)$-th Hermite functions. As a result, the Vlasov equation — although variable coefficient — may be integrated at a cost which is *linear* in the Hermite truncation N, versus a quadratic dependence when Hermite functions are applied to the Navier-Stokes equations.

The irony is: Although the close connection between simple models and Hermite functions motivated these numerical studies in equatorial fluid dynamics and plasma physics, there are serious problems with straightforward application of the Hermite Galerkin method. In equatorial oceanography, the difficulty is that for many important problems such as Yoshida's 1959 model of an equatorial jet and Moore's theory for reflection from an eastern boundary, the flow decays *algebraically* with latitude so that the Hermite coefficients decay algebraically, too. This slow convergence can be fixed by applying sum acceleration methods (Boyd and Moore, 1986, and Holvorcem, 1992).

Similarly, early calculations in plasma physics were bedeviled by filamentation in the velocity coordinate, which necessitated $N >> 100$ to compute the solution for large times. This was solved by a filter introduced by Klimas(1987) and Klimas and Farrell(1994), who used a tensor product Fourier basis instead of a Fourier-Hermite scheme. Holloway(1996a,b) and Schumer and Holloway(1998) showed that the convergence rate is greatly improved by scaling the Hermite functions by a constant. (This is a good strategy for *general* applications, too, as suggested by Boyd(1984a) and Tang(1993).) The scaled-and-filtered Hermite algorithm is very effective for the Vlasov equation as shown by Holloway and Schumer. The irony is that both modifications pull the numerical solution away from the unfiltered and unscaled analytical approximation which motivated the choice of Hermite functions in the first place.

17.5   Laguerre Functions: Basis for the Semi-Infinite Interval

The Laguerre functions are a complete orthogonal basis for the domain $x \in [0, \infty]$. They are close cousins of the Hermite functions and have a similar form: the product of a decaying

Table 17.3: A Selected Bibliography of Laguerre Function Theory & Applications

References	Comments
My Name Is Legion	Many applications in quantum chemistry
Schwartz(1963)	Effects of singularities on rate-of-convergence
Francis(1972)	Vertical structure in weather forecasting model using L_n, not Laguerre function, as basis. Showed this scheme needs prohibitively short timestep
Hoskins(1973)	Comment on Francis; Legendre basis requires a constraint which Laguerre does not
Gottlieb&Orszag(1977)	Wave equation and heat equation
Maday&Pernaud-Thomas &Vandeven (1985)	Theory; hyperbolic PDEs
Phillips&Davies(1988)	Semi-infinite spectral elements
Mavriplis(1989)	Legendre spectral elements coupled with one semi-infinite element with Laguerre
Coulaud&Funaro &Kavian(1990)	Elliptic PDEs in exterior domains
Funaro (1990b,1992b)	Theory
Boyd (1992c)	Fast Multipole Methods provide fast Laguerre transform
Iranzo&Falqués (1992)	Comparisons with rational Chebyshev TL_n Good examples & attention to practical details
DeVeronico & Funaro & Reali(1994)	Spectral elements for wave propagation with Laguerre basis for the end elements, which extend to infinity
Khabibrakhmanov&Summers(1998)	Generalized Laguerre for nonlinear equations
Black(1998)	spectral element alternative to Laguerre: employs the mapping of Boyd(1987b) from semi-infinite domain into finite interval

exponential with a polynomial:

$$\phi_n(x) \equiv \exp(-x/2)\, L_n(x), n = 0, 1, \ldots \qquad (17.32)$$

where $L_n(x)$ is the n-th Laguerre polynomial. Three-term recurrence relations for evaluating both the basis functions and their derivatives are given in Appendix A.

If a function $f(x)$ decays exponentially as $|x| \rightarrow \infty$, then the Laguerre series for f will usually converge exponentially fast within the largest parabola, with focus at the origin, which is free of singularities of $f(x)$. However, it seems likely that very rapidly growing or decaying functions, such as an exponential-of-an-exponential, may have series that converge only on the positive real x-axis. However, we have not seen any investigation of this in the literature.

We will not discuss Laguerre functions in detail because (i) their theory is very similar to Hermite functions and (ii) there have been few applications of Laguerre functions outside of quantum mechanics as indicated by the brevity of Table 17.3. Within quantum chemistry, however, there have been lots and lots of applications.

The reason is that the Laguerre functions are the exact radial factors for the eigenfunctions of the hydrogen atom. In the Linear Combinations of Atomic Orbitals (LCAO) method, basis functions are constructed by using hydrogenic orbitals centered on the first atom plus hydrogenic orbitals centered on the second atom and so on. Thus, Laguerre functions (multiplied by spherical harmonics) are the building blocks for many Galerkin computations of molecular structures as explained in Chapter 3. The hydrogen s-orbitals, which were employed to approximate the H_2^+ ion in Chapter 3, are actually just the lowest Laguerre functions, $\phi_0(r/[a_0/2])$ where a_0 is the Bohr radius of the hydrogen atom.

Just as for Hermite functions, the Laguerre basis always implicitly contains a scale factor – in this case, the Bohr radius.

Iranzo and Falqués (1992) give detailed comparisons between Laguerre functions and the rational Chebyshev functions for the semi-infinite interval (Sec. 11 below). Some theory is given in Funaro's (1992) book and his two earlier articles listed in Table 17.3.

Mavriplis (1989) is unusual in employing Laguerre polynomials – without the exponential – as the basis. This unorthodoxy makes it easier to match the Laguerre series with Legendre polynomial series in a spectral element (domain decomposition) computation. However, it is has the disadvantage that the *absolute* error of her solution is unbounded as $x \rightarrow \infty$.

This seems disastrous, but really is not. The true is that no standard basis can resolve an infinite number of oscillations with a finite number of basis functions. The Hermite or Laguerre function or rational Chebyshev series for a function like $\exp(-x)\cos(x)$ must inevitably have large *relative* error at infinity because the function continues to oscillate for all x whereas a truncated sum of basis functions must have only a finite number of zeros, and will therefore vary monotonically for some sufficiently large x. However, the *absolute* error is small everywhere because both the exact solution and the numerical approximation (except for Mavriplis' scheme) are exponentially small where the relative error is not small.

The Laguerre polynomial series is guilty of only making this breakdown in relative error, which is hidden in other methods, obvious through a large absolute error. This breakdown is also present, though implicit, with all the other standard basis sets.

The only way out is to add special basis functions to a standard series as explained in Sec. 13.

17.6   Mapping Methods: New Basis Sets Through a Change of Coordinates

By using a transformation that maps an infinite interval into a finite domain, it is possible to generate a great variety of new basis sets for the infinite interval that are the images under the change-of-coordinate of Chebyshev polynomials or Fourier series. This strategy is merely a generalization of the particular map $x = \cos(t)$, which transforms the cosine functions into the Chebyshev polynomials.

An infinite variety of maps is possible, but one can make some useful points by considering three particular classes of mappings, each illustrated by an example:

1. "Logarithmic maps":

$$y = \text{arctanh}(x), \qquad x \in [-1, 1] \tag{17.33}$$

2. "Algebraic maps"

$$y = \frac{L x}{\sqrt{1 - x^2}}, \qquad x \in [-1, 1] \tag{17.34}$$

3. "Exponential maps"

$$y = \sinh(Lt), \qquad t \in [-\pi, \pi] \tag{17.35}$$

where $y \in [-\infty, \infty]$. The names for these families of maps are chosen by how rapidly y increases with $x \to \pm 1$.

Grosch & Orszag(1977) and Boyd (1982a) were big boosters of algebraic mappings. These lead to the rational Chebyshev functions, described in the next seven sections. Their strength is that they are "minimalist" mappings, creating a not-very-violent change from one coordinate to the other, and they have optimal properties in the limit $N \to \infty$ for fixed L.

Nevertheless, the other two families of mappings are useful, too. The logarithmic mappings have the disadvantage that well-behaved functions on the infinite interval are turned into nasty functions on $x \in [-1, 1]$. For example, the mapping $y = \text{arctanh}(x)$ transforms

$$\text{sech}^\alpha(y) \to (1 - x^2)^\alpha \tag{17.36}$$

The Hermite and rational Chebyshev series for the hyperbolic secant function converge exponentially fast for any positive α, even non-integral α. However, the function $(1 - x^2)^\alpha$ has branch points at both endpoints unless α is an integer.

The arctanh mapping is useful almost entirely in the *reverse* direction, transforming functions with *endpoint singularities* at the ends of a finite interval into well-behaved functions on the infinite interval. One can then apply a standard infinite interval basis to obtain very accurate solutions to singular problems (Chap. 16, Secs. 5).

There is one exception: in the theory of solitary waves and of shocks, there are many perturbation series which involve only powers of the hyperbolic secant and/or tangent. The arctanh mapping (with $L = 1$) transforms these polynomials in hyperbolic functions of y into ordinary polynomials in x, making it easy to compute these expansions to high order (Boyd, 1995a, 95j).

The exponential maps go so far to the opposite extreme that they are inferior to the algebraic maps in the limit $N \to \infty$. Cloot and Weideman(1990, 1992), Weideman and Cloot(1990) and Boyd (1994b) have shown that the Weideman and Cloot map (17.35) is nevertheless extremely efficient in real world situations because the asymptotic limit is approached very, very slowly with N. We will illustrate the effectiveness of the Weideman-Cloot change-of-coordinate later in the chapter. First, though, we shall turn to algebraic mappings where there has been much wider experience.

17.7   Algebraically Mapped Chebyshev Polynomials: $TB_n(y)$

These basis functions are defined on the interval $y \in [-\infty, \infty]$ by

$$TB_n(y) \equiv T_n(x) \equiv \cos(nt) \qquad (17.37)$$

where the coordinates are related via

$$y = \frac{Lx}{\sqrt{1-x^2}} \qquad ; \qquad x = \frac{y}{\sqrt{L^2+y^2}} \qquad (17.38a)$$

$$y = L\cot(t) \qquad ; \qquad t = \text{arccot}(y/L) \qquad (17.38b)$$

The first 11 $TB_n(y)$ are listed in Table 17.5. As one can see from the table or from the right member of (17.38a), the $TB_{2n}(y)$ are *rational* functions which are *symmetric* about $y = 0$. The odd degree $TB_{2n+1}(y)$ are antisymmetric and are in the form of a rational function divided by a left-over factor of $\sqrt{L^2+y^2}$. We shall refer to all these basis functions as the "rational Chebyshev functions on an infinite interval" even though the odd degree members of the basis are not rational. Fig. 17.4 illustrates the first four odd $TB_n(y)$. Note that the map parameter merely changes the y-scale of the wave (the oscillations become narrower when L increases) without altering the shape, so these graphs for $L = 1$ apply for all L if we replace y by (y/L).

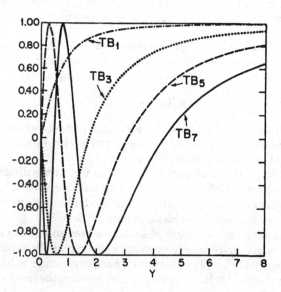

Figure 17.4: Graphs of the first four rational Chebyshev functions on $x \in [-\infty, \infty]$. Because the odd degree $TB_n(y)$ are all antisymmetric about $y = 0$, they are illustrated only for positive y. All asymptote to 1 as $y \to \infty$.

Table 17.4: A Selected Bibliography of Rational Chebyshev Functions

References	Comments
Grosch&Orszag(1977)	Algebraic map: infinite interval in y to $x \in [-1, 1]$ $T_n(x)$ basis, equivalent to rational Chebyshev in y
Boyd (1982a)	Steepest descent asymptotics for spectral coefficients
Christov (1982)	Rational functions related to TB_n; application to rational solution of Burgers-like model
Cain&Ferziger&Reynolds(1984)	Map infinite interval to finite through $y = L\cot(x)$ and apply Fourier in x; equivalent to $TB_n(y)$
Boyd (1985a)	Change-of-coordinate to detour around interior singularity into the complex plane
Boyd (1987a)	Theory and asymptotics for TB_n
Boyd (1987b)	Theory & examples for semi-infinite interval: TL_n
Boyd (1987c)	Quadrature on infinite and semi-infinite intervals
Lin & Pierrehumbert (1988)	Tensor product of $TL_n(z) \otimes TB_m(y)$ for two-dimensional baroclinic instability
Boyd (1990a)	Theory; relation of TB_n to other rational basis sets
Christov&Bekyarov(1990), Bekyarov&Christov(1991)	Nonlinear eigenvalue (soliton); basis is cousin of TB_n
Boyd (1990b,1991a,1991d, 1991e,1995b,1995j)	TB_n/radiation function basis for nonlocal solitons
Boyd (1990d)	TB_n/radiation function basis for quantum scattering (continuous spectrum)
Boyd (1991d,1995a)	TB_n for nonlinear eigenvalue (soliton)
Weideman&Cloot(1990)	Comparisons with the Weideman-Cloot sinh-mapping
Cloot (1991)	Comparisons with solution-adaptive mapping
Cloot&Weideman (1992)	Comparisons with the Weideman-Cloot sinh-mapping; adaptive algorithm to vary the map parameter with time
Liu&Liu&Tang(1992,1994)	TB_n to solve nonlinear boundary value problems for heteroclinic (shock-like) & homoclinic (soliton-like) solutions
Falqués&Iranzo(1992)	Edge waves in the ocean with TL_n
Boyd (1993)	Algebraic manipulation language computations
Chen (1993)	Eigenvalues (hydrodynamic stability) using TL_n
Weideman(1994a,1994b,1995a)	Error function series, good for complex z, using basis $\{(L + iz)/(L - iz)\}^n$, $n = -\infty, \infty$
Weideman(1995b)	Hilbert Transform computation via $\{(L + iz)/(L - iz)\}^n$ Compares well with alternative Fourier algorithm.
Boyd (1996c)	Legendre, quantum and tidal equations; traps and snares in eigencalculations
Gill&Sneddon(1995,1996a,1996b)	Complex-plane maps (revisited) for eigenfunctions singular on or near interior of (real) computational domain
Sneddon (1996)	Complex-plane mappings for a semi-infinite interval
Yang&Akylas(1996)	TB_n/radiation function basis for nonlocal solitons
Matsushima&Marcus(1997)	Polar coordinates: radial basis is rational functions which are images of associated Legendre functions to defeat the "pole problem"

Table 17.5: Rational Chebyshev functions for the infinite interval: $TB_n(y)$.

(For map parameter $L = 1$).

n	$TB_n(y)$
	[Symmetric about $y = 0$]
0	1
2	$(y^2 - 1)/(y^2 + 1)$
4	$(y^4 - 6y^2 + 1)/(y^2 + 1)^2$
6	$(y^6 - 15y^4 + 15y^2 - 1)/(y^2 + 1)^3$
8	$(y^8 - 28y^6 + 70y^4 - 28y^2 + 1)/(y^2 + 1)^4$
10	$(y^{10} - 45y^8 + 210y^6 - 210y^4 + 45y^2 + 1)/(y^2 + 1)^5$
	[Antisymmetric about $y = 0$]
1	$y/(y^2 + 1)^{1/2}$
3	$y(y^2 - 3)/(y^2 + 1)^{3/2}$
5	$y(y^4 - 10y^2 + 5)/(y^2 + 1)^{5/2}$
7	$y(y^6 - 21y^4 + 35y^2 - 7)/(y^2 + 1)^{7/2}$
9	$y(y^8 - 36y^6 + 126y^4 - 84y^2 + 9)/(y^2 + 1)^{9/2}$

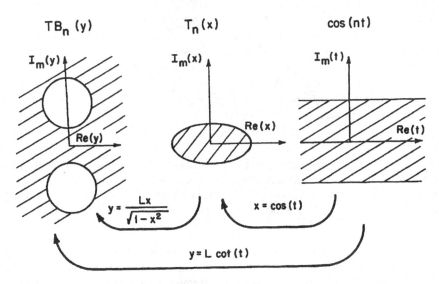

Figure 17.5: Schematic showing the mapping relationships between the rational functions $TB_n(y)$, the Chebyshev polynomials $T_n(x)$, and the Fourier functions $\cos(nt)$. The large arrows indicate the mappings that transform one set of basis functions into another. Each series converges within the cross-hatched region; the mappings transform the boundaries of one region into those of another.

Fig. 17.5 shows the mappings between y, x, and t and the relationships between the basis functions in each coordinate and their domains of convergence. The series converges within the largest shaded area that does not contain a singularity of the function being expanded.

The rational Chebyshev functions are orthogonal on $[-\infty, \infty]$ with the weight function (for $L = 1$) of $1/(1 + y^2)$. However, the easiest way to program is to use the trigonometric representation. The necessary derivative conversion formulas are given in Appendix E.

The earliest use of these functions (without a special notation) is Boyd (1982a). However, Grosch & Orszag (1977) *implicitly* used the same basis by mapping to the interval $x \in [-1, 1]$ and applying ordinary Chebyshev polynomials in x. The most complete treatment is Boyd (1987a).

These orthogonal rational functions have many useful properties. For example, if the differential equation has *polynomial* or *rational* coefficients, one can show [by converting the equation from y to the trigonometric argument t and using trigonometric identities] that the corresponding Galerkin matrix will be banded. A fuller discussion is given in Boyd (1987a).

The rate of convergence of the $TB_n(y)$ series is normally exponential but subgeometric. In the same way that the convergence domain of both sinc and Hermite expansions is a symmetric strip parallel to the real y-axis, the domain of convergence for the rational Chebyshev functions is the *exterior* of a bipolar coordinate surface in the complex y-plane with foci at $\pm iL$ as shown in the left panel of Fig. 17.5. In words, this means that if $f(y)$ is *rational* with no singularity at ∞, then the $TB_n(y)$ expansion of $f(y)$ will converge geometrically. The closer the poles of $f(y)$ are to those of the mapping at $\pm iL$, the more rapid the convergence.

In practice, geometric convergence is rare because the solutions of most differential

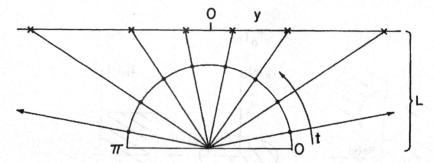

Figure 17.6: Graphical construction, devised by Mark Storz, of the interpolation points for the rational Chebyshev functions which are orthogonal on $y \in [-\infty, \infty]$. The evenly spaced grid points on the semicircle of unit radius (dots) become the roots of the $TB_n(y)$ ($\times$'s) on the line labelled "y". L is the map parameter

equations are singular at infinity. This limits the convergence domain of the $TB_n(y)$ to the real y-axis and reduces the convergence rate from geometric to subgeometric. Nonetheless, the rational Chebyshev functions are still an extremely useful basic set, easy to program and just as efficient as the Hermite and sinc expansions in almost all cases.

Fig. 17.6 is Storz' graphical construction of the interpolation grid. Note that — in contrast to Chebyshev polynomials — the smallest distance between adjacent grid points is $O(1/N)$. This implies that the explicit time-stepping restrictions will be no more severe for these orthogonal rational functions than when finite differences replace them as proved by Weideman (1992).

He shows that rational Chebyshev differentiation matrices have good properties: like those of Hermite functions, the Galerkin matrices are banded and the first derivative is skew-symmetric while the second derivative is symmetric. The corresponding collocation matrices are full matrices without symmetry. However, Weideman argues that the pseudospectral differentiation matrices are "asymptotically normal" in the sense that determining timestep limits from the eigenvalues of the differentiation matrices is legitimate.

In contrast, the Legendre polynomial differentiation matrices are highly non-normal. The concept of "pseudoeigenvalues" is needed to realistically estimate timesteps (Trefethen and Trummer, 1987, Trefethen, 1988, Reddy and Trefethen, 1990, and Trefethen et al., 1993).

Weideman (1992, 1994a, 1994b, 1995a, 1995b) is but one of several authors who find it convenient to work with complex-valued forms of the rational Chebyshev functions. Christov (1982) and Christov and Bekyarov (1990), for example, use linear combinations of the functions

$$\rho_n \equiv \frac{1}{\sqrt{\pi}} \frac{(iy-1)^n}{(iy+1)^{n+1}}, \qquad n = 0, \pm 1, \pm 2, \ldots \qquad (17.39)$$

These functions were introduced by Norbert Weiner as the Fourier transforms of Laguerre functions (Higgins, 1977). Boyd (1990a) shows that with the change-of-coordinate $y = \cot(t)$, these complex-valued functions are simply the terms of a Fourier series in t, or the difference of two such functions, and thus are mathematically equivalent to the TB_n defined here or the SB_n which will be defined in the section after next. One may use whatever form of the rational Chebyshev functions is convenient; we prefer the TB_n basis defined above.

Matsushima and Marcus (1997) have introduced an interesting variant: rational functions which are the images of associated Legendre functions rather than Chebyshev polynomials. Although this sacrifices the applicability of the Fast Fourier Transform, the rational associated Legendre functions, when used as radial basis functions in polar coordinates, have no "pole problem". This means that when using an explicit time-marching algorithm, one can use a time step an order of magnitude larger than with a rational Chebyshev basis. For boundary value or eigenproblems, or when the time-marching method is implicit, this advantage disappears, so the rational associated Legendre functions are useful for one restricted but important niche: explicit time-integrations of unbounded domains in polar coordinates.

17.8 Behavioral versus Numerical Boundary Conditions

In Chap. 6, Sec. 3, we briefly discussed two kinds of boundary conditions: "behavioral" and "numerical". When a problem is periodic, the boundary conditions are behavioral, that is to say, we require that the solution has the behavior of spatial periodicity. We cannot specify in advance what values $u(y)$ will take at the ends of the interval. In contrast, when we solve a non-periodic boundary value problem, we usually must explicitly impose numerical constraints such as $u(-1) = \alpha$, $u(1) = \beta$. Behavioral boundary conditions, like what are called "natural" conditions in finite element theory, do not require any modifications to the basis set. The sines and cosines of the Fourier series, for example, automatically and individually have the behavior of periodicity. In contrast, numerical boundary conditions are "essential", to again borrow a label from finite element theory: They must be imposed as *explicit* constraints.

Boundary conditions at infinity may usually be treated as *either* "behavioral" or "numerical". This flexibility is useful because there are advantages and drawbacks to both approaches. We will describe each in turn.

Although a general theorem is lacking, it is usually true that the *behavioral* condition of *boundedness* at infinity is sufficient to uniquely determine the solution. The reason is that the differential equation is *usually singular* at *infinity* so that only one of the many linearly independent solutions is bounded at that point. If we use an unmodified series of $TB_n(y)$ as the basis, the series will automatically converge to that solution which is finite at infinity in the same way that a trigonometric series will automatically converge to that solution which is periodic.

It is rather ironic, but, because we can ignore the boundary conditions in setting up the pseudospectral or Galerkin's matrix, the singularity of the differential equation at infinity actually makes the problem *easier* instead of harder, as one might have expected. The same thing is usually true of endpoint singularities when the interval is finite. The Chebyshev polynomials and the Associated Legendre functions, for example, both solve second order Sturm-Liouville differential equations which are singular at $x = \pm 1$. The ODE

$$\sqrt{1 - x^2} \left[\sqrt{1 - x^2}\, u_x \right]_x + \lambda u = 0 \tag{17.40}$$

has solutions which are analytic at ± 1 only when (i) $\lambda = n^2$ and (ii) $u(x)$ is $T_n(x)$. If one blindly applies a Chebyshev basis to (17.40) — without explicitly imposing boundary conditions — the eigenvalues of the pseudospectral or Galerkin's matrix will nonetheless converge to n^2 and the corresponding matrix eigenfunctions will have all but one element equal to zero.

Gottlieb and Orszag (1977, pg. 32) point out that it is precisely such singular Sturm-Liouville problems that furnish the best basis sets: Because we need not impose boundary conditions on the solutions of (17.40) [except the implicit one of being analytic at the endpoints], the Chebyshev polynomials will give geometric convergence for all functions independent of their boundary behavior, so long as $f(x)$ is analytic at the endpoints. When the eigenfunctions of a Sturm-Liouville problem have to satisfy specific numerical boundary conditions such as $u(\pm 1) = 0$, series of these eigenfunctions converge geometrically only for equally specific and special classes of functions.

Rule-of-Thumb 15 (Behavioral Boundary Conditions at Infinity) *If the desired solution of a differential equation is the only solution which is bounded at infinity, then one may usually obtain exponential accuracy by using an unmodified series of the rational functions* $TB_n(y)$.

If this doesn't work, or doesn't work well, one can always explicitly impose the constraint that $u(y)$ *vanishes at infinity, but usually this is an unnecessary waste of programming hours.*

The best way to explain this rule-of-thumb is by means of a specific example. The differential equation (Boyd, 1987b)

$$y\,u_{yy} + (y+1)\,u_y + \lambda u = 0 \qquad\qquad y \in [0, \infty] \qquad (17.41)$$

is defined on the semi-infinite interval $[0, \infty]$, and therefore strictly belongs in Sec. 12, but it illustrates behavioral boundary conditions very well. The problem is singular both at the origin and at infinity. Eq. (17.41) has solutions which are analytic at both endpoints only when (i) $\lambda = n$ where n is an integer ≥ 0 and (ii) $u(y)$ is the $(n-1)$-st Laguerre function. When we expand $u(y)$ as a series of the rational functions $TL_n(y)$ without the imposition of boundary conditions, it works — but just barely. With $N = 40$, the lowest numerical eigenfunction agrees with $u(y) = \exp(-y)$, the exact answer, only to a couple of decimal places. The correct eigenvalue ($\lambda = 1$) is mirrored by a pair of complex conjugate eigenvalues with real parts approximately equal to 1 and imaginary parts on the order of 0.01. For so large an N, this is astonishingly poor accuracy.

The rub is that the other linearly independent solution of (17.41) when λ is a positive integer does blow up as $|y| \to \infty$ — but only as a power of y rather than exponentially. The remedy is to make a change-of-unknown to a new variable $w(y)$ defined by

$$w = \exp(y/2)\,u(y) \qquad\qquad (17.42)$$

which is the solution of

$$y\,w_{yy} + w_y + \left[-\frac{1}{2} - \frac{1}{4}y + \lambda \right] w = 0 \qquad\qquad (17.43)$$

The exact eigenvalues are unchanged by the transformation except for the disappearance of $\lambda = 0$, which corresponds to the trivial eigenfunction $u(y) \equiv 1$. After the change-of-unknown (17.42), $N = 40$ gives 12 accurate eigenvalues, and these are all real. From (17.42), we see that the offending second solutions of (17.43) blow up exponentially, rather than algebraically, as $|y| \to \infty$, and the unmodified spectral series has a much easier time separating the true eigenfunctions from the unbounded solutions.

We see why the rule-of-thumb is not a theorem; it still applies to (17.41) in principle, but is extremely inefficient for this problem without the change-of-unknown. Eq. (17.41) combines two types of behavioral boundary conditions in a single example: the solution must be bounded at both endpoints even though one is finite and the other at ∞.

There seems to be no simple theorem that covers all eventualities. Gottlieb and Orszag (1977, pg. 152–153) discuss a Sturm-Liouville eigenproblem whose solutions are the $J_7(y)$ Bessel functions. The boundary condition at $y = 0$ is that $u(y)$ be analytic in spite of the fact that the differential equation is singular there. Blindly applying Chebyshev polynomials gives an approximation that converges on the true eigenvalue as N increases; indeed, with $N = 26$, the approximation is exact to five decimal places! The only problem is that when we impose the boundary conditions $u(0) = u_y(0) = 0$ with the same N, we obtain *eleven* decimal places.[2]

The conclusion is that in coping with behavioral boundary conditions, one should be flexible, prepared to modify the basis set or explicitly impose constraints if need be. "Usually"[3], however, treating boundary conditions at infinity as "behavioral" or "natural" is successful. Boyd (1987a, b) gives a dozen examples; except for (17.41), using the unconstrained $TB_n(y)$ or $TL_n(y)$ gave many decimal places of accuracy with a small number of basis functions.

Nevertheless, it is possible to treat infinity as a *numerical* boundary condition by imposing

$$u(\infty) = 0 \tag{17.44}$$

because the solutions to most problems decay as $y \to \infty$. The programming is more complex because the basis must be modified. On the other hand, if the solution is zero at one or both endpoints, we can reduce the size of the basis set without a loss of accuracy by switching to new basis functions which also vanish at the boundaries.

Indeed, if $u(x)$ decays exponentially fast, then it is true that

$$\frac{d^k u}{dy^k}(\infty) = 0 \tag{17.45}$$

for *any* finite k. Thus, we have the theoretical flexibility to impose an *arbitrary* number of endpoint zeros on the basis.

In practice, this freedom should be used cautiously. If all the basis functions have high order zeros at the endpoints, they will all be almost indistinguishable from zero — and each other — over a large part of the interval. This may give numerical ill-conditioning. Merilees (1973a) shows that Robert's suggested basis functions for the sphere — an ordinary cosine series multiplied by $\sin^m(\theta)$ to mimic the m-th order zero of the corresponding spherical harmonic — is disastrously ill-conditioned for large m. Furthermore, imposition of high order zeros complicates the programming.

Nonetheless, Boyd (1988f) obtained good accuracy for the Flierl-Petviashvili monopole, discussed in Appendix D, by applying the basis

$$\phi_{2n}(y) \equiv TB_{2n}(y) - 1 \tag{17.46}$$

which vanishes as $1/y^2$ as $|y| \to \infty$. The analytical one-basis function approximation was sufficiently accurate to serve as a good first guess for the Newton-Kantorovich iteration. The $N=1$ Chebyshev approximation — $u(y) \equiv$ a constant — was considerably less accurate!

17.9   Expansions for Functions Which Decay Algebraically with y or Asymptote to a Constant

The rational function

$$f(y) \equiv \frac{1}{1 + y^2} \tag{17.47}$$

---

[2]The singularity of the differential equation at the origin actually forces the first *six* derivatives of $J_7(y)$ to vanish at the origin, but it is not necessary to impose all these constraints to obtain very high accuracy.

[3]A pure mathematician would loathe the word "usually", but as Briggs *et al.* (1981) note, "numerical analysts are the academic world's greatest plumbers."

is rather harmless, but both Hermite series and sinc expansions fail miserably in approximating it. The trouble with the former is that the Hermite functions all decay exponentially (like Gaussians) for very large y. One can show that the transition from oscillation to decay occurs for the n-th Hermite function at the "turning points" given by

$$y_t = \pm\sqrt{2n+1} \tag{17.48}$$

Now $f(100)$ is still as large as $1/10,000$, so it follows that to obtain four decimal place accuracy for (17.47), we need to have a Hermite basis large enough so that at least some of the Hermite functions have turning points at $|y| = 100$ or larger. Eq. (17.48) implies that we need at least 5,000 Hermite functions to achieve this — a ridiculously large number. As explained in Sec. 3, the Hermite series of a function which decays algebraically with $|y|$ will have coefficients a_n that decay algebraically with n.

The sinc expansion is similarly in trouble: the coefficients of the Whittaker cardinal series are $f(jh)$, and for a slowly decaying $f(y)$ and a reasonably small grid spacing h, it follows that $f(jh)$ will not be small until j is huge — in other words, we must keep a preposterously large number of sinc functions in the series to obtain even moderate accuracy.

The orthogonal rational basis functions, however, have no such problem. Inspecting Table 17.5, we see that

$$f(y) \equiv \frac{1}{2}\{TB_0(y) - TB_2(y)\} \tag{17.49}$$

if $L = 1$. Even if we choose a different map parameter, the change of variable $y = L\cot(t)$ shows that

$$f(y[t]) = \frac{1 - \cos^2(t)}{1 + (L^2 - 1)\cos^2(t)} \tag{17.50}$$

which has no singularities for any real t and therefore has a geometrically convergent Fourier cosine series. The $TB_n(y)$ expansion of $f(y)$ has the same coefficients as the Fourier series (in t) of $f(y[t])$, so it follows that it, too, must converge geometrically for arbitrary, real L.

Therefore, the RATIONAL CHEBYSHEV FUNCTIONS ARE THE BASIS SET OF CHOICE FOR $f(y)$ THAT DECAY *ALGEBRAICALLY* RATHER THAN EXPONENTIALLY WITH y, OR WHICH ASYMPTOTE TO A CONSTANT AS $y \to \infty$. There are limits; if $f(y)$ blows up, even linearly with y, as $y \to \infty$, then its image under the mapping $y \to t$ will be infinite at $t = 0$, and neither the Fourier series nor the $TB_n(y)$ expansion will even be defined. As long as $f(y)$ is finite, however, then an exponentially convergent series of rational Chebyshev functions is at least a possibility.

We have to be content with the vague phrase "is a possibility" because we need more specific information about the asymptotic behavior of the function. If we consider the function

$$g(y) \equiv \frac{1}{(1 + y^2)^{1/3}} \tag{17.51}$$

then since

$$y = L\cot(t) \approx L/t \qquad t \ll 1 \tag{17.52}$$

it follows that image of (17.51) under the mapping (17.52) will behave like

$$g(y[t]) \approx t^{2/3} \qquad t \ll 1 \tag{17.53}$$

Table 17.6: Examples of functions which asymptote to a constant or decay algebraically with y. These illustrate the four classes of functions that have both symmetry with respect to $y = 0$ (denoted by S for symmetric and A for antisymmetric in the column labeled "Code") and also have asymptotic expansions which contain only even or only odd powers of $1/y$ (indicated by E or O in the "Code" column). The third and fourth columns give the mathematical forms of these symmetries. The rightmost column indicates the restricted basis set that is sufficient to represent all $u(y)$ that fall into this symmetry class.

$u(y[t])$	$u(y)$	Asymptotic Form	Parity with respect to $y=0$	Code	Basis Set
$\cos(t)$	$\dfrac{y}{\sqrt{1+y^2}}$	$\sim \dfrac{y}{|y|} + O\left(\dfrac{1}{y^2}\right)$	$u(y) = -u(-y)$	A & E	TB_{2n+1}
$\cos(2t)$	$\dfrac{y^2 - 1}{y^2 + 1}$	$\sim 1 + O\left(\dfrac{1}{y^2}\right)$	$u(y) = u(-y)$	S & E	TB_{2n}
$\sin(t)$	$\dfrac{1}{\sqrt{1+y^2}}$	$\sim \dfrac{1}{|y|} + O\left(\dfrac{1}{|y|^3}\right)$	$u(y) = u(-y)$	S & O	Odd sines SB_{2n}
$\sin(2t)$	$\dfrac{2y}{1+y^2}$	$\sim \dfrac{2}{y} + O\left(\dfrac{1}{y^3}\right)$	$u(y) = -u(-y)$	A & O	Even sines SB_{2n+1}

and the Fourier and $TB_n(y)$ series will converge algebraically ($a_n \sim O(1/n^{5/3})$) because of the branch point at $t = 0$.

There are a couple of remedies. One, discussed in Boyd (1987a), is to examine the *general* Fourier series in the trigonometric coordinate t. If the asymptotic behavior of $f(y)$ may be represented as a series of the inverse *odd* powers of y, then the sine series will converge exponentially fast. The images of the sines under the inverse mapping define a new basis set, $SB_n(y) \equiv \sin\{(n+1)[\text{arccot}(y/L)]\}$ for $n = 0, 1, 2, \ldots$.[4] These functions, which are rational in y, are *never needed* except to approximate a function which is *decaying algebraically* as $|y| \to \infty$. Even then, they may be unnecessary if $f(y)$ may represented as series of inverse *even* powers of y for $|y| \gg 1$. Table 17.6 summarizes the cases.

The $SB_n(y; L)$ can be computed directly by introducing $Y \equiv y/L$ and then applying the three-term recurrence

$$SB_0 = \frac{1}{\sqrt{1+Y^2}}, \qquad SB_1 = \frac{2Y}{1+Y^2}, \qquad SB_{n+1} = 2\frac{Y}{\sqrt{1+Y^2}}SB_n - SB_{n-1} \qquad (17.54)$$

The second option is to apply a mapping. For example, the "bad" function (17.51) has the geometrically converging series

$$\frac{1}{(1+y^2)^{1/3}} = \sum_{n=0}^{\infty} a_{2n}\, TB_{2n}(y^{1/3}) \qquad (17.55)$$

which is equivalent to mapping the problem from y to the trigonometric coordinate t using $y = L \cot(t^3)$.

---

[4]The indexing convention for the SB has been adjusted from the first edition of this book so that SB_0 is the lowest basis function and the even degree SB are symmetric with respect to $y = 0$.

Similarly, a change-of-coordinate will redeem Hermite and sinc series if the mapping transforms *algebraic* decay into *exponential* decay as noted by Eggert, Jarratt, and Lund (1987). An example on the semi-finite interval:

$$y = \exp(x) - 1 \qquad\qquad y \ \& \ x \in [0, \infty] \tag{17.56}$$

Decay as $O(1/y)$, for example, becomes decay as $O(\exp[-x])$.

Nevertheless, if one is forced to accept the burden of a mapping, it is far simpler to solve the transformed problem using a Fourier series rather than something more complicated like Hermite or sinc expansions. Consequently, the earlier assertion remains true: the best way to approximate algebraically decaying functions is to use a rational Chebyshev series, which is a Fourier series with a change-of-coordinate. However, the possibility of applying sinc and Hermite functions even to "bad" cases like algebraic decay is again a reminder of the tremendous power of mappings.

17.10   Numerical Examples for the Rational Chebyshev Functions

In oceanography, the steady north-south current of the "Yoshida jet" solves

$$v_{yy} - y^2 v = y \tag{17.57}$$

For large y,

$$v(y) \sim -\frac{1}{y} \tag{17.58}$$

This slow, algebraic decay with $|y|$ destroys both Hermite series and sinc expansions. Because $v(y)$ is an antisymmetric function with *odd* powers of $1/y$, Table 17.6 implies that the $TB_n(y)$ will fail, too. However, the *mapping* from y to t will still succeed provided that we use a *sine* series instead of the usual cosine series to solve the transformed differential equation. Fig. 17.7 shows the exact solution and the 2-point, 3-point, and 4-point collocation approximations. The $N = 6$ series listed in the table is indistinguishable from the exact solution to within the thickness of the curve.

Table 17.7 lists the coefficients for various numbers of interpolation points. As always, the error is the sum of two contributions. The truncation error is the neglect of all coefficients a_n with $n > N$; the missing coefficients are represented by dashes. In addition, those coefficients that are calculated differ from those of the exact solution. As we read a row from left to right, however, we can see this "discretization error" decrease as each coefficient converges to its exact value as N increases.

Note that because $v(y)$ is antisymmetric, only odd basis functions ($\sin(t[y])$, $\sin(3t[y])$, $\sin(5t[y])$, etc.) were used in the expansion. All the collocation points were on the interior of $y \in [0, \infty]$ (instead of $[-\infty, \infty]$) for the same reason.

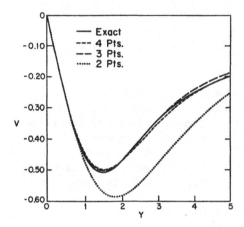

Figure 17.7: The latitudinal velocity $v(y)$ for the "Yoshida jet". The map parameter $L = 3$. SB_n were used instead of the more common TB_n to capture the very slow $1/y$ decay of $v(y)$ with spectral accuracy.

Table 17.7: The coefficients of the spectral series for the Yoshida jet $v(y)$ as computed using various numbers of collocation points with a basis of odd SB functions.

n	2 pts.	3 pts.	4 pts.	7 pts.	21 pts.
1	-0.47769	-0.388976	-0.394428	-0.395329	-0.395324
3	0.199475	0.179639	0.176129	0.179237	0.179224
5	— —	-0.050409	-0.04471	-0.051163	-0.051147
7	— —	— —	0.001399	0.002620	0.002651
9	— —	— —	— —	0.003856	0.003740
11	— —	— —	— —	-0.000541	-0.000592
13	— —	— —	— —	-0.000452	-0.000446
15	— —	— —	— —	— —	0.000060

The second example is

$$u_{yy} + \left[n(n+1)\,\mathrm{sech}^2(y) + 1\right]u = P_n(\tanh[y]) \tag{17.59}$$

where n is an integer ≥ 0 and $P_n(x)$ is the n-th Legendre polynomial.

Eq. (17.59) is merely Legendre's differential equation transformed from $x \in [-1, 1]$ to $y \in [-\infty, \infty]$ via $y = \mathrm{arctanh}(x)$. On the finite interval, the differential equation is singular at both endpoints. The boundary conditions are that the solution must be analytic at $x = \pm 1$ in spite of the singularities. The Legendre polynomials are the only solutions that satisfy this requirement, and only when n is an integer.

The solutions of the transformed equation (17.59) asymptote to the constant 1 as $y \to \infty$ for all n and to either 1 (even n) or -1 (odd n) as $y \to -\infty$. The coefficients of the Hermite series for such a function would decay as $1/n^{1/4}$ at the fastest. However, a rational Chebyshev expansion is extremely effective. The boundary conditions are natural (behavioral) both before and after the arctanh-mapping, so it is not necessary to impose any constraints on the $TB_n(y)$.

Fig. 17.8 compares the exact solution (solid line) with the sixteen-point pseudospectral approximation for the case $n = 12$. Because $P_{13}(\tanh[y])$ is symmetric about $y = 0$, only symmetric basis functions were used, and the sixteen-point approximation included $T_{30}(y)$. The approximation for $N = 38$ (20 collocation points, all with $y \geq 0$), was indistinguishable from the exact solution to within the thickness of the curve.

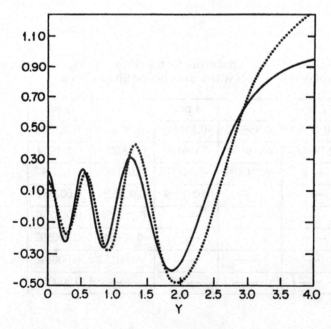

Figure 17.8: Solid curve: $P_{12}(\tanh[y])$ where $P_{12}(x)$ is the Legendre polynomial. Dotted curve: numerical approximation by solving the transformed Legendre differential equation using 16 symmetric basis functions, i. e. $TB_0(y)$, $TB_2(y)$, $TB_4(y)$, $\ldots$, $TB_{30}(y)$.

17.11   Rational Chebyshev Functions on $y \in [0, \infty]$: the $TL_n(y)$

These basis functions are defined by

$$TL_n(y) \equiv T_n(x) = \cos(nt) \tag{17.60}$$

where the argument of the rational Chebyshev functions, y, is related to the argument of the Chebyshev polynomials, x, and the argument of the trigonometric functions, t, via

$$y = \frac{L(1+x)}{1-x} \quad\leftrightarrow\quad x = \frac{y-L}{y+L} \tag{17.61}$$

$$y = L\cot^2\left(\frac{t}{2}\right) \quad\leftrightarrow\quad t = 2\operatorname{arccot}\left(\sqrt{\frac{y}{L}}\right) \tag{17.62}$$

The first few functions are given explicitly in Table 17.8. The pseudospectral interpolation points are

$$y = L\cot^2\left(\frac{t_i}{2}\right) \quad\leftrightarrow\quad t_i \equiv \frac{(2i-1)\pi}{2N+2} \quad i = 1, \ldots, N+1 \tag{17.63}$$

Table 17.8: Rational Chebyshev functions for the semi-infinite interval: $TL_n(y)$.

n	$TL_n(y; L)$
0	1
1	$(y - L)/(y + L)$
2	$(y^2 - 6yL + L^2)/(y + L)^2$
3	$(y - L)(y^2 - 14yL + L^2)/(y + L)^3$
4	$(y^4 - 28Ly^3 + 70L^2y^2 - 28L^3y + L^4)/(y + L)^4$

We will discuss these functions only briefly because their characteristics are very similar to those of the rational Chebyshev functions on $y \in [-\infty, \infty]$ and Boyd (1987b) gives a thorough treatment. If $f(y)$ is rational with no singularity at ∞, then the $TL_n(y)$ series will have geometric convergence. Usually, however, the solution will be singular at infinity so that convergence will be *subgeometric*.

If the differential equation has polynomial or rational coefficients (in y), the Galerkin matrix will be banded.

If $f(y)$ asymptotes to a constant or decays algebraically with y as $y \to \infty$, then the $TL_n(y)$ series may still have an exponential rate of convergence. A series of Laguerre functions, however, will converge very poorly unless $f(y)$ decays exponentially for $y \gg 1$. In a modest difference from the $TB_n(y)$ case, the map (17.62) shows that a cosine series in t will give rapid convergence if $f(y)$ has an asymptotic series as $y \to \infty$ in inverse negative powers of y. The sines are needed only to represent *half-integral* powers of y.

Boyd (1982b) offers guidelines for optimizing the map parameter L, but some trial-and-error is inevitable. (The simplest criterion is: choose L to roughly equal the scale of variation of the solution.) Fortunately, the accuracy is usually quite insensitive to L so long as it is of the same order-of-magnitude as the optimum value. In the next section, we offer an illustration.

17.12 Numerical Examples for Mapped Chebyshev Methods on a Semi-Infinite Interval

Example #1: Laguerre Eigenvalue Problem

$$u_{yy} + \left[\frac{1}{y} - \frac{1}{4} - \frac{\lambda}{y}\right] u = 0 \tag{17.64a}$$

$$u(y) \text{ analytic and bounded at both } y = 0 \text{ and } y = \infty \tag{17.64b}$$

where λ is the eigenvalue. The exact solution is

$$u = \exp\left(-\frac{y}{2}\right) y\, L_n^1(y) \tag{17.65a}$$

$$\lambda \equiv n, \quad n \text{ an integer} \geq 0 \tag{17.65b}$$

where $L_n^1(y)$ denotes the first order Laguerre polynomial of degree n. Since the boundary conditions are "behavioral" at both endpoints, unmodified $TL_n(y)$ are a good basis. The $(N + 1) \times (N + 1)$ pseudospectral matrix always has $(N+1)$ eigenvalues, but the $(N+1)$-st eigenfunction of the original differential equation is always oscillating too rapidly to be resolved by interpolation using only $(N+1)$ points. Consequently, we face the difficulty — inevitable for any eigenvalue problem, whether on a finite or unbounded interval — that the lowest few eigenvalues of the matrix are good approximations to those of the differential equation whereas the larger matrix eigenvalues are numerical artifacts.

Fig. 17.9 shows the number of "good" eigenvalues computed with 41 collocation points as a function of the map parameter L where a "good" eigenvalue is arbitarily defined as one within 0.05 of the exact eigenvalue (17.65b). Because N is large and the eigenfunctions are entire, the theory of Boyd (1982a) argues that L should be fairly large, and the graph confirms it. ($L_{optimum} = 32$.) It is reassuring, however, that taking L too large or small by a factor of 2 still yields at least $N/3$ "good" eigenvalues. It is striking that the lowest two eigenvalues are accurately calculated for L anywhere between 1 and 512, a range of three orders-of-magnitude!

Example #2: Global Expansion of the K-Bessel Function, i. e.

$$f(r) \equiv r\, K_1(r) \qquad \text{on} \qquad r \in [0, \infty] \tag{17.66}$$

where $K_1(r)$ is the usual imaginary Bessel function. Library software for evaluating this function normally employs (i) a power series representation with logarithmic terms for small r and (ii) an asymptotic series in inverse powers of r for $r \gg 1$. Our goal is much more ambitious: to use mapping to obtain a *single* expansion that is accurate for *all* $r \geq 0$.

The Bessel function K_1 has a simple pole at the origin, but the factor of r in (17.66) removes this so that $f(r)$ is finite. However, no multiplicative factor can remove the $\log(r)$, which itself multiplies an infinite series in r. Fortunately, since $f(r)$ is bounded at the origin, we can use the exponential mapping trick introduced by Stenger. Since $K_1(r)$ decays as $\exp(-r)$ for large r, it is convenient to use a mapping which is linear in r for large r since we do not have any unusual problem at ∞. Therefore, we first apply the change of coordinate

$$r = \text{arcsinh}(e^y) \tag{17.67}$$

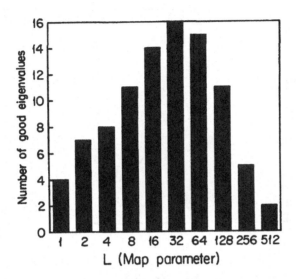

Figure 17.9: The number of "good" eigenvalues of the Laguerre equation as a function of the map parameter L. A "good" eigenvalue is arbitrarily defined as one whose absolute error is less than 0.05. All runs used 41 collocation points and the basis functions $TL_0(y), \ldots, TL_{40}(y)$ on the interval $y \in [0, \infty]$.

to transform to $y \in [-\infty, \infty]$ and then apply the $TB_n(y)$. Although it is a little confusing to use the Chebyshev rational functions on the *infinite* interval in solving a problem which was originally defined on a *semi-infinite* domain, this example does illustrate the power of mappings.

The end result, after conversion back to the original coordinate r, is an approximation of the form

$$
\begin{aligned}
r\,K_1(r) \;=\; & 0.534 - 0.6\,TB_1(\log[\sinh(r)]) - 0.068\,TB_2(\log[\sinh(r)]) \qquad (17.68) \\
+ \; & 0.125\,TB_3(\log[\sinh(r)]) + 0.032\,TB_4(\log[\sinh(r)]) \\
- \; & 0.032\,TB_5(\log[\sinh(r)]) + \text{terms smaller than } 0.01
\end{aligned}
$$

Because we are resolving a logarithmic singularity at $r = 0$, a good choice of L must be much smaller than for the previous example. Fig. 17.10 shows the result with map parameter $L = 4$: the logarithm of the error to base 10 is graphed against N, the truncation.

The approximation is almost too good. If the series converged geometrically, then the curve would approximate a straight line. The actual graph is a fairly good approximation to a straight line — but one can rigorously prove that the rate of convergence must be *subgeometric* because $r\,K_1(r)$ is singular at both endpoints. A careful look shows that the curve is beginning to flatten out a bit near the right edge of the graph; the slope would tend to zero from above if we extended the graph to sufficiently large N. This would be rather foolish, however, because $N = 24$ already gives six decimal place accuracy, and within the range illustrated, the subgeometrically convergent series does a remarkably good job of imitating the straight line rate-of-decrease-of-error that is the spoor of geometric convergence on a log-linear plot.

This example serves to remind us that theoretical ideas about orders and rates of convergence are almost always *asymptotic* estimates. In this case, the concept of "subgeomet-

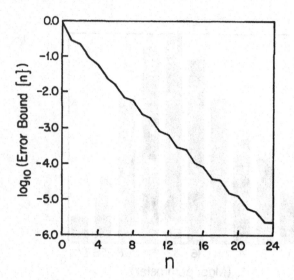

Figure 17.10: The base-10 logarithm of the error bound as a function of n, the degree of the highest $TB_n(y)$, in the representation of the function $f(r) \equiv r\,K_1(r)$ on $r \in [0, \infty]$. To resolve the logarithmic singularity at the origin, r is mapped into the infinite interval $y \in [-\infty, \infty]$ via $r = \operatorname{arcsinh}(\exp[y])$ and the result is expanded using the rational Chebyshev functions in the new coordinate y. The n-th error bound is the sum of the absolute values of all coefficients a_m with $m > n$; this is a slight overestimate of the actual maximum pointwise error on $r \in [0, \infty]$.

ric" convergence gives a misleading picture of how rapidly the error decreases for small and moderate N. Boyd (1978a, Table VIII) offers another amusing illustration.

17.13　Functions Which Oscillate Without Exponential Decay at Infinity

Despite the effectiveness of the mappings and basis sets illustrated above and in the journal articles mentioned earlier, there are still classes of problems where further research is badly needed. When a function $f(y)$ oscillates as $y \to \infty$, there is no difficulty as long as the function also decays *exponentially* fast with y. It is unnecessary for the basis functions to

Table 17.9: A Bibliography of Spectral Solutions to Oscillate-at-Infinity Problems

References	Comments
Boyd (1987b)	Theory & TL_n expansions of amplitude and phase for J_0 Bessel function
Boyd (1990b,1991a,1991d, 1991e,1995b,1995j)	TB_n/radiation function basis for nonlocal solitons
Boyd (1990d)	TB_n/radiation function basis for quantum scattering (continuous spectrum)
Yang&Akylas(1996)	Weakly nonlocal solitary waves

resolve thousands and thousands of wavelengths since the function is negligibly small outside a region that includes only a few alternations in sign.

When $f(y)$ decays algebraically with y, the $TB_n(y)$ and $TL_n(y)$ still converge rapidly as long as either (i) the function does not oscillate or (ii) the frequency of the oscillations dies out as $y \to \infty$. However, if the function decays slowly, but the rate of oscillation does not, then a conventional expansion is in big trouble.

A prototype is the Bessel function

$$J_0(y) \sim \sqrt{\frac{2}{\pi y}} \left\{ \left[1 - \frac{9}{128\,y^2} + O\left(y^{-4}\right) \right] \cos\left(y - \frac{\pi}{4}\right) \right.$$

$$\left. + \left[-\frac{1}{8y} + O\left(y^{-3}\right) \right] \sin\left(y - \frac{\pi}{4}\right) \right\} \qquad (17.69)$$

for $y \gg 1$. It is not possible for a finite number of basis functions to resolve an infinite number of oscillations. Because $J_0(y)$ decays in magnitude only as $1/\sqrt{y}$, if we fail to resolve the oscillation between $y = 200\pi$ and $y = 202\pi$, for example, we will make an error of $O(1/\sqrt{200\pi}) = 0.04$. To accurately approximate the first hundred wavelengths, however, would take several hundred $TL_n(y)$. So many basis functions for only 4% accuracy!

Library software avoids this problem by using two approximations: the asymptotic expansion (17.69) for large y and a power series for small y. It is not possible to extend the large y expansion to $y = 0$ since the series in $1/y$ are divergent. If we wish to obtain a "global" approximation, that is to say, a single expansion accurate for all of $y \in [0, \infty]$, we

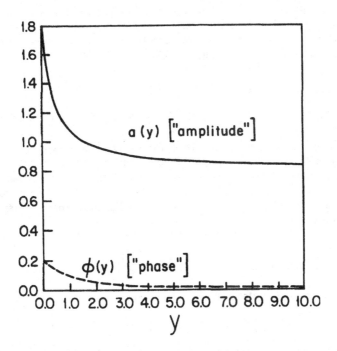

Figure 17.11: The functions $a(y)$ and $\phi(y)$ in the approximation
$$\sqrt{1 + y}\, J_0(y) \approx a(y)\cos(y - \pi/4 - \phi[y])$$
which is *uniformly* accurate for $y \in [0, \infty]$.

must mimic the asymptotic expansion (17.69). Boyd (1987b) assumed the *double* series

$$\sqrt{1+y}\, J_0(y) \approx \cos\left(y - \frac{\pi}{4}\right) \sum_{n=0}^{M} a_n \, TL_n(y) + \sin\left(y - \frac{\pi}{4}\right) \sum_{n=0}^{N} b_n \, TL_n(y) \qquad (17.70)$$

and computed the coefficients by interpolation at $(M + N + 2)$ points, spaced as in (17.63). With $M = 15$ and $N = 5$, the maximum absolute error was less than 0.00001 everywhere on the entire semi-infinite interval.

In some respects, there is nothing mysterious about this: the expansion (17.70) is a faithful mirror of the asymptotic series and the $TL_n(y)$ converge rapidly for functions with asymptotic series in negative powers of y. However, the success is also rather surprising (and encouraging) because (17.70) is a blind leap off a cliff. There is no convergence theory whatsoever for basis sets composed of mixtures of sines and cosines multiplied by the $TL_n(y)$. Since the $\cos(y - \pi/4)$ and $\sin(y - \pi/4)$ factors are still rapidly oscillating for large y while the rational Chebyshev functions have each smoothly asymptoted to 1, there is no reason to suppose that a distribution of points that is optimum for the $TL_n(y)$ alone will capture the much more complicated (and rapidly varying) behavior of the rational-Chebyshev-multiplied-by-a-cosine used to compute the coefficients in (17.70). We can make a strong case, via the WKB approximation from whence (17.69) came, that $J_0(y)$ has an accurate approximation in the form of (17.70). We have no theory, however, to indicate precisely how to compute those coefficients.

In point of fact, the simple-minded interpolation is somewhat ill-conditioned, and it was not possible to obtain more than five decimal places of accuracy when computing in sixteen decimal place arithmetic. A Galerkin-type procedure (although it, too, was somewhat ill-conditioned) achieved 2×10^{-7} accuracy. Perhaps the ill-conditioning could be removed by using the Euler summation method to more accurately evaluate the Galerkin's integrals, which have highly oscillatory integrands. Fig. 17.11 shows the "amplitude" $a(y)$ and "phase" $\phi(y)$ when elementary trigonometric identities are used to rewrite (17.70) as

$$\sqrt{1+y}\, J_0(y) = a(y) \cos\left(y - \frac{\pi}{4} - \phi[y]\right) \qquad (17.71)$$

Similar difficulties arise in quantum scattering (Boyd, 1990d), in which the incoming, transmitted, and reflected waves all propagate even at spatial infinity, and in the theory of weakly nonlocal solitary waves, which decay from a large central core to small amplitude oscillations of constant amplitude (Boyd, 1995b, 1998b). An alternative strategy, similar in spirit to the method described above but different in detail, is to use a standard rational Chebyshev basis (or whatever) and augment it with one or two special "radiation functions", which have the correct oscillations-at-infinity built into them. Thisis discussed in the Chap. 19, Secs. 4 and 5.

17.14 Weideman-Cloot Sinh Mapping

Weideman and Cloot(1990) introduced the mapping

$$y = \sinh(Lt), \qquad y \in [-\infty, \infty] \ \& \ t \in [-\pi, \pi] \qquad (17.72)$$

They apply a Fourier pseudospectral method in the new coordinate t. Since the image of $t \in [-\pi, \pi]$ is $y \in [-y_{max}, y_{max}]$ where y_{max} is finite, their scheme combines both mapping and domain truncation in a single algorithm. Weideman and Cloot(1990), Cloot(1991), Cloot and Weideman (1992) and Boyd (1994b) offer both theoretical and empirical evidence that this Fourier-domain-truncation-with-mapping is as effective as the rational Chebyshev basis, or better, in many applications.

The crucial point is that because $y_{max} \equiv \sinh(L\pi)$ grows *exponentially* fast with L, very small increases in L produce large decreases in the domain truncation error $E_{DT} \equiv |u(y_{max})|$. There is also a series truncation error $E_S(N)$ that arises because the Fourier series in t must be truncated. Since the Fourier coefficients decrease geometrically for most functions, it follows that $\log|E_S|$ is roughly $O(N)$ for *fixed* L. To keep the domain truncation error as small as the series truncation error, however, we must increase L with N so that $\log|E_{DT}|$ is also $O(N)$. However, because the sinh-map increases the size of the computational domain *exponentially* fast with L, it follows that increasing L *logarithmically* with N will ensure that the domain truncation error falls off geometrically with N.

The bad news is that increasing L will move the singularities of $u(y[t])$ closer to the real t-axis. As explained in Chapter 2, the pole or branch point whose location has the smallest imaginary part controls the rate of convergence of the Fourier series. If the convergence-limiting singularity is at $t = t_s$, then $\log|E_S(N)| \sim -|\Im(t_s)| \, N$. It follows that a logarithmic increase of L with N will cause a logarithmic decrease in $|\Im(t_s)|$ so that $\log|E_S|$ and also the total error (series plus domain truncation) must decrease no faster than

$$\log|E| \sim O\left(\frac{N}{\log(N)}\right) \tag{17.73}$$

Boyd (1994b) dubs this "quasi-geometric" convergence because it fails to be truly geometric only because of the logarithm.

Cloot and Weideman (1990) show further that if we optimize L so that the domain truncation and series truncation errors are roughly the same and decrease as rapidly as possible in lockstep at $N \to \infty$, then the optimum L has a remarkably simple form for a large class of functions For example, suppose that $u(y)$ is such that

$$E_{DT} \sim \exp(-A y_{max}^r), \qquad y_{max} \gg 1 \tag{17.74}$$

and has a convergence-limiting singularity (that is, the pole or branch point whose location has the smallest imaginary part of all the singularities of the function) at $y = y_s$. Applying the mapping and then simplifying under the assumption that $L \gg 1$ gives,

$$\log(E_{DT}) \sim -A \, 2^{-r} \exp(r\pi L) \tag{17.75}$$

$$\log(E_S) \sim -N \frac{|\Im(y_s)|}{L} \tag{17.76}$$

The sum $E_{DT} + E_S$ is minimized as $N \to \infty$ by

$$L_{opt} \sim \frac{1}{\pi} \log(N^{1/r}) + O(\log(\log(N))) \tag{17.77}$$

In contrast, rational Chebyshev series have subgeometric convergence. For a function like $\operatorname{sech}(Ay)$, the asymptotic analysis of Boyd(1982b) shows $L_{opt} \sim 1.07 N^{1/3}/A$ and $\log|E_S| \sim -1.47 N^{2/3}$, which is subgeometric convergence with exponential index of convergence $2/3$. The optimum map parameter depends upon A, the width parameter of $u(y)$, whereas the optimum L for the Cloot-Weideman mapping depends only upon N.

Fig. 17.12 shows that for $\operatorname{sech}(y)$, the Weideman-Cloot sinh-mapped Fourier algorithm gives much smaller errors for the same number of degrees of freedom. However, it also shows a problem. The asymptotic estimate for the best map parameter for the rational Chebyshev basis, $L \sim 3.8$, is fairly close to the bottom of the curve of the actual maximum

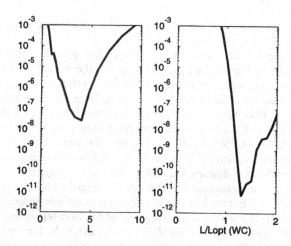

Figure 17.12: Errors in the rational Chebyshev series (left panel) and the Weideman-Cloot algorithm for the function $u(y) = \text{sech}(y)$ for various map parameters L (left graph) or domain parameters L (right). $L_{opt} \equiv (1/\pi)\log(N)$ is Cloot and Weideman's own estimate of the best L. Twenty-two rational Chebyshev functions were used for the left panel; the same number of cosine functions for the right graph.

pointwise error versus L, and the error in the TB series is not very sensitive to L anyway. However, $L = 1$ in the Weideman-Cloot method gives only one-third the number of correct digits obtainable by using L only 30% larger; the Weideman-Cloot method is much more sensitive to L.

This sensitivity difference between mapping and domain truncation was noted by Boyd (1982b). With domain truncation, the total error is the sum of two terms which grow or decay differently with L: the domain truncation error E_{DT} decreases with L while the error in the N-term series increases with L. The total error has the sharp pointed shape of the letter "V" on a graph of error versus L. For *entire* functions, the rational Chebyshev basis is better because there is only the series error, and the graph of error versus L is flat at the minimum.

Unfortunately, when $u(y)$ has a singularity, the error for the TB basis is also the sum of two unrelated terms; one is the series truncation error associated with the singularity, the other depends on how rapidly $u(y)$ decays with $|y|$ for large $|y|$. Consequently, the curve on the left of Fig. 17.12 is V-shaped. However, the penalty for underestimating the optimum L is less drastic for the rational Chebyshev basis than for the sinh-map.

One difficulty in estimating L for the sinh-mapping is that the relative error in the leading order approximation to L_{opt} is $\log(|\Im(y_s)| \pi r 2^r/A) / \log(N)$, which is rather large unless N is huge. Indeed, although the leading order estimate is technically independent of A, the width of $\text{sech}(Ay)$, it is actually a good idea to make a preliminary linear change-of-coordinate so that $A \sim O(1)$ in the new coordinate. Otherwise, the simple approximation $L_{opt} \sim (1/\pi)\log(N)$ may be useless until N is ridiculously large.

Fig. 17.13 illustrates the spectral coefficients for the two methods. The map parameters were deliberately chosen to be optimal (empirically) for $N = 44$, but the coefficients are graphed to much larger N. For both methods, the coefficients decrease roughly geometrically until N is slightly larger than that for which these map parameters are optimal, and then both series flatten out. The Fourier series runs into a wall and becomes very flat because the coefficients for $N > 50$ are those of Gibbs' Phenomenon, controlled by the

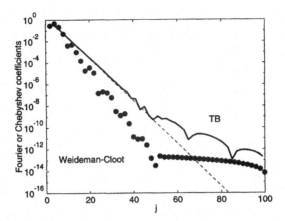

Figure 17.13: Absolute values of spectral coefficients for the rational Chebyshev series (solid) and the Cloot-Weideman algorithm (disks) for the function $f(y) = \mathrm{sech}(y)$. The dashed line is a fit to the leading coefficients of the TB series.

jump discontinuity at $t = \pi$ whose magnitude is the domain truncation error. The rational Chebyshev series flattens out, as shown by comparison with the dashed dividing line, which was chosen to match the slope of the coefficients a_j for $j < 40$. However, the error continues to decrease exponentially fast, albeit at a subgeometric rate. The curves provide some useful empirical guidance for optimizing L as expressed by the following:

Rule-of-Thumb 16 (OPTIMIZING INFINITE INTERVAL MAP PARAMETER)
Plot the coefficients a_j versus degree on a log-linear plot. If the graph abruptly flattens at some N, then this implies that L is SMALL *for the given N, and one should increase L until the flattening is postponed to $j = N$.*

Fig. 17.14, which is identical to Fig. 17.12 except for a different $u(y)$ and N, illustrates how difficult it is make rigid rules about the superiority of one method over another. For the function $u(y) = \mathrm{sech}(y)\cos(3y)$, the smallest errors attainable for a given N are roughly the same for each choice of basis. One would probably be more comfortable using the rational Chebyshev basis because it is less sensitive to the choice of the map parameter.

For solving model time evolution equations, Weideman and Cloot (1990) concluded that the sinh-map gave the smallest errors and was most efficient when the width of the solution changes little with time. However, the rational Chebyshev basis was superior when the width of the solution varied greatly with time. Both authors have continued to employ both schemes in their own later work.

Their own inability to choose one over the other reiterates a key theme: on the infinite and semi-infinite intervals, there are a lot of good choices, but no perfect choice.

17.15 Summary

When the target solution $u(y)$ decays exponentially as the coordinate tends to infinity, there are many good spectrally-accurate algorithms. Sinc functions are the simplest, Hermite and Laguerre functions have the closest connection with the physics for some classes of problems, rational Chebyshev functions and the Weideman-Cloot sinh-mapped Fourier

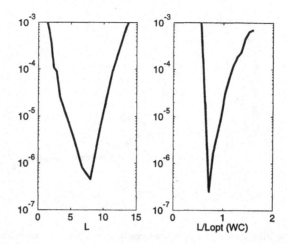

Figure 17.14: Errors in the rational Chebyshev series (left panel) and the Weideman-Cloot algorithm for the function $u(y) = \text{sech}(y)\cos(3y)$ for various map parameters L or domain parameters L/L_{opt} where $L_{opt} \equiv (1/\pi)\log(N)$ is the Cloot and Weideman's own estimate of the best L. Twenty-two rational Chebyshev functions were used for the left panel; the same number of cosine functions for the right graph.

scheme exploit the Fast Fourier Transform. We have not made a general recommendation about the "best" basis because there is no compelling reason to choose one basis over another that applies to all cases.

When the function $u(y)$ decays as an inverse power of y, then rational Chebyshev functions are the clear favorite. This basis is equivalent to making a change of coordinate, such as $y = L\cot(t)$ for the infinite interval, and then applying a Fourier basis in the new coordinate t. However, one must analyze the asymptotics of the solution carefully to ensure that the function $u(y[t])$ is well-behaved in t. If $u(y)$ decays slowly but as some fractional power of y, then one can still obtain spectral accuracy. However, one must modify the cotangent mapping so that $u(y)$ is infinitely differentiable everywhere in the new coordinate t.

Another class of nasty problems is when a function defined on a semi-infinite interval is singular at one endpoint or on or near the interior of the interval. The singularities can be resolved by combining the ideas of this chapter with those of the previous chapter. For an endpoint singularity, for example, one can map the semi-infinite interval to $t \in [0, \pi]$ using a change-of-coordinate that clusters the grid points with a spacing that decreases exponentially near the singular endpoint $y = 0$ as illustrated by the series for K_1 Bessel function above. Similarly, one can choose the map to detour around a singularity on the interior of the interval whether the interval is finite or unbounded.

When $u(y)$ decays to an oscillation rather than to zero as $|y| \to \infty$, the best strategy is to use special basis functions, added to a standard infinite interval basis, which match the oscillations, leaving the standard basis to approximate the part of the function which decays without oscillation. We have illustrated this strategy using the amplitude-and-phase approximations to the J_0 Bessel function. However, the same idea works well in quantum scattering and computations of weakly nonlocal solitary waves.

For all basis sets on an unbounded interval, there is a scale factor or map parameter that (if chosen properly) matches the width of the basis functions to the width of the solution. One must use much narrower basis functions to compute the structure of a molecule than to compute a solitary wave approximation to the Great Red Spot of Jupiter, which is 40,000 km across.

For Whittaker cardinal functions, the scaling factor is the grid spacing. When Chebyshev polynomials are mapped from $[-1, 1]$ to $[-\infty, \infty]$ to give the rational basis functions denoted by $TB_n(y)$, the scaling factor is the map parameter L. The Hermite functions do not *explicitly* contain a free parameter, but we always have the option of performing the expansion in terms of $\psi_n(\alpha y)$ instead of $\psi_n(y)$ where α is an arbitrary scaling factor. The proper choice of α is just as important to maximizing the efficiency of the Hermite series as is a good choice of L for mapped Chebyshev polynomials. When the domain is unbounded, there is always, either explicitly or implicitly, a scale parameter to be chosen.

Chapter 18

Spherical, Cylindrical, Toroidal & Elliptical Geometry

"The book of nature is written in the characters of geometry"
— Galileo Galilei

18.1   Introduction

Cylindrical and spherical geometry require special theory, special grids, and special basis functions. There is no easy road to compute flows on the surface of a sphere or solve a partial differential equation on the interior of a disk.

The heart of the problem is that at the origin of a polar coordinate system, or at the north and south poles of the surface of a sphere, the lines of constant polar angle or constant longitude converge in a single point. This convergence-of-meridians has two unfortunate consequences. First, there is a severe time-stepping limit, the so-called "pole problem". Second, the differential equation is almost invariably singular at this convergence point even when the solution is everywhere smooth and infinitely differentiable.

In this chapter, we hope to initiate the reader into the mysteries of disk and sphere. In the first part, we discuss polar coordinates. In the second, we concentrate on latitude and longitude, the coordinates of the surface of a sphere. As explained in the next section, polar, cylindrical, spherical, and toroidal coordinates are very closely related: there is really only one species of "pole problem", manifesting itself as a disease in four different coordinate systems.

Spectral algorithms for all these coordinates have been developed. The numerical technology has become so mature that spectral methods for weather forecasting and climate modelling have conquered a global empire embracing more countries than that of Caesar or Alexander. Nevertheless, as the century ends and massively-parallel computing gathers momentum, spherical harmonic models are under siege. As we describe the techniques, we shall narrate both the virtues and faults of each spectral and non-spectral option and let the reader be his own prophet.

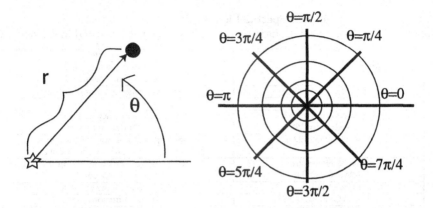

Figure 18.1: Left: Definitions of radius r and angle θ for a typical point (black disk) in polar coordinates. Right: In polar coordinates, the contours of constant radius r are concentric circles. The contours of constant angle θ are semi-infinite rays through the origin (thick lines).

18.2 Polar Coordinates and Their Relationship to Cylindrical, Spherical and Toroidal Coordinates

In polar coordinates, a point is specified by the radius r, which is the distance from the origin to point, and the angle θ, as illustrated in the left side of Fig. 18.1. The Cartesian coordinates x and y are connected to polar coordinates through the relation

$$x = r \cos(\theta), \qquad y = r \sin(\theta) \tag{18.1}$$

The contours of constant radius r are concentric circles centered on the origin. The isolines of polar angle θ are semi-infinite rays all converging at the origin as shown on the right of Fig. 18.1.

Cylindrical coordinates are obtained by adding a third coordinate z which measures distance along a vertical axis perpendicular to the plane defined by polar coordinates. The isosurfaces of constant r are cylinders. The rays of constant θ now converge at every point on the vertical axis. Each plane of constant z is a two-dimensional surface where points are located by polar coordinates.

Toroidal coordinates are obtained by wrapping the cylinder around into the shape of a bagel or doughnut. The surfaces of constant r are now tori, and the coordinate z, now renamed the "toroidal angle" λ (Fig. 18.3), is always cyclic.

In spherical coordinates, the radial coordinate is now the distance from the origin to a point in three dimensions rather than two, and the surfaces of constant radius are spheres. Latitude and longitude, as on any geographical map, locate points on the surface of a given sphere. (Actually, in scientific applications, it is customary to replace latitude by "colatitude", which is latitude minus $\pi/2$, but this is merely a convention that does not alter the pole problem.)

A "polar cap" is a small region centered on either the north or the south pole, bounded by colatitude $\theta = \theta_0$ where θ_0 is small. The polar cap is a slightly puckered disk. As $\theta_0 \to 0$, that is, as the polar cap becomes smaller and smaller, the polar cap becomes flatter and flatter. In the limit, the convergence of meridians (lines of constant longitude) at the

Table 18.1: Spectral Methods in a Disk or Cylinder
Note: Exterior flows and solutions in an unbounded domain are listed in the next table.

References	Comments
Mercier & Raugel (1982)	Theoretical justification for pole conditions in mixed spectral/finite element [in French]
Orszag&Patera (1983)	Robert-type basis in polar coordinates
Tan(1985)	3D Poisson equation in cylinder
Randriamampianina& Bontoux&Roux(1987)	flows with buoyancy and rotation in a cylinder (in French)
Lewis&Bellan(1990)	Symmetry conditions for scalars & vectors across $r = 0$ in cylindrical coordinates
Eisen&Heinrichs &Witsch(1991)	mapped unit disk onto rectangle and imposed pole conditions
Bouaoudia&Marcus (1991)	Robert basis (now obsolete)
Pulicani&Ouazzani(1991)	3D flow and diffusion in cylinder
Huang &Sloan(1993b)	singularity analysis; good discussion of finite difference preconditioning Bessel eigenproblem & Poisson eq. in a disk
Launaye *et al.*(1994)	axisymmetric 2D crystal growth in a cylinder
Van Kemenade&Deville(1994b)	spectral elements; non-Newtonian flow
Godon&Shaviv(1993,1995) Godon(1995,1996abc,1997a,b) Godon*et al.*(1995)	Chebyshev in r; Fourier in θ astrophysics problems accretion disk boundary layers
Priymak(1995),Priymak&Miyazaki(1998)	Turbulence
Fornberg(1995)	Chebyshev/Fourier basis with strong polar filtering
Matsushima and Marcus(1995)	One-Sided Jacobi basis
Raspo& Ouazzani & Peyret(1994,1996)	Multidomain scheme for axisymmetric flow within a cylinder
Shen(1997)	fast algorithms for Poisson eq. and related operators in disk or cylinder
Verkley(1997a,1997b)	One-sided Jacobi basis; hydrodynamics
Matsushima&Marcus(1997)	unbounded vortical flows, polar coordinates
Lopez&Shen(1998)	Navier-Stokes equations with fast algorithms for the semi-implicit time marching

north pole is indistinguishable from the convergence of the rays of constant θ at the origin in polar coordinates.

Thus, the special difficulties caused by the convergence of isolines of an angular coordinate at a single point are *identical* in all four coordinate systems.

18.3 Polar Coordinates: Apparent Singularity at the Origin and Boundary Condition at the Origin

The second consequence of the convergence of the meridians is that differential equations are *singular* at $r = 0$. For example, the Laplace operator is

$$\nabla^2 \equiv \frac{\partial^2}{\partial r^2} + \frac{1}{r}\frac{\partial}{\partial r} + \frac{1}{r^2}\frac{\partial^2}{\partial \theta^2} \tag{18.2}$$

Thus, the coefficients of the first radial derivative and second polar derivative are both singular at $r = 0$. And yet, when the Laplacian is written in Cartesian coordinates, it is constant coefficient:

$$\nabla^2 \equiv \frac{\partial^2}{\partial x^2} + \frac{\partial^2}{\partial y^2} \tag{18.3}$$

A simple function like $f(x,y) = x^2 + y^2$ has a Laplacian which is a constant. What could be less singular!

The resolution of the paradox is that the singularities of the Laplace operator, when written in polar coordinates, are only "apparent". The solution to a differential equation is usually smooth and infinitely differentiable at the pole.

To avoid evaluating differential equation coefficients where they are infinite, the spectral grid usually excludes the origin. This is only a trivial complication, however; the interesting issue is: How does one impose boundary conditions at the origin?

For finite difference methods, one must normally impose explicit, numerical boundary conditions, based on careful analysis of the behavior of the solution near the origin, at the grid point nearest $r = 0$. For spectral methods, however, the singularity actually *simplifies* the algorithm. To a spectral method, the boundary condition at the origin is *behavioral* rather than *numerical*: the correct behavior is that the solution should be analytic at the origin even though the coefficients of the differential equation are not. Since the terms of a spectral series are individually analytic, it follows that a Chebyshev or similar series in radius automatically satisfies the boundary condition at $r = 0$, and no additional explicit conditions are necessary. If the differential equation were not singular at $r = 0$, it would be necessary (for a second order equation) to impose an explicit numerical boundary condition at the origin. Thus, the singularity of the coordinate has ironically *simplified* the use of a spectral method.

For best results, one must respect the parity symmetry at the origin which is explained in the next two sections. However, one only needs to constrain the spectral series through a numerical boundary condition at the outer boundary of the disk. If one applies polar coordinates to an infinite domain, then both boundary conditions are behavioral, and no conditions need to be explicitly imposed on the spectral series.

18.4 Polar Coordinates: Parity Theorem

Theorem 35 (Polar Coordinates: Parity in Radius) *In polar coordinates where r is radius and θ is angle, expand an arbitary function as a Fourier series in θ:*

$$f(r, \theta) = \sum_{m=0}^{\infty} f_m(r) \cos(m\theta) + g_m(r) \sin(m\theta) \tag{18.4}$$

Suppose that $f(r, \theta)$ is a SCALAR *or the z-velocity in cylindrical coordinates or is the product of r with the radial or tangential components of a vector in polar coordinates. Then if the function is analytic at $r = 0$, continuity of the function and its derivatives demands the following:*

(i) $f_m(r)$ and $g_m(r)$ have m-th order zeros at $r = 0$.

(ii) If m is EVEN, *then $f_m(r)$ and $g_m(r)$ are both* SYMMETRIC *about $r = 0$ and their power series contain only* EVEN *powers of r.*

(iii) If m is ODD, *then $f_m(r)$ and $g_m(r)$ are both* ANTISYMMETRIC *about $r = 0$ and their power series contain only* ODD *powers of r.*

Proof: A geometric argument for spherical coordinates is given in Sec. 18.8 below. This visual argument applies also to polar coordinates because latitude and longitude become a local polar coordinate system in a small neighborhood of the north and south poles. Here, we present a different line of proof, couched as an informal, heuristic argument. Because there seems to have been much confusion about this theorem in the literature and multiple rederivations of its conclusions, we give this argument in some detail, but the rest of this section can be skipped without loss of continuity. A careful, rigorous proof is given in Eisen, Heinrichs &Witsch(1991).

First, note that any function $f(x, y)$ which is analytic at and near $x = y = 0$ can be locally approximated, to any desired accuracy, by a bivariate polynomial of sufficiently high order. (Indeed, a double Taylor series will suffice). Substituting $x = r \cos(\theta)$ and $y = r \sin(\theta)$, it follows that a bivariate polynomial in $r \cos(\theta)$ and $r \sin(\theta)$ will suffice as a proxy for $f(r, \theta)$ to any desired accuracy. The monomials that appear in this polynomial approximation have properties described by the following two lemmas.

Lemma 1 (Polar Wavenumbers of Cartesian Powers) *Let "polar wavenumber m" denote the Fourier terms $\cos(m\theta)$ or $\sin(m\theta)$. When a monomial*

$$M_{jk} \equiv x^j y^k \tag{18.5}$$

in powers of the Cartesian coordinates x and y is converted into polar coordinates and expanded as a Fourier series in the polar angle θ, polar wavenumber m will only appear in the expansion of the monomial if the total degree of the monomial is greater than or equal to m, that is, if

$$d \equiv j + k \geq m \tag{18.6}$$

Proof: A monomial of total degree d can be written, expressing the sine and cosine as complex exponentials, in the form

$$x^j y^k \equiv r^{j+k} \frac{(-i)^k}{2^d} \prod_{n=1}^{d} \left(\exp(i\theta) \pm \exp(-i\theta) \right) \tag{18.7}$$

The crucial fact is that each exponential in the product is proportional to polar wavenumber one, that is, to $\exp(\pm i\theta)$. To obtain a term of wavenumber 7, for example, for example, it is necessary to multiply such wavenumber one terms together at least 7 times. This requires that the number of factors in the product is at least 7. Since the number of factors is the total degree d, it follows that the total degree of the monomial must be at least 7. The same reasoning applies to arbitrary wavenumber m.

The factor of r^d in front of the product in Eq. (18.7) then immediately implies the following.

Lemma 2 (Radial Degree of Cartesian Powers) *A Cartesian monomial $x^j y^k$ is proportional to r^d where $d \equiv j + k$ is the total degree of the monomial.*

Because of Lemma 1, it follows that a Cartesian monomial will contain terms like $\cos(m\theta)$ or $\sin(m\theta)$ only if it is proportional to r^d where $d \geq m$.

Together, these lemmas show that the terms proportional to either $\cos(m\theta)$ or $\sin(m\theta)$ must also be proportional to r^m, and thus the coefficients of these terms in the Fourier series of $f(r, \theta)$ must have an m-th order zero at $r = 0$.

The parity of the Fourier coefficients with respect to $r = 0$ follows from a third lemma.

Lemma 3 (Monomial Parity) *A monomial $x^j y^k$, when expanded as a Fourier series in polar angle θ, is a trigonometric polynomial of degree $j + k$ which contains contains only those polar wavenumbers of the same parity as the degree $d = j+k$. That is, if d is odd, then the Fourier series of $x^j y^k$ contains only $\{\cos(d\theta), \sin(d\theta), \cos([d-2]\theta), \sin([d-2]\theta), \cos([d-4]\theta), \sin([d-4]\theta), \dots, \}$.*

Odd polar wavenumbers m are thus multiplied only by odd powers of r, and even powers only by even.

Proof: The product

$$x^j y^k \equiv r^{j+k} \frac{(-i)^k}{2^d} \prod_{n=1}^{d} \left(\exp(i\theta) \pm \exp(-i\theta) \right), \tag{18.8}$$

can be expanded into terms of the form $\exp(in\theta)$ where n depends on how many factors of $\exp(i\theta)$ and how many factors of its complex conjugate are combined in a given term. The term of highest polar wavenumber is $\exp(id\theta)$ when all d exponentials $\exp(i\theta)$ are combined. The crucial point is that when $\exp(i\theta)$ is replaced by its complex conjugate $\exp(-i\theta)$, which is the other term inside each factor in the product, the degree j is lowered by *two*. Thus, all wavenumbers j in the expansion of the product will be of the same parity, either even or odd, as the total degree d.

The radial dependence of the monomial $x^j y^k$ is r^d where d is the sum of j and k. If d is even, then r^d is symmetric with respect to $r = 0$. (We have already seen that if d is even, then all the Fourier coefficients of the monomial are of even polar wavenumber, too.). Similarly, if d is odd, then its radial factor r^d is antisymmetric (i. e., odd parity) with respect to $r = 0$. Thus odd powers of r are paired only with $\cos(m\theta)$ and $\sin(m\theta)$ where m is odd and even powers of r are paired only with even m.

The parity theorem has major implications for the choice of grids and basis functions as explained in the next section.

18.5 Radial Basis Sets and Radial Grids

On a finite domain in radius, which can be always be normalized to $r \in [0, 1]$ by a linear change-of-coordinate, there are several options (where m denotes the angular wavenumber):

1. Bessel Functions: $J_m(j_{m,j}\, r) \, \cos(m\theta), \sin(m\theta)$

2. Polar Robert Functions: $r^m \, T_j(r)$, $j = 0, 1, 2, \ldots$

3. Shifted Chebyshev Polynomials of Linear Argument:
   $T_j(2r - 1)$, $j = 0, 1, 2, \ldots$

4. Shifted Chebyshev Polynomials of Quadratic Argument:
   $T_j(2r^2 - 1)$, $(m$ even$)$, $r\, T_j(2r^2 - 1)$, $(m$ odd$), j = 0, 1, 2, \ldots$

5. Unshifted Chebyshev Polynomials of Appropriate Parity:
   $T_{2j}(r) \cos(2m\theta), T_{2j-1}(r) \cos([2m - 1]\theta)$, $j = 0, 1, 2, \ldots$

6. One-Sided Jacobi Polynomials:
   $r^m \, P_j^{0,|m|}(2r^2 - 1) \, \cos(m\theta), \sin(m\theta)$

The "cylindrical harmonics" are

$$\mathcal{J}_n^m(r, \theta) \equiv J_m(j_{m,n}\, r) \left\{ \begin{array}{l} \cos(m\theta) \\ \sin(m\theta) \end{array} \right. \tag{18.9}$$

where there are two basis functions, one proportional to $\cos(m\theta)$ and the other to $\sin(m\theta)$, for all pairs of integers (m, n) with $m > 0$ and where $j_{m,n}$ denotes the n-th zero of the m-th Bessel function $J_m(r)$. The cylindrical harmonics are popular in engineering because these are the eigenfunctions of the Laplace operator in polar or cylindrical coordinates, which made it easy in the old pre-computer days to apply a Galerkin method. However, as noted by Gottlieb and Orszag(1977), the Bessel series usually converges only at an algebraic rate. (An analysis of the precise rate of convergence is given in Boyd and Flyer(1999).)

Bouaoudia and Marcus(1991) developed a fast transform for the polar "Robert" functions, which mimic the spherical basis invented by Robert(1966):

$$\phi_{m,n}(r, \theta) = r^m T_n(r) \left\{ \begin{array}{l} \cos(m\theta) \\ \sin(m\theta) \end{array} \right. \tag{18.10}$$

where n is restricted to have the same parity (i. e., even or odd) as m. Unfortunately, Merilees(1973a) showed that the analogous spherical basis is horribly ill-conditioned and therefore useless except at very low resolution. The problem is that for large m and small n, the basis functions are zero over almost all the interval because of the r^m factor; note that $r^m = \exp(m \log(r)) \approx \exp(-m(1 - r))$ for $r \approx 1$, and thus decays exponentially fast away from the boundary at $r = 1$. In the narrow region around $r = 1$ where the small n functions are non-zero, T_0, T_2 and other low degree basis functions vary so slowly that $\phi_{m,0} \approx \phi_{m,2}$, destroying the linear independence of the functions in the basis. Marcus abandoned the polar Robert functions in his later work, and instead went to One-sided Jacobi polynomials as explained below.

The third bad option is to expand $f_m(r)$ as a series of Shifted-Chebyshev polynomials with a linear argument:

$$T_j^*(r) \equiv T_j(2r - 1), \qquad r \in [0, 1] \tag{18.11}$$

where the asterisk is the usual notation for the Shifted-Chebyshev polynomials. The reason that this option is bad is that the Shifted-Chebyshev grid has points clustered near both $r = 0$ and $r = 1$. However, the disk bounded by $r = \rho$ has an area which is only the fraction ρ^2 of the area of the unit disk. Near the origin, points are separated by $O(1/N^2)$. It follows that the high density of points near the origin is giving high resolution of only a tiny, $O(1/N^4)$-in-area portion of the disk. Unless the physics imposes near-singularities or high gradients near the origin, clustering grid points at $r = 0$ is a bad idea. And even in this special situation of small-scale features near the origin, it is conceptually better to regard the use of Shifted-Chebyshev polynomials as "mapping to resolve large gradients near the origin" rather than "Shifted-Chebyshev polynomials are a good *general* representation for radial dependence".

Fig. 18.2 compares the Shifted-Chebyshev series (dashed) with two other options. The shallow slope of the coefficients of the Shifted-Chebyshev series shows that one needs roughly twice as many terms to obtain a given degree of accuracy with the Shifted-Chebyshev polynomials as with the alternatives.

The Shifted-Chebyshev polynomials of quadratic argument and the Parity-Restricted Chebyshev polynomials, which are also shown in the figure, both work well, but generate only a single curve on the graph. The reason is that these two options, although seemingly very different, are in fact the same because of the identity

$$T_j(2r^2 - 1) \equiv T_{2j}(r), \qquad \forall j, r \tag{18.12}$$

The grid with parity uses $r_j^{parity} = \cos(t_j/2)$ whereas the shifted grid with quadratic argument has $r_j = \{(1 + \cos(t_j))/2\}^{1/2}$ where the t_j are an evenly spaced grid on the interval $t \in [0, \pi]$ for Lobatto, Chebyshev-roots, or Radau grids. (The Radau grid includes $r = 1$ but not $r = 0$.) The trigonometric identity

$$\cos\left(\frac{t}{2}\right) = \sqrt{\frac{1 + \cos(t)}{2}} \tag{18.13}$$

shows that

$$r_j^{parity} = r_j, \qquad \forall j \tag{18.14}$$

Thus, both the basis functions and the interpolation grid are identical for these two methods. The distance between adjacent points close to origin is $O(1/N)$.

Nevertheless, when there is a large number M of points in the polar angle θ and N points in radius, the distance between adjacent gridpoints on the circle of radius r_1, the smallest radial gridpoint, will be $O(1/(MN))$. Since the explicit time-stepping limit is roughly the amount of time required for advection or diffusion from one grid point to the next, it follows that the Parity-Restricted Chebyshev polynomials will give a time step limit which is $O(1/(MN))$ for advective stability and $O(M^{-2}N^{-2})$ for diffusion. Ouch! On the sphere, this motivates the choice of spherical harmonics as the basis, which eliminates this "pole problem" of a very short time-step. A similarly-motivated basis for the disk is described in the next section.

18.5.1 One-Sided Jacobi Basis for the Radial Coordinate

Matsushima and Marcus(1995) and Verkley(1997a,1997b) independently proposed One-Sided Jacobi polynomials, multiplied by r^m:

$$\phi_n^m(r, \theta) \equiv W_n^m(r) \begin{cases} \cos(m\theta) \\ \sin(m\theta) \end{cases} \tag{18.15}$$

where in Verkley's notation

$$W_n^m(r) \equiv r^m \, P_{(n-m)/2}^{0,m}(2r^2 - 1), \qquad m = 0, 1, 2, \dots; \qquad n = m, m+2, m+4, \dots \tag{18.16}$$

where $P_k^{0,m}(s)$ is the Jacobi polynomial of degree k of order $(0, m)$ in its argument s. (Note that because $s \equiv 2r^2 - 1$, the Jacobi part of the basis function is of degree $(n - m)$ in r.) This basis function explicitly enforces all the constraints of the parity theorem; each basis function is symmetric in r for even m and odd in r for odd m. The factor of r^m ensures that each basis function has an m-th order zero at the origin. Verkley(1997a, Table 1) provides the explicit form of these basis functions for small n and m.

The Jacobi polynomials are orthogonal so that

$$\int_{-1}^{1} P_k^{\alpha,\beta}(s) P_{k'}^{\alpha,\beta}(s)(1 - s)^\alpha (1 + s)^\beta ds = 0, \qquad k \neq k' \tag{18.17}$$

Changing variables from s to r and setting $\alpha = 0, \beta = m$ gives

$$\int_0^1 P_k^{0,m}(2r^2 - 1) P_{k'}^{0,m}(2r^2 - 1) r^{2m} dr = 0, \qquad k \neq k' \tag{18.18}$$

or in other words,

$$\int_0^1 W_n^m(r) W_{n'}^m(r) \, r \, dr = 0, \qquad n \neq n' \tag{18.19}$$

These Jacobi polynomials of order $(0, m)$ are "one-sided" in the sense that the weight function in Eq.(18.18) diminishes rapidly (as r^{2m}) as the origin is approached, but varies only slightly near the outer edge of the disk. If the Jacobi polynomials oscillated uniformly in r like the Robert functions, they would be nearly-dependent and ill-conditioned like the Robert functions. The orthogonality constraint, however, forces the polynomials of different degree to be as independent as possible. This requires that the polynomials oscillate mostly near $r = 1$: the roots of the basis functions move closer and closer to the outer boundary for fixed degree as the order m increases.

The one-sideness is the magic. The basis functions of high wavenumber m are the villians that describe the fast tangential advection or diffusion that limits the timestep. However, the radial parts of these basis functions, if one-sided Jacobi polynomials, have only negligible amplitude close to the origin where the grid points are close. The rapid advection or diffusion is suppressed, and a much longer timestep is possible.

As with spherical harmonics, discussed later, there are two options for truncating the basis: "rectangular" and "triangular". A rectangular truncation employs the same number of radial basis functions for each angular wavenumber m. It is the easiest to program, but has fallen from grace in spherical coordinates. The triangular truncation decreases the number of radial basis functions by m with each increase in m until the highest wavenumber has but a single radial basis function. It has been shown for spherical harmonics that this gives the most uniform resolution (and allows the longest timestep) of any truncation that includes a given maximum zonal wavenumber M. This property of "equiareal resolution", defined formally below in Sec. 18.13, has not been proved for the One-Sided Jacobi basis, but it seems plausible that a triangular truncation is preferable here, too.

The usual strategy of choosing the roots of the polynomials as the interpolation points does not work here because the zeros of the $P_k^{(0,m)}$ are *different* for each polar wavenumber m. To take transforms in θ, we need the unknowns defined on a single set of radial grid points. Therefore, one is forced to abandon a pseudospectral strategy and instead apply a Galerkin method. The integration points are the Legendre-Radau grid which includes $r = 1$ but not $r = 0$. To avoid quadrature errors, it is necessary to use more grid points than unknowns, but the transform from radial grid point values to coefficients can be still be expressed as the multiplication of a vector of grid point values or coefficients by a matrix, albeit now a rectangular matrix.

These basis functions, unlike the spherical harmonics they mimic, are not eigenfunctions of Laplace operator. However, both Matsushima and Marcus(1995) and Verkley(1997a) derive recurrence relations which allow the Laplacian to be inverted by solving a pentadiagonal matrix problem. Similarly, derivatives can be evaluated at an operation count directly proportional to the number of spectral coefficients in the truncation. Overall, the efficiency of this basis is roughly the same as for spherical harmonics. Indeed, the algorithmic connection with spherical harmonics is so close that Verkley's shallow-water solver for the disk is merely a rewrite of a previous code that solved the same hydrodynamic equations on the surface of a sphere via spherical harmonics.

Unfortunately, the Jacobi basis shares the vices of spherical harmonics. First, there is no Fast Fourier Transform for the Jacobi polynomials, so all radial transforms from grid point values to spectral coefficients must be done by Matrix Multiplication Transform (MMT). This costs $O(N^2)$ for each wavenumber m versus only $O(N \log_2(N))$ for an FFT. (Transforms in θ are still done by FFT, thank goodness.)

Second, the messy recurrence relations and basis functions require some investment of both learning time and programming time. (We may hope that someday the necessary subroutines will become widely available in software libraries, and then this objection will disappear).

There is one profound difference from spherical harmonics, however. The One-sided Jacobi basis has the usual high density of grid points near $r = 1$ where grid points are separated in r by $O(1/N^2)$. With a triangular truncation so that the maximum polar wavenumber M is equal to N, the tangential distance between grid points on the outermost ring is

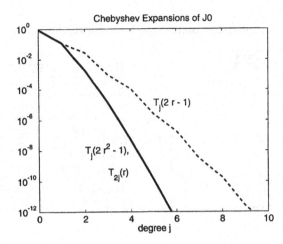

Figure 18.2: Chebyshev expansions of the Bessel function $J_0(r)$. The slowly-convergent series (dashed) is the expansion in Shifted-Chebyshev polynomials: $T_j^*(r) \equiv T_j(2r - 1)$. The two good techniques are (i) an expansion in Shifted-Chebyshev polynomials with a QUADRATIC argument $T_j(r^2) = T_j(2r^2 - 1)$ and (ii) an expansion in the EVEN Chebyshev polynomials without a shift: $T_{2j}(r)$. These two options generate only a single curve because (i) and (ii) are in fact IDENTICAL because of the identity $T_j(2r^2 - 1) = T_{2j}(r)$.

also $O(1/N^2)$. This implies that the One-sided Jacobi basis has not completely solved the difficulty of a restrictive timestep; indeed, if the time scales for tangential advection or diffusion near the origin are similar to the time scales for radial advection or diffusion near the boundary, then the switch from a Chebyshev basis to the Jacobi basis will not increase the time step at all.

Nevertheless, this basis has worked well in Matsushima and Marcus(1995) and Verkley (1997a,b).

18.5.2 Boundary Value & Eigenvalue Problems on a Disk

When there is no timestep, the pressure to use One-Sided Jacobi basis functions in radius is greatly reduced. Orszag(1974) and Boyd(1978c) have shown that satisfying all the pole conditions is not terribly important to numerical efficiency; a regular Chebyshev basis seems to work just fine. (Their papers actually solve problems on the sphere, but because the polar caps are well-approximated by small disks, their conclusions should apply to polar coordinates, too.) Matsushima and Marcus(1995) show by solving the eigenproblem whose exact solution is the J_m Bessel function that the Jacobi basis can reach a given error tolerance with only half as many coefficients as the Parity-Restricted Chebyshev polynomials when $m = 50$. However, in a Fourier-Chebyshev series in two dimensions, the Fourier series in m usually converges rapidly, so the Chebyshev basis is converging slowly only for wavenumbers whose amplitude is negligible anyway.

Our recommendation for boundary and eigenvalue problems is to try a Chebyshev basis first. If the speed of the resulting (rather easy-to-program) code is unsatisfactory, the One-Sided Jacobi basis can be used as a fallback with the Chebyshev program used to check the newer and much more intricate Jacobi code.

Table 18.2: Cylindrical or Polar Coordinates: Unbounded Domain or External to a Cylinder

References	Comments
Zebib(1987a)	Stability of flow past a cylinder
Don&Gottlieb(1990)	unsteady flow past a cylinder
Deane & Kevrekidis &Karniadakis & Orszag(1991)	spectral elements; low order basis of empirical orthogonal eigenfunctions
Mayer&Powell(1992)	Eigenvalue calculation of stability of trailing vortex through domain truncation
Mittal & Balachandar(1996)	flow past elliptic cylinder in elliptic coordinates good discussion of boundary conditions, blending inflow & outflow
Blackburn&Henderson(1996)	flow past a vibrating cyclinder
Matsushima&Marcus(1997)	Unbounded domain including the origin; 2D and 3D vortex flows

18.5.3   Unbounded Domains Including the Origin in Cylindrical Coordinates

Matsushima&Marcus(1997) have extended their earlier work by applying an algebraic change-of-coordinate in r to map an unbounded domain into a disk. They then apply the One-Sided Jacobi basis described above. Like the rational basis TB_j for the infinite interval described in Chapter 17, the Matsushima-Marcus basis functions are rational functions of $r \in [0, \infty]$.

The alternative is the even TB functions for even polar wavenumber m and the $TB_{2j-1}(r)$ for odd m: a Parity-Restricted Rational Chebyshev basis. The simpler functions are preferable for boundary and eigenvalue problems, but the Matsushima-Marcus basis may give faster convergence and allow a longer timestep.

Flows exterior to a cylinder in an unbounded domain have been the subject of much study, both analytical and numerical. Exterior flows are free of the pole problem, but imposing boundary conditions on a flow at large radius can be tricky. We have therefore collected some illustrations in Table 18.2.

18.6   Annular Domains

Flow in an annulus that does not include the origin is free from the pole problem. We recommend a Fourier basis in the angle θ and a standard Chebyshev basis in the radial coordinate r. In three dimensions, one should use a Chebyshev basis in the axial coordinate z unless periodic boundary conditions in z are imposed, in which case a Fourier basis is greatly preferable.

There is little difference between annular flow and channel flow except that cylindrical coordinates introduce an r-dependent "metric" factor into the Laplace operator, etc., whereas the Laplace operator has constant coefficients in Cartesian coordinates. The usual methods for separable partial differential equations, discussed in Chapter 15, Sec. 11, still apply.

For most practical purposes, however, an annular channel is simply a straight channel which is periodic in the downstream direction. A standard Fourier/Chebyshev basis is best. Some representative works are listed in Table 18.3.

Table 18.3: Annular Flows

References	Comments
Marcus(1984a,1990)	Taylor-Couette flow
Le Quéré &Pecheux(1989,1990)	axisymmetric convection; bifurcations
Chaouche(1990a) Chaouche *et al.*(1990)	axisymmetric flows; influence matrix method
Randriamampianina(1994)	3D flow; vorticity-vector-potential

18.7   Spherical Coordinates: An Overview

The pattern of latitude and longitude lines on the surface of the sphere is, near the poles, the same as that of a polar coordinate system in a disk. So, it is hardly surprising that there are many similarities between cylindrical coordinates and spherical coordinates. However, there are important differences.

One is that a sphere has two poles rather than one. A second, more important difference is that the surface of a sphere has *no boundaries*. This has the profound consequence that spherical basis sets automatically and individually satisfy the *behavioral* boundary conditions on the sphere. In contrast, it is necessary to impose *numerical* boundary conditions at the boundary circle of the disk.

A third difference is that for the sphere, there is an "obvious" basis set, the spherical harmonics, which are nearly ideal. Spherical harmonics give equiareal resolution, exponential convergence, and trivial inversion of the Laplace operator, which is the eigenoperator for these functions. In contrast, the obvious basis for the disk, the cylindrical harmonics, are Bessel functions of radius with only a poor, algebraic rate of convergence. One can match most of the good properties of spherical harmonics by using the One-Sided Jacobi basis for the disk, but these are not eigenfunctions of the Laplace operator.

The fourth difference is that the number of simulations in a cylinder is fairly modest compared to the large body of work that has been done in spherical coordinates for weather forecasting, climate modelling, mantle convection and stellar flows. For this reason, we shall discuss spectral methods on the sphere in great detail.

18.8   The Parity Factor for Scalars: Sphere versus Torus

Topologically, the sphere is a two-dimensional manifold of genus zero while the torus ("doughnut") is of genus one. Ironically, however, it is easier to compute on the surface of a torus than the surface of a sphere.

Science fiction writers have described toroidal planets (Boyd, 1981c, 1984c) and plasma physicists solve flows in tori to model the ring-shaped fusion generators known as tokamaks (Schnack *et al.*, 1984). In this section, we will discuss a two-dimensional basis for the surface of a torus. However, one may define a three-dimensional orthogonal coordinate system ("toroidal coordinates", Morse and Feshbach, 1953) in which the third coordinate is radial distance from the surface of a torus to its centerline.

Fig. 18.3 illustrates toroidal geometry. To stress the analogue with the sphere, we use λ, which is longitude-like, for the "toroidal" angle and θ, which is analogous to colatitude, for the "poloidal" angle. All physical solutions must be periodic with a period of 2π in both coordinates. As shown in the lower half of the figure, it is trivial to "flatten" the torus by making a slit through a meridian, bending the cut torus into a cylinder, and then making a second cut parallel to the "equator" to unwrap the cylinder into a rectangle.

(a)

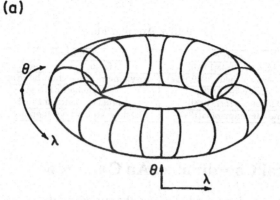

(b)

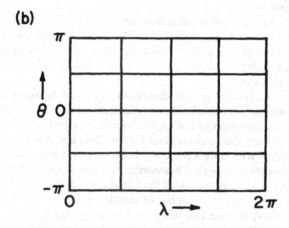

Figure 18.3: Toroidal coordinates. (a) The torus. (b) The torus developed into a rectangle.

Because the geometry is periodic in both coordinates, the natural basis set on the torus is a double Fourier series, i. e. any function on the surface of the torus can be expanded as

$$f(\lambda, \theta) = \sum_{m,n}^{\infty} a_{mn}^{cc} \cos(m\lambda)\cos(n\theta) + \sum_{m,n}^{\infty} a_{mn}^{cs}\cos(m\lambda)\sin(n\theta) \qquad (18.20)$$

$$+ \sum_{m,n}^{\infty} a_{mn}^{sc}\sin(m\lambda)\cos(n\theta) + \sum_{m,n}^{\infty} a_{mn}^{ss}\sin(m\lambda)\sin(n\theta)$$

If $f(\lambda, \theta)$ is free of singularities on the torus, then the series (18.20) will converge exponentially. Schnack, Baxter, and Caramana (1984) is a good illustration of a code (in plasma physics) that employs a two-dimensional Fourier series in toroidal coordinates.

The sphere is more complex because it is *not* a surface that can be unwrapped or "developed" into a cylinder. Its geometry is fundamentally different from that of the periodic rectangle shown in Fig. 18.3b. To expand a scalar function on the sphere, we need only *half* the terms in the general double Fourier series shown above. To explain which half, however, requires us to look down on the north pole of the sphere and flatten the "polar cap" region into a disk with a local polar coordinate system as shown in Fig. 18.4.

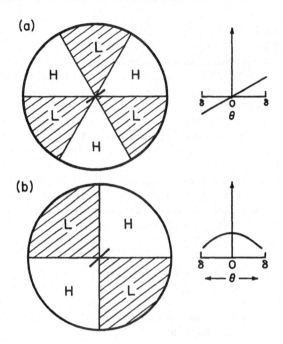

Figure 18.4: Polar projections showing the positive (H) and negative (L; cross-hatched) regions for a term $g_m(\theta) \sin(m\lambda)$ in the Fourier expansion of a function. The one-dimensional graphs on the right show how $g_m(\theta)$ must vary along the short, pole-crossing line segments marked on the polar plots.

(a) Zonal wavenumber m is *odd*. The pole-crossing line segment has one end where $\sin(m\lambda)$ is negative and the other where $\sin(m\lambda)$ is positive. To avoid a discontinuity, $g_m(\theta)$, must be antisymmetric about the pole.

(b) Zonal wavenumber m is *even*, so $\sin(m\lambda)$ is positive everywhere along the pole-crossing segment. To avoid a discontinuity, $g_m(\theta)$ must be symmetric about the pole.

Note that $f(\lambda, \theta)$ can be written in the form

$$f(\lambda, \theta) = \sum_{m=0}^{\infty} f_m(\theta) \cos(m\lambda) + \sum_{m=1}^{\infty} g_m(\theta) \sin(m\lambda) \qquad (18.21)$$

The two diagrams in Fig. 18.4 schematically illustrate two typical terms in the sums: one for m odd (top) and the other for m even (bottom). Let us follow $g_m(\theta) \sin(m\lambda)$ along a meridian over the pole.

Fig. 18.4a shows that when m is *odd*, the component $g_m(\theta) \sin(m\lambda)$ must always *change sign* as the pole is crossed. This sign change depends only on the fact that the zonal wavenumber m is odd, and is quite independent of the details of $g_m(\theta)$ and also of whether $m = 1, 3, 5, \ldots$ so long as m is odd. This implies that $g_m(\theta) \sin(m\lambda)$ must have a *jump discontinuity* at the pole unless

$$g_m(0) = g_m(\pi) = 0 \qquad [m \text{ odd}] \qquad (18.22)$$

The same diagram applies to $f_m(\theta) \cos(m\lambda)$ by rotating it through $\pi/2$ degrees, so the same reasoning implies that $f_m(\theta)$ must have zeros at both poles for all odd longitudinal wavenumbers m.

This argument can be extended to higher latitudinal derivatives to show that ALL EVEN derivatives of $f_m(\theta)$ and $g_m(\theta)$ must vanish at both poles. The Fourier series for these functions could in principle include both $\sin(n\theta)$ and $\cos(n\theta)$ terms, but whereas the sine terms individually have all even derivatives zero at the poles, the sum of the cosine terms would be constrained by an infinite number of constraints. Reexpansion of a spherical harmonic series into double Fourier series shows that the trivial solution to these constraints is in fact the only solution: the colatitude cosine coefficients are always zero when the longitudinal wavenumber m is odd. The series (18.20) (or one of its derivatives) will have a *discontinuity* in $f(\lambda, \theta)$ unless $f_m(\theta)$ and $g_m(\theta)$ are the sums of Fourier *sine* series in θ when the *zonal wavenumber m is odd*.

When m is even, Fig. 18.4b shows that the function must have the same value on a given meridian on either side of the pole. The analytical proof is that if we compare two points that are at equal distances from the poles, but on opposite sides, $\theta = \delta$ for both points, but λ differs by π. When m is even, $\sin(m\lambda) = \sin(m[\lambda + \pi])$, so

$$g_m(\delta) \sin(m\lambda) = g_m(\delta) \sin(m[\lambda + \pi]) \quad \text{for any } \delta, \lambda \; [m = 0, 2, 4, \ldots] \tag{18.23}$$

as shown in the meridional slice graphed on the right side of Fig. 18.4b. However, if this function is symmetric with respect to the pole on the line segment marked in the figure, it follows that its derivative must be antisymmetric. (Recall from Chapter 8 that differentiation is a parity-reversing operation.) However, an antisymmetric function is zero at the point of symmetry, in this case, the pole.

It follows that

$$\frac{d\,g_m}{d\theta}(0) = \frac{d\,g_m}{d\theta}(\pi) = 0 \qquad [m \text{ even}] \tag{18.24}$$

Extending this argument to higher derivatives shows that all the sine coefficients must be zero when m is even.[1]

Thus, a *scalar* function $f(\lambda, \theta)$ that has no singularities on the sphere may be expanded as an exponentially convergent Fourier series of the special form

$$f(\lambda, \theta) \;=\; \sum_{\substack{m=0, 2, 4, \ldots \\ n=0}}^{\infty} \{ a_{mn}^c \cos(m\lambda) + a_{mn}^s \sin(m\lambda) \} \cos(n\theta)$$

$$+ \sum_{\substack{m=1, 3, 5, \ldots \\ n=1}}^{\infty} \{ b_{mn}^c \cos(m\lambda) + b_{mn}^s \sin(m\lambda) \} \sin(n\theta) \tag{18.25}$$

The identity

$$\sin(\theta) \cos(n\,\theta) = \frac{1}{2} \{\sin([n-1]\theta) + \sin([n+1]\theta)\} \tag{18.26}$$

shows that equivalently, one can replace $\sin(n\theta)$ by a basis whose elements are $\{\sin(\theta)$ $\cos(n\theta)\}$. [Exercise for the reader: Derive the relationship between the Fourier coefficients for these two alternative forms of the sine series.] For historical reasons, this $\sin(\theta)$ -times-a-cosine series representation is often used to replace $\sin(n\theta)$ in the terms in the second line of (18.25). This extracted factor of $\sin(\theta)$ for m *odd* is the "parity factor".

---

[1]The argument that zero derivatives to all orders implies zero coefficients is not rigorous because a C^∞ function such as $\exp(-1/[\theta(\pi - \theta)])$ has zero derivatives to all orders at the poles. However, such functions can be represented on $\theta \in [0, \pi]$ by *either* a cosine series or sine series with a subgeometric but exponential rate of convergence. Thus it is true, even when such exceptions are considered, that one needs only latitudinal Fourier terms of a single parity to approximate $f_m(\theta)$ and $g_m(\theta)$.

We added the restriction "for a scalar function" in stating (18.25) because the arguments for a *vector* function such as the wind velocities are more complicated as explained in the next section. The conclusions are similar, however: to represent the components of a vector, one needs only half a general Fourier series, and which components in θ must be kept is different for even and odd zonal wavenumber.

18.9 Parity II: Horizontal Velocities & Other Vector Components

Computing the spherical harmonic expansions of vectors on the sphere is a little tricky. As explained in Orszag (1974) the three *Cartesian* velocity components, u_x, u_y, and u_z, all transform like scalars, that is, may all be expanded directly in spherical harmonics[2]. The same is true for the *radial* velocity u_r in *spherical* components. (A meteorologist would call u_r the "vertical" component.) However, the horizontal velocity components $u_\lambda \equiv d\lambda/dt$ and $u_\theta \equiv d\theta/dt$ transform differently. As noted by Robert (1966), u_λ and u_θ should be expanded as $1/\sin(\theta)$ times a spherical harmonic series.

For this reason, most spherical harmonics models replace the zonal and meridional velocities by

$$U \; \equiv \; u_\lambda \sin(\theta) = \sum_{m,n}^{\infty} u_{mn} Y_n^m(\lambda, \theta) \qquad \text{[modified east-west wind]} \qquad (18.27)$$

$$V \; \equiv \; u_\theta \sin(\theta) = \sum_{m,n}^{\infty} v_{mn} Y_n^m(\lambda, \theta) \qquad \text{[modified north-south wind]} \qquad (18.28)$$

Orszag (1974) gives an analytical proof of (18.27) and (18.28). We instead will offer a graphical justification.

Fig. 18.5 illustrates a polar cap view of a typical flow generated by a *Cartesian* velocity component which is function of spherical components with even wavenumber — in this particular case, a flow which is a constant, independent of all three spatial coordinates. When we decompose this flow into spherical velocity components as shown in the two lower panels, we find that the zonal velocity and meridional velocity components are both of *odd* zonal wavenumber. For this example of wavenumber one velocities, each has a single nodal meridian, shown by the dotted line, where the field is zero. This interchange of even wavenumber for odd wavenumber and vice versa when we decompose a Cartesian vector component into spherical vector components explains why the parity factor for u_λ and u_θ for odd m is just what we expect for a scalar quantity for even m — none.

When we inspect a polar view of an odd wavenumber portion of a scalar quantity as in Fig. 18.4a, we observe that it changes sign as we cross the poles while moving on a meridian. We argued that the colatitude dependence of the scalar had to be antisymmetric about $\theta = 0$ — that is, like $\sin(\theta)$ — so the scalar would not have a jump discontinuity at the poles. Why does this argument fail for u_λ and u_θ for odd m? The answer is that these spherical vector components actually *are discontinuous* at the poles!

If we look again at the upper left of Fig. 18.5, we note that what is interpreted as a westward flow in the top of the disk is an eastward flow in the bottom half (opposite hemisphere). The wind velocity, however, does not go to zero at the pole — the magnitude and direction of the flow are independent of coordinate. If we embed a fixed Cartesian

---

[2]Warning: in this section, subscripts will be used to denote a particular component of a vector and not differentiation with respect to that coordinate.

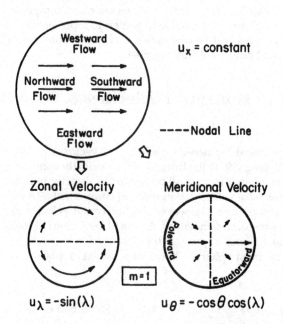

Figure 18.5: Polar cap view of a flow parallel to the Cartesion x-axis, chosen to be representative of even zonal wavenumber. In this case, the zonal wavenumber is zero — the flow is independent of position.

When decomposed into vector components, however, both the zonal and meridional velocity have *odd* zonal wavenumber, in this case, $m = 1$. To avoid a discontinuity at the pole, both spherical velocity components must be *symmetric* about the pole even though a *scalar* quantity of *odd* wavenumber would have to be *antisymmetric*.

coordinate system in the globe so that the origin is at the north pole, then along the y-axis the north-south flow is zero and

$$|u_\lambda| = |u_x| \qquad \text{for } \lambda = \pm \frac{\pi}{2}, \text{ all } \theta \quad [\text{y-axis}] \qquad (18.29)$$

so that the magnitude of the zonal velocity is independent of θ along the whole meridian. The only way that this constancy of magnitude can be reconciled with the change of sign is if u_λ has a jump discontinuity as the pole is crossed:

$$u_\lambda = \begin{cases} -u_x & \lambda = \dfrac{\pi}{2} \\ u_x & \lambda = -\dfrac{\pi}{2} \end{cases} \qquad \text{all } \theta \quad [\text{y-axis}] \qquad (18.30)$$

This in turn requires that we omit the $\sin(\theta)$ parity that is needed for well-behaved quantities of odd zonal wavenumber (such as scalars).

It is important to note, however, that this discontinuity is caused by a change in mathematical interpretation rather than a physical variation in the flow itself. In Fig. 18.5, the wind as measured by an anemometer never changes either magnitude or direction. What does change is that we interpret the same wind as an eastward wind on one side of the pole but as a westward current on the other. Thus, the prescription against a physical discontinuity remains in force, and nothing in this section should be (mis)interpreted as implying otherwise.

The meridional velocity component has a similar jump discontinuity on the x-axis (and every other meridian except the y-axis where $u_\theta \equiv 0$). The modified velocity components U and V in contrast, are continuous across the poles — another reason for preferring them to u_λ and u_θ. However, the radial velocity is *not* discontinuous as we cross the pole: up away from the center of the earth is still up.

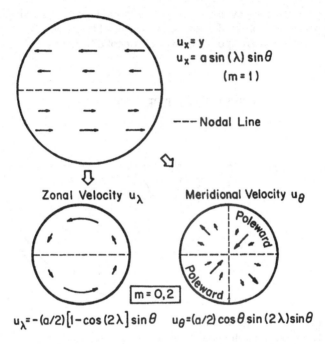

$$u_x = y$$
$$u_x = a \sin(\lambda) \sin\theta$$
$$(m = 1)$$

---- Nodal Line

Zonal Velocity u_λ Meridional Velocity u_θ

$m = 0, 2$

$$u_\lambda = -(a/2)\left[1 - \cos(2\lambda)\right]\sin\theta \qquad u_\theta = (a/2)\cos\theta\sin(2\lambda)\sin\theta$$

Figure 18.6: Same as previous figure except that the Cartesian velocity is now a function of *odd* zonal wavenumber, in this case, $m = 1$. The spherical velocity components are the sums of terms of *even* zonal wavenumber. To avoid polar discontinuities, the velocity components must *vanish* at the poles. A *scalar* quantity of similar zonal wavenumber would be symmetric with respect to the poles.

Fig. 18.6 illustrates the situation when the spherical velocity components are of *even* zonal wavenumber. For the case shown, the zonal velocity is eastward over the whole globe. The meridional velocity alternates in sign, but if we follow a meridian over the pole, we find that a poleward flow is a poleward flow on both sides of the pole. However, this in turn requires that u_θ decrease smoothly to zero as the pole is approached along any meridian. This in turn requires the parity factor, $\sin(\theta)$. This may be seen in the analytical expressions for u_θ and u_θ at the bottom of Fig. 18.6.

Swarztrauber (1981, 1993) presents some good illustrations of discontinuities in horizontal wind fields. His proposed remedy, *vector* spherical harmonics, will be discussed in Sec. 18.22.

18.10   The Pole Problem: Spherical Coordinates

The Courant-Friedrichs-Levy criterion for the stability of *explicit* time-stepping algorithms is that

$$\Delta t < \frac{\Delta x}{c} \qquad (18.31)$$

where c is the speed of the *fastest* waves allowed by the differential equations being solved and Δx is the *smallest* spatial grid interval. One reason why nearly-uniform finite difference or finite elements grids are popular is that Δx is roughly the same everywhere so that the time-step limit is not unduly expensive.

On a sphere, the simplest tactic is to use latitude and longitude as coordinates and apply an even grid spacing in λ and θ. Denoting the radius of the sphere by a, the rub is that the distance Δx between two grid points on a circle of colatitude θ is

$$\Delta x = a \, \sin(\theta) \, \Delta \lambda \qquad (18.32)$$

which tends to zero at the poles as shown graphically in Fig. 18.7.

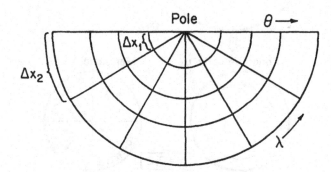

Figure 18.7: A grid with uniform spacing in latitude and longitude as viewed from Polaris, the Pole Star. The meridians (lines of constant longitude) all converge at the pole. Consequently, $\Delta x \to 0$ as the pole is approached, even though $\Delta \lambda$ is constant. The very small grid spacing near the pole is the "pole problem": One must use a very, very short time step, or the numerical flow will violate the Courant-Friedrichs-Levy criterion and become unstable.

The result is that a time step which is stable and reasonable for tropical and temperate latitudes — typically 10 minutes for a fully explicit, "primitive equations" numerical weather prediction model — will give instability near the poles. As the run continues, this localized instability will gradually spread equatorward like a cancer until the whole globe is consumed with numerical error.

One remedy is an artificially high viscosity near the poles. Heavy damping destroys the accuracy of the model at high latitudes, but it is known that high latitudes have little physical influence on lower latitudes, and few complaints about inaccurate local forecasts have come from polar bears.

The alternative is to attack the root of the problem, which is that, close to the poles, the resolution in longitude is needlessly high. Why use very small Δx near the poles when the accuracy of the model as a whole is constrained by the much poorer resolution everywhere else? The high zonal wavenumber components of the solution are the only ones that become unstable. Therefore, the preferred methods for solving the "pole problem" are based on *lowering zonal resolution near the poles*.

One approach, used in the GFDL-Princeton General Circulation Mode, is to selectively *delete* grid points from the latitude-longitude grid at high latitudes. This is messy to program. Worse still, the grid becomes irregular and the resulting uncentered finite difference approximations are not very accurate.

A second alternative is a zonal-scale-selective filtering at high latitudes. This is merely a variant of the artificial polar damping technique, of course, but a scale-selective dissipation is physically reasonable in a way that a scale-independent filtering is not. The reason is that the model cannot resolve the very short longitudinal scales of the damped components except near the poles. There is no reason to mourn the damping of zonal wavenumber 30 at 85 degrees of latitude where its associated zonal scale is only 20 km when the longitudinal resolution at the equator is as coarse as 200 km.

The third alternative is to use a spectral or pseudospectral method with spherical harmonics as the basis set. Because of the close connection between the spherical harmonics and the geometry of the sphere, *this* basis set plays no favorites with respect to latitude: the spherical harmonics have the property of "equiareal resolution" on the sphere as explained in Sec. 18.13 below.

To be sure, the spherical harmonics are more complicated to program than simple finite differences on a *uniform* latitude-longitude grid. But if one wants to survive the cancerous instabilities of the "pole problem", one is *forced* to use something more complicated than a uniform λ-θ grid.

18.11 Spherical Harmonics: Introduction

Spherical harmonics fix the pole problem, contain the proper "parity factors", and offer exponential convergence for functions that are infinitely differentiable on the sphere. Unfortunately, spherical harmonics are more complicated than any of the other basis sets in common use because they are two-dimensional. We will later discuss icosahedral grids and near-uniform triangularizations to emphasize that this difficulty is unavoidable: there is no way to solve problems on the sphere that does not require some complicated programming and careful attention to detail.

The spherical harmonics are

$$Y_n^m(\lambda, \theta) \equiv e^{im\lambda} P_n^m(\theta) \qquad \begin{array}{c} m, \ n \ \text{non-negative integers} \\ \text{such that } n \geq m \end{array} \qquad (18.33)$$

$$Y_n^{-m}(\lambda, \theta) \equiv e^{-im\lambda} P_n^m(\theta) (-1)^m$$

Thus, in longitude, the spherical harmonics are merely a Fourier series. Eq. (18.33) displays the Fourier series written in the form of complex exponentials; some authors prefer ordinary sines and cosines in λ, but the difference is only one of convention.

The functions $P_n^m(\theta)$, the "associated Legendre functions" of order [zonal wavenumber] m and degree n, are where all the complications lie. They are defined by

$$P_n^m(\theta) \equiv \sin^m(\theta) C_{n-m}^m(\cos[\theta]) \qquad (18.34)$$

where the $C_n^m(\cos[\theta])$ are the "Gegenbauer polynomials"[3]. The subscript denotes the degree of the polynomial; the coefficients of the polynomials are different for each different zonal wavenumber m.

---

[3]Warning: The superscript m used in this section is related to the superscript m' of Appendix A via $m=m'-1/2$.

The factor $\sin^m(\theta)$ in (18.34) has a three-fold significance. First, recall that $\sin^{2k}(\theta) \equiv (1-\cos^2[\theta])^k$ = a finite Fourier cosine series with $2k$ terms. Thus, all the associated Legendre functions with m even may be written in terms of a Fourier cosine series. However, when m is odd, one factor of $\sin(\theta)$ cannot be converted into a cosine polynomial, and [combining $\sin(\theta)$ with the cosine polynomial via the usual trigonometric identities] $P_n^m(\theta)$ is a (finite) Fourier sine series. Thus, $\sin^m(\theta)$ is the "parity factor" discussed earlier in the chapter; it is automatically built into the spherical harmonics.

The second significance of the sine factor is that it implies that the spherical harmonics have an m-th order zero at the poles — a multiple root in *colatitude* whose degree is determined by the wavenumber in *longitude*. It is this high-order zero which enables spherical harmonics to avoid the pole problem: the basis functions of high zonal wavenumber m would have wastefully high longitudinal resolution at high latitudes except that the $\sin^m(\theta)$ factor forces these spherical harmonics to have negligible amplitude near the poles.

The third consequence of the $\sin^m(\theta)$ factor is that modes with large m and $(n-m) \ll m$ have little amplitude outside a narrow band centered on the equator. This justifies the Hermite function asymptotic approximation discussed later.

The definition of the "degree" n in (18.34) seems rather peculiar in that the degree of the Gegenbauer polynomial is $(n - m)$; it would seem much more natural to define the degree of the spherical harmonic to equal that of the polynomial. However, the convention shown in (18.34) has become universally adopted because this definition of degree-of-harmonic simplifies almost all other descriptions of the associated Legendre functions.

The most important of these is truncation: what cutoffs on m and n are most efficient? As will be explained below, one can prove that the spherical harmonics give equal resolution of all areas of the globe if and only if the truncation is *triangular*, that is, the retained basis functions satisfy the inequalities

$$|m| \le N \quad \text{and} \quad n \le N \qquad \text{"triangular truncation"} \qquad (18.35)$$

for some integer N as illustrated in Table 18.4. A triangular truncation up to and including wavenumber N is usually denoted by the shorthand of "T" plus the numerical value of N. A "TN" truncation retains a total of $(N + 1)^2$ spherical harmonics.

The Gegenbauer polynomials satisfy an orthogonality relation that can be expressed in two illuminating ways:

$$\int_0^\pi C_k^m(\cos[\theta])\, C_j^m(\cos[\theta]) \sin^{2m}(\theta)\, d\theta \;\; = \;\; 0 \quad \text{if } j \ne k \qquad (18.36)$$

$$\int_{-1}^1 C_k^m(\mu)\, C_j^m(\mu)\, (1 - \mu^2)^m\, d\mu \;\; = \;\; 0 \quad \text{if } j \ne k \qquad (18.37)$$

First, note that both forms of orthogonality apply only to different polynomials of the *same* zonal wavenumber m. With the spherical harmonics, one is not dealing with a single set of orthogonal polynomials, but rather with a countable infinity of sets.

Second, (18.36) shows a fourth role for the factor of $\sin^m(\theta)$ in (18.34): when two associated Legendre functions of the same m are multiplied together, they collectively supply the $\sin^{2m}(\theta)$ weight factor in (18.36). Thus, the associated Legendre functions of the same order m are orthogonal on $\theta \in [0, \pi]$ with a weight function of unity.

When written in terms of $\mu = \cos(\theta)$, the Gegenbauer polynomials are ordinary (rather than trigonometric) polynomials. Eq. (18.37) shows that the Gegenbauer family includes the Chebyshev and Legendre polynomials as the special cases $m = -1/2$ and $m = 0$. Like other orthogonal polynomials, the Gegenbauer polynomials and their derivatives may be

Table 18.4: An illustration of allowed combinations (m, n) for the spherical harmonics Y_n^m. The X's show the modes that retained in a triangular truncation with a cutoff at $N = 3$ ("T3"); the O's show the modes that are added when the cutoff is increased to $N = 5$ ("T5"). The total number of modes is $(N + 1)^2$ in a TN truncation.

$m \rightarrow$	-5	-4	-3	-2	-1	0	1	2	3	4	5
n										
$\downarrow$										
0						X					
1					X	X	X				
2				X	X	X	X	X			
3			X	X	X	X	X	X	X		
4		O	O	O	O	O	O	O	O	O	
5	O	O	O	O	O	O	O	O	O	O	O

evaluated for arbitrary m and n via three-term recurrence relations. Handbooks on mathematical functions are stuffed with such identities, so we shall here focus on the qualitative properties of the basis functions.

One important property is that as m increases, the roots of the n-th degree polynomial move closer and closer to the origin (equator). Since basis functions give good resolution where they are oscillating and poor resolution where they vary slowly and monotonically, it follows that the high order Gegenbauer polynomials give good resolution near the equator and poor resolution near the poles — exactly what is needed to avoid the pole problem and give uniform resolution over the whole globe. [When we consider a triangular truncation of harmonics, the small m functions resolve the small polar cap region; the large areas near the equator are resolved through the combined efforts of *all* the harmonics.]

Recall that an N-term orthogonal polynomial series minimizes the mean square of the error where the error is weighted by the weighting function in the orthogonality integral. The weight function $(1 - \mu^2)^m$ tolerates large error near the poles for the sake of higher accuracy near the origin.

Another consequence of this stress on the interior is that when the Gegenbauer polynomials are normalized so that the integrals in (18.36) and (18.37) are unity when $j = k$, one finds that

$$\max_{x \in [-1, 1]} |\tilde{C}_k^m(x)| \sim O\left(k^{m+1/2}\right) \tag{18.38}$$

where the tilde over C_k^m denotes the normalized Gegenbauer polynomials. In other words, Eq. 18.36 implies that the maximum values of a normalized polynomial are large compared to its $O(1)$ average value on the interval $x \in [-1, 1]$. This implies that the amplitude of the polynomial's oscillations is highly nonuniform (in contrast to the Chebyshev polynomials, whose maxima and minima are all the same). The amplitude increases rapidly as one approaches either pole. However, these oscillations of the polynomials are tempered by the decay of the $\sin^m(\theta)$ factor as we move away from the equator so that the associated Legendre function rather resembles a Hermite function — that is, to say, there is a band around

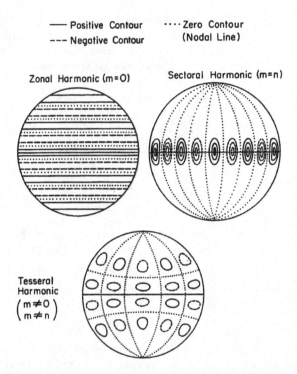

Figure 18.8: Schematic of typical spherical harmonics. The "zonal" harmonics have zero zonal wavenumber and their latitudinal structure is that of the ordinary Legendre polynomials $P_n(\cos\theta)$ where θ is colatitude. The "sectoral" harmonics are the modes which have no latitude circles as nodal lines. The "sectoral" harmonics of high zonal wavenumber are equatorially trapped and have negligible amplitude outside the tropics. The "tesseral" harmonics are the general case.

the equator where it oscillates with many (roughly) equal crests and troughs; outside this band, $P_n^m(\cos[\theta])$ simply decays monotonically to zero. (This band is very narrow if $n \approx m$ and $m \gg 1$ but extends almost all the way to poles if $(n - m)$ is not small in comparison to m.) For the special case of the so-called "zonal harmonics", $m = 0$, the band of oscillations is global and the high latitude "decay zones" do not exist. Fig. 18.8 schematically illustrates typical spherical harmonics.

18.12 Legendre Transforms and Other Sorrows

18.12.1 FFT in Longitude/MMT in Latitude

In longitude, spherical harmonics are just a Fourier series. The optimum longitudinal grid is therefore evenly spaced. Since two-dimensional transforms are most efficiently performed as a nested sequence of one-dimensional partial sums or transforms, irregardless of how these one-dimensional sums are evaluated, the longitudinal partial sums are invariably evaluated via the FFT. In latitude, however, it is necessary to use Gaussian quadrature, which is slower [$O(N^2)$ versus $O(N \log_2 N)$].

The optimum Gaussian quadrature abscissas for a given wavenumber are the roots of an m-th order Gegenbauer polynomial. However, because we must calculate transforms in both λ and θ, it is necessary to use the same points in colatitude for all m. Thus, the spherical grid is the direct product of an evenly spaced Fourier grid in longitude with the quadrature abscissas for Gauss-Legendre integration. (These grid points in colatitude are the roots of a Legendre polynomial, and must be recalculated whenever the resolution is changed.)

Orszag (1974) makes the comment " ... the goal of numerical simulations is accurate reproduction of physical hydrodynamics; it is not clear, at least to the author, that quadratic conservation properties [and other means of suppressing aliasing at the expense of additional operation] have very much to do with the attainment of that goal." Chen(1993) has performed numerical experiments suggesting that dealiasing is simply a waste of computer resources if the wavenumbers near N are appropriately filtered. However, his conclusion is controversial in the sense that models in Eulerian coordinates require such strong filtering near the aliasing limit that the upper third of the wavenumber spectrum is largely destroyed as it would be in a dealiased scheme. Most spherical harmonics models, especially those using Eulerian rather than semi-Lagrangian advective schemes, are de-aliased.

However, the semi-Lagrangian treatment of advection has allowed explicit dealiasing to be dropped from the current ECMWF operational forecasting model. The nonlinear advective terms are never directly evaluated in a semi-Lagrangian method; instead, fluid is advected through the method of characteristics and the resulting fields are interpolated to the pseudospectral grid. Aliasing is not completely eliminated by semi-Lagrangian advection, but it is greatly weakened, and the inherent dissipation is sufficient to suppress aliasing instability. Nevertheless, we shall describe de-aliased models in what follows.

With a total of $2N + 1$ degrees of freedom in longitude, we need at least $2N + 1$ grid points in λ, but it is customary to use $3N + 1$ points so as to eliminate aliasing according to the 3/2's rule. In colatitude, each spherical harmonic in a triangular truncation is a polynomial in $\cos(\theta)$ [or such a polynomial multiplied by $\sin(\theta)$] of degree at most N. For $m = \pm N$, the degree of the polynomial is furnished entirely by the $\sin^{|m|}(\theta)$ factor; for $m = 0$, it is the degree of the Gegenbauer polynomial, but for all m, the degree is at most N. In expanding the quadratically nonlinear terms, we must evaluate integrals of the product of three harmonics, so the total degree of the integrands we must perform is never larger than $3N$. It follows that we need a minimum of $(3/2)N$ grid points in colatitude to evaluate these integrals exactly and thus again avoid aliasing.

18.12.2 Substitutes and Accelerators for the MMT

Spherical harmonics have one great fault: slow transforms for Associated Legendre functions in latitude. This vice has has inspired a quest for alternatives which has lasted more than three decades and counting. In longitude, the spherical harmonics are just a Fourier series. All grid-to-coefficient and coefficient-to-grid transforms for spherical harmonics (and alternatives to them) employ a Fourier series in longitude and the Fast Fourier Transform in this coordinate. It is only the transform in latitude which is troublesome because the FFT is inapplicable to Associated Legendre series.

The reason that the Legendre transform is important is that spectral algorithms require repeated jumps from grid point values to coefficients and back again at each time step. At the end of the twentieth century, spherical harmonic weather forecasting models use the Matrix Multiplication Transform (MMT) to jump from Legendre coefficients to grid point values and back again. For each zonal wavenumber m, this multiplication of a vector by a dense matrix requires $O(N^2)$ operations. It follows that the cost is very large compared to the $O(N \log_2(N))$ expense of an FFT. For many years, there has been a feeling that eventually N would grow so large that spherical harmonics models would become prohibitively expensive, forcing a switch to finite elements or other non-spectral algorithms.

Reality is different from expectation. The highest resolution forecasting code at the time of writing (1998) is the T639 model of the European Centre for Medium-Range Weather Forecasting (ECMWF), which is being tested but will not be operational for another couple of years. Although $N = 639$, the Legendre transforms still consume only 19% of the running time of the model. How is this possible? The short answer is that even though there is no algorithm as nifty as the FFT for Legendre series, there are several tricks for accelerating the MMT which have so far saved spherical harmonics from obsolesence:

1. Parity Matrix Multiplication Transform (PMMT)

2. The Vectorizability and Parallelizability of Matrix Multiplication

3. Reduced Polar Grid

4. Complicated Physics and Chemistry in Weather Forecasting Models

5. Schuster-Dilts Triangular Matrix Acceleration

6. Generalized Fast Fourier Transforms

18.12.3   Parity and Legendre Transforms

Because the Associated Legendre functions are all either symmetric or antisymmetric with respect to the equator, a transformation from grid point values to N spectral coefficients (or the reverse) can always be split into two subproblems of size $N/2$. Thus, after first applying the FFT to transform from grid point values in latitude and longtitude to the grid point values of zonal wavenumber m, a set of grid point values $f_m(\theta)$ can be combined into components symmetric and antisymmetric with respect to the equator (where colatitude ($\theta = $ latitude $- \pi/2$) by taking sums and differences:

$$f_m^S(\theta) \equiv \frac{1}{2}\left(f_m(\theta) + f_m(\pi - \theta)\right), \quad f_m^A(\theta) \equiv \frac{1}{2}\left(f_m(\theta) - f_m(\pi - \theta)\right), \quad \theta \in [0, \pi/2] \quad (18.39)$$

Each collection of grid point values of definite parity can then be transformed by a matrix multiplication into the corresponding even or odd Legendre coefficients. Each rectangular transformation matrix has only half the rows and half the columns of the original, non-parity-exploiting transformation, but there are two smaller transformations to be taken. If we ignore the $O(N)$ cost of taking sums and differences compared to the $O(N^2)$ of the matrix-vector multiplications of the transformations themselves, parity saves exactly a factor of two. Symmetry is discussed in greater detail in Chapter 8.

18.12.4   Hurrah for Matrix/Vector Multiplication

The multiplication of a vector by a matrix is one of the most fundamental operations of linear algebra. Because of this, it is an operation which is often coded as a blazingly fast assembly language program in software libraries. An assembly language routine will typically speed up the matrix/vector multiplication by a factor of two.

Furthermore, supercomputers with vector-processing hardware, such as the Cray 1 and its arithmurgical descendants, process the matrix/vector multiplication very efficiently. Indeed, it is not an exaggeration to say that the vector hardware is built primarily to do this one operation as efficiently as possible. Because this operation requires no conditional statements, branching, or any other overhead, Legendre transforms can be computed on vector hardware at very close to the machine's optimum floating point execution rate. In contrast, the sparse matrix operations of finite differences require a lot of starting-and-stopping: one cannot exploit the holes in a sparse matrix without working very hard to inform the hardware of where the holes are. Because of this, when finite difference and spherical harmonics codes are compared in terms of flops (that is, floating point operations per second), the spherical harmonics program usually has a much higher flop rate, much closer to the machine's theoretical optimum, than a finite difference code. Thus, although the spherical harmonic code may require a much larger number of floating point operations for a given N than a finite difference or finite element code, the wall-clock "grind time" may be about the same.

Similarly, the multiplication of a dense matrix by a vector is a speedy process on a massively parallel machine. A matrix/vector multiply is actually just a set of independent vector-vector multiplications, one for each row. Furthermore, the MMT matrix elements are computed in a preprocessing step. It follows that on a massively parallel machine, it is easy to subdivide the matrix/vector multiply among processors by assigning one or more rows to each processor. To initiate the parallel computation, it is only necessary to broadcast the vector of grid point values to each processor simultaneously, and all floating point operations can then be executed in parallel. Again, because the matrix/vector multiplication is the fundamental operation of linear algebra, there are very good assembly language routines and carefully designed algorithms for each type of hardware.

The conclusion is that in high performance computing, Legendre transforms-by-matrix are much more efficient and less costly than operation counts would suggest. In a computer science course, one may only be asked to add up multiplications and additions, but in the real world, it is important to look at the clock.

18.12.5 Reduced Grid and Other Tricks

Hortal and Simmons(1991) and Courtier and Naughton (1994) showed that one could reduce the cost of a spectral model by systematically deleting points from the tensor product grid near the poles with little loss of accuracy. The reductions are not large (perhaps 30%), but to paraphrase the late U. S. Senator Dirksen: "A billion multiplications here, a billion additions there — pretty soon you're talking about real savings."

Swarztrauber(1996) has carefully compared a large number of different formulations of spherical harmonic methods to show that by proper organization, one can minimize the number of Legendre transforms that are needed on each step of the algorithm.

18.12.6 Schuster-Dilts Triangular Matrix Acceleration

Schuster(1903) discovered that the dense matrix/vector multiplication of the usual Gaussian-quadrature PMMT could be replaced by multiplication by a *triangular* matrix to reduce the cost by a factor of two. Dilts(1985) independently reinvented this technique. The first part of the algorithm is a two-dimensional FFT to transform from a grid which is evenly spaced in both latitude and longtitude to the coefficients of a two-dimensional Fourier series. The second step is to multiply the Fourier coefficients by an upper triangular matrix to convert the Fourier coefficients to spherical harmonic coefficients.

Because of the familiar identity $T_j(\cos(\theta)) \equiv \cos(j\theta)$, the Fourier series in latitude can be equally well be regarded as a Chebyshev series in the variable $x \equiv \cos(\theta)$ where θ is colatitude. This interpretation is conceptually useful because it shows that the function of the triangular matrix is to convert from one polynomial basis (Chebyshev) to another polynomial basis (associated Legendre). These conversion is done one zonal wavenumber at a time. To simplify the discussion, we shall illustrate only the case of zonal wavenumber $m = 0$ (ordinary Legendre polynomials).

One important point is that the transformation respects parity. Thus, the even degree Fourier coefficients in latitude transform only to even degree spherical harmonics, and similarly odd transforms only to odd.

First, split the the θ-dependent coefficient of the $m = 0$ Fourier series into components symmetric and antisymmetric with respect to the equator as in Eq.(18.39). Let the symmetric part have the Chebyshev polynomial expansion

$$f_m^S(x) = \sum_{j=0}^{3} b_{2j} T_{2j}(x) \tag{18.40}$$

$$= b_0 + b_2(2x^2 - 1) + b_4(8x^4 - 8x^2 + 1) + b_6(32x^4 - 48x^4 + 18x^2 - 1)$$

which is equivalent to a cosine series in θ where $x = \cos(\theta)$ where θ is colatitude. (For simplicity, we take only four terms, but the method can be extended to arbitrary order.) Our goal is to convert this to the corresponding Legendre series:

$$f_m^S(x) = \sum_{j=0}^{3} a_{2j} P_{2j}(x)$$

$$= a_0 + a_2 \left\{ \frac{3}{2}x^2 - \frac{1}{2} \right\} + a_4 \left\{ \frac{35}{8}x^4 - \frac{15}{4}x^2 + \frac{3}{8} \right\} \tag{18.41}$$

$$+ a_6 \left\{ \frac{231}{16}x^6 - \frac{315}{16}x^4 + \frac{105}{16}x^2 - \frac{5}{16} \right\}$$

The expansion of the j-th Chebyshev polynomial in terms of Legendre polynomials has the coefficients $(T_j, P_k)/(P_k, P_k)$ where (p, q) denotes the unweighted integral of $p(x) q(x)$ from -1 to 1. Substituting this into the Chebyshev series and collecting Legendre terms, one finds

$$\vec{a} = \vec{M}\vec{b} \tag{18.42}$$

where $\vec{a}$ and $\vec{b}$ are column vectors containing the Legendre and Chebyshev coefficients, respectively, and where the elements of the square matrix $\vec{M}$ are

$$M_{ij} \equiv \int_{-1}^{1} dx\, T_{2j-2}(x) P_{2i-2}(x) \left/ \int_{-1}^{1} dx\, P_{2i-2}(x) P_{2i-2}(x) \right., \qquad i, j = 1, 2, \dots \tag{18.43}$$

Because the expansion of a polynomial of degree j requires only the Legendre polynomials of degree j and lower, the matrix $\vec{M}$ is upper triangular.

For example, if the expansions are truncated at sixth degree, only P_6 in the Legendre expansion and T_6 in the Chebyshev series contain a term proportional to x^6. It follows that $32x^6 b_6 = (231/16)x^6 a_6$ or in other words $M_{44} = 512/231$. Similarly, only two terms in each series contain terms proportional to x^4. This implies $8b_4 - 48b_6 = (35/8)a_4 - (315/16)a_6$. Since a_6 is already known, this can be rewritten as $a_4 = (8/35)\{8b_4 - (48/11)b_6\}$. Similarly, a_2 can be computed as a weighted sum of b_2, b_4 and b_6 and a_0 as a weighted sum of four coefficients. The necessary multipliers are the upper 4×4 block of the matrix M:

$$
\begin{vmatrix} a_0 \\ a_2 \\ a_4 \\ a_6 \end{vmatrix} = \begin{vmatrix} 1 & -\frac{1}{3} & -\frac{1}{15} & -\frac{1}{35} \\ 0 & \frac{4}{3} & -\frac{16}{21} & -\frac{4}{21} \\ 0 & 0 & \frac{64}{35} & -\frac{384}{385} \\ 0 & 0 & 0 & \frac{512}{231} \end{vmatrix} \begin{vmatrix} b_0 \\ b_2 \\ b_4 \\ b_6 \end{vmatrix}
\tag{18.44}
$$

Similar matrices give the transform from the Chebyshev (latitudinal Fourier) series for the antisymmetric $m = 0$ Legendre polynomials and for the associated Legendre functions of all higher m. The savings is only about a factor of two; note that the cost of the FFT in latitude is negligible compared to the cost of the triangular matrix/vector multiply in the limit $N \to \infty$. However, this is an adequate reward for implementing a very simple algorithm. (Healy, Rockmore, Kostelec and Moore, 1998, amusingly call this the "semi-naive" algorithm precisely because it is so simple.)

Curiously, this algorithm has been never been used in an operational forecasting or climate code. One reason is that one must store the elements of a different triangular matrix for each equatorial parity and each m, which adds up to a lot of storage. A more significant reason is that the Schuster-Dilts algorithm requires grid point values at evenly spaced points in θ, which is a Chebyshev grid in $x = \cos(\theta)$. The usual PMMT algorithm uses points on the Legendre grid. Because a typical forecasting or climate code is $O(50, 000)$ lines of code, it is not trivial to rewrite a model to accept a different grid, even though the two grids are very similar and most of the chemistry and hydrologic cycle, etc., are evaluated point-by-point.

18.12.7 Generalized FFT: Multipoles and All That

As reviewed in Chapter 10, generalizations of the Fast Fourier Transform have been developed which replace the $O(N^2)$ cost of a Legendre transform (for a given zonal wavenumber) by a cost proportional to $O(N \log_2(N))$ operations — the same as for the Fast Fourier Transform. Orszag's (1986) method exploits the three-term recurrence relation satisfied by the Associated Legendre Functions. Boyd(1992c) and Dutt&Rokhlin(1993, 1995) independently observed that Fast Multipole Methods could also be applied. Recently, Rockmore *et al.*(1998) have developed a very sophisticated algorithm specifically for spherical harmonics.

The bad news is that these algorithms have a huge proportionality constant compared to the FFT. As a result, none of the generalized FFTs has ever been used in an operational climate or weather forecasting model. However, as forecasting resolutions climb above T400, these algorithms may become faster than the PMMT, and perhaps in time replace it.

18.12.8 Summary

The relative slowness of the PMMT (Parity-Exploiting Matrix Multiplication Transform) for transforming the latitudinal basis functions is the great inefficiency-maker for spherical harmonics algorithms. However, clever tricks, as catalogued above, have kept the spherical harmonics from obsolescence.

Table 18.5: Legendre Transforms Bibliography

References	Comments
Schuster(1903)	Triangular matrix multiplication transform
Dilts (1985)	Modern reinvention of Schuster's scheme
Brown (1985)	Fast spherical harmonic transform
Orszag (1986)	Fast transform for any basis satisfying 3-term recurrence relation
Elowitz&Hill &Duvall (1989)	Compares fast transforms of Dilts (1985),Brown (1985) but conclusion is false
Alpert&Rokhlin(1991)	Fast Legendre-to-Chebyshev transform; does not generalize to associated Legendre
Boyd (1992c)	Showed FMM could be applied to non-Chebyshev basis
Dutt&Rokhlin (1993,1995)	FMM nonequispaced FFT
Jakob-Chien&Alpert(1997) Yarvin&Rokhlin(1998)	Grid-to-spherical-harmonics-to-grid projective filter for longitude/latitude time-dependent double Fourier series
Foster&Worley(1997)	Comparison of parallel spherical harmonic transforms
Healy&Rockmore& Kostelec&Moore(1999)	Fast spherical harmonic transform; freeware at www.cs.dartmouth.edu/ geelong/sphere
Swarztrauber&Spotz(2000) Spotz&Swarztrauber(2000)	"Weighted Orthogonal Complement" algorithm reduces storage by $O(N)$; faster than alternatives because algorithm mostly stays in the on-chip cache

18.13   Equiareal Resolution and the Addition Theorem

Definition 40 (Equiareal Resolution) *A numerical algorithm which has the property that its numerical characteristics are invariant to a rotation of the north pole of the coordinate system so that features of a given size are resolved equally well or badly regardless of whether they are located at the poles, equator, or anywhere in between.*

A so-called "triangular truncation" of a spherical harmonic basis has this property because of the following theorem and its corollaries.

Theorem 36 (ADDITION THEOREM:) *Let (λ', θ') denote longitude and latitude as measured relative to a set of coordinate axes rotated with respect to the original axes. Let (λ, θ) denote the angles measured relative to the original, unrotated coordinate system. Then*

$$Y_n^m(\lambda', \theta') = \sum_{m'=-n}^{n} D_{mm'}^{(n)}(R) Y_n^{m'}(\lambda, \theta) \tag{18.45}$$

where the coefficients $D_{mm'}^{(n)}$ are functions of the rotation angle.
COROLLARY 1: The spherical harmonics of degree n form a (2n+1)-dimensional representation of the continuous rotation group.
COROLLARY 2: A triangular truncation of spherical harmonics, that is, keeping only those harmonics such that

$$n \leq N \qquad all \ m \tag{18.46}$$

gives equal resolution to equal areas on the globe, regardless of where those areas are located.

PROOF: Eq. (18.45) is a classical result discussed in most quantum mechanics texts such as Merzbacher (1970).

The first corollary is merely a way of restating the theorem in the jargon of group theory. When we rotate the pattern that is a particular spherical harmonic Y_n^m, we create a new function of latitude and longitude. Like all such functions, this may be expanded in a spherical harmonics series, but there is no obvious reason why these series should not contain an infinite number of terms. The theorem shows, however, that the series contains at most $(2n+1)$ terms, and all the non-zero harmonics have the *same degree* n as the function that is rotated. Thus, the spherical harmonics of degree n form a *closed subset* under rotation through *arbitrary* angles — and this is what is required to form a "representation of the rotation group".

The collection of spherical harmonics that are retained in a "triangular truncation" therefore are closed under rotation, too. This implies the property of equiareal resolution. In the words of Orszag (1974), "The basic mathematical reason is that, under arbitrary rotations, expansions truncated at harmonics of degree N remain truncated at degree N so that the resolution of such series must be uniform over the sphere."

18.14 Variable Resolution Spherical Harmonics Models

"Limited-area" weather forecasting models offer the advantages of very high resolution over a small portion of the earth's surface — higher than would be affordable if this resolution were extended over the entire globe. Most national weather services run limited-area models targeted at the country that pays for them. Limited-area models are also used to track tropical hurricanes, which have such small scales that it is difficult for uniform resolution global models to track them accurately.

One obvious strategy is to employ a dense but uniform grid over a small portion of the globe and specify inflow-outflow conditions at the sides, usually from climate data or from interpolation of a global model. Such non-global limited-area codes are in wide use, but there are difficulties. One is that specifying lateral boundary conditions turns out be very hard; rather elaborate blending procedures are necessary to avoid corruption of the high-resolution data within the domain by the low-resolution data at the boundaries. Moreover, as models incorporate more and more physics and chemistry, it has become increasingly painful to maintain two completely separate models, one global and one limited-area, using different numerical schemes, tuning parameters and so on.

An alternative is to use the global model as the limited-area model, too. This can be done by a smooth change of coordinates that maps the surface of the sphere into itself. In physical space, the transformed grid has a high density of points over the region of interest, but decreases to lower and lower density as one moves away from the target region. No artificial sidewall boundary conditions are needed because there are no sidewalls. A single model can be used for both global and regional forecasting by switching on or off a few metric factors in the evaluation of derivatives.

Schmidt(1977, 1982) proposed a conformal sphere-to-sphere mapping which he tested successfully in a simple code. With refinements, this has been adopted by Météo-France for its primary operational weather forecasting code ("Arpege") as described by Courtier & Geleyn(1988), Courtier *et al.*(1991), and Déqué&Piedelievre(1995). Hardiker(1997) has shown that such mappings are equally effective for tracking hurricanes.

These variable-resolution models have been sufficiently successful that the desire to combine global and regional models into a single code is not a major threat to continued use of spherical harmonics for weather forecasting. Rather, the big Thing-That-Goes-Bump-in-the-Night is the switch to massively parallel machines, which may or may not be happy doing PMMTs for very large N.

Table 18.6: Variable Resolution Spherical Harmonic Models

References	Comments
Schmidt(1977,1982)	Conformal mapping to give high local resolution to track cyclones
Courtier&Geleyn(1988)	Variable resolution (mapped) weather prediction
Courtier et al.(1991)	Arpege project at Météo-France: global weather model with high resolution in Europe
Déqué&Piedelievre(1995)	Experience with Arpege variable resolution weather model
Hardiker(1997)	Conformal mapping for variable resolution

18.15   Spherical Harmonics and Physics

The spherical harmonics are the eigenfunctions of the two-dimensional Laplacian operator

$$\nabla^2 Y_n^m = -n(n+1) Y_n^m \tag{18.47}$$

where

$$\nabla^2 \equiv \frac{\partial^2}{\partial \theta^2} + \cot(\theta) \frac{\partial}{\partial \theta} + \left[\frac{1}{\sin^2(\theta)} \right] \frac{\partial^2}{\partial \lambda^2} \tag{18.48}$$

Because of this, the spherical harmonics are fundamental solutions to many problems in physics.

In geophysics, for example, Haurwitz (1940) showed that the streamfunction for linear, barotropic Rossby waves was proportional to a spherical harmonic, that is to say, the spherical harmonics are the quasi-geostrophic normal modes of the earth's atmosphere. Similarly, Longuet-Higgins (1968) has shown that for gravity waves in the barotropic limit, the velocity potential is proportional to a spherical harmonic. In both cases, elementary identities show that the velocities and other quantities may be written as pairs of spherical harmonics.

Barrett (1958) observed that since the (purely westward) phase velocity is

$$c_{\text{phase}} = -\frac{2\Omega}{n(n+1)} \qquad \text{[Rossby-Haurwitz waves]} \tag{18.49}$$

one may synthesize a uniformly propagating disturbance from an arbitrary sum of spherical harmonics of the same degree n. In particular, an "eccentric" spherical harmonic, that is, a spherical harmonic rotated so that its pole at some arbitrary latitude θ_p, is a steadily propagating Rossby wave because of the addition theorem of Sec. 18.13. In more recent times, such rotated spherical harmonics have been the basis for constructing nonlinear modons in spherical geometry as explained by Tribbia(1984b).

One could multiply these examples with dozens from other fields (Morse and Feshbach, 1953). This close and intimate connection between the spherical harmonics and the physics, as well as their good numerical properties, have helped to make spherical harmonics popular.

18.16   Asymptotic Approximations I: Polar-Cap and Bessel Functions

At high latitudes,

$$\cos(\theta) \approx 1 \qquad \& \qquad \sin(\theta) \approx \theta \qquad\qquad \theta \ll 1 \tag{18.50}$$

and the horizontal Laplacian operator becomes

$$\nabla^2 \approx \frac{\partial^2}{\partial\theta^2} + \frac{1}{\theta}\frac{\partial}{\partial\theta} + \frac{1}{\theta^2}\frac{\partial^2}{\partial\lambda^2} \tag{18.51}$$

which is identical in form with the Laplacian in plane polar coordinates if we identify θ with radius r and λ with the polar angle. For a given zonal wavenumber m, the second λ-derivative in (18.51) becomes multiplication by $(-m^2)$ and the eigenequation $\nabla^2 Y_n^m = -n(n+1)$ becomes Bessel's equation:

$$\left(\frac{d^2}{d\theta^2} + \frac{1}{\theta}\frac{d}{d\theta} + \left\{ k^2 - \frac{m^2}{\theta^2} \right\} \right) J_m(k\theta) = 0 \tag{18.52}$$

where

$$k \equiv \sqrt{n(n+1)} \tag{18.53}$$

A more heuristic (but equally correct) way of justifying this planar, polar coordinate approximation is to simply look down on a globe from above one of the poles. As one moves closer and closer to the poles, the sphere flattens into a plane, and the meridians form a network of radial lines.

Since this "polar-cap" approximation becomes exact near the pole, the Bessel functions must be consistent with the known behavior of the spherical harmonics at the pole. We note that $J_m(r)$ has an m-th order zero at $r = 0$ — mimicking the m-th order zero of Y_n^m at the pole. If we expand Y_n^m as a power series in θ, we find that the expansion contains only every other power of θ; even powers of θ when m is even and odd powers of θ (because of the "parity factor", $\sin(\theta)$) when m is odd. Similarly, the expansion of $J_m(r)$ is in *alternating* powers of r; since the expansion begins with r^m, the powers are the *odd powers* of r when m is *odd* and the *even* powers of r when m is *even*.

The Bessel function (for $m > 0$) rises monotonically to a turning point at $r \approx m$ and then oscillates for larger values of its argument. Fig. 18.9 illustrates three typical Bessel functions. The turning colatitude is

$$\theta_t \approx \frac{m}{\sqrt{n(n+1)}} \qquad \text{["turning colatitude"]} \tag{18.54}$$

When m is roughly equal to n (recall that n cannot be smaller than m), then the predicted turning latitude is $\theta_t \approx 1$, that is, only about 30 degrees from the equator. The polar cap approximation is not accurate that close to the equator, so what the Bessel approximation tells us in this case is simply that the spherical harmonic has most of its amplitude at low latitudes (where we shall use a different approximation given in the next section). Near the pole, the Bessel approximation is still valid — but the behavior of both $J_m(k\theta)$ and Y_n^m is dominated by the m-th order zero at the pole, so the polar cap approximation does not tell us anything we did not already know.

When $n \gg m$, however, the predicted turning point is close to the poles and the Bessel approximation is more useful. There is some region equatorward of θ_t where the Bessel approximation is legitimate, and we can use the asymptotic approximation

$$J_m\left([n(n+1)]^{\frac{1}{2}}\theta\right) \sim \left[\frac{4}{\pi^2 n(n+1)\theta^2} \right]^{\frac{1}{4}} \cos\left\{ [n(n+1)]^{\frac{1}{2}}\theta - \frac{(2m+1)\pi}{4} \right\} \quad \theta \gg \theta_t \tag{18.55}$$

Eq. (18.55) shows explicitly that a spherical harmonic oscillates equatorward of its turning latitude. It also shows that the amplitude of the oscillation *decreases* in the direction of the equator (as $1/\sqrt{\theta}$) — the maximum of the harmonic is just equatorward of the turning latitude.

The "polar-plane" approximation, which invented by B. Haurwitz, has been systematically developed for geophysical applications (Bridger and Stevens, 1980).

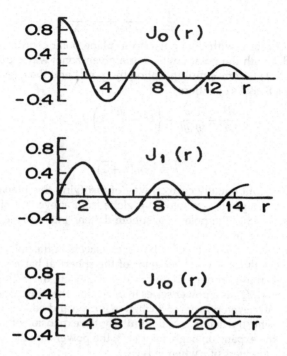

Figure 18.9: Three Bessel functions, illustrating the behavior of spherical harmonics near the pole, $r = 0$. Note that $J_{10}(r)$, which is representative of high degree harmonics, has a "turning latitude" at about $r = 10$: the function is exponentially decaying towards the pole for small r, and oscillatory for larger r. ($r \equiv \sqrt{n(n+1)}\,\theta$ increases toward the equator.)

18.17 Asymptotic Approximations, II: High Zonal Wavenumber & Hermite Functions

Abramowitz & Stegun (1965) give the asymptotic approximation

$$P_{m+n'}^m[\cos(\theta)] \quad \sim \quad q \exp(-\tfrac{1}{2} m\phi^2)\, H_{n'}(\sqrt{m}\,\phi) \tag{18.56}$$

$$m \to \infty, \; n' \text{ fixed}, \; \phi \sim O\left(\frac{1}{\sqrt{m}}\right)$$

where q is a constant, ϕ is latitude [not colatitude] and $H_{n'}$ is the usual Hermite polynomial. This approximation is complementary to that of the previous section. First, the "polar cap" approximation is accurate only for *high* latitudes while (18.56) is limited to a band around the equator. Second, the Bessel approximation becomes more useful when $n \gg m$ whereas the Hermite formula is most accurate when $m \gg n'$, that is, in the limit of large zonal wavenumber for fixed n' where n' is the number of zeros of the harmonic (excluding those at the poles).

With a triangular truncation at degree N, there are but two harmonics with zonal wavenumber N, and their common latitudinal structure is, according to (18.56), given by the Gaussian, $\exp(-\sqrt{N}\,\phi^2)$. Now the highest zonal wavenumber would impose the

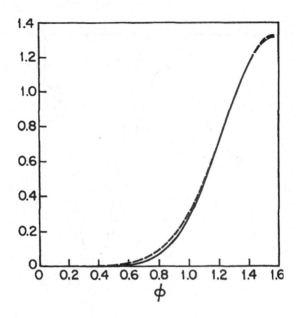

Figure 18.10: A comparison of the exact associated Legendre function, $P_9^9(\phi)$ [solid] with its asymtotic approximation, the Hermite function $\exp(-[9/2]\,\phi^2)$ [dashed]. ϕ is colatitude & $\phi = \pi/2$ is the equator.

strictest requirement on the time step if we expanded the unknowns in a double cosine series (instead of spherical harmonics)– but zonal wavenumber N does not translate into a very small longitudinal grid interval near the poles for the spherical harmonics because the Gaussian has an exponentially small amplitude near the pole. The equatorial confinement embodied in (18.56) is how the "pole problem" is avoided when we use spherical harmonics to integrate time-dependent equations.

Figs. 18.10 and 18.11 show that the Hermite approximation is not uniform in degree n (or n'); as n increases for fixed zonal wavenumber, the harmonic becomes wider and wider in latitude and (18.56) becomes less and less accurate. However, inspecting Fig. 18.11 more carefully, we note that even as the approximation becomes *numerically* less accurate, it remains *qualitatively* correct in the sense that the harmonic oscillates close to the equator and then decays monotonically towards the poles on the far side of the turning latitudes. [One may in fact obtain a very accurate approximation to P_{13}^9 merely by replacing $\sqrt{m}$ in (18.56) by 3.33 to narrow the Hermite function.]

One obvious question is: what is the *physical* reason for this equatorial confinement of spherical harmonics with $(n - m) \ll m$? Fig. 18.12 shows a ray-tracing argument that makes the trapping at least plausible. Because of the spherical geometry, a ray that leaves the equator at say a 45 degree angle (that is, moving in a north-east direction) will *not* reach the poles if it moves in a straight line on the sphere, that is, if the ray moves as a geodesic. When the zonal wavenumber is large in comparison to the latitudinal wavenumber, i. e. $m \gg (n - m)$, then the equivalent ray is almost parallel to the equator and the turning latitude is only a short distance from the equator.

To be sure, this argument is only heuristic. We have not attempted to justify ray-tracing. Indeed, the spherical harmonics often appear in electrostatic and gravitational problems where no rays are evident. However, this refraction-by-geometry is a real and physically important effect in many wave problems. Boyd (1985b) gives a geophysical discussion.

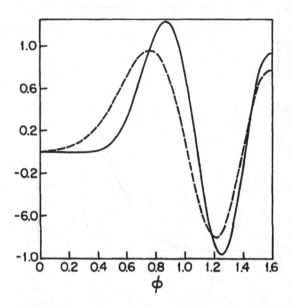

Figure 18.11: A comparison of the exact $P_{13}^9(\phi)$ [solid] with its asymptotic approximation, $\exp[-(9/2)\phi^2]\,H_4(3\phi)$, which is shown as the dashed curve. For a given zonal wavenumber [$m = 9$ for both this figure and its predecessor], the approximation worsens with increasing degree, but remains qualitatively accurate even for fairly large degree.

18.18   Software: Spherical Harmonics

John Adams and Paul N. Swarztrauber of the National Center for Atmospheric Research have written a very comprehensive software library to compute spherical harmonics, their derivatives, transforms from grid point values to spectral coefficients, interpolation from an evenly spaced latitudinal grid to the Gaussian grid employed by forecasting models, and computation and manipulation of vector spherical harmonics. SPHEREPACK 3.0 is now available at **http://www.scd.ucar.edu/css/software/spherepack**. A shallow water model, complete with a full suite of test cases, and a three-dimensional global circulation model are also available from NCAR, complete with documenation.

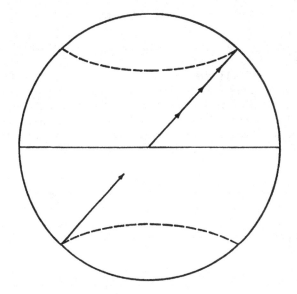

Figure 18.12: Schematic of the behavior of rays trapped on the surface of a sphere. In the absence of other refractive mechanisms, the rays follow geodesics, that is, great circle trajectories. The illustrated ray cannot propagate poleward of the dotted circles: it is latitudinally confined.

18.19   Semi-Implicit Spherical Harmonic Methods for the Shallow Water Wave Equations

Real forecasting models have three spatial dimensions, but since the radial (vertical) coordinate is irrelevant to spherical harmonics, we will discuss the simpler case of the shallow water wave equations, which are two-dimensional in space, based on the classic paper of Bourke (1972).

The shallow water equations may be written in the form

$$\vec{V}_t = -(\zeta + f)\vec{k} \times \vec{V} - \nabla\left(\phi' + \frac{1}{2}\vec{V} \cdot \vec{V}\right) \tag{18.57a}$$

$$\phi'_t = -\nabla \cdot (\phi'\vec{V}) - \Phi D \tag{18.57b}$$

where ζ is the vorticity [strictly, the vertical component of the vorticity vector], f is the Coriolis parameter, $\vec{k}$ is a unit vector in the vertical direction, $\vec{V}$ is the horizontal velocity vector, and D is the divergence of $\vec{V}$. The height field has been split into a mean part, Φ, and deviations from the mean, ϕ', because these parts will be treated differently in the time-marching algorithm.

The major problem in Eq. (18.57) is that its free oscillations are composed of two sets of waves with wildly disparate time scales: (i) slow, low-frequency Rossby motions and (ii) high frequency gravity waves. Existing data networks cannot adequately resolve the gravity waves, and these high frequency motions have little effect on weather or climate. From the perspective of global weather modelling, the gravity waves are simply noise. Unfortunately, this noise shortens the maximum stable time step for an *explicit* time differencing scheme from about 60 minutes to 10 minutes.

For this reason, so-called "semi-implicit" time marching schemes (Chapter 12) have become very popular in meteorology. A small set of terms which are essential to the (linear) propagation of gravity waves are treated implicitly, but most of the linear terms and all the nonlinear terms are handled explicitly. This makes it possible to increase the time step by a factor of six with no significant loss of accuracy and without the iterations that would be needed in a fully implicit code.

One major reason why spherical harmonics have carried the field in meteorology is that these expansions allow very simple and efficient semi-implicit time-stepping algorithms. The two crucial tricks are to (i) decompose the horizontal velocity into the contributions of a scalar streamfunction ψ and scalar velocity potential χ via

$$\vec{V} \equiv \vec{k} \times \nabla\psi + \nabla\chi \tag{18.58}$$

and (ii) replace the horizontal momentum equations by a vorticity equation and a divergence equation, which are obtained by taking the curl and divergence of (18.57a), respectively:

$$\zeta_t = -\nabla \cdot \left[(\zeta + f)\vec{V}\right] \qquad\qquad \text{[vorticity eq.]} \quad (18.59)$$

$$D_t = \vec{k} \cdot \nabla \times \left[(\zeta + f)\vec{V}\right] - \nabla\left(\phi' + \frac{1}{2}\vec{V} \cdot \vec{V}\right) \qquad \text{[divergence eq.]} \quad (18.60)$$

This vorticity/divergence form is useful in part because of the exact relations

$$\zeta = \nabla^2\psi \qquad\qquad \& \qquad\qquad D = \nabla^2\chi \tag{18.61}$$

Since the spherical harmonics are the eigenfunctions of the Laplacian, (18.61) is equivalent to the statement that the spherical harmonic coefficients of ζ and D are directly proportional to those of ψ and χ, respectively. Thus, the vorticity and divergence equations plus the "height" equation (18.59) form a closed set of three equations in three unknowns: ψ, χ, and ϕ'. Furthermore, all the unknowns are scalars and may be expanded directly in spherical harmonics.

The second virtue of the vorticity-streamfunction procedure is that a semi-implicit algorithm for the shallow water set requires implicit treatment of only (i) the $\nabla^2 \phi'$ term in the divergence equation and (ii) the $\nabla^2 \chi (= D)$ term in the height equation. With a finite difference model, one would have to solve two Poisson equations at every time step because ϕ' and χ appear only in the form of the Laplacian operator acting upon these unknowns. With a spectral model using spherical harmonics (but only spherical harmonics!), we do not have to invert the Laplacian because the spherical harmonics are its eigenfunctions: the Laplacian applied to a spherical harmonic is equivalent to multiplying that harmonic by $-n(n+1)$ where n is the degree of the harmonic.

This in turn leads to the third trick: applying Galerkin's method instead of the pseudospectral algorithm. Normally, the latter is simpler and easier to program, but a pseudospectral application of spherical harmonics would require inverting large matrices at every time step.

The fourth trick is to introduce the modified velocity components $U(\equiv u_\lambda \sin[\theta])$ and $V(\equiv u_\theta \sin[\theta])$; although not unknowns, they are useful auxiliary quantities for evaluating the nonlinear terms. If we expand

$$\psi = \sum_{m=-N}^{N} \sum_{n=|m|}^{N} \psi_{mn} Y_n^m \tag{18.62}$$

and similarly for the other fields, identities and recurrence relations for the spherical harmonics and (18.58) show that the coefficients of U and V are

$$U_{mn} = (n-1) D_{mn} \psi_{m,n-1} - (n+2) D_{m,n+1} \psi_{m,n+1} + im \chi_{mn} \tag{18.63}$$

$$V_{mn} = -(n-1) D_{mn} \chi_{m,n-1} + (n+2) D_{m,n+1} \chi_{m,n+1} + im \psi_{mn} \tag{18.64}$$

where

$$D_{mn} \equiv \sqrt{\frac{n^2 - m^2}{4 n^2 - 1}} \tag{18.65}$$

The fifth trick is the pseudospectral strategy of evaluating the nonlinear terms in gridpoint space rather than by forming convolution sums or inner products of the spherical harmonic coefficients. First, the series for $\nabla^2 \psi$, ϕ', U, and V are evaluated on a latitude and longitude grid. [Note that the coefficients for the $\nabla^2 \psi$ are those for ψ multiplied by $-n(n+1)$.] Then the five needed nonlinear products are formed: $U \nabla^2 \psi$, $V \nabla^2 \psi$, $U \phi'$, $V \phi'$, and $[(U^2 + V^2)/2]$ by multiplying the gridpoint values of the factors. Finally, these products are reexpanded in spherical harmonics.

In a fully pseudospectral treatment, we omit the step of expanding the nonlinear products in basis functions and instead advance the calculation in time on the grid point values. However, here it is better to advance the spectral coefficients in time. The reason is that the *linear* terms in the shallow water wave equations — including the critical terms that we wish to treat implicitly — involve the Laplacian operating on an unknown. If we use the spherical harmonics rather than the equivalent cardinal functions as the unknowns, then the Laplacians are reduced to factors of $-n(n+1)$ multiplying the coefficients.

The end result is to convert the shallow water wave equations to the set of ordinary differential equations in time:

$$-n(n+1)\frac{d\psi_{mn}}{dt} = \frac{1}{2}\left\{\langle imA_m, P_n^m\rangle - \left\langle B_m, \frac{dP_n^m}{d\mu}\right\rangle\right\} \tag{18.66a}$$

$$+ \left[n(n-1)D_{mn}\chi_{m,n-1} + (n+1)(n+2)D_{m,n+1}\chi_{m,n+1} - V_{mn}\right]$$

$$-n(n+1)\frac{d\chi_{mn}}{dt} = \frac{1}{2}\left\{\langle imB_m, P_n^m\rangle + \left\langle A_m, \frac{dP_n^m}{d\mu}\right\rangle\right\} \tag{18.66b}$$

$$- \left[n(n-1)D_{mn}\psi_{m,n-1} + (n+1)(n+2)D_{m,n+1}\psi_{m,n+1} + U_{mn}\right]$$

$$+ n(n+1)\left[\frac{1}{2}\langle E_m, P_n\rangle + \phi_{mn}\right]$$

$$\frac{d\phi_{mn}}{dt} = -\frac{1}{2}\langle imC_m, P_n^m\rangle - \left\langle D_m, \frac{dP_n^m}{d\mu}\right\rangle + \Phi n(n+1)\chi_{mn} \tag{18.66c}$$

where $\mu = \cos(\theta)$, and $A_m(\mu)$, $B_m(\mu)$, $C_m(\mu)$, $D_m(\mu)$, and $E_m(\mu)$ are the coefficients of the longitudinal Fourier series of the nonlinear terms: $U\nabla^2\psi$, $V\nabla^2\psi$, $U\phi'$, $V\phi'$, and $(U^2 + V^2)/2$, respectively, and the inner product is

$$\langle a(\mu), b(\mu)\rangle \equiv \int_{-1}^{1} \frac{a(\mu)\,b(\mu)}{1 - \mu^2}\,d\mu \tag{18.67}$$

The inner products are evaluated by Gaussian quadrature using the model grid points, which are the roots of the Legendre polynomials, $P_n(\mu)$.

There are many variants to the semi-implicit time-stepping, but the important point is that all terms are treated explicitly in (18.66) except for $\phi_{mn}(t)$ in the divergence equation, (18.66b), and $\chi_{mn}(t)$ in the height equation, (18.66c). More details are given in Bourke (1972, 1974) and Bourke et al. (1977). A well-documented and intensively tested shallow water code is available from the National Center for Atmospheric Research. The same algorithm is used in the three-dimensional climate model, the CCM3, which is publicly available at http://neit.cgd.ucar.edu/cms/ccm3/.

The shallow water wave equations are inviscid, but it is trivial to add viscous damping. The viscosity operator is the Laplacian and the spherical harmonics are its eigenfunctions.

18.20   Fronts and Topography: Smoothing/Filters

18.20.1   Fronts and Topography

Fluid flows spontaneously develop narrow, curved regions of very large gradients. These regions are known as "fronts" in geophysics and "shocks" in aerospace and plasma physics. The frontal zones narrow to jump discontinuities in the limit of zero viscosity. For physically reasonable viscosities, which are very small, the frontal zones are much narrower than the separation between grid points. Therefore, it is necessary to apply some kind of filtering to prevent computational catastrophe.

Another curse is peculiar to meteorology. The topography of the earth's surface varies on a wide variety of scales including scales far below that which can be resolved by a global model. Worse still, the topography has a discontinuous first derivative at the coasts where the non-zero slope of land abruptly gives way to the zero slope of the oceans. (Of course, the ocean is continually rippled by waves, but on the scale of a global forecasting model,

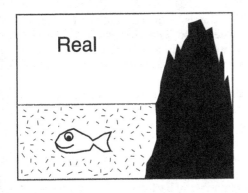 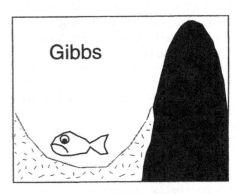

Figure 18.13: Left: The real world. The lower boundary for an atmospheric model is the jagged surface of the land (black) plus the flat surface of the ocean (confetti-shading). The boundary has a discontinuous slope at the coast. Right: "Gibbs" world. When the lower boundary of the atmosphere is approximated by a finite sum of spherical harmonics, large valleys are created in the ocean. The frowning fish had better learn to breathe air!

these wind-induced ripples are negligible.) It follows that if the topography is expanded as a series of spherical harmonics, the expansion will have only an algebraic rate of convergence with large Gibbs ripples. (These ripples are sometimes called "spectral ringing".) Even at high resolution, the partial sums of the topography may have undulating valleys one hundred meters deep in the oceans, as illustrated in the cartoon Fig. 18.13.

Both fronts and topography require smoothing or filtering of spectral series. For fronts, the filtering is applied at every time step as an artificial dissipation. Since front problems are intimately connected with spectral blocking, aliasing instability, and other difficulties described at length in Chapter 11, we shall concentrate on the difficulty peculiar to meteorology, which is the highly irregular lower boundary of the fluid. The topography needs to be filtered only once in a pre-processing step. However, the mechanics of making good filters is similar for both once-only and every-step filtering.

18.20.2   Mechanics of Filtering

If $Q(\lambda, \theta)$ denotes a function on the surface of a sphere with spherical harmonic coefficients Q_n^m, then its filtrate Q_F can be represented in spectral space as

$$Q_F(\lambda, \theta) \equiv \sum_{n=0}^{\infty} \sum_{m=-n}^{n} \sigma_{nm} Q_n^m Y_n^m(\lambda, \theta) \tag{18.68}$$

where the σ_{nm} are the filter coefficients. Q_F can also be represented in physical space as an integral:

$$Q_F(\lambda, \theta) \equiv \int_0^{2\pi} d\lambda' \int_0^{\pi} \sin(\theta')d\theta'\, w(\lambda', \theta')\, Q(\lambda', \theta') \tag{18.69}$$

where the primed coordinates denote latitude and longitude in a coordinate system that has been *rotated* so that the point (λ, θ) is the north pole of the rotated coordinate system.

The reason for this apparently complicated tactic of a different coordinate system inside the integral for each point is justified by a theorem proved by Sardeshmuhk and Hoskins

(1984): if the weight function $w(\lambda', \theta')$ is "isotropic", that is, a function only of the distance from the point where Q is evaluated inside the integral to the point where Q_F is being computed, then w is a function only of θ' in the rotated coordinate system. Furthermore, if the filter is isotropic, they prove that the weights σ_{nm} must be a function of n only, that is, must be the same for all spherical harmonics of the same subscript.

Sardeshmukh and Hoskins note that *truncation* of a spectral series is just a special case of filtering in which all the filter weights are either one or zero. The triangular truncation of order N, usually denoted as "TN", is the isotropic filter

$$\sigma_{nm} = \left\{ \begin{array}{cc} 1 & n \le N \\ 0 & \text{otherwise} \end{array} \right. \quad \text{[triangular truncation]} \tag{18.70}$$

The rhomboidal truncation, which keeps the same number of basis functions for each zonal wavenumber [superscript] m, is *not* isotropic — another reason why it has fallen from grace.

For an isotropic filter, the weight function varies only with θ' in the integral representation of Q_F. It can therefore be expanded as a one-dimensional series of ordinary Legendre polynomials (Sardeshmukh and Hoskins, 1984):

$$w(\theta') = \frac{1}{4\pi} \sum_{n=0}^{\infty} \sigma_n (2n+1) P_n(\cos(\theta')) \tag{18.71}$$

In the limit that all $\sigma_n = 1$, this is just the Legendre series for the Dirac delta function. When the series is truncated, the weight function is a finite Legendre approximation to the delta-function (Fig. 18.14).

Unfortunately, this approximation for finite N is rather ugly. It resembles a Bessel function with a maximum at the origin and oscillations that decrease very slowly with radius. Thus, values of Q far from the point (λ, θ) have a strong influence on the values of Q_F at that point. A weight function that decays *rapidly* and *monontonically* is much more reasonable than one with wiggles and slow decay, as produced by truncation.

In this sense, truncating a series is a rather stupid idea: a filtered approximation, if the filter weights are chosen properly, is much more reasonable. But how are the weights to be chosen?

18.20.3 Spherical splines

Truncation of a spectral series generates an approximation which is the best possible in a least-squares sense. The reason that the truncated series is unsatisfactory is that the approximation is very wiggly. Ocean valleys are both ugly and unphysical!

The remedy is to compute an approximation which minimizes a "cost" function that not only penalizes least-square error, but also excessive wiggles. Spherical splines of order k are a weighted sum of spherical harmonics which minimizes the cost function

$$\mathcal{J} \equiv \int_{\Omega} d\Omega \left(Q(\lambda, \theta) - Q_N(\lambda, \theta) \right)^2 + \alpha \int_{\Omega} d\Omega \left(\nabla^{2k} Q_N \right)^2 \tag{18.72}$$

where Ω denotes the surface of the sphere and the approximation Q_N is

$$Q_N = \sum_{n=0}^{N} \sum_{m=-n}^{n} a_n^m Y_n^m(\lambda, \theta) \tag{18.73}$$

When the parameter $\alpha = 0$, the cost function is the usual mean-square error and Q_N is the unfiltered truncation of the series. As $\alpha > 0$ increases, the cost function depends more and more on the smoothness of the approximation since wiggles imply large coefficients for high degree spherical harmonics. Recalling that

$$\nabla^2 Y_n^m = -n(n+1)Y_n^m \tag{18.74}$$

it follows that high degree coefficients contribute disproportionately to the second term in the cost function. The cost can only be minimized by reducing the wiggles, that is, by reducing the large n coefficients, even though this slightly increases the mean-square error. One can prove that the minimum of the cost function is rigorously given by

$$a_n^m = \frac{b_n^m}{1 + \alpha n^{2k}(n+1)^{2k}} \tag{18.75}$$

where the b_n^m are the usual unfiltered spherical harmonic coefficients.

Lindberg and Broccoli(1996) have applied spherical splines to topography. Wahba(1990) reviews the general theory. The method can be generalized by replacing a single power of the Laplacian in the second term by a weighted sum of Laplacians to make the weights in Eq.(18.75) almost anything that one pleases. Almost any reasonable choice — Lindberg and Broccoli set $k = 1$ so that the smoothness penalty was the mean-square norm of the Laplacian — will produce a smoother approximation than truncation.

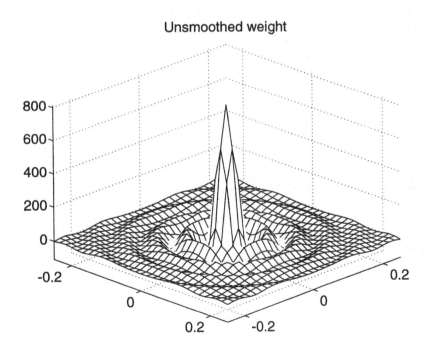

Figure 18.14: Mesh plot of the weight function for an unfiltered truncation of a spherical harmonics series. ($N = 100$.)

18.20.4   Filter Order

It is convenient to describe the properties of a filter in terms of a continuous function σ such that the discrete weights are given by σ for arguments ranging between zero and one:

$$\sigma_n = \sigma\left(\frac{n}{N+1}\right), \qquad n = 0, 1, 2, \ldots, N \tag{18.76}$$

Definition 41 (Filter Order) *A filter is said to be of* ORDER *"p" if*

$$\sigma\left(\frac{j}{N}\right) \sim O\left(\frac{1}{N^p}\right), \qquad N \to \infty, \; fixed \; j \tag{18.77}$$

or equivalently, if $1 - \sigma$ *is an analytic function that has a* $(p-1)$-st *order zero at the origin.*

The concept of the order is important for two reasons. First, if the function is infinitely differentiable so that its spectral series is converging exponentially fast, then it is both necessary and sufficient for the filter order to be p so that the errors produced by the filtering are $O(1/N^p)$ where N is the truncation. Put another way, it is desirable to use a filter of high order to avoid throwing away the advantages of a high order method when computing smooth solutions. Second, when the solution is *not* smooth and its unfiltered series converges only at an algebraic rate, it has been shown that, at least in one dimension, one can recover the true solution to within $O(1/N^p)$ by using a filter of order p at any fixed distance away from the singularity of the solution (Vandeven, 1991).

The bad news is that the proportionality constant in Vandeven's theorem is proportional to the p-th power of distance from the singularity, so that filtering, at least filtering through weighted partial sums, is unable to avoid $O(1)$ errors in the immediate neighborhood of the front or shock. This is not too surprising because a slight shift in the location of a jump discontinuity implies that a small region that should be on the low side of the jump is placed on the high side, creating a pointwise error equal to the magnitude of the jump. Away from the discontinuity where the solution is smooth, however, we can filter the slowly convergent series to recover the function to arbitrarily high order — if and only if we use a high order filter.

Vandeven shows that to be truly p-th order, a filter $\sigma(\zeta)$ must have also have a $(p-1)$-st order zero at $\zeta = 1$. Spherical smoothing splines, even for large order k, *technically* flunk these conditions. So also does the exponential filter favored by Hoskins(1980),

$$\sigma^{(exponential)}(\zeta) \equiv \exp(-\text{constant}(n(n+1))^k \, \zeta) \tag{18.78}$$

If, however the filter weights are exponentially small for the high n coefficients, the $\zeta = 1$ conditions are *approximately* satisfied. Spherical splines (for high powers k of the Laplace operator) and the exponential filter thus work as high order filters *in practice*.

Vandeven himself devised a filter which satisfied all the conditions of his theorem exactly. Unfortunately, his "ultimate" filter is an incomplete beta function. Boyd(1996c) showed that there is a close relationship between Vandeven's filter and the two hundred-year old series acceleration method of Euler. He also demonstrated that both can be approximated quite accurately by an analytical function that satisfies Vandeven's conditions exactly, the "Erfc-Log" filter:

$$\sigma_{Erfc-Log}(\theta; p) \equiv \frac{1}{2}\text{erfc}\left\{2p^{1/2}\,\bar{\theta}\sqrt{\frac{-\log(1 - 4\bar{\theta}^2)}{4\bar{\theta}^2}}\right\} \tag{18.79}$$

where

$$\bar{\theta} \equiv |\theta| - \frac{1}{2} \tag{18.80}$$

18.20.5 Filtering with Spatially-Variable Order

Boyd(1995e) shows that the optimum order p for a function with a discontinuity is an increasing function of distance from the singularity. This is consistent with the philosophy of Flux-Corrected Transport algorithms for flows with shocks: low order method around the shock, a higher order method farther away. For a Fourier series with a spatial period of 2π, the optimal order is

$$\boxed{p_{optimum}(x) = 1 + N\,\frac{d(x)}{2\pi}} \qquad (18.81)$$

where $d(x)$ is the distance from point x to the singularity. (This distance is calculated by interpreting the interval $x \in [0, 2\pi]$ as a circle.)

As a one-dimensional example, consider the "sawtooth" function defined by

$$\text{Sw}(x) \equiv \begin{cases} x/\pi & x \in [-\pi, \pi] \\ \text{Sw}(x + 2\pi m) & |x| \geq \pi, m = integer \end{cases} \qquad (18.82)$$

which has the very slowly converging Fourier series

$$\text{Sw}(x) \equiv \frac{2}{\pi} \sum_{j=1}^{\infty} \frac{(-1)^{j+1}}{j} \sin(jx) \qquad \forall\, x \qquad (18.83)$$

Fig. 18.15 shows filtering of both fixed order and spatially-varying order. With 100 terms, the accuracy is greater than $1/100{,}000$ over 90% of the spatial interval. Fig. 18.16 is a zoom plot showing a small portion of the spatial interval. It is indeed possible to recover spectral accuracy outside of a small neighborhood of the discontinuity. The best filter is of spatially-varying order and is of high order away from the discontinuity.

18.20.6 Topographic Filtering in Meteorology

A wide variety of filters have been employed to smooth topography for spherical harmonics models. Because they are high order filters, the exponential filter of Hoskins(1980) and Sardeshmukh and Hoskins(1984) and the spherical splines of Lindberg and Broccoli(1996) look good, at least when the tunable order parameter is sufficiently large. Bouteloup(1995) employed an optimization approximation very similar to spherical splines, but with more elaborate smoothness penalties. In contrast, Navarra, Stern and Miyakoda's (1994) Cesaro filter is a low order filter and therefore tends to smooth even the largest spatial scales too much. However, Navarra *et al.* nevertheless obtained a smoothed topography that was much more satisfactory than the truncated, unfiltered series, which emphasizes that for a function with a discontinuity or discontinuous slope, almost *any* filtering is better than none.

The spurious ocean valleys have been especially annoying to numerical modellers because darn it, one really ought to be able to represent a flat ocean surface as a flat surface. A child with a ruler and a pencil can draw an approximation much better than the truncated spherical harmonics series! This suggested to Navarra, Stern and Miyakoda(1994), Bouteloup(1995) and Lindberg and Broccoli(1996) that it would be desirable to employ a non-isotropic filter which is strong over water, and weak over land. Lindberg and Broccoli, for example, restricted the integral of the Laplacian to water only so that the approximation was penalized for oscillations over water, where it should be flat, but not over land, where the oscillations might represent the real hills and valleys.

This differential land/water filter is sound strategy. However, distance from the singularity — for topography, it is distance from the coast where the slope of the atmospheric boundary is discontinuous — also should control the order p of the filter.

Table 18.7: Filtering and Smoothing on the Sphere

References	Comments
Hoskins(1980)	exponential filter
Sardeshmukh&Hoskins(1984)	Isotropy Theorem; exponential filter
Hogan&Rosmond(1991)	comparison of several filters; applied Lanczos filter to Navy forecast model
Navarra&Stern &Miyakoda(1994)	2D isotropic filter (Cesaro); 2D physical space filter applied to oceans only
Bouteloup(1995)	optimization method with multiple penalty constraints
Lindberg&Broccoli(1996)	spherical spline; orography & precipitation
Gelb(1997) Gelb&Navarra(1998)	Gegenbauer regularization, restricted to discontinuities parallel to either latitude or longitude lines

Clearly, there is a need for further experimentation and creativity. Gelb (1997) and Gelb and Navarra(1998) used an ingenious regularization procedure which replaces a high order spherical harmonic series by a low order Gegenbauer polynomial series. Unfortunately, their method requires that the discontinuities lie on the walls of a rectangle or a union of rectangles. In its present state, the Gegenbauer method is not sufficiently general to accomodate real topography. However, this strategy, which can be applied to other spectral series, is still under rapid development and perhaps its present deficiencies will be overcome. In any event, it shows that the algorithm developers have not yet run out of ideas.

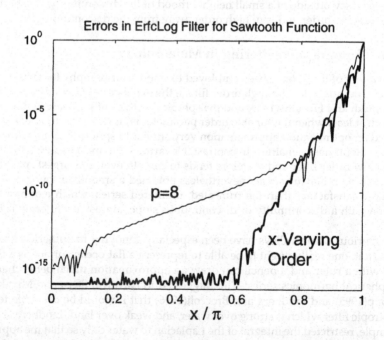

Figure 18.15: Errors in ErfcLog-filtered, 100-sine approximations to the piecewise linear or "sawtooth" function. Thin line: fixed filter order $p = 8$. Thick line: optimal, spatially-varying p.

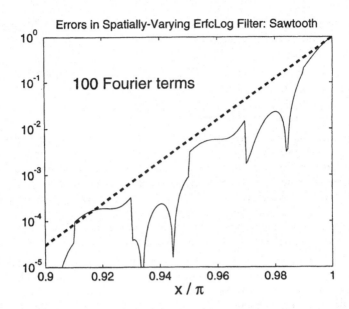

Figure 18.16: Zoom plot of the errors in ErfcLog-filtered (one hundred term) sine series for the sawtooth function. The discontinuity of $Sw(x)$ is at $x = \pi$ (right edge of the graph). The thin solid curve shows the error when the order p is varied with distance from the discontinuity in an optimal way. The thick dashed curve is an upper bound or "envelope" for the errors, showing (on this log-linear plot) that the error falls exponentially with distance from the singularity at $x = \pi$.

18.21 Resolution of Spectral Models

Filtering and dealiasing raise an obvious question: How small are the features that a spectral model can resolve with a given N (Laprise, 1992, and Lander and Hoskins, 1997)?

The central argument of Lander and Hoskins is that a graph of spectral coefficients versus degree n can be divided into three regions. (For a Fourier series, n is wavenumber; for spherical harmonics, n is the subscript of the harmonic Y_n^m.) These regions are:

1. "Unresolved scales" $< ----- >$ "Dealiasing pad"

$$N < n < \frac{3}{2} N \qquad (18.84)$$

2. "Unbelievable scales" $< ----- >$ "Discretization-addled" wavenumbers

$$N_P < n < N \qquad (18.85)$$

3. "Believable scales"

$$n \leq N_P \qquad (18.86)$$

The "dealiasing pad" or what Lander and Hoskins call simply the "unresolved scales" arise because the usual weather forecasting model has roughly three-halves as many points in each of latitude and longitude as the truncation N. This allows the integrals of the

Galerkin method to be evaluated without error and is consistent with the Orszag Two-Thirds Rule for removing aliasing error. However, these wavenumbers with $n > N$ are not retained in the truncation.

Lander and Hoskins' "unbelievable scales" are those wavenumbers which are kept in the trucation, but are untrustworthy. Unfortunately, spectral coefficients near the truncation N are corrupted by a wide variety of errors. The schematic illustrates three such error-mechanisms, labelled in italic fonts.

First, if a time-dependent hydrodynamics model is not strongly damped, it tends to develop accumulate noise near the truncation limit, causing the graph of spectral coefficients to develop an upward curl, as shown by one of the dashed curves in Fig. 18.17. Second, if the model incorporates a strong artificial viscosity, then the spectral coefficients may be damped to unrealistically small amplitude as shown by the downward curling dashed curve. If damping is chosen correctly, the computed coefficients will curl neither up nor down, but will be of roughly the correct order of magnitude. However, " correct magnitude" is as unsatisfactory to the arithmurgist as to the brain surgeon.

In addition to these sources of error, there is a third coefficient-addling influence which is always present, even for one-dimensional boundary value problems: the "discretization error" E_D defined by Definition 9 in Chapter 2 and illustrated in Tables 3.1 and 3.2. This error arises because the coefficients that we compute are always influenced by the coefficients that are not computed. Consequently, all the calculated coefficients have *absolute* errors which are the order of magnitude of the truncation error E_T, which is the magnitude of the uncomputed coefficients for $n > N$. Because the low degree coefficients are large compared to the truncation error, their relative error is small and these spectral coefficients are "believable". The coefficients near the truncation limit, however, are of the same magnitude as the discretization error E_D, and therefore have such large *relative* errors as to be "unbelievable".

Lander and Hoskins suggest two strategies. One is to filter the wind and pressure, etc., that are input to the physics parameterizations to include only the "believable" scales. This will increase cost, but reduce noise.

Their second strategy is that the physical parameterizations of chemistry, hydrologic cycle and so on should be computed only on a coarse grid restricted to these believable scales. (This explains the notation for the boundary of these wavenumber, N_P — "P" for parameterizations and also "P" for physics.) Instead, it is the usual practice to compute the chemistry and physics on the entire grid. Although the ozone concentration on such a fine grid could be in principle be expanded with harmonics up to $N = (3/2)N$, only coefficients in the triangular truncation TN are ever calculated. Since the non-hydrodynamic calculations account for at least half the total running time, one could gain much by restricting chemistry and physics to the "believable" scales only.

One drawback of Lander and Hoskins' second strategy is the need to interpolate to and from the coarse grid [where the parameterizations are evaluated] to the fine grid [used to compute nonlinear products]. However, the parameterizations are so expensive that it seems likely that there would be significant savings even so. Furthermore, the fine scales of the ozone, water vapor and so on force the dynamics at "unbelievable" scales; the high resolution parameterizations may actually be reducing accuracy! However, neither of their suggestions has been tested in a real code.

One important practical issue is: What is N_P, the highest believable wavenumber? Lander and Hoskins discuss five choices. Their preference is for a length scale r_G which is defined in terms of the "point spread function". [discrete approximation to the Dirac delta-function] as the distance from its global maximum to the first minimum . This is equivalent to

$$N_P \approx 0.63N \approx \frac{(3/2)}{2.4} N \qquad (18.87)$$

This would reduce the total number of points for the non-hydrodynamic parameterizations by a factor of $5.76 = 2.4^2$.

However, Lander and Hoskins hedge, and suggest that sometimes a larger N_P may be desirable for safety's sake. The truth is that "believability" is likely to be highly problem-dependent and model-dependent. Nevertheless, it is only by understanding that there are three regions in the wavenumber spectrum that we can hope to make rational choices in model architecture, filters and so on.

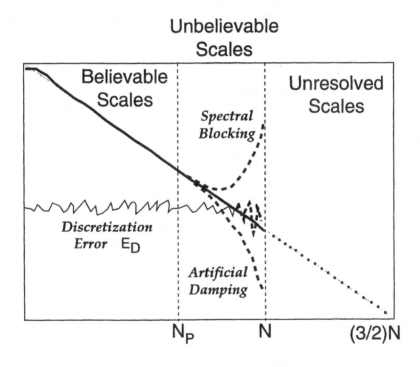

Figure 18.17: Schematic of Fourier or spherical harmonics coefficients versus degree N. The thick solid line shows the exact coefficients for $n \leq N$; the dotted line shows the exact coefficients which are neglected when the series is truncated at degree N. The "believable" coefficients are accurate; the length scales of the corresponding spherical harmonics define the "believable scales". The "unbelievable" scales are corrupted by numerical difficulties. Three villians (labelled in italics) are illustrated by thick dashed curves, which show how the computed coefficients are distorted by each pathology. The absolute errors for the discretization error are roughly the same for all coefficients in the truncation (thin jagged line), but the *relative* errors are large only for $n \approx N$.

18.22   Vector Basis Functions: Vector Spherical Harmonics & Hough Functions

When solving a *system* of equations for a *vector* of unknown fields, there is no reason why one cannot use vector basis functions. The eigenmodes of the linearized system will normally be vectors themselves. If the modes differ drastically — for example, the gravity wave oscillations of the shallow water wave equations have very high frequencies in comparison to the Rossby modes — than it may be possible to greatly improve numerical efficiency by choosing vector basis functions that mimic the important modes. When we expand all the unknowns in the shallow water or primitive equations in spherical harmonics, it is not possible to discriminate in any way between the gravity and Rossby modes. One *can* discriminate using vector basis functions.

Several alternatives to ordinary (scalar) spherical harmonics have been discussed. Swarztrauber (1981, 1993) and Swarztrauber and Kasahara (1985) have described two different varieties of "vector spherical harmonics". The basic strategy is to transform the spherical components to the corresponding Cartesian components (which are always continuous) and then expand in (scalar) spherical harmonics. It is convenient to label Swarztrauber's new basis set the "vector spherical harmonics" even though he takes pain to note that this term has been previously applied to a different set: the eigenfunctions of the *vector* Helmholtz equation. Swarztrauber has written library software ("SPHEREPACK") to compute his new basis functions and has applied them to Laplace's tidal equations, which are the linearized shallow water wave equations.

Shen and Wang(1999) have recently proposed a new vector spherical harmonic/Legendre algorithm for the primitive equations. The theoretical analysis and numerical experiments are promising, but their algorithm has not yet been used for meteorological applications.

The second approach, suggested by T. W. Flattery for data analysis but developed for time-integration by A. Kasahara, is to use the vector eigenfunctions of Laplace's tidal equations as the basis set. In three dimensions, separation-of-variables shows that after the vertical dependence has been eliminated, the two-dimensional equations are simply Laplace's tidal equations with a different equivalent depth parameter for each vertical mode. Thus, although the basis set is two-dimensional, these functions of latitude and longitude become the three-dimensional normal modes when multiplied by the proper vertical structure function.

The motive for this choice of basis set is that these so-called Hough functions are intimately connected with the physics. In particular, to filter gravity waves, which are "noise" insofar as the forecast is concerned, one may simply omit the gravity wave Hough modes from the basis set.

There are alternatives for eliminating the gravity waves; the most famous is the quasigeostrophic approximation, which drops terms from the shallow water wave equations such that the simplified partial differential equations have only Rossby waves as their free modes of oscillation. Unfortunately, the errors in quasi-geostrophy (and its generalizations) are large at low latitudes. Furthermore, deleting terms from the differential equation is an all-or-nothing approach: the only choice is between models that allow all gravity modes or none at all.

In contrast, Hough models allow selectivity. There are a few gravity waves with relatively low frequencies which can be retained in a trucation of the Hough basis, but are always excluded by the quasi-geostrophic approximation.

One obvious difficulty is that the Hough functions are the eigenmodes of the *linearized* equations, but the dynamics of the global atmosphere is strongly *nonlinear*. One may well ask: Doesn't the nonlinearity produce such strong mode-mode coupling, such strong distortions of the normal modes themselves, as to invalidate the physical interpretation of the Hough modes as the gravitational and Rossby oscillations of the atmosphere?

The answer to this question is: Yes, but not so as to diminish the usefulness of Hough functions as a basis set. The reason is that because of the great differences between Rossby and gravity waves, the couplings produced by nonlinearity are primarily between modes in the same class — Rossby with Rossby, gravity with gravity. What is important for selectively filtering gravity waves is not whether a particular Rossby Hough function closely resembles a free oscillation as observed in the real atmosphere. Rather, what matters is that even with nonlinearity, dissipation, and other complications, the Rossby Hough functions are, to a good approximation, a complete set for expanding the low-frequency, weakly divergent motions that dominate weather and climate and this set is orthogonal to the gravity Hough functions. The Rossby-to-gravity coupling is weak.

Kasahara (1977, 1978) has clearly demonstrated the feasibility of integrating the hydrodynamic equations using Hough functions. It is important to note that one can resolve a slow flow with roughly one-third the number of degrees of freedom of a conventional spherical harmonics model since there are two gravity modes for every Rossby Hough function. However, like Swarztrauber's equally feasible vector basis set, Kasahara's Hough functions have not yet caught on, except for the middle atmosphere model of Callaghan *et al.*(1999).

There is a technical reason for this: when the number of vertical levels is large, the eigenvalues for the higher vertical modes (what meteorologists call the "equivalent depths" of the modes) become very small. This disrupts the clear separation between Rossby and gravity waves which exists for the lowest few vertical modes and is the motive for choosing a Hough basis in the first place.

18.23 Radial/Vertical Coordinate: Spectral or Non-Spectral?

18.23.1 Basis for Axial Coordinate in Cylindrical Coordinates

For cylindrical coordinates, the z-axis is *not* cyclic, so Chebyshev polynomials (if the cylinder is bounded in z) or rational Chebyshev functions (if the cylinder is unbounded) are best. Thus, a cylindrical basis is usually double Chebyshev (in r and z) and Fourier in the angle θ.

18.23.2 Axial Basis in Toroidal Coordinates

Toroidal coordinates are similar to cylindrical except that the vertical axis is closed upon itself to form a circle instead of an infinite line. Because this closure-into-a-circle forces the "toroidal angle" λ to be cyclic, one should always use an ordinary Fourier series in this coordinate. Thus, a toroidal basis is Fourier in θ, Fourier in λ, and Chebyshev in the third coordinate, which is radial distance from the center of the torus.

18.23.3 Vertical/Radial Basis in Spherical Coordinates

Although spherical harmonics have been extremely popular in atmospheric science, such models almost invariably use finite difference approximations in the vertical. It is not that the idea of a vertical series has been tried and found wanting, but rather it has not been tried. Francis(1972) and Hoskins(1973) explored Laguerre functions (for an atmosphere of semi-infinite depth), but found that explicit time step limits would be very severe. Machenhauer and Daley(1972, 1974) and Eliassen and Machenauer (1974) experimented with Legendre polynomials in depth, but had to use *ad hoc* modifications; this work was never published in a journal.

Kasahara and Puri(1981) and Mizzi, Tribbia and Curry(1995) tried the eigenfuctions of the so-called "vertical structure equation" as the basis functions in height. Mizzi *et al.* found that the rate of convergence was lamentably slow compared to the expansion of test solutions in Chebyshev or Legendre polynomials. This is hardly surprising: using normal modes for a problem with numerical boundary conditions is akin to using a Fourier basis for a non-periodic problem and gives the same slow, algebraic rate of convergence. (Indeed, for idealized temperature profiles, the normal modes *are* sine functions.)

And there the matter has rested except for Shen and Wang(1999). In contrast, Marcus and Tuckerman (1987a, b), Mamun and Tuckerman (1995) and Glatzmaier(1984, 1988) have computed the flows in a rotating spherical annulus using Chebyshev polynomials in radius. Glatzmaier's algorithm was so successful that it has been used for studies of stellar dynamos, mantle convection, and the atmosphere of a gas giant planet totalling at least thirty-five articles and book chapters. Because Glatzmaier's algorithm has been used so widely, it is worth briefly describing here.

18.24   Stellar Convection in a Spherical Annulus: Glatzmaier (1984)

This model simulates stellar convection by solving the anelastic equations of motion for a spherical shell heated from below. The inner wall is at 0.53 of the solar radius and the outer wall at 0.93 of the solar radius so that there are seven scale heights within the shell.

The latitude-longitude basis was spherical harmonics in a triangular T31 truncation. The radial basis was Chebyshev polynomials up to and including $T_{16}(r)$.

The time-integration scheme was semi-implicit. The explicit Adams-Bashforth scheme was applied to all terms except diffusion, which was treated implicitly with the Crank-Nicholson scheme. Without magnetic forces, the cost was about 1.0 sec per timestep on a CRAY-1.

The implicit Crank-Nicholson scheme does not require splitting because the spherical harmonics are *eigenfunctions* of the *horizontal part* of the *diffusion* operator. Because of this, it is only necessary to solve a *one-dimensional* boundary value problem in *radius only* to implement the Crank-Nicholson method. Ordinarily, this would require storing the LU decomposition of $O(L^2)$ Chebyshev collocation matrices, one for each horizontal basis function. However, the eigenvalue of the diffusion equation is $-n(n+1)$, independent of the zonal wavenumber m. Consequently, Glatzmaier is only obliged to perform $(N+1)$ LU decompositions in a preprocessing stage and then store these thirty-two 17 x 17 matrices throughout the run. The cost of the backsolves for the Crank-Nicholson treatment of diffusion is $O(L^2 N^2)$ whereas the number of gridpoints is only $O(L^2 N)$, so we have lost a factor of N relative to a finite difference method. This will be only a small part of the overall work – this is a *very* complicated set of equations because of the magnetic and Coriolis forces – unless N is a good deal larger than 17.

The smallness of N is a tribute to the accuracy of pseudospectral methods. The density varies by a factor of $\exp(-7)$ over the shell,[4] and there are viscous boundary layers at both the top and the bottom, but seventeen grid points in radius suffice. The model does not use a variable grid because the built-in quadratic stretching of the Chebyshev grid is sufficient.

---

[4]This is the physical meaning of the statement that the shell is "seven scale heights" thick.

Glatzmaier reports that a test calculation with $N = 65$ differed from the results for $N = 17$ by only 2 % after 300 time steps.

Although meteorological spherical harmonics codes are almost always dealiased, the stellar convection model is *aliased*; no 2/3's rule here! Glatzmaier states the position of that large group of modellers who ignore aliasing: "An unaliased method, which would require significantly more computer time and memory, would ensure that the error due to the truncation of the harmonics is normal [i. e., orthogonal] to the retained set of harmonics. However, a truncation error would still exist... If the two methods [aliased and unaliased] produce solutions that are not negligibly different, both solutions are bad. Only when enough harmonics are retained so that both methods produce negligibly different solutions are the solutions good".

The incompressibility constraint is enforced by decomposing the three dimensional velocity into

$$\vec{v} = \vec{\nabla} \times \vec{\nabla} \times (W\hat{r}) + \vec{\nabla} \times (Z\hat{r}) \qquad (18.88)$$

where $\hat{r}$ is a unit vector in the radial direction and $(\vec{\nabla}\times)$ denotes the curl. Through elementary vector identities, one may verify that the divergence of (18.88) is zero. The functions W and Z are *scalar*, and generate the "poloidal" and "toroidal" components of the flow. The incompressibility has reduced the number of scalar functions needed to specify the flow from three to two, and similarly for the magnetic field, which is also nondivergent (unless the Sun has magnetic monopoles!) This poloidal/toroidal decomposition is the three-dimensional, spherical analogue of the usual vorticity streamfunction formulation, but the pressure is *not* eliminated and remains one of the six unknowns: W, Z, the two similar magnetic functions, p, and the entropy s.

The continued presence of the pressure as an unknown creates a problem. At the rigid boundaries, the constraints of no normal flow and no stress imply *four* conditions on W, which satisfies only a *second* order equation in r. However, the pressure p also satisfies a second order equation, and there are no intrinsic boundary conditions on p. Glatzmaier therefore treats part of the radial diffusion for W explicitly by including the terms that depend on p. He then demands that p have whatever values at the boundary are necessary to make the explicit terms for W equal to zero so that W remains consistent with the four boundary conditions upon it.

This sounds awkward when expressed in words; it is not a lot of laughs for the programmer either. The pressure is a perpetual nuisance in models which lack a prognostic equation for p. Pressure gradients appear in the governing equation, just as gradients of the velocity do, but we have physical boundary conditions only on $\vec{v}$. The pressure is a passive quantity which varies so that the incompressibility constraint is always satisfied, and we have to live with this awkwardness.

18.25   Non-Tensor Grids: Icosahedral Grids and the Radiolaria

Spherical harmonics, in which every part of the globe directly influences every other part, have always alarmed proponents of parallel computing because of the need to share lots of grid point values between different processors of a massively parallel machine. This has not become a difficulty as early as expected because even at T639 resolution, preliminary tests have shown that Legendre transforms cost only 19% of the overall CPU time using 48 vector processors. (The model at T213 resolution runs quite efficiently on 500 processors of a Cray T3E.) Nevertheless, there has been a search for finite difference and finite element alternatives to spherical harmonics. Unfortunately, the globe imposes special difficulties even on low-order methods: there is no free lunch on the sphere.

One obvious alternative to pseudospectral methods is to simply lay down an evenly spaced grid or triangulation of the sphere and use finite differences or finite elements. Unfortunately, an evenly spaced grid and a triangulation composed of identical triangles are both impossible except at very low resolution.

The choice of an evenly spaced latitude-longitude grid creates a severe pole problem for all non-spherical harmonic schemes, both low order differences and high order spectral methods, unless (i) the time-stepping is *fully* implicit, which is prohibitively expensive or (ii) strong filtering is applied near the poles.

The other grid option is to abandon a tensor grid and instead simply spread grid points as uniformly as possible over the sphere. As reviewed by Saff & Kuijlaars(1997), this turns out to be very difficult.

The problem was identified by Leonard Euler in the middle of the eighteenth century, but it first became important in science when marine biologists on H. M. S. *Challenger* dredged up samples of microscopic ocean life a century ago. Fig. 18.18 shows the "radiolaria", *Aulonia hexagona*. At first glance, the siliceous skeleton appears composed of an even net of hexagons. One might imagine that one could circumscribe the skeleton with a globe and subdivide each hexagon into six equilateral triangles to obtain a uniform triangulation of the sphere, or compute finite differences directly on the hexagonal grid. Unfortunately, the appearance of a uniform net of hexagons is an *illusion*. W. D'Arcy Thompson (1917) shows that every such hexagonal array must have 12 pentagons inserted in the lattice.

The vertices of the five Platonic solids, in contrast, are uniformly spaced. However, the icosahedron, with twelve vertices and twenty equilateral triangles, is the Platonic solid with the largest number of sides. One can hardly hope to model the weather well with just twelve grid points!

The best one can hope for is to create an *almost* uniform triangulation. Baumgardner and Frederickson (1985) show how one can use the icosahedron as a starting point for covering the globe with little triangles. Each of the twenty sections is bounded by a triangle of identical size, but the triangles within each icosahedral face are of different sizes.

Table 18.8: Gridpoint Methods on the Sphere Bibliography

References	Comments
Taylor (1995)	proves spherical harmonics interpolation problem is insoluble; expansion must be computed by integration with extra points
Sloan(1995)	interpolation and "hyperinterpolation" on sphere
Baumgardner&Frederickson(1985)	Icosahedral/triangular grid
Heikes&Randall(1995a,b)	Twisted icosahedral grid
Ronchi&Iacono & Paolucci(1997)	New finite differences ("cubed sphere") compared with spherical harmonic code
Rančić&Purser Mesinger(1996)	Expanded cube grid: comparisons of gnomonic and conformal coordinates
Purser&Rancic(1997)	Finite difference grid based on conformal octagon
Swarztrauber&Williamson &Drake(1997)	icosahedral grid; 3D Cartesian method projected onto the surface of a sphere
Thuburn(1997)	icosahedral-hexahedral finite difference grid
Saff & Kuijlaars(1997)	Popular review of the mathematical difficulties in computing nearly-uniform grids on the sphere
Stuhne&Peltier(1999)	icosahedral grid

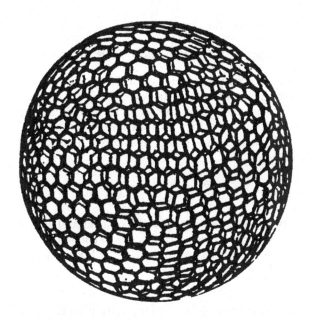

Figure 18.18: The skeleton of *aulonia hexagona*. After *Growth and Form* by W. D'Arcy Thompson (1917). Although the lattice *appears* to be composed entirely of hexagons, Thompson showed that there must be at least a dozen pentagons, too. It is impossible to embed the sphere in a lattice of identical polygons when the number of polygons is greater than 20.

These nearly-uniform grids work well — as evident in the picture of the radiolaria, the degree of non-uniformity is so small as to be almost invisible when the number of grid points is large. The rub is that the grid nevertheless *is* non-uniform and this leaves two problems: (i) programming is much more difficult than for a uniform grid and (ii) the accuracy of finite differences deteriorates because the grid is not entirely regular. In addition, recent models have been bedeviled by stability problems and spurious excitation of longitudinal wavenumber five disturbances. Heikes and Randall(1995a,b) used a "twisted" icosahedral grid which respects the parity symmetry with respect to the equator, but at the expense of more complication. Algorithms that replace the icosahedron with another Platonic solid, such as a cube or octagon, have been tried, too (Table 18.8).

All of these variants work, but all are messy and most have required some tweaking. These faults of almost-uniform grid methods are some consolation for the intricacies of spherical harmonics.

18.26 Spectral Alternatives to Spherical Harmonics, I: Robert Functions

The Fast Fourier Transform (FFT) converts grid point values of a function to its Fourier coefficients in only $O(N \log_2 N)$ operations. Since the spherical harmonics are an ordinary Fourier series in longitude, we can apply the FFT in λ, but not to the Gegenbauer polynomials in θ. This inspired serious efforts in the late 60's and early 70's to find substitutes for the latitudinal basis functions.

André Robert suggested the basis functions

$$\sin^{|m|}(\theta) \cos(n\theta) \qquad\qquad [\text{"modified Robert functions"}] \qquad (18.89)$$

These basis functions reproduce the polar behavior of the spherical harmonics. The high order zeros for large wavenumber m eliminate the "pole problem"; one can integrate using (18.89) with a fairly large time step and no special tricks. Merilees (1973a) pointed out the fatal flaw of the "Robert" functions: they are horribly ill-conditioned. The problem is that for large m, the $\sin^{|m|}(\theta)$ factor is zero except in a narrow boundary layer around the equator. Within this layer

$$\cos(n\theta) \approx 1 - \frac{n^2\theta^2}{2} \approx 1 \qquad\qquad \text{if } |\theta| \ll 1 \qquad (18.90)$$

so that the lowest few Robert functions are graphically indistinguishable for large m. This translates into large round-off errors when a function is expanded in terms of these (almost) linearly dependent functions. The result is that the Robert functions have never been employed for large-scale numerical models.

18.27   Spectral Alternatives to Spherical Harmonics, II: Parity-Modified Fourier Series

Merilees (1973b) and Orszag (1974) have discussed parity-modified Fourier series in which the latitudinal dependence of the zonal wavenumber m component of a scalar function is expanded in the form

$$f_m(\theta) = \sin^s(\theta) \sum_{n=0}^{\infty} a_n \cos(n\theta) \qquad (18.91)$$

where

$$s \equiv \begin{cases} 0 & m \text{ even} \\ 1 & n \text{ odd} \end{cases} \qquad (18.92)$$

The pseudospectral grid points are

$$\theta_i \equiv \frac{\pi(2i-1)}{2N} \qquad\qquad i = 1, \dots, N, \qquad (18.93)$$

that is, an evenly spaced grid with all points on the *interior* of $\theta \in [0, \pi]$. This implies that we may freely divide out $\sin^s(\theta)$ and then apply the FFT (a Fast Cosine Transform, actually). The coordinate singularities of the differential operators at $\theta = 0$ and π cause no major problems because we do not need to evaluate any quantities at the poles.

One obvious concern is that the spherical harmonics (and all well-behaved functions on the sphere) have m-th order zeros at the poles whereas the modified Fourier basis functions in (18.91) have at most a first order root. However, to quote Orszag (1974): "The detailed behavior expressed by the $\sin^{|m|}(\theta)$ factor is usually of little direct interest in a numerical simulation ... The above argument suggests that surface harmonic-series contain much information on the behavior of functions near the poles that is not of primary interest." Boyd (1978b) offers an amusing illustration of Orszag's point by applying the pseudospectral method with 60 basis functions to calculate the eigenvalues of the two-dimensional

Table 18.9: Spectral Alternatives to Spherical Harmonics: Bibliography
Note: "Double Fourier" denotes a Fourier basis in both latitude and longitude.

References	Comments
Robert (1966)	"Robert" function basis, but these are very ill-conditioned
Merilees (1973a)	Ill-conditioning of "Robert" basis $sin^m(\theta)\cos(n\theta)$
Merilees (1974)	Double Fourier series with high latitude filtering
Orszag (1974)	Comparison of spherical harmonics with double Fourier series
Boyd (1978b)	Comparison of Chebyshev, Fourier & Associated Legendre for boundary value and eigenvalue problems
Yee(1981)	Poisson equation on sphere; double Fourier basis
Spotz&Taylor & Swarztrauber(1998)	high-latitude-filtered double Fourier series
Shen(1999)	double Fourier basis
Cheong(2000, 2001)	double Fourier series

Laplacian, i. e. the spherical harmonics themselves, for $m = 49$. Although the 60 term approximation may have at most 61 zeros (counting multiplicity and the two roots of $\sin(\theta)$) and the spherical harmonics have $98 + (n - m)$ roots, the lowest twenty eigenvalues are calculated to within an accuracy of 1 part in 70 million.

For so-called "jury" problems, that is to say, for boundary value and eigenvalue problems, Boyd (1978b) argues that the parity-modified Fourier series is the best basis set. It is much easier to program than spherical harmonics and the accuracy does not seem poorer.

For time-dependent or "marching" problems, however, the "pole" problem returns in full force. Orszag (1974) has shown that the time-step limits for realistic calculations will be quite severe. He suggested several remedies. First, one may apply a polar filtering every ten or twenty time steps. Second, one may split the convective terms into (i) a surface harmonic of degree 2 that gives the flow over the poles and (ii) the rest of the convective terms and then treat (i) implicitly; the bulk of the terms, i. e. (ii), may be treated explicitly without instability because (ii) does not involve flow over the pole. Third, treating viscous terms implicitly — almost always necessary with spectral and pseudospectral algorithms — tends to stabilize the convective terms, too, at least for a Backwards-Euler or other implicitly-dissipative scheme.

Even so, longitude/latitude double Fourier series have been little used for time-dependent calculations in the quarter-century after Orszag's article. Part of the reason is inertia; as explained earlier, the relatively slow cost of associated Legendre transforms has not been a major liability for spectral models even at the highest resolution employed in weather forecasting at the dawn of the third millenium, $T639$.

In recent years, however, double Fourier series have made a comeback in the articles of Shen(1999), Cheong(2000, 2001) and Spotz, Taylor and Swarztrauber(1998). Spotz *et al.*, for example, revived an idea of Merilees(1974) with an improved filter, and showed that latitudinal Fourier series can be both fast and accurate when the polar filtering is strong enough.

18.28 Grid-to-Spherical-Harmonics-to-Grid Projective Filter for Time-Dependent Latitudinal Fourier Series

The ultimate polar filter is to transform grid point values on a tensor product latitude/longitude grid to a triangular truncation of spherical harmonics and then sum the spectral series to obtain a different set of grid point values. The output of the fore-and-back transform will inherit the equiareal resolution property of spherical harmonics. The pole problem is completely suppressed.

This combination of transforms has been described as both a "spherical filter" and a "projection" in the literature; we shall call it a "projective filter". The reason for the terminological confusion is that the projective filter is different from other filters in that its purpose is not to provide artificial viscosity or suppress small-scale noise and aliasing instability. Rather, the grid-to-spectral half of the projective filter is a non-dissipative transform. It is, in the usual parlance of functional analysis, a "projection" of the function $f(\lambda, \theta)$ onto the subspace spanned by the spherical harmonics. The spectral-to-grid half of the transform is merely a summation of the truncated series. The grid-to-grid transform is more than a projection (because it is a summation, too), but in some sense less than a filter (because the only filtering is a *truncation*).

The weakness of a latitude/longitude Fourier method is that projective filtering has a cost which is the same order of magnitude as an Associated Legendre transform. However, because only the unknowns themselves must be projectively filtered — no *derivatives* need be transformed — double Fourier algorithms have the potential of extending the lifetime of global spectral methods on the sphere for at least another decade.

The method which Spotz and Swarztrauber (2000) label both "naive" and "standard" is to compute both halves of the projective filter through separate Parity Matrix Multiplication Transforms (PMMT) in latitude. (All transformations in longitude are performed by the Fast Fourier Transform (FFT), which convert grid point values into functions $f_m(\theta)$ which describe how the coefficient of $\cos(m\lambda)$ or $\sin(m\lambda)$ varies with colatitude.) The adjective "Parity" means that grid point values of $f(\theta)$ are first split into components that are symmetric and antisymmetric with respect to the equator $\theta = \pi/2$ through $f^S \equiv \{ f(\theta) + f(\pi - \theta) \}$ and $f^A \equiv \{ f(\theta) - f(\pi - \theta) \}$. Each component is then projectively filtered separately; the even and odd parity filtrates are recombined by adding their grid point values in the northern hemisphere and subtracting the filtered values of f^A from the filtrate of f^S in the southern hemisphere.

Yarvin and Rokhlin(1998) noted that one could reduce the cost by 25 % by combining the forward and backward transform matrices into a single $N \times N$ matrix for small m while performing the forward and backward transforms separately for large m. (Note that the forward and backward transform matrices are rectangular with smaller dimension $(N - m)$; for the extreme of $m = N$, the transform matrices are only vectors; two vector-vector multiplies are obviously cheaper than an $N \times N$ matrix-vector multiply.)

Swarztrauber and Spotz (2000) developed the "Weighted Orthogonal Complement" projective filter which has only half the cost of the "naive"/"standard" method and only two-thirds the expense of Yarvin and Rokhlin's "direct" method. Better yet, they prove that the WOC matrices can be written as subsets of a couple of "master" matrices, reducing storage for transform/projection matrices from $O(N^3)$ to only $O(N^2)$.

For the shallow water wave equations, nine Legendre transforms are needed at each time step with a spherical harmonic basis. With a double Fourier expansion and the WOC projective filter, the transform cost is reduced to the equivalent of only three Legendre transforms.

Jakob-Chien and Alpert(1997) developed a projective filter based on Fast Multipole Method (FMM) ideas; this was improved by Yarvin and Rokhlin(1998). As with other multipole transforms, the FFM projective filter is faster than the WOC or any MMT-type method in the limit $N \to \infty$, but the cross-over point where the FMM scheme is faster is likely to be at very large N.

Spotz and Swarztrauber(2000) carefully compare all these projective filters for various N, both with and without the application of the Two-Thirds Rule. (Dealiasing truncation reduces the advantages of the newer MMT schemes relative to the "naive" method.) They

Table 18.10: Projective Filters Bibliography

References	Comments
Jakob-Chien&Alpert(1997) Yarvin&Rokhlin(1998)	Grid-to-spherical-harmonics-to-grid projective filter for longitude/latitude time-dependent double Fourier series
Swarztrauber&Spotz(2000) Spotz&Swarztrauber(2000)	"Weighted Orthogonal Complement" algorithm reduces storage by $O(N)$; faster than alternatives because algorithm mostly stays in the on-chip cache
Cheong(2001)	Approximate projective filter (hyperviscosity) which combines damping with projection

find that their WOC algorithm is faster than the "direct" and "standard" PMMT and the FMM at least up to $N = 200$. The WOC method is moderately faster than formal operation counts would suggest; the very low storage of the WOC leads to more efficient utilization of the on-chip cache memory of current machines. The WOC method also parallelizes very well.

Cheong(2001) employs an alternative: Polar filtering by a combination of fourth order and sixth order hyperviscosity. Recall that Y_n^m is an eigenfunction of the two-dimensional Laplace operator on the surface of the sphere with the eigenvalue $\lambda_n = -n(n+1)$. It follows that a dissipation which is proportional to a high power of the Laplacian can be tuned to strongly damp all spherical harmonics with $n > N$ while having only a slight effect on harmonics retained within a triangular truncation. The Fourier-Galerkin discretization of the Laplace operator is a tridiagonal matrix, so the filter is very fast.

In contrast to the Spotz-Swarztrauber and FMM methods, Cheong's hyperviscosity is not an exact projection onto spherical harmonics. Since most time-dependent models require a little scale-dependent dissipation for noise-suppression anyway, an approximate projective filter is satisfactory. However, there has not been a detailed comparison of the relative speed and accuracy of exact projective filters versus approximate filters that damp and project simultaneously.

Swarztrauber and Spotz (2000) note, "The goal of an $O(N^2 \log N)$ harmonic spectral method remains elusive; however, the "projection" method provides a new avenue of research. Perhaps the development of a fast projection [projective filter] will prove to be easier than the development of a fast harmonic transform." Sufficient progress has already been made to encourage the development of double Fourier methods on the sphere. However, as of 2000, latitude/longitude Fourier series have not yet been applied to an operational forecasting or climate model.

18.29 Spectral Elements on the Sphere

Another alternative to global spectral methods is domain decomposition. A moderately high order Legendre series is used in each subdomain, and then the pieces are stitched together as in finite elements to obtain a global approximation. The great virtue of spectral elements is that each subdomain can be assigned to a different processor of a multiprocessor machine. The articles listed in Table 18.11 show that the algorithm parallelizes very well, and seems promising for both atmospheric and oceanic modelling.

Table 18.11: Spectral Elements on the Sphere Bibliography

References	Comments
Ma(1992,1993a,b)	Ocean basin models
Iskandarani&Haidvogel &Boyd(1995)	global ocean model
Levin&Iskandarani &Haidvogel(1997)	ocean model
Taylor&Tribbia & Iskandarani(1997)	atmospheric model; shallow water equations
Curchitser&Iskandarani & Haidvogel(1998), Haidvogel&Curchitser &Iskandarani &Hughes &Taylor(1997)	Ocean modelling on a variety of massively parallel computers

18.30 Spherical Harmonics Besieged

Although spherical harmonics are the discretization of choice for the ECMWF forecasting model, the NCAR Community Climate model, and Glatzmaier's stellar convection and mantle flow simulations at the turn of the milllenium, this spectral basis is under attack on two fronts. The first assault is from massively parallel (MP) computing. The worry is that global basis sets of any kind, and spherical harmonics with their slow Legendre transforms in particular, will become lamentably inefficient when the number of processors is huge. As a result, a number of national weather services have switched to finite element or other local, low order algorithms. Some American research groups, such as those at NASA, employ finite difference codes almost exclusively.

Table 18.12: Comparisons of Finite Difference with Spectral Models: Bibliography

References	Comments
Doron&Hollingsworth &Hoskins&Simmons(1974)	Weather forecasting
Hoskins&Simmons(1975)	Comparison of spherical harmonic/differences in z with 3D finite difference weather forecasting code
Simmons&Hoskins(1975)	Spectral vs. finite-differences: growing baroclinic wave
Merilees et al.(1977)	FD with polar filtering compared with pseudospectral
Browning&Hack &Swarztrauber(1988)	finite differences of 2d, 4th, and 6th order compared with spherical harmonics
Held&Suarez (1994)	2d/4th finite difference; climate model
Gustafsson&McDonald(1996)	HIRLAM grid point vs. spectral semi-Lagrangian; models are similar in accuracy and efficiency
Fornberg&Merrill(1997)	Double Fourier vs. spherical harmonics vs. 2d & 4th order differences for a very smooth solution
Spotz&Taylor & Swarztrauber(1998)	Comparisons of compact 4th order differences with spherical harmonics and filtered double Fourier

So far, however, the MP/low order assault has not yet carried the fort although some of the outworks have fallen. In the first place, ingenious coding and the inherent efficiency of matrix multiplication has made spectral codes run fast even on machines with moderate parallelism (forty-eight processors). (See, for example, Drake et al.(1995) and other articles in a special issue of *Parallel Computing*, vol. 21, number 10.) Because the architecture of twenty-first century machines is still in a state of flux, it is impossible to say whether the spherical harmonics are doomed or whether ingenuity will continue to make them competitive. Second, the new finite order methods on the sphere have a shorter history than spherical harmonics, and consequently many schemes are still nagged by stability problems and also by their low accuracy for smooth, large scale structures that are much better resolved by spectral codes.

Most of the studies in Table 18.12 which have compared finite difference and spectral codes have found the latter were better. However, Held and Suarez(1994) have claimed that a low order difference model is just as accurate and runs faster; Gustafsson and McDonald(1996) report a tie. The meteorological community is applying more sophisticated benchmark tests including both smooth and discontinuous solutions, so these controversies will eventually be resolved. In the meantime, spherical harmonics models soldier on.

The second assault is the replacement of spherical harmonics by an alternative high order method. Spectral elements, which allow the work to be split among many processors without sacrificing high accuracy, have great promise (Table 18.11. However, only two groups, one each in meteorology and oceanography, are developing such models. Two-dimensional Fourier series are very promising, too (Spotz, Taylor and Swarztrauber, 1998, and Fornberg and Merrill, 1997, Shen, 1999, and Cheong, 2000). However, work to date has been confined to the shallow water wave equations — two-dimensional, no clouds, no chemistry. It seems likely that one or both of these alternatives will someday be used for operational forecasting, but *when* is very murky.

The third assault is from conservative, shock-resolving van Leer/Roe schemes, which have become very popular in aerodynamics. The Lin and Rood algorithm (1996) has gained a wide following. However, experience with such low-order-but-conservative schemes is still limited.

The only conclusion we can make at the turn of the twenty-first century is: the siege continues.

18.31 Elliptic and Elliptic Cylinder Coordinates

Two-dimensional elliptic coordinates (μ, η) are defined by

$$x = \cosh(\mu) \cos(\eta) \qquad \& \qquad y = -\sinh(\mu) \sin(\eta) \qquad (18.94)$$

where x and y are the usual Cartesian coordinates in the plane. The reason for the name "elliptic" is that the surfaces of constant μ are all ellipses with foci at $x = \pm 1$ along the x-axis as illustrated in Fig. 2-16 of Chapter 2. At large distances from these foci, the ellipses degenerate into circles and elliptical coordinates become ordinary polar coordinates. In this limit, $\mu \approx \log(r)$ where r is the polar radius and $\eta \approx \theta$, the polar angle.

Elliptic cylinder coordinates add a vertical coordinate z which is perpendicular to the x-y plane. These three-dimensional coordinates generalize elliptic coordinates in the same way that cylindrical coordinates extend polar coordinates to three dimensions. Since the z-axis is straight, one should use either Chebyshev polynomials or rational Chebyshev functions for z, just as for cylindrical coordinates.

Because the quasi-angular coordinate η is cyclic, all functions are periodic in η, which implies a Fourier basis in η. However, the coordinate singularity implies that μ-dependent Fourier coefficients have definite parity, just as for spherical and polar coordinates. However, there are intriguing differences because the singular coordinate $\mu = 0$ is a singular *line* (the whole interval $x \in [-1, 1]$) rather than a single point.

Theorem 37 (ELLIPTICAL COORDINATES: PARITY in QUASI-RADIAL COORD.)

If $F(\mu, \eta)$ is an analytic scalar function which is expanded as

$$F(\mu, \eta) = \sum_{m=0}^{\infty} f_m(\mu) \cos(m\eta) + g_m(\mu) \sin(m\eta) \tag{18.95}$$

then continuity of $F(\mu, \eta)$ and all its derivatives at $\mu = 0$ implies:

(i) $f_m(\mu)$ is symmetric *with respect to $\mu = 0$ for all m*

(ii) $g_m(\mu)$ *is* antisymmetric *for all m*

Thus, one should expand the coefficients of $\cos(m\eta)$ as even Chebyshev polynomials in μ while the sine terms should be represented as a sum of odd polynomials only.

PROOF: All functions in two dimensions can be approximated to arbitrary accuracy as polynomials in the Cartesian coordinates x and y. This in turn implies that any well-behaved function can be written as a sum of the powers

$$x^j y^k = \cos^j(\eta) \sin^k(\eta) \cosh^j(\mu) \sinh^k(\mu) \tag{18.96}$$

The parity of $x^j y^k$ with respect to both η and μ is controlled entirely by k. If k is even, then $\sin^k(\eta)$ and $\sinh^k(\mu)$ are both symmetric functions; if k is odd, then (18.96) is also odd in both elliptic coordinates. Now a function which is symmetric in η may be expanded as a cosine series in η. The fact that even parity in $\eta \longleftrightarrow$ even parity in μ implies that all the cosine coefficients must be symmetric in μ, which is proposition (i). Similar reasoning implies (ii).

One could alternatively prove the theorem by showing that it is necessary for continuity of $F(\mu, \eta)$ and all its derivative across the coordinate singularity at $\mu = 0$. (By singularity, we mean that differential equations in elliptical coordinates have coefficients that are singular at $\mu = 0$; the *solution* is analytic on that line segment.)

Boyd and Ma(1990) and Mittal and Balachandar(1996) are fluid mechanics applications of spectral methods in elliptic coordinates, Fourier in η and Chebyshev polynomials in μ.

18.32 Summary

The end-of-the-century state of the art is a mixture of the known and unknown. The known includes: coping with coordinate singularities and implementing stable, fast algorithms in spherical, cylindrical or polar coordinates and in some related coordinate systems; these methods run well on coarse-grained parallel computers with sixteen or fewer processors.

One unknown is: Can the Legendre transform be accelerated to keep spherical harmonics or One-Sided Jacobi polynomials competitive at very large N? Another unknown is: Can spherical harmonics schemes and related algorithms be efficiently adapted to massively-parallel machines? Will spherical harmonics be replaced by latitude-and-longitude Fourier series or spectral elements? Will spectral methods of all kinds be routed by low order finite difference or finite element methods for weather forecasting and climate modelling?

In the areas of this chapter, the Age of Exploration is still not over.

Table 18.13: Reviews, Books, and Model Descriptions

References	Comments
Bourke&McAvaney &Puri&Thurling(1977)	Comprehensive review, describing spherical harmonics code which is the basis for ECMWF forecast and NCAR climate models
Haltiner&Williams(1980)	one chapter on spherical harmonic methods
Sela (1980, 1995)	Reviews of National Meteorological Center (USA) forecast model
Daley(1981)	Review concentrating on normal mode initialization for weather prediction models on sphere
Gordon&Stern(1982)	Description of GFDL global circulation model
Kanamitsu et al.(1983)	Description of the operational model of the Japanese Meteorological Agency
Boer et al.(1984)	Review of Canadian Climate Centre model
Jarraud &Baede (1985)	lengthy review of atmospheric codes
Daley(1991)	Monograph about data analysis, initialization and other issues for weather prediction
Hogan&Rosmond(1991)	U. S. Navy global operational spectral code
Jakob-Chien & &Hack&Williamson(1995)	seven benchmark problems for shallow water eqs.; fronts, waves, Gibbs' wiggles

Chapter 19

Special Tricks

"If it works once, it's a trick. If it works twice, it's a method. If it works three times, it's a law."
— Source Unknown

"Card tricks are as important in numerical analysis as in stage magic."
— J. P. Boyd

19.1   Introduction

In this chapter, we give brief, self-contained descriptions of a number of ideas that are useful even if they are applicable only under special circumstances. These special tricks can be safely skipped by the reader who is interested only in the major themes of spectral methods. Nonetheless, the assertion that "numerical analysis is as much an art as a science" has become a proverb. Much of the reason is precisely because special technques are the best way of solving certain exotic problems; card tricks, if we may so dub the ideas illustrated in this chapter, are as important in computation as in conjuration.

Besides the special algorithms collected in this chapter, we have discussed a number of useful tricks elsewhere including the following.

First, one can use nonlinear degrees-of-freedom, such as width or shape parameters, in solving differential equations. Recall that when the numerical solution is approximated by a series, $u_N(x)$ is a *linear* function of the unknown series coefficients, $a_0, \ldots, a_N$. But it need not always be so. *Nonlinear* unknowns were used with great success in the H_2^+ ion (Chapt. 3, Sec. 7) and the Korteweg-deVries soliton (Appendix G of the first edition of this book). (An especially good candidate as a nonlinear degree-of-freedom is a *width* parameter that matches the scale of the basis functions to the width of the solution.)

Second, the standard methods of Chapt. 17 for solving problems on an infinite interval will fail unless the oscillations in $u(x)$ decay exponentially as $|x| \to \infty$. (This condition may be met either because $u(x)$ itself decays exponentially, with or without oscillations, or because the oscillations rapidly decay even though $u(x)$ asymptotes to a constant or its mean value decreases algebraically with $|x|$.) If $u(x)$ oscillates with slowly decaying amplitude as $|x| \to \infty$, then all standard methods fail because the rational Chebyshev

442

functions (or whatever) are being asked to faithfully mimic an infinite number of crests and troughs. Chapt. 17, Sec. 13, shows that one may solve this problem, at least for the Bessel functions $J_n(x)$, by computing *two* simultaneous series: one for the amplitude and one for the phase of the oscillation.

The Bessel double expansion is a particular case of a very general strategy: modifying a standard basis set by using special functions that match the pathological behavior of the solution. Other examples are described in Secs. 5, 6, and 7 below.

Third, when the solution is singular or nearly singular on the interior of an interval, one can *detour* around the branch point by integrating on an arc in the complex plane. This device, very useful for computing weakly unstable eigenvalues in hydrodynamic instability theory, is described in Chapt. 7, Sec. 11.

Fourth, when $u(x)$ has regions of steep gradient, one can use a change-of-coordinate to cluster grid points in the frontal region. Chapt. 13 describes a wide variety of real-valued mappings; transformations for spatially periodic problems are in Sec. 7 and adaptive grids are explained in Sec. 9.

These ideas and those explained below are only a sampler. No doubt the reader will invent his own variations.

19.2 Sideband Truncation

Normally, we truncate a spectral expansion by including all basis functions up to a certain cutoff, beginning with $\phi_0(x)$. For special problems, however, this may be inefficient. The large eigenfunctions of Mathieu's equation furnish a good example. These solve

$$u_{xx} + [\lambda - 2q\cos(2x)]u = 0 \tag{19.1}$$

subject to the condition of periodicity with period 2π with λ as the eigenvalue and q as a constant parameter.

When $q = 0$, the eigenvalues and eigenfunctions are

$$\lambda_n = n^2 \quad ; \quad u_0 = 1; \quad u_n(x) = \sin(nx) \quad \text{or} \quad \cos(nx) \tag{19.2}$$

For non-zero q, the unperturbed solutions are still good approximations to those of the full Mathieu problem, (19.1), whenever $n^2 \gg q$. If $n = 15$, for example, the spectral coefficient with the largest amplitude will not be a_0 or a_1, but a_{15}. If we use the most drastic possible truncation — one term! — then $\sin(15\,x)$ or $\cos(15\,x)$ must be that term.

One can show by using the trigonometric identity

$$\cos(2x)\cos(nx) = 0.5\left\{\cos[(n+2)x] + \cos[(n-2)x]\right\} \tag{19.3}$$

(and its analogue for $\sin(nx)$) that the perturbation, $2q\cos(2x)$, will only couple Fourier terms whose degrees differ by two and are of the same type, either sines or cosines. The eigenfunctions of Mathieu's equation therefore fall into four separate classes. The eigenfunctions of each class can be computed independently of the eigenfunctions of the other three classes.

These four groups of eigenfunctions illustrate the double-parity symmetry of Fourier functions (Chapter 8, Sec. 2). Each eigenfunction has definite parity both with respect to $x = 0$ and with respect to $x = \pi/2$. The little table below lists the four classes in the conventional notation. The middle column gives the double-parity symmetry of each class with the parity with respect to $x = 0$ listed first. The third column lists the Fourier functions that suffice to represent all modes in a given class; each of these basis functions has the same double symmetry as the Mathieu functions.

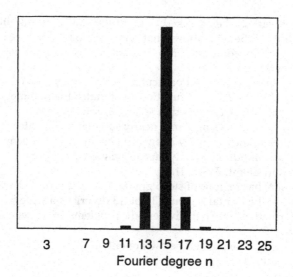

Figure 19.1: Absolute values of the Fourier coefficients of the Mathieu function, $ce_{15}(x)$, when the parameter $q = 10$. This is the motive for the "sideband truncation": only five of the coefficients are non-negligible, but they are *not* the coefficients of the lowest five Fourier terms.

$ce_{2n}(x)$	...	Even-Even symmetry	...	$\{1, \cos(2x), \cos(4x) \dots\}$
$ce_{2n+1}(x)$	...	Even-Odd symmetry	...	$\{\cos(x), \cos(3x), \cos(5x) \dots\}$
$se_{2n}(x)$	...	Odd-Odd symmetry	...	$\{\sin(2x), \sin(4x), \sin(6x) \dots\}$
$se_{2n+1}(x)$	...	Odd-Even symmetry	...	$\{\sin(x), \sin(3x), \sin(5x) \dots\}$

For small q, one can show via perturbation theory that $a_{n\pm2}$ is $O(q)$, $a_{n\pm4}$ is $O(q^2)$, $a_{n\pm6}$ is $O(q^3)$ and so on. The coefficient $a_{n\pm4}$ is only $O(q^2)$ because (19.3) shows that the perturbation cannot couple these components directly to $\cos(nx)$, but only to the much smaller components, $\cos[(n \pm 2)x]$.

Fig. 19.1 illustrates these remarks by plotting the absolute value of the Fourier cosine coefficients for $ce_{15}(x)$ for $q = 10$ (Lowan *et al.*, 1951). We do not need 1 or $\cos(2x)$ or $\cos(4x)$ in our basis set (for this eigenvalue and this value of q) because their coefficients are very, very small. The important basis functions are those which are oscillating on nearly the same scale as the dominant basis function, $\cos(nx)$.

Definition 42 (SIDEBAND TRUNCATION)

When the basis set is restricted to basis functions of the form $\phi_{n\pm m}(x)$ where $m \ll n$, this is said to be a "SIDEBAND TRUNCATION".

The basis function $\phi_n(x)$, normally the function with the largest coefficient, is the "fundamental" and the other basis functions in the truncation are the "SIDEBANDS".

When we apply Galerkin's method with a sideband truncation to five basis functions, the resulting matrix problem is of the form

$$\begin{vmatrix} [\lambda - (n-4)^2] & -q & 0 & 0 & 0 \\ 0 & [\lambda - (n-2)^2] & -q & 0 & 0 \\ 0 & -q & [\lambda - n^2] & -q & 0 \\ 0 & 0 & -q & [\lambda - (n+2)^2] & -q \\ 0 & 0 & 0 & -q & [\lambda - (n+4)^2] \end{vmatrix} \begin{vmatrix} a_{n-4} \\ a_{n-2} \\ a_n \\ a_{n+2} \\ a_{n+4} \end{vmatrix} = \vec{0}$$

$$(19.4)$$

Eq. (19.4) applies to all four classes of basis functions provided that $n \geq 7$. The condition for a nontrivial solution is that the determinant of the 5×5 matrix in (19.4) should equal 0, which gives a quintic equation to determine λ. If we truncate to just the fundamental plus two sidebands, then the "secular determinant" is that of the inner 3×3 block in (19.4) which is enclosed by dots.

The polynomial equations that determine λ are

$$\begin{aligned} P_3(\lambda; q, n) &\equiv \lambda^3 - (3n^2 + 8)\lambda^2 + (3n^4 - 2q^2 + 16)\lambda \\ &+ 8q^2 - 16n^2 + 2n^2q^2 + 8n^4 - n^6 \end{aligned} \qquad (19.5)$$

plus a similar equation for the 5×5 determinant (not displayed). Since we are interested in the mode whose unperturbed eigenvalue is n^2 and since the eigenvalue is changed only a little bit by the perturbation provided that $n^2 \gg q$, we do not need to apply a general polynomial equation solver to $P_3(\lambda)$ and $P_5(\lambda)$. Instead, we can apply Newton's iteration once to obtain an accurate *analytical* solution in the form

$$\lambda = \lambda_0 + \delta \qquad (19.6)$$

where

$$\lambda_0 \equiv -n^2 \qquad (19.7)$$

The 3×3 truncation gives

$$\delta_3 = -\frac{P_3(\lambda_0)}{dP_3(\lambda_0)/d\lambda} \qquad (19.8)$$

$$= \frac{4q^2}{q^2 + 8n^2 - 8} \qquad (19.9)$$

and similarly[1]

$$\delta_5 = \frac{32q^4 + 512q^2n^2 - 1024q^2}{3q^4 - 896q^2 + 64n^2q^2 + 1024n^4 - 5120n^2 + 4096} \qquad (19.10)$$

Fig. 19.2 compares $\delta_3(q)$ and $\delta_5(q)$ with the exact correction to the eigenvalue for $n = 15$. Although we have included no cosines lower than $\cos(11x)$, the 5×5 sideband truncation is almost indistinguishable from the exact eigenvalue correction for $q \leq 25$.

When q/n^2 is large, the perturbation is so strong that the non-negligible sidebands extend all the way to the lowest mode, thus implicitly reverting to a normal spectral expansion. "Sideband truncation" has only a limited range of applicability.

---

[1]The determinants and the Newton's iteration were evaluated using the algebraic manipulation language REDUCE

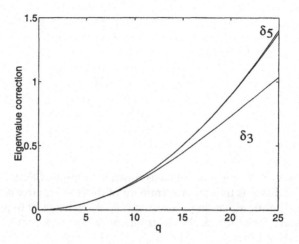

Figure 19.2: A comparison of the exact eigenvalue correction, $\delta(q)$, for the mode ce_{15} with the three-basis-function and five-basis-function approximations computed in the text. The five-term approximation, $\delta_5(q)$, is almost indistinguishable from the exact $\delta(q)$.

Nevertheless, there are other physical problems where this trick is useful. An example would be the direct (as opposed to perturbative) calculation of "envelope solitons" (Boyd, 1983a, b, c). There are many interesting phenomena which can be modelled by a fast "carrier" wave modulated by a slowly-varying "envelope" or amplitude; a sideband truncation is appropriate for all of them.

19.3   Special Basis Functions, I: Corner Singularities

When the computational domain has sharp angles, the PDE solution is usually weakly singular at such angles or corners. As noted in Chap. 16, mapping methods are often very effective in dealing with corner singularities because the solution is bounded even at the corners. However, mappings rob resolution from the interior of the domain to squeeze more grid points near the walls. A better option, at least in theory, is to subtract the branch points from the solution so that the Chebyshev series is only required to represent the non-singular part of $u(x, y)$.

Schultz, Lee, and Boyd (1989) describe an illuminating example: the wall-driven incompressible flow in a two-dimensional rectangular cavity. The velocity is discontinuous at the upper corners where the moving wall meets two stationary boundaries. The result is that if one defines a local polar coordinate system (r_1, θ_1) centered on the upper left-hand corner, the flow is singular there with the strongest branch point taking the form

$$\psi_{s1} = \frac{r_1 \left[\theta_1 (\cos\theta_1 - \frac{\pi}{2}\sin\theta_1) - \left(\frac{\pi}{2}\right)^2 \sin\theta_1 \right]}{\frac{\pi^2}{4} - 1} \tag{19.11}$$

plus a similar form for the other upper corner.

When the streamfunction is partitioned as

$$\psi(x, y) = \psi_a(x, y) + \psi_s(x, y) \tag{19.12}$$

where $\psi_s(x, y)$ includes the two singular terms of the form of (19.11) and $\psi_a(x, y)$ is expanded as a 30×30 two-dimensional Chebyshev series, the streamfunction for zero Reynolds' number is accurate to at least 1 part in 10,000,000,000, relative to its maximum. The singularity-subtraction has added at least six decimal places of accuracy for this truncation.

The remarkable thing about this accuracy is that the driven cavity flow has weaker singularities at the lower corners, too. Moffatt (1964) showed that as the corners are approached, the flow becomes more and more linear because the velocities tend to zero (relative to the walls). This near-corner linearity allowed him to show that the flow consists of an infinite sequence of eddies of rapidly diminishing size and intensity – the so-called "Moffatt eddies". Because the singular term (19.11) is driven directly by the wall, it is completely known. However, the "Moffatt eddies" are driven by the large eddy in the center of the cavity, so the "Moffatt eddies" are known only to within multiplicative constants which must be determined by matching to the interior flow.

This problem illustrates two general themes. The first is that one can generally determine the form of the singular basis functions, as in (19.11), by a *linear* asymptotic analysis even when the problem as a whole is strongly *nonlinear*.

The second theme is that *weak* singularities may often be *ignored*. The "Moffatt eddies" are described by functions with branch points, but neither ψ nor its first couple of derivatives are unbounded. The result is that one can obtain ten decimal place accuracy without applying special tricks to these corner eddies.

Which corner branch points can be ignored and which should be represented by singular basis functions depends on (i) the target accuracy and (ii) the strength of the nonlinearity.

If the goal is thirty decimal place accuracy, for example, one could greatly improve efficiency by supplementing the Chebyshev basis by special functions in the form of Moffatt eddies of undetermined strength. The greater the target accuracy, the more singular functions that should be included in the spectral basis set.

The strength of the nonlinearity is important, too. Boyd (1986) solved another problem, the two-dimensional Bratu equation, which has corner singularities. He showed that as the amplitude becomes larger, the solution series converges more and more slowly. The corner singularities become less and less important in this same limit.

Schultz, Lee and Boyd (1989) show that the same is true of the driven cavity flow. As the Reynolds number increases, the flow develops interior fronts. These regions of steep gradients are smoothed by viscosity so that the flow is analytic everywhere in these frontal zones — in contrast to the corner eddies, which are genuinely pathological at the corners. Nevertheless, the Chebyshev series converges more and more slowly as the internal fronts become steeper. Unless the target accuracy is greater than ten decimal places, these internal fronts are a greater numerical challenge than resolving the Moffatt eddies in the corners.

For many problems, the spectral coefficients are the sum of two or more independent contributions. The corner singularities, because their contributions decay only algebraically with n, must dominate in the limit $n \rightarrow \infty$. However, if the corner branch points are weak and the internal fronts are strong, the exponentially decaying contributions from the complex poles or branch points associated with the fronts may be larger than the contributions of the corner singularities for all $n < n_{\text{cross-over}}$. For example, consider a Chebyshev coefficient which depends on n as

$$a_n \sim 10 \exp\left(-\frac{n}{3}\right) + \frac{0.000001}{n^5} \tag{19.13}$$

A pure mathematician would be content with the statement that $a_n \sim 0.000001/n^5$ in the limit $n \rightarrow \infty$. The numerical analyst is called to a higher standard: the exponentially-decaying term in (19.13) is larger than the second, algebraically decaying term for all $n < 120$.

Boyd (1987e, 1988d) gives other vivid examples of this competition between competing numerical influences. Each problem exhibits a "cross-over truncation" where the slope of the error as a function of N dramatically changes. Boyd (1988d) is a one-dimensional application of singular basis functions to resolve logarithmic endpoint singularities.

The driven cavity flow and the Bratu equation both exhibit a similar tug-of-war between branch points of different strengths and different nearness-to-the-square. When the corner singularities win, singular basis functions are extremely helpful.

19.4   Special Basis Functions, II: Wave Scattering

The scattering of waves incident upon a reflector is another problem where special radiation functions are very useful (Boyd, 1990d). To illustrate this idea, we solve the quantum mechanics problem of the scattering of a sine wave by a potential well in one dimension.

The physical background is described in Morse and Feshbach (1953). The Schrödinger equation for a potential energy $V(x)$ is, letting k^2 denote the energy,

$$\psi_{xx} + \left[k^2 - V(x)\right]\psi = 0 \tag{19.14}$$

It is implicitly assumed that $V(x) \to 0$ as $|x| \to \infty$ so that the asymptotic solution is a plane wave of wavenumber k.

The scattering problem is to calculate the transmitted and reflected waves when a plane wave of unit amplitude is incident from the left. The wavefunction ψ has the asymptotic behavior

$$\psi \sim e^{ikx} + \alpha(k)\,e^{-ikx} \qquad\qquad x \to -\infty \tag{19.15a}$$
$$\psi \sim \beta(k)\,e^{ikx} \qquad\qquad x \to \infty \tag{19.15b}$$

where the first term in (19.15a) is the incident wave (the forcing), the second term is the reflected wave, and (19.15b) is the transmitted wave. The complex coefficients $\alpha(k)$ and $\beta(k)$ are unknowns; the primary goal is to calculate the reflection coefficient $\mathcal{R}$ defined by

$$\mathcal{R} \equiv |\alpha|^2 \tag{19.16}$$

If $V(x)$ decays exponentially fast as $|x| \to \infty$, then $\psi(x)$ tends to its asymptotic forms (19.15) exponentially fast, too. The rational Chebyshev functions, $TB_n(x)$, are a natural and exponentially convergent basis for representing the difference between ψ and its asymptotic forms.

Unfortunately, the rational Chebyshev functions cannot represent the asymptotic sinusoidal waves themselves. Because the sine waves do not decay in amplitude, a small absolute error on the infinite interval is possible if and only if a moderate number of rational functions can faithfully represent an *infinite* number of crests and troughs. This is too much to ask of rational Chebyshev functions, sinc functions, Hermite functions, or any standard basis set.

However, one can retrieve spectral accuracy for the entire problem by supplementing the rational Chebyshev functions by a pair of special "radiation" functions.

First, define two real functions, $C(x)$ and $S(x)$, by the requirement that each solves (19.14) and has the boundary behavior

$$C(x) \;\to\; \cos(kx) \qquad\qquad \text{as }\; x \to -\infty \tag{19.17a}$$
$$S(x) \;\to\; \sin(kx) \qquad\qquad \text{as }\; x \to -\infty \tag{19.17b}$$

Since these are linearly independent, the general solution of (19.14) is a linear combination of these two functions. Comparing (19.17) with (19.15a) shows that

$$\psi = (1 + \alpha) C(x) + i (1 - \alpha) S(x) \tag{19.18}$$

To compute α and β, we need the asymptotic behavior of $C(x)$ and $S(x)$ as $x \to \infty$. Since $V(x) \to 0$ in this limit (by assumption), it follows that their most general asymptotic behavior must take the form

$$C(x) \sim \gamma_1 \cos(kx) + \gamma_2 \sin(kx) \qquad\qquad x \to \infty \tag{19.19a}$$

$$S(x) \sim \sigma_1 \cos(kx) + \sigma_2 \sin(kx) \qquad\qquad x \to \infty \tag{19.19b}$$

Once we have calculated the four parameters $(\gamma_1, \gamma_2, \sigma_1, \sigma_1)$, we can match (19.19) to (19.15b) to determine the four unknowns, the real and imaginary parts of α and β.

To compute $C(x)$ and $S(x)$, we write

$$C(x) = \cos(kx) + \tilde{C}(x) \qquad\qquad \& \qquad\qquad S(x) = \sin(kx) + \tilde{S}(x) \tag{19.20}$$

$$\tilde{C}(x) = \sum_{n=0}^{N-3} a_n \, TB_n(x) + a_{N-2} \, H(x) \, \cos(kx) + a_{N-1} \, H(x) \, \sin(kx) \tag{19.21}$$

where

$$H(x) \equiv \frac{1}{2} (1 + \tanh(x)) \tag{19.22}$$

is a smoothed approximation to a step function. The expansion for $\tilde{S}(x)$ is identical in form with that for $\tilde{C}(x)$. The two "radiation basis functions" have the asymptotic behavior

$$H(x) \cos(kx) \to \begin{cases} 0 & \text{as } x \to -\infty \\ \cos(kx) & \text{as } x \to \infty \end{cases} \tag{19.23}$$

and similarly for the other function. Because these basis functions are "one-sided", they vanish as $x \to -\infty$ and are irrelevant to the asymptotic behavior of $C(x)$ in this limit. In the opposite limit, $x \to \infty$, one finds that as N increases,

$$a_{N-2} \to \gamma_1 - 1 \qquad\qquad \& \qquad\qquad a_{N-1} \to \gamma_2 \qquad\qquad N \to \infty \tag{19.24}$$

Similarly, the two radiation coefficients for $\tilde{S}(x)$ converge to σ_1 and $\sigma_2 - 1$. Thus, the coefficients of the radiation functions give the needed constants in the asymptotic approximations for $C(x)$ and $S(x)$.

Substituting (19.20) into (19.14) shows that

$$\tilde{C}_{xx} + \left[k^2 - V(x)\right] \tilde{C} = V(x) \cos(kx) \tag{19.25}$$

The differential equation for $\tilde{S}(x)$ is identical except that $\cos(kx)$ on the right side of (19.25) must be replaced by $\sin(kx)$. To solve (19.25), we apply the pseudospectral method with the usual N interpolation points for the rational Chebyshev basis:

$$x_i \equiv L \cot\left(\frac{\pi \, [2i - 1]}{2N}\right) \qquad\qquad i = 1, \dots, N \tag{19.26}$$

where L is a constant map parameter (=2 in the example below). Because the problems for $\tilde{C}(x)$ and $\tilde{S}(x)$ differ only in the inhomogeneous term, it is necessary to compute and LU-factorize only a single square matrix whose elements are

$$H_{ij} \equiv \phi_{j,xx}(x_i) + \left\{ k^2 - V(x_i) \right\} \phi_j(x_i) \tag{19.27}$$

where the $\{\phi_j(x)\}$ are the basis functions defined above: rational Chebyshev functions plus two special radiation functions The factorized matrix equation must be solved twice (Appendix B) with different inhomogeneous terms:

$$f_i \equiv V(x_i)\cos(kx_i), \quad [\tilde{C}] \qquad g_i \equiv V(x_i)\cos(kx_i), \quad [\tilde{S}] \tag{19.28}$$

Matching (19.19) to (19.15b) shows that α is the solution of the 2×2 real system

$$\begin{vmatrix} (\gamma_1 + \sigma_2) & (\sigma_1 - \gamma_2) \\ (\gamma_2 - \sigma_1) & (\gamma_1 + \sigma_2) \end{vmatrix} \begin{vmatrix} \mathrm{Re}(\alpha) \\ \mathrm{Im}(\alpha) \end{vmatrix} = \begin{vmatrix} (\sigma_2 - \gamma_1) \\ -(\sigma_1 + \gamma_2) \end{vmatrix} \tag{19.29}$$

The rate-determining step is to invert a single $N \times N$ square matrix: $O([2/3]N^3)$ multiplications and additions.

Table 19.1 shows the results for the particular case

$$V(x) = -v \, \mathrm{sech}^2(x) \tag{19.30}$$

for $v = 1$ and various wavenumbers k. The exact reflection coefficient for this potential is

$$\mathcal{R} = \frac{1 + \cos(2\pi\sqrt{v + 1/4})}{\cosh(2\pi k) + \cos(2\pi\sqrt{v + 1/4})} \tag{19.31}$$

Because of the hyperbolic function in (19.31), the reflection coefficient is *exponentially small* in $1/k$. This is a general feature of scattering from a potential well, and not something special to a sech^2 potential. Because of this, one must solve the Schroedinger equation to *very high accuracy* to compute even a *crude* approximation to $\mathcal{R}$ for *large* k. Otherwise, the tiny reflection coefficient will be swamped by numerical error.

Thus, spectral methods – with spectral accuracy – are very valuable for scattering problems.

19.5   Special Basis Functions, III: Weakly Nonlocal Solitary Waves

This idea of special "radiation basis functions" can be applied to many other types of wave problems where the solution asymptotes to a sinusoidal wave rather than to zero. Boyd (1991a, 1990b, 1991d, 1991e, 1995b, 1995h, 1995j) describes applications to "non-local solitary waves" and other nonlinear waves.

19.6   Root-Finding by Means of Chebyshev Polynomials: Lanczos Economization and Ioakimidis' Method

The standard numerical strategy for solving $f(x) = 0$ is to *approximate* $f(x)$ by something simpler which can be solved explicitly, usually a *local* approximation in the vicinity of a "first guess" for the root. In Newton's method, the approximation is a linear Taylor series

Table 19.1: The exact and numerical reflection coefficients $\mathcal{R}$ for the sech^2 potential as computed using 48 rational Chebyshev functions and two radiation functions.

k	$R_{numerical}$	R_{exact}	Absolute Error
0.3	0.423097	0.423097	2.7E−8
0.6	0.0774332	0.0774332	2.6E−8
0.9	0.0121005	0.0121005	−3.1E−8
1.2	0.00184535	0.00184535	−9.5E−10
1.5	0.000280391	0.000280376	1.5E−8
1.8	0.000042562	0.000042576	−1.3E−8
2.1	0.000006471	0.000006465	6.1E−9
2.4	0.000000978	0.000000982	−3.2E−9
2.7	0.000000151	0.000000149	1.5E−9
3.0	0.000000022	0.000000023	−8.6E−10

expansion about $x = x_i$. In the secant method, the approximation is the linear polynomial that interpolates $f(x)$ at $x = x_i$ and x_{i-1}. The Cauchy and Muller methods are the same except that the approximations are the quadratic Taylor series and quadratic interpolating polynomial. Halley's scheme uses the [1, 1] Padé approximant, which is that ratio of one linear polynomial over another whose Taylor expansion matches that of $f(x)$ through quadratic terms. Shafer suggests using higher order Padé approximants: the [2,3] gives a fifth-order method, but requires only solving a quadratic (the numerator) to find the root.

An obvious generalization is to use Chebyshev polynomials if the root can be localized in a narrow region. C. Lanczos used a simple argument to localize the root of a cubic polynomial within a finite interval, converted the polynomial into a Chebyshev series on the interval, and then solved the $N = 2$ truncation of the Chebyshev series, which is a quadratic polynomial.

Boyd (1995f) applied this strategy of "Lanczos economization"[2] to nonlinear eigenvalue problems. The goal is to find the root of the determinant of a system of linear equation when the matrix elements depend nonlinearly on the eigenparameter x. The cost of evaluating $f(x)$ is very expensive if the dimension of the matrix is large – $O(10^6)$ for a 100 x 100 matrix. It therefore is useful to evaluate the determinant on an interval in x and then compute the roots by finding those of the Chebyshev approximant.

The most reliable methods for finding all roots on an interval unfortunately require frequent evaluations of $f(x)$. One such robust strategy is to compute $f(x)$ on a large number of points (say one thousand!), look for sign changes, and then refine the roots thus located

---

[2]"Lanczos economization" is usually applied in a somewhat narrower sense to the process of converting a truncated power series to a Chebyshev series of the same order, truncating to the lowest acceptable degree, and then converting back to an ordinary polynomial to obtain an approximation that is cheaper to evaluate and more uniform than the higher-degree polynomial from whence it came. The root-finding algorithm described here is similar in the sense that a function which is expensive to evaluate is replaced by a truncated Chebyshev series that is much cheaper.

by Newton's method. This would be hideously expensive if we worked directly with $f(x)$. An N-term Chebyshev approximant can be evaluated in $O(N)$ operations, however, so that one can graph the expansion at a thousand points for perhaps less cost than evaluating the original function just once.

Lanczos economization can be applied to find roots of functions of two or even three unknowns, but unfortunately it is unworkable – unless the degree of the Chebyshev interpolant is very low – in higher dimensions. The other Chebyshev-based rootfinding technique, due to Ioakimidis , is unfortunately to limited to a small number of unknowns also. Nevertheless, it is an illuminating and clever application of Chebyshev ideas.

Suppose that $f(x)$ has only a single root on $x \in [a, b]$. Let ρ denote the (unknown!) location of the root. The integral

$$I \equiv \int_a^b \frac{x - \rho}{f(x)} \frac{1}{\sqrt{1 - x^2}} \, dx \tag{19.32}$$

can be integrated with exponential accuracy by a Chebyshev quadrature because the zero in $f(x)$ cancels that in the numerator. It follows that the results of Gauss-Chebyshev and Chebyshev-Lobatto quadrature must differ by an amount exponentially small in N

$$I_{Gauss}(N) = I_{Lobatto} + O(\exp(-qN)) \tag{19.33}$$

for some positive constant q (neglecting powers of N and other algebraic factors of N multiplying the exponential) where

$$I_{Gauss}(N) \equiv \sum_{j=1}^N w_j^G \frac{(x_j^G - \rho)}{f(x_j^G)} \tag{19.34}$$

and similarly for the Lobatto quadrature, which uses the "endpoints-and-extrema" grid instead of the "interior" grid where the w_j are the quadrature weights. Neglecting the exponential error terms, we can rearrange the terms in the equation $I_{Gauss} = I_{Lobatto}$ to obtain

$$\rho \left\{ \sum_{j=1}^N \left(\frac{w_j^G}{f(x_j^G)} - \frac{w_j^L}{f(x_j^L)} \right) \right\} = \left\{ \sum_{j=1}^N \left(x_j^G \frac{w_j^G}{f(x_j^G)} - x_j^L \frac{w_j^L}{f(x_j^L)} \right) \right\} \tag{19.35}$$

Dividing by the sum on the left gives an explicit formula for the root ρ as a ratio of weighted values of $f(x)$. For Chebyshev quadrature with the usual Chebyshev weight function in the integral Eq.(19.32), the weights are all identical except for the two endpoints, and the quadrature points for both grids combined are the images of an evenly spaced grid under the cosine mapping. This greatly simplifies the final approximation to the root to

$$\rho = \left\{ \sum_{j=0}^{2N} {}'' (-1)^j \frac{x_j}{f(x_j)} \right\} \Big/ \left\{ \sum_{j=0}^{2N} {}'' (-1)^j \frac{1}{f(x_j)} \right\} \tag{19.36}$$

where the double prime on the sums denotes that the first and last terms in both the numerator and denominator should be multiplied by $(1/2)$ and

$$x_j = \frac{1}{2} \left\{ (a + b) - (a - b) \cos \left(\frac{j\,\pi}{2N} \right) \right\} \tag{19.37}$$

Fig. 19.3 shows that for a typical example, the convergence of the approximation with N is indeed geometric. Ioakimidis (unpublished preprint) has shown that the idea can be generalized to find the roots of a pair of equations in two unknowns.

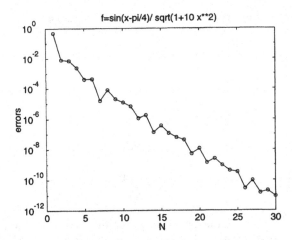

Figure 19.3: Absolute value of the absolute error in approximating the root nearest the origin of the function $\sin(x - \pi/4)/(1 + 10\,x^2)^{1/2}$ using the quadrature interval [-1, 1] in Ioakimidis' non-iterative Chebyshev root-finding algorithm. The number of evaluations of $f(x)$ is $2N + 1$ where N is the order of the method.

19.7 Spectrally-Accurate Algorithms for the Hilbert Transform

The Hilbert transform of a function $f(x)$ is defined as the Principal Value (PV) integral

$$H\{f\}(y) \equiv \frac{1}{\pi} PV \int_{-\infty}^{\infty} \frac{f(x)}{x - y} dy \qquad (19.38)$$

If $f(x) \in L_2(-\infty, \infty)$, then the Hilbert transform is in the function space L_2 also. The Hilbert transform is important in signal processing, optics, and water waves. For example, the famous Benjamin-Ono equation, which has solitary wave solutions, is

$$u_t + uu_x + H\{u_{xx}\} = 0 \qquad (19.39)$$

For functions that decay exponentially fast in $|x|$, there are at least three Hilbert transform algorithms whose accuracy increases exponentially but subgeometrically with N, the number of degrees of freedom:

1. sinc quadrature (Stenger(1981); first proposed by Kress and Martensen):

$$H\{f\}(y) \approx \frac{h}{\pi} \sum_{j=-N}^{N} f(jh) \frac{\cos(\pi[y - jh]/h) - 1}{y - jh} \qquad (19.40)$$

   where h is the grid spacing

2. Fourier method (Weideman and James, 1992):

$$H\{f\}(y) = F^{-1}\{i\ \mathrm{sgn}(k)F\{f\}(k)\} \qquad (19.41)$$

   where F denotes the Fourier transform and F^{-1} is the inverse transform.

3. Rational Chebyshev series (Weideman, 1995b, Weideman and James, 1992))

$$H\{f\}(y) = \sum_{j=-\infty}^{\infty} i\,\text{sgn}(j)a_j\rho_j(y) \tag{19.42}$$

where

$$f(y) = \sum_{j=-\infty}^{\infty} a_j\rho_j(y) \tag{19.43}$$

$$\rho_j \equiv \frac{(1+ix)^j}{(1-ix)^{j+1}} \tag{19.44}$$

The Fourier algorithm is the easiest to program. In words, the procedure is to take the Fast Fourier Transform of $f(x)$, multiply wavenumber k by $\text{sgn}(k)$, and then take the inverse Fast Fourier Transform of the result.

19.8   Spectrally-Accurate Quadrature Methods

19.8.1   Introduction: Gaussian and Clenshaw-Curtis Quadrature

The "quadrature problem" is to evaluate an integral such as

$$I \equiv \int_a^b f(x)w(x)dx \tag{19.45}$$

by an approximation of the form

$$I \approx \sum_{j=1}^{N} w_j\,f(x_j) \tag{19.46}$$

Here, $w(x) \geq 0$ is a user-chosen weight function and $f(x)$ is a smooth but otherwise arbitrary function. The "quadrature weights" w_j are invariably *different* from the values of $w(x)$ evaluated at the "quadrature abscissas" x_j. However, both quadrature weights and abscissas are *independent* of $f(x)$, and depend only on the interval $[a,b]$ and the weight function $w(x)$.

To obtain spectral accuracy, the underlying strategy of all algorithms discussed here is to expand $f(x)$ as a truncated spectral series and then integrate term-by-term, analytically.

For each weight function $w(x)$ and choice of interval, $x \in [a,b]$, there is a unique choice of basis set such that the spectral integration yields a "Gaussian quadrature" or "Gauss-Jacobi integration" scheme. As has already been discussed in Chapter 4, Sec. 3, when $w \equiv 1$ and $x \in [-1,1]$, for example, the ordinary Legendre polynomials are the basis which gives a Gaussian quadrature.

The special advantage of Gaussian quadrature is that the N-point formula is exact when $f(x)$ is a polynomial of degree $(2N+1)$ or less that. That is, Gaussian quadrature is exact (ignoring roundoff) when the degree of the polynomial is *twice* the number of quadrature points. Non-Gaussian quadratures are also possible, such as that obtained by using the Chebyshev roots as abscissas for a weight function $w(x) = 1$. The disadvantage of a non-Gaussian scheme is that such a quadrature is exact only for polynomials of degree $(N-1)$, that is, only about half the degree for which Gaussian quadrature is error-free.

Clenshaw and Curtis(1960) pointed out, however, that non-Gaussian schemes could have some advantages over Gaussian quadrature. First, the Gaussian abscissas are the roots of transcendental polynomials, which must be computed or looked up in tables. In contrast, the roots and quadrature weights for Chebyshev polynomials are given by simple analytical formulas. Second, for all Gaussian quadratures, when the integral is computed twice using different N to check for accuracy, the evaluations of $f(x)$ for the first computation are completely useless for the second computation. The reason is that for Gaussian integration formulas, the abscissas for a given N never equal the abscissas of a different N (except for the endpoints). This is unfortunate when the integrand $f(x)$ is expensive to evaluate. In contrast, for certain Chebyshev polynomial schemes, previous evaluations of $f(x)$ can be recycled when N is doubled. This inexpensive adaptive quadrature schemes which double N until two successive approximations to the integral agree to within a user-specified tolerance, or satisfy other criteria for accuracy explained below.

Thus, "Clenshaw-Curtis" quadratures have the twin virtues of simplicity and easy and inexpensive adaptivity.

19.8.2 Clenshaw-Curtis Adaptivity

Clenshaw and Curtis(1960) noted that the Chebyshev-Lobatto grid of $(N + 2)$ points

$$x_j = \cos\left(\frac{\pi j}{N + 1}\right), \qquad j = 0, 1, \ldots, (N + 1) \tag{19.47}$$

has the property that when $N + 1$ is doubled, all points on the old grid are also part of the new, higher resolution grid. This allows inexpensive computation of the same integral for multiple resolutions because it is only necessary to evaluate $f(x)$ on the finest grid, and this provides all the evaluations of f needed for the coarser grids, too.

The Chebyshev polynomials are the images of cosine functions under the mapping $x = \cos(t)$. Thus, the Clenshaw-Curtis grid is an evenly spaced grid in the trigonometric coordinate t. Although also evenly spaced in t, the usual Chebyshev "interior" or "roots" grid does *not* have the property that the points at small N are contained in points of higher N, so Clenshaw and Curtis rejected it in favor of the Lobatto, endpoints-and-extrema grid.

Automatic adaptivity also requires an estimate for the error E_N. Several possibilities have been described in the literature, such as Clenshaw and Curtis' own estimate: the maximum of the absolute value of the three highest degree coefficients of the Chebyshev series of the indefinite integral of f. However, the simplest and most conservative is

$$E_N^{estimate} \equiv |I_N - I_{N/2}| \tag{19.48}$$

where I_N and $I_{N/2}$ denote the $(N + 2)$-point and $(N + 1)/2 + 1$-point approximations to the integral I. In words, when $N + 1$ is doubled, the error is almost certainly less than the difference between two successive estimates of the integral, provided that this difference is small. Gentleman(1972a), who discusses and compares several estimates, says, "Considerable experience with the subroutine CQUAD (Gentleman, 1972c), however, indicates that this simple estimate [Eq.(19.48)] appears unusually realistic here [for quadrature]. ... The naive estimate also has the advantage that when the integral is being computed to essentially full-word accuracy, a reasonable indication of the representational rounding error in the answer is frequently given. The alternative error estimates ... ignore this — often to one's embarrassment."

We agree whole-heartedly. Computers have improved greatly in the 26 years since his article was published, and there is little excuse for trading reliability for a slightly smaller value of the maximum value of N.

His comment on rounding error is a reminder that spectral methods are so accurate that machine precision will limit accuracy rather than theoretical rates of convergence. The difference between I_N and $I_{N/2}$ cannot decrease below ten or one hundred times the machine epsilon, that is, the minimum error of a single floating point operation on a given species of computer (usually about 10^{-16} on most contemporary machines). In contrast, a theoretical estimate can be smaller than machine-epsilon, which is meaningless unless one calculates using multiple precision arithmetic.

The strategy for adaptive quadrature is simple: pick a starting value of N, compute an approximation, double $N + 1$, compute another approximation, compare the difference to the desired error tolerance, and then double $N + 1$ again and again until two successive approximations differ by a satisfactorily tiny amount.

19.8.3   Mechanics

Boyd(1987c) has given a general treatment for Curtis-Clenshaw quadrature that embraces the infinite and semi-infinite intervals as well as $x \in [-1, 1]$. The first step is to transform the interval in x to the trigonometric coordinate t. The second step is to approximate the integral by an evenly spaced quadrature formula on the interval $t \in [0, \pi]$. The quadrature weights are simply the integrals of the trigonometric cardinal functios with the "metric" factor that results from the change of coordinates. The results may be summarized as follows:

1. $x \in [-1, 1]$: The transformation of the integral is

$$x \;=\; \cos(t) \tag{19.49}$$

$$I \;=\; \int_{-1}^{1} f(x)dx \;=\; \int_{0}^{\pi} f(\cos(t)) \sin(t)\,dt \tag{19.50}$$

The quadrature approximation is

$$I_N \;\equiv\; \sum_{j=1}^{N} w_j\, f(\cos(t_j)) \tag{19.51}$$

$$t_j \;\equiv\; \pi j/(N+1), \qquad j = 1, 2, \ldots, N \tag{19.52}$$

$$w_j \;\equiv\; \sin(t_j)\frac{2}{N+1}\sum_{m=1}^{N} \sin(mt_j)\,[1 - \cos(m\pi)]/m \tag{19.53}$$

2. $x \in [0, \infty]$: The transformation of the integral is (Boyd, 1987b, and Chapter 17), using the same map as yields the TL rational Chebyshev basis:

$$x \;=\; L\cot^2(t/2), \qquad L \text{ is a user-choosable constant} \tag{19.54}$$

$$I \;=\; \int_{0}^{\infty} f(x)dx \;=\; \int_{0}^{\pi} f(L\cot^2(t/2))\, 2L\,\frac{\sin(t)}{[1 - \cos(t)]^2}\,dt \tag{19.55}$$

The quadrature is

$$I_N \;\equiv\; \sum_{j=1}^{N} w_j\, f(L\cot^2(t_j/2)) \tag{19.56}$$

$$t_j \;\equiv\; \pi j/(N+1), \qquad j = 1, 2, \ldots, N \tag{19.57}$$

$$w_j \;\equiv\; 2L\frac{\sin(t_j)}{[1 - \cos(t_j)]^2}\frac{2}{N+1}\sum_{m=1}^{N} \sin(mt_j)\,\frac{[1 - \cos(m\pi)]}{m} \tag{19.58}$$

3. $x \in [-\infty, \infty]$: The transformation of the integral is (Boyd, 1987b, and Chapter 17), using the same map as yields the TB rational Chebyshev basis:

$$x \; = \; L \cot(t), \qquad L \text{ is a user-choosable constant} \qquad (19.59)$$

$$I \; = \; \int_0^\infty f(x)dx \; = \; \int_0^\pi f(L\cot(t)) \, 2L \, \frac{1}{\sin^2(t)} \, dt \qquad (19.60)$$

The quadrature is

$$I_N \; \equiv \; \sum_{j=1}^N w_j \, f(L\cot(t_j)) \qquad (19.61)$$

$$t_j \; \equiv \; \pi j/(N+1), \qquad j = 0, 1, \dots, N+1 \qquad (19.62)$$

$$w_j \; \equiv \; \begin{cases} L\pi/[\sin^2(t_j)(N+1)], & 0 < j < (N+1) \\ L\pi/[\sin^2(t_j)(2N+2)], & j = 0 \text{ and } j = N+1 \end{cases} \qquad (19.63)$$

Because the transformed integrals have integrands which are *antisymmetric* with respect to $t = 0$ for the finite and semi-infinite intervals, Boyd expanded the product of f with the metric factor in terms of *sine* cardinal functions. Thus, the endpoint are *omitted* for these two cases so that each uses only the N interior points of the $(N+2)$-point Lobatto grid.

Alternatively, one may expand only f as a Chebyshev or Fourier series and then evaluate integrals of the products of the cardinal functions with the metric factor, which is $\sin(t)$ for the finite interval. This requires two additional evaluations of f, but raises the degree of f for which the approximation is exact only by one. The quadrature weights given by Fraser and Wilson (1966) as their Eq. (4.3).

Clenshaw and Curtis (1960) and Gentleman(1972a, b, c) prefer to compute the Chebyshev series for $f(x)$ first by a Fast Fourier transform and then integrate term-by-term through a recurrence formula. This is an $O(N \log_2(N))$ procedure whereas the cost of evaluating the weights through Boyd's method is $O(N^2)$. Since these weights need only be computed once for a given N, however, we recommend Boyd's procedure.

19.8.4   Integration of Periodic Functions and the Trapezoidal Rule

Every student of numerical analysis learns the trapezoidal rule, and learns that the error is $O(1/N^2)$ where N is the number of grid points. Indeed, this estimate of only second order accuracy is correct for *non-periodic* functions. However, when the integrand $f(x)$ is *periodic* and *infinitely differentiable* for all real x and the integration interval coincides in length with the spatial period P, the trapezoidal rule becomes *spectrally accurate*.

Theorem 38 (TRAPEZOIDAL RULE ERROR for PERIODIC INTEGRANDS)
trapezoidal rule approximation to the integral of a function $f(x)$ on the interval $x \in [-P/2, P/2]$ is

$$\int_{-P/2}^{P/2} f(x)dx \equiv I_N = h \left\{ \frac{f(-P/2) + f(P/2)}{2} + \sum_{k=-(N-1)}^{N-1} f(-P/2 + hk) \right\} \qquad (19.64)$$

$h \equiv P/N$ *is the grid spacing. If $f(x)$ has the Fourier series*

$$f(x) \equiv \frac{1}{2}\alpha_0 + \sum_{n=1}^\infty \alpha_n \cos\left(n\frac{2\pi}{P}x\right) + \sum_{n=1}^\infty \beta_n \sin\left(n\frac{2\pi}{P}x\right), \qquad x \in [-P/2, P/2], \quad (19.65)$$

then the error in the trapezoidal rule is given without approximation as

$$E_N \equiv \int_{-P/2}^{P/2} f(x)dx - I_N = -P\{\alpha_N + \alpha_{2N} + \alpha_{3N} + \dots\} \qquad (19.66)$$

If $f(x)$ is non-periodic, then the Fourier coefficients decrease as $O(1/N^2)$ and the trapezoidal rule is only second-order accurate, as taught in elementary numerical analysis texts.

If $f(x)$ is periodic with period P and analytic on the interval $x \in [-P/2, P/2]$, then the Fourier coefficients are exponentially small *in N and the trapezoidal rule is* spectrally accurate.

Proof: The error series follows trivially from Theorem 19 because the integral is also the constant in the Fourier series. The rest of the theorem follows from the Fourier convergence theory for non-periodic and periodic functions described in Chap. 2. Q. E. D.

It is important to note that since the trapezoidal rule uses an evenly spaced grid, the Clenshaw-Curtis adaptivity still applies.

19.8.5   Infinite Intervals and the Trapezoidal Rule

If $f(x)$ is a function that decays exponentially fast with $|x|$, then integrals on the infinite interval can be performed with spectral accuracy by applying the trapezoidal rule on a large but finite interval through "domain truncation" (Sec 17.2). The approximation is the same as Eq.(19.64),

$$\int_{-\infty}^{\infty} f(x)dx \approx I_N \equiv h\left\{ \frac{f(-P/2) + f(P/2)}{2} + \sum_{k=1}^{N-1} f\left(-P/2 + hk\right) \right\} \qquad (19.67)$$

where

$$|I - I_N| \sim \max_{|x|>L}\left(|f(x)|\right) + O(\alpha_N) \qquad (19.68)$$

where α_N is again the N-th Fourier coefficient when $f(x)$ is expanded as a Fourier series on the interval $x \in [-P/2, P/2]$. Note that because $f(x)$ is exponentially small at the ends of the (large!) interval $x \in [-P/2, P/2]$, it is quite unnecessary for $f(x)$ to be periodic, and the discontinuity in the Fourier series at the ends of the interval will be $O(f(P/2))$ and therefore exponentially small if we have chosen a sufficiently large P.

Because of its simplicity, the trapezoidal rule is usually preferable to the TB-integration formula (19.61). However, the latter can be sometimes applied to integrands that decay only as an inverse power of x whereas the trapezoidal rule is always restricted to exponentially decaying f. Its accuracy is also more sensitive to the choice of P than is the TB-integration rule to the choice of map parameter L (Boyd, 1982a). However, because numerical integration is fairly cheap task, this sensitivity of the trapezoidal rule is rarely a problem.

19.8.6   Singular Integrands

Integrals with integrable singularities can be computed to spectral accuracy by making a change of coordinates. For example, suppose that the function is singular at both ends of the interval $x \in [-1, 1]$. Make the change of variable

$$x = \tanh(z/L) \qquad (19.69)$$

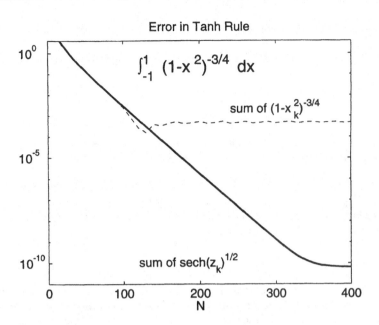

Figure 19.4: An illustration of the exponential convergence of the tanh-rule where N is the number of grid points. The integrand $(1 - x^2)^{-3/4}$ is singular at both endpoints. After the change of coordinates to $x = \tanh(x/L)$ where L is arbitrarily chosen to be one here, the integral is converted to a nonsingular integral on $z \in [-\infty, \infty]$. The "tanh rule" is simply the trapezoidal rule, applied on a large but finite interval in z. The tanh rule is sensitive to roundoff. When the integrand is expressed in terms of $\text{sech}(z)$, one can obtain better than ten decimal place accuracy (thick, solid curve). However, when the integrand is computed in its original form $(1-x^2)^{-3/4}$, the tanh rule gives only three correct digits (dashed curve). The difficulty with $(1-x^2)^{-3/4}$ is that near the endpoints where $x^2 \approx 1$, this is the fractional power of the small difference of two much larger terms, generating large roundoff error. (Note that the tanh rule has an exponentially high density of points, i. e., lots of points, in these endpoint regions of strong cancellation.)

where L is a user-choosable scaling factor. Noting that $dx/dz = (d/dz)\tanh(z/L) = (1/L)\text{sech}^2(z/L)$ it follows that

$$I \equiv \int_{-1}^{1} f(x)\,dx = \frac{1}{L} \int_{-\infty}^{\infty} f\left(\tanh(z/L)\right)\,\text{sech}^2(z/L)\,dz \qquad (19.70)$$

If $f(x)$ blows up more slowly than a first order pole as $x \to 1$, that is, $f(x)$ grows no faster than $(1 - x)^{-\alpha}$ for some $\alpha < 1$ and similarly at the other endpoint, then (i) the original integral in x is finite and (ii) the transformed integral converges because the sech^2 factor, which arises from the change of coordinates, decays faster than $f\left(\tanh(z/L)\right)$ grows as $|z| \to \infty$.

To recover an exponential rate of convergence, the transformed integral can be approximated by any spectrally-accurate quadrature for the infinite interval. The simplest is the trapezoidal rule, which becomes

$$I_N \equiv \frac{h}{L} \left\{ \frac{f(\tanh(-P/(2L))) + f(\tanh(P/(2L)))}{2} \operatorname{sech}^2(P/(2L)) \right.$$

$$+ \left. \sum_{k=1}^{N-1} f\left(\tanh(-P/(2L) + hk/L)\right) \operatorname{sech}^2\left(-P/(2L) + hk/L\right) \right\} \qquad (19.71)$$

where $h \equiv P/N$ is the grid spacing in the new coordinate z and where P is chosen sufficiently large that the integrand is negligible at $z = P/(2L)$. This formula is usually called the "tanh" rule for obvious reasons.

When the integral is singular at a point $x = a$ in the interior of the integration interval, one should split the integral into two and apply the tanh rule separately to each piece. If there is only a single singularity, one can replace the tanh mapping by a similar exponential mapping onto a semi-infinite integral.

19.8.7   Sets and Solitaries

Sometimes, one needs to repeatedly integrate a large set of integrals which fall into a common class. For this case, the most efficient procedure is to estimate how large N must be by computing a few members of the class for several different N. One may then use the smallest acceptable N for all the remaining integrals. Since most computations are performed for fixed N, Gaussian quadrature is the cheapest way to integrate a whole set of integrals.

When one needs to integrate a single, solitary integral, then an adaptive procedure like the Curtis-Clenshaw strategy is more efficient. The non-Gaussian spectral quadrature schemes described above are then very effective.

Chapter 20

Symbolic Calculations: Spectral Methods in Algebraic Manipulation Languages

"[A computer] can arrange and combine its numerical quantities exactly as if they were letters or any other general sysmbols; and in fact it might be possible to bring out its results in algebraic notation, were provisions made accordingly."

– Augusta Ada Byron, Countess of Lovelace (1844)

"Originally, the calculation [of the Bondi metric problem in general relativity] had required something like six months by hand ... the computer calculation, using REDUCE2, required only 90 seconds on a CDC-7600 ... and corrected six errors in the published manual derivation."

– R. A. d'Inverno (1976)

20.1   Introduction

Computer algebra has had a strange and checkered history. The notion of manipulating symbols instead of merely numbers is as old as the concept of the computer itself as shown by the quote from the Countess of Lovelace. Algebraic manipulation (AM) languages such as REDUCE and MACSYMA have been available for more than twenty years. And yet the spread of applications of algebraic manipulation has been, to use the punning words of a 1979 reviewer, "a case of creeping flow". Two decade later, the flow is still creeping, but with a noticeably smaller viscosity.

Several problems have retarded the growth of "AM" languages. First, symbolic calculations are best done *interactively* with a live human watching the stream of formulas materialize on the screen, and intervening as needed with substitution and simplification rules. Symbolic manipulation was therefore a rather and slow painful business on screenless, teletype terminals. Second, AM programs are terrible memory hogs; many research projects were prematurely truncated through running out of storage. Third, symbolic manipulation is inherently an order-of-magnitude more expensive than numerical manipulation: a number is a number is a number (with apologies to Gertrude Stein), but a symbolic variable must be tagged with all sorts of baggage to indicate precisely what the symbol symbolizes. Any bold soul who wants to push the state of the art in calculating perturbation series will sadly turn away from Maple and Mathematica to slog through the agonies of writing a FORTRAN program to do the same job – the numerical program will be ten times as long, but it may run 100 times faster and require only a tenth the storage to reach a given perturbation order.

Paradoxically, however, it is as certain as tomorrow's sunrise that algebraic manipulation will triumph in the end. The scientist of the twenty-first century will use AM everyday, and will no more be able to imagine life without Maple or Mathematica or their ilk than his counterpart of the 30's could imagine designing without a slide rule. It seems unlikely, after twenty years of waiting, that algebraic manipulation will ever become the *major* computational tool for more than a tiny fraction of scientific projects. As more and more engineers have AM on their desks, however, algebraic manipulation will likely become an indispensable tool for small, "scratchpad" calculations of the kind we do every day.

It follows that symbolic spectral codes will always remain few compared to spectral models that manipulate only numbers. Nevertheless, there are several reasons why spectral algorithms and algebraic manipulation languages are an especially favorable pairing.

First, interpolatory and non-interpolatory spectral methods both manipulate *polynomials* and *trigonometric* functions, which are precisely what REDUCE, Maple, and Mathematica multiply, divide, differentiate, and integrate most easily. Second, spectral methods have always been used at *low* order to compute *analytical* solutions to *simple* problems (Finlayson, 1973). Computer algebra greatly extends the "analytic range" of spectral methods to more complex problems and higher resolution. Third, a major use of symbolic manipulation has been to compute perturbation series. In many applications, the perturbation series is a spectral series (Rand, 1984; Rand and Armbruster, 1987; Mills, 1987; Haupt and Boyd, 1988, 1991).

In the remainder of this chapter, we discuss the striking differences between numerical and symbolic spectral methods. When calculating by hand or by REDUCE, one must apply very different rules to choose the most efficient method, the optimum strategy. Indeed, many of the precepts for numerical calculations must be reversed or inverted when the number of degrees of freedom is small and the goal is not a table or a graph but an algebraic expression. The next two sections discuss such "inversions". The fourth section is a catalogue of the little tricks and devices that are important in writing efficient programs. Sec. 5 describes several examples. The final section is a summary of the current state-of-the-art and lists several open problems.

20.2 Strategy

Table 20.1 summarizes four strategies for symbolic calculations. All spectral methods substitute a series with unknown coefficients a_j into the differential equation to obtain the so-called residual function, $R(x; a_0, a_1, \ldots, a_{N-1})$ where N is the number of basis functions in the approximation. If the trial solution $u_N(x)$ were exact, then the residual would be zero for all x. With a truncated spectral approximation, the best we can do (barring special cases) is to make the residual as small as possible in some sense.

The most widely used discretization strategy for numerical applications is known variously as "collocation", "orthogonal collocation", "discrete ordinates" or the "pseudospectral" method. One obtains N equations in N unknowns by demanding that the residual

Table 20.1: Precepts for spectral methods: Numerical ($N >> 1$) versus Symbolic (N small)

SYMBOLIC [$N \sim O(3)$]	NUMERICAL [$N >> 1$]
Galerkin or Methods of Moments	Collocation
Legendre or Gegenbauer or Powers	Chebyshev polynomials

Polynomialization
    (i)Fourier and rational Chebyshev $\Longrightarrow$ Chebyshev
    (ii)Transcendental data and forcing $\Longrightarrow$ Polynomial
	Unnecessary

Rationalization
    (i)Collocation at rational points
 	
    (ii)Replace irrational numbers
    &floating point numbers by rational numbers
	(i) Collocation at roots
	of T_N, which are irrational
	(ii) Unnecessary

function should be zero at each of N interpolation points. In contrast, the Galerkin method demands that the residual should be orthogonal to each of the first N basis functions. The accuracy difference between the two discretization strategies is negligible for large N, so collocation is usually preferred because it requires fewer computations and less programming (Chapter 4).

For small N, however, Galerkin's method may be two or three times more accurate than collocation. To put it another way, the rule of thumb quoted in Chapter 3, Sec. 9, is that the Galerkin solution with $N = m$ is usually as accurate as the pseudospectral solution with $N = m + 1$ or $m + 2$, provided that the solution is sufficiently smooth that a small N is sufficient to resolve it. Since the complexity of a symbolic answer usually grows exponentially with N, we must reverse the usual precept, and suggest that for calculations in Maple or REDUCE, etc., one should use Galerkin's method wherever possible.

Unfortunately, when a problem is cursed with an exponential nonlinearity as for the Bratu equation, $u_{xx} + \exp(u) = 0$, Galerkin's method may be completely unfeasible. However, it may be still be possible to get a simple but accurate answer by using collocation, as in Boyd (1986c), Finlayson (1973) and our Example Two.

In floating point computations, it is almost always necessary, unless N is very small, to represent both the basis functions and the weight functions as orthogonal polynomials. Because algebraic manipulation languages compute in exact, rational arithmetic wherever possible and also because N is very small, one may use the equivalent powers of x as both basis functions and as the Galerkin test functions, in which case Galerkin's method is usually called the "method of moments". Numerical ill-conditioning, which is a perennial worry in high resolution number-crunching, is almost never a worry in symbolic manipulation.

A second issue is the choice of the basis functions. As explained in Chapter 2, the Chebyshev and Legendre polynomials are part of a large family ("Gegenbauer polynomials") which are orthogonal on the interval [-1, 1] with weight function $w(x) = (1 - x^2)^\alpha$. The Chebyshev polynomials are the special case $\alpha = -1/2$; the Legendre are $\alpha = 0$, i. e., no weighting at all. The Chebyshev weighting, heavily biased towards the endpoints, is optimum for expanding *general* functions; the Legendre error is smaller than the Chebyshev over most of the interval, but is much larger (roughly by $O(N^{1/2})$) near the endpoints as illustrated in Sec. 13 of Chapter 2.

The solution to a boundary value problem is not arbitrary, however; its boundary behavior is strongly constrained by the boundary conditions. Therefore – especially for an approximation with a very small number of degrees of freedom – it is preferable to replace Chebyshev polynomials by Legendre polynomials or perhaps even Gegenbauer polynomials of larger α (Finlayson, 1973). If α is a positive integer, there is the further benefit that the Galerkin integrals are simpler because of the absence of the square root of the Chebyshev weight, increasing the probability that the integration routines, built-in to all major algebraic languages, will be able to perform the integrations analytically.

The third strategic suggestion has no counterpart in numerical calculations: It is to convert all transcendental functions to polynomials. A Fourier cosine series, for example, can be converted into a Chebyshev series by making the change of variable $t = \arccos(x)$ and using the identity $T_n(x) = \cos(nt)$ for all n. The rational Chebyshev functions, which are a basis for infinite and semi-infinite domains (Chapter 17), are also just Chebyshev polynomials after a change of variable. Similarly, powers of $\mathrm{sech}(x)$, which often arise in soliton theory, can be "polynomialized" by defining $z = \tanh(x)$ and applying the identity $\mathrm{sech}^2(x) = 1 - \tanh^2(x) = 1 - z^2$. Lastly, transcendental coefficients of a differential equation can be approximated by polynomials. The error in

$$\cos([\pi/2]x) \approx 1 - 0.4967(\pi x/2)^2 + 0.03705(\pi x/2)^4, \tag{20.1}$$

for example, is less 0.0009 on $x \in [-1, 1]$.

It is often convenient to delay "polynomialization" until an intermediate stage. Maple, for example, can differentiate and multiply trigonometric polynomials and then convert them to a standard Fourier series through the "combine(expression,trig)" command. However, it has no command to pick off the coefficients of a Fourier series, so polynomialization through a command like "subs(z=cos(x),z**2=cos(2*x), z**3=cos(3*x), expression)" is essential to exploiting the "coeff" command, which does return a particular coefficient of a polynomial.

Algebraic manipulation languages are more efficient at calculations with integers and rational numbers than floating point numbers, so the fourth strategy is "rationalization", that is, converting floating point numbers to rational numbers. The usual collocation points, for example, are the irrational roots or extrema of the $(N + 1)$-st basis function. When collocation, as opposed to the preferred Galerkin method, is necessary, it is usually better to approximate an irrational collocation point such as $(1/5)^{1/2} = 0.44721\ldots$ by a rational number such as $4/9 = 0.444\ldots$. For small N, the error is small. Other floating point numbers in the boundary conditions, differential equation, etc., should usually be "rationalized" too.

Arnold (1983) shows that for any irrational number μ, there exists a sequence of rational approximations, p/q, which converge to μ. The error of a given member of the sequence is less than the reciprocal of the square of the denominator, i. e.,

$$|\mu - p/q| < 1/q^2 \tag{20.2}$$

The well-known approximation $\pi \approx 355/113$, which has a relative error of less than 10^{-6}, was long thought to be very special. What (20.2) shows, in words, is that for *any* μ, there exists an approximation with a three-digit denominator and six-digit accuracy. The sequence of optimum rational approximations can be computed by an algorithm known variously as the "Euclidean algorithm", "algorithm of continued fractions", or "the algorithm of stretching the noses" (Arnold, 1983). The floating point coefficients in (20.1) can be approximated by 75/151 and 1/27, for example, with errors of only 0.00011 and 0.000014, respectively.

There are two major exceptions to "rationalization". First, when the answer depends only a single irrational such as $2^{1/2}$, it may be more efficient to carry a symbolic variable ROOT2 through the remainder of the algebra than to replace it by a rational number. The second is that rationalization may be unnecessary if the later stages of the computation have to be done numerically, as often happens. It is not, alas, unusual to see hundred digit numbers when a long calculation is performed in exact arithmetic ("integer swell") as shown in Table 20.5 below. The precepts in Table 20.1 should be interpreted as suggestions and guidelines rather than commandments. Let common sense be the highest law!

Table 20.2: Maple program for Example One: Linear BVP with Boundary Layer

```
u:= (1-x*x) * (a0 + a1*x*x + a2*x**4 + a3*x**6);
R4:=tau*tau*diff(u,x,x) - u + 1;
eq0:=integrate(R4,x=-1..1); eq1:=integrate(R4*x*x, x=-1..1);
eq2:= integrate(R4*x**4, x=-1..1); eq3:= integrate(R4*x**6, x=-1..1);
solve( eq0, eq1, eq2, eq3, a0, a1, a2, a3 );
```

A couple of other important pieces of advice have been omitted from the table because they are not inversions of usual practice. First, exploiting parity and other symmetries (Chapter 8) is always highly advantageous in numerical calculations. The sort of simple, idealized problems that are amenable to symbolic spectral methods are precisely the kind of problem that are most likely to have symmetries. Therefore, the standard precept: Look for Symmetry! is redoubled for symbolic computations.

Second, one may impose boundary conditions through either "boundary-bordering" or "basis recombination" as explained in Chapter 6. In symbolic calculations, the latter is strongly preferred because incorporating the boundary conditions into the basis functions may halve the number of degrees of freedom, and enormously reduce the complexity of the algebraic formula which is the answer. Furthermore, it is particularly easy to incorporate homogeneous boundary conditions into a symbolic answer: for homogeneous Dirichlet conditions: Multiply the trial solution by $(1 - x^2)$.

20.3 Examples

All examples were programmed in REDUCE or Maple. The choice of languages has no particular significance except that both are widely used and the author knows them.

Example One: Linear ODE BVP (Carrier & Pearson, 1968)

$$\epsilon^2 u_{xx} - u = -1; \qquad u(-1) = u(1) = 0 \tag{20.3}$$

where ϵ is a constant. The exact solution is

$$u(x) = 1 - \cosh(x/\epsilon)/\cosh(1/\epsilon) \tag{20.4}$$

Because $u(x)$ is symmetric, we shall assume an expansion in even powers of x. Further, we incorporate the homogeneous boundary condition into the series. Arbitrarily, we choose $N = 4$:

$$u_4(x) = (1 - x^2)\left\{a_0 + a_2 x^2 + a_4 x^4 + a_3 x^6\right\} \tag{20.5}$$

Substituting (20.5) into (20.3) gives the residual function

$$\begin{aligned} R_4 &\equiv \epsilon u_{4,xx} - u_4 + 1 \\ &= 1 - \left\{[1 + 2\epsilon^2]a_0 - 2\epsilon^2 a_2\right\} + \left\{a_0 - [1 + 12\epsilon^2]a_1 + 12\epsilon^2 a_2\right\}x^2 \\ &\quad + \left\{a_1 - [1 + 30\epsilon^2]a_2 + 30\epsilon^2 a_3\right\} + \left\{a_2 - [1 + 56\epsilon^2]\epsilon^2 a_3\right\}x^6 + a_3 x^8 \end{aligned} \tag{20.6}$$

Because $u(x)$ has a boundary layer of width $O(\epsilon)$ – all the variation of $u(x)$ is close to the endpoints – we shall not include a factor of $(1 - x^2)$ in the Galerkin conditions, but instead demand that the unweighted integrals of R_4 with each of the first four even powers of x should be zero:

$$\int_{-1}^{1} x^{2j} R_4(x; a_0, a_1, a_2, a_3)dx = 0, \qquad j = 0, 1, 2, 3 \tag{20.7}$$

Table 20.3: The maximum pointwise errors in the four-term Legendre approximation as a function of the parameter ϵ for Example One: Linear BVP with Boundary Layer

ϵ	L_∞ error
1/20	0.081
3/40	0.030
1/10	0.012
3/20	0.0046
1/5	0.00040
1/4	0.000099
3/10	0.000028
4/10	0.0000033
1/2	0.00000056
3/4	1.7E-8
1	1.3E-9

This gives four equations in four unknowns. This can be cast as a matrix equation of the form H A = F and solved in REDUCE by the single statement A:=(1/H)*F. A REDUCE program for general 2d order BVPs is Table 17-1 of the first edition of this book (Boyd, 1989a); the equivalent Maple program is Table 20.2;.

The solution is

$$a0 \;=\; 3(43243200\epsilon^6 + 2162160\epsilon^4 + 27720\epsilon^2 + 31)/D(\epsilon)$$

$$a1 \;=\; 33(327600\epsilon^4 + 41)/D(\epsilon)$$

$$a2 \;=\; 429(840\epsilon^2 - 13)/D(\epsilon)$$

$$a3 \;=\; 6435/D(\epsilon) \tag{20.8}$$

$$D(\epsilon) = 128(2027025\epsilon^8 + 945945\epsilon^6 + 51975\epsilon^4 + 630\epsilon^2 + 1) \tag{20.9}$$

Observe that the residual function R_4 – and therefore the elements of the matrix equation – are quadratic in ϵ. The coefficients of the spectral series a_j are rational functions of ϵ with numerator and denominator polynomials of degree no higher than the degree of elements of H multiplied by N, that is $2 \cdot 4 = 8$. This rational dependence upon a parameter is not special to this equation, but is a consequence of the following.

Theorem 39 (MATRICES WHOSE ELEMENTS DEPEND on a PARAMETER)   *Let the elements of a matrix H of dimension N be polynomials in a parameter ϵ of degree at most k and let A denote the column vector which solves $HA = F$ where F is a column vector whose elements are polynomials in ϵ of degree at most m. Then (i) $\det(H)$ is a polynomial in ϵ of degree at most Nk. (ii) The elements of A are rational functions of ϵ. The degree of the numerator is $k(N-1) + m$; the degree of the denominator is kN.*

The proof is a trivial consequence of Cramer's Rule for solving matrix equations (Vinograde, 1967, pg. 60) and the definition of a determinant (op. cit., pg. 169). Q. E. D.

The boundary layer is so narrow for $\epsilon = 1/10$ that the half-width of the layer is only 0.07, that is, 50% of the drop of $u(x)$ from its maximum to zero occurs between $x = 0.93$ and $x = 1$. Nevertheless, the maximum pointwise error of the four-term Legendre-Galerkin approximation is only 0.017. Table 20.3 shows that the error is no worse than 8% even for ϵ as small as 1/20.

Like all spectral approximations, $u_4(x; \epsilon)$ is highly uniform in the *coordinate* x, but this does not imply uniformity in the *parameters*. In this example, the error is highly non-uniform in the parameter ϵ.

Example Two: The Quartic Oscillator of Quantum Mechanics
The special challenge of this eigenvalue problem is that it is posed on an infinite interval.

$$u_{yy} + (E - y^4)u = 0, \qquad y \in [-\infty, \infty] \qquad (20.10)$$

with the boundary condition

$$|u| \to 0 \qquad \text{as } |y| \to \infty \qquad (20.11)$$

where E is the eigenvalue, the energy of the quantum bound state. The rational Chebyshev functions described in Boyd (1987a) and in Chapter 17 are a good basis set for the infinite interval. To "polynomialize" these rational functions and simplify the integrals, make the change of variable

$$y = Lx/\sqrt{1 - x^2} \qquad (20.12)$$

where L is a user-choosable constant ("map parameter"). By using the chain rule or Table E-5 here, this change-of-coordinate transforms the problem into

$$(1 - x^2)^4 \left\{ (1 - x^2)u_{xx} - 3xu_x \right\} /L^2 + \left\{ (1 - x^2)^2 E - L^4 x^4 \right\} u = 0 \qquad (20.13)$$

$$x \in [-1, 1], \qquad u(\pm 1) = 0 \qquad (20.14)$$

The rational Chebyshev functions are converted into ordinary Chebyshev polynomials by this transformation, so both the infinite interval and rational basis functions are swept under the rug by the mapping (20.12).

The solution is not very sensitive to the map parameter L as long as it roughly matches the scale of the basis functions to the width of $u(x)$; we somewhat arbitrarily set $L = 2$.

One can prove that the eigenfunctions of (20.10) are all symmetric (for even mode number) or antisymmetric (for odd mode number). To compute the lowest few symmetric modes, we set

$$u = (1 - x^2)\{a_1 + a_2 x^2 + a_3 x^4 + a_4 x^6 + a_5 x^8\} \qquad (20.15)$$

Table 20.4: Maple program for Example Two: Quartic Oscillator

```
Q:=1-x*x; L:=2; # L=map parameter;
u:=(1-x*x)*(a1+a2*x*x+a3*x**4+a4*x**6+a5*x**8);
R:=Q**4 *(Q*diff(u,x,x)-3*x*diff(u,x) )/(L*L) + (lamb*Q*Q - L**4 * x**4 )*u;
# E is a Maple reserved symbol, so we use "lamb" for the eigenvalue
# To use Gegenbauer polynomials of first order instead of Legendre,
# multiply the integrand in the loop by (1-x*x);
for ii from 1 by 1 to 5 do eq.ii:=integrate(x**(2*ii-2)*R,x=-1..1); od:
with(linalg); J:=matrix(5,5);
for ii from 1 by 1 to 5 do for j from 1 by 1 to 5 do
J[ii,j]:=coeff(eq.ii,a.j,1);  od;  od;
secdet:=simplify(det(J)); # secdet=D(E)
Eigenvalue:=fsolve(secdet,lamb);
```

Table 20.5: Errors for Example Two: Quartic Oscillator

	Exact	$w(x) = 1$: Legendre		$w(x) = (1-x^2)$: Gegenbauer	
n	E_{exact}	$E_{numerical}$	Error	$E_{numerical}$	Error
0	1.060	1.065	0.005	1.061	0.001
2	7.456	7.655	0.199	7.423	-0.033
4	16.26	17.38	1.12	17.17	0.91
6	26.528	64.32	Huge	42.53	Huge
8	37.92	715.66	Huge	330.7	Huge

We converted the differential equation into a 5-dimensional matrix problem by demanding that the integral of the residual with the test functions $\{1, x^2, x^4, x^6, x^8\}$ should be 0. Since the matrix equation is homogeneous, it has a solution only if its determinant $D(E) = 0$. The vanishing of this "secular determinant", to borrow the usual terminology of quantum mechanics, determines the eigenvalue E.

Table 20.4 is a short Maple program which produces

$$D(E) = 1 - 1.143E + 0.203E^2 - 0.0102E^3 + 0.000124E^4 - 0.000000153E^5 \qquad (20.16)$$

after the coefficients have been converted to floating point numbers. The exact determinant, given in Boyd (1993), illustrates "integer swell": the coefficients are ratios of as many as 29 digits divided by 35 digits. Table 20.5 lists the roots, the exact eigenvalues (from Bender and Orszag, 1978, pg. 523), and the absolute errors. As is usual in numerical solutions to eigenvalue problems, the lowest numerical eigenvalue is very accurate, the eigenvalues up to roughly $N/2$ are of decreasing accuracy, and the upper few eigenvalues of the $N = 5$ matrix bear no useful resemblance to those of the differential equation. The accuracy of the lowest root is quite good for such a low order approximation. For comparison, we redid the problem with the inner product weight of $(1 - x^2)$ instead of 1. As suggested in the previous section, there is a striking improvement in accuracy, at least for the lowest two modes.

Example Three: Branching (Bifurcation) for the Fifth-Order KdV Equation

The twist in this problem (Boyd, 1986a) is that the equation depends nonlinearly on the eigenparameter a. This is a serious complication for conventional software, but computer algebra problems are indifferent as to whether the matrix is linear or nonlinear in a.

The problem is to compute a solution to the homogeneous, linear equation

$$-\Delta_{xxxxx} - c\Delta_x + \{U(x; a)\Delta\}_x = 0 \qquad (20.17)$$

$$\Delta(x) = \Delta(x + 2\pi) \qquad \forall x \qquad (20.18)$$

where $U(x; a)$ is the "cnoidal wave" of the Fifth-Order Korteweg-deVries equation (FKdV), that is, a spatially periodic solution to the nonlinear eigenvalue problem

$$-U_{xxxxx} - cU_x + UU_x = 0 \qquad (20.19)$$

$$U(x) = U(x + \pi) \qquad \text{for all } x \qquad (20.20)$$

where $c(a)$ is the eigenvalue and the one-parameter family of cnoidal waves is parameterized by a, the amplitude of $\cos(2x)$ in the Fourier series for $U(x; a)$. The physical background is rather complicated (Boyd, 1986a), but the salient facts are these. The cnoidal waves which solve (20.19) have a period of π, are symmetric with respect to $x = 0$, and exist for all real values of a. For sufficiently large amplitudes, (20.19) also has "bicnoidal" solutions which have a period of 2π – double that of the cnoidal waves – and two peaks flanking the origin. For some value of a, the bicnoidal waves are a subharmonic bifurcation from the cnoidal wave. Away from the bifurcation, the linearized equation for small perturbations to the cnoidal wave, (20.17), has no solution except the trivial one. At the bifurcation point, however, (20.17) has a nontrivial eigensolution because the solution to (20.19) is no longer unique. The cnoidal and bicnoidal solutions coexist at the bifurcation; near but not actually at the bifurcation point, the bicnoidal wave is approximately described by $u_{bicnoidal}(x) \approx U(x; a) + g\Delta(x)$ for some constant g where $\Delta(x)$ is the eigensolution (with zero eigenvalue) of (20.17).

To determine the bifurcation point, we must first compute $U(x; a)$ and then substitute the result into (20.17). When the linearized equation is discretized by a spectral method, the resulting matrix equation will have a nontrivial solution if and only if the determinant $\Delta(a) = 0$. Thus, the bifurcation point $a = a_{bifurcation}$ can be found by computing the roots of the determinant $\Delta(a)$.

To compute the cnoidal wave, we could apply the spectral method to (20.19) followed by one or two Newton's iteration. It turns out, however, that the bicnoidal wave bifurcates at such a small amplitude that the cnoidal wave is adequately approximated at the bifurcation point by the first two terms of its perturbation series ("Stokes' expansion") (Boyd, 1986a):

$$U(x; a) \approx a \cos(2x) + [a^2/960] \cos(4x) \tag{20.21}$$

$$c(a) \approx -16 \left\{1 - a^2/30720\right\} \tag{20.22}$$

Exploiting the known symmetry of $\Delta(x)$, we assume a Fourier series and choose $N = 4$:

$$\Delta(x) \approx a_1 \cos(x) + a_2 \cos(2x) + a_3 \cos(3x) + a_4 \cos(4x) \tag{20.23}$$

Because the equation contains only *odd* derivatives of Δ, all the cosines are turned into sines. Therefore, we demand that the residual should be orthogonal to the first four sine functions, which converts (20.17) into the 4 x 4 matrix system

$$\begin{vmatrix} -28800 - 960a + a^2 & 0 & -960a - a^2 & 0 \\ 0 & 0 & 0 & -1920a \\ -960a - a^2 & 0 & 124800 + a^2 & 0 \\ 0 & -960a & 0 & 460800 + a^2 \end{vmatrix} \begin{vmatrix} a_1 \\ a_2 \\ a_3 \\ a_4 \end{vmatrix} = \begin{vmatrix} 0 \\ 0 \\ 0 \\ 0 \end{vmatrix} \tag{20.24}$$

In a standard eigenvalue problem, the matrix elements are all linear functions of the eigenparameter a, but here the elements are quadratic in a. Nevertheless, taking the determinant of (20.24) gives a polynomial in a which can be solved for the bifurcation point:

$$\Delta(a) = a^2 \left(3a^3 + 860a^2 + 124800a + 3744000\right) \tag{20.25}$$

The only nonzero root is $a_{bifurcation} = -39.10$; the exact value is -39.37, an error of only 0.7%.

The other real root, $a = 0$, is the limit point of the cnoidal wave, legitimate but irrelevant and trivial.

The checkerboard of zeros in the Galerkin matrix implies that even and odd terms in the cosine series are uncoupled; we could have obtained the same answer by keeping only the $\cos(x)$ and $\cos(3x)$ terms in (20.23) as done in Boyd (1986a). Overlooking a symmetry will never give an incorrect answer, but it will give an unduly complicated and inaccurate one for a given N. We would have obtained a similar checkerboard of zeros for Example Three if we had not assumed a symmetric basis set to compute symmetric eigenmodes only.

Example Four: Reduction of a Partial Differential Equation to a Set of Ordinary Differential Equations: An Unorthodox Derivation of the Korteweg-deVries Equation

Spectral methods can be applied to reduce partial differential equations directly to an algebraic system for the spectral coefficients. Sometimes, as illustrated by this example, it is more useful to apply the spectral method only in one coordinate, reducing the problem to a system of ordinary differential equations in the remaining dimension. The ODE system can often be solved analytically, sometimes even when it is nonlinear as here.

To illustrate the possibilites, we analyze the travelling-wave solutions of the so-called "Ageostrophic Equatorial Wave" or "AEW" equation (Boyd, 1991b):

$$u_{xx} + u_{yy} - y^2 u - u/c - yu^2/c = 0, \qquad x, y \in [-\infty, \infty] \qquad (20.26)$$

$$|u(x, y)| \to 0 \ \text{as} \ |y| \to \infty \qquad (20.27)$$

and various boundary conditions in x. The phase speed c is the eigenvalue In the limit of infinitesimal amplitude, this nonlinear eigenvalue problem has solutions of the form

$$u(x, y) = \cos(kx) \ \psi_n(y), \qquad c = -1/(2n + 1 + k^2) \qquad (20.28)$$

where $\psi_n(y)$ is the n-th Hermite function. This suggests that the Hermite functions, which individually satisfy the latitudinal boundary condition of decay, would be a good spectral basis set.

The Hermite functions have definite parity with respect to $y = 0$, the equator, and one can show that the nonlinear term does not disrupt this symmetry. Therefore, to look for general, nonlinear solutions which are antisymmetric with respect to $y = 0$, it is sufficient to assume an expansion in the Hermite functions of odd n only:

$$u \approx A_1(x)2y \exp(-0.5y^2) + A_3(x) \left\{8y^3 - 12y\right\} \exp(-0.5y^2) \qquad (20.29)$$

Table 20.6: Maple Listing for Example Four: Reduction of a Partial Differential Equation

```
u:=exp(-y*y/2)*(a1(x)*2*y + a3(x)*(8*y*y*y-12*y));
R:=diff(u,x,x)+diff(u,y,y)-y*y*u-u/c-y*u*u/c;
# Next substitution is necessary because the presence of x-operators;
# will defeat the y-integration routine. This is a workaround.;
R:=subs(diff(a1(x),x,x)=A1xx,diff(a3(x),x,x)=A3xx,a1(x)=A1,a3(x)=A3,R);
integrand1:=R*2*y*exp(-y*y/2);
integrand3:=R*(8*y*y*y-12*y)*exp(-y*y/2);
# More workarounds. First, we apply Maple simplification &;
# expansion operators. Then, we use a special Maple command to do;
# the integrations term by term;
integrand1:=expand(simplify(integrand1)); integrand3:=expand(simplify(integrand3));
eq1:=0: eq3:=0:
for piece in integrand1 do eq1:=eq1+integrate(piece,y=0..infinity) od:
for piece in integrand3 do eq3:=eq3+integrate(piece,y=0..infinity) od:
eq1:=expand(eq1); eq3:=expand(eq3);
# Pick off the coefficient of A1xx, A3xx, and call them div1, div3;
div1:=coeff(eq1,A1xx,1); div3:=coeff(eq3,A3xx,1);
eq1:=expand(simplify(eq1/div1)); eq3:=expand(simplify(eq3/div3));
eq1f:=evalf(eq1); eq3f:=evalf(eq3);
```

and similarly at higher order. By applying the usual Galerkin method, we obtain (for $N = 2$) the set of two coupled nonlinear equations in two unknowns:

$$A_{1,xx} \quad -(3 + \tfrac{1}{c})A_1 \quad -\tfrac{1}{c}1.09A_1^2 - \tfrac{1}{c}1.45A_1A_3 - \tfrac{1}{c}19.8A_3^2 = 0$$

$$A_{3,xx} \quad -(7 + \tfrac{1}{c})A_3 \quad -\tfrac{1}{c}5.93A_3^2 - \tfrac{1}{c}1.65A_1A_3 - \tfrac{1}{c}0.0302A_1^2 = 0 \qquad (20.30)$$

The Maple listing is Table 20.6.

This example demonstrates one of the strengths of algebraic manipulation: Correction of errors in hand calculations. The set (20.30) was originally given in Boyd (1989e), but with the coefficient of A_1A_3 in the second equation equal to 0.82 instead of the correct value shown above. The error was found only when the derivation was repeated using computer algebra.

When (20.30) is further approximated by setting one or the other of the unknowns to zero, the result is the travelling-wave form of the Korteweg-deVries equation. The general solution of this $N = 1$ approximation is an elliptic function; limiting cases are infinitesimal cosine waves (which recovers (20.28)) and solitary waves.

When both modes are retained, the result is the system (20.30) which cannot be analytically solved. However, "maximally simplified" models like the two-mode system are often very useful in illuminating complicated behavior. E. N. Lorenz employed this philosophy in modelling two-dimensional convection by a system of three ordinary differential equations which had been previously derived by B. A. Saltzman. Lorenz' discovery of chaos in this system, which now bears his name, was a great milestone in the theory of chaotic dynamics, strange attractors, and fractals. Similarly, Boyd (1989e) shows that a two-mode model is the smallest which displays the "weakly nonlocal" solitary waves of the full AEW equation.

The single-mode approximation, which is just the KdV equation, is usually obtained by a very different route which employs the singular perturbation technique known as the "method of multiple scales" (Bender and Orszag, 1978, Boyd, 1980c). The end result, though, is identical with applying the spectral method with a single basis function. The perturbative analysis is really just a systematic way to answer the question: When is a one-mode approximation justified? The answer is that when the solitary wave is very wide with a length scale which is $O(1/\epsilon)$ and an amplitude of $O(\epsilon^2)$ where $\epsilon \ll 1$ is a small parameter, then the nonlinear terms coupling $A_1(x)$ with $A_3(x)$ will be small. The self-interaction term cannot be neglected, however, if the corresponding linear term is also small, i. e., of $O(\epsilon^4)$. This happens when the phase speed differs from the $k = 0$ ("long wave") limit of the linear phase speed by an amount of $O(\epsilon^2)$. It is also necessary to use a spectral basis whose functions have the structure of the infinitesimal waves, i. e., Hermite functions for this example.

The formal justification for small ϵ is heartening, but experience with the method of multiple scales (and few-mode models) is that both are often qualitatively accurate even when $\epsilon \sim O(1)$, that is, far outside their region of formal accuracy (Bender and Orszag, 1978, Finlayson, 1973). The symbolic spectral method is a very quick and general alternative to the perturbation theory.

The only caveat is one illustrated by all the examples above: To get the most out of the spectral method (or any other solution algorithm), it is important to understand the physics. Does $u(x)$ have a boundary layer? Is it smooth so that small N will be okay, or does it have a complicated structure that will require very large N (and probably make symbolic calculations unfeasible)? Is some choice of basis functions, such as Hermite functions for the AEW equation, particularly well-matched to the solutions? Computer algebra and numerical calculations are not alternatives to analysis but merely its extension.

20.4   Summary and Open Problems

There has been considerable progress in marrying spectral methods with algebraic manipulation languages as summarized here and in the review, (Boyd, 1993). None of the guidelines should be interpreted too rigidly; a couple of examples deliberately broke some of the precepts of earlier sections to emphasize the need for common sense.

The major open problem is coping with nonlinearity. The examples above show some success, but only for fairly easy problems. Perturbation expansions or a single Newton iteration may be essential in obtaining an answer which is simple enough to be useful.

Fox and Parker (1968) summarize a large body of hand calculations which exploited recurrence relations of the Chebyshev polynomials instead of Galerkin's method. These recurrences significantly improve accuracy for small N. This line of attack has been continued by Ortiz and his students as reviewed in Chapter 21. The Galerkin method is simpler and more general, so we have omitted these alternative Chebyshev algorithms from this discussion. However, these recurrences lend themselves well to symbolic computations.

Algebraic manipulation is still dominated by a "perturbation-and-power series" mentality. However, there is much more to symbolic mathematics than perturbation series or the derivation of exact algebraic relationships. The main algorithms of this work, spectral methods and Newton's iteration, are still badly underutilized in symbolic calculations. Perhaps some readers will change this.

Chapter 21

The Tau-Method

"I never, never want to be a pioneer ... It's always best to come in second, when you can look at all the mistakes the pioneers made — and then take advantage of them."
— Seymour Cray

21.1 Introduction

The tau-method is both an algorithm and a philosophy. It was invented by Cornelius Lanc-zos in the same (1938) paper that gave the world the pseudospectral method.

As an algorithm, the tau-method is a synonym for expanding the residual function as a series of Chebyshev polynomials and then applying the boundary conditions as side constraints. Thus, it is indistinguishable from what we have earlier described as Galerkin's method with "boundary bordering". The formal definition is the following.

Definition 43 (Tau-Method) *A mean-weighted residual method is a TAU-METHOD if the "test functions" are the Chebyshev polynomials and the inner product is the usual Chebyshev integral inner product.*

For example, to solve a second order ODE with $u(-1) = u(1) = 0$, the τ-method would impose the constraints

$$(T_m, R(x; a_0, \ldots, a_N)) = 0 \qquad m = 0, \ldots, N - 2 \qquad (21.1)$$

$$\sum_{n=0}^{N} a_n T_n(\pm 1) = 0 \qquad (21.2)$$

where the $\{a_n\}$ are the Chebyshev coefficients of $u(x)$ and $R(x; a_0, \ldots, a_N)$ is, as usual, the residual function obtained by substituting the truncated Chebyshev series into the differential equation.

In contrast, Canuto *et al.* (1988) prefer to apply the label "Galerkin" only to that mean weighted residual method that chooses the "test" functions to be basis functions that satisfy the homogeneous boundary conditions. For example,

$$(\phi_m, R(x; b_2, \ldots b_N)) = 0 \qquad m = 2, \ldots, N \qquad (21.3)$$

473

where the basis functions are

$$\phi_{2n}(x) \equiv T_{2n}(x) - 1 \qquad ; \qquad \phi_{2n+1}(x) \equiv T_{2n+1}(x) - x \tag{21.4}$$

These two methods do give slightly different answers because the two highest-degree constraints in (21.3) involve the inner product of the residual with $T_{N-1}(x)$ and $T_N(x)$ whereas the highest Chebyshev polynomial multiplying the residual in the tau-method inner product (21.1) is only $T_{N-2}(x)$. For this reason, this terminological distinction is popular in the literature (Gottlieb and Orszag, 1977, for example). However, the accuracy differences between the "tau" and "Galerkin" methods are likely to be negligible.

At the end of the chapter, we will return to issues of nomenclature.

First, however, we shall discuss the *philosophy* of the τ-method. Lanczos recognized that there are two ways of attacking a differential equation. The obvious approach is to compute an approximate solution to the exact, unmodified differential equation. The second is to compute the *exact* solution to a *modification* of the original differential equation. If the "modification" is small, then the solution to the modified problem will be a good approximation to that of the original problem.

This second strategy — to solve *approximate* equations exactly — is the *philosophy* of the τ-method. In the next section, we shall apply this tactic to approximate a rational function. In Sec. 3, we shall extend this philosophy to differential equations. Finally, in Sec. 4, we shall discuss Lanczos' "canonical polynomials", which have been the jumping-off point for a long series of papers by E. L. Ortiz and his collaborators.

21.2 τ-Approximation for a Rational Function

Suppose the goal is to approximate a rational function $f(x)$ by a polynomial f_N of degree N. Let

$$f(x) = P_p(x)/Q_q(x) \tag{21.5}$$

where $P(x)$ and $Q(x)$ are polynomials of degree p and q, respectively. Substituting $f \to f_N$ and multiplying through by $Q(x)$ gives

$$f_N(x)\, Q(x) = P(x) \tag{21.6}$$

which has only polynomials on either side of the equal sign. At first glance, it appears that we could compute the $(N + 1)$ coefficients of $f_N(x)$ merely by matching powers of x. However, the degree of $f_N(x)Q(x)$ is $(N + q)$; thus, we have more conditions on the coefficients of $f_N(x)$ than we have unknowns to satisfy them.

Lanczos observed that there is an *exact, polynomial* solution to the perturbed problem

$$f_N(x)\, Q(x) = P(x) + \epsilon(x) \tag{21.7}$$

where $\epsilon(x)$ is a polynomial of degree $(N + q)$ with q undetermined coefficients. Eq. (21.7) has polynomials of degree $(N + q)$ on both sides with a total of $(N + q + 1)$ unknowns. Matching the powers of x in (21.7) gives a set of $(N + q + 1)$ linear equations to determine the unknown coefficients of $f_N(x)$ and $\epsilon(x)$.

The secret to success is to choose $\epsilon(x)$ — more accurately, to choose $N + 1$ of the $(N + q + 1)$ coefficients of $\epsilon(x)$ — such that the perturbation is small in some appropriate sense. The obvious choice is to set all $\epsilon_j = 0$ for $j \le N$. In this case, $f_N(x)$ is the usual power series approximation to $f(x)$. The trouble with this choice is that $\epsilon(x)$ is highly non-uniform; because it is $O(x^{N+1})$ for small $|x|$, this "power series perturbation" is very, very small near the origin and then grows rapidly for larger x. The result is that (i) the accuracy of $f_N(x)$ is similarly non-uniform (good for small $|x|$ and increasingly bad as $|x|$ increases) and (ii) the error may be huge. Indeed, if $|x|$ is larger than the absolute value of any of the roots of the denominator, $Q(x)$, then the approximation (which is just a truncated power series) will diverge as $N \to \infty$ even if $f(x)$ is bounded and smooth for all real x.

Lanczos' second key observation is that the Chebyshev polynomials oscillate as uniformly as possible on their canonical interval, [-1, 1]. It follows that if we define

$$\epsilon(x) = \sum_{j=N+1}^{N+q} \tau_j T_j(x) \qquad \text{["Lanczos perturbation"]} \qquad (21.8)$$

the perturbation will *spatially uniform* in magnitude on $x \in [-1, 1]$ and so will the error $|f(x) - f_N(x)|$.

Of course, the error may still be large if the coefficients τ_j are large. However, observe that the coefficients of $\epsilon(x)$ are simply those of the smooth function $f_N(x) Q(x) - P(x)$. We saw in Chapter 2 that the coefficients of any well-behaved function fall off exponentially fast with N (for sufficiently large N), so that it follows that the τ_j will be exponentially small, too, at least for $N \gg 1$.

If we write

$$f_N(x) \equiv \sum_{n=0}^{N} f_n T_n(x) \qquad (21.9)$$

then the f_n are determined by solving the linear equations

$$\sum_{i=0}^{N} (T_i, T_j Q) f_j = (T_i, P) \qquad i = 0, \dots, N \qquad (21.10)$$

where

$$\tau_j = (T_j, P - f_N Q) \qquad j = N+1, \dots, (N+q) \qquad (21.11)$$

Several points are important. First, it is *not* necessary to explicitly compute the τ-coefficients in order to determine the approximation $f_N(x)$; (21.11) is a second, separate step that can be applied only after $f_N(x)$ has already been computed.

Second, the τ-approximation is not simply the Chebyshev expansion of $f(x)$; the solution of the linear system (21.10), whose matrix elements are inner products of *polynomials*, is usually different from the coefficient integrals of the Chebyshev series of $f(x)$, which are the inner products of the Chebyshev polynomials with the *rational* function $f(x)$. As $N \to \infty$, of course, the differences between the τ-approximation, the Chebyshev series, and the pseudospectral interpolant decrease exponentially fast with N.

Third, (21.10) is *not* the way Lanczos himself performed the calculation. In his time B. C. (Before Computers), it was more convenient to represent $f_N(x)$ as an ordinary polynomial, expand the Chebyshev polynomials in $\epsilon(x)$ as powers of x, and then match the powers of x. This saved work, but it required *simultaneously* determining both $f_N(x)$ and $\epsilon(x)$. Lanczos' variant obscures the fact that it is possible to determine $f_N(x)$ *independently* of the perturbation $\epsilon(x)$ if one represents all quantities as Chebyshev polynomials.

Fourth, the τ-coefficients are useful for *a posterior* error analysis because

$$f(x) - f_N(x) = \frac{\epsilon(x)}{Q(x)} \leq \frac{1}{\min |Q(x)|} \sum_{j=N+1}^{N+q} |\tau| \qquad (21.12)$$

if we can find a lower bound on $Q(x)$ on $x \in [-1, 1]$. In practice, this error analysis, although highly praised in Fox & Parker (1968), is not useful in the age of microcomputers: it is easier to simply evaluate the difference between $f(x)$ and $f_N(x)$ for a large number of points and take the maximum.

Furthermore, the τ-method is not normally the method of choice even for generating a Chebyshev approximation to a given $f(x)$. Numerically evaluating the inner products of $f(x)$ with the polynomials, i. e. the usual Chebyshev series, is easy and avoids the costly inversion of a matrix. The τ-method is only useful in conjuction with algebraic manipulation methods for *small* N where one wants an approximation with rational coefficients or where $f(x)$ may contain symbolic parameters, making (21.11) more attractive.

Nevertheless, Lanczos' philosophy of exactly solving a perturbed problem is useful in other areas of applied mathematics. The technique that evolved from that philosophy is still useful even today for solving differential equations.

21.3   Differential Equations

It is not possible to solve

$$u_x + u = 0 \qquad\qquad \& \qquad\qquad u(-1) = 1 \qquad\qquad (21.13)$$

exactly as a polynomial. The residual function $R(x; a_0, \dots, a_N)$ is a polynomial of degree N, too, but the initial condition provides one constraint, leaving only N free coefficients in $u_N(x)$ to force the $(N + 1)$ coefficients of $R(x)$ to vanish. Lanczos observed that the modified problem

$$v_x + v = \tau\, T_N(x) \qquad\qquad \& \qquad\qquad v(-1) = 1 \qquad\qquad (21.14)$$

does have a unique solution. As before, if we take the inner product of (21.14) with each of the first N Chebyshev polynomials and reserve the first row to impose the boundary condition, we obtain an $(N + 1) \times (N + 1)$ system to determine the coefficients of the Chebyshev series for $v(x)$ — *independent* of τ. We shall not bother to write down this matrix equation because it is identical with "Galerkin's method with boundary bordering" as defined in Chapter 3.

This example is solved in Fox & Parker by expanding $T_N(x)$ as an ordinary polynomial and matching powers of x. This is more efficient for hand calculation even though it ignores the orthogonality of the Chebyshev polynomials and makes it necessary to determine τ *simultaneously* with the coefficients of the power series representation of $v(x)$. We obtain the same answer either way; the important point is that the perturbation on the R. H. S. in (21.14) is a Chebyshev polynomial, which guarantees spectral accuracy. In contrast, perturbing (21.13) by a forcing proportional to x^N would give a solution which would be the first $(N + 1)$ terms of the power series of $\exp(-x - 1)$, the exact solution.

Just as for the rational function, the error can be bounded by a function proportion to τ. Just as for the rational function, this error analysis is usually not worth the bother.

For more complicated differential equations, the same principle applies except that it may be necessary to use many τ terms or even an infinite series.

21.4   Canonical Polynomials

Lanczos observed that one strategy for bypassing the cost of inverting an $(N + 1)$ by $(N + 1)$ matrix is to use what he dubbed the "canonical" polynomials. Note that for (21.13),

Table 21.1: A Selected Bibliography of Tau Methods

References	Comments
Lanczos(1938)	Invention of the tau method
Wragg(1966)	Lanczos-tau method for Stefan problem
Ortiz(1969)	
Namasivayam&Ortiz(1981)	Best approximation and PDEs via tau method
Ortiz&Samara(1981,1983)	Operational approach: ODEs, eigenvalue eqs.
Liu&Ortiz&Pun(1984)	Steklov's PDE eigenvalue problem
Ortiz&Pun(1985)	Nonlinear PDEs
Liu&Ortiz(1986)	PDEs through the tau-lines method
Ito&Teglas(1986,1987)	Legendre-tau for functional differential equations and linear quadratic optimal control
Ortiz(1987)	Review: singular PDEs and tau method
da Silva(1987)	Review
Ortiz&Pham Ngoc Dinh(1987)	Recursions for nonlinear PDEs
Liu&Ortiz(1987a)	High order ODE eigenvalue problems; Orr-Sommerfeld Eq.
Liu&Ortiz(1987b)	Nonlinear dependence of linear ODE on eigenparameter
Liu&Ortiz(1989)	Functional-differential eigenvalue problems
Hosseini Ali Abadi & Ortiz(1991b)	Non-uniform space-time elements for tau simulation of solitons
Khajah&Ortiz(1992)	functional equations
Namasivayam&Ortiz(1993)	Error analysis for tau method
Khajah&Ortiz(1993)	Rational approximations via tau method
El-Daou & Ortiz & Samara(1993)	Unified approach to tau-method and Chebyshev expansions
El-Daou & Ortiz(1993)	Error analysis
El-Daou & Ortiz(1994a)	Recursive collocation with canonical polynomials
El-Daou & Ortiz(1994b)	Collocation/tau; weighting subspaces
Ortiz(1994)	Review
Dongarra&Straughan &Walker(1996)	Eigenvalue calculations
Straughan&Walker(1996)	Compound matrix and tau for eigenproblems in porous convection
Siyyam&Syam(1997)	Poisson equation
Froes Bunchaft(1997)	Extensions to theory of canonical polynomials
El-Daou&Ortiz(1997)	Existence & stability proofs for singular perturbation problems, independent of perturbation parameter ϵ

it is trivial to obtain the solution when the R. H. S. is a power of x. These "canonical polynomials" $p_j(x)$ are defined as the solutions of

$$p_{j,x} + p_j = x^j \qquad ; \qquad p_j(-1) = 1 \qquad (21.15)$$

Since each is the exact, power series solution to a simple problem, they can be computed recursively.

Now $T_N(x)$ is a sum of powers of x, so it follows that the τ-solution of 21.14) is a sum of the canonical polynomials. If we write

$$T_N(x) = \sum_{j=0}^{N} t_{N,j}\, x^j \qquad (21.16)$$

then

$$v(x) = \tau \sum_{j=0}^{N} t_{N,j}\, p_j(x) \qquad (21.17)$$

We then choose τ so that the boundary condition is satisfied — in this case, this requires $\tau = 1/\sum t_{N,j}$ — and $v(x)$ is completely determined. Since each $p_j(x)$ may be computed in $O(N)$ operations via recursion and since there are N polynomials, the total cost is $O(N^2)$ — a huge savings over the $O(N^3)$ cost of matrix inversion.

The method of canonical polynomials is less efficient when one needs several τ terms instead of a single Chebyshev polynomial as sufficed for (21.13). Furthermore, the canonical polynomials are *problem-dependent*, and thus must be recomputed from scratch for each differential equation.

Nevertheless, E. L. Ortiz and his collaborators have solved many problems including nonlinear partial differential equations by combining the τ-method with the technique of "canonical polynomials". He and his group have made many refinements to extend the range of Lanczos' ideas. The relevant articles are described in the bibliography. However, the canonical polynomials method has never been popular outside of his group.

21.5   Nomenclature

We have not used this nomenclature of "tau-method" earlier in the book because from a programmer's viewpoint, the distinction is between "basis recombination" (=Galerkin's) and "boundary bordering" (=tau-method). The accuracy difference between (21.1) and (21.3) is usually negligible. Furthermore, these same options of "basis recombination" and "boundary bordering" also exist for collocation methods (although in this case, the accuracy difference is not merely small but zero). Lastly, when the test functions are different from the basis functions, the label preferred by the finite element community is "Petrov-Galerkin" rather than "tau-method".

Chapter 22

Domain Decomposition Methods

"We stress that spectral domain decomposition methods are a recent and rapidly evolving subject. The interested reader is advised to keep abreast of the literature."
— Canuto, Hussaini, Quarteroni, and Zang (1988)

22.1 Introduction

The historical trend in both finite difference and finite element methods has been the replacement of second order schemes by higher order formulas. Indeed, finite elements have fissioned into "h-type" and "p-type" strategies. The former improve accuracy by reducing the grid spacing while p, the order of the method, is kept fixed. In contrast, "p-type" codes partition the domain into a few large pieces (fixed h) and refine the solution by increasing the degree p of the polynomial within each element.

In the last few years, the trend in spectral methods has been in the opposite direction: to replace a global approximation by local polynomials defined only in part of the domain. Such piecewise spectral methods are almost indistinguishable from p-type finite elements.

These techniques are variously called "global elements", "spectral elements", "spectral substructuring", and a variety of other names. Since "domain decomposition pseudospectral and Chebyshev-Galerkin methods" is rather a jawbreaker, we shall use "spectral elements" as a shorthand for all the various algorithms in this family.

There are several motives for "domain decomposition" or "substructuring". One is that spectral elements convert differential equations into sparse rather than dense matrices which are cheaper to invert. A second is that in complicated geometry, it may be difficult or impossible to map the domain into a rectangle or a disk without introducing artificial singularities or boundary layers into the transformed solution. Mapping the domain into multiple rectangles and/or disks is much easier. A third reason is that mapping into sectorial elements can eliminate "corner singularities".

It seems improbable that spectral elements will ever completely chase global expansions from the field, any more than higher order methods have eliminated second order calculations. Nevertheless, they have become a major part of spectral algorithms and, as noted in the quote above, an important research frontier.

479

22.2   Notation

Ω will denote the total domain, $\partial\Omega$ its boundary, and $\Omega_j, j = 1, \ldots, M$, will denote its M subdomains. N_j will denote the degree of the polynomial used to approximate u on the j-th subdomain, and $N \equiv N_1 + N_2 + \cdots N_M$ is the total number of degrees of freedom on the global domain. The numerical solution will be denoted by u^N while u_j will indicate the *restriction* of the numerical solution to the j-th subdomain. In spectral elements, we approximate $u(x)$ by a collection of separate approximations which are each valid only on a particular subdomain and are undefined elsewhere. Thus, $u^N(x)$ is the collection of polynomials while $u_j(x)$ is a single polynomial defined only on Ω_j.

22.3   Connecting the Subdomains: Patching

Definition 44 (PATCHING) *Subdomains are connected by "PATCHING" if the solutions in different elements are matched along their common boundary by requiring that $u(x)$ and a finite number of its derivatives are equal along the interface.*

The number of matching conditions is the number that is necessary to give a unique solution in each subdomain. As a rule-of-thumb, two matching conditions are equivalent to a single numerical boundary condition. Thus, one must match both $u(x)$ and du/dx at an interface for a second order differential equation.

For ordinary differential equations and for elliptic partial differential equations, patching is usually straightforward. For example, suppose the goal is to solve a second order ordinary differential equation on the interval $x \in [-1, 1]$ by splitting the segment into two subintervals: $[-1, d]$ & $[d, 1]$. If one knew $u(d)$ [in addition to the usual Dirichlet boundary conditions, $u(-1) = \alpha$ and $u(1) = \beta$], then the original problem would be equivalent to two completely separate and distinct boundary value problems, one on each subinterval, which could be solved independently of one another.

Unfortunately, we are rarely given three boundary conditions for a second order problem! However, by demanding that both u and du/dx are continuous at $x = d$, we obtain two interface conditions which, together with the Dirichlet conditions at $x = \pm 1$, give a total of four constraints. This is precisely what is needed to uniquely determine the solution of *two* otherwise independent boundary value problems of second order.

To be specific, let Ω denote the interval $x \in [a, b]$ and

$$L u \equiv q_2(x)\, u_{xx} + q_1(x)\, u_x + q_0(x)\, u = f(x) \qquad \text{in } \Omega \qquad (22.1)$$

$$u(a) = \alpha \qquad\qquad \& \qquad\qquad u(b) = \beta \qquad (22.2)$$

The two-element subdomain solution may be mathematically expressed as

$$L u_1 = f \qquad \text{in } \Omega_1 \qquad \& \qquad u_1(a) = \alpha \qquad (22.3)$$

and

$$L u_2 = f \qquad \text{in } \Omega_2 \qquad \& \qquad u_2(b) = \beta \qquad (22.4)$$

with the interface conditions

$$u_1(d) = u_2(d) \qquad (22.5a)$$

$$u_{1,x}(d) = u_{2,x}(d) \qquad (22.5b)$$

where the subscript "x" denotes differentiation with respect to x.

Patching for *hyperbolic* equations is more complex. The simplest such equation is

$$u_t + c u_x = 0 \qquad [\text{"One-Dimensional Advection Eq."}] \qquad (22.6)$$

on $x \in [a, b]$ where $(c > 0)$

$$u(a, t) = u_L(t) \qquad (22.7)$$

$$u(x, 0) = u_0(x) \qquad (22.8)$$

which has the exact solution

$$u(x, t) = \begin{cases} u_0(x - ct) & x > (a + ct) \\ u_L\left(t - \dfrac{x - a}{c}\right) & x \le (a + ct) \end{cases} \qquad (22.9)$$

Thus, $u(x, t)$ at a given point x_0 is influenced only by events "upstream", that is, by the forcing and initial condition for $x < x_0$. The solution at $x = x_0$ is unaffected by downstream values of the initial condition.

The numerical strategy which conforms to the mathematics of the exact solution is to solve the wave equation independently on each element, one at a time, beginning at the left boundary. The computed solution at the right of the j-th element provides the left-hand boundary forcing for the $(j+1)$-st element.

If instead one uses an *averaged* interface condition such as

$$\frac{\partial u_N}{\partial t} + \frac{c}{2}\frac{\partial u_j}{\partial x} + \frac{c}{2}\frac{\partial u_{j+1}}{\partial x} = 0 \qquad \text{at } x = d_j \qquad (22.10)$$

the calculation may be unstable, at least if the number of grid points in element j is greater than that in $(j+1)$. Eq. (22.10) would seem to be stupid almost beyond belief since it both increases the computational work and misrepresents the mathematics of the original problem. One would never want to apply (22.10) to a *single* scalar equation like (22.6). However, when we solve a *system* of equations with waves travelling both to the left and the right, it may be difficult to apply anything better than (22.10).

The best approach, as reviewed by Canuto *et al.* (1988) and Kopriva (1986, 1987, 1989a, 1998, 1999), is to diagonalize the operator of the system of differential equations — either exactly or approximately — so as to separate right-running and left-running waves. One may then separately apply "upwind" conditions on each wave component. This is still very much a research frontier.

22.4 The Weak Coupling of Elemental Solutions: the Key to Efficiency

An important theme is that solutions on different subdomains are connected only through the *interface matching conditions,* (22.5). Almost all spectral element methods exploit this weak element-to-element coupling.

To illustrate how and why, consider a linear, second order different equation, i. e. (22.1) and (22.2). Instead of just two subintervals as in the previous section, however, divide the interval $x \in [a, b]$ into M subdomains. Let d_j denote the boundary between element $(j$-1$)$ and element j and define

$$U_j \equiv u(d_j) \qquad (22.11)$$

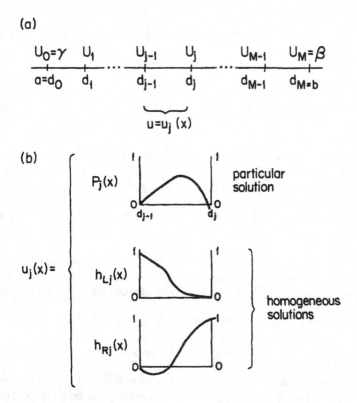

Figure 22.1: Schematic of the interval $[a, b]$ split into subdomains. The element boundaries are denoted by d_j; the values of $u(x)$ at these boundaries are denoted by U_j. The solution on the j-th element, $u_j(x)$, is the sum of a particular integral $p_j(x)$ plus the weighted contributions of the homogeneous solutions $h_{Lj}(x)$ and $h_{Rj}(x)$. The schematic illustrates the boundary conditions satisfied by each component.

The condition of continuity of $u(x)$ at the domain boundaries may be enforced implicitly by demanding that both $u_{j-1}(x)$ and $u_j(x)$ equal U_j at $x = d_j$. These definitions are illustrated in Fig. 22.1.

The solution on the j-th element, $u_j(x)$, can always be written as the sum of a particular integral $p_j(x)$ plus two homogeneous solutions. One always has the freedom to choose the particular integral so that it vanishes at both elemental boundaries. One may similarly choose the homogeneous solutions so that one — call it $h_{Lj}(x)$ — is equal to one at the left domain boundary and zero at the other while $h_{Rj}(x)$ is non-zero at the right boundary but not at the left wall. If the differential equation is $L u = f$, these three components are defined by

$$L p_j(x) = f \quad \& \quad p_j(d_{j-1}) = p_j(d_j) = 0 \quad \text{[particular integral]} \tag{22.12}$$

$$L h_{Lj} = 0 \quad \& \quad h_{Lj}(d_{j-1}) = 1 \quad \& \quad h_{Lj}(d_j) = 0 \tag{22.13a}$$
$$\text{[homogeneous solutions]}$$
$$L h_{Rj} = 0 \quad \& \quad h_{Rj}(d_{j-1}) = 0 \quad \& \quad h_{Rj}(d_j) = 1 \tag{22.13b}$$

These components are schematically shown in Fig. 22.1.

It follows that the *general* solution to the differential equation can be written — without approximation — as

$$u_j(x) = p_j(x) + U_{j-1}\,h_{Lj}(x) + U_j\,h_{Rj}(x) \qquad (22.14)$$

This decomposition is useful because one can numerically compute all the particular integrals and homogeneous solutions intra-elementally — that is, calculate each $p_j(x)$ or $h_{Lj}(x)$ or $h_{Rj}(x)$ *independently* of the solutions on all the other elements. The boundary value problems defined by (22.12) and (22.13) are completely uncoupled.

Unfortunately, the elemental solutions $u_j(x)$ are not completely specified because (22.14) contains $(M+1)$ unknown parameters, U_j, the values of $u(x)$ at the domain boundaries. The two end values are determined by the boundary conditions for the original problem:

$$u(a) = \alpha \quad \rightarrow \quad U_0 = \alpha \qquad \& \qquad u(b) = \beta \quad \rightarrow \quad U_M = \beta \qquad (22.15)$$

The remaining domain boundary values of $u(x)$ are determined by the requirement of continuity of the first derivative at each of the $(M-1)$ interior domain walls. This gives, using primes to denote differentiation,

$$h'_{Lj}\,U_{j-1} + \left[h'_{Rj} - h'_{L,j+1} \right] U_j - h'_{R,j+1}\,U_{j+1} = p'_{j+1} - p'_j \qquad j = 1, \ldots, M-1 \qquad (22.16)$$

where all functions are evaluated at $x = d_j$. The reason that this matrix system is tridiagonal, soluble in only $O(8M)$ multiplications and additions, is that only four different homogeneous solutions, two for element $(j-1)$ and two for element j, enter the continuity condition at $x = d_j$. There are only three unknowns in each row of (22.16) because $U_j [\equiv u(d_j)]$ is the coefficient of both $h_{Rj}(x)$ and $h_{L,j+1}(x)$.

Thus, the cost of coupling elements together via continuity of $u(x)$ and du/dx is trivial. The real work is in computing the particular and homogenous solutions on each element. This may be done by whatever method — indeed, by whatever subroutine — one would have employed to solve the boundary value problem by a single Chebyshev series on the whole interval.

To compare the global expansion with the subdomain procedure, assume that the global series has N degrees of freedom while N/M spectral coefficients are used on each subdomain. If the cost to solve a boundary value problem with ν Chebyshev coefficients is $O(\nu^3)$ operations and $O(\nu^2)$ storage — true if the pseudospectral method is combined with Gaussian elimination — then the relative costs are

$$\text{Global:} \qquad O(N^3) \quad \text{ops.} \quad \& \qquad O(N^2) \quad \text{store} \qquad (22.17a)$$

$$\text{Subdomain:} \quad M\,O([N/M]^3) \text{ ops.} \quad \& \quad M\,O([N/M]^2) \text{ store} \qquad (22.17b)$$

The spectral element method is cheaper by a factor of

$$1/M^2 \text{ operations} \qquad \& \qquad 1/M \text{ storage} \qquad (22.18)$$

These are enormous factors if $M \gg 1$! Of course, to retain the high accuracy of spectral elements, one must keep $N/M \geq 6$ (or so). Furthermore, more efficient methods for solving the boundary value problems, such as the preconditioned iterations discussed earlier, would reduce the competitive advantages of the subdomain strategy versus the use of a single global expansion. Nevertheless, even if (22.18) is a little optimistic, it still dramatizes the potential of "spectral substructuring".

The key to efficiency is that most of the work is reduced to solving boundary value problems intra-elementally, i. e. (22.12) & (22.13). If we assumed M different expansions on each of the M elements and then lumped all the collocation, boundary, and continuity conditions into a single matrix of dimension N, the subdomain method would be just as expensive as using a single global expansion — and not as accurate. An efficient spectral element code is one that solves the global problem in two stages. The first is to solve many uncoupled smaller problems within each element. The second stage is to couple the elemental solutions together into a single, continuous global solution.

22.5   Variational Principles

The heart of the variational formulation is to multiply the differential equation residual by a test function and then *integrate-by-parts* so that a term like $\phi_i(x)\,\phi_{j,xx}(x)$ becomes transformed into the product of first derivatives. When the basis functions are very low order polynomials, this trick is important. For example, a popular finite element basis is composed of piecewise linear polynomials, the so-called "tent" or "chapeau" functions. Since the second derivative of these functions is identically zero, it is not possible to solve second order equations with "tent" functions by collocation. After integration-by-parts, however, the piecewise linear polynomials give second order accuracy.

For example, the self-adjoint boundary value problem

$$- [p(x)\,u_x]_x + q(x)\,u = f(x) \qquad\qquad u(-1) = u(1) = 0 \qquad\qquad (22.19)$$

is equivalent to minimizing the functional

$$I(v) \equiv \int_{-1}^{1} \left\{ p(x)\,[v_x]^2 + q(x)\,v^2 - 2\,v\,f(x) \right\}\,dx \qquad\qquad (22.20)$$

over the class of functions $v(x)$ such that $v(-1) = v(1) = 0$. In practice, (22.20) is minimized over a finite dimensional space of basis functions, $\{\phi_j,\ j = 1,\ \ldots,\ N\}$. The coefficients of the series for $u(x)$ are determined by solving the matrix equation $\vec{\vec{H}}\vec{a} = \vec{F}$ where

$$H_{ij} \equiv (\phi_{i,x},\,p\,\phi_{j,x}) + (\phi_i,\,q\,\phi_j) \qquad\qquad (22.21)$$

$$F_i \equiv (\phi_i,\,f) \qquad\qquad (22.22)$$

$$(r,\,s) \equiv \int_{-1}^{1} r(x)\,s(x)\,dx \qquad\qquad (22.23)$$

Although (22.21) is justified by the calculus of variations, it is obvious that (22.21) is identical with the usual Galerkin matrix element except for use of the integration-by-parts

$$\int_{-1}^{1} \phi_i \left(- [p\,\phi_{j,\,x}]_x \right)\,dx \quad \longrightarrow \quad \int_{-1}^{1} \phi_{i,\,x}\,p\,\phi_{j,\,x} \qquad\qquad (22.24)$$

(The boundary term, $(\phi_i(1)\,p(1)\,\phi_{j,\,x}(1) - \phi_i(-1)\,p(-1)\,\phi_{j,\,x}(-1))$, is zero because the basis functions satisfy the boundary conditions (22.19).)

As noted above, the variational form is essential to successful use of piecewise functions of first degree to solve differential equations of second order. As the degree of the polynomials increases, however, the advantage of the variational form becomes smaller and smaller. This is the reason that the integrated-by-parts form (22.21) is rarely used with global spectral methods; in comparison to the standard Galerkin procedure, the variational matrix is not worth the bother.

The domain decomposition/variational principle strategy allows considerable flexibility. Delves and Hall (1979) add additional terms to the variational functional (22.20) ("global elements") so that the basis functions need not even be continous at the element boundaries. As the resolution is increased, however, the discontinuities in u decrease rapidly, and the order-of-magnitude of these jumps is never larger than that of the maximum pointwise error.

Patera (1984), who dubbed his method "spectral elements", takes a middle ground. His basis functions belong to C^0, that is, are continuous. However, continuity of the first and higher derivatives at inter-element boundaries is only approximate.

Phillips and Davies (1988), who also use the term "spectral elements", employ basis functions that are at least C^1. Their basis explicitly satisfies all boundary conditions and also obeys the element-interface conditions that are required to specify a unique solution to the (undiscretized) problem.

The higher the degree of continuity, the smaller the matrix which discretizes the differential equation. With Delves' and Hall's "global elements", the basis has extra degrees of freedom — the freedom to be discontinuous on inter-element boundaries — which are explicitly eliminated in Phillips' and Davies' method. Furthermore, the variational functional is simpler when the basis functions are C^1 because the functional lacks the extra terms that are needed to enforce continuity of $u(x)$ if the basis functions are not continuous. Thus, in principle, using basis functions with continuous first derivatives would be optimum. In practice, computing and manipulating basis functions in C^1 may be very difficult if the geometry is complicated, so the choice of the degree of continuity of the basis set must be left to individual discretion. However, the C^0 basis has become the most popular (Karniadakis and Sherwin, 1999).

22.6 Choice of Basis & Grid: Cardinal versus Orthogonal Polynomial, Chebyshev versus Legendre, Interior versus Extrema-and-Endpoints Grid

Phillips and Davies (1988) use ordinary Chebyshev polynomials as the intra-element basis, but it is far more common to use the Chebyshev or Legendre *cardinal* functions. The reasons are similar to those that have made the pseudospectral method popular for global spectral methods. The issue of $T_n(x)$ versus cardinal function is not very important for boundary value problems if the matrix problem is solved by direct methods, but — as true of global algorithms — it is of paramount important when *time-integration* or *iteration* is applied.

In older books on finite elements such as Strang and Fix (1973), only a non-cardinal basis was used. The drawback is that because the piecewise polynomials are not orthogonal, the time derivatives for the differential equation $du/dt = f(u, x, t)$ were multiplied, in discretized form, by the so-called "mass matrix" $\vec{\vec{M}}$. That is, the finite element method converted a partial differential equation into the coupled system of ordinary differential equations

$$\vec{\vec{M}} \frac{d\vec{a}}{dt} = \dots \tag{22.25}$$

where the ellipsis ($\dots$) denotes an unspecified right-hand side and where the elements of

$\vec{M}$ are

$$M_{ij} \equiv (\phi_i, \phi_j) \qquad \text{"mass matrix"} \qquad (22.26)$$

In a non-cardinal basis, finite element methods are always implicit in the sense that one must backsolve using the LU-factorization of the "mass matrix" at every time step.

If one shifts to a cardinal function basis and applies collocation instead of integration, however, the new mass matrix is diagonal. This pseudospectral approach is essential for efficient, explicit time-integration, just as for global Chebyshev methods.

"Spectral elements" began with a Chebyshev basis, but Legendre polynomials have become popular, too. One reason is that the usual drawbacks of Legendre polynomials — collocation points that are not known in closed form, poorer accuracy near the boundaries, and so on — are less important for polynomials of moderate degree than when N is large. A more compelling reason is that the variational principle employs inner products with a weight function of unity. Because the Chebyshev polynomials are orthogonal with respect to a weight function of $1/\sqrt{1-x^2}$, it is more awkward to use them instead of Legendre polynomials while still remaining faithful to the variational principle.

Either basis, however, works just fine; the difference between them seems to have more to do with mathematical tidiness than with efficiency. The variational formalism is popular, but collocation with either Chebyshev or Legendre polynomials is successful, too.

As noted earlier, there are two standard collocation grids for Chebyshev polynomials: (i) the "roots" or "Gauss-Chebyshev" grid, whose points are the zeros of $T_N(x)$ and (ii) the "extrema-and-endpoints" or "Gauss-Chebyshev-Lobatto" grid, which includes the roots of the *derivative* of $T_N(x)$ plus the endpoints ± 1. As shown in Chapter 2, both have equally good theoretical properties and the choice between them is usually a matter of preference.

With spectral elements, however, the "extrema-and-endpoints" or "Lobatto" grid is much more popular than the roots grid. The reason is that one must explicitly match solutions along inter-element boundaries. In the variational formalism, it is awkward to impose continuity on the basis sets, apply the isoparameteric mapping described in Sec. 10, and so on if none of the grid points coincides with the element walls. However, Schumack, Schultz, and Boyd (1989) have found that the roots grid is just as accurate as the Lobatto grid for computing Stokes' flows.

Nevertheless, the Legendre basis with Lobatto grid has become the canonical choice for spectral elements. The cardinal functions, derivative matrix and grid points are all given in Appendix F, Sec. 10.

22.7 Patching versus Variational Formalism

Canuto *et al.* (1988) show that the differences between the patching & variational approaches to spectral elements are small. Indeed, the variational principle can be expressed in terms of patching and vice versa if the two formalisms are appropriately generalized.

For example, the variational method is equivalent to patching if the condition of continuity of du/dx at inter-element boundaries, (22.5b), is replaced by

$$u_{1,x}(d) - u_{2,x}(d) + w_0\, R_1(d) + w_{N_2}\, R_2(d) \qquad (22.27)$$

where R_1 and R_2 are the differential equation residuals in elements 1 and 2 and where w_0 and w_{N_2} are the first and last weights of the Gauss-Lobatto quadrature. Since the quadrature weights are $O(1/N_i^2)$ and the residuals are very small if the numerical solution is accurate, it follows that (22.27) is only a small perturbation of (22.5b). Further, the magnitude of the residual terms in (22.27) is always the same order of magnitude as the error in the numerical solution: exponentially small.

Thus, whether one applies (22.5b) or (22.27) is irrelevant to accuracy. It is obvious, however, that (22.27) is harder to program and also harder to explain and justify. In practice, one would always either apply patching in the standard form (22.5b) or apply the variational principle as explained in Sec. 5. The only use of (22.27) is conceptual: it shows that patching is very *close* to the variational formalism without being *identical*.

In a similar way, the variational method can mimic patching if the variational principle is slightly modified. Again, however, the modifications do not raise or lower the order of the method. There is no compelling reason for choosing patching over the variational approach or vice versa; it is a matter of preference and taste.

22.8   Matrix Inversion

As reviewed by Canuto *et al.* (1988), linear algebra methods used with spectral elements include:

(i) Alternating Schwarz method
(ii) Static condensation
(iii) Influence matrix method
(iv) Conjugate gradient and
(v) Multigrid.

The Schwarz algorithm defines *overlapping* elements and then *iterates* to match the solutions in different subdomains. Canuto & Funaro (1988) report great success and Schwarz iteration has become popular with finite difference methods, too.

"Static condensation" is a two-stage application of Gaussian elimination. On a given subdomain, the numerical solution has both internal and external degrees of freedom. The external degrees of freedom are those which can be computed only simultaneously with the corresponding quantities for all the other elements. For the second order ordinary differential equation of Sec. 4, for example, the general solution can be written as

$$u_j(x) = p_j(x) + U_{j-1} \, h_{Lj}(x) + U_j \, h_{Rj}(x) \qquad\qquad x \in [d_{j-1}, \, d_j] \qquad (22.28)$$

where $U_j \equiv u(x_j)$ and the particular solution $p_j(x)$ and the homogeneous solutions, $h_{Lj}(x)$ & $h_{Rj}(x)$, satisfy (22.12) and (22.13). The internal degrees of freedom are the polynomial coefficients of $p_j(x)$, $h_{Lj}(x)$ & $h_{Rj}(x)$. These coefficients are "internal" because one can compute these three functions on the j-th element without using any information outside the j-th element. This is possible because these three functions are arbitrarily assigned boundary values of either 1 or 0.

The external degrees of freedom are the values of $u_j(x)$ on the boundaries of the element, U_j. These can be determined only by matching the elemental solutions across the global domain to form a single continuous solution.

The first stage of static condensation is to eliminate the *internal* degrees of freedom. The second stage is to solve a reduced matrix system (i. e. (22.16)) for the external degrees of freedom. The dimension of this "condensed" system is small in comparison to N, the total number of degrees of freedom, resulting in huge savings.

The influence matrix method, which is the theme of the next section, is very similar to static condensation. Both are *direct* methods that perform as much of the work as possible within each element before solving a matrix problem of moderate size to link the piecewise polynomials together.

The conjugate gradient and multigrid algorithms are *iterations*. However, both share a common philosophy with the direct methods: exploit the weak inter-element coupling through heavy use of intra-element operations. As emphasized in Sec. 4, this strategy is fundamental to all spectral element procedures.

Both multigrid and a conjugate gradient-like algorithm have already been discussed in an earlier chapter. However, Rønquist & Patera (1987b) and Maday & Munoz (1988) have shown that the incomplete-factorizations employed with *global* Chebyshev multigrid are unnecessary for Legendre spectral elements. Instead, a simple pointwise Richardson's iteration (which these authors label with Jacobi's name) is extremely effective, at least for simple problems.

It is usual to abandon iteration when multigrid has descended to a grid with only a small number of points and solve the matrix problem on the coarsest grid via Gaussian elimination. In Rønquist & Patera (1987b), the multigrid/Richardson's iteration polyalgorithm is applied only *within* each element. The direct-solution-on-the-coarsest-grid is the step at which the various elemental solutions are matched together. Thus, the non-iterative part of multigrid is identical with the second stage of static condensation or the influence matrix algorithm described below.

22.9   The Influence Matrix Method

If we knew the values of $u(x, y)$ on the walls of an element, then the problem on that subdomain would be completely self-contained. If we had a subroutine that solved the differential equation using a *global* spectral series, we could call this subprogram once and be done.

The difficulty is that when one or more of the walls of an element is an *internal* boundary, the values of $u(x, y)$ on that wall are *unknown*. Nevertheless, if we discretize the boundary, we can calculate the *influence* of each boundary point on the rest of the solution. This calculation of "influence" requires nothing more nor less than calling the subroutine to solve the Dirichlet problem on the j-th element.

Let the elliptic equation be

$$Lu = f \tag{22.29}$$

For simplicity, assume homogeneous Dirichlet conditions. (If the boundary conditions are inhomogeneous, we may always modify u and f to recast the problem so that the boundary conditions are homogeneous as described in Chapter 6, Secs. 5 and 6). The "discrete boundary Green's functions on the j-th element", which we shall call the "elemental Green's functions" for short, are defined as the pseudospectral solutions of

$$L\gamma_i = 0 \qquad \text{on} \qquad (x, y) \in \Omega_j \tag{22.30a}$$

$$\gamma_i(x_k, y_k) = \delta_{ik} \qquad\qquad (x_k, y_k) \in \partial\Omega_j \tag{22.30b}$$

In words, each function $\gamma_i(x, y)$ vanishes at all but one of the points on the boundary of the element. Strictly speaking, we should add subscripts "j" to both the boundary Green's functions and to the boundary point labels to specify the j-th element, but we have suppressed these for simplicity. We shall append the element index below wherever needed to avoid confusion.

In addition, it is usually necessary to cover the domain with curved quadrilaterals and then map each into the unit square as described in the next section. After such a transformation, L, f, Ω_j, and $\partial\Omega_j$ should have tildes to denote that they have been altered by the mapping. For simplicity, such tildes have been omitted. The mapping is important, however, and forms the theme of the next section.

To enforce continuity of $u(x)$ at the inter-element boundaries, it is helpful to assemble the elemental Green's functions into global Green's functions that display the *total* influence of a given boundary point. Let K denote an index that runs over *all* interior boundary

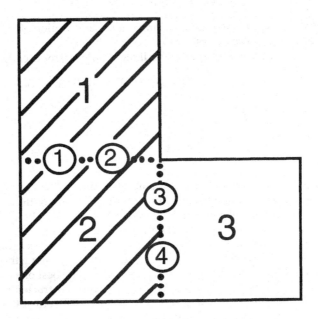

Figure 22.2: Schematic of an L-shaped domain which has been divided into three square elements. The boundaries between elements are shown by dotted lines. The grid points on the *interior* of these intra-element boundaries are shown by numbers in circles. We need four "boundary elemental Green's functions". The Green's function $G_2(x, y)$ is equal to one at boundary grid point 2 and is zero at each of the other three circled points. G_2 is also zero at all grid points on the boundaries of the L-shaped domain (not marked). The cross-hatching shows the two elements where $G_2(x, y)$ is non-zero.

points and let $J(K)$ and $J'(K)$ denote the elements whose common wall contains the point labelled by K. Fig. 22.2 schematically illustrates this system for an L-shaped domain partitioned into three elements. The index K runs from 1 to 4. The point with $K = 2$, for example, lies on the boundary between elements 1 and 2, so that $J(2) = 1$ and $J'(2) = 2$.

Further, let $i(K)$ and $i'(K)$ denote the label of the K-th point in the local number system on the elements $j = J(K)$ and $j = J'(K)$. Then, taking $j(x, y)$ to denote the number of the element which contains the point (x, y),

$$
G_K(x, y) \equiv \begin{cases} \gamma_{i(K), J(K)} & j = J(K) \\ 0 & j \neq J(K), J'(K) \\ \gamma_{i'(K), J'(K)} & j = J'(K) \end{cases} \tag{22.31}
$$

At corner points where *four* domains come together, the definition of G_K must be generalized to include four non-zero pieces.

We also must compute a single particular integral $p_j(x, y)$ for each element; these solve

$$
L p_j = f \qquad x, y \in \Omega_j \quad \& \quad p_j = 0 \quad \text{on} \quad \partial\Omega_j \tag{22.32}
$$

Then

$$
u(x, y) = \sum_{K=1}^{K_{\max}} u_K \, G_K(x, y) + \sum_{j=1}^{M} p_j(x, y) \tag{22.33}
$$

where $K_{\max}$ is the total number of interior boundary points (=4 in Fig. 22.2), M is the number of subdomains, and u_K denotes the value of $u(x, y)$ at the K-th boundary point. Eq. (22.33) is the two-dimensional generalization of (22.28); $h_{Lj}(x)$ and $h_{Rj}(x)$ are the elemental boundary Green's functions for that one-dimensional example.

The usual patching conditions are that both u and its derivatives should be continuous at all boundary points. (Strictly speaking, one wants continuity along the whole interval of each inter-element boundary, but with any discretization, the best one can do is to impose such conditions only at the grid points.) The $G_K(x, y)$ are constructed to be continuous at the boundary points. The particular integrals $p_j(x, y)$ are all zero at the boundary points and therefore are continuous also. Thus, the only non-trivial conditions are those on the derivatives.

Imposing continuity of the *normal* derivative, that is, the derivative in the direction perpendicular to the wall, gives us one condition at each of the $K_{\max}$ interior boundary points. These continuity conditions give a matrix problem of dimension $K_{\max}$ to determine the free parameters in (22.33), which are the u_K. Note that because $K_{\max} \ll N$, the dimension of this element-coupling matrix system is relatively small.

Naturally, one wants continuity of the *tangential* derivative, too. However, continuity of u implies that the numerical solutions in two neighboring elements agree at each of the boundary points along their common wall. It follows that if we expand u_i and u_j as one-dimensional functions of the coordinate tangential to the wall, their interpolating polynomials will be identical. It follows that the derivatives tangential to the wall are forced to match because u_i and u_j match along the wall. (For simplicity, this argument assumed that the number of boundary points is the same for each element, but this assumption may be relaxed without altering the conclusion: the jump in the tangential derivative across the wall is either zero or spectrally small.)

At a corner point, this argument fails. It seems that one would need to explicitly impose continuity of the derivatives with respect to both x and y, which would give more conditions than unknowns u_K, creating an overdetermined matrix system. It turns out, however, that it is sufficient to impose continuity of *either* derivative at a corner point; the jump in the other derivative will be spectrally small (Canuto *et al.*, 1988, pg. 454).

Once the u_K have been found, the R. H. S. of (22.33) is completely known. The matrix systems which must be solved to determine the solution within each element are rather small: the dimension is N_j, the number of grid points that lie in the interior (and not on the boundaries) of the j-th element. In most practical applications, $N_j \approx 50$–100 (Rønquist & Patera, 1987b).

22.10 Two-Dimensional Mappings & Sectorial Elements

Spectral elements, like finite elements, may accomodate very irregular or complicated geometry through mapping. Each element is square in the computational coefficients (r, s), but may be a highly distorted quadrilateral in physical space where the coordinates are (x, y).

The mapping mechanics are a straightforward (if messy) extension of the one-dimensional transformations of Chapter 16. One writes

$$x = F(r, s) \qquad \& \qquad y = G(r, s) \qquad r, s \in [-1, 1] \tag{22.34}$$

and then applies the chain rule to transform derivatives with respect to the physical coordinates x and y into derivatives with respect to the computational coordinates (Table E.9, Appendix E).

The simplest map is that which transforms the rectangle $[a_x, b_x] \times [a_y, b_y]$ in physical space into the unit square:

$$x = \frac{1}{2} \left[(b_x + a_x) + (b_x - a_x) r \right] \qquad \& \qquad y = \frac{1}{2} \left[(b_y + a_y) + (b_y - a_y) s \right] \tag{22.35}$$

However, one is free to use analytical maps of almost any form desired.

When the geometry is so complex so that a closed-form mapping for each element cannot be found, the usual strategy is to employ a so-called "isoparametric mapping". The mapping functions $F(r, s)$ and $G(r, s)$ are approximated by double Chebyshev series of the *same order* as will be used for the solution $u(x, y)$, i. e.

$$x = \sum_{i=0}^{N} \sum_{j=0}^{N} X_{ij} C_i(r) C_j(s) \tag{22.36}$$

where the C_i are the usual cardinal functions on the "Lobatto" grid (Appendix F, Sec. 6), plus a similar expansion for $y(r, s)$.

The earliest strategies for coping with curving boundaries were quite crude. A circle might be approximated by an octagon, or a curving boundary replaced by a zigzag of line segments that alternately paralleled the x and y axis so that the computational domain was a union of rectangles. As higher order finite elements were developed, such crudeness was no longer adequate: the poor approximation to the boundary would erase the high accuracy inherent in the basis set. At the other extreme, using an eighth order polynomial for the map functions while employing only a fourth order basis for $u(x, y)$ is also stupid; the numerical solution cannot resolve the fine structure in the mapping functions. The isoparametric strategy is the common-sense tactic of approximating all parts of the problem — solution & mapping functions — by polynomials of the same order.

Generating numerical grids is a complex topic in and of itself; (Thompson, Warsi, and Mastin, 1985). However, when the geometry is not complex, or when the number of elements is sufficiently large, it is not difficult to devise a systematic strategy for mapping the whole domain, one curvilinear quadrilateral at a time.

The "Global Element School" (Delves & Hall, 1979, Delves & Phillips, 1980, Kermode, McKerrell & Delves, 1985, Delves *et al.*, 1986, and McKerrell, 1988) have creatively used mapping to neutralize corner singularities. The basic strategy, which is borrowed from Wait, is to map a *sectorial* or *triangular* element in physical space into the unit square in the computational coordinates r and s using a map which is designed to remove a specific type of singularity.

To explain this strategy, it is convenient to use the polar coordinates (in physical space)

$$\rho \equiv \sqrt{x^2 + y^2} \qquad \& \qquad \theta = \arctan\left(\frac{y}{x}\right) \qquad (22.37)$$

where $\rho = 0$ is the location of the boundary corner where $u(x, y)$ is singular. To apply the method, one *must* be able to characterize the branch point using local analysis. Assume, for example, that the singularity is of the form

$$u(x, y) = [\text{constant}] \; \rho^w f(\theta) + \text{less singular functions} \qquad (22.38)$$

for some non-integral constant w.

When the element in (x, y) is the sector

$$\rho \in [0, \alpha] \qquad \& \qquad \theta \in [0, \beta] \qquad (22.39)$$

where α is the radius of the sector and β is the sector's angular width, then the mapping which (i) transforms the sector into the unit square and (ii) neutralizes a singularity of the form of (22.38) is

$$\rho = \alpha \left[\frac{1+r}{2}\right]^{1/w} \qquad (22.40a)$$

$$\theta = \frac{\beta}{2}(1+s) \qquad (22.40b)$$

The original Cartesian coordinates are given by $x = \rho(r)\cos(\theta[s])$ and $y = \rho(r)\sin(\theta[s])$. The polar form is very useful, however, because it shows that ρ^w becomes a *linear* function of the computational coordinate r, and thus is *nonsingular* in the $(r\text{-}s)$ plane.

Similarly, one can choose a triangular element in the $x\text{-}y$ (physical) plane. To define the element, specify sides of length A and B which intersect at the origin with an angle β between them, oriented so that the side of length A is parallel to the x-axis. Then a "desingularizing" mapping is

$$\rho = \left[\frac{1+r}{2}\right]^{1/w} \sqrt{\frac{(1-s)^2 B^2 + (1+s)^2 A^2 + 2(1-s^2)AB\cos(\beta)}{4}} \qquad (22.41a)$$

$$\theta = \arctan\left\{\frac{B\sin(\beta)(1-s)}{B\cos(\beta)(1-s) + A(1+s)}\right\} \qquad (22.41b)$$

Neither (22.40) nor (22.41) is unique; many other choices of mappings are possible. The point is that *if* one knows the degree of the corner branch and *if* one is willing to accept the extra programming complexity of additional element shapes and mappings, spectral elements can eliminate corner singularities.

22.11 Prospectus

The geometries of nature and engineering are both complicated, full of arcs and spirals and web-like networks of beams and curving cylinders. It is far easier to fit a complicated shape by an assemblage of many triangles or curved quadrilaterals than by a single deformed square. It is for this reason that finite elements have become so popular.

However, as computers grow in speed and power, second order methods become less and less satisfactory. After all, Sir Robert Southwell and his colleagues de G. Allen and Vaisey were applying fourth order methods with desktop calculators half a century ago.

The future of science and engineering is likely to be domain decomposition strategies that use polynomials of medium order — sixth to eighth degree, say — on each element.

Nevertheless, the fundamentals of spectral series are independent of N, the degree of the polynomial, and also independent of whether the domain of the polynomial is the whole domain of the problem, or merely a subdomain. Like Newtonian mechanics, spectral methods will surely endure rapid, perhaps even revolutionary change. Like Newtonian mechanics, what has been described in these twenty-two chapters will not be replaced, but merely extended.

Chapter 23

Books and Reviews

"Aun Aprendo" — "I am still learning"

—— Francisco de Goya (1746-1828), written on a sketch drawn in his old age

Table 23.1: Books and Reviews on Spectral Methods

Books		
Bernardi&Maday(1992)	250 pp.	In French; mathematical theory
Boyd(1989,2000)	665 pp.	Text & Encyclopedia; elementary mathematical level
Canuto, Hussaini, Quarteroni&Zang(1987)	500 pp.	Half algorithms and practice, half theory Available in paperback and hardcover
Delves&Freeman(1981)	275 pp.	Theory of Chebyshev-Galerkin methods and block-and-diagonal ("Delves-Freeman") matrix iterations
Finlayson(1972)	410 pp.	Low order pseudospectral & Galerkin lots of examples from fluids and chemical engineering
Fornberg(1996)	200 pp.	Classical pseudospectral; very practical Available in paperback and hardcover
Fox&Parker(1968)	200 pp.	Very readable, but its Galerkin-by-recurrence and differential-eq.-through-double-integration are now rare
Funaro, D. (1992)	300 pp.	Variational and approximation theory plus algorithms; one chapter on domain decomposition
Funaro, D. (1997)	210 pp.	Spectral elements; readable description of the variational theory; many numerical examples
Gottlieb & Orszag(1977)	150 pp.	Mix of examples and stability theory
Guo(1998)	200 pp.	Proofs; highly mathematical
Karniadakis&Sherwin(1999)	448 pp.	Quadrilateral and triangular spectral elements
Mercier(1989)	200 pp.	Translation, without updating, of earlier book in French

Review Articles		
Bert&Malik(1996)	28 pp.	Pseudospectral as high order finite differences under the name "differential quadrature"; mechanics applications
Delves(1976)	13 pp.	Integral and differential equations; Chebyshev basis block-and-diagonal (Delves-Freeman) matrix iterations
Fornberg&Sloan(1994)	65 pp.	Chebyshev & Fourier single-domain
Givi&Madnia(1993)	44 pp.	Spectral methods in combustion
Gottlieb,Hussaini&Orszag(1984)	54 pp.	Single-domain pseudospectral
Hussaini,Kopriva&Patera(1989)	31 pp.	Chebyshev & Fourier single-domain
Jarraud&Baede(1985)	41 pp.	Spectral methods in numerical weather prediction
Maday&Patera(1987)	73 pp.	Spectral elements; fluids
Robson & Prytz(1993)	31 pp.	Pseudospectral as the "discrete ordinates" method physics applications
Fischer&Patera(1994)	44 pp.	Spectral elements on massively parallel machines

Appendix A

A Bestiary of Basis Functions

"Talk with M. Hermite: he never invokes a concrete image; yet you soon perceive that the most abstract entities are for him like living creatures."
— Henri Poincaré

A.1 Trigonometric Basis Functions: Fourier Series

Applications: All problems with *periodic* boundary conditions

Interval: $x \in [-\pi, \pi]$ or $x \in [0, 2\pi]$

(Because of the periodicity, these two intervals are equivalent.)

General Fourier Series:

$$f(x) = a_0 + \sum_{n=1}^{\infty} a_n \cos(nx) + \sum_{n=1}^{\infty} b_n \sin(nx) \tag{A.1}$$

$$
\begin{aligned}
a_0 &= (1/2\pi) \int_{-\pi}^{\pi} f(x)dx \\
a_n &= (1/\pi) \int_{-\pi}^{\pi} f(x) \cos(nx)dx \\
b_n &= (1/\pi) \int_{-\pi}^{\pi} f(x) \sin(nx)dx
\end{aligned}
\tag{A.2}
$$

495

If $f(x)$ is symmetric about $x = 0$, one needs only the cosine terms, giving a Fourier cosine series. If $f(x)$ is antisymmetric about the origin, that is, $f(x) = -f(-x)$ for all x, then one needs only the sine terms. One should halve the computational interval to avoid applying redundant collocation conditions, i. e.

Interval: $x \in [0, \pi]$ or $x \in \left[-\dfrac{\pi}{2}, \dfrac{\pi}{2}\right]$ Fourier Cosine Series
                                                                             or Fourier Sine Series

If $f(x)$ has definite symmetry with respect to both $x = 0$ and $x = \pi/2$, then one needs only the even or odd terms of a Fourier cosine or sine series as explained in Chapter 8. For these "quarter-Fourier series" in which one uses only the cosine terms with even n, or the sine terms of even n, then

Interval: $x \in [0, \pi/2]$ "Quarter-Wave" Series with Definite Parity
                                             with Respect to Both $x = 0$ and $x = \pi/2$

Collocation Points:

$$
x_i = \begin{cases} \pi i/N & i = 1, \ldots, 2N & \text{(General)} \\ \pi i/N & i = 1, \ldots, N & \text{(Symmetric or Antisymmetric)} \\ \pi i/(2N) & i = 1, \ldots, N & \text{(Quarter-Wave [Double Symmetry])} \end{cases}
$$

(A.3)

A.2 Chebyshev Polynomials: $T_n(x)$

Applications: Any problem whatsoever.

Weight:
$$\rho(x) \equiv \frac{1}{\sqrt{1 - x^2}} \tag{A.4}$$

Inner Product:
$$\int_{-1}^{1} \frac{T_m T_n}{\sqrt{1 - x^2}} \, dx = \begin{cases} 0 & m \neq n \\ \pi & m = n = 0 \\ \pi/2 & m = n \neq 0 \end{cases} \tag{A.5}$$

Standardization:
$$T_n(\cos[t]) = \cos(nt) \tag{A.6}$$

Inequality:
$$|T_n(x)| \leq 1 \qquad \text{for all } n, \text{ all on } [-1, 1] \tag{A.7}$$

Three-Term Recurrence:
$$T_0 \equiv 1 \qquad ; \qquad T_1(x) \equiv x \tag{A.8}$$
$$T_{n+1}(x) = 2 x \, T_n(x) - T_{n-1}(x) \qquad n \geq 1 \tag{A.9}$$

Differentiation rules:

(i) Use the definition of $T_n(x)$ in terms of the cosine, (A.6), and Table E–2 [RECOMMENDED because of its simplicity]. For example,

$$\frac{dT_n}{dx} = \frac{n \sin(nt)}{\sin(t)} \tag{A.10}$$

$$\frac{d^2 T_n}{dx^2} = -\frac{n^2 \cos(nt)}{\sin^2(t)} + \frac{n \cos(t) \sin(nt)}{\sin^3(t)} \tag{A.11}$$

where $t = \arccos(x)$.

(ii) Use the relationship between the Chebyshev and Gegenbauer polynomials, which is

$$\frac{d^m}{dx^m} T_n(x) = n \, 2^{m-1} \, (m-1)! \, C_{n-m}^{(m)}(x) \tag{A.12}$$

$$\left. \frac{d^p T_n}{dx^p} \right|_{x = \pm 1} = (\pm 1)^{n+p} \prod_{k=0}^{p-1} \frac{n^2 - k^2}{2k + 1} \tag{A.13}$$

(iii) Compute the Chebyshev series for the derivative from that for $u(x)$ itself. Let

$$\frac{d^q u}{dx^q} = \sum_{k}^{N} a_k^{(q)} T_k(x) \tag{A.14}$$

so that the superscript "q" denotes the coefficients of the q-th derivative. These may be computed from the Chebyshev coefficients of the $(q-1)$-st derivative by the recurrence relation (in descending order)

$$a_N^{(q)} = a_{N-1}^{(q)} = 0 \tag{A.15}$$

$$a_{k-1}^{(q)} = \frac{1}{c_{k-1}}\left\{2k\,a_k^{(q-1)} + a_{k+1}^{(q)}\right\}, \qquad k = N-1, N-2, N-3, \dots, 1$$

where $c_k = 2$ if $k = 0$ and $c_k = 1$ for $k > 0$.

Notes:

$$(\text{i})\frac{d}{dx}T_n(x) = nU_{n-1}(x) \tag{A.16}$$

Higher order differentiation is mildly ill-conditioned as discussed by Breuer and Everson (1992), Huang and Sloan (1992), Merryfield and Shizgal (1993) and Bayliss, Class and Matkowsky (1994). Heinrichs (1991) has suggested a partial remedy: Basis recombination.

Integration:

$$\int^x dy\, T_n(y) = \left\{\frac{T_{n+1}(x)}{2(n+1)} + \frac{T_{n-1}(x)}{2(n-1)}\right\}, \qquad n \geq 2 \tag{A.17}$$

Flyer (1998) gives asymptotic upper bounds for the coefficients in the Chebyshev series for the iterated integral of arbitrary order as well as a discussion of integration identities in general.

Karageorghis (1988a, b) has analyzed the Chebyshev coefficients of the general order derivative and of its moments.

Collocation Points: either of the two choices

$$x_i = \cos\left[\frac{(2i-1)\pi}{2N}\right] \quad i = 1, \dots, N \qquad \begin{array}{c}\text{["Roots" or}\\ \text{"Gauss-Chebyshev"]}\end{array} \tag{A.18}$$

$$x_i = \cos\left[\frac{\pi i}{N-1}\right] \quad i = 0, \dots, N-1 \qquad \begin{array}{c}\text{["Extrema-plus-Endpoints"}\\ \text{or "Gauss-Lobatto"]}\end{array} \tag{A.19}$$

Differential Equation (Sturm-Liouville Problem)

$$(1-x^2)\frac{d^2 T_n}{dx^2} - x\frac{dT_n}{dx} + n^2 T_n = 0, \qquad n = 0, 1, 2, \dots \tag{A.20}$$

A.3   Chebyshev Polynomials of the Second Kind: $U_n(x)$

Applications: Proportional to first derivative of Chebyshev polynomials of the first kind. Rarely used as a basis set.

Interval: $x \in [-1, 1]$

Weight:

$$\rho(x) \equiv \sqrt{1 - x^2} \tag{A.21}$$

Inner Product:

$$\int_{-1}^{1} U_m(x)\, U_n(x)\, \sqrt{1 - x^2}\, dx = \left(\frac{\pi}{2}\right) \delta_{mn} \tag{A.22}$$

Standardization:

$$U_n(1) = n + 1 \tag{A.23}$$

Trigonometric Form:

$$U_n(\cos\theta) \equiv \frac{\sin[(n + 1)\theta]}{\sin\theta} \tag{A.24}$$

Inequality:

$$|U_{n+1}(x)| \leq n + 1 \qquad \text{for all } n, \text{ all } x \text{ on } [-1, 1] \tag{A.25}$$

Three-Term Recurrence and Starting Values:

$$U_0(x) \equiv 1 \qquad ; \qquad U_1(x) \equiv 2x \tag{A.26}$$
$$U_{n+1}(x) = 2x\, U_n(x) - U_{n-1}(x) \qquad n \geq 1 \tag{A.27}$$

Differentiation Rules:

    (i) Use trigonometric form (A.24)
    (ii) Use relationship between Chebyshev & Gegenbauer polynomials

$$\frac{d^m}{dx^m} U_n(x) = 2^m\, m!\; C_{n-m}^{(m+1)}(x) \tag{A.28}$$

Collocation Points:

$$x_i = -\cos\left[\frac{i\pi}{N + 1}\right] \qquad ; \qquad i = 1, 2, \ldots, N \tag{A.29}$$

A.4   Legendre Polynomials: $P_n(x)$

Applications: Spectral elements; alternative to Chebyshev polynomials for non-periodic problems; also the axisymmetric spherical harmonics for problems in spherical geometry.

Interval: $x \in [-1, 1]$ Weight: $\rho(x) \equiv 1$

Inner Product:

$$\int_{-1}^{1} P_m \, P_n \, dx = \frac{2}{2n+1} \delta_{mn} \tag{A.30}$$

Endpoint Values:

$$P_n(\pm 1) = (\pm 1)^n, \qquad dP_n/dx(\pm 1) = (\pm 1)^{n-1} n(n+1)/2 \tag{A.31}$$

Inequalities:

$$|P_n(x)| \leq 1 \qquad |dP_n/dx| \leq n(n+1)/2 \quad \text{for all } n, \text{ all } x \in [-1, 1] \tag{A.32}$$

Nonuniformity:

$$P_n(0) = \frac{1 \cdot 3 \cdots (n-1)}{2 \cdot 4 \cdots (n)} \sim \sqrt{2/(\pi n)}, \qquad n \text{ even} \tag{A.33}$$

Asymptotic Approximation as $n \to \infty$ for fixed t:

$$P_n(\cos(t)) \sim \sqrt{\frac{2}{n\pi \sin(t)}} \sin\left\{(n+1/2)t + \frac{\pi}{4}\right\} + O(n^{-3/2}) \tag{A.34}$$

Three-Term Recurrence and Starting Values:

$$P_0 \equiv 1, \qquad\qquad\qquad\qquad P_1(x) = x \tag{A.35}$$

$$(n+1) P_{n+1}(x) = (2n+1) x P_n(x) - n P_{n-1}(x) \tag{A.36}$$

Differentiation:

Use relationship between Legendre and Gegenbauer Polynomials

$$\frac{d^m}{dx^m} P_n(x) = 1 \cdot 3 \cdot 5 \cdots (2m-1) \, C_{n-m}^{(m+1/2)}(x) \tag{A.37}$$

Phillips(1988) gives the Legendre coefficients of a general-order derivative of an infinitely differentiable function.

Collocation Points (Gaussian Quadrature Abscissas):

Not known in closed form for general N; see pgs. 916–919 of Abramowitz & Stegun(1965), *NBS Handbook*. Analytical expressions for up to 9-point grids are given in Appendix F.

If

$$f(x) \equiv \sum_{n=0}^{\infty} a_n P_n(x), \qquad x f(x) \equiv \sum_{n=0}^{\infty} b_n P_n(x) \qquad \text{(A.38)}$$

$$\frac{df}{dx} \equiv \sum_{n=0}^{\infty} a_n^{(1)} P_n(x), \qquad \frac{d^2 f}{dx^2} \equiv \sum_{n=0}^{\infty} a_n^{(2)} P_n(x) \qquad \text{(A.39)}$$

then the coefficients of the derived series are given in terms of the coefficients of $f(x)$ itself as

Multiplication by x:

$$b_n = \frac{n}{2n-1} a_{n-1} + \frac{n+1}{2n+3} a_{n+1}, \qquad n \geq 1 \quad [b_0 = 0] \qquad \text{(A.40)}$$

First Derivative:

$$a_n^{(1)} = (2n+1) \sum_{p=n+1, p+n \,\text{odd}}^{\infty} a_p \qquad \text{(A.41)}$$

Second Derivative:

$$a_n^{(2)} = (n+1/2) \sum_{p=n+2, p+n \,\text{even}}^{\infty} \{p(p+1) - n(n+1)\} a_p \qquad \text{(A.42)}$$

A.5   Gegenbauer Polynomials

Applications: Spherical geometry as the latitudinal factors for the spherical harmonics, Gottlieb *et al* regularization of Gibbs' Phenomenon

Alternative name: "Ultraspherical" polynomials

Interval: $x \in [-1, 1]$

Weight Function:

$$\rho(x) = \left(1 - x^2\right)^{m-1/2} \tag{A.43}$$

Inner Product:

$$\int_{-1}^{1} C_n^{(m)} \, C_n^{(m)} \, (1 - x^2)^{m-1/2} \, dx = \frac{\pi \, 2^{1-2m} \, \Gamma(n + 2m)}{n! \, (n + m) \, \Gamma(m)^2} \equiv h_n^m \tag{A.44}$$

Expansion Coefficients of a Function $f(x)$

$$f(x) = \sum_{j=0}^{\infty} a_j^m \, C_j^m(x) \tag{A.45}$$

$$a_j^m \equiv \frac{1}{h_j^m} \int_{-1}^{1} f(x) \, C_j^m(x) \, (1 - x^2)^{m-1/2} \, dx \tag{A.46}$$

where h_j^m is defined by A.44 and where the Gegenbauer polynomials are not orthonormal, but standardized as in A.49.

Expansion of Complex Exponential

$$\exp(\, i \, \pi \, \omega \, x) = \sum_{j=0}^{\infty} f_j^m \, C_j^m(x) \tag{A.47}$$

$$f_j^m = i^j \, \Gamma(m) \left(\frac{2}{\pi \, \omega}\right)^m (j + m) \, J_{j+m}(\, \pi \, \omega \,) \tag{A.48}$$

Standardization:

$$C_n^{(m)}(1) = \frac{\Gamma(n + 2m)}{n! \, \Gamma(2m)} \tag{A.49}$$

Asymptotics of Endpoint Values

$$C_n^n(1) \sim \sqrt{\frac{1}{3\,\pi\,n}} \left(\frac{27}{4}\right)^n, \qquad n \to \infty \qquad [m = n] \tag{A.50}$$

$$C_n^m(1) \;\sim\; (n+1)^{2m-1} \frac{1}{\sqrt{\pi}} \left(\frac{1}{m}\right)^{2m-1/2} \exp\{2\,m\,[1 - \log(2)\,]\,\},$$
$$\text{fixed } m, n \to \infty \tag{A.51}$$

Inequality:

$$|C_n^{(m)}(x)| \le C_n^{(m)}(1) \qquad \text{for } m > 1/2,\ x \in [-1,\,1] \tag{A.52}$$

Uniformity Ratio

$$\rho_{2n}(m) \equiv \frac{C_{2n}^m(1)}{|C_{2n}^m(0)|} = \frac{\Gamma(2n+2m)\,n!\,\Gamma(m)}{(2n)!\,\Gamma(2m)\,\Gamma(n+m)} \tag{A.53}$$

$$\rho_{2n}(2n) \equiv \frac{C_{2n}^{2n}(1)}{|C_{2n}^{2n}(0)|} = \frac{\Gamma(6n)\,n!}{2n\,\Gamma(4n)\,\Gamma(3n)} \tag{A.54}$$

Three-Term Recurrence:

$$C_0^{(m)} \equiv 1 \qquad\qquad ; \qquad\qquad C_1^{(m)}(x) \equiv 2\,m\,x \tag{A.55}$$

$$(n+1)\,C_{n+1}^{(m)}(x) = 2\,(n+m)\,x\,C_n^{(m)}(x) - (n+2m-1)\,C_{n-1}^{(m)}(x) \tag{A.56}$$

Note: the recurrence relation is weakly unstable for $m > 0$, and the instability increases as m increases. Consequently, library software usually calculates the spherical harmonics using a different, more complicated recurrence relation (see book by Belousov, 1962). However, our own numerical experiments, perturbing the recurrence relation by $\epsilon\rho(n)$ where $\rho(n)$ is a different random number between 0 and 1 for each m, have shown the maximum error for all x and all polynomial degrees < 100 is only $263\,\epsilon$ for $m = 24.5$ and 1247ϵ for $m = 49.5$. Nehrkorn (1990) found the same through his more detailed experiments. This suggests that the recurrence relation is actually quite satisfactory if a relative error as large as 1.E-10 is acceptable.

Differentiation Rule:

$$\frac{d}{dx}C_n^{(m)}(x) = 2\,m\,C_{n-1}^{(m+1)}(x) \tag{A.57}$$

Phillips and Karageorghis (1990) derive formulas for the coefficients of integrated expansions of Gegenbauer polynomials.

Collocations Points (Gaussian Quadrature Abscissas):

See Davis & Rabinowitz, *Methods of Numerical Integration*. Canuto *et al.* (1988) give a short FORTRAN subroutine to compute the Lobatto grid for general order in their Appendix C.

Special Cases:

$$T_n(x) = \left(\frac{n}{2}\right) C_n^{(0)}(x) \qquad\qquad \text{[Chebyshev]} \tag{A.58}$$

$$P_n(x) = C_n^{(1/2)}(x) \qquad\qquad\qquad \text{[Legendre]} \tag{A.59}$$

$$U_n(x) = C_n^{(1)}(x) \qquad\qquad \text{[Chebyshev of the 2d Kind]} \tag{A.60}$$

Gegenbauer Polynomials: Explicit Form

$$
\begin{aligned}
C_0^m &= 1; \qquad C_1^m(x) = 2m\, x \\
C_2^m &= 2m(1+m)\, x^2 - m \\
C_3^m &= \left\{\frac{8}{3}m + 4m^2 + \frac{4}{3}m^3\right\} x^3 - 2m(1+m)x \\
C_4^m &= \left(\frac{2}{3}m^4 + 4m^3 + \frac{22}{3}m^2 + 4m\right) x^4 - (2m^3 + 6m^2 + 4m)x^2 + \frac{1}{2}m(1+m) \\
C_5^m &= \left(\frac{4}{15}m^5 + \frac{8}{3}m^4 + \frac{28}{3}m^3 + \frac{40}{3}m^2 + \frac{32}{5}\right) x^5 \\
&\quad - \left(\frac{4}{3}m^4 + 8m^3 + \frac{44}{3}m^2 + 8m\right) x^3 + (m^3 + 3m^2 + 2m)x \tag{A.61}
\end{aligned}
$$

Endpoint values

$$
\begin{aligned}
C_0^m(1) &= 1; \qquad C_1^m(1) = 2m \\
C_2^m(1) &= m + 2m^2; \qquad C_3^m(1) = \frac{4}{3}m^3 + 2m^2 + \frac{2}{3}m \\
C_4^m(1) &= \frac{2}{3}m^4 + 2m^3 + \frac{11}{6}m^2 + \frac{1}{2}m \\
C_5^m(1) &= \frac{4}{15}m^5 + \frac{4}{3}m^4 + \frac{7}{3}m^3 + \frac{5}{3}m^2 + \frac{2}{5}m \tag{A.62}
\end{aligned}
$$

A.6   Hermite Polynomials: $H_n(x)$

Applications: doubly infinite interval; $f(x) \to 0$ as $|x| \to \infty$

Basis Functions:

$$\phi_n(x) \equiv \exp[-x^2/2]\, H_n(x) \tag{A.63}$$

Interval

$$x \in [-\infty,\, \infty] \tag{A.64}$$

Weight:

$$\rho(x) \equiv \exp[-x^2] \tag{A.65}$$

Inner Product:

$$\int_{-\infty}^{\infty} e^{-x^2}\, H_m(x)\, H_n(x)\, dx = \delta_{mn}\, \sqrt{\pi}\, 2^n n! \tag{A.66}$$

Standardization:

$$H_n(x) = 2^n\, x^n + \cdots \tag{A.67}$$

Three-Term Recurrence and Starting Values:

$$H_0 \equiv 1 \qquad ; \qquad H_1(x) \equiv 2\,x \tag{A.68}$$

$$H_{n+1}(x) = 2\,x\, H_n(x) - 2\,n\, H_{n-1}(x) \tag{A.69}$$

Differentiation Rules:

$$\frac{dH_n}{dx} \equiv 2n\,H_{n-1}(x) \qquad\qquad \text{all } n \qquad\qquad (A.70)$$

Collocation Points (Gaussian Abscissas): Abramowitz & Stegun(1965), *NBS Handbook*, pg. 924 , Davis and Rabinowitz, *Numerical Integration*.

The orthonormal Hermite functions ψ_n such that $\int_{-\infty}^{\infty} \psi_n(x)^2 dx = 1$ for all n may be computed by

$$\psi_0 \equiv \pi^{-1/4} \exp(-(1/2)x^2) \quad ; \quad \psi_1(x) \equiv \sqrt{2}\, x\, \psi_0 \qquad\qquad (A.71)$$

$$\psi_{n+1}(x) = \sqrt{\tfrac{2}{n+1}}\, x\, \psi_n(x) - \sqrt{\tfrac{n}{n+1}}\, \psi_{n-1}(x) \qquad\qquad (A.72)$$

and the derivatives $\psi_{n,x}$ from

$$\psi_{0,x} \equiv -x\psi_0 \quad ; \quad \psi_{1,x}(x) \equiv -x\,\psi_1 + \sqrt{2}x\, \psi_0 \qquad\qquad (A.73)$$

$$\psi_{n+1,x} = -x\,\psi_{n+1} + \sqrt{2\,n}\,\psi_n \qquad\qquad (A.74)$$

The orthonormal Hermite functions are better-conditioned than their unnormalized counterparts.

Inequality:

$$|\psi_n(x)| = \leq 0.816 \ \forall x \qquad\qquad (A.75)$$

Expansion of a Gaussian:

$$\exp(-x^2) = \sum_{m=0}^{\infty} a_{2m}\, \psi_{2m}(x) \qquad\qquad (A.76)$$

$$a_2 = (-1)^m \frac{\pi^{1/4}\sqrt{2/3}\,\sqrt{2m!}}{6^m\, m!} \qquad\qquad (A.77)$$

$$\sim (-1)^m\, 0.9709835 \frac{1}{3^m\,(2m)^{1/4}}, \qquad m \to \infty \qquad\qquad (A.78)$$

A.7 Rational Chebyshev Functions on an Infinite Interval: $TB_n(y)$

Applications: Doubly infinite interval

Interval: $y \in [-\infty, \infty]$

Let L denote a constant map parameter, user-chosen to match width of the function to be expanded. (Some experimentation may be necessary.) Then

Chebyshev definition:

$$TB_n(y) = T_n \left(\frac{y}{\sqrt{L^2 + y^2}} \right) \qquad \text{(A.79)}$$

Trigonometric definition:

$$
\begin{aligned}
TB_n(y) &= \cos \left(n \arccot \left(\frac{y}{L} \right) \right) & \text{(A.80)} \\
&= \cos(nt) & \text{(A.81)}
\end{aligned}
$$

where

$$y = L \cot(t) \qquad \longleftrightarrow \qquad t = \arccot \left(\frac{y}{L} \right) \qquad \text{(A.82)}$$

Weight:

$$\rho(y) \equiv \frac{L}{L^2 + y^2} \qquad \text{(A.83)}$$

Inner Product:

$$\int_{-\infty}^{\infty} TB_m(y) \, TB_n(y) \left[\frac{L}{L^2 + y^2} \right] dy = \begin{cases} \pi & m = n = 0 \\ 0 & m \neq n \\ \pi/2 & m = n > 1 \end{cases} \qquad \text{(A.84)}$$

Asymptotic Behavior:

$$TB_n(y) \sim 1 \quad \text{as } |y| \to \infty \quad \text{for all } n \qquad \text{(A.85)}$$

(Note that this basis set may usually be applied without modification to compute solutions that tend to 0 as $|y| \to \infty$.)

Derivatives:

Apply the trigonometric form (A.81) in combination with Table E–4.

References: Boyd (1987a).

A.8   Laguerre Polynomials: $L_n(x)$

Applications: Semi-Infinite interval; $f(x) \to 0$ as $x \to \infty$

Laguerre functions (the basis set):

$$\phi_n(x) \equiv \exp[-x/2]\, L_n(x) \tag{A.86}$$

Interval: $x \in [0, \infty]$

Weight:

$$\rho(x) \equiv \exp[-x/2] \tag{A.87}$$

Inner Product:

$$\int_0^\infty e^{-x}\, L_m(x)\, L_n(x)\, dx = \delta_{mn} \tag{A.88}$$

Standardization:

$$L_n(x) = \frac{(-1)^n}{n!}\, x^n + \cdots \tag{A.89}$$

Inequality:

$$|\phi_n(x)| = |\exp(-x/2)\, L_n(x)| \le 1 \tag{A.90}$$

Three-Term Recurrence and Starting Values:

$$L_0(x) = 1 \qquad ; \qquad L_1(x) = -x + 1 \tag{A.91}$$

$$(n+1)\, L_{n+1}(x) = \{(2n+1) - x\}\, L_n(x) - n\, L_{n-1}(x) \tag{A.92}$$

Differential Relations:

$$\frac{d^m}{dx^m} L_{n+m}(x) = (-1)^m\, L_n^{(m)}(x) \tag{A.93}$$

where the $L_n^{(m)}(x)$ are the generalized Laguerre polynomials. These can be computed by using the recurrence

$$L_0^{(m)}(x) \equiv 1 \qquad ; \qquad L_1^{(m)}(x) \equiv 1 + m - x \tag{A.94}$$

$$(n+1)\, L_{n+1}^{(m)}(x) = \{(2n+m+1) - x\}\, L_n^{(m)}(x) - (n+m)\, L_{n-1}^{(m)}(x) \tag{A.95}$$

Collocation Points:

Abramowitz & Stegun(1965), *NBS Handbook*, pg. 923.

A.9 Rational Chebyshev Functions on Semi-Infinite Interval: $TL_n(y)$

Interval: $y \in [0, \infty]$

(Generalization: $r \in [r_0, \infty]$ by substituting $y = r - r_0$ in all formulas below.) Let L denote a constant map parameter, user-chosen to match width of the function to be expanded. (Some experimentation may be necessary.) Then

Chebyshev definition:

$$TL_n(y) \equiv T_n\left(\frac{y - L}{y + L}\right) \tag{A.96}$$

Trigonometric definition:

$$TL_n(y) \quad \equiv \quad \cos\left[2\,n\arccot\left(\sqrt{\frac{y}{L}}\right)\right] \tag{A.97}$$

$$\equiv \quad \cos(nt) \tag{A.98}$$

where

$$y \equiv L\cot^2\left(\frac{t}{2}\right) \quad \longleftrightarrow \quad t = 2\arccot\left(\sqrt{\frac{y}{L}}\right) \tag{A.99}$$

Weight function:

$$\rho(y) \equiv \frac{1}{y + L}\sqrt{\frac{L}{y}} \tag{A.100}$$

$$\int_0^\infty TL_m(y;\,L)\,TL_n(y;\,L)\left\{\frac{1}{y + L}\sqrt{\frac{L}{y}}\right\}\,dy = \begin{cases} \pi & m = n = 0 \\ 0 & m \neq n \\ \pi/2 & m = n > 0 \end{cases} \tag{A.101}$$

Derivatives:

Apply trigonometric form (A.98) in Table E–6.

References: Boyd (1987b).

Table A.1: Flow Chart on the Choice of Basis Functions

If	Basis Set is
$f(x)$ is periodic	Fourier series
$f(x)$ is periodic & symmetric about $x = 0$	Fourier cosine
$f(x)$ is periodic & antisymmetric about $x = 0$	Fourier sine
$x \in [a, b]$ & $f(x)$ is non-periodic	Chebyshev polys. Legendre polys.
$y \in [0, \infty]$ & $f(y)$ decays exponentially as $y \to \infty$	$TL_n(y)$ Laguerre functions
$y \in [0, \infty]$ & $f(y)$ has asymptotic series in inverse powers of y	$TL_n(y)$ only
$y \in [-\infty, \infty]$ & $f(y)$ decays exponentially as $y \to \infty$	$TB_n(y)$ sinc functions or Hermite functions
$y \in [-\infty, \infty]$ & $f(y)$ has asymptotic series in inverse powers of y	$TB_n(y)$ only
$f(\lambda, \theta)$ is a function of latitude and longitude	spherical harmonics

A.10 Graphs of Convergence Domains in the Complex Plane

Regions of convergence of various basis sets in the complex plane along with the expansion interval appropriate to that basis. The cross-hatching indicates the region *within* which the series converges. If $f(x)$ denotes the function being expanded, then there is usually a pole or branch point of $f(x)$ on the curve bounding the region of convergence. In all cases, the function $f(x)$ has no singularities *inside* the convergence region; the *size* of the region of convergence is thus controlled by the poles and branch points of the function which is being approximated.

The *shapes* of the curves that bound that convergence regions, however, are determined entirely by the *basis* set. Thus, a Chebyshev series always converges within a certain ellipse while diverging outside that ellipse, but whether the ellipse is large or small depends on the singularities of $f(x)$, and which one is closest to the interval $x \in [-1, 1]$.

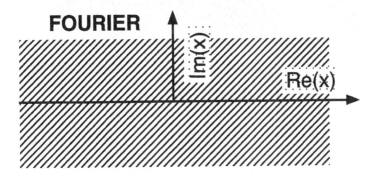

Figure A.1: The convergence boundary for a Fourier series is the pair of straight lines, $\text{Im}(x) = $ constant

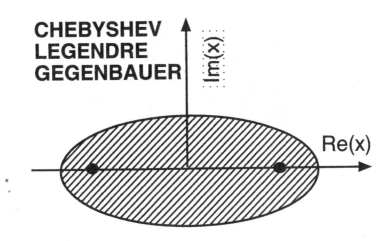

Figure A.2: Chebyshev, Legendre, and Gegenbauer polynomials of all orders: ellipse with foci at $x = \pm 1$. The foci are marked by black disks.

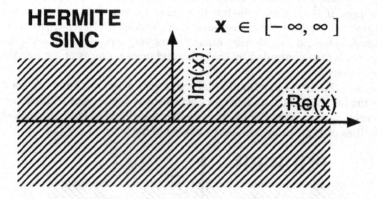

Figure A.3: Hermite & sinc functions: straight lines, $\mathrm{Im}(x)$ = constant

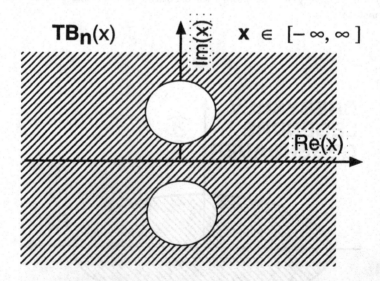

Figure A.4: $TB_n(x)$: pair of circles, the contours of "bipolar" coordinates in the complex x-plane

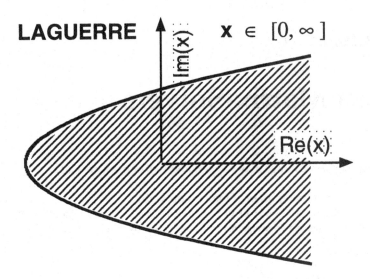

Figure A.5: Laguerre functions: parabola with focus at $x = 0$

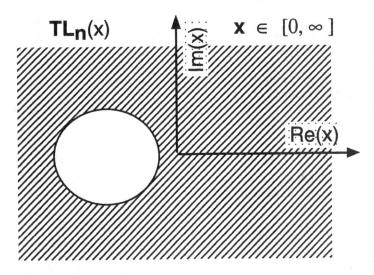

Figure A.6: $TL_n(x)$: quartic curve with no standard name

Appendix B

Direct Matrix-Solvers

"When iterating, three things may happen and two of them are bad."

— J. P. Boyd (paraphrase of football proverb)

In football, the three fates of a forward pass are "completion" (good), "incomplete" (bad) and "interception" (very bad). Similarly, an iteration may converge, may converge too slowly to be useful, or diverge.

B.1 Matrix Factorizations

The most efficient "direct" way to solve a matrix equation is to first simplify the square matrix to a special factored form and then perform a "backsolve". A "direct" method is a non-iterative algorithm; iterative matrix-solvers are discussed in Chapter 15 above.

The three most useful factorizations are (i) the product of two triangular matrices $\vec{L}$ and $\vec{U}$, usually called the "LU" factorization (or "Crout reduction") (ii) the product of a "lower triangular" matrix $\vec{L}$ and its transpose ["Choleski factorization"], possible only for a *symmetric* matrix and (iii) the product of an "orthogonal" matrix $\vec{Q}$ with an upper triangular matrix $\vec{R}$, commonly called the "QR" decomposition. The relative costs and merits of these three methods are given Table B.1.

Very robust library software is available for all these algorithms, so the user need never write his own. However, a little knowledge of matrix properties is important in choosing the most efficient subroutine.

Triangular matrices and the LU and Cholesky factorizations are defined by following.

Table B.1: Cost of Matrix-Solving Methods for Dense (Full) Matrices
Note: In the cost estimates, N is the size of the (dense) matrix which is factored. The algorithms all require equal numbers of additions/subtractions and multiplications/division; the numbers are the sums of all operations.

Name	Factorization Cost	Backsolve Cost	Comments
LU	$\frac{2}{3}N^3$	$2N^2$	Default method
Cholesky	$\frac{1}{3}N^3$	$2N^2$	Cheaper than LU, but only applies to symmetric matrices
QR	$\frac{4}{3}N^3$	$4N^2$	Good for ill-conditioned matrices, but twice as costly as LU

Definition 45 (TRIANGULAR MATRIX) *A matrix $\vec{\vec{A}}$ is said to be in LOWER TRIANGULAR or UPPER TRIANGULAR form, respectively, if*

$$A_{ij} = 0 \qquad \text{whenever} \quad j > i \qquad \text{[LOWER TRIANGULAR]} \qquad \text{(B.1)}$$

$$A_{ij} = 0 \qquad \text{whenever} \quad i < j \qquad \text{[UPPER TRIANGULAR]} \qquad \text{(B.2)}$$

In other words, on one side, either the upper or lower, of the diagonal row of elements A_{ii}, a triangular matrix $\vec{\vec{A}}$ has nothing but elements whose values are zero.

Theorem 40 (*LU* DECOMPOSITION OF A MATRIX) *If its determinant is non-zero, then an otherwise arbitrary matrix $\vec{\vec{A}}$ can always be factored into the product of a lower triangular matrix $\vec{\vec{L}}$ and an upper triangular matrix $\vec{\vec{U}}$. For 4×4 matrices, for example,*

$$\begin{vmatrix} l_{11} & 0 & 0 & 0 \\ l_{21} & l_{22} & 0 & 0 \\ l_{31} & l_{32} & l_{33} & 0 \\ l_{41} & l_{42} & l_{43} & l_{44} \end{vmatrix} \begin{vmatrix} 1 & u_{12} & u_{13} & u_{14} \\ 0 & 1 & u_{23} & u_{24} \\ 0 & 0 & 1 & u_{34} \\ 0 & 0 & 0 & 1 \end{vmatrix} = \begin{vmatrix} a_{11} & a_{12} & a_{13} & a_{14} \\ a_{21} & a_{22} & a_{23} & a_{24} \\ a_{31} & a_{32} & a_{33} & a_{34} \\ a_{41} & a_{42} & a_{43} & a_{44} \end{vmatrix} \qquad \text{(B.3)}$$

If the matrix $\vec{\vec{A}}$ is REAL, SYMMETRIC, and POSITIVE DEFINITE, then it can factored into

$$\vec{\vec{L}}\,\vec{\vec{L}}^{T} = \vec{\vec{A}} \qquad \text{[Cholesky decomposition]} \qquad \text{(B.4)}$$

where superscript T denotes the transpose. The Cholesky decomposition needs only half the number of operations and half the storage of the standard LU decomposition.

PROOF: The elements of $\vec{\vec{L}}$ and $\vec{\vec{U}}$ [$u_{ii} \equiv 1$] are given explicitly by

$$l_{ij} = a_{ij} - \sum_{k=1}^{j-1} l_{ik}\, u_{kj} \qquad\qquad i \geq j \qquad \text{(B.5a)}$$

$$u_{ij} = \frac{1}{l_{ii}} \left\{ a_{ij} - \sum_{k=1}^{i-1} l_{ik}\, u_{kj} \right\} \qquad\qquad i < j \qquad \text{(B.5b)}$$

which is nothing more than a restatement of the fact that the elements of $\vec{\vec{A}}$ are given by the matrix product of $\vec{\vec{L}}$ and $\vec{\vec{U}}$. The elements are computed in the order

$$l_{i1}, \; u_{1j}; \; l_{i2}, \; u_{2j}; \; \ldots, \; l_{i,N-1}, \; u_{N-1,j}; \; l_{NN},$$

that is, we compute one column of $\vec{\vec{L}}$, then the corresponding row of $\vec{\vec{U}}$, and return to evaluate the next column of $\vec{\vec{L}}$ and so on. Q. E. D.

The proof ignores pivoting, which is rarely necessary for matrices arising from spectral or finite element discretizations. Good reviews of Gaussian elimination and factorization are given in Conte and de Boor (1980) and Forsythe and Moler (1967).

The matrices which are the Chebyshev or Legendre pseudospectral discretizations of derivatives may be somewhat ill-conditioned, that is, generate a lot of round-off error in the LU method, when N is large (> 100) and/or the order of the derivatives is high (third derivatives or higher). One remedy is to use a different factorization

Theorem 41 (QR Factorization)

- *An arbitrary matrix $\vec{\vec{A}}$, even if rectangular or singular, can always be factored as the product of an orthogonal matrix $\vec{\vec{Q}}$ multiplied on the right by a triangular matrix $\vec{\vec{R}}$. (A matrix is "orthogonal" or "unitary" if its transpose is also its inverse.)*

- *The matrix $\vec{\vec{Q}}$ is (implicitly) the product of a series of rank-one orthogonal matrices ("Householder transformations"). Because all transformations are unitary, QR factorization is much more resistant to accumulation of roundoff error than the LU factorization.*

- *It is expensive to explicitly compute $\vec{\vec{Q}}$. To solve $\vec{\vec{A}}\vec{x} = \vec{b}$, one alternative is to apply the Householder transformations simultaneously to both $\vec{\vec{A}}$ and $\vec{b}$ to transform the matrix into $\vec{\vec{R}}$ and the vector into $\vec{\vec{Q}}^{T}\vec{b}$. The solution $\vec{x}$ is then computed by backsolving the triangular system $\vec{\vec{R}}\vec{x} = \vec{b}$.*

- *Alternatively, one can use the "semi-normal equations": one multiplication of a vector by a square matrix followed by two triangular backsolves:*

$$\vec{\vec{R}}^{T}\left\{\vec{\vec{R}}\vec{x}\right\} = \vec{\vec{A}}^{T}\vec{b} \tag{B.6}$$

This has less numerical stability (to roundoff) than the "Q-and-R" or "simultaneous transformation" approach, so the semi-normal backsolve is usually repeated with $\vec{b}$ replaced by $\vec{b} - \vec{\vec{A}}\vec{x}$, the residual of the linear system, to provide one step of "iterative refinement".

- *$\vec{\vec{R}}$ is a Cholesky factor of $\vec{\vec{A}}^{T}\vec{\vec{A}}$, that is*

$$\vec{\vec{R}}^{T}\vec{\vec{R}} = \vec{\vec{A}}^{T}\vec{\vec{A}} \tag{B.7}$$

This implies that the QR factorization doubles the bandwidth of a banded matrix; the ratio of (QR cost)/(LU cost) is even bigger for sparse matrices than the ratio of two for dense matrices.

- *If $\vec{\vec{A}}$ contains even a single full row, $\vec{\vec{R}}$ will be a dense matrix.*

- *For overdetermined systems, the "normal equations" $\vec{\vec{A}}^{T}\vec{\vec{A}}\vec{x} = \vec{\vec{A}}^{T}\vec{b}$ and the QR factorization will both give bounded solutions if the columns of $\vec{\vec{A}}$ are all linearly independent ("full column rank"). However, accumulation of roundoff error is often disastrous in the normal equations; QR factorization is much more stable.*

- *If the matrix is intrinsically singular, and not merely apparently singular because of accumulated roundoff, $\vec{\vec{R}}$ will have one or more zero diagonal elements and the QR algorithm will fail.*

QR may be a useful fallback algorithm when LU software crashes with a message like "Matrix is nearly singular". However, my personal experience is that 95% of the occurrences of an ill-conditioned pseudospectral matrix arise because of a coding error. Dennis and Schnabel (1983) give a brief and readable treatment of QR.

Definition 46 (SPARSE & DENSE) *A matrix $\vec{\vec{A}}$ is said to be SPARSE if most of its matrix elements are zero. A matrix whose elements are all (or mostly) non-zero is said to be DENSE or FULL.*

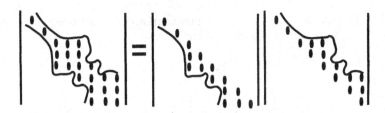

Figure B.1: Schematic of sparse matrix with disks representing nonzero matrix elements. The heavy curves are the "skyline" which bounds the non-zero elements; it is (by definition) symmetric with respect to the diagonal of the matrix. The right half of the figure shows that the lower and upper triangular factors of a matrix have the same "skyline".

Spectral methods usually generate dense matrices (alas!). The costs of the factorizations for such dense matrices are given in Table B.1. However, domain decomposition generates sparse matrices (with a few dense blocks). Galerkin discretizations of special differential equations, (usually constant coefficient and linear) , also generate sparse matrices (Chapter 15, Sec. 10). The LU and Cholesky factorizations can then be modified to omit operations on the zero elements, thereby greatly reducing cost.

Because matrix sparsity occurs in many patterns and can generate enormous savings, a huge literature on "sparse matrix" methods has arisen. We cannot possibly do it justice in an appendix, but must instead refer the reader to a good book on numerical linear algebra. However, some patterns are so common that it is worthwhile to briefly describe them here.

The most general sparse matrix software which can be explained in simple terms is a "skyline solver". Kikuchi (1986) gives both FORTRAN and BASIC codes and also a good explanation of the rather complicated "skylines" that can arise from finite element and spectral element discretizations. The concept of the "skyline", which is defined formally below , is important because it is *preserved* by LU factorization as also illustrated in the graph.

Definition 47 (SKYLINE of a MATRIX) *Let all the nonzero elements of a matrix $\vec{\vec{A}}$ satisfy the inequality*

$$A_{ij}, A_{ji} = 0 \ \forall |i - j| \geq \sigma(i)$$

where $\sigma(i)$ is chosen to be as small as possible for each i. Then $\sigma(i)$ is the "SKYLINE" of the matrix $\vec{\vec{A}}$.

The crucial point is that if the matrix is sparse, many of the operations above involve only elements known to be zero. The cost of the factorization can be greatly reduced by replacing the upper limits on the sums by the skyline function $\sigma(i)$.

The factorization allows the original problem

$$\vec{\vec{A}} \vec{x} = \vec{f} \tag{B.8}$$

to be written without approximation as

$$\vec{\vec{L}} \vec{g} = \vec{f} \tag{B.9}$$

$$\vec{\vec{U}} \vec{x} = \vec{g} \tag{B.10}$$

The "backsolve" step is to compute $\vec{x}$ from these triangular matrix equations in $O(2N^2)$ operations through the recursions ("backsolve" steps)

$$g_1 = \frac{1}{l_{11}} f_1 \tag{B.11}$$

$$g_i = \frac{1}{l_{ii}} \left\{ f_i - \sum_{j=1}^{i-1} l_{ij} g_j \right\} \qquad i = 2, 3, \ldots, N \tag{B.12}$$

$$x_N = g_N \tag{B.13}$$

$$x_i = g_i - \sum_{j=i+1}^{N} u_{ij} x_j \qquad i = (N-1), (N-2), \ldots, 2 \tag{B.14}$$

Like the LU factorization, the backsolve step can also be greatly reduced in cost by exploiting the sparsity of the matrix $\vec{A}$: eliminating operations on elements known *a priori* to be zero.

In many cases, a general skyline solver is inefficient because the matrix has very simple patterns of sparsity. We describe three useful special cases of sparse matrices in the next four sections.

B.2   Banded Matrix

Definition 48 (BANDWIDTH OF A MATRIX) *A matrix $\vec{\vec{A}}$ is said to be of BANDWIDTH $(2m+1)$ if its elements satisfy the equation*

$$a_{ij} = 0 \qquad \textit{for all} \ \ |i - j| > m \tag{B.15}$$

The special case $m = 1$ has at most three non-zero elements in each row (or column) and is called "TRIDIAGONAL". The special case $m = 2$ has at most five non-zero elements in each row and is called "PENTADIAGONAL".

Theorem 42 (LU DECOMPOSITION: BANDED MATRIX) *The LU decomposition of a matrix of bandwidth $(2m+1)$ has factors which are of the same bandwidth, $(2m+1)$, in addition to being triangular.*

The cost of performing an LU decomposition is

$$\text{\# multiplications} = N(m^2 + m) \tag{B.16}$$
$$\text{\# additions} = Nm^2 \tag{B.17}$$

Eqs. (B.5) through (B.13) still apply; we merely truncate the sums to reflect the bandwidth of $\vec{\vec{A}}$, $\vec{\vec{L}}$, and $\vec{\vec{U}}$.

A number of special cases are collected in Table B.2.

PROOF: The bandwidth preserving property can be shown directly from (B.5).

Table B.2: Operation Counts for Gaussian Elimination.

There are two steps in solving a matrix equation: (i) the LU factorization of the square matrix $\vec{A}$, which must be repeated whenever $\vec{A}$ is altered and (ii) the backsolve, which must be repeated for each different column vector $\vec{f}$. The costs for each step are shown separately below because it is common (in semi-implicit time stepping algorithms, for instance) to perform only a single LU factorization followed by many backsolves (one per time step). Below, we list the costs for general m under the simplifying assumptions that (i) $N \gg m$ where N is the dimension of the matrix and (ii) multiplications and divisions are equally costly and additions and subtractions are equally costly. Warning: all entries assume pivoting is unnecessary.

(a) Full matrix

LU factorization:	$N^3/3$ mults.	&	$N^3/3$ adds.
Backsolve:	N^2 mults.	&	N^2 adds.

(b) Banded matrix of bandwidth m

LU factorization:	$N(m^2 + m)$ mults.	&	Nm^2 adds.
Backsolve:	$N(2m + 1)$ mults.	&	$N(2m)$ adds.

(c) Special bandwidths.

The first number listed in each pair is the number of multiplications or divisions; the second is the number of additions and subtractions.

m	Name	LU total	Backsolve	LU+Backsolve
1	[Tridiagonal]	2N & N	3N & 2N	5 N mult. & 3 N add.
2	[Pentadiagonal]	6N & 4N	5N & 4N	11 N mult. & 8 N add.
3	[Heptadiagonal]	12N & 9N	7N & 6N	19 N mult. & 15 N add.
4	[Enneadiagonal]	20N & 16N	9N & 8N	29 N mult. & 24 N add.
5	[Hendecadiagonal]	30N & 25N	11N & 10N	41 N mult. & 35 N add.
6	[Tridecadiagonal]	42N & 36N	13N & 12N	55 N mult. & 48 N add.

B.3   Matrix-of-Matrices Theorem

Theorem 43 (MATRICES WHOSE ELEMENTS ARE MATRICES)

All the ordinary rules of matrix multiplication and factorization, including the LU decomposition for full and banded matrices, still apply when the elements of the matrix are themselves matrices, provided one is careful about the non-commutativity of matrix multiplication.

PROOF: Matrix operations (addition, subtraction, multiplication, and division) obey exactly the same rules as those for ordinary numbers except that division must be interpreted as multiplication by the inverse matrix and multiplication and division are not commutative. It follows that formulas like (B.5) – (B.13) will remain true when the a_{ij}, l_{ij}, and u_{ij} are matrices so long as we do not reverse the order of any factors. The division by l_{ii} in (B.11) was written as left multiplication by the inverse of l_{ii} precisely for this reason. Q. E. D.

This matrix-of-matrices concept has many interesting applications. In the next two sections, we shall discuss a couple.

B.4   Block-Banded Elimination: the "Lindzen-Kuo" Algorithm

When the usual second order finite differences are applied to a second order boundary value problem in two dimensions[1], the result is a matrix of dimension $(MN) \times (MN)$. When the differential equation is everywhere elliptic and sufficiently well-behaved, the matrix problem may be quite efficiently solvable through Successive Over-Relaxation or some other iterative method. However, the iterations usually diverge if the matrix is not positive definite, and this can happen even for a harmless Helmholtz equation like

$$\triangle u + K u = f \tag{B.18}$$

for constant K where K is sufficiently large and positive. (However, multi-grid iterations may work even for an indefinite matrix if Gaussian elimination is used on the coarsest grid.)

An alternative is to apply Gaussian elimination (Lindzen and Kuo, 1969) in such a way as to exploit the "block-banded" structure of the matrix. By "block-banded" we mean that the matrix is tridiagonal if we group square blocks of dimension $M \times M$ together. Alternatively, we can simply treat the "block-banded" matrix as an ordinary banded matrix of bandwidth $3M$, but it is conceptually simpler and algorithmically more efficient to use the block structure instead.

The history of the Lindzen-Kuo note is rather illuminating. Although the method is often called the "Lindzen-Kuo" method in geophysics since most applications in that area trace their ancestry back to their note, band LU factorization is a very obvious idea. They themselves eventually found an earlier reference in Richtmyer (1957) but chose to write a note anyway because most numerical analysis texts did not discuss Gaussian elimination even in the context of one-dimensional boundary value problems, preferring "shooting" instead.

---

[1]The same is true in three dimensions, but we will discuss only two dimensions for simplicity — and because the matrices are usually too large for this method to be practical in three dimensions.

The reason for this is not that block-banded elimination doesn't work, but rather that it is difficult to *prove* that it does. Lindzen and Kuo found empirically, however, that block-banded elimination is very robust. It worked for "stiff" one-dimensional problems with many alternating regions of exponential and oscillatory behavior, applied with as many as 2000 grid points. It was equally successful in solving two-dimensional boundary value problems of mixed elliptic-hyperbolic type. I have used the method with great success myself, including cases in which one dimension was treated with finite differences and the other with Chebyshev polynomials.

Second order finite differences give, for a one dimensional boundary value problem, an ordinary tridiagonal matrix. For a two-dimensional BVP, these methods generate a *block tridiagonal* matrix, that is to say, a matrix which is in the same form as a tridiagonal matrix of dimension N except that each "element" is itself a matrix of dimension M where M is the number of grid points in the second coordinate.

The beauty of the "matrix-of-matrices" concept is that it implies we can solve both one- and two-dimensional problems by exactly the *same* algorithm. Since tridiagonal and block tridiagonal matrices are so important in applications — in particular, they are generated by Chebyshev methods in a number of important special cases — it is useful to write down explicit formulas.

Since there are no more than three non-zero elements or blocks in each matrix, it is helpful to denote each by a symbol with a single subscript. The LU factorization is

$$
\begin{vmatrix}
\alpha_1 & 0 & 0 & 0 \\
b_2 & \alpha_2 & 0 & 0 \\
0 & b_3 & \alpha_3 & 0 \\
0 & 0 & b_4 & \alpha_4
\end{vmatrix}
\begin{vmatrix}
1 & \delta_1 & 0 & 0 \\
0 & 1 & \delta_2 & 0 \\
0 & 0 & 1 & \delta_3 \\
0 & 0 & 0 & 1
\end{vmatrix}
=
\begin{vmatrix}
a_1 & c_1 & 0 & 0 \\
b_2 & a_2 & c_2 & 0 \\
0 & b_3 & a_3 & c_3 \\
0 & 0 & b_4 & a_4
\end{vmatrix}
\tag{B.19}
$$

for the 4×4 case. Note that the "subdiagonal" elements b_i are identical with the "subdiagonal" of L, so we only have to compute two elements (or $M \times M$ matrices) for each of the N rows of $\vec{\vec{L}}$ and $\vec{\vec{U}}$.

The solution of the tridiagonal problem

$$
\vec{\vec{A}} \vec{x} = \vec{f}
\tag{B.20}
$$

is then

$$
\alpha_1 = a_1
\tag{B.21}
$$

$$
g_1 = (\alpha_1)^{-1} f_1
\tag{B.22}
$$

$$
\delta_i = \alpha_i^{-1} c_i \qquad\qquad i = 1, \ldots, N-1
\tag{B.23}
$$

$$
\alpha_{i+1} = a_{i+1} - b_{i+1} \delta_i
\tag{B.24}
$$

$$
g_{i+1} = (\alpha_{i+1})^{-1} [f_{i+1} - b_{i+1} g_i]
\tag{B.25}
$$

and the backward recurrence

$$
x_N = g_N
\tag{B.26}
$$

$$
x_j = g_j - \delta_j x_{j+1} \qquad j = N-1, N-2, \ldots, 1
$$

When this is applied to *block* tridiagonal matrices, we have a problem: the *inverse* of a sparse matrix is generally a *full* matrix. Consequently, even if the block matrices a_i, b_i, and c_i are themselves tridiagonal, as would be true with finite differences in both directions, all the δ_i and all the α_i matrices except α_1 are *full* matrices. For this reason, a *mixed pseudospectral-finite difference* algorithm is very appealing (Boyd, 1978c); we sacrifice nothing by using collocation with M points since the matrices generated during Gaussian elimination will be full anyway, regardless of whether a_i, b_i, and c_i are sparse (finite difference) or full (pseudospectral).

Above, we segregated the LU factorization on the left and the steps for the backsolve on the right because if we are only going to compute a single solution for a given matrix $\vec{\vec{A}}$, it is most efficient to compute the back-solve intermediate, $\vec{g}$, on the same forward pass in which we convert the α_i. That way, we only need to store the δ_i matrices at a cost of (NM^2) memory cells. This method is simpler even for a one dimensional problem since the complete solution is computed with just two DO loops.

To compute many solutions with the same matrix $\vec{\vec{A}}$ but different $\vec{f}$ — the normal situation in an implicit time-stepping code — then one would perform the steps on the left only once, and solve for $\vec{g}$ and $\vec{x}$ in a separate subroutine. Unfortunately, this needs the LU factors of the α_i in addition to the δ_i, requiring $2NM^2$ storage.

The rate-determining step is solving $\alpha_i \delta_i = c_i$, which requires Gaussian elimination on a total of N matrices at a cost of $O(NM^3)$. This is obviously much more expensive than an optimized iterative method, which is $O(NM \log M)$. Each back-solve requires $O(NM^2)$ operations.

It goes without saying that block-elimination is a method of last resort; iterative methods are always much faster when they can be applied. However, geophysical wave equations are often of mixed elliptic-hyperbolic or lack the property of positive definiteness even when everywhere elliptic. The great strength of the block LU algorithm is that it is very robust. It *always* gives a solution even for those ghastly equations of mixed type[2]. It is very successful in one-dimensional problems where "shooting" or even "parallel shooting" would fail.

Geophysical applications include Lindzen (1970), Lindzen & Hong (1974), Forbes and Garrett (1976), and Schoeberl & Geller (1977).

B.5   Block and "Bordered" Matrices: the Faddeev-Faddeeva Factorization

When Galerkin's method with Chebyshev polynomials is applied to

$$u_{xx} + K\,u = f(x) \tag{B.27}$$

where K is constant, the result (after a little algebra) is almost a tridiagonal matrix (Chapter 15, Sec. 10). (Actually, *two* tridiagonal matrices since we would normally split $f(x)$ into symmetric and antisymmetric parts (Chapter 8) because the alternative is solving a pentadiagonal matrix of twice the dimension and half its elements equal to zero.) Although most rows of one of these two Galerkin matrices do contain only three non-zero elements, we must use one row of each matrix to impose the boundary condition, and this row has all non-zero elements.

There are two ways to deal with this. First, one can use the formulas for the general LU decomposition to show that the property of having just one full row can be preserved by the LU factorization. Consequently, one can apply (B.5)–(B.13) in such a way that all the

---

[2]At least for geophysical problems with so-called "inertial" or "critical" latitudes.

sums are taken over just two elements *except* for one row where the sum runs over all N elements.

The second way is "matrix-bordering" (Faddeev & Faddeeva, 1963, Strang & Fix, 1973). To solve

$$\vec{\vec{A}}\,\vec{x} = \vec{f}, \tag{B.28}$$

write $\vec{\vec{A}}$ in the block form

$$\vec{\vec{A}} = \begin{vmatrix} \vec{\vec{A}}_{11} & \vec{\vec{A}}_{12} \\ \vec{\vec{A}}_{21} & \vec{\vec{A}}_{22} \end{vmatrix} \tag{B.29}$$

where the $\vec{\vec{A}}_{ij}$ are themselves matrices. The blocks can be of different sizes provided they follow the pattern shown below where m is an integer such that $1 \le m < N$

$$\begin{aligned} &\vec{\vec{A}}_{11} : m \times m &&\vec{\vec{A}}_{12} : m \times (N - m) \\ &\vec{\vec{A}}_{21} : (N - m) \times m &&\vec{\vec{A}}_{22} : (N - m) \times (N - m) \end{aligned} \tag{B.30}$$

For example, if $\vec{\vec{A}}$ is tridiagonal except for a full row at the bottom, we can pick $\vec{\vec{A}}_{11}$ to be the $(N - 1) \times (N - 1)$ tridiagonal matrix. Then the dimensions of the blocks are

$$\begin{aligned} &\vec{\vec{A}}_{11} : (N - 1) \times (N - 1) &&\vec{\vec{A}}_{12} : (N - 1) \times 1 \\ &\vec{\vec{A}}_{21} : 1 \times (N - 1) &&\vec{\vec{A}}_{22} : 1 \times 1 \quad \text{(scalar!)} \end{aligned} \tag{B.31}$$

$\vec{\vec{A}}$ can be written as the product of matrices $\vec{\vec{L}}$ and $\vec{\vec{U}}$ of block form, too:

$$\begin{vmatrix} \vec{\vec{L}}_{11} & 0 \\ \vec{\vec{L}}_{21} & \vec{\vec{L}}_{22} \end{vmatrix} \begin{vmatrix} \vec{\vec{U}}_{11} & \vec{\vec{U}}_{12} \\ \vec{0} & \vec{\vec{U}}_{22} \end{vmatrix} = \begin{vmatrix} \vec{\vec{A}}_{11} & \vec{\vec{A}}_{12} \\ \vec{\vec{A}}_{21} & \vec{\vec{A}}_{22} \end{vmatrix} \tag{B.32}$$

where both $\vec{\vec{L}}$ & $\vec{\vec{U}}$ are triangular and

$$\vec{\vec{L}}_{11} \vec{\vec{U}}_{11} = \vec{\vec{A}}_{11} \tag{B.33}$$

$$\vec{\vec{L}}_{21} \vec{\vec{U}}_{11} = \vec{\vec{A}}_{21} \tag{B.34}$$

$$\vec{\vec{L}}_{11} \vec{\vec{U}}_{12} = \vec{\vec{A}}_{12} \tag{B.35}$$

$$\vec{\vec{L}}_{22} \vec{\vec{U}}_{22} = \vec{\vec{A}}_{22} - \vec{\vec{L}}_{21} \vec{\vec{U}}_{12} \tag{B.36}$$

Optionally, one can force the diagonal blocks of $\vec{\vec{U}}$ to be identity matrices, but this would save no labor. To solve (B.35), we must compute the LU factorization of $\vec{\vec{A}}_{11}$ anyway. Consequently, the program is simpler and no less efficient if we choose $\vec{\vec{U}}_{11}$ to be one of the usual LU factors of $\vec{\vec{A}}_{11}$. Taking the transpose of (B.34) gives

$$\vec{\vec{U}}_{11}^{T} \vec{\vec{L}}_{21}^{T} = \vec{\vec{A}}_{21}^{T} \tag{B.37}$$

which can be easily solved for $\vec{L}_{21}$ since the transpose of $\vec{U}_{11}$ is a lower triangular matrix. We then can solve (B.33)–(B.36) in sequence to obtain the block factorization of $\vec{A}$.

The crucial point is that if $\vec{A}_{11}$ is banded, then $\vec{L}_{11}$ and $\vec{U}_{11}$ will be banded also. Thus, if $\vec{A}_{11}$ is tridiagonal, as in Orszag and Kells(1980), we can find its factors in $O(N)$ operations. The remaining steps cost $O(N)$ operations for the same reason. The fact that $\vec{A}_{21}$ is a full row implies that $\vec{L}_{21}$ will be full also, but the extra expense is only $O(N)$.

L. M. Delves has developed spectral iterations which require inverting matrices with an $m \times m$ *full* block for $\vec{A}_{11}$ but with zeros in all elements outside this block except along the main diagonal (Delves and Freeman, 1981). The work is $O((2/3)m^3 + kN \log_2 N)$ operations per iteration; if $m \ll N$ and the number of iterations is not too large, this is much faster than solving the full $N \times N$ matrix problem via Gaussian elimination. At present, Delves' spectral/block iteration method is the fastest for solving *integral* equations via Chebyshev polynomials. It can be applied to differential equations, too. Details are given in the monograph by Delves and Freeman (1981) and Chapter 15.

B.6   Cyclic Banded Matrices (Periodic Boundary Conditions)

For periodic boundary conditions, finite difference methods — useful as pseudospectral pre-conditioners — generate so-called "cyclic" banded matrices. Outside the diagonal band, the only non-zero elements are those in a small triangular region in the upper right corner. Temperton (1975) and Navon (1987) describe matrix factorizations for cyclic matrices. However, these are unnecessary for periodic problems with parity: the symmetries give rise to ordinary banded banded matrices — very helpful when using finite differences to precondition an iteration with a Fourier cosine or Chebyshev pseudospectral

B.7   Parting shots

- Always use a direct, rather than an iterative method, if you can afford it. Direct methods almost always work; iterative schemes sometimes fail.

- Iterative methods are considerably less robust for real-world problems than for the idealized problems in the textbooks of mathematicians.

- Hotellier proved in 1944 that Gaussian elimination (LU factorization) could magnify roundoff error by 4^N where N is the size of the matrix. This would seem to make it impossible to solve matrix problems with $N > 40$ even in seventeen decimal place arithmetic! However, this worst case is exceedingly rare in practice. In numerical analysis, there is sometimes an enormous gap between what can be proven and what actually happens.

- Clever matrix-of-a-matrix methods have lowered the asymptotic cost of solving a dense matrix from $O(N^3)$ operations to $O(N^{2.376})$. L. N. Trefethen has a $100 bet with P. Alfeld that the exponent will lowered to $O(N^{2+\epsilon})$ for arbitrarily small ϵ by 2006. However, these new algorithms have huge proportionality constants. Consequently, Gaussian elimination is the best unless N is really, really huge. A numerical analyst's concept of efficiency is the $N \to \infty$ limit, which may have little to do with engineering reality.

Trefethen and Bau(1997) give more information about roundoff and asymptotically cheap algorithms.

Karageorghis and Paprzycki(1996, 1998), and earlier papers by Karageorghis and collaborators, describe efficient direct methods even for the somewhat complicated matrices generated by spectral domain decomposition.

Appendix C

The Newton-Kantorovich Method for Nonlinear Boundary & Eigenvalue Problems

"Nature and Nature's laws were hid in night. God said: *Let Newton be*! and all was light."
— Alexander Pope (1688–1744)

" ... with his prism and silent face ... a mind forever voyaging through strange seas of thought, alone."
— William Wordsworth (of Sir Isaac Newton)

C.1 Introduction

Most algorithms for solving nonlinear equations or systems of equations are variations of Newton's method. When we apply spectral, finite difference, or finite element approximations to a *nonlinear differential* equation, the result is a *nonlinear* system of *algebraic* equations equal in number to the total number of degrees of freedom, regardless of whether the degrees of freedom are the grid point values of the solution, $\{u(x_i)\}$, or the coefficients of the spectral series for $u(x)$, $\{a_n\}$. We must then apply Newton's method — or something equivalent to it — to iterate from a first guess until the numerical solution is sufficiently accurate.

There are several practical difficulties. First, we need a first guess — this is a universal problem, and no choice of iteration scheme can eliminate it. Second, Newton's method requires the elements of the "Jacobian matrix" of the system of equations. If we write the system as

$$F_i(a_0, a_1, \ldots, a_N) = 0 \qquad\qquad i = 0, \ldots, N \qquad\qquad (C.1)$$

then the elements of the Jacobian matrix $\vec{\vec{J}}$ are

$$J_{ij} \equiv \partial F_i / \partial a_j \qquad\qquad i, j = 0, \ldots, N \qquad\qquad (C.2)$$

526

It is usually messy and time-consuming to calculate the Jacobian, which must (in a strict Newton's method) be recomputed at each iteration. Third, each iteration requires *inverting* the Jacobian. If we let $\vec{a}^{(i)}$ denote the vector of the spectral coefficients at the i-th iteration, then

$$\vec{a}^{(i+1)} = \vec{a}^{(i)} - \vec{\vec{J}}^{-1} \vec{F} \tag{C.3}$$

where $\vec{F}$ is the vector whose elements are the equations of the system and where both $\vec{F}$ and $\vec{\vec{J}}$ are evaluated using $\vec{a} = \vec{a}^{(i)}$. This may be the most expensive step of all since the Jacobian matrix — at least with pseudospectral methods — is usually a full matrix, so it costs $O(N^3)$ operations to determine each iterate from its predecessor.

Resolving the first difficulty requires the "continuation" method discussed below, but we can often reduce the magnitude of the second and third problems by applying Newton's method directly to the original *differential* equation, and then applying the spectral method to convert the resulting iterative sequence of *linear* differential equations into the corresponding linear matrix problem. When Newton's method is applied directly to a differential equation — rather than to algebraic equations — it is often called the "Newton-Kantorovich" method.

The justification for the ordinary Newton's method is a Taylor expansion followed by linearization. If we wish to solve $f(x) = 0$, then assuming we have a good guess for a root, $x \approx x^{(i)}$, we can Taylor expand $f(x)$ to obtain

$$f(x) = f(x^{(i)}) + f_x(x^{(i)}) \left[x - x^{(i)} \right] + O\left(\left[x - x^{(i)} \right]^2 \right) \tag{C.4}$$

If our guess is sufficiently good, we can ignore the quadratic terms and treat (C.4) as a *linear* equation for x,

$$x^{(i+1)} = x^{(i)} - \frac{f(x^{(i)})}{f_x\left(x^{(i)}\right)} \tag{C.5}$$

This is identical with (C.3) except for the extension from one unknown to many.

To illustrate the generalization known as the Newton-Kantorovich method, consider the first order differential equation

$$u_x = F(x, u[x]) \tag{C.6}$$

Suppose we have an iterate, $u^{(i)}(x)$, which is a good approximation to the true solution $u(x)$. Taylor expand (C.6) about this *function*:

$$u_x = F\left(x, u^{(i)}[x] \right) + F_u\left(x, u^{(i)}[x] \right) \left[u(x) - u^{(i)}(x) \right] + O\left(\left[u - u^{(i)} \right]^2 \right) \tag{C.7}$$

where F_u denotes the derivative of the function $F(x, u)$ with respect to u. This implies that the *linear* differential equation for the next iterate is

$$u_x^{(i+1)} - \{F_u\}u^{(i+1)} = F - F_u\, u^{(i)} \tag{C.8}$$

where F and F_u are both evaluated with $u = u^{(i)}$. Equivalently, we can solve a differential equation for the *correction* to the iterate by defining

$$u^{(i+1)}(x) \equiv u^{(i)}(x) + \triangle(x) \tag{C.9}$$

Then $\triangle(x)$ is the solution of

$$\triangle_x - \{F_u\}\triangle = \left\{F - u_x^{(i)}\right\} \tag{C.10}$$

Since Taylor expansions about a *function* instead of a point may be unfamiliar, a brief explanation is in order. Technically, operations on functionals, which we may loosely define as "functions of a function", are described by a branch of mathematics called "functional analysis". The most elementary and readable introduction is a little paperback by W. W. Sawyer (1978).

The punchline is this: the linear term in a generalized Taylor series is actually a *linear operator*. In (C.3), the operator is multiplication by the Jacobian matrix. In (C.10), the operator is the differential operator that is applied to $\triangle(x)$ to generate the L. H. S. of the equation. In either case, this linear operator is known as a "Frechet derivative". The general definition is the following.

Definition 49 (FRECHET DERIVATIVE) *Given a general nonlinear operator $\mathcal{N}(u)$, which may include derivatives or partial derivatives, its FRECHET DIFFERENTIAL is defined by*

$$\mathcal{N}_u\triangle \equiv \lim_{\epsilon \to 0} \frac{\mathcal{N}(u + \epsilon\,\triangle(x)) - \mathcal{N}(u)}{\epsilon} \tag{C.11}$$

which is equivalent to

$$\mathcal{N}_u\triangle \equiv \left.\frac{\partial \mathcal{N}(u + \epsilon\triangle)}{\partial \epsilon}\right|_{\epsilon=0} \tag{C.12}$$

The numerical result, $\mathcal{N}_u\triangle$, is the FRECHET DIFFERENTIAL of the operator $\mathcal{N}$ in the direction $\triangle$ while $\mathcal{N}_u$ is the FRECHET DERIVATIVE and is a linear operator.

This definition is an algorithm for computing the first order Taylor term. We merely multiply $\triangle(x)$ by a dummy variable ϵ and then take the partial derivative of $\mathcal{N}(u, x; \epsilon)$ with respect to ϵ in the usual way, that is, with u and x kept fixed. The dummy variable ϵ takes us away from the complicated world of functional analysis into the far simpler realm of ordinary calculus.

The need for the "direction $\triangle(x)$" in the definition of the Frechet differential can be seen by expanding $u(x)$ in a spectral series. The operator equation $\mathcal{N}(u) = 0$ becomes a *system* of nonlinear algebraic equations. When a function depends upon more than one unknown, we cannot specify "the" derivative: we have to specify the *direction*. The Frechet differential is the "directional derivative" of ordinary multi-variable calculus.

What is important about the Frechet derivative is that it gives the linear term in our generalized Taylor expansion: for any operator $\mathcal{N}(u)$,

$$\mathcal{N}(u + \triangle) = \mathcal{N}(u) + \mathcal{N}_u(u)\triangle + O(\triangle^2) \quad \text{[Generalized Taylor Series]} \tag{C.13}$$

For the special case in which $\mathcal{N}(u)$ is a system of algebraic equations,

$$\vec{F}\left(\vec{a} + \vec{\triangle}\right) = \vec{F}(\vec{a}) + \vec{\vec{J}}\vec{\triangle} + O\left(\vec{\triangle}^2\right) \tag{C.14}$$

the Frechet derivative is the Jacobian matrix $\vec{\vec{J}}$ of the system of functions $\vec{F}(\vec{a})$.

With this definition in hand, we can easily create a Newton-Kantorovich iteration for almost any equation, and we shall give many examples later. An obvious question is: Why bother? After all, when we apply the pseudospectral method (or finite difference or finite element algorithms) to the linear differential equation that we must solve at each step of the Newton-Kantorovich iteration, we are *implicitly* computing the Jacobian of the system of equations that we would have obtained by applying the same numerical method first, and then Newton's method afterwards.

There are two answers. First, it is often easier to write down the matrix elements in the Newton-Kantorovich formalism. If we already have a subroutine for solving linear differential equations, for example, we merely embed this in an iteration loop and write subroutines defining $F(x, u)$ and $F_u(x, u)$. It is easy to change parameters or adapt the code to quite different equations merely by altering the subroutines for $F(x, u)$, etc.

The second answer is that posing the problem as a sequence of linear differential equations implies that we also have the *theoretical* machinery for such equations at our disposal. For example, we can apply the iterative methods of Chapter 15, which solve the matrix problems generated by differential equations much more cheaply than Gaussian elimination.

In principle, we could use all these same tricks even if we applied Newton's method second instead of first — after all, we are always in the business of solving differential equations, and reversing the order cannot change that. *Conceptually*, however, it is much simpler to have the differential equation clearly in view.

C.2 Examples

The second order ordinary differential equation

$$u_{xx} = F(x, u, u_x) \tag{C.15}$$

generates the iteration

$$\triangle_{xx} - F_{u_x}\triangle_x - F_u\triangle = F - u_{xx}^{(i)} \tag{C.16}$$

where F and all its derivatives are evaluated with $u = u^{(i)}$ and $u_x = u_x^{(i)}$, and where $\triangle(x)$, as in the previous section, is the *correction* to the i-th iterate:

$$u^{(i+1)}(x) \equiv u^{(i)}(x) + \triangle(x) \tag{C.17}$$

The Frechet derivative is more complicated here than for a first order equation because $u(x)$ appears as both the second and third arguments of the three-variable function $F(x, u, u_x)$, but the principle is unchanged.

For instance, let

$$F(x, u, u_x) \equiv \alpha(u_x)^2 + \exp(x\,u[x]) \tag{C.18}$$

The needed derivatives are

$$F_u(x, u, u_x) = x \exp(x\,u) \tag{C.19}$$

$$F_{u_x}(x, u, u_x) = 2\,\alpha\,u_x \tag{C.20}$$

We can also extend the idea to equations in which the highest derivative appears nonlinearly. Norton (1964) gives the example

$$u\,u_{xx} + A(u_x)^2 + B\,u = B\left[20 + \frac{1}{12}\sin(\pi x)\right] \tag{C.21}$$

which, despite its rather bizarre appearance, arose in analyzing ocean waves. We first define $\mathcal{N}(u)$ to equal all the terms in (C.21). Next, make the substitution $u \to u + \epsilon\Delta$ and collect powers of ϵ. We find

$$\mathcal{N}(u + \epsilon\Delta) = \left\{ u\,u_{xx} + A(u_x)^2 + B\,u - B\left[20 + \tfrac{1}{12}\sin(\pi x)\right]\right\} \tag{C.22}$$

$$+\epsilon\left\{ u\,\Delta_{xx} + u_{xx}\,\Delta + 2\,A\,u_x\,\Delta_x + B\,\Delta\right\} + \epsilon^2\left\{\Delta\,\Delta_{xx} + A\,(\Delta_x)^2\right\}$$

To evaluate the Frechet derivative, we subtract the first line of the R. H. S. of (C.22) — the term independent of ϵ — and then divide by ϵ. When we take the limit $\epsilon \to 0$, the term in ϵ^2 disappears as we could have anticipated directly from its form: it is *nonlinear* in Δ, and the whole point of the iteration is to create a *linear* differential equation for Δ. Thus, the Frechet derivative is equal to the *linear* term in the expansion of $\mathcal{N}(u + \epsilon\Delta)$ in powers of ϵ, and this is a *general theorem*. Our generalized Taylor series is then

$$\mathcal{N}(u + \Delta) \approx \mathcal{N}(u) + \mathcal{N}_u(u)\,\Delta \tag{C.23}$$

and equating this to zero gives

$$\mathcal{N}_u(u)\,\Delta = -\mathcal{N}(u) \tag{C.24}$$

which is

$$u^{(i)}\,\Delta_{xx} + 2\,A\,u_x^{(i)}\,\Delta_x + (u_{xx}^{(i)} + B)\,\Delta \tag{C.25}$$

$$= -\left\{ u^{(i)}\,u_{xx}^{(i)} + A(u_x^{(i)})^2 + B\,u^{(i)} - B\left[20 + \frac{1}{12}\sin(\pi x)\right]\right\}$$

Inspecting (C.25), we see that the R. H. S. is simply the ϵ-independent term in the power series expansion of $\mathcal{N}(u + \epsilon\Delta)$. The L. H. S. is simply the term in the same series which is *linear* in ϵ. Once we have recognized this, we can bypass some of the intermediate steps and proceed immediately from (C.22) to (C.25). The intermediate equations, however, are still *conceptually* important because they remind us that the iteration is simply a generalized form of Newton's method.

We have not discussed boundary conditions, but these are usually obvious. If the first guess satisfies the boundary condition, then we can safely impose homogeneous boundary conditions on $\Delta(x)$ at all iterations. However, we must be careful to impose conditions that are consistent with those on $u(x)$. If we have Dirichlet conditions, $u(-1) = \alpha$, $u(1) = \beta$, then we would set $\Delta(\pm 1) = 0$. With Neuman conditions on $u(x)$, $du/dx(-1) = \alpha$ and $du/dx(1) = \beta$, then we would impose $d\Delta/dx(\pm 1) = 0$. Finally, periodic boundary conditions, i. e. using a Fourier series, should be imposed on $\Delta(x)$ if the problem specifies such conditions for $u(x)$.

The Newton-Kantorovich method applies equally well to partial differential equations and to integral equations. For example,

$$u_{xx} + u_{yy} + \sqrt{u_x^2 + u_y^2} = 0 \tag{C.26}$$

generates the iteration

$$\Delta_{xx} + \Delta_{yy} + \frac{u_x\,\Delta_x + u_y\,\Delta_y}{\sqrt{\left(u_x^{(i)}\right)^2 + \left(u_y^{(i)}\right)^2}} \tag{C.27}$$

$$- \left\{ u_{xx}^{(i)} + u_{yy}^{(i)} + \sqrt{\left(u_x^{(i)}\right)^2 + \left(u_y^{(i)}\right)^2}\right\}$$

We can create an iteration even for the most bizarre equations, but the issues of whether solutions exist for nonlinear equations, or whether they are smooth even if they do exist, are very difficult — as difficult as the Newton-Kantorovich method is elementary.

C.3 Eigenvalue Problems

When the equation is nonlinear, the distinction between eigenvalue and boundary value problems may be negligible. A linear boundary value problem has a solution for arbitrary forcing, but a linear eigenvalue problem has a solution only for particular, *discrete* values of the eigenparameter. It is rather different when the eigenvalue problem is nonlinear. For example, the periodic solutions of the Korteweg-deVries equation ("cnoidal waves") and the solitary waves (also called "solitons") of the same equation satisfy the differential equation

$$u_{xxx} + (u - c)\, u_x = 0 \qquad \text{[Korteweg-deVries Eq.]} \qquad (\text{C.28})$$

where c is the eigenvalue. The KDV equation can be explicitly solved in terms of the so-called elliptic cosine function, and this makes it possible to prove that solutions exist for a *continuous* range of c:

$$u(-\infty) = u(\infty) = 0 \qquad : \qquad \text{Solitary waves exist for any } c > 0 \qquad (\text{C.29})$$

$$u(x) = u(x + 2\pi) \qquad : \qquad \text{Cnoidal waves exist for any } c > -1. \qquad (\text{C.30})$$

What makes it possible for the eigenparameter to have a continuous range is that for a *nonlinear* differential equation, the relative magnitude of different terms changes with the *amplitude* of the solution. For the Korteweg-deVries problem, the wave amplitude for either of the two cases is a unique, monotonically increasing function of c.

For other problems in this class, the wave amplitude need not be a monotonic function of a nor is the solution necessarily unique. For example, if we replace the third degree term in (C.28) by a fifth derivative, one can show (Boyd, 1986b) that in addition to the usual solitons with a single peak, there are also solitons with two peaks, three peaks, etc. (The multi-peaked solitons are actually bound states of the single-peaked solitons.)

The existence of solutions for a *continuous* range of the eigenparameter, instead of merely a set of *discrete* values of c, is a common property of nonlinear equations. It follows that to compute a solution to (C.28), for example, one can simply choose an almost-arbitrary value of c — "almost-arbitrary" means it must be larger than the cutoffs shown in (C.29) and (C.30) — and then solve it by the Newton-Kantorovich method as one would any other *boundary* value problem. The equation for the correction $\Delta(x)$ to $u^{(i)}(x)$ is

$$\Delta_{xxx} + (u^{(i)} - c)\, \Delta_x + u_x^{(i)}\, \Delta = -u_{xxx}^{(i)} + (u^{(i)} - c)\, u_x^{(i)} \qquad (\text{C.31})$$

The only possible complication in solving (C.31) is that the operator on the L. H. S. has two eigensolutions with zero eigenvalue. In a word, (C.31) does not have a unique solution. The physical reason for this is that the solitons of the Korteweg-deVries equation form a two-parameter family.

The first degree of freedom is translational invariance: if $u(x)$ is a solution, then so is $u(x + \phi)$ for arbitrary constant ϕ. If we Taylor expand $u(x + \phi)$ and keep only the first term so as to be consistent with the linearization inherent in (C.31), then

$$u(x + \phi) = u(x) + \phi \frac{du}{dx} + O(\phi^2) \qquad (\text{C.32})$$

must be a solution of the linearized Korteweg-deVries equation for arbitrary ϕ. It follows that the solution of (C.31) is not unique, but is determined only to within addition of an arbitrary amount of

$$\Delta_1(x) \equiv \frac{du}{dx} \tag{C.33}$$

which therefore is an eigenfunction (with zero eigenvalue) of the linear operator of (C.31).

Fortunately, the solitons are symmetric about their peak. The derivative of a symmetric function is antisymmetric. It follows that the column vector containing the spectral coefficients of du/dx is not a linear eigenfunction of the discretized form of (C.31) when the basis set is restricted to *only symmetric* basis functions (either cosines, in the periodic case, or even degree rational Chebyshev functions on the infinite interval).

The second degree of freedom is that if $u(x)$ is a solution of (C.28), then so is

$$v(x) = -\epsilon + (1 + \epsilon) \, u \left(\sqrt{1 + \epsilon} \, x \right) \tag{C.34}$$

for arbitrary ϵ. (In technical terms, the KdV equation has a continuous Lie group symmetry, in this case, a "dilational" symmetry.) Again, we must Taylor expand $v(x)$ so as to remain consistent with the linearization inherent in (C.31). We find that the other eigenfunction [the $O(\epsilon)$ term in (C.34)] is

$$\Delta_2(x) \equiv -1 + u(x) + \frac{x}{2} \frac{du}{dx} \tag{C.35}$$

We can exclude $\Delta_2(x)$ on the infinite interval by imposing the boundary condition $u(\pm\infty) = 0$. Adding a term proportional to $\Delta_2(x)$, although consistent with the differential equation (C.31), would violate the boundary condition because $\Delta_2(x)$ asymptotes to a non-zero constant at infinity. In a similar way, we can remove the non-uniqueness for the periodic case by imposing the condition that the average of $u(x)$ over the interval is 0. The simplest way to implement this condition is to omit the constant from the basis set, using only the set $\{\cos(nx)\}$ with $n > 0$ to approximate the solution.

Once we have computed the soliton or cnoidal wave for a given c, we can then take that single solution and generate a two-parameter family of solutions by exploiting these two Lie group symmetries. Thus, the translational and dilational symmetries are both curse and a blessing. The symmetries complicate the calculations by generating two eigenmodes with zero eigenvalues. However, the symmetries are also a blessing because they reduce the number of parameters by two.

For some differential equations, solutions exist only for *discrete* values of the eigenparameter. An example is van der Pol's equation:

$$u_{tt} + u = \Gamma \, u_t \left(1 - u^2 \right) \qquad \text{[Van der Pol's Eq.]} \tag{C.36}$$

If we replaced the term in the () by a constant, then (C.36) would describe a harmonic oscillator which was damped or excited depending on the *sign* of the constant. Because this "constant" in (C.36) is actually a function of $u(x)$, we see that the solution of the nonlinear equation is self-excited when u is small and self-damping when u is large. When (C.36) is given an arbitary initial condition, the solution eventually settles down to a nonlinear *limit cycle* in which the damping and self-excitation balance out over the course of one period of the cycle to create a permanent, self-sustaining oscillation.

We can solve for the period and structure of the limit cycle as a nonlinear eigenvalue problem by adopting the following device. Let ω denote the (unknown) frequency of the limit cycle. If we make the change of variable

$$x \equiv \omega t, \tag{C.37}$$

then the limit cycle will automatically be periodic with a period of 2π and we can compute a solution by assuming a Fourier series in x^1. The problem is then

$$\omega^2 u_{xx} + u = \Gamma \omega u_x (1 - u^2) \tag{C.38}$$

In contrast to the undamped Korteweg-deVries equation, it is *not* possible to pick an arbitrary value of ω and find a solution. The reason is that the tug-of-war between the damping and the self-excitation normally has only a single solution for a given value of the parameter Γ. It is not possible to have arbitrarily large amplitudes for u, for example, because the equation would then be damped over almost the whole cycle. If we started the calculation with a very large value of $u(t = 0)$, we would find that $u(t)$ would quickly decay into a limit cycle of moderate amplitude and a *unique* frequency ω. Consequently, just as in a linear eigenvalue problem, we must vary both $u(x)$ and the eigenparameter ω until we have obtained a unique solution.

Van der Pol's equation happens to also have the property of translational invariance. In this respect, it is like the Korteweg-deVries equation: its solution is not unique, but rather is a continuous family, and the linearized differential equation of the Newton-Kantorovich has an eigenvalue with zero eigenfunction. Thus, the matrix which is the pseudospectral discretization of the linearized ODE will be almost singular [*exactly* singular in the limit $N \to \infty$] and the iteration will diverge unless we add additional constraints or conditions to restrict the solution to a unique member of the continuous family of solutions.

Translational invariance may be destroyed by imposing a phase condition such as

$$u_x(x = 0) = 0 \qquad \text{[Phase Condition]} \tag{C.39}$$

which demands that the origin be either a maximum or a minimum of $u(x)$. For the KdV solutions, it is possible to impose (C.39) *implicitly* by restricting the basis set to functions which are *symmetric* about $x = 0$.

However, because of the presence of both even and odd order derivatives in van der Pol's equation, its solution is not symmetric about any point and we cannot impose a phase condition by restricting the basis set; we have to use both sines and cosines. To obtain a unique solution, we must therefore impose the phase condition (C.39) *explicitly*.

Since a solution does not exist for arbitary ω, but only for one particular frequency, we must retain ω as an additional unknown. Thus, if we apply the pseudospectral method with N coefficients and N interpolation points, we must solve an $(N+1) \times (N+1)$ matrix equation at each iteration. The $(N+1)$-st row of the matrix is the spectral form of (C.39); the $(N+1)$-st unknown is $\delta\omega$, the correction to the frequency at the i-th iteration.

The Newton-Kantorovich iteration is a modest generalization of that used for earlier examples. We make the simultaneous replacements

$$u(x) \quad \longrightarrow \quad u(x) + \epsilon \triangle(x) \tag{C.40a}$$

$$\omega \quad \longrightarrow \quad \omega + \epsilon (\delta \omega) \tag{C.40b}$$

in (C.38), collect powers of ϵ, and then the iteration, as for the other examples, consists of equating the first order terms (divided by ϵ) to the negative of the zeroth order terms. In particular, if we write (C.38) in the symbolic form

$$u_{xx} = F(x, u, u_x, \omega) \tag{C.41}$$

then the linearized differential equation is

$$\triangle_{xx} - F_{u_x} \triangle_x - F_u \triangle - F_\omega (\delta \omega) = - \left\{ u_{xx}^{(i)} - F \right\} \tag{C.42}$$

---

[1] Note that time-periodic solutions, usually with the frequency as an eigenparameter, are an exception to the usual rule that spectral methods are applied only to spatial coordinates.

where F and all its derivatives are evaluated with $u = u^{(i)}$ and $\omega = \omega^{(i)}$. The only extra term is the derivative with respect to ω. We expand $\triangle(x)$ as a Fourier series with a total of N terms, demand that (C.42) is satisfied at N interpolation points, add the spectral form of the phase condition, and then solve the linear matrix equation for the column vector whose first N elements are the N spectral coefficients of $\triangle(x)$ and whose $(N+1)$-st element is $(\delta\omega)$. We then add the corrections to $u^{(i)}$ and $\omega^{(i)}$ and repeat the iteration until it has converged.

It should be clear, as stressed at the end of the preceding chapter, that the Newton-Kantorovich method is always applicable and the iteration can always be derived by very elementary methods. ("Elementary" is a euphemism for freshman calculus! Functional analysis is *not* needed). However, the differences between the Korteweg-deVries and van der Pol cases, the need to impose phase conditions in some problems but not others, the possibility that the eigenparameter may assume a continuous range of values or just one — all in all, they show that the subject of nonlinear differential equations is very complicated. It is important to thoroughly understand the particular problem you are trying to solve because the range of possibilities — no solution, a solution only for a discrete value of a parameter, solution for a continuous range of values of the parameter — is rather large.

The only sure statement is this: given a sufficiently good first guess, the Newton-Kantorovich method will always compute a solution of the indicated form if it exists. If the solution does not exist, or you have left out phase conditions so that the solution of the numerical problem is not unique, the iteration will (probably!) diverge, or the pseudospectral matrix for the iteration differential equation will be a singular matrix.

C.4   Summary

SUCCESS: Given a sufficiently good first guess, the Newton-Kantorovich method will converge. Furthermore, the asymptotic convergence will be *quadratic* in the sense that the magnitude of the correction $\triangle^{(i+1)}$ will be roughly the *square* of the magnitude of $\triangle^{(i)}$ for sufficiently large i.

Notes: (i) The quadratic convergence does not occur until the iteration is close to the final solution (ii) The corrections will not decrease indefinitely, but will "stall out" at some very small magnitude which is determined by the roundoff of the computer — typically $O(10^{-12}$ to $10^{-14})$. (iii) The iteration is self-correcting, so it is possible to have small errors in solving the linearized ODE while still obtaining the correct answer. The signature of such errors is *geometric* rather than *quadratic* convergence. (iv) The iteration does not converge to the exact solution to the differential equation, but only to the exact solution of the N-basis function discretization of the differential equation. Remember to vary N and compare!

FAILURE: Newton's method may diverge because of the following:
(i) Insufficiently accurate first guess.
(ii) No solution exists for the nonlinear problem.
(iii) A family of solutions, parameterized by a continuous variable, exists, rather than a single, unique solution.
(iv) Programming error.

Notes: (i) Close to a "limit" or "bifurcation" point as discussed in Appendix D, a *very* accurate first guess is needed. (ii) Small perturbations may sometimes drastically modify nonlinear solutions; perturbing the KdV equation (C.28) so as to make the coefficients vary with x will destroy the translational invariance, and reduce an infinite number of solutions to either one or none, depending upon whether one can find a soliton such that the first

Table C.1: Nonlinear Boundary Value Problems Solved by Spectral Methods: A Select Bibliography

References	Comments
Boyd(1986c)	FKdV bicnoidal; Fourier pseudospectral
Yang&Keller(1986)	Fourier/finite difference algorithm for pipe flow; good use of pseudoarclength continuation to follow multiple branches
Van Dooren&Janssen(1989)	period-doubling for forced Duffing equation Fourier Galerkin
Haupt&Boyd(1988, 1991)	FKdV cnoidal waves; KdV double cnoidal wave
Boyd(1989b)	review; many examples
Boyd & Haupt(1991)	Fourier pseudospectral
Heinrichs(1992d)	quasi-linear elliptic problems
Boyd(1995h)	multiple precision; finite difference preconditioned nonlinear Richardson's iteration; nonlocal solitons
Mamun&Tuckerman(1995)	good discussion of how a time-marching code can be easily modified to a Nonlinear Richardson's iteration nonlinear steady flow between concentric spheres
Yang&Akylas(1996)	weakly nonlocal capillary-gravity solitons
Boyd (1997b)	Camassa-Holm "peakons"; TB pseudospectral
Boyd (1998c)	monograph on solitons; 3 chapters on spectral methods
Nicholls(1998)	water waves; Fourier in 2 horizontal coords.
Schultz&Vanden-Broeck & Jiang&Perlin(1998)	spectral boundary integral; Newton continuation; water waves with surface tension

order perturbation is orthogonal to the eigenfunctions. (iii) Too many solutions is as bad as too few: one must add "side" conditions like (C.39) until a unique solution is obtained. (iv) Because Newton's iteration is self-correcting, a code may pass tests for some parameters — and then fail in a different parameter range. Geometric convergence, that is, $\| \triangle^{(i+1)} \| / \| \triangle^{(i)} \| \sim \rho$ for some constant $\rho < 1$, indicates that the Jacobian matrix is inconsistent with the N equations which are being solved. (This is true if the iteration is a "honest" Newton's method with recomputation of the linearized differential equation at each step; if the same linearized ODE is used at every iteration to save money, then geometric convergence is the expected price for this modification of Newton's method.)

Appendix D

The Continuation Method

"Creeps in this petty pace from day to day"
— W. Shakespeare, *Macbeth*

D.1 Introduction

Iterative methods have the vice that they must be initialized with an approximate solution. The "continuation" method is a strategy for systematically constructing a first guess and tracing a branch of solutions.

The basic idea is very simple. Suppose that a problem depends upon a parameter α. Suppose further that we know the solution u_0 for $\alpha = 0$. We can then compute $u(\alpha = \delta)$ by applying the Newton or Newton-Kantorovich iteration with $u^{(0)}(\delta) = u_0$, provided that δ is sufficiently small. Similarly, we can set $u^{(0)}(2\,\delta) = u(\delta)$, and so on: at each step, we use the solution for $u(n\,\delta)$ as the first guess for calculating $u([n+1]\,\delta)$.

The better our guess, the faster the iteration will converge. The basic scheme can be refined by using the linear extrapolation:

$$u^{(0)}([n+1]\,\delta) = u(n\,\delta) + \delta\,\{u(n\,\delta) - u([n-1]\,\delta)\} \tag{D.1}$$

In summary,

$$[\text{Analytical solution}] \quad \longrightarrow \quad u(\alpha = 0) \tag{D.2}$$

$$u(0) \quad \longrightarrow \quad u^{(0)}(\delta) \quad \xrightarrow{\text{Newton iteration}} \quad u(\delta) \tag{D.3}$$

$$\left\{ \begin{array}{c} u([n-1]\,\delta) \\ u(n\,\delta) \end{array} \right\} \xrightarrow{\text{Linear Extrapolation}} u^{(0)}([n+1]\,\delta) \xrightarrow{\text{Newton's iter.}} u([n+1]\,\delta) \tag{D.4}$$

In words, we have the following.

536

Definition 50 (CONTINUATION METHOD) *This procedure solves the equation*

$$\mathcal{N}(u; \alpha) = 0 \tag{D.5}$$

for u by marching in small steps in α. $\mathcal{N}$ is the nonlinear operator and α is the parameter. The march begins with a known solution — typically, a linear solution which can be found analytically or numerically — and applies an iteration to compute u at each value of α, USING PREVIOUSLY COMPUTED VALUES OF u FOR SMALLER α TO EXTRAPOLATE A FIRST GUESS FOR THE SOLUTION AT THE CURRENT α.

$\mathcal{N}$ may include derivatives or integrations, and may represent a system of nonlinear equations. The unknown u may be the solution to either an algebraic equation or system of equations, or the solution to a differential equation or system of such equations.

Strictly speaking, "continuation method" should be "continuation methods" in Definition 50 because the basic idea of parameter-marching to obtain the first guess can be applied with a wide variety of iteration and extrapolation schemes. We can always replace Newton's method by the secant method, for example; since the latter has a smaller radius of convergence than Newton's, the price we must pay is a smaller maximum step in α.

Similarly, to define better extrapolation formulas, we can use either ODE integration schemes or higher-order Lagrangian interpolation in α — very carefully since interpolation may diverge as the degree of the interpolating polynomial becomes large, especially when *extrapolating*.

The ODE methods are based on differentiating $\mathcal{N}(u[\alpha], \alpha)$ with respect to α to obtain

$$\mathcal{N}_u(u, \alpha) u_\alpha + \mathcal{N}_\alpha = 0 \tag{D.6}$$

For the special case that $\mathcal{N}(u, \alpha)$ is a single algebraic equation, (D.6) is the first order ordinary differential equation

$$\frac{du}{d\alpha} = -\frac{\partial \mathcal{N}(u, \alpha)/\partial \alpha}{\partial \mathcal{N}(u, \alpha)/\partial u} \qquad \text{["Davidenko equation"]} \tag{D.7}$$

Given a single initial condition such as $u(\alpha = 0) = u_0$, we can apply our favorite Runge-Kutta or predictor-corrector method in α to trace the solution branch.

This differential-equation-in-the-parameter is named after Davidenko who first observed that nonlinear algebraic equations could be solved by integrating an ODE or system of ODE's in the parameter. When extended to systems of M nonlinear equations in M unknowns, u and $\mathcal{N}$ become M-dimensional vectors and $\partial \mathcal{N}/\partial u$ becomes the Jacobian matrix whose elements are given by $\partial \mathcal{N}_i/\partial u_j$. Integrating (D.7) is then relatively expensive because one must compute $du/d\alpha$ by Gaussian elimination.

In principle, Newton's method can be completely replaced by the integration of the Davidenko equation. However, the errors in a Runge-Kutta scheme will gradually accumulate as we march, so it is a good idea to periodically apply Newton's method so as to reinitialize the march with a $u(\alpha)$ that is accurate to full machine precision.

D.2 Examples

Let the goal be to solve the polynomial equation

$$P(x; \alpha) = 0 \tag{D.8}$$

where α is a parameter. Let x_0 be a root which is known for $\alpha = 0$. We can trace this solution branch $x(\alpha)$ emanating from the point $(x_0, 0)$ by solving the ODE initial-value problem

$$\frac{dx}{d\alpha} = -\frac{\partial P(x; \alpha)/\partial \alpha}{\partial P(x; \alpha)/\partial x} \tag{D.9a}$$

$$x(0) = x_0 \tag{D.9b}$$

For example,

$$P(x; \alpha) \equiv \alpha x^2 - 2x + 1 \quad \longrightarrow \tag{D.10}$$

$$\partial P/\partial \alpha = x^2 \tag{D.11}$$

$$\partial P/\partial x = 2\alpha x - 2 \tag{D.12}$$

As $\alpha \to 0$, this collapses into the linear equation $(-2x + 1) = 0$, which has the unique solution $x = 1/2$. We can trace $x(\alpha)$ away from $\alpha = 0$ by solving

$$\frac{dx}{d\alpha} = -\frac{x^2}{2(\alpha x - 1)} \tag{D.13a}$$

$$x(0) = \frac{1}{2} \tag{D.13b}$$

Polynomials have multiple solutions, but we can apply the same method to trace all the roots. For small α, one can easily show that

$$x_2(\alpha) \approx \frac{2}{\alpha} \qquad\qquad \alpha \ll 1 \tag{D.14}$$

which can be used to initialize the *same* differential equation, (D.13a), for small α to trace the second solution for all α for which this root is real.

Two difficulties arise. First, what happens when two roots of the polynomial coincide? The answer is that the simple marching procedure will probably fail. There are good strategies for dealing with "limit points" and "bifurcation points", explained in later sections. The second problem is to obtain the first guess for continuation, and we turn to this next.

D.3   Initialization Strategies

To apply continuation, we need a solution for $\alpha = 0$ or equivalently, a first guess which is sufficiently close so that the Newton-Kantorovich method will convert the guess into a solution. Strategies for obtaining this needed solution or first guess include the following:

(i) a linear or perturbative solution
(ii) invention of an artificial marching parameter and
(iii) low-order collocation.

The first strategy may be illustrated by cnoidal waves of the so-called FKdV equation. The ODE and boundary conditions are

$$-u_{xxxxx} + (u - c)u_x = 0 \tag{D.15}$$

$$u(x + 2\pi) = u(x) \tag{D.16}$$

where c is the phase speed. For small amplitude, the nonlinear term can be ignored and solving the linearized equation gives

$$u(x) \approx a\cos(x) \qquad ; \qquad c \approx -1, \qquad a << 1 \qquad \text{(D.17)}$$

which can be used to initialize the march. The most obvious marching parameter is the phase speed c, but we can also define the coefficient of $\cos(x)$ in the Fourier expansion of $u(x)$ to be the amplitude a, and march in a. Fig. D.1 shows $c(a)$.

Since dc/da varies much more rapidly for small amplitude than for large, it is efficient to employ a variable step-size in the marching parameter a. One could, for example, monitor the number of Newton's iterations that are required at each value of a. When it falls below a cutoff, one could automatically double the step-size. When the number of Newton's iterations is large, one may automatically halve the step-size. Library ODE-solvers, such as might be applied to the Davidenko equation, use similar strategies.

Unfortunately, in the vicinity of a limit point, $du/d\alpha \rightarrow \infty$, so no variable step-size strategy will work. However, there are other remedies as explained in the next section.

The second strategy of inventing an artificial continuation parameter may be applied to solve a polynomial equation

$$R(x) = 0 \qquad \text{(D.18)}$$

for all its N roots. There is an infinite variety of choices, but Wasserstrom (1973) defines

$$P(x; \alpha) \equiv \alpha R(x) + (1 - \alpha)[x^N - 1] \qquad \text{(D.19)}$$

When $\alpha = 0$, the roots of $P(x; \alpha)$ are the N roots of unity. When $\alpha = 1$, $P(x; 1) = R(x)$. When α is a *physical* parameter, we usually graph $x(\alpha)$. When α is an artificial marching parameter, however, the intermediate values of $x(\alpha)$ are simply thrown away. Fig. D.2

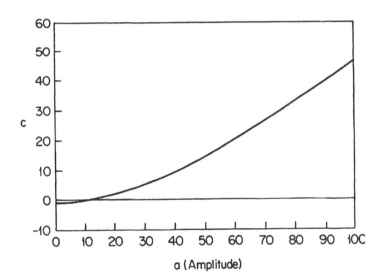

Figure D.1: Phase speed c versus amplitude a for the cnoidal wave of the fifth-degree KdV equation, taken from Boyd (1986b). For small amplitude, the differential equation is linear and $c \approx -1$ while $u(x) \approx a\cos(x)$. This linear solution can be used to initialize the continuation method, which can then march to arbitrarily large amplitude in small steps.

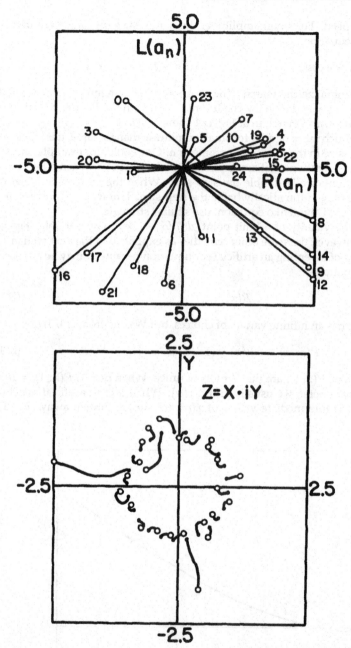

Figure D.2: An illustration of the continuation method for a polynomial of degree 25.
(a) Trajectories of the coefficients of the polynomials as functions of the marching parameter α.
(b) Trajectories of the roots of the polynomial.
This example and graphs are taken from Wasserstrom (1973).

shows how the roots vary with α for a 25-th degree polynomial taken from Wasserstrom (1973).

Is this a good way to solve polynomial equations? If there is a single solution of physical interest and if α is a physical parameter, then continuation is very efficient. When roots $x(\alpha)$ collide as α varies, the continuation is defeated. However, for *complex* trajectories, this is very unlikely. The technical jargon is that (D.19) is a "probability-one homotopy" (Li, 1987, and Morgan, 1987). What this means is that if one picks the coefficients of $R(x)$ as random *complex-valued* numbers, the probability of root-collision is infinitesimal.

Unfortunately, the situation is different when the trajectories are real. If $R(x)$ has any complex roots, then continuation from a polynomial $P(x; \alpha)$ which has only real roots will inevitably fail because pairs of roots must merge before separating as a complex conjugate pair. Thus, one needs a complex-valued embedding strategy like Wasserstrom's. Furthermore, *near*-collisions will cause problems if the roots pass sufficiently close, and the probability of such near-collisions is *not zero* — although it is very small.

For a single polynomial equation, the simplest strategy is the Traub-Jenkins package which is available in most software libraries. For *systems* of polynomial equations, however, little library code exists and the continuation strategy is quite attractive as described in the book by Morgan (1987).

For a differential equation such as

$$W_{rr} + \frac{1}{r}W_r - W - W^2 = 0 \qquad W \to 0 \text{ as } |r| \to \infty \qquad (D.20)$$

which is known as the "Petviashvili monopole vortex" problem in geophysical fluid dynamics, linearization fails because the linearized problem has only the solutions $K_0(r)$, which is singular at the origin, and $I_0(r)$, which is unbounded for large r. This problem is "essentially nonlinear" in the sense that the nonlinear term is necessary to remove the singularity at the origin.

The physical background of this problem shows that the solution qualitatively resembles $\exp(-r)$. One way to exploit this is to define an artificial forcing function to be the residual of the trial solution $\exp(-r)$, that is,

$$f(r) \equiv \left[\frac{\partial^2}{\partial r^2} + \frac{1}{r}\frac{\partial}{\partial r} - 1 - e^{-r}\right]e^{-r} \qquad (D.21)$$

If we solve perturbed problem

$$u_{rr} + \frac{1}{r}u_r - u - u^2 = (1 - \alpha)f(r) \qquad (D.22)$$

then Newton's iteration will converge for $\alpha = 0$ because $\exp(-r)$ is, *by construction*, the exact solution. As we march in small steps in α, $u(r; \alpha)$ is slowly deformed from $\exp(-r)(\alpha = 0)$ to the Petviashvili monopole, $W(r)(\alpha = 1)$.

This method of the "artificial forcing function" has a broad range of usefulness. It helps to choose a trial solution that qualitatively resembles the true solution because $u(r; \alpha)$ varies less between $\alpha = 0$ and $\alpha = 1$, but this is by no means mandatory.

A variant is to choose the forcing function for simplicity such as a constant δ:

$$u_{rr} + \frac{1}{r}u_r - u - u^2 = \delta \qquad (D.23)$$

For small δ, the perturbed differential equation has *two* solutions: (i) the monopole (large amplitude) and (ii) the artificial (small amplitude) solution which may be calculated by ignoring the nonlinear term, solving the resulting linear BVP, and then using the linear solution to initialize Newton's iteration. If we are lucky, these two solution branches, $u_1(r; \delta)$ and $u_2(r; \delta)$ will join at a limit point, $\delta = \delta_{limit}$. Because of the critical point, marching in δ will never "turn the corner" around the limit point onto the upper branch; this tactic of a constant forcing function is practical only when combined with "pseudoarclength continuation", which is briefly described below. Glowinski, Keller, and Reinhart (1985) have successfully applied this method to a problem very closely related to (D.20).

The third numerical strategy for obtaining a first guess is to compute a crude solution using a very low order approximation which can be obtained in closed form. For example, if one can invent a two-parameter guess, one can then make a contour plot of the residual when this trial solution is substituted into the differential equation, and pick the parameters that minimize the residual. A more direct tactic in this same spirit is to apply low order collocation.

For example, the solution of (D.20) is (a) symmetric[1] and (b) vanishes as $|r| \to \infty$. It follows that if we solve this using the orthogonal rational Chebyshev functions $TB_n(r)$, discussed in Chapter 17, then we need only the symmetric (even n) basis functions. Further, we can create new basis functions which individually satisfy the boundary conditions at infinity by taking linear combinations of the $TB_n(r)$. In particular, choosing $\phi_0(r) \equiv TB_2(r) - TB_0(r)$ and truncating the basis to just this single function gives the approximation

$$W(r) \approx a_0 \frac{2}{1 + [y/L]^2} \tag{D.24}$$

where L is the user-choosable map parameter associated with the $TB_n(r)$. Choosing $L = 2$, which is known to be reasonable from calculating the expansion of functions that decay as $\exp(-r)$, and applying the one-point pseudospectral algorithm gives a *linear* equation whose solution is $a_0 = -1$. The resulting approximation has a maximum error of about 16% of the maximum of the monopole, but it *is* sufficiently close so that Newton's iteration with many basis functions will converge. Boyd (1986c) and Finlayson (1973) give other examples.

It goes without saying that this approach is more risky than the slow, systematic march of continuation in an artificial parameter. It is also much cheaper in machine time in the sense that if it works, low-order collocation may obviate the need for continuation, at least in non-physical parameters. There is also the satisfaction of "making friends with the function", to use Tai-Tsun Wu's poetical phrase: brute force numerical computation is often no better at giving insight into the physics or the structure of the solution than television is at giving insight into real human behavior.

D.4   Limit Points

Definition 51 (LIMIT POINT) *If $\mathcal{N}(u; \alpha) = 0$ is a single nonlinear equation with a solution branch $u(\alpha)$, then a LIMIT POINT occurs wherever*

$$\mathcal{N}_u(u[\alpha]; \alpha) = 0 \qquad ; \qquad \mathcal{N}_\alpha(u[\alpha]; \alpha) \neq 0 \tag{D.25}$$

The Davidenko equation implies that an equivalent definition is

$$\frac{du}{d\alpha} \to \infty \qquad as \qquad \alpha \to \alpha_{limit} \tag{D.26}$$

---

[1]In cylindrical coordinates, the r-dependent coefficients of the terms in the Fourier series in polar angle θ are symmetric about $r = 0$ for even wavenumber (including the axisymmetric term) and the coefficients of odd wavenumber are antisymmetric in r as explained in Chapter 18.

Two distinct solution branches $u_1(\alpha)$ and $u_2(\alpha)$ meet at a limit point, and exist only on one side of the limit point. That is, if $u_1(\alpha)$ and $u_2(\alpha)$ exist for $\alpha \leq \alpha_{limit}$, then they do not exist (as real solutions) for $\alpha > \alpha_{limit}$.

Newton's iteration must fail at and near a limit point because the iteration is attracted equally to both branches, and responds by converging to neither.

When $\mathcal{N}(u; \alpha)$ is a system of equations, (D.26) still applies. The interpretation of (D.25) is that (i) the determinant of the Jacobian matrix $\mathcal{N}_u$ is zero and (ii) the column vector $\mathcal{N}_\alpha$ does not lie in the range of $\mathcal{N}_u$ so that the matrix equation $\mathcal{N}_u \, du/d\alpha = -\mathcal{N}_\alpha$ has no bounded solution.

As an example, the simple algebraic equation

$$(u - 1/2)^2 + \alpha^2 = 1 \tag{D.27}$$

has limit points at $\alpha = \pm 1$ where $u = 1/2$. This is merely the equation of a circle in the u-α plane as shown in Fig. D.3, but nonetheless it is impossible to trace the entire curve in a single pass by marching in α. By using a variable step-size, we can come as close to the limit point as we want, but we can never turn the corner and march from the lower branch onto the upper branch using small steps in either α or u.

However, it is trivial to write the equation of the circle in (D.28) in parametric form as

$$u = \cos(t) + 1/2 \qquad ; \qquad \alpha = \sin(t) \tag{D.28}$$

Both u and α are continuous, single-valued functions of t so that we can trace both branches of the solution in one continuous path if we march in t. The new continuation parameter t is the *arc-length* along the solution curve.

It is usually impossible to *analytically* define the arclength as in (D.28), but H. B. Keller and various collaborators(Keller, 1977, Decker and Keller, 1980, Chan, 1984, and Glowinski,

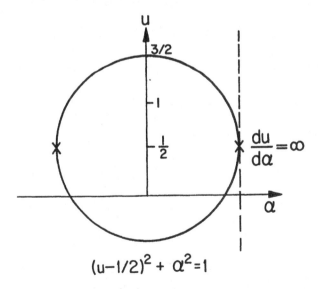

$$(u-1/2)^2 + \alpha^2 = 1$$

Figure D.3: Solution curve $u(\alpha)$ for a quadratic equation. The two *limit* points are marked by $\times$'s. There, as indicated on the right, the solution curve is tangent to a vertical line (dashed) and $du/d\alpha = \infty$. For $-1 < \alpha < 1$, there are two solutions for a given α. At the limit points, these upper and lower branches meet. Unable to decide whether to converge to the upper or lower branch, Newton's iteration diverges at the limit points.

Keller and Reinhart, 1985, and Keller, 1992) have developed a "pseudo-arclength continuation method" which can follow a solution smoothly around a limit point. The "globally convergent homotopy" methods reviewed by Li (1987) are very similar. We postpone a discussion until Sec. 6.

D.5 Bifurcation points

As in the previous section, the definition below and the examples are restricted to a *single* equation in a *single* unknown.

Definition 52 (BIFURCATION POINT) *This is a point in parameter space where two branches cross. The formal definition for a single equation $\mathcal{N}(u; \alpha) = 0$ is that a bifurcation point occurs whenever*

$$\mathcal{N}_u(u[\alpha]; \alpha) = 0 \qquad \& \qquad \mathcal{N}_\alpha(u[\alpha]; \alpha) = 0, \qquad (D.29)$$

which differs from the definition of a limit point in that both *derivatives are 0 instead of just one.*

Newton's method fails in the vicinity of a bifurcation point; the iteration literally cannot decide which branch of the two branches to converge to.

For a system of equations $\mathcal{N}(u) = 0$ where u is a vector of unknowns, (D.29) become the conditions that (i) the Jacobian matrix $\mathcal{N}_u$ has a vanishing determinant and (ii) $\mathcal{N}_\alpha$ is within the range of $\mathcal{N}_u$ so that the matrix equation $\mathcal{N}_u du/d\alpha = -\mathcal{N}_\alpha$ has bounded solution even though the Jacobian matrix is singular.

An example is the quadratic equation

$$u^2 - 2u - \alpha(\alpha - 2) = 0 \qquad (D.30)$$

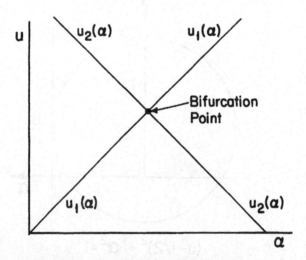

Figure D.4: Same as Fig. D.3 except that the quadratic has been altered so that its solution branches have a *bifurcation* point instead of a pair of limit points. At a bifurcation point, two branches cross. As at a limit point, Newton's iteration cannot converge to either one but instead diverges. Unlike a limit point, however, all branches have a finite slope $du/d\alpha$ at a bifurcation point. A limit point is one where branches *merge*; a bifurcation point is one where branches *cross*.

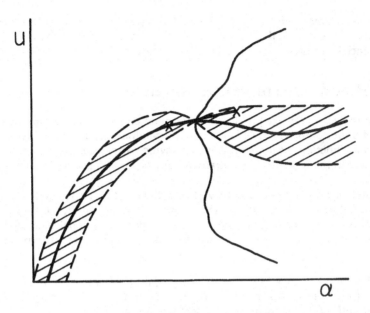

Figure D.5: Schematic of "shooting a bifurcation point". Newton's iteration converges when the first guess is within the cross-hatched region. Unfortunately, this convergence region shrinks to zero width at the bifurcation point. However, linear extrapolation from the left "x" will give a first guess as indicated by the right "x". This lies within the convergence zone on the far side of the bifurcation point. Newton's iteration will converge from a guess of the point marked by the right "x", and one may then apply continuation to trace out the whole solution branch to the right of the bifurcation point. The only complication is the issue of how far to proceed along the straight line when extrapolating beyond the bifurcation point; some trial-and-error may be needed.

whose two roots are

$$u_1(\alpha) = \alpha \qquad ; \qquad u_2(\alpha) = 2 - \alpha \qquad (D.31)$$

The bifurcation point is located at $\alpha = 1$ where $u_1 = u_2$ and the two branches cross. The ODE in α for (D.30) is

$$\frac{du}{d\alpha} = \frac{\alpha - 1}{u - 1} \qquad (D.32)$$

At the limit point, the denominator of (D.32) tends to 0 — but so does the numerator. The slope $du/d\alpha$ is always finite. Consequently, although we cannot compute $u(\alpha)$ at the bifurcation point, we can calculate either branch without using a small step-size even at points very *near* $\alpha = \alpha_{\text{bifurcation}}$.

Fig. D.5 shows how it is possible to, in Keller's words, "shoot the bifurcation point". The radius of convergence of Newton's method (in u) shrinks to 0 at the bifurcation point, but expands on either side. Consequently, if we use a rather *large* step in α, we can jump from the region of convergence on one side of the bifurcation point to the corresponding region on the other side. We do not need tricks like "pseudoarclength continuation" near a bifurcation point.

A harder task is to shift to the *other* branch, that is, to jump from $u_1(\alpha)$ to $u_2(\alpha)$. By using a local Taylor approximation near the bifurcation point (all the needed terms can be computed numerically), we can obtain a first guess for points on the other branch very close to the bifurcation. All one needs is a single point on $u_2(\alpha)$; one may then trace the rest of the branch by ordinary continuation. Seydel (1988) and Keller (1992) are good references.

D.6 Pseudoarclength Continuation

As noted in Sec. 4, a major problem for continuation in a physical parameter α is the existence of limit points where $du/d\alpha \to \infty$. In pseudoarclength continuation, both u and α are taken as functions of a new parameter s which approximates arclength along the solution curve in the u-α plane. The gradient $(du/ds, d\alpha/ds)$ is always finite so there are no infinities.

The arclength is defined by approximating the solution curve $u(\alpha)$ by many short line segments. In the limit that the segments become infinitely short, the sum of the length of the segments between two points on the curve is the arclength between those points.

Mathematically, suppose that the solution curve is specified parametrically, i. e.

$$u = u(t) \qquad \& \qquad \alpha = \alpha(t) \tag{D.33}$$

Let Δt denote the change in t between the endpoints of one of the infinitesimal line segments that we use to measure arclength. Let Δu and $\Delta \alpha$ denote changes in u and α between the ends of the segment. By the Pythagorean theorem,

$$(\Delta s)^2 = (\Delta u)^2 + (\Delta \alpha)^2 \tag{D.34}$$

which, dividing by Δt and then taking the limit, is

$$\left(\frac{ds}{dt}\right)^2 = \left(\frac{du}{dt}\right)^2 + \left(\frac{d\alpha}{dt}\right)^2 \tag{D.35}$$

Our goal is to parameterize the solution curve in terms of the arclength so that $t \equiv s$. This implies that the L. H. S. of (D.35) is one, and this in turn gives a constraint which can be applied to "inflate" our system of M unknowns, $u(\alpha)$, into a system of $M+1$ unknowns $(u(s), \alpha(s))$ by giving us the $(M+1)$-st nonlinear equation:

$$\left(\frac{du}{ds}\right)^2 + \left(\frac{d\alpha}{ds}\right)^2 = 1 \qquad \text{["Arclength Constraint"]} \tag{D.36}$$

For simplicity, we have written u as a scalar, but the only change if u is a vector is that $(du/ds)^2 = (du_1/ds)^2 + (du_2/ds)^2 + \ldots$, i. e. is the usual scalar product of the vector du/ds with itself.

There is one modest complication: (D.36) involves the s-derivatives of u and α rather than these quantities themselves. Strictly speaking, when u is a vector of dimension M, the system $\mathcal{N}(u; \alpha) = 0$ plus (D.36) is a system of $(M+1)$ equations in $(2M+2)$ unknowns! However, H. Keller, who invented this technique, pointed out that we are not interested in the arclength for its own sake, but rather as a computational device for evading limit points. It follows that it is legitimate to *approximate* the s-derivatives in (D.36) in any way such that $u(s)$ & $\alpha(s)$ become the unknowns. When such approximations are made, s is no longer exactly equal to the arclength, but is merely an approximation to it. It is for this reason that this family of algorithms is called "*pseudo*arclength" continuation.

Many approximations are possible. The simplest is a forward finite difference approximation to (some of) the derivatives in (D.36), which gives

$$S_1(u, \alpha; s) = 0 \tag{D.37a}$$

$$S_1 \equiv \frac{du(s_0)}{ds}[u(s_0 + \delta s) - u(s_0)] + \frac{d\alpha(s_0)}{ds}[\alpha(s_0 + \delta s) - \alpha(s_0)] - \delta s \tag{D.37b}$$

where s_0 is a constant which is normally the value of the arclength at the previous step of the continuation, and $\delta s \equiv s - s_0$.

An alternative is useful in overcoming a drawback of standard pseudoarclength continuation: Because s is the marching parameter and α is an unknown, there is no simple way to compute $u(\alpha)$ at a given "target" value of α. The problem is that we do not know which value of s corresponds to the target α. However, if we generalize (D.37) to

$$S_\theta \equiv \theta\frac{du}{ds}[u(s_0 + \delta s) - u(s_0)] + (2 - \theta)\frac{d\alpha}{ds}[\alpha(s_0 + \delta s) - \alpha(s_0)] - \delta s \qquad (D.38)$$

then $S_\theta(\theta = 1) = S_1$. However, when $\theta = 0$, (D.38) becomes an approximation to

$$\left(\frac{d\alpha}{ds}\right)^2 = 1 \qquad\qquad [\theta = 0], \qquad (D.39)$$

that is, s becomes identical with α. We have switched back to α as the marching parameter, and it is trivial to march to a desired "target" value of α.

Thus, it is possible to switch from ordinary continuation (in α) to pseudosarclength continuation in s and back again merely by changing θ. In the vicinity of a limit point, of course, taking $\theta = 0$ will lead to disaster, but very close to a limit point, we have no hope of marching to a target α anyway. There may not even be a solution if $\alpha_{target} > \alpha_{limit}$. Away from a limit point, however, we can apply pseudoarclength continuation to march close to a target value of α and then switch to $\theta = 0$ to march directly to α_{target} on the final step.

In any event, Newton's iteration for the arclength continuation requires solving the $(M + 1) \times (M + 1)$ matrix equation which in block form is

$$\left|\begin{array}{cc} \mathcal{N}_u & \mathcal{N}_\alpha \\ S_u & S_\alpha \end{array}\right| \left|\begin{array}{c} \Delta u \\ \Delta\alpha \end{array}\right| = \left|\begin{array}{c} -\mathcal{N}(u; \alpha) \\ -S(u; \alpha) \end{array}\right| \qquad (D.40)$$

When u is an M-dimensional vector, $\mathcal{N}_u$ is the $M \times M$ Jacobian matrix whose elements are $(\mathcal{N}_u)_{ij} \equiv \partial\mathcal{N}_i/\partial u_j$. Similarly, $\mathcal{N}_\alpha$ is the M-dimensional column vector whose elements are $(\mathcal{N}_\alpha)_i \equiv \partial\mathcal{N}_i/\partial\alpha$, S_u is the M-dimensional row vector with $(S_u)_j \equiv \partial S/\partial u_j$, and S_α is a scalar.

The one minor technical complication is that elements of S require the s-derivatives of u and α. Fortunately, we can obtain them in two steps. First, solve the system

$$\left|\begin{array}{cc} \mathcal{N}_u & \mathcal{N}_\alpha \\ 0 & 1 \end{array}\right| \left|\begin{array}{c} du/dt \\ d\alpha/dt \end{array}\right| = \left|\begin{array}{c} 0 \\ 1 \end{array}\right| \qquad (D.41)$$

which modifies only the R. H. S. and the last row of (D.40). Second, rescale du/dt and $d\alpha/dt$ via

$$\left(\frac{du}{ds}, \frac{d\alpha}{ds}\right)^T = \frac{(du/dt, d\alpha/dt)^T}{\sqrt{[du/dt]^2 + 1}} \qquad (D.42)$$

The justification for (D.41)–(D.42) is that along the solution curves, $\mathcal{N}(u[t], \alpha[t]) = 0$ where t is any parameter that parameterizes the solution curve. It follows that $d\mathcal{N}/dt \equiv 0$, but this is merely the top M rows of (D.41). This condition, $d\mathcal{N}/dt = 0$, gives only M constraints on $(M+1)$ unknowns, so to obtain a system with a unique solution, we arbitrarily demand that $d\alpha/dt = 1$, which is expressed by the bottom row of (D.41). For small displacements along the solution curve, the parameter t must be proportional to s so that the t-derivatives of u and α are proportional to the desired s-derivatives. We can determine the proportionality constant by recalling that the length of the vector $(du/ds, d\alpha/ds)^T$ is unity, which implies (D.42).

Thus, the overall algorithm may be summarized as:

1. Generate a first guess for u & α. [Predictor step]

2. Apply Newton's method to compute u & α by repeatedly solving the $(M+1)\times(M+1)$ matrix problem (D.40). [Corrector step]

3. Modify the last row of the square matrix and the R. H. S. of (D.40) to give (D.41). Solve (D.41) and apply the rescaling (D.42) to compute du/ds and $d\alpha/ds$.

4. Increase $s \to s + \delta s$ and return to 1.

When $s = 0$, the first guess is usually $\alpha = \alpha_0$, $u = u_0(\alpha_0)$ where u_0 is a known analytical solution (often a linear solution). For larger s, the simplest predictor is

$$u(s_0 + \delta s) \approx u(s_0) + \frac{du(s_0)}{ds}\,\delta s \quad \& \quad \alpha(s_0 + \delta s) \approx \alpha(s_0) + \frac{d\alpha(s_0)}{ds}\,\delta s \qquad \text{(D.43)}$$

but many alternatives are possible. Bank & Chan (1986) point out that a good predictor can greatly reduce the computation time by reducing the number of Newton iterations and apply a rather *ad hoc* but effective procedure that requires solving two simultaneous equations in two unknowns. Generating a good first guess is still a subject of active research, but (D.43) is safe and reliable.

If the first guess for $s = 0$ is exact, one may skip the Newton's iteration for this one parameter value. If, however, the march is initialized with an approximate solution, then (D.40) cannot be applied because it requires s-derivatives which we do not yet know. The remedy is to modify (D.40) by replacing the last row of by of the square matrix by $(0\ 1)^T$ and setting the last element of the R. H. S. equal to 0. This modified matrix problem is then the classical Newton's iteration with u as the unknown and α fixed; once $u(\alpha = \alpha_0)$ has been refined to the desired degree of accuracy, one may compute the corresponding s-derivatives in step 3 and solve (D.40) without modification at all later points on the solution curve.

The extra computation required by pseudoarclength continuation versus the classic Newton's method is very minor, especially when $M \gg 1$. By solving (D.40) and (D.41) via the 2×2 block decomposition which is described in Appendix B, one may solve (D.41) for the s-derivatives without having to recompute the full LU decomposition of a large matrix; the factorization of the Jacobian matrix $\mathcal{N}_u$, which is the costly step, can be applied to both (D.40) and (D.41).

The block-factorization fails near limit points because the Jacobian matrix $\mathcal{N}_u$ is singular at a limit point. This is not a serious difficulty in practice, however. Near the limit point, the solution varies with s proportional to the singular (zero eigenvalue) eigenfunction of the Jacobian anyway, so the 2×2 block solution of (D.41) will give the correct s-derivatives even very near the limit point.

Unfortunately the accuracy of the Newton's step (D.40) is degraded if the block factorization is used too near a limit point. The remedy is to switch to standard Gaussian elimination for the entire inflated system. The beauty of the pseudoarclength continuation method is precisely that the inflated, $(M+1)$-dimensional matrix system is *not* singular even when the M-dimensional Jacobian matrix is. Bank & Chan (1986) describe a method for obtaining accurate solutions via block-factorization for (D.40), but it would take us too far afield to describe their technique in detail.

It is important to note that pseudoarclength continuation eliminates only *limit* points. At bifurcation points, continuation still has difficulties. However, we can "shoot" the bifurcation points as described in the previous section. It is comforting, however, that Keller has found that "shooting" tends to be easier with arclength continuation than with standard continuation.

Seydel (1988) is a good treatment of continuation, limit points, and bifurcation points which assumes only an undergraduate mathematical background: ordinary differential equations, partial derivatives, and matrix algebra. Keller (1992) is an outstanding mix of review and research papers at a higher mathematical level.

Appendix E

Change-of-Coordinate Derivative Transformations

"A Chebyshev polynomial is a Fourier cosine in disguise"
  — J. Boyd

In many problems, it is useful to replace the original coordinate y by a new variable x:

$$y = f(x) \tag{E.1}$$

One common application is to convert a Chebyshev series in y into a Fourier cosine series in x, in which case $f(x) \equiv \cos(x)$. In this appendix, we collect (mostly unpublished) tables that are useful in transforming derivatives. The first is for general $f(x)$; the later tables are for a variety of important special mappings. For example, the ODE

$$a_2(y)\, u_{yy} + a_1(y)\, u_y + a_0(y)\, u = g(y) \tag{E.2}$$

becomes

$$a_2(f[x])\frac{f_1(x)\, u_{xx} - f_2(x)\, u_x}{f_1(x)^3} + a_1(f[x])\frac{u_x}{f_1(x)} + a_0(f[x]) = g(f[x]) \tag{E.3}$$

where, in a notation used throughout the tables,

$$f_n(x) \equiv \frac{d^n f}{dx^n} \tag{E.4}$$

Transformation of Orthogonality Integral

$$\int_0^{\pi} \cos(mx) \cos(nx)\, dx \equiv \int_{f(0)}^{f(\pi)} \frac{\phi_m(y)\, \phi_n(y)}{f_1\left(f^{-1}[y]\right)}\, dy; \tag{E.5}$$

Table E.1: Transformation of derivatives under a general mapping $y = f(x)$.

$$u_{n,y} = \frac{1}{f_1^{2n+1}} \{b_{n,n}\, u_{n,x} + b_{n,n-1}\, u_{n-1,x} + \cdots\}$$

where f_n is the n-th derivative of the map $f(x)$; the $b_{n,k}$ are below.

$$u_y = \frac{1}{f_1(x)}\, u_x$$

$$u_{yy} = \frac{1}{f_1(x)^3} \{f_1\, u_{xx} - f_2\, u_x\}$$

u_{yyy} $\qquad [\times 1/f_1^5]$

u_x	u_{xx}	u_{xxx}
$-f_3 f_1 + 3 f_2^2$	$-3 f_2 f_1$	f_1^2

u_{yyyy} $\qquad [\times 1/f_1^7]$

u_x	u_{xx}	u_{xxx}	u_{xxxx}
$-f_4 f_1^2 + 10 f_3 f_2 f_1 - 15 f_2^3$	$-4 f_3 f_1^2 + 15 f_2^2 f_1$	$-6 f_2 f_1^2$	f_1^3

u_{yyyyy} $\qquad [\times 1/f_1^9]$

u_x	u_{xx}	u_{xxx}	u_{xxxx}	u_{xxxxx}
$-f_5 f_1^3 + 15 f_4 f_2 f_1^2$ $+10 f_3^2 f_1^2 - 105 f_3 f_2^2 f_1$ $+105 f_2^4$	$-5 f_4 f_1^3$ $+60 f_3 f_2 f_1^2$ $-105 f_2^3 f_1$	$-10 f_3 f_1^3$ $+45 f_2^2 f_1^2$	$-10 f_2 f_1^3$	f_1^4

Table E.1: Transformation of derivatives under a general mapping $y = f(x)$ [CONTINUED]

$$u_{yyyyyy} \qquad [\times 1/f_1^{11}]$$

u_x	u_{xx}	u_{xxx}	u_{xxxx}	u_{xxxxx}
$+1260\, f_3\, f_2^3\, f_1$ $-280\, f_3^2\, f_2\, f_1^2$ $-210\, f_4\, f_2^2\, f_1^2$ $+35\, f_4\, f_3\, f_1^3$ $+21\, f_5\, f_2\, f_1^3$ $-945\, f_2^5 - f_6\, f_1^4$	$-840\, f_3\, f_2^2\, f_1^2$ $+945\, f_2^4\, f_1$ $+70\, f_3^2\, f_1^3$ $+105\, f_4\, f_2\, f_1^3$ $-6\, f_5\, f_1^4$	$+210\, f_3\, f_2\, f_1^3$ $-420\, f_2^3\, f_1^2$ $-15\, f_4\, f_1^4$	$+105\, f_2^2\, f_1^3$ $-20\, f_3\, f_1^4$	$-15\, f_2\, f_1^4$

u_{xxxxxx}
f_1^5

Table E.2: Transformations of derivatives for the mapping $y = \cos(x)$ which converts a Chebyshev series in $T_n(y)$ into a Fourier cosine series in $\cos(nx)$.

$$u_y = -\frac{1}{\sin(x)}\, u_x$$

$$u_{yy} = \frac{1}{\sin^3(x)} \left\{ \sin(x)\, u_{xx} - \cos(x)\, u_x \right\}$$

$u_{yyy} \qquad \left[\times 1/\sin^5(x) \right]$

u_x	u_{xx}	u_{xxx}
$-\sin^2 -3\cos^2$	$+3\cos\sin$	$-\sin^2$

$u_{yyyy} \qquad \left[\times 1/\sin^7(x) \right]$

u_x	u_{xx}	u_{xxx}	u_{xxxx}
$-15\cos^3 -9\cos\sin^2$	$+4\sin^3 +15\cos^2\sin$	$-6\cos\sin^2$	$\sin^3$

$u_{yyyyy} \qquad \left[\times 1/\sin^9(x) \right]$

u_x	u_{xx}	u_{xxx}	u_{xxxx}	u_{xxxxx}
$-9\sin^4 -105\cos^4$ $-90\cos^2\sin^2$	$+105\cos^3\sin$ $+55\cos\sin^3$	$-10\sin^4$ $-45\cos^2\sin^2$	$+10\cos\sin^3$	$-\sin^4$

$u_{yyyyyy} \qquad \left[\times 1/\sin^{11}(x) \right]$

u_x	u_{xx}	u_{xxx}	u_{xxxx}	u_{xxxxx}
$-225\cos\sin^4$ $-1050\cos^3\sin^2$ $-945\cos^5$	$+64\sin^5$ $+735\cos^2\sin^3$ $+945\cos^4\sin$	$-420\cos^3\sin^2$ $-195\cos\sin^4$	$105\cos^2\sin^3$ $+20\sin^5$	$-15\cos\sin^4$ $\dfrac{u_{xxxxxx}}{\sin^5}$

Table E.3: Transformations of derivatives for the mapping $y = \arccos(x)$, which converts a Fourier cosine series in y into a Chebyshev series in x. (This is the inverse of the mapping of Table E.2.) To simplify the tables, we used the auxiliary parameter: $Q(x) \equiv 1 - x^2$

$$u_y = -\sqrt{Q}\, u_x$$

$$u_{yy} = Q\, u_{xx} - x\, u_x$$

$$u_{yyy} = \sqrt{Q}\, \{ -Q\, u_{xxx} + 3\, x\, u_{xx} + u_x \}$$

$$u_{yyyy} = Q^2\, u_{xxxx} - 6\, Q\, x\, u_{xxx} + (3 - 7\, Q)\, u_{xx} + x\, u_x$$

$$u_{yyyyy} = \sqrt{Q}\, \{ -Q^2\, u_{xxxxx} + 10\, Q\, x\, u_{xxxx} + (25\, Q - 15)\, u_{xxx} - 15\, x\, u_{xx} - u_x \}$$

$$u_{yyyyyy} = Q^3\, u_{xxxxxx} - 15\, x\, Q^2\, u_{xxxxx} + (45\, Q - 65\, Q^2)\, u_{xxxx}$$
$$+ (90\, x\, Q - 15\, x)\, u_{xxx} + (31\, Q - 15)\, u_{xx} - x\, u_x$$

Table E.4: Transformations of derivatives for the mapping $y = L\cot(x)$ which converts a rational-Chebyshev series in $TB_n(y)$ into a Fourier cosine series in $\cos(nx)$, $y \in [-\infty, \infty]$ and $x \in [0, \pi]$. L is a constant, the "map parameter".

$$u_y = -\frac{\sin^2(x)}{L}\, u_x$$

$$u_{yy} = \frac{\sin^3(x)}{L^2}\,\{\sin(x)\,u_{xx} + 2\,\cos(x)\,u_x\}$$

u_{yyy} $\quad\left[\times\sin^4(x)/L^3\right]$

u_x	u_{xx}	u_{xxx}
$8\sin^2 - 6$	$-6\cos\sin$	$-\sin^2$

u_{yyyy} $\quad\left[\times\sin^5(x)/L^4\right]$

u_x	u_{xx}	u_{xxx}	u_{xxxx}
$24\cos - 48\cos\sin^2$	$36\sin - 44\sin^3$	$12\cos\sin^2$	$\sin^3$

u_{yyyyy} $\quad\left[\times\sin^6(x)/L^5\right]$

u_x	u_{xx}	u_{xxx}	u_{xxxx}	u_{xxxxx}
$-384\sin^4$ $+480\sin^2 - 120$	$400\cos\sin^3$ $-240\cos\sin$	$140\sin^4$ $-120\sin^2$	$-20\cos\sin^3$	$-\sin^4$

u_{yyyyyy} $\quad\left[\times\sin^7/L^6\right]$

u_x	u_{xx}	u_{xxx}	u_{xxxx}	u_{xxxxx}	u_{xxxxxx}
$3840\cos\sin^4$ $-3840\cos\sin^2$ $+720\cos$	$1800\sin$ $-6000\sin^3$ $+4384\sin^5$	$1200\cos\sin^2$ $-1800\cos\sin^4$	$300\sin^3$ $-340\sin^5$	$30\cos\sin^4$	$\sin^5$

$[\times\sin^q/L^p]$ denotes that all entries in the box below must be multiplied by this factor.

Table E.5: Transformations of derivatives for the mapping $y = Lx/\sqrt{1-x^2}$ where L is a constant map parameter. This transformation converts a series of $TB_n(y)$ into a Chebyshev series in x, that is, $TB_n(y) = T_n(x)$, $\forall n$, $y \in [-\infty, \infty]$, $x \in [-1, 1]$. Defining the auxiliary parameter $Q(x) \equiv 1 - x^2$ greatly simplifies the tables.

$$u_y = \sqrt{Q}\,Q\,u_x/L$$

$$u_{yy} = Q^2\left\{Q\,u_{xx} - 3\,x\,u_x\right\}/L^2$$

$$u_{yyy} = \sqrt{Q}\,Q^2\left\{Q^2\,u_{xxx} - 9\,x\,Q\,u_{xx} + (12 - 15\,Q)\,u_x\right\}/L^3$$

$$u_{yyyy} = Q^3\left\{Q^3\,u_{xxxx} - 18\,x\,Q^2\,u_{xxx} + (75\,Q - 87\,Q^2)\,u_{xx}\right.$$
$$\left. + (105\,x\,Q - 60\,x)\,u_x\right\}/L^4$$

$$u_{yyyyy} = \sqrt{Q}\,Q^3\left\{Q^4\,u_{xxxxx} - 30\,x\,Q^3\,u_{xxxx} + (255\,Q^2 - 285\,Q^3)\,u_{xxx}\right.$$
$$\left. + x\,Q(975\,Q - 660)\,u_{xx} + (360 - 1260\,Q + 945\,Q^2)\,u_x\right\}/L^5$$

$$u_{yyyyyy} = Q^4\left\{Q^5\,u_{xxxxxx} - 45\,x\,Q^4\,u_{xxxxx} + Q^3(645 - 705\,Q)\,u_{xxxx}\right.$$
$$+ x\,Q^2(4680\,Q - 3465)\,u_{xxx} + Q\,(6300 - 18585\,Q + 12645\,Q^2)\,u_{xx}$$
$$\left. - x\,(2520 - 11340\,Q + 10395\,Q^2)\,u_x\right\}/L^6$$

Table E.6: Transformations of derivatives for the mapping $y = L \cot^2(x/2)$ which converts a rational-Chebyshev series in $TL_n(y)$ into a Fourier cosine series in $\cos(nx)$. L is a constant, the "map parameter". Note that all sines and cosines in the table have arguments of $(x/2)$, not x.

Note: the $[\times \sin^q /\{2^n L^p \cos^r\}]$ denotes that all entries in the box below must be multiplied by this common factor.

$$u_y = -\{\sin^3(x/2)/[L\,\cos(x/2)]\}\,u_x$$

$$u_{yy} = \frac{\sin^5(x/2)}{2\,L^2\cos^3(x/2)}\left\{2\cos\left(\frac{x}{2}\right)\sin\left(\frac{x}{2}\right)u_{xx} + \left[3 - 2\sin^2\left(\frac{x}{2}\right)\right]u_x\right\}$$

u_{yyy} $\quad[\times \sin^7(x/2)/\{4\,L^3\cos^5(x/2)\}]$

u_x	u_{xx}	u_{xxx}
$-8\sin^4 +20\sin^2 -15$	$12\cos\sin^3 -18\cos\sin$	$4\sin^4 -4\sin^2$

u_{yyyy} $\quad[\times \sin^9 /\{8\,L^4\cos^7\}]$

u_x	u_{xx}	u_{xxx}	u_{xxxx}
$-48\sin^6 +168\sin^4$ $-210\sin^2 +105$	$88\cos\sin^5 -232\cos\sin^3$ $+174\cos\sin$	$48\sin^6 -120\sin^4$ $+72\sin^2$	$-8\cos\sin^5$ $+8\cos\sin^3$

u_{yyyyy} $\quad[\times \sin^{11} /\{16\,L^5\cos^9\}]$

u_x	u_{xx}	u_{xxx}	u_{xxxx}	u_{xxxxx}
$-384\sin^8 +1728\sin^6$ $-3024\sin^4$ $+2520\sin^2 -945$	$800\cos\sin^7$ $-2960\cos\sin^5$ $+3900\cos\sin^3$ $-1950\cos\sin$	$560\sin^8$ $-2080\sin^6$ $+2660\sin^4$ $-1140\sin^2$	$-160\cos\sin^7$ $+400\cos\sin^5$ $-240\cos\sin^3$	$-16\sin^8$ $+32\sin^6$ $-16\sin^4$

Table E.7: Transformations of derivatives for the mapping $y = L(1+x)/(1-x) + r_0$ where $y \in [r_0, \infty]$ and $x \in [-1, 1]$. This is similar to the map of Table E.6 except that it takes the semi-infinite interval in y to the standard interval for Chebyshev polynomials (rather than that for Fourier series). Defining the auxiliary parameter $Q(x) \equiv x^2 - 2x + 1$ greatly simplifies the tables. Note that the translation of the origin given by r_0 disappears when the mapping is differentiated, and therefore all entries below are *independent* of r_0.

$$u_y = u_x \frac{Q}{2L}$$

$$u_{yy} = \{Q\, u_{xx} + 2\,(x-1)\, u_x\} \frac{Q}{4L^2}$$

$$u_{yyy} = \{Q\, u_{xxx} + 6\,(x-1)\, u_{xx} + 6\, u_x\} \frac{Q^2}{8L^3}$$

$$u_{yyyy} = \{Q^2\, u_{xxxx} + 12\,(x-1)\, Q\, u_{xxx} + 36\, Q\, u_{xx} + 24\,(x-1)\, u_x\} \frac{Q^2}{16L^4}$$

$$u_{yyyyy} = \{Q^2 u_{xxxxx} + 20\,(x-1)\, Q\, u_{xxxx} + 120\, Q\, u_{xxx} + 240\,(x-1)\, u_{xx}$$
$$+ 120\, u_x\} \frac{Q^3}{32L^5}$$

$$u_{yyyyyy} = \{Q^3\, u_{xxxxxx} + 30\,(x-1)\, Q^2\, u_{xxxxx} + 300\, Q^2\, u_{xxxx}$$
$$+ 1200\,(x-1)\, Q\, u_{xxx} + 1800\, Q\, u_{xx} + 720\,(x-1)\, u_x\} \frac{Q^3}{64L^6}$$

Table E.8: Transformation of derivatives for the mapping $y = L \operatorname{arctanh}(x)$ which converts polynomials in $\operatorname{sech}(y/L)$ and $\tanh(y/L)$ into ordinary polynomials in x. In particular, $\tanh(y/L) = x$ & $\operatorname{sech}^2(y/L) = 1 - x^2$ The auxiliary parameter is $Q(x) \equiv 1 - x^2$

$$u_y = u_x \frac{Q}{L}$$

$$u_{yy} = \{Q\,u_{xx} - 2\,x\,u_x\} \frac{Q}{L^2}$$

$$u_{yyy} = \{Q^2\,u_{xxx} - 6\,x\,Q\,u_{xx} + (4 - 6\,Q)\,u_x\} \frac{Q}{L^3}$$

$$u_{yyyy} = \{Q^3\,u_{xxxx} - 12\,x\,Q^2\,u_{xxx} + Q\,(28 - 36\,Q)\,u_{xx} + x\,(24\,Q - 8)\,u_x\} \frac{Q}{L^4}$$

$$u_{5y} = \{Q^4\,u_{xxxxx} - 20\,x\,Q^3\,u_{xxxx} + Q^2\,(100 - 120\,Q)\,u_{xxx}$$
$$+ x\,Q\,(240\,Q - 120)\,u_{xx} + (16 - 120\,Q + 120\,Q^2)\,u_x\} \frac{Q}{L^5}$$

$$u_{6y} = \{Q^5\,u_{xxxxxx} - 30\,x\,Q^4\,u_{xxxxx} + Q^3\,(260 - 300\,Q)\,u_{xxxx}$$
$$+ x\,Q^2\,(1200\,Q - 720)\,u_{xxx} + Q\,(496 - 2160\,Q + 1800\,Q^2)\,u_{xx}$$
$$- x\,(32 - 480\,Q + 720\,Q^2)\,u_x\} \frac{Q}{L^6}$$

Table E.9: Transformation of derivatives under a general mapping $x = f(r), y = g(s)$

$$J \equiv f_r g_s - f_s g_r \tag{E.6}$$

$$\frac{\partial u}{\partial x} = \frac{g_s u_r - g_r u_s}{J}, \qquad \frac{\partial u}{\partial y} = \frac{-f_s u_r + f_r u_s}{J} \tag{E.7}$$

$$
\begin{aligned}
u_{xx} &= \frac{g_s g_s u_{rr} - 2 g_r g_s u_{rs} + g_r g_r u_{ss}}{J^2} \\
&\quad - u_y \frac{g_s g_s g_{rr} - 2 g_r g_s g_{rs} + g_r g_r g_{ss}}{J^2} \\
&\quad - u_x \frac{g_s g_s f_{rr} - 2 g_r g_s f_{ss} + g_r g_r f_{ss}}{J^2}
\end{aligned} \tag{E.8}
$$

$$
\begin{aligned}
u_{yy} &= \frac{f_s f_s u_{rr} - 2 f_r f_s u_{rs} + f_r f_r * u_{ss}}{J^2} \\
&\quad - u_y \frac{f_s f_s g_{rr} - 2 f_r f_s g_{rs} + f_r f_r g_{ss}}{J^2} \\
&\quad - u_x \frac{f_s f_s f_{rr} - 2 f_r f_s f_{ss} + f_r f_r f_{ss}}{J^2}
\end{aligned} \tag{E.9}
$$

$$J_r = f_{rr} g_s + f_r g_{rs} - f_{rs} g_r - f_s g_{rr};, \qquad J_s = f_{rs} g_s + f_r g_{ss} - f_{ss} g_r - f_s g_{rs}; \tag{E.10}$$

$$
\begin{aligned}
u_{xy} &= \frac{((f_r g_s + f_s g_r) u_{rs} - f_r g_r u_{ss} - f_s g_s u_{rr})}{J^2} \\
&\quad + \left\{ \frac{(f_r g_{ss} - f_s g_{rs})}{J^2} + \frac{(f_s g_s J_r - f_r g_s J_s)}{J^3} \right\} u_r \\
&\quad + \left\{ \frac{(f_s g_{rr} - f_r g_{rs})}{J^2} + \frac{(f_r g_r J_s - f_s g_r J_r)}{J^3} \right\} u_s
\end{aligned} \tag{E.11}
$$

Appendix F

Cardinal Functions

[of the Whittaker cardinal function, sinc(x)]: "a function of royal blood ... whose distinguished properties separate it from its bourgeois brethren."
— Sir Edmund Whittaker (1915)

F.1   Introduction

The cardinal functions $C_j(x)$ for a given type of interpolation (trigonometric, polynomial, etc.) and for a set of interpolation points x_i are defined by the requirement that

$$C_j(x_i) = \delta_{ij} \qquad\qquad i, j = 1, \ldots, N \qquad\qquad \text{(F.1)}$$

where δ_{ij} is the usual Kronecker delta symbol defined by

$$\delta_{ij} \equiv \begin{cases} 1 & i = j \\ 0 & i \ne j \end{cases} \qquad \text{[Kronecker delta]} \qquad\qquad \text{(F.2)}$$

There is no universal terminology; other authors refer to the cardinal functions as the "cardinal basis", "Lagrange basis", or "the fundamental polynomials of interpolation".

$\vec{\delta_k}$ denotes the matrix of the k-th derivative at the interpolation points:

$$(\delta_k)_{ij} \equiv \left. \frac{d^k C_j}{dx^k} \right|_{x = x_i} \qquad\qquad \text{(F.3)}$$

The tables use the auxiliary column vectors defined by:

$$c_j \equiv \begin{cases} 2 & j = \pm N \\ 1 & \text{otherwise} \end{cases} \qquad \& \qquad p_j \equiv \begin{cases} 2 & j = 0 \text{ or } N \\ 1 & \text{otherwise} \end{cases} \qquad\qquad \text{(F.4)}$$

F.2   General Fourier Series: Endpoint Grid

Taken from Gottlieb, Hussaini, and Orszag(1984).

Grid points:

$$x_i = \frac{\pi i}{N} \qquad\qquad i = 0, \ldots, 2N - 1 \tag{F.5}$$

Cardinal Funcs.:

$$C_j(x) \;\equiv\; \frac{1}{2N}\, \sin[N(x - x_j)]\, \cot[0.5(x - x_j)] \tag{F.6}$$

$$\equiv\; \frac{1}{2N} \sum_{k=-N}^{N} \frac{1}{c_k}\, \exp[ik(x - x_j)] \tag{F.7}$$

1st Derivative:

$$(\delta_1)_{ij} \equiv \begin{cases} 0 & i = j \\ 0.5(-1)^{i-j} \cot[0.5(x_i - x_j)] & i \neq j \end{cases} \tag{F.8}$$

2d Derivative:

$$(\delta_2)_{ij} \equiv \begin{cases} -(1 + 2N^2)/6 & i = j \\ 0.5(-1)^{i-j+1}/\sin^2[0.5(x_i - x_j)] & i \neq j \end{cases} \tag{F.9}$$

F.3   Fourier Cosine Series: Endpoint Grid

$$x_i = \frac{\pi i}{N} \qquad i = 0, \dots, N \qquad [(N+1) \text{ degrees of freedom}] \qquad \text{(F.10)}$$

$$C_j^{\cos}(x) \equiv \begin{cases} (-1)^{j+1} \dfrac{\sin(x) \sin(Nx)}{p_j \, N[\cos(x) - \cos(x_j)]} \\[2ex] \dfrac{2}{N p_j} \displaystyle\sum_{m=0}^{N} \dfrac{1}{p_m} \cos(mx_j) \cos(mx) \end{cases} \qquad \text{(F.11)}$$

where $p_j = 2$ if $j = 0$ or N and $p_j = 1$ if $j = 1, \dots, N-1$.

1st Deriv.: $(N+1) \times (N+1)$ matrix

$$\left. \frac{dC_j^{\cos}}{dx} \right|_{x=x_i} \equiv (\delta_1^{\cos})_{ij} \equiv \begin{cases} (\delta_1)_{ij} & j = 0 \text{ or } N \\[1ex] (\delta_1)_{ij} + (\delta_1)_{i,2N-j} & \text{otherwise} \end{cases} \qquad \text{(F.12)}$$

where $\vec{\delta_1}$ is the first derivative matrix for a general Fourier series (F.8).

Alternative 1st Derivative:

$$(\delta_1^{\cos})_{ij} \equiv \begin{cases} 0 & i = 0 \text{ or } i = N, \text{ all } j \\[1ex] 0.5 \cot(x_j) & i = j; \ j = 1, \dots, N-1 \\[1ex] \dfrac{(-1)^{j+1} \sin(x_i) \cos(Nx_i)}{p_j[\cos(x_i) - \cos(x_j)]} & i \neq j, \, 0, \, N \end{cases} \qquad \text{(F.13)}$$

2d Derivative:

$$\vec{\delta_2}^{\cos} \equiv \vec{\delta_1}^{\sin} \, \vec{\delta_1}^{\cos} \qquad \text{(F.14)}$$

where $\vec{\delta_1}^{\sin}$ is the first derivative matrix for a Fourier *sine* series, defined in the next subsection.

Quadrature formula: If $f(x) = f(-x)$, then

$$\int_0^\pi f(x)\, dx \approx \sum_{i=0}^N w_i\, f(x_i) \tag{F.15}$$

where the x_i are given by (F.10) and where the weights are

$$w_i \equiv \begin{cases} \pi/N & i = 1, 2, \ldots, N-1 \\ \pi/(2N) & i = 0 \text{ or } i = N \end{cases} \tag{F.16}$$

These weights are identical with those of the usual *trapezoidal* rule, but the accuracy for this special case of a *periodic, symmetric* function is *exponentially accurate*. The quadrature is exact if $f(x)$ is a trigonometric cosine polynomial of degree at most N (where the number of interpolation points is $N + 1$).

F.4   Fourier Sine Series: Endpoint Grid

$$x_i = \frac{\pi i}{N} \qquad i = 1, \dots, N-1 \qquad [(N\text{-}1) \text{ degrees of freedom}] \qquad (\text{F.17})$$

Note: by symmetry, a Fourier sine series must vanish at $x = 0$ & π, so there are only $(N-1)$ degrees of freedom. We can define the derivative matrix as one of dimension $(N+1) \times (N+1)$ for use in computing the even derivatives of a cosine series [by extending (F.17) to include $i = 0$ and $i = N$], but in applying collocation to compute a solution in the form of a sine series, we should use only the interior points.

Cardinal Funcs:

$$C_j^{\text{sin}}(x) \equiv \frac{(-1)^{j+1}\sin(x_j)\sin(Nx)}{N[\cos(x) - \cos(x_j)]} \qquad j = 1, \dots, N-1 \qquad (\text{F.18a})$$

$$C_j^{\text{sin}}(x) \equiv \frac{2}{N}\sum_{m=1}^{N-1}\sin(mx_j)\sin(mx) \qquad j = 1, \dots, N-1 \qquad (\text{F.18b})$$

1st Deriv.: $(N+1) \times (N+1)$ matrix

$$\left(\delta_1^{\text{sin}}\right)_{ij} \equiv \begin{cases} 0 & j = 0 \text{ or } N \\ (\delta_1)_{ij} - (\delta_1)_{i,2N-j} & \text{otherwise} \end{cases} \qquad (\text{F.19})$$

where δ_1 is the first derivative matrix for a general Fourier series, defined in (F.8), and where c_j is defined by (F.4). An equivalent definition of the sine derivative is

Alternative 1st Deriv:

$$\left(\delta_1^{\text{sin}}\right)_{ij} \equiv \begin{cases} 0 & j = 0 \text{ or } N \\ -0.5\cot(x_j) & i = j \\ \dfrac{(-1)^{i+j+1}\sin(x_j)}{\cos(x_i) - \cos(x_j)} & \text{otherwise} \end{cases} \qquad (\text{F.20})$$

2d Deriv:

$$\vec{\delta}_2^{\text{sin}} = \vec{\delta}_1^{\text{cos}}\,\vec{\delta}_1^{\text{sin}} \qquad (\text{F.21})$$

Exponentially Accurate Quadrature:

If $f(x) = -f(-x)$, then

$$\int_0^\pi f(x)\, dx = \sum_{i=1} w_i\, f(x_i) \qquad\qquad (F.22)$$

where the x_i are defined by (F.17) and the weights are given by

$$w_i \equiv \frac{2}{N} \sum_{m=1}^{N-1} \frac{1}{m}\, \sin(m x_i)[1 - \cos(m\pi)] \qquad\qquad (F.23)$$

This quadrature formula is exact if $f(x)$ is a trigonometric sine polynomial of degree $(N-1)$ or less; note that $(N-1)$ is the actual number of quadrature abscissas.

F.5 Cosine Cardinal Functions on the Interior (Rectangle Rule) Grid

Previously unpublished.

Grid points:

$$x_i = \frac{(2i-1)\,\pi}{2N} \qquad \text{[N degrees of freedom]} \tag{F.24}$$

$$C_j^{\cos}(x) \equiv (-1)^{j+1} \frac{\cos(Nx)\sin(x_j)}{N[\cos(x)-\cos(x_j)]} \tag{F.25}$$

$$= \frac{2}{N}\sum_{m=0}^{N} \frac{1}{p_m}\cos(mx_j)\cos(mx) \tag{F.26}$$

where the $1/p_m$ term means that the $m = 0$ and $m = N$ terms in the sum should be divided by $(1/2)$.

Quadrature Rule:

If $f(x)$ is a periodic but symmetric function with a period of 2π, then

$$\int_0^{\pi} f(x)\,dx \approx \sum_{j=1}^{N} w_j\,f(x_j) \tag{F.27}$$

where the quadrature weights are given by

$$w_j \equiv \frac{\pi}{N} \qquad \text{["Rectangle Rule"]} \tag{F.28}$$

independent of j. Although the "rectangle rule" is rather crude $[O(1/N^2)]$ accuracy for integrating *non-periodic* integrals, it is has an error that decreases *exponentially* fast with N when applied to an integrand which is (i) periodic and (ii) symmetric about the origin.

F.6   Sine Cardinal Functions on the Interior (Rectangle Rule) Grid

Previously unpublished.

Grid points:

$$x_i = \frac{(2i-1)\pi}{2N} \qquad [N \text{ degrees of freedom}] \tag{F.29}$$

$$C_j^{\sin}(x) \equiv \sin(Nx_j)\cos(Nx)\sin(x)/(N[\cos(x) - \cos(x_j)]) \tag{F.30a}$$

$$= \frac{2}{N}\sum_{m=1}^{N}\frac{1}{c_m}\sin(mx_j)\sin(mx) \tag{F.30b}$$

where the $1/c_m$ term means that the $m = N$ term in the sum should be divided by $(1/2)$.

Quadrature Rule:

If $f(x)$ is a periodic but antisymmetric function with a period of 2π, then

$$\int_0^\pi f(x)\,dx \approx \sum_{j=1}^{N} w_j\, f(x_j) \tag{F.31}$$

where the quadrature weights are given by

$$w_j \equiv \frac{2}{N^2}\sin(Nx_j)\sin^2\left(\frac{N\pi}{2}\right) + \frac{4}{N}\sum_{m=1}^{N-1}\frac{1}{m}\sin(mx_j)\sin^2\left(\frac{m\pi}{2}\right) \tag{F.32}$$

F.7   Sinc(x): **Whittaker cardinal function**

These are employed to represent functions on the interval $x \in [-\infty, \infty]$. It is necessary to simultaneously (i) decrease the spacing h between neighboring interpolation points and (ii) increase the number of sinc functions retained in the expansion to obtain higher accuracy for a given $f(x)$. The function being expanded must decay as $|x| \to \infty$. The formulas are taken from Stenger(1981).

Grid points:

$$x_i = h\,i \qquad\qquad i = 0, \pm 1, \pm 2, \ldots \tag{F.33}$$

Cardinal Functions:

$$C_j(x) \equiv \operatorname{sinc}\left(\frac{x - jh}{h}\right) \tag{F.34}$$

where

$$\operatorname{sinc}(x) \equiv \frac{\sin(\pi x)}{\pi x} \tag{F.35}$$

1st Derivative:

$$h\left(\delta_1^{\text{sinc}}\right)_{i,i+n} = \begin{cases} 0 & n = 0 \\ (-1)^{n+1}/n & n \neq 0 \end{cases} \tag{F.36}$$

2d Derivative:

$$h^2\left(\delta_2^{\text{sinc}}\right)_{i,i+n} = \begin{cases} -\pi^2/3 & n = 0 \\ -2(-1)^n/n^2 & n \neq 0 \end{cases} \tag{F.37}$$

$$h^3\left(\delta_3^{\text{sinc}}\right)_{i,i+n} = \begin{cases} 0 & n = 0 \\ (-1)^{n+1}\left[6/n^3 - \pi^2/n\right] & n \neq 0 \end{cases} \tag{F.38}$$

$$h^4\left(\delta_4^{\text{sinc}}\right)_{i,i+n} = \begin{cases} \pi^4/5 & n = 0 \\ 4(-1)^n[\pi^2/n^2 - 6/n^4] & n \neq 0 \end{cases} \tag{F.39}$$

$$h^5\left(\delta_5^{\text{sinc}}\right)_{i,i+n} = \begin{cases} 0 & n = 0 \\ (-1)^{n+1}[\pi^4/n - 20\pi^2/n^3 + 120/n^5] & n \neq 0 \end{cases} \tag{F.40}$$

$$h^6\left(\delta_6^{\text{sinc}}\right)_{i,i+n} = \begin{cases} -\pi^2/7 & n = 0 \\ -6(-1)^n[\pi^4/n^2 - 20\pi^2/n^4 + 120/n^6] & n \neq 0 \end{cases} \tag{F.41}$$

Fourier representation:

$$\operatorname{sinc}(x) \equiv \frac{1}{2\pi}\int_{-\pi}^{\pi} \exp(\pm ikx)\,dk \tag{F.42}$$

F.8   Chebyshev Polynomials: Extrema & Endpoints ("Gauss-Lobatto") Grid

Taken from Gottlieb, Hussaini, and Orszag(1984).

Grid points:

$$x_i = \cos\left(\frac{\pi i}{N}\right) \qquad\qquad i = 0, \ldots, N \qquad\qquad (F.43)$$

Cardinal Funcs:

$$C_j(x) \equiv (-1)^{j+1}\frac{(1-x^2)}{c_j N^2 (x-x_j)}\frac{dT_N(x)}{dx} \qquad (F.44a)$$

$$= \frac{2}{N p_j}\sum_{m=0}^{N}\frac{1}{p_m}T_m(x_j)\,T_m(x) \qquad (F.44b)$$

where $p_j = 2$ if $j = 0$ or N and $p_j = 1$ if $j = 1, \ldots, N-1$; $c_j = 1$ if $|j| < N$; $c_{\pm N} = 2$.

1st Derivative:

$$\left(\delta_1^{\text{Cheb}}\right)_{ij} \equiv \frac{dC_j}{dx}\Bigg|_{x=x_i} = \begin{cases} (1+2N^2)/6 & i = j = 0 \\ -(1+2N^2)/6 & i = j = N \\ -x_j/[2(1-x_j^2)] & i = j;\ 0 < j < N \\ (-1)^{i+j}p_i/[p_j(x_i - x_j)] & i \neq j \end{cases} \qquad (F.45)$$

where $p_0 = p_N = 2$, $p_j = 1$ otherwise.

Higher Derivative:

$$\delta_k^{\text{Cheb}} = \left(\delta_1^{\text{Cheb}}\right)^k \qquad\qquad (F.46)$$

where the exponent k denotes the usual matrix multiplication of k copies of the first derivative matrix.

Note: The first derivative matrix is not antisymmetric (with respect to the interchange of i and j) and the second derivative matrix is not symmetric. This makes it difficult to prove stability theorems for time-integration methods (Gottlieb et al., 1984).

F.9 Chebyshev Polynomials: Interior or "Roots" Grid

Grid Points:

$$x_i = \cos\left(\pi \frac{2i-1}{2N}\right) \qquad i = 1, \ldots, N \qquad \text{(F.47)}$$

Cardinal Functions:

$$C_j(x) \equiv \frac{T_N(x)}{T_N'(x_j)} [x - x_j] \qquad \text{(F.48)}$$

$$\equiv \frac{\cos(Nt)\sin(t_j)}{N\sin(Nt_j)[\cos(t) - \cos(t_j)]} \qquad \text{(F.49)}$$

where the T_N' is the x-derivative of $T_N(x)$ and where in the trigonometric form, $t = \arccos(x)$ and $t_j = \arccos(x_j)$ where the branch is chosen so that $0 \le t \le \pi$.

1st Derivative:

$$\left.\frac{dC_j}{dx}\right|_{x=x_i} \equiv (\delta_1)_{ij} = \begin{cases} 0.5\, x_j/(1-x_j^2) & i = j \\ (-1)^{i+j} \dfrac{\sqrt{(1-x_j^2)/(1-x_i^2)}}{x_i - x_j} & i \ne j \end{cases} \qquad \text{(F.50)}$$

Second Derivative:

$$\left.\frac{d^2C_j}{dx^2}\right|_{x=x_i} \equiv (\delta_2)_{ij} = \begin{cases} \dfrac{x_j^2}{(1-x_j^2)^2} - \dfrac{N^2 - 1}{3(1-x_j^2)} & i = j \\ (\delta_1)_{ij}\left\{\dfrac{x_i}{1-x_i^2} - \dfrac{2}{x_i - x_j}\right\} & i \ne j \end{cases} \qquad \text{(F.51)}$$

F.10   Legendre Polynomials: Extrema & Endpoints Grid ("Gaus Lobatto")

These data are from Gottlieb, Hussaini, & Orszag(1984).

Grid points:

$$x_0 = -1 \quad \& \quad x_N = 1 \quad \& \quad \text{the } (N\text{-}1) \text{ roots of } dP_N/dx \qquad (F.52)$$

Cardinal Functions:

$$C_j(x) \equiv \frac{-(1-x^2)}{N(N+1)P_N(x_j)(x-x_j)} \frac{dP_N(x)}{dx} \qquad (F.53)$$

Quadrature weights:

$$w_j \equiv \frac{2}{N(N+1)\{P_N(x_j)\}^2} \qquad (F.54)$$

First Derivative Matrix:

$$\left(\delta_1^{\text{Leg}}\right)_{ij} = \begin{cases} (1/4)N(N+1) & i = j = 0 \\ -(1/4)N(N+1) & i = j = N \\ 0 & i = j \ \& \ 0 < j < N \\ P_N(x_i)/[P_N(x_j)(x_i - x_j)] & i \neq j \end{cases} \qquad (F.55)$$

Four-Point Interpolation

$$P_3 = -\frac{3}{2}x + \frac{5}{2}x^3, \qquad \frac{dP_3}{dx} = -\frac{3}{2} + \frac{15}{2}x^2 \qquad (F.56)$$

Table F.1: Grid points and weights for 4-point Legendre-Lobatto

x_j		w_j	
± 1	-	1/6	0.166666666666667
± 0.447213595499958	$\pm\sqrt{1/5}$	5/6	0.833333333333333

Five-Point Interpolation

$$P_4 = \frac{3}{8} - \frac{15}{4}x^2 + \frac{35}{8}x^4, \qquad \frac{dP_4}{dx} = -\frac{15}{2}x + \frac{35}{2}x^3 \tag{F.57}$$

Table F.2: Grid points and weights for 5-point Legendre-Lobatto

x_j		w_j	
±1	-	1/10	0.1
±0.654653670707977	$\pm\sqrt{3/7}$	49/90	0.544444444444444
0	-	32/45	0.711111111111111

Six-Point Interpolation

$$P_5 = \frac{15}{8}x - \frac{35}{4}x^3 + \frac{63}{8}x^5; \qquad \frac{dP_5}{dx} = \frac{15}{8} - \frac{105}{4}x^2 + \frac{315}{8}x^4 \tag{F.58}$$

Table F.3: Grid points and weights for 6-point Legendre-Lobatto

Notes: Variants for weights: $w(\pm0.765) = 9/(20 + 10/\sqrt{7})$ and $w(\pm0.285) = 9/(20 - 10/\sqrt{7})$.

x_j		w_j	
±1	-	1/15	0.066666666666667
±0.765055323929464	$\pm\sqrt{(7 + 2\sqrt{7})/21}$	$\frac{1}{15\{P_5(x_j)\}^2}$	0.378474956297847
±0.285231516480645	$\pm\sqrt{(7 - 2\sqrt{7})/21}$	$\frac{1}{15\{P_5(x_j)\}^2}$	0.554858377035486

Seven-Point Interpolation

$$P_6 = -\frac{5}{16} + \frac{105}{16}x^2 - \frac{315}{16}x^4 + \frac{231}{16}x^6, \quad \frac{dP_6}{dx} = \frac{105}{8}x - \frac{315}{4}x^3 + \frac{693}{8}x^5 \tag{F.59}$$

Table F.4: Grid points and weights for 7-point Legendre-Lobatto

Notes: Variants for weights: $w(\pm0.8302) = 43923/(175\left\{3 + 7\sqrt{15}\right\}^2)$ and
$w(\pm0.468) = 43923/(175\left\{-3 + 7\sqrt{15}\right\}^2)$.

x_j		w_j	
±1	-	1/21	0.047619047619047
±0.830223389627857	$\pm\sqrt{(15 + 2\sqrt{15})/33}$	$\frac{1}{21\{P_6(x_j)\}^2}$	0.276826047361566
±0.468848793470714	$\pm\sqrt{(15 - 2\sqrt{15})/33}$	$\frac{1}{21\{P_6(x_j)\}^2}$	0.431745381209862
0	-	$\frac{256}{525}$	0.487619047619048

Eight-Point Interpolation

$$P_7 = -\frac{35}{16}x + \frac{315}{16}x^3 - \frac{693}{16}x^5 + \frac{429}{16}x^7 \tag{F.60}$$

$$\frac{dP_7}{dx} = -\frac{35}{16} + \frac{945}{16}x^2 - \frac{3465}{16}x^4 + \frac{3003}{16}x^6 \tag{F.61}$$

Table F.5: Grid points and weights for 8-point Legendre-Lobatto

Notes: $r = 320\sqrt{55}/265837$ and $\phi = \arccos(\sqrt{55}/30)$. x_j in the expression for the weights denotes the corresponding grid point.

x_j		w_j	
±1	-	1/28	0.035714285714285
±0.871740148509606	$\pm\sqrt{2r^{1/3}\cos\left(\frac{\phi}{3} + \frac{5}{13}\right)}$	$\frac{1}{28\{P_7(x_j)\}^2}$	0.210704227143506
±0.591700181433142	$\pm\sqrt{2r^{1/3}\cos\left(\frac{\phi}{3} + \frac{5}{13} + \frac{4\pi}{3}\right)}$	$\frac{1}{28\{P_7(x_j)\}^2}$	0.341122692483504
±0.209299217902479	$\pm\sqrt{2r^{1/3}\cos\left(\frac{\phi}{3} + \frac{5}{13} + \frac{2\pi}{3}\right)}$	$\frac{1}{28\{P_7(x_j)\}^2}$	0.412458794658704

Nine-Point Interpolation

$$P_8 = \frac{35}{128} - \frac{315}{32}x^2 + \frac{3465}{64}x^4 - \frac{3003}{32}x^6 + \frac{6435}{128}x^8 \tag{F.62}$$

$$\frac{dP_8}{dx} = -\frac{315}{16}x + \frac{3465}{16}x^3 - \frac{9009}{16}x^5 + \frac{6435}{16}x^7 \tag{F.63}$$

Table F.6: Grid points and weights for 9-point Legendre-Lobatto

Notes: $r = 448\sqrt{91}/570375$ and $\phi = \arccos(\sqrt{91}/154)$. x_j in the expression for the weights denotes the corresponding grid point.

x_j		w_j	
±1	-	1/36	0.027777777777777
±0.899757995411460	$\pm\sqrt{2r^{1/3}\cos\left(\frac{\phi}{3} + \frac{7}{15}\right)}$	$\frac{1}{36\{P_8(x_j)\}^2}$	0.165495361560805
±0.677186279510737	$\pm\sqrt{2r^{1/3}\cos\left(\frac{\phi}{3} + \frac{7}{15} + \frac{4\pi}{3}\right)}$	$\frac{1}{36\{P_8(x_j)\}^2}$	0.274538712500161
±0.363117463826178	$\pm\sqrt{2r^{1/3}\cos\left(\frac{\phi}{3} + \frac{7}{15} + \frac{2\pi}{3}\right)}$	$\frac{1}{36\{P_8(x_j)\}^2}$	0.346428510973046
0	-	$\frac{4096}{11025}$	0.371519274376417

Canuto, Hussaini, Quarteroni and Zang (1988), Appendix C, give a short FORTRAN program to compute the Legendre-Lobatto points for general degree.

Appendix G

Transformation of Derivative Boundary Conditions

"Pseudospectral algorithms are simply N-th order finite differences in disguise."
— J. P. Boyd

One modest complication in applying boundary conditions imposed on the *derivatives* of the solution is that the mapping functions which relate the derivatives in the original coordinate y to those of the new coordinate x are usually *singular* at the *endpoints*. However, it is straightforward to obtain the correct conditions via l'Hopital's Rule, which is equivalent to making power series expansions about the endpoint and then taking $x \to$ endpoint.

To illustrate, consider mapping that converts Chebyshev polynomials in y into cosines in x.

$$y = \cos(x) \qquad y \in [-1, 1] \quad \& \quad x \in [0, \pi] \qquad \text{(G.1)}$$

The method, however, can be applied to any of the mappings discussed above.

Suppose that the boundary conditions are

$$u_y(1) = \alpha \qquad \& \qquad u_{yy}(1) = \beta \qquad \text{(G.2)}$$

Now from Table E.2,

$$u_y = -\frac{u_x}{\sin(x)} \qquad \text{(G.3)}$$

The complication is that as $y \to 1$, $x \to 0$, and the denominator of (G.3) is 0.

However, $u(y)$ is mapped into $u(\cos[x])$ which is always *symmetric* about both $x = 0$ and $x = \pi$. This in turn implies that *all* the *odd derivatives* of $u(y[x])$ with respect to x must *vanish* at both endpoints. Thus, the R. H. S. of (G.3) is finite because both its numerator and denominator vanish. To calculate the limit at the endpoint, write the Taylor expansions

$$u_x(x) \approx u_x(0) + x\,u_{xx}(0) + O(x^2) \qquad\qquad |x| \ll 1 \qquad \text{(G.4)}$$

$$\approx x\,u_{xx}(0) + O(x^3) \qquad \text{(G.5)}$$

since all the odd derivatives vanish. Since

$$\sin(x) \approx x - \frac{x^3}{6} + O(x^5) \qquad \text{(G.6)}$$

one finds

$$u_y(y) \approx -x\, u_{xx}(0)/x + O(x^2) \qquad\qquad |x| \ll 1 \qquad\qquad (G.7)$$

which, after cancelling the common power of x, implies that the correct boundary condition is

$$u_y(1) = -u_{xx}(0) \qquad\qquad (G.8)$$

Thus, THE CORRECT WAY TO IMPOSE THE BOUNDARY CONDITION ON THE *FIRST* DERIVATIVE WITH RESPECT TO y IS TO IMPOSE IT ON THE *SECOND* DERIVATIVE WITH RESPECT TO x, i. e.

$$u_{xx}(0) = -\alpha = -u_y(1) \qquad\qquad (G.9)$$

Similarly, for the second derivative, note

$$u_{yy} = \frac{1}{\sin^2(x)} u_{xx} - \frac{\cos(x)}{\sin^3(x)} u_x \qquad\qquad (G.10)$$

Via Taylor expansions,

$$u_{yy} \approx \left\{ \frac{1}{x^2} + \frac{1}{3} \right\} \left\{ u_{xx}(0) + \frac{x^2}{2} u_{xxxx}(0) \right\} - \left\{ \frac{1}{x^3} + O(x) \right\} \left\{ x\, u_{xx}(0) + \frac{x^3}{6} u_{xxxx}(0) \right\}$$

$$(G.11)$$

$$u_{yy}(1) \equiv \frac{1}{3} \left\{ u_{xx}(0) + u_{xxxx}(0) \right\} \qquad\qquad (G.12)$$

which means that on the transformed problem in x, we should impose the boundary condition that

$$u_{xxxx}(x = 0) = 3\beta + \alpha = 3\, u_{yy}(y = 1) + u_y(y = 1) \qquad\qquad (G.13)$$

Glossary

"Like Adam, naming the birds and the flowers"
  — J. J. Sylvester   [of his invention of many mathematical terms]

ALGEBRAIC CONVERGENCE: If the ALGEBRAIC INDEX OF CONVERGENCE for a series is finite, then the series has "algebraic convergence". If $a_n \sim 1/n^k$, $n \to \infty$ for some finite k, then the series has "algebraic convergence".

ALGEBRAIC INDEX OF CONVERGENCE k: This index k is the largest number for which

$$\lim_{n \to \infty} |a_n| \, n^k < \infty$$

where the a_n are the series coefficients. The usual proof that a given function has algebraic index of convergence k is to apply k integrations-by-parts to the coefficient integrals. If this process of integration-by-parts can be repeated indefinitely so that the equation is true for arbitrarily large k, then the convergence is "infinite order" or "exponential". If the coefficients $a_n \sim 1/n^k$, $n \to \infty$, then k is the "algebraic index of convergence".

ANTISYMMETRIC: If a function is such that $f(x) = -f(x)$ for all x, it is said to be "ANTISYMMETRIC with respect to the origin" or to be of "odd PARITY".

ARITHMURGY: Synonym for "number-crunching". (From the Greek $\alpha\rho\theta\mu o\sigma$, "number", and $-\epsilon\rho\gamma o\sigma$, "working".)

ASSUMPTION OF EQUAL ERRORS: This empirical principle, which cannot be rigorously proved but is supported by strong heuristic arguments and by practical experience, states the "discretization error" and the "truncation error" are of the same order-of-magnitude. Whenever it is true, one can estimate the total error in a spectral calculation merely by estimating the truncation error, which is usually much easier to estimate than the discretization error. (Chap. 2.)

ASYMPTOTIC RATE OF GEOMETRIC CONVERGENCE: If a series has GEOMETRIC CONVERGENCE, that is, if

$$|a_n| < \exp(-n\mu) \qquad \text{all } n,$$

then the "asymptotic rate of geometric convergence" is the largest μ for which the above bound is true. The asymptotic rate of convergence is undefined for series with algebraic or subgeometric convergence.

577

BANDED (MATRIX): A label applied to a matrix whose elements A_{ij} are all zero except for diagonal bands around the main diagonal, that is, $A_{ij} = 0$ if $|i - j| > m$ for some m. (Appendix B.)

BANDWIDTH (of a MATRIX): If the elements $A_{ij} = 0$ unless $|i - j| \leq m$, then $(2m + 1)$ is the BANDWIDTH of the BANDED MATRIX.

BASIS FUNCTIONS: The members of a BASIS SET. Examples of basis functions are the Chebyshev polynomials and the Hermite functions (and in general all the classes of functions described in Appendix A).

BASIS RECOMBINATION: A strategy for satisfying numerical boundary conditions in which the original basis set, such as Chebyshev polynomials, is replaced by a new basis composed of linear combinations of the original basis functions such that each member of the new basis *individually* satisfies the boundary conditions. The alternative strategy is BOUNDARY BORDERING. (Chap. 6, Sec. 4.)

BASIS SET: The collection of functions which are used to approximate the solution of a differential equation. The Fourier functions $\{1, \cos(nx), \sin(nx)$ for $n = 1, 2, \ldots\}$ and the Chebyshev polynomials $\{T_n(x), n = 0, 1, \ldots\}$ are two examples of basis sets.

BEHAVIORAL BOUNDARY CONDITION: A boundary condition that imposes a certain behavior on the solution rather specifying a numerical constraint. Examples of behavioral boundary conditions include: (i) periodicity with a certain period L (ii) boundedness and infinite differentiability at a point where the coefficients of the differential equation are singular. It is usually possible to satisfy such conditions by proper choice of basis function. For example, the sines and cosines of a Fourier series are periodic, so no further action is needed to enforce periodicity on the solution.

BIFURCATION POINT: A point where two branches of a solution $u(\alpha)$ cross. Synonym is "crossing point". For both $\alpha > \alpha_{\text{limit}}$ and $\alpha < \alpha_{\text{limit}}$, two different solutions exist which meet at the bifurcation point. Newton's method fails at a bifurcation point, but it is possible to "shoot the bifurcation point" or switch branches as explained in Appendix D. (When "bifurcation" is used in a broader sense, a "crossing point" is also called a "trans-critical bifurcation".)

BLOCKING, SPECTRAL: See SPECTRAL BLOCKING.

BOUNDARY BORDERING : A strategy for satisfying boundary conditions in which these are imposed as explicit constraints in the matrix problem which is solved for the spectral coefficients. Some of the collocation or Galerkin conditions on the residual are replaced by a "border" of rows which impose the constraints. The major alternative is BASIS RECOMBINATION. (Chap. 6.)

CHOLESKY FACTORIZATION: Factorization of a SYMMETRIC matrix into the product LL^T where L is lower triangular and L^T is its transpose; Cholesky factorization requires only half as many operations as the usual LU factorization. (Appendix B.)

COLLOCATION: Adjective for labelling methods which determine the series coefficients by demanding that the residual function be zero on a grid of points. In this book, a synonym for PSEUDOSPECTRAL and for METHOD OF SELECTED POINTS. Elsewhere in the literature, this adjective is also applied to certain finite element methods.

COLLOCATION POINTS: the grid of points where the residual function must vanish. Synonym for INTERPOLATION POINTS.

COMPATIBILITY CONDITIONS: A countably infinite set of constraints on the initial conditions of a time-dependent partial differential equation which are necessary for the solution to be analytic everywhere in the space-time domain. Example: an incompressible flow will be strongly singular at $t = 0$ if the initial condition is divergent at some point in the fluid, and will have discontinuities (vortex sheets) at rigid walls if the initial velocity does not satisfy the no-slip boundary condition of vanishing at the walls. If the initial flow does not satisfy additional constraints, the flow will be singular, but more weakly in the sense of having more derivatives than if the nondivergence and no-slip constraints are violated. The constraints are different from each partial differential equation.

COMPLETENESS: A basis set is "complete" for a given class of functions if all functions within the class can be represented to arbitrarily high accuracy as a sum of a sufficiently large number of basis functions.

CROUT REDUCTION: Solution of a matrix equation by direct LU FACTORIZATION.

DARBOUX'S PRINCIPLE: Theorem that for a function $f(x)$, the asymptotic form of the spectral coefficients a_n as $n \to \infty$ are controlled by the singularities (poles, branch points, etc.) of $f(x)$. (Theorem 1.)

DENSE MATRIX: A matrix whose elements are mostly nonzero so that the zero elements do not offer any structure which can be exploited to reduce the cost of LU factorization. A synonym for FULL MATRIX; the opposite of a SPARSE MATRIX.

DIFFERENTIAL QUADRATURE/DIFFERENTIAL CUBATURE: Almost a synonym for "PSEUDOSPECTRAL". The DQ method, like the pseudospectral, uses difference formulas, based on either trigonometric or polynomial interpolation, in which every point on the grid (or every point on a coordinate line in multiple dimensions) contributes to the approximation at every point on the grid (or coordinate line). The difference from the pseudospectral method is that the distribution of points is completely *arbitrary* in theory. (In practice, the roots of Chebyshev polynomials are commonly used, making DQ identical with Chebyshev pseudospectral.) Differential Cubature is a multi-dimensional scheme which does not use a tensor product basis, but instead approximates derivatives using the grid point values at every point of the two-dimensional or three-dimensional grid. (This is expensive!).

DIRECT MATRIX METHOD: A non-iterative algorithm for solving a matrix equation, as in the phrase "fast direct methods". Examples include Gaussian elimination, Crout reduction, Cholesky factorization, LU factorization, skyline solver, and static condensation (Appendix B).

DISCRETE ORDINATES METHOD: Synonym for "PSEUDOSPECTRAL" or "COLLO-CATION" method. (This term is most common among physicists and chemists.)

DISCRETIZATION ERROR: This is the difference between the first $(N+1)$ coefficients of the exact solution and the corresponding coefficients as computed using a spectral or pseudospectral algorithm with $(N+1)$ basis functions. It is distinct from (but usually the same order-of-magnitude as) the "truncation error". (Def. 9.)

DOMAIN TRUNCATION: A method of solving problems on unbounded domains by replacing the interval $y \in [-\infty, \infty]$ by $y \in [-L, L]$. If $|u(\pm L)|$ decays exponentially with L, then it is possible to obtain solutions of arbitrary accuracy by choosing L sufficiently large. (Chap. 17.)

ENVELOPE OF THE COEFFICIENTS: A smooth, monotonically decreasing curve which is a tight bound on oscillatory Chebyshev or Fourier coefficients $\{a_n\}$ in the sense that the absolute value of the coefficients is arbitrarily close to the envelope infinitely often as $n \to \infty$. (Borrowed from wave theory, where the "envelope of a wave packet" has the identical meaning.)

ENVELOPE OF THE INTERPOLATION ERROR: The envelope ρ is defined by

$$\rho \equiv \frac{E_I(x; N)}{\Phi_N(x)}$$

where $E_I(x; N)$ is the interpolation error and $\Phi_N(x)$ is the function whose roots are the interpolation grid. (Boyd, 1990c.)

EQUAL ERRORS, ASSUMPTION OF: This empirical principle, unprovable but supported by strong heuristic arguments and practical experience, states that the "DISCRETIZATION ERROR" and "TRUNCATION ERROR" and "INTERPOLATION ERROR" are the same order of magnitude. (Chap. 2.)

ESSENTIAL BOUNDARY CONDITION: A boundary condition is "essential" if it must be explicitly imposed on the approximation to the solution of a differential equation; usually, essential boundary conditions are NUMERICAL BOUNDARY CONDITIONS, that is, require the solution u or its derivative to equal a certain number or function on the boundary.

EXPLICIT (TIME-MARCHING): A time-integration scheme in which the solution at the next time level is given by an explicit formula that does NOT require solving a boundary value problem. (See also IMPLICITLY-IMPLICIT.) (Chap. 9.)

EXPONENTIAL CONVERGENCE: A spectral series possesses the property of "exponential convergence" if the error decreases faster than any finite inverse power of N as N, the number of terms in the truncated series, increases. Typically, the series coefficients decrease as $O(\exp[-pn^r])$ for some positive constants p and r. A synonym for "infinite order convergence".

EXPONENTIAL INDEX OF CONVERGENCE: If the series coefficients a_n satisfy

$$a_n \sim O[s \exp(-q\, n^r)], \qquad n \to \infty$$

where s, q, and r are constants, then r is the "exponential convergence index". Alternative definitions are given in Chap. 2 as Def. 4.

FFT: Abbreviation for FAST FOURIER TRANSFORM. (Chap. 10.)

FINITE ELEMENT METHOD: A class of algorithms in which the domain is divided into subdomains and the solution is approximated by a different polynomial on each subdomain. At high order, these piecewise approximations must be constructed by the same strategies as for global spectral methods, and are therefore called SPECTRAL ELEMENTS or "h-p" finite elements.

FOLD POINT: Synonym for LIMIT POINT.

GEOMETRIC CONVERGENCE: A series whose coefficients are decreasing

$$a_n \sim \alpha(n)p^n \leftrightarrow \alpha(n)\exp[-n\,|\log(p)|]) \qquad\qquad n \to \infty \qquad |p| < 1$$

has the property of "geometric convergence" where $\alpha(n)$ denotes a function that varies algebraically, rather than exponentially, with n such as a power of n. The reason for the name is that the terms of a geometrically convergent series can always be bounded by those of a geometric series, that is, by the terms of the power series expansion of $\alpha/(\beta + x)$ for some α and β where these are constants. (All convergent power series have geometric convergence. All Chebyshev series for functions which have no singularities on $x \in [-1, 1]$ (including the endpoints) also have geometric convergence).

GIBBS OSCILLATIONS: When a function with a discontinuity or a discontinuity in a derivative of some order is approximated by a spectral series truncated at N terms, the sum of the truncated series oscillates with a wavelength which is $O(1/N)$. These oscillations are also called "SPECTRAL RINGING".

GIBBS PHENOMENON: When a function with a discontinuity is approximated by a spectral series truncated at N terms, the sum of the truncated series overshoots the true value of the function at the edge of the discontinuity by a fixed amount (approximately 0.089 times the height of the jump), independent of N. GIBBS PHENOMENON also includes GIBBS OSCILLATIONS.

GLOBAL ELEMENT: (Now rare.) Sometimes a synonym for SPECTRAL ELEMENTS, often used in a narrower sense to label elements which are discontinuous at inter-element boundaries.

h-p FINITE ELEMENT: See SPECTRAL ELEMENT.

IMBRICATE SERIES: An infinite series for a *spatially periodic* function which is the superposition of an infinite number of evenly spaced identical copies of a "pattern" function $A(x)$. It may be shown that all periodic functions have imbricate series in addition to their Fourier expansions, and often the imbricate series converge faster. Imbricate series may be generalized to an arbitrary number of dimensions.

IMPLICIT (TIME-MARCHING): A time-integration scheme in which it is necessary to solve a boundary-value problem to compute the solution at the new time level. The Crank-Nicholson method is an example. (Chap. 12.)

IMPLICITLY-IMPLICIT: A label for a time-dependent partial differential equation in which the time derivative is multiplied by a differential operator L so that it is necessary to solve a boundary value problem at every time step even to apply an explicit time-marching scheme. (Chap. 9.)

INFINITE ORDER CONVERGENCE: A spectral series possesses the property of "infinite order convergence" if the error decreases faster than any finite inverse power of N as N, the number of terms in the truncated series, increases. A synonym for "exponential convergence".

INTERPOLANT: An approximation $P_{N-1}(x)$ whose free parameters or coefficients are chosen by the requirement that

$$f(x_i) = P_{N-1}(x_i) \qquad\qquad i = 1, \ldots, N$$

at a set of N grid points. The process of computing such an approximation is INTERPOLATION.

ISOLA: For a nonlinear equation that depends upon two parameters, α and τ, an "isola" is a one-parameter family of solutions, parameterized by some quantity s, such that $(\alpha(s), \tau(s))$ is a simple closed curve in the $\alpha - \tau$ plane. (From Latin "insula", "island", through Italian "isolato", "isolate".)

JURY PROBLEM: Numerical slang for a boundary value problem in one or more dimensions. The reason for the name is that when discretized by any reasonable difference or spectral scheme, the solution must be computed on the entire grid *simultaneously*. All the boundary and interior values render a "verdict" on the solution at each point. See also MARCHING PROBLEM.

LANCZOS-TAU METHOD: Technique for solving differential equations which computes an *exact* solution to an *approximate* differential equation. (The differential equation is perturbed in such a way that an exact solution of the perturbed problem is possible.) This philosophy is opposite to the usual strategy of computing an approximate solution to the original, exact differential equation. (Chap. 21.)

LIMIT POINT: A point where a solution $u(\alpha)$ of a nonlinear equation curves back so that there are two solutions for α on one side of $\alpha = \alpha_{\text{limit}}$ and no solutions on the other side of the limit point. As the limit point is approached, $du/d\alpha \to \infty$. Special methods ["pseudoarclength continuation" or "globally convergent homotopy"] are needed to "turn the corner" and march from the lower branch through the limit point onto the upper branch or vice versa. Synonyms are "FOLD POINT", "TURNING POINT", and "SADDLE-NODE BIFURCATION". (Appendix D.)

MARCHING PROBLEM: Arithmurgical slang for a time-dependent problem. It is not necessary to compute the solution through the entire space-time plane at once. Instead, one can march from one time level to the next, computing over all space at fixed t independently of the unknown future. See also JURY PROBLEM.

METHOD OF SELECTED POINTS: Synonym for PSEUDOSPECTRAL & ORTHOGONAL COLLOCATION. (Now rare.)

MMT: MATRIX MULTIPLICATION TRANSFORM: This is N-point interpolation using N-point Gaussian quadrature or summation of an N-point spectral series at each of the N grid points. In either direction, the transform is computed by multiplying an N-dimensional column vector by an N-dimensional square matrix. (PMMT is the same except that the transform is split into two multiplications by $(N/2)$-dimensional square matrices to exploit parity. (Chap. 10.)

NATURAL BOUNDARY CONDITION: A boundary condition is "natural" if it need not be explicitly imposed, but rather is implicitly satisfied by the choice of basis functions. Examples include periodicity (implicit in Fourier basis) and analyticity at an endpoint where the coefficients of the differential equation are singular (implicit in Chebyshev basis).

NON-INTERPOLATING: (i) Adjective used to label that family of spectral methods which do not employ an auxiliary interpolation grid to compute the series coefficients. Galerkin's method and the Lanczos-tau method are examples (ii) Label for a semi-Lagrangian time-marching scheme invented by Ritchie (Chap. 14.)

NUMERICAL BOUNDARY CONDITION: A constraint such as $u(-1) = -0.5$ which involves a number. It is always necessary to modify either the basis set or the Galerkin or pseudospectral matrix to enforce such conditions.

ORTHOGONAL COLLOCATION: A collocation method in which the grid points are the roots or extrema of the basis functions. Synonym for PSEUDOSPECTRAL.

PARITY: A symmetry of functions such that either $f(x) = f(-x)$ for all x ("SYMMETRIC with respect to the origin") or $f(x) = -f(-x)$ for all x ("ANTISYMMETRIC with respect to the origin". A symmetric function is said to be of "even parity" while an antisymmetric function is said to be of "odd parity".

PERIODIC: A function $f(x)$ is "periodic" with period L if and only if

$$f(x + L) = f(x)$$

for all x.

PMMT: PARITY MATRIX MULTIPLICATION: Same as the MMT except parity is exploited by taking sums and differences of the input vector to compute its symmetric and antisymmetric parts and each is transformed separately by a matrix-vector multiplication. The two transforms are combined to give the transform of the original unsymmetric series or set of grid point values. (Chap. 10.)

PRECONDITIONING: A technique for accelerating the convergence of an iteration for solving $\Lambda x - f$ by changing the iteration matrix to $H^{-1}\Lambda$. The matrix H is the "preconditioning matrix" and is chosen to be an approximation to Λ (in the sense of having *approximately* the same eigenvalues), but is much less expensive to invert than Λ. (Chap. 15.)

PSEUDOSPECTRAL: an algorithm which uses an interpolation grid to determine the coefficients of a spectral series. Synonyms are ORTHOGONAL COLLOCATION & METHOD OF SELECTED POINTS.

RESIDUAL FUNCTION: When an approximate solution u_N is substituted into a differential, integral, or matrix equation, the result is the RESIDUAL function, usually denoted $R(x; a_0, a_1, \ldots, a_N)$. The residual function would be identically zero if the approximate solution were exact. (Chap. 1.)

RULE OF THREE NAMES: Every term in this glossary has at least two synonyms.

SEMI-IMPLICIT (TIME-MARCHING): a time-integration method that treats some terms implicitly and others explicitly. Such algorithms are very common in hydrodynamics where diffusion is treated implicitly but the nonlinear advection is explicit; this avoids inverting large systems of *nonlinear* algebraic equations at each time step.

SEMI-LAGRANGIAN: A family of time-marching algorithms for fluid mechanics which use the method of characteristics to integrate the advection terms while employing conventional strategies for the rest. Excellent shock resolution; now widely employed in numerical weather forecasting. (Chap. 14.)

SEPARABLE: An adjective applied to a partial differential (PDE) equation which denotes that the PDE can be split into a sequence of ordinary differential equations including Sturm-Liouville eigenproblems. The procedure for reducing a PDE to ODEs is the METHOD of SEPARATION of VARIABLES.

SKEW-SYMMETRIC: A matrix $\vec{A}$ is SKEW-SYMMETRIC if and only if its tranpose is its negative, that is, $\vec{A}^T = -\vec{A}$. All eigenvalues of a skew-symmetric matrix are pure imaginary or zero. Similarly, an operator is skew-symmetric if its adjoint is its negative; it, too, has pure imaginary eigenvalues.

SKYLINE SOLVER: An algorithm for the LU FACTORIZATION of a SPARSE matrix which omits nonzero operations. It is a generalization of a BANDED matrix solver because the "skyline" allows for different numbers of nonzero elements in different rows (Appendix B).

SPARSE MATRIX: A matrix whose elements are mostly zero. The opposite of a DENSE or FULL MATRIX.

SPECTRAL: (i) A catch-all term for all methods (including pseudospectral techiques) which expand the unknown as a series of global, infinitely differentiable expansion functions. (ii) In a narrower sense, it denotes those algorithms that use only the expansion functions and their integrals and recurrence relations (as opposed to pseudospectral methods, which need the values of the functions on a grid of interpolation points). [Definition (ii) is not used in this book.]

SPECTRAL BLOCKING: The (spurious) accumulation of energy at or near the smallest resolved scales. Common in nonlinear hydrodynamics; also arises when the CFL limit is violated only for the shortest resolvable scales. The signature of spectral blocking is that the $\log |a_n|$ rises instead of falls with n near $n = N$, the truncation limit. The usual remedy is to add a little dissipation which is highly scale-selective, damping only wavenumbers or coefficients close to the truncation limit. (Chap. 11.)

SPECTRAL ELEMENT METHOD: A technique in which the computational domain is subdivided into several subintervals or blocks and then separate spectral series are employed on each interval. It differs from the SUBDOMAIN method only by employing a variational principle and integration-by-parts to more closely parallel finite element algorithms.

SPECTRAL RINGING: Spurious, small scale oscillations which appear when a function with a discontinuity or discontinuous slope is approximated by a truncated spectral series; synonym for GIBBS' OSCILLATIONS.

SPONGE LAYER: To absorb waves near the ends of a finite interval $y \in [-L, L]$ when domain truncation is used to simulate an infinite spatial interval, a large artificial viscosity is often added in the neighborhood of $y = \pm L$. The regions of large damping are called "sponge layers" because they absorb the waves so that the center of the domain is uncorrupted by spurious waves reflected from $y = \pm L$.

STATIC CONDENSATION: A direct method for solving the matrix equations which result from finite element or spectral element discretizations. As a first step, the degrees of freedom internal to each element are eliminated in favor of the grid point values on subdomain walls. This generates a dense matrix of much smaller size than the original sparse matrix, which is then solved by standard LU factorization.

SUBDOMAIN SPECTRAL METHOD: Instead of solving the problem by using a *single* series which is valid over the whole domain, the computational region is split into several smaller regions or "subdomains". A separate Chebyshev or Legendre expansion is then defined on each subdomain, and the expansions are matched together at the subdomain boundaries. There are many variants of this strategy, depending upon whether a variational principle is or is not used and also upon how the algorithm enforces continuity from one subdomain to the next. See also SPECTRAL ELEMENTS. (Chap. 22.)

SUBGEOMETRIC CONVERGENCE: There are three equivalent definitions. (i) The series converges exponentially fast with n, but too slowly for the coefficients to be bounded in magnitude by $c \exp(-pn)$ for any positive constants c and p. (ii) The index of exponential convergence $r < 1$. [$r = 1$ is "geometric convergence".] (iii)

$$\lim_{n \to \infty} \log(|a_n|)/n = 0$$

Subgeometrically-converging series are the expansions of functions which are singular but infinitely differentiable at some point (or points) on the expansion interval (often at infinity). In mathematical jargon, such weakly singular functions "are in C^∞, but not C^Ω."

SUPERGEOMETRIC CONVERGENCE: If the coefficients can be bounded by

$$|a_n| \le O\{s \exp[-(n/j) \log n]\}$$

for sufficiently large n and some constants s and j, then the convergence rate is "supergeometric". This is possible only when the function being expanded is an "entire" function, that is, one which has no singularities anywhere in the complex plane except at ∞.

SUPERIOR LIMIT: For a sequence $\{a_n\}$, the superior limit or supremum limit is written $\lim \sup\{a_n\}$ and denotes the lower bound of the almost upper bounds of the sequence. (A number is an almost upper bound for a sequence if only a finite number of members of the sequence exceed the "almost upper bound" in value.) Strictly speaking, definitions of convergence rates should be expressed in terms of superior limits, rather than ordinary limits, to allow for oscillations and zeros in the sequence as $n \to \infty$. A synonym is "supremum limit".

SYMMETRIC: (i) [Of a matrix]: $A_{ij} = A_{ji}$. (Appendix B.) The longer but more precise term "CENTROSYMMETRIC" is sometimes used as a synonym. (ii) [Of a function $f(x)$]: $f(x) = f(-x)$ for all x. (Chap. 8.)

TAU-METHOD: A mean weighted residual method for which the weighting functions are the Chebyshev polynomials. Also called the LANCZOS TAU-METHOD. (Chapt. 21.)

TENSOR PRODUCT BASIS: A multi-dimensional basis whose elements are the products of one-dimensional basis elements. In two dimensions,

$$\Phi_{mn}(x, y) \equiv \phi_m(x) \, \psi_n(y)$$

TENSOR PRODUCT GRID: A multi-dimensional grid whose MN points are formed from the corresponding one-dimensional grids:

$$\underline{x}_{ij} \equiv (x_i, y_j) \qquad i = 1, \dots, M \quad \& \quad j = 1, \dots, N$$

TRUNCATION ERROR: The error made by neglecting all coefficients a_n in the spectral series such that $n > N$ for some "truncation" N.

WEAKLY NONLOCAL SOLITARY WAVE: A steadily-translating, finite amplitude wave that decays to a small amplitude oscillation (rather to zero) as one moves away from the core of the disturbance.

Index

Bibliography

Abadi, M. H. A. and Ortiz, E. L.: 1991, A tau method based on non-uniform space-time elements for the numerical simulation of solitons, *Computers Math. Applic.* **22**(9), 7–19.

Abarbanel, S. and Gottlieb, D.: 1985, Information content in spectral calculations, *in* E. M. Murman and S. S. Abarbanel (eds), *Progress and Supercomputing in Computational Fluid Dynamics*, number 6 in *Proc. U. S.-Israel Workshop*, Birkhauser, Boston, pp. 345–356.

Abarbanel, S., Gottlieb, D. and Carpenter, M. H.: 1996, On the removal of boundary errors caused by Runge-Kutta integration of nonlinear partial differential equations, *SIAM Journal of Scientific Computing* **17**(3), 777–782. Not spectral, but essential in optimizing Runge-Kutta methods with Chebyshev or other high order spatial discretizations.

Abarbanel, S., Gottlieb, D. and Tadmor, E.: 1986, Spectral methods for discontinuous problems, *in* K. W. Morton and M. J. Baines (eds), *Numerical Methods for Fluid Dynamics, II*, Oxford University Press, Oxford, pp. 129–153.

Abe, K. and Inoue, O.: 1980, Fourier expansion solution of the Korteweg-deVries equation, *Journal of Computational Physics* **34**, 202–210.

Abramowitz, M. and Stegun, I. A.: 1965, *Handbook of Mathematical Functions*, Dover, New York.

Abril-Raymundo, M. R. and García-Archilla, B.: 2000, Approximation properties of a mapped Chebyshev method, *Applied Numerical Mathematics* **32**(2), 119–136. Kosloff/Tal-Ezer mapping is analyzed through both theory and numerical experiments; it proves to have a poor *asymptotic* rate of convergence, but only for very large N.

Alpert, B. A. and Rokhlin, V.: 1991, A fast algorithm for the evaluation of Legendre expansions, *SIAM Journal of Scientific Computing* **12**, 158–179. Provides a fast Legendre-to-Chebyshev transform.

Amon, C. H. and Patera, A. T.: 1989, Numerical calculation of stable three-dimensional tertiary states in grooved-channel flow, *Physics of Fluids A* **1**, 2005–2009.

Anagnostou, G., Maday, Y. and Patera, A. T.: 1991, Numerical simulation of incompressible fluid flows, *Concurrency-Practice and Experience* **3**(6), 667–685. Review: high order operator-integration factors methods, sliding mesh spectral mortar-element, and parallel computing.

Anderson, C. and Dahleh, M. D.: 1996, Rapid computation of the discrete Fourier transform, *SIAM Journal of Scientific Computing* **17**, 913–919. Off-grid interpolation to irregularly spaced points by using local Taylor expansions at a cost of about 8 FFTs.

Anderson, D. L. T.: 1973, An ocean model using parabolic cylinder functions, World Meteorological Organization, Geneva. Model of equatorial ocean where Hermite functions are the linearized normal modes; discontinued.

Aoyagi, A.: 1995, Nonlinear leapfrog instability for Fornberg's pattern, *Journal of Computational Physics* **120**, 316–322. Aliasing instability.

Arakawa, A.: 1966, Computational design for long-term numerical integration of the equations of fluid motion: Two dimensional incompressible flow, *Journal of Computational Physics* **1**, 119–143. Energy-conserving finite differences schemes.

Armstrong, T. P.: 1967, Numerical studies of nonlinear Vlasov equation, *Physics of Fluids* **10**, 1269–1280. Fourier-Galerkin in the space dimension, Hermite-Galerkin in the velocity coordinate. Slow convergence of the Hermite series, a difficulty not fully overcome until variable scaling introduced in Holloway, 1996a,b.

Arnold, V. I.: 1983, *Geometrical Methods in the Theory of Ordinary Differential Equations*, Springer-Verlag, New York, chapter 17, pp. 110–111. The algorithm of "counting the noses" to approximate irrational numbers by rational fractions.

Augenbaum, J. M.: 1989, An adaptive pseudospectral method for discontinuous problems, *Applied Numerical Mathematics* **5**, 459–480.

Augenbaum, J. M.: 1990, Multidomain adaptive pseudospectral methods for acoustic wave propagation in discontinuous media, *in* D. Lee, A. Cakmak and R. Vichnevetsky (eds), *Computational Acoustics-Seismo-Ocean Acoustics and Modeling*, number 3 in *Computational Acoustics*, Elsevier/North-Holland, Amsterdam, pp. 19–40.

Augenbaum, J. M.: 1992, The pseudospectral method for limited-area elastic wave calculations, *in* W. E. Fitzgibbon and M. F. Wheeler (eds), *Computational Methods in Geosciences*, Frontiers in Applied Mathematics, SIAM, Philadelphia. Chebyshev method needs only about half the resolution in each coordinate compared to fourth order differences.

Averbuch, A., Ioffe, L., Israeli, M. and Vozovoi, L.: 1998a, Two-dimensional parallel solver for the solution of navier-stokes equations with constant and variable coefficients using ADI on cells, *Parallel Comput.* **24**(5–6), 673–699. Local Fourier basis for nonperiodic problems, here with domain decomposition.

Averbuch, A., Israeli, M. and Vozovoi, L.: 1995, Parallel implementation of nonlinear evolution problems using parabolic domain decomposition, *Parallel Comput.* **21**(7), 1151–1183. Fourier method for nonperiodic boundary conditions; multidomain.

Averbuch, A., Israeli, M. and Vozovoi, L.: 1998b, A fast poisson solver of arbitrary order accuracy in rectangular regions, *SIAM J. Sci. Comput.* **19**(3), 933–952. Fourier method for nonperiodic boundary conditions with subtractions to handle Gibbs' Phenomenon and corner singularities.

Averbuch, A., Vozovoi, L. and Israeli, M.: 1997, On a fast direct elliptic solver by a modified fourier method, *Numer. Algorithms* **15**(3–4), 287–313. Fourier method for nonperiodic boundary conditions with subtractions to handle Gibbs' Phenomenon and corner singularities.

Babolian, E. and Delves, L. M.: 1979, An augmented Galerkin method for first kind Fredholm equations, *Journal of the Institute for Mathematics and its Applications* **24**, 151–174. Galerkin basis is Chebyshev polynomials.

Babolian, E. and Delves, L. M.: 1981, A fast Galerkin method for linear integro-differential equations, *IMA Journal of Numerical Analysis* **1**, 193–213. Chebyshev series.

Baer, F.: 1977, Adjustment of initial conditions required to suppress gravity oscillations in non-linear flows, *Contributions to Atmospheric Physics* **50**, 350–366.

Baer, F. and Tribbia, J. J.: 1977, On complete filtering of gravity modes through nonlinear initialization, *Monthly Weather Review* **105**, 1536–1539.

Bain, M.: 1978, Convergence theorems for Hermite functions and Jacobi polynomials, *Journal of the Institute for Mathematics and its Applications* **21**, 379–386.

Baker, Jr., G. A.: 1975, *Essentials of Padé Approximants*, Academic Press, New York. 220 pp.

Baker, M. D., Süli, E. and Ware, A. F.: 1992, Stability and convergence of the spectral Lagrange-Galerkin method for periodic and non-periodic convection-dominated diffusion problems, *Technical Report 92/19*, Oxford University Computing Laboratory, Numerical Analysis Group,Wolfson Bldg., Parks Rd., Oxford, England OX1 3QD.

Banerjee, K.: 1978, Hermite function solution of quantum anharmonic oscillator, *Proceedings Of The Royal Society of London, Series A* **364**, 264.

Banerjee, K., Bhatnagar, S. P., Choudhury, V. and Kanwal, S. S.: 1978, *Proceedings of the Royal Society of London A*. Hermite functions $\psi_n(\alpha y)$ with empirically-optimized scale factor α.

Bank, R. E. and Chan, T. F.: 1986, PLTMGC: A multi-grid continuation program for parameterized nonlinear elliptic systems, *SIAM Journal of Scientific and Statistical Computing* **7**, 540–559.

Barrett, E. W.: 1958, Eccentric circumpolar vortices in a barotropic atmosphere, *Tellus* **10**, 395–400. Spherical harmonics.

Basdevant, C., Deville, M., Haldenvvang, P., Lacroix, J., Orlandi, D., Patera, A., Peyret, R. and Quazzani, J.: 1986, Spectral and finite difference solutions of Burgers' equation, *Comput. Fluids* **14**, 23–41.

Bates, J. R.: 1984, An efficient semi-Lagrangian and alternating-direction implicit method for integrating the shallow-water equations, *Monthly Weather Review* **112**, 2033–2047.

Bates, J. R. and McDonald, A.: 1982, Multiply-upstream, semi-Lagrangian advective schemes: Analysis and application to a multilevel primitive equation model, *Monthly Weather Review* **110**, 1831–1842.

Bates, J. R., Semazzi, F. H. M., Higgins, R. W. and Barros, S. R. M.: 1990, Integration of the shallow-water equations on the sphere using a vector semi-Lagrangian scheme with a multigrid solver, *Monthly Weather Review* **118**, 1615–1627.

Baumgardner, J. R. and Frederickson, P. O.: 1985, Icosahedral discretization of the two-sphere, *SIAM Journal of Numerical Analysis* **22**, 1107–1115.

Bayliss, A. and Matkowsky, B.: 1987, Fronts, relaxation oscillations, and period-doubling in solid fuel combustion, *Journal of Computational Physics* **71**, 147–168. Adaptive pseudospectral grid.

Bayliss, A. and Matkowsky, B. J.: 1992, Nonlinear dynamics of cellular flames, *SIAM Journal of Applied Mathematics* **52**(2), 396–415. Chebyshev and Fourier pseudospectral combined with adaptive shifted arctan map and first order splitting time integration.

Bayliss, A. and Turkel, E.: 1992, Mappings and accuracy for Chebyshev pseudospectral approximations, *Journal of Computational Physics* **101**, 349–359.

Bayliss, A., Belytschko, T., Kulkarni, M. and Lott-Crumpler, D. A.: 1994, On the dynamics and role of imperfections for localization in thermo-viscoplastic materials, *Modelling and Simulation of Materials in Science and Engineering* **2**, 941–964. Adaptive Chebyshev pseudospectral method to compute adiabatic shear bands via time-marching with shifted arctan mapping.

Bayliss, A., Garbey, M. and Matkowsky, B. J.: 1995, Adaptive pseudo-spectral domain decomposition and the approximation of multiple layers, *Journal of Computational Physics* **119**, 132–141.

Bayliss, A., Gottlieb, D., Matkowsky, B. J. and Minkoff, M.: 1989, An adaptive pseudospectral method for reaction-diffusion problems, *Journal of Computational Physics* **81**, 421–443.

Bayliss, A., Kuske, R. and Matkowsky, B. J.: 1990, A two-dimensional adaptive pseudospectral method, *Journal of Computational Physics* **91**, 174–196. Chebyshev polynomials and one-dimensional arctan/tan mapping with map parameters that are allowed to vary with the other spatial coordinate.

Belousov, S. L.: 1962, *Tables of Normalized Associate Legendre Polynomials*, Macmillan, New York.

Ben Belgacem, F. and Maday, Y.: 1997, The mortar element method for three dimensional finite elements, *Rairo-Math. Model Numer.* **31**(2), 289–302.

Bender, C. M. and Orszag, S. A.: 1978, *Advanced Mathematical Methods for Scientists and Engineers*, McGraw-Hill, New York. 594 pp.

Bermejo, R.: 1990, On the equivalence of semi-Lagrangian and particle-in-cell finite-element methods, *Monthly Weather Review* **118**, 979–987. Not spectral.

Bermejo, R. and Staniforth, A.: 1992, The conversion of semi-Lagrangian advection schemes to quasi-monotone schemes, *Monthly Weather Review* **120**, 2622–2632.

Bernardi, C. and Maday, Y.: 1992, *Approximations spectrales de problémes aux limites elliptiques*, Springer-Verlag France, Paris, France. 250 pp. In French; highly mathematical.

Bert, C. W. and Malik, M.: 1996, Differential quadrature method in computational mechanics: A review, *Applied Mechanics Reviews* **49**, 1–28. Pseudospectral method implemented as N-th order finite difference scheme.

Bialecki, B.: 1989, Sinc-type approximations in h^1 norm with application to boundary value problems, *Journal of Computational and Applied Mathematics* **25**, 289–303.

Bialecki, B.: 1991, Sinc-collocation methods for two-point boundary value problems, *IMA Journal of Numerical Analysis* **11**, 357–375.

Birkhoff, G. and Fix, G.: 1970, Variational methods for eigenvalue problems, *Numerical Solution of Field Problems in Continuum Mechanics*, Vol. 2, SIAM-AMS Proceedings of the American Mathematical Society, Providence, pp. 111–151. Hermite function and Fourier solution of eigenvalue problems.

Birkhoff, G. and Lynch, R. E.: 1984, *Numerical Solution of Elliptic Problems*, Society for Industrial and Applied Mathematics, Philadelphia. 300 pp.

Black, K.: 1998, Spectral elements on infinite domains, *SIAM J. Sci. Comput.* **19**(5), 1667–1681. Employs the change-of-coordinate of Boyd(1987b).

Blackburn, H. and Henderson, R. D.: 1996, Lock-in behavior in simulation vortex-induced simulation, *Experimental Thermal and Fluid Science* **12**, 184–189. Spectral element model of vibrations of a circular cylinder, induced by vibrations due to vortices in the surrounding fluid, using stiffly stable (Backward Differentiation) time integration. The grid is fixed to the (moving) cylinder and equations solved in this accelerating reference frame. ODEs for the motion of the cylinder are solved simultaneously with the hydrodynamic PDEs.

Blaisdell, G. A., T.Spyropoulos, E. and Qin, J. H.: 1996, The effect of the formulation of nonlinear terms on aliasing errors in spectral methods, *Applied Numerical Mathematics* **21**, 207–219. Show through both Burgers' equation and turbulence models that the skew-symmetric form of advection is more robust than the alternatives.

Blue, J. L. and Gummel, H. K.: 1970, Rational approximations to matrix exponential for systems of stiff differential equations, *Journal of Computational Physics* **5**, 70–83. Not spectral.

Boer, G. J., McFarlane, N. A., Laprise, R., Henderson, J. D. and Blanchet, J.-P.: 1984, The Canadian Climate Centre spectral atmospheric general circulation model, *Atmosphere-Ocean* **22**, 397–429.

Boffi, D. and Funaro, D.: 1994, An alternative approach to the analysis and the approximation of the Navier-Stokes equations, *Journal of Scientific Computing* **9**(1), 16.

Boomkamp, P. A. M., Boersma, B. J., Miesen, R. H. M. and Beijnon, G. V.: 1997, A Chebyshev collocation method for solving two-phase flow stability problems, *J. Comput. Phys.* **132**, 191–200. Collocation/QZ algorithm for eigenvalue problems, using a three-subdomain strategy.

Böttcher, M.: 1996, A semi-Lagrangian advection scheme with modified exponential splines, *Monthly Weather Review* **124**, 716. Not spectral.

Bouaoudi and Marcus, P. S.: 1991, Fast and accurate spectral treatment of coordinate singularities, *Journal of Computational Physics* **96**, 217–223. Polar and spherical coordinates using Robert functions. In later work, the authors abandoned this approach because this basis is highly ill-conditioned.

Boukir, K., Maday, Y. and Métivet, B.: 1997, A high order characteristics method for the incompressible Navier-Stokes equations, *Internat. J. Numer. Meth. Fluids* **25**(12), 1421–1454.

Bourke, W.: 1972, An efficient, one-level primitive-equation spectral model, *Monthly Weather Review* **100**, 683–689. Spherical harmonics.

Bourke, W.: 1974, A multi-level spectral model. I. Formulation and hemispheric integrations, *Monthly Weather Review* **102**, 687–701.

Bourke, W., McAvaney, K., Puri, K. and Thurling, R.: 1977, *Global modelling of atmospheric flow by spectral methods*, Vol. 17 of *Methods in Computational Physics*, Academic Press, New York, pp. 267–334.

Bouteloup, Y.: 1995, Improvement of the spectral representation of the earth topography, *Mon. Wea. Rev.* **123**, 1560–1573.

Bowers, K. L. and Lund, J.: 1987, Numerical solution of singular Poisson equations by the sinc-Galerkin method, *SIAM Journal of Numerical Analysis* **24**, 36–51.

Boyd, J. P.: 1978a, A Chebyshev polynomial method for computing analytic solutions to eigenvalue problems with application to the anharmonic oscillator, *Journal of Mathematical Physics* **19**, 1445–1456.

Boyd, J. P.: 1978b, The choice of spectral functions on a sphere for boundary and eigenvalue problems: A comparison of Chebyshev, Fourier and associated Legendre expansions, *Monthly Weather Review* **106**, 1184–1191.

Boyd, J. P.: 1978c, Spectral and pseudospectral methods for eigenvalue and nonseparable boundary value problems, *Monthly Weather Review* **106**, 1192–1203.

Boyd, J. P.: 1980a, The nonlinear equatorial Kelvin wave, *Journal of Physical Oceanography* **10**, 1–11.

Boyd, J. P.: 1980b, The rate of convergence of Hermite function series, *Mathematics of Computation* **35**, 1309–1316.

Boyd, J. P.: 1980c, Equatorial solitary waves, Part I: Rossby solitons, *Journal of Physical Oceanography* **10**, 1699–1718.

Boyd, J. P.: 1981a, A Sturm-Liouville eigenproblem with an interior pole, *Journal of Physical Oceanography* **22**, 1575–1590. Background on waves with critical points; nothing on spectral methods.

Boyd, J. P.: 1981b, The rate of convergence of Chebyshev polynomials for functions which have asymptotic power series about one endpoint, *Journal of Physical Oceanography* **37**, 189–196.

Boyd, J. P.: 1981c, The Moonbow, *Isaac Asimov's SF Mag.* **5**, 18–37. Properties of a toroidal world.

Boyd, J. P.: 1982a, The optimization of convergence for Chebyshev polynomial methods in an unbounded domain, *Journal of Computational Physics* **45**, 43–79. Infinite and semi-infinite intervals; guidelines for choosing the map parameter or domain size L.

Boyd, J. P.: 1982b, The effects of meridional shear on planetary waves, Part I: Nonsingular profiles, *Journal of the Atmospheric Sciences* **39**, 756–769.

Boyd, J. P.: 1982c, The effects of meridional shear on planetary waves, Part II: Critical latitudes, *Journal of the Atmospheric Sciences* **39**, 770–790. First application of cubic-plus-linear mapping with spectral methods. The detour procedure of Boyd (1985a) is better in this context.

Boyd, J. P.: 1982d, A Chebyshev polynomial rate-of-convergence theorem for Stieltjes functions, *Mathematics of Computation* 39, 201–206.

Boyd, J. P.: 1982e, Theta functions, Gaussian series, and spatially periodic solutions of the Korteweg-de Vries equation, *Journal of Mathematical Physics* 23, 375–387.

Boyd, J. P.: 1983a, Equatorial solitary waves, Part II: Envelope solitons, *Journal of Physical Oceanography* 13, 428–449. This and the next two papers use Hermite series to solve linear, separable equations in perturbation theory for nonlinear waves.

Boyd, J. P.: 1983b, Long wave/short wave resonance in equatorial waves, *Journal of Physical Oceanography* 13, 450–458.

Boyd, J. P.: 1983c, Second harmonic resonance for equatorial waves, *Journal of Physical Oceanography* 13, 459–466.

Boyd, J. P.: 1983d, The continuous spectrum of linear Couette flow with the beta effect, *Journal of the Atmospheric Sciences* 40, 2304–2308.

Boyd, J. P.: 1984a, The asymptotic coefficients of Hermite series, *Journal of Computational Physics* 54, 382–410.

Boyd, J. P.: 1984b, Equatorial solitary waves, Part IV: Kelvin solitons in a shear flow, *Dynamics of Atmospheres and Oceans* 8, 173–184.

Boyd, J. P.: 1984c, Cnoidal waves as exact sums of repeated solitary waves: New series for elliptic functions, *SIAM Journal of Applied Mathematics* 44, 952–955. Imbricate series for nonlinear waves.

Boyd, J. P.: 1984d, The double cnoidal wave of the Korteweg-de Vries equation: An overview, *Journal of the Mathematical Physics* 25, 3390–3401.

Boyd, J. P.: 1984e, Perturbation theory for the double cnoidal wave of the Korteweg-de Vries equation, *Journal of the Mathematical Physics* 25, 3402–3414.

Boyd, J. P.: 1984f, The special modular transformation for the polycnoidal waves of the Korteweg-de Vries equation, *Journal of the Mathematical Physics* 25, 3390–3401.

Boyd, J. P.: 1984g, Earthflight, *in* S. Shwarz (ed.), *Habitats*, DAW, New York, pp. 201–218. Toroidal planet.

Boyd, J. P.: 1985a, Complex coordinate methods for hydrodynamic instabilities and Sturm-Liouville problems with an interior singularity, *Journal of Computational Physics* 57, 454–471.

Boyd, J. P.: 1985b, Equatorial solitary waves, Part 3: Modons, *Journal of Physical Oceanography* 15, 46–54.

Boyd, J. P.: 1985c, An analytical and numerical study of the two-dimensional Bratu equation, *Journal of Scientific Computing* 1, 183–206. Nonlinear eigenvalue problem with 8-fold symmetry.

Boyd, J. P.: 1985d, Barotropic equatorial waves: The non-uniformity of the equatorial beta-plane, *Journal of the Atmospheric Sciences* 42, 1965–1967.

Boyd, J. P.: 1986a, Solitons from sine waves: analytical and numerical methods for non-integrable solitary and cnoidal waves, *Physica D* 21, 227–246. Fourier pseudospectral with continuation and the Newton-Kantorovich iteration.

Boyd, J. P.: 1986b, Polynomial series versus sinc expansions for functions with corner or endpoint singularities, *Journal of Computational Physics* **64**, 266–269.

Boyd, J. P.: 1987a, Exponentially convergent Fourier/Chebyshev quadrature schemes on bounded and infinite intervals, *Journal of Scientific Computing* **2**, 99–109.

Boyd, J. P.: 1987b, Spectral methods using rational basis functions on an infinite interval, *Journal of Computational Physics* **69**, 112–142.

Boyd, J. P.: 1987c, Orthogonal rational functions on a semi-infinite interval, *Journal of Computational Physics* **70**, 63–88.

Boyd, J. P.: 1987d, Generalized solitary and cnoidal waves, *in* G. Brantstator, J. J. Tribbia and R. Madden (eds), *NCAR Colloquium on Low Frequency Variability in the Atmosphere*, National Center for Atmospheric Research, Boulder, Colorado, pp. 717–722. Numerical calculation of the exponentially small wings of the ϕ^4 breather.

Boyd, J. P.: 1988a, Chebyshev domain truncation is inferior to Fourier domain truncation for solving problems on an infinite interval, *Journal of Scientific Computing* **3**, 109–120.

Boyd, J. P.: 1988b, An analytical solution for a nonlinear differential equation with logarithmic decay, *Advances in Applied Mathematics* **9**, 358–363.

Boyd, J. P.: 1988c, The superiority of Fourier domain truncation to Chebyshev domain truncation for solving problems on an infinite interval, *Journal of Scientific Computing* **3**, 109–120.

Boyd, J. P.: 1989a, *Chebyshev and Fourier Spectral Methods*, Springer-Verlag, New York. 792 pp.

Boyd, J. P.: 1989b, New directions in solitons and nonlinear periodic waves: Polycnoidal waves, imbricated solitons, weakly non-local solitary waves and numerical boundary value algorithms, *in* T.-Y. Wu and J. W. Hutchinson (eds), *Advances in Applied Mechanics*, number 27 in *Advances in Applied Mechanics*, Academic Press, New York, pp. 1–82.

Boyd, J. P.: 1989c, Periodic solutions generated by superposition of solitary waves for the quarticly nonlinear Korteweg-de Vries equation, *ZAMP* **40**, 940–944. Imbrication of solitary wave generates good *approximate* periodic solutions.

Boyd, J. P.: 1989d, The asymptotic Chebyshev coefficients for functions with logarithmic endpoint singularities, *Applied Mathematics and Computation* **29**, 49–67.

Boyd, J. P.: 1989e, Non-local equatorial solitary waves, *in* J. C. J. Nihoul and B. M. Jamart (eds), *Mesoscale/Synoptic Coherent Structures in Geophysical Turbulence: Proc. 20th Liege Coll. on Hydrodynamics*, Elsevier, Amsterdam, pp. 103–112.

Boyd, J. P.: 1990a, The orthogonal rational functions of Higgins and Christov and Chebyshev polynomials, *Journal of Approximation Theory* **61**, 98–103.

Boyd, J. P.: 1990b, A numerical calculation of a weakly non-local solitary wave: the ϕ^4 breather, *Nonlinearity* **3**, 177–195.

Boyd, J. P.: 1990c, The envelope of the error for Chebyshev and Fourier interpolation, *Journal of Scientific Computing* **5**, 311–363.

Boyd, J. P.: 1990d, A Chebyshev/radiation function pseudospectral method for wave scattering, *Computers in Physics* **4**, 83–85.

Boyd, J. P.: 1991a, A comparison of numerical and analytical methods for the reduced wave equation with multiple spatial scales, *Applied Numerical Mathematics* **7**, 453–479.

Boyd, J. P.: 1991b, Monopolar and dipolar vortex solitons in two space dimensions, *Wave Motion* **57**, 223–243.

Boyd, J. P.: 1991c, Nonlinear equatorial waves, *in* A. R. Osborne (ed.), *Nonlinear Topics of Ocean Physics: Fermi Summer School, Course LIX*, North-Holland, Amsterdam, pp. 51–97.

Boyd, J. P.: 1991d, Weakly nonlocal solitary waves, *in* A. R. Osborne (ed.), *Nonlinear Topics of Ocean Physics: Fermi Summer School, Course LIX*, North-Holland, Amsterdam, pp. 527–556.

Boyd, J. P.: 1991e, Weakly nonlocal solitons for capillary-gravity waves: Fifth-degree Korteweg-de Vries equation, *Physica D* **48**, 129–146.

Boyd, J. P.: 1991f, Sum-accelerated pseudospectral methods: The Euler-accelerated sinc algorithm, *Applied Numerical Mathematics* **7**, 287–296.

Boyd, J. P.: 1992a, The arctan/tan and Kepler-Burger mappings for periodic solutions with a shock, front, or internal boundary layer, *Journal of Computational Physics* **98**, 181–193. Numerical trick which is useful for solitary waves and cnoidal waves.

Boyd, J. P.: 1992b, The energy spectrum of fronts: The time evolution of shocks in Burgers' equation, *Journal of the Atmospheric Sciences* **49**, 128–139.

Boyd, J. P.: 1992c, Multipole expansions and pseudospectral cardinal functions: A new generalization of the Fast Fourier Transform, *Journal of Computational Physics* **102**, 184–186.

Boyd, J. P.: 1992d, A fast algorithm for Chebyshev and Fourier interpolation onto an irregular grid, *Journal of Computational Physics* **103**, 243–257.

Boyd, J. P.: 1992e, Defeating the Runge phenomenon for equispaced polynomial interpolation via Tikhonov regularization, *Applied Mathematics Letters* **5**, 57–59.

Boyd, J. P.: 1993, Chebyshev and Legendre spectral methods in algebraic manipulation languages, *Journal of Symbolic Computing* **16**, 377–399.

Boyd, J. P.: 1994a, Hyperviscous shock layers and diffusion zones: Monotonicity, spectral viscosity, and pseudospectral methods for high order differential equations, *Journal of Scientific Computing* **9**, 81–106.

Boyd, J. P.: 1994b, The rate of convergence of Fourier coefficients for entire functions of infinite order with application to the Weideman-Cloot sinh-mapping for pseudospectral computations on an infinite interval, *Journal of Computational Physics* **110**, 360–372.

Boyd, J. P.: 1994c, The slow manifold of a five mode model, *Journal of the Atmospheric Sciences* **51**, 1057–1064.

Boyd, J. P.: 1994d, Nonlocal modons on the beta-plane, *Geophysical and Astrophysical Fluid Dynamics* **75**, 163–182.

Boyd, J. P.: 1994e, Time-marching on the slow manifold: The relationship between the nonlinear Galerkin method and implicit timestepping algorithms, *Applied Mathematics Letters* **7**, 95–99.

Boyd, J. P.: 1994f, Sum-accelerated pseudospectral methods: Finite differences and sech-weighted differences, *Computer Methods in Applied Mechanics and Engineering* **116**, 1–11. Typos: in the second line of the theorem, "derivative of x" should be "derivative of u; in (4.3), $\exp(ijk)$ should be $\exp(ijK)$.

Boyd, J. P.: 1995a, Weakly nonlocal envelope solitary waves: Numerical calculations for the Klein-Gordon (ϕ^4) equation, *Wave Motion* **21**, 311–330.

Boyd, J. P.: 1995b, A hyperasymptotic perturbative method for computing the radiation coefficient for weakly nonlocal solitary waves, *Journal of Computational Physics* **120**, 15–32.

Boyd, J. P.: 1995c, Eight definitions of the slow manifold: Seiches, pseudoseiches and exponential smallness, *Dynamics of Atmospheres and Oceans* **22**, 49–75.

Boyd, J. P.: 1995d, A lag-averaged generalization of Euler's method for accelerating series, *Applied Mathematics and Computation* **72**, 146–166.

Boyd, J. P.: 1995e, A Chebyshev polynomial interval-searching method ("Lanczos economization") for solving a nonlinear equation with application to the nonlinear eigenvalue problem, *Journal of Computational Physics* **118**, 1–8.

Boyd, J. P.: 1995f, Multiple precision pseudospectral computations of the radiation coefficient for weakly nonlocal solitary waves: Fifth-Order Korteweg-deVries equation, *Computers in Physics* **9**, 324–334.

Boyd, J. P.: 1996a, Asymptotic Chebyshev coefficients for two functions with very rapidly or very slowly divergent power series about one endpoint, *Applied Mathematics Letters* **9**(2), 11–15.

Boyd, J. P.: 1996b, Traps and snares in eigenvalue calculations with application to pseudospectral computations of ocean tides in a basin bounded by meridians, *Journal of Computational Physics* **126**, 11–20. Corrigendum, **136**, no. 1, 227-228 (1997).

Boyd, J. P.: 1996c, Numerical computations of a nearly singular nonlinear equation: Weakly nonlocal bound states of solitons for the Fifth-Order Korteweg-deVries equation, *Journal of Computational Physics* **124**, 55–70.

Boyd, J. P.: 1996d, The Erfc-Log filter and the asymptotics of the Vandeven and Euler sequence accelerations, *in* A. V. Ilin and L. R. Scott (eds), *Proceedings of the Third International Conference on Spectral and High Order Methods*, Houston Journal of Mathematics, Houston, Texas, pp. 267–276.

Boyd, J. P.: 1997a, Padé approximant algorithm for solving nonlinear ODE boundary value broblems on an unbounded domain, *Computers and Physics* **11**(3), 299–303.

Boyd, J. P.: 1997b, Pseudospectral/Delves-Freeman computations of the radiation coefficient for weakly nonlocal solitary waves of the Third Order Nonlinear Schroedinger Equation and their relation to hyperasymptotic perturbation theory, *Journal of Computational Physics* **138**, 665–694.

Boyd, J. P.: 1997c, The periodic generalization of Camassa-Holm "peakons": An exact superposition of solitary waves, *Applied Mathematics and Computation* **81**(2), 173–187. Classical solitons.

Boyd, J. P.: 1997d, Construction of Lighthill's unitary functions: The imbricate series of unity, *Applied Mathematics and Computation* **86**, 1–10.

Boyd, J. P.: 1998a, Radiative decay of weakly nonlocal solitary waves, *Wave Motion* **27**, 211–221.

Boyd, J. P.: 1998b, *Weakly Nonlocal Solitary Waves and Beyond-All-Orders Asymptotics: Generalized Solitons and Hyperasymptotic Perturbation Theory*, Vol. 442 of *Mathematics and Its Applications*, Kluwer, Amsterdam. 608 pp. Three chapters on spectral methods, solving spectral-discretized nonlinear equations, and special methods for nonlocal solitary waves and other phenomena that radiate to spatial infinity.

Boyd, J. P.: 1998c, High order models for the nonlinear shallow water wave equations on the equatorial beta-plane with application to Kelvin wave frontogenesis, *Dynamics of Atmospheres and Oceans* **28**(2), 69–91.

Boyd, J. P.: 1998d, Two comments on filtering, *J. Comput. Phys.* **143**(1), 283–288. Shows how to apply filters or sum acceleration methods to spectral series so as to preserve the boundary conditions. Also explains why additional boundary conditions are not needed: high order filtering operators can be interpreted as powers of the Legendre or Chebyshev differential operator, which is singular at the boundaries.

Boyd, J. P.: 1998e, Global approximations to the principal real-valued branch of the Lambert W-function, *Appl. Math. Lett.* **11**(6), 27–31.

Boyd, J. P.: 1999a, A numerical comparison of seven grids for polynomial interpolation on the interval, *Comput. Math. Appl.* **38**(3–4), 35–50.

Boyd, J. P.: 1999b, The Blasius function in the complex plane, *J. Experimental Math.* **8**(4), 381–394. Rational Chebyshev expansions TL for the similarity solution to the boundary layer flow over a flat plate.

Boyd, J. P.: 2000, *Chebyshev and Fourier Spectral Methods*, 2d edn, Dover, Mineola, New York. 665 pp. Heavily revised and updated second edition of Boyd(1989).

Boyd, J. P.: 2003, *Essays on Chebyshev and Fourier Spectral Methods*. To appear.

Boyd, J. P. and Christidis, Z. D.: 1982, Low wavenumber instability on the equatorial beta-plane, *Geophysical Research Letters* **9**, 769–772.

Boyd, J. P. and Christidis, Z. D.: 1983, Instability on the equatorial beta-plane, *in* J. Nihoul (ed.), *Hydrodynamics of the Equatorial Ocean*, Elsevier, Amsterdam, pp. 339–351.

Boyd, J. P. and Christidis, Z. D.: 1987, The continuous spectrum of equatorial Rossby waves in a shear flow, *Dynamics of Atmospheres and Oceans* **11**, 139–151.

Boyd, J. P. and Flyer, N.: 1999, Compatibility conditions for time-dependent partial differential equations and the the rate of convergence of Chebyshev and Fourier spectral methods, *Comput. Meths. Appl. Mech. Engrg.* **175**(3–4), 281–309.

Boyd, J. P. and Haupt, S. E.: 1991, Polycnoidal waves: Spatially periodic generalizations of multiple solitary waves, *in* A. R. Osborne (ed.), *Nonlinear Topics of Ocean Physics: Fermi Summer School, Course LIX*, North-Holland, Amsterdam, pp. 827–856.

Boyd, J. P. and Ma, H.: 1990, Numerical study of elliptical modons by a spectral method, *Journal of Fluid Mechanics* **221**, 597–611.

Boyd, J. P. and Natarov, A.: 1998, A Sturm-Liouville eigenproblem of the Fourth Kind: A critical latitude with equatorial trapping, *Stud. Appl. Math.* **101**, 433–455. Rational TB calculation of logarithmically singular eigenproblems along an infinite interval parallel to the real axis, but shifted from it by a constant.

Boyd, J. P. and Tan, B.: 1998, Vortex crystals and non-existence of non-axisymmetric solitary waves in the Flierl-Petviashvili equation, *Chaos, Solitons and Fractals* **9**, 2007–2021. Double Fourier algorithm for a generalized, two-dimensional quasi-geostrophic equation.

Boyd, J. P. and Tan, B.: 1999, Composite bound states of wide and narrow envelope solitons in the Coupled Schroedinger equations through matched asymptotic expansions, *Nonlinearity* **12**, 1449–1469.

Brachet, M. E., Meiron, D. I., Orszag, S. A., Nickel, B. G., Morf, R. H. and Frisch, U.: 1983, Small-scale structure of the Taylor-Green vortex, *Journal of Fluid Mechanics* **130**, 411–452. Three-dimensional Fourier computations with exploitation of 64-fold symmetry.

Brandt, A., Fulton, S. R. and Taylor, G. D.: 1985, Improved spectral multigrid methods for periodic elliptic problems, *Journal of Computational Physics* **58**, 96–112.

Braverman, E., Israeli, M., Averbuch, A. and Vozovoi, L.: 1998, A fast 3D Poisson solver of arbitrary order accuracy, *J. Comput. Phys.* **144**(1), 109–136. Fourier spectral method for nonperiodic boundary conditions with subtractions to improve convergence and remedy corner singularities.

Brenier, B., Roux, B. and Bontoux, P.: 1986, Comparaison des méthodes tau-Chebyshev et Galerkin dans l'étude de stabilité des mouvements de convection naturelle. Problème des valeurs propres parasites, *Journal de Méchanique théorique et appliquée* **5**, 95–119.

Bridger, A. F. C. and Stevens, D. E.: 1980, Long atmospheric waves and the polar-plane approximation to the earth's spherical geometry, *Journal of the Atmospheric Sciences* **37**, 534–544.

Briggs, W. L.: 1987, *A Multigrid Tutorial*, SIAM, Philadelphia. 88 pp.

Briggs, W. L. and Henson, V. E.: 1995, *The DFT: An Owner's Manual for the Discrete Fourier Transform*, Society for Industrial and Applied Mathematics, Philadelphia.

Briggs, W. L., Newell, A. C. and Sarie, T.: 1981, The mechanism by which many partial difference equations destabilize, *in* H. Haken (ed.), *Chaos and Order in Nature*, Springer-Verlag, New York, pp. 269–273. Aliasing instability.

Brown, D. L. and Minion, M. L.: 1995, Performance of under-resolved two-dimensional flow simulations, *Journal of Computational Physics* **122**, 165–183.

Brown, J. D., Chu, M. T., Ellison, D. C. and Plemmons, R. J. (eds): 1994, *Proceedings of the Cornelius Lanczos International Centenary Conference*, Society for Industrial and Applied Mathematics, Philadelphia. Collection; many articles on spectral methods.

Browning, G. L., Hack, J. J. and Swarztrauber, P. N.: 1988, A comparison of three numerical methods for solving differential equations on the sphere, *Monthly Weather Review* **117**, 1058–1075. Spherical harmonics versus fourth and sixth order finite differences.

Cai, W., Gottlieb, D. and Harten, A.: 1992a, Cell-averaging Chebyshev methods for hyperbolic problems, *Comput. & Math. Applics.* **24**, 37–49.

Cai, W., Gottlieb, D. and Shu, C.: 1989, Essentially nonoscillatory spectral Fourier methods for shock wave calculation, *Mathematics of Computation* **52**(186), 389–410.

Cai, W., Gottlieb, D. and Shu, C. W.: 1992b, On one-sided filters for spectral Fourier approximation of discontinuous functions, *SIAM Journal of Numerical Analysis* **29**, 905–916.

Cain, A. B., Ferziger, J. H. and Reynolds, W. C.: 1984, Discrete orthogonal function expansions for non-uniform grids using the Fast Fourier transform, *Journal of Computational Physics* **56**, 272–286. Use Fourier series with a mapping which is equivalent to rational Chebyshev.

Calahan, D. A.: 1967, Numerical solution of linear systems with widely separated time constants, *Proceedings IEEE* **55**, 2016–2017. Not spectral; higher order Padé time marching scheme.

Callaghan, P., Fusco, A., Francis, G. and Salby, M.: 1999, A Hough spectral model for three-dimensional studies of the middle atmosphere, *J. Atmos. Sci.* **56**, 1461–1480. Basis uses three-dimensional eigenfunctions of the primitive equations.

Canuto, C. and Funaro, D.: 1988, The Schwarz algorithm for spectral methods, *SIAM Journal of Numerical Analysis* **25**, 24–40.

Canuto, C. and Quarteroni, A.: 1985, Preconditioned minimal residual methods for Chebyshev spectral calculations, *Journal of Computational Physics* **60**, 315–337.

Canuto, C., Hussaini, M. Y., Quarteroni, A. and Zang, T. A.: 1988, *Spectral Methods for Fluid Dynamics*, Springer-Verlag, New York. Classic text, 556 pp. Very comprehensive and readable, now in paperback.

Carcione, J. M.: 1994, Boundary conditions for wave propagation problems, *in* C. Bernardi and Y. Maday (eds), *Analysis, Algorithms and Applications of Spectral and High Order Methods for Partial Differential Equations*, Selected Papers from the International Conference on Spectral and High Order Methods (ICOSAHOM '92), Le Corum, Montpellier, France, 22-26 June 1992, North-Holland, Amsterdam, pp. 457–468. Also in Finite Elements in Analysis and Design, vo. 16, pp. 317-327. Compares standard Chebyshev grid with the stretched Kosloff/Tal-Ezer grid.

Carcione, J. M.: 1996, A 2-D Chebyshev differential operator for the elastic wave equation, *Computer Methods in Applied Mechanics and Engineering* **130**, 33–45. Adaptive grid mapping using the Kosloff/Tal-Ezer map (for longer time step), the Augenbaum map (to resolve narrow features) and different maps on different subdomains for maximum flexibility.

Carpenter, M. H.: 1996, Spectral methods on arbitrary grids, *Journal of Computational Physics* **129**(1), 74–86. Differentiation is performed using one grid of points while the equation is collocated on a different grid. Generalization of Don and Gottlieb(1994).

Carpenter, R. L., Droegemeier, K. K., Woodward, P. R. and Hane, C. E.: 1990, Application of the piecewise parabolic method (PPM) to meteorological modeling, *Monthly Weather Review* **118**, 586–612. Not spectral.

Carrier, G. F. and Pearson, C. E.: 1968, *Ordinary Differential Equations*, Blaisdell, Waltham, MA. 229 pp; not spectral.

Carse, G. A. and Urquhart, J.: 1914, Harmonic analysis, *in* E. M. Horsburgh (ed.), *Modern Instruments and Methods of Calculation*, G. Bell and Sons, in cooperation with the Royal Society of Edinburgh, London. 300 pp. Reviews "Runge grouping", which is a variant of the FFT that was widely used in the early 20th century and requires only real arithmetic. This algorithm was lost in mid-century until rediscovered, in complex form, by Cooley and Tukey. The article also describes the Runge-FFT computing forms devised by E. T. Whittaker [today, we would call them "spreadsheets"] and the mechanical harmonic analyzer of Michaelson and Stratton (1898) which empirically discovered the Gibbs' phenomenon.

Chan, T. and Kerkhoven, T.: 1985, Fourier methods with extended stability intervals for the Korteweg-deVries equation, *SIAM Journal of Numerical Analysis* **22**, 441–454.

Chan, T. F.: 1984, Newton-like pseudo-arclength methods for computing simple turning points, *SIAM Journal of Scientific and Statistical Computing* **5**, 135–148. Not spectral, but useful for tracking branches of the solutions to nonlinear equations as a parameter is varied.

Chaouche, A. M.: 1990, A collocation method basedon an influence matrix technique for axisymmetric flows in an annulus, *Rech. Aérosp.* **1990-5**, 1–13.

Chaouche, A., Randriamampianina, A. and Bontoux, P.: 1990, A collocation method based on an influence matrix technique for axisymmetric flows in an annulus, *Computer Methods in Applied Mechanics and Engineering* **80**, 237–244.

Chapman, S. and Lindzen, R. S.: 1970, *Atmospheric Tides*, D. Reidel, Dordrecht, Holland. 200 pp., computes Hough functions by solving eigenvalue problem for Laplace's tidal equations by spherical harmonic expanions.

Chen, H. B.: 1993a, On the instability of a full non-parallel flow — Kovasznay flow, *International Journal for Numerical Methods in Fluids* **17**, 731–754. Eigenvalue study using rational Chebyshev functions TL_n for the semi-infinite interval.

Chen, S., Doolen, G. D., Kraichnan, R. H. and She, Z.-S.: 1993, On statistical correlation between velocity increments and locally averaged dissipation in homogeneous turbulence, *Physics of Fluids A* **5**, 458–463.

Chen, X.-S.: 1993b, The aliased and dealiased spectral models of the shallow-water equations, *Monthly Weather Review* **121**, 834–852. Solves the KdV equation and also the spherical shallow water wave equations for both very smooth initial conditions (Rossby-Haurwitz waves) and realistic conditions (observational data from FGGE). In all cases, the aliased models are no better than aliased models with the same number of points after filtering (the dealiased code uses $(3/2)N$ points to compute N modes); the aliased models are always considerably worse than dealiased models when compared on the basis of the same number of collocation points (the dealiased code has a smaller number of modes than the aliased code after the dealiasing filtering is applied). One of the dealiased methods is novel in that it uses the Walsh Hadamard transform instead of the usual FFT. The WT transform method is a little cheaper than the FFT for the same accuracy, but all of the dealiased codes are 1.7 to 2 times more expensive than the aliasing codes with the same number of grid points. but unfortunately no more accurate.

Cheong, H. B.: 2000, Double fourier series on a sphere: Applications to elliptic and vorticity equations, *J. Comput. Phys.* **157**(1), 327–349.

Cheong, H. B.: 2001, Application of double fourier series to shallow water equations on a sphere, *J. Comput. Phys.* Submitted.

Christov, C. I.: 1982, A complete orthonormal system in $L^2(-\infty, \infty)$ space, *SIAM Journal of Applied Mathematics* **42**, 1337–1344.

Christov, C. I. and Bekyarov, K. L.: 1990, A Fourier-series method for solving soliton problems, *SIAM Journal of Scientific and Statistical Computing* **11**, 631–647. Rational functions.

Chu, M. T.: 1988, On the continuous realization of iterative processes, *SIAM Review* **30**, 375–387. Differential equations in pseudotime as models for Newton's and other iterations.

Cividini, A. and Zampieri, E.: 1997, Nonlinear stress analysis problems by spectral collocation methods, *Comput. Methods Appl. Mech. Engrg.* **145**, 185–201.

Clenshaw, C. W.: 1957, The numerical solution of linear differential equations in Chebyshev series, *Proceedings of the Cambridge Philosophical Society* **53**, 134–149.

Clenshaw, C. W. and Curtis, A. R.: 1960, A method for numerical integration on an automatic computer, *Numerische Mathematik* **2**, 197–205. Chebyshev polynomial quadrature scheme.

Clercx, H. J. H.: 1997, A spectral solver for the Navier-Stokes equations in the velocity-vorticity formulation for flows with two nonperiodic directions, *J. Comput. Phys.* **137**, 186–211.

Cloot, A.: 1991, Equidistributing mapping and spectral method for the computation on unbounded domains, *Applied Mathematics Letters* **4**, 23–27.

Cloot, A. and Weideman, J. A. C.: 1990, Spectral methods and mappings for evolution equations on the infinite line, *Computer Methods in Applied Mechanics and Engineering* **80**, 467–481.

Cloot, A. and Weideman, J. A. C.: 1992, An adaptive algorithm for spectral computations on unbounded domains, *Journal of Computational Physics* **102**, 398–406.

Cloot, A., Herbst, B. M. and Weideman, J. A. C.: 1990, A numerical study of the cubic-quintic Schrodinger equation, *Journal of Computational Physics* **86**, 127–146.

Concus, P. and Golub, G. H.: 1973, Use of fast direct methods for the efficient numerical solution of nonseparable elliptic equations, *SIAM Journal of Numerical Analysis*. Not spectral, but shows how a fast direct method, either spectral or otherwise, can be used to efficiently solve nonseparable elliptic equations.

Cooley, J. W. and Tukey, J. W.: 1965, An algorithm for the machine calculation of complex Fourier series, *Mathematics of Computation* **19**, 297–301.

Corral, R. and Jiménez, J.: 1995, Fourier/Chebyshev methods for the incompressible Navier-Stokes equations in infinite domains, *J. Comput. Phys.* **121**, 261.

Côté, J., Gravel, S. and Staniforth, A.: 1990, Improving variable-resolution finite-element semi-Lagrangian integration schemes by pseudostaggering, *Monthly Weather Review* **118**, 2718–2731.

Côté, J., Gravel, S. and Staniforth, A.: 1995, A generalized family of schemes that eliminates the spurious resonant response of semi-Lagrangian schemes to orographic forcing, *Monthly Weather Review* **123**, 3605.

Coulson, C. A.: 1961, *Valence*, Oxford University Press, New York. 250 pp.

Courtier, P. and Geleyn, J.-F.: 1988, A global numerical weather prediction model with variable resolution: Application to the shallow-water equations, *Quarterly Journal of Royal Meteorological Society* **114**, 1321–1346.

Courtier, P. and Naughton, M.: 1994, A polar problem in the reduced Gaussian grid, *Quart. J. Roy. Met. Soc.* **120**, 1389–1407.

Courtier, P., Freydier, C., Geleyn, J. F., Rabier, F. and Rochas, M.: 1991, The Arpege project at Météo-France., *Proceedings of Numerical Methods in Atmospheric Models*, European Center for Medium Range Forecasting, European Center for Medium Range Forecasting, Shinfield Park, Reading, United Kingdom, pp. 193–231.

Coutsias, E. A., Hagstrom, T. and Torres, D.: 1996, An efficient spectral method for ordinary differential equations with rational function coefficients, *Mathematics of Computation* **65**, 611–635. Obtained banded matrix representations of one-dimensional differential operators for all standard polynomial basis sets, assuming that the coefficients of the operators are restricted to polynomials or rational functions. These banded matrices can be inverted in $O(N)$ operations.

Couzy, W. and Deville, M.: 1994, Iterative solution technique for spectral-element pressure operators at high Reynolds number, *in* S. Wagner et al. (eds), *Proceedings of the Second European Computational Fluid Dynamics Conference*, Stuttgart, Germany, pp. 613–618.

Craik, A. D. D.: 1985, *Wave Interactions and Fluid Flows*, Cambridge University Press, New York. Not spectral, but a good description of weakly nonlinear waves and resonant triad and four-wave interactions.

Curchitser, E. N., Iskandarani, M. and Haidvogel, D. B.: 1998, A spectral element solution of the shallow-water equations on multiprocessor computers, *J. Atmos. Oceanic Technology* **15**(2), 510–521.

da Silva, M. R.: 1987, A quick survey of recent developments and applications of the tau-method, *in* E. L. Ortiz (ed.), *Numerical Approximations of P. D. E., Part III*, North-Holland, Amsterdam, pp. 297–308.

Daley, R.: 1980, The development of efficient time integration schemes using model normal modes, *Monthly Weather Review* **108**, 100–110.

Daley, R.: 1981, Normal mode initialization, *Reviews of Geophysics and Space Physics* **19**(3), 450–468. Review.

Daley, R.: 1991, *Atmospheric Data Analysis*, Cambridge University Press, New York.

Daley, R. and Bourassu, Y.: 1978, Rhomboidal versus triangular spherical harmonic truncation: Some verification statistics, *Atmosphere-Ocean* **16**, 187–196.

Davis, P. J.: 1975, *Interpolation and Approximation*, Dover Publications, New York. 200 pp.

Davis, P. J. and Rabinowitz, P.: 1984, *Methods of Numerical Integration*, 3rd edn, Academic Press, Boston. 612pp.

Dawkins, P. T., Dunbar, S. R. and Douglass, R. W.: 1998, The origin and nature of spurious eigenvalues in the spectral tau method, *J. Comput. Phys.* **147**(2), 441–462. Theoretical paper proving the existence of two spurious eigenvalues for the exemplary problem $u_{xxxx} = \lambda u_{xx}$ which are larger than N^4. The cure and numerical experiments are also described.

de Veronico, M. C., Funaro, D. and Reali, G. C.: 1994, A novel numerical technique to investigate nonlinear guided waves: approximation of the Nonlinear Schroedinger equation by nonperiodic pseudospectral methods, *Numerical Methods for Partial Differential Equations* **10**(6), 667–675. Domain decomposition with Legendre polynomials on interior domains and Laguerre polynomials on the exterior elements, which extend to infinity.

de Vries, R. W. and Zandbergen, P. J.: 1989, The numerical solution of the biharmonic equation, using a spectral multigrid method, *in* W. F. Ballhaus and M. Y. Hussaini (eds), *Advances in Fluid Dynamics*, Springer-Verlag, New York, pp. 25–35.

Deane, A. E., Kevrekidis, I. G., Karniadakis, G. E. and Orszag, S. A.: 1991, Low-dimensional modeling for complex geometry flows: Application to grooved channels and circular cylinders, *Physics of Fluids A* **3**, 2337–2354. Spectral elements; also derivation and application of a four-mode model using "empirical eigenfunctions" as the basis (also known as "proper orthogonal decomposition".

Debussche, A., Dubois, T. and Temam, R.: 1995, The Nonlinear Galerkin method: A multiscale method applied to the simulation of homogeneous turbulent flows, *Theoretical and Computational Fluid Dynamics* **7**, 279–315.

Decker, D. W. and Keller, H. B.: 1980, Path following near bifurcation, *Communications in Pure and Applied Mathematics* **34**, 149–175.

Delves, L. M.: 1976, Expansion methods, *in* G. Hall and J. M. Watts (eds), *Modern Numerical Methods for O. D. E.s*, Clarendon Press, Oxford University Press, Oxford, pp. 269–281. REVIEW.

Delves, L. M.: 1977a, A fast method for the solution of Fredholm integral equations, *Journal of the Institute for Mathematics and its Applications* **20**, 173–184.

Delves, L. M.: 1977b, On the solution of the linear equation arising from Galerkin methods, *Journal of the Institute for Mathematics and its Applications* **20**, 163–171.

Delves, L. M.: 1977c, A linear equation solver for Galerkin and least squares methods, *J. Comp.* **20**, 371–374.

Delves, L. M., Abd-Elal, L. F. and Hendry, J. A.: 1979, A fast Galerkin algorithm for the solution of Fredholm integral equations, *Journal of the Institute for Mathematics and its Applications* **23**, 139–166.

Delves, L. M., Abd-Elal, L. F. and Hendry, J. A.: 1981a, A set of modules for the solution of integral equations, *Comp. J.* **24**, 184–190.

Delves, L. M. and Freeman, T. N.: 1981, *Analysis of Global Expansion Methods: Weakly Asymptotically Diagonal Systems*, Academic Press, New York. 275 pp. Mostly theory with only a handful of elementary examples, but the preconditioned "Delves-Freeman" iteration is very interesting.

Delves, L. M. and Hall, C. A.: 1979, An implicit matching procedure for global element calculations, *Journal of the Institute for Mathematics and its Applications* **23**, 223–234.

Delves, L. M. and Mead, K. O.: 1971, On the convergence rates of variational methods. I. Asymptotically diagonal systems, *Mathematics of Computation* **25**, 699–716. Theory only.

Delves, L. M. and Phillips, C.: 1980, A fast implementation of the global element method, *Journal of the Institute for Mathematics and its Applications* **25**, 177–197.

Delves, L. M., McKerrell, A. and Henry, J. A.: 1981b, A note on Chebyshev methods for the solution of partial differential equations, *Journal of Computational Physics* **41**, 444–452.

Delves, L. M., McKerrell, A. and Peters, S. A.: 1986, Performance of GEM2 on the ELLPACK problem population, *International Journal for Numerical Methods in Engineering* **23**, 229–238.

Demaret, P. and Deville, M. O.: 1989, Chebyshev pseudospectral solution of the Stokes equation using finite element preconditioning, *Journal of Computational Physics* **83**, 463–484.

Demaret, P. and Deville, M. O.: 1991, Chebyshev collocations solutions of the Navier-Stokes equations using multi-domain decomposition and finite element preconditioning, *Journal of Computational Physics* **95**, 359–386.

Demaret, P., Deville, M. O. and Schneidesch, C.: 1989, Thermal convection solutions by Chebyshev pseudospectral multi-domain decomposition and finite element preconditioning, *Applied Numerical Mathematics* **6**, 107–121. Nonlinear steady flows through nested iterations: outer Newton/inner Richardson.

Déqué Piedelievre, M.: 1995, High resolution climate model over Europe, *Climate Dynamics* **11**, 321–339.

Dettori, L., Gottlieb, D. and Témam, R.: 1995, Nonlinear Galerkin method: the two-level Fourier collocation case, *Journal of Scientific Computing* **10**, 371–.

Deville, M.: 1990, Chebyshev collocation solutions of flow problems, *in* C. Canuto and A. Quarteroni (eds), *Spectral and High Order Methods for Partial Differential Equations: Proceedings of the ICOSAHOM '89 Conference in Como, Italy*, North-Holland/Elsevier, Amsterdam, pp. 27–38. Also in Comput. Meths. Appl. Mech. Engrg., vol. 80.

Deville, M. and Labrosse, G.: 1982, An algorithm for the evaluation of multi-dimensional (direct and inverse) discrete Chebyshev transform, *Journal of Computational and Applied Mathematics* **8**, 293–304.

Deville, M. and Mund, E.: 1984, On a mixed one step/Chebyshev pseudospectral technique for the integration of parabolic problems using finite element preconditioning, *in* C. Brezinski, A. Draux, A. P. Magnus, P. Maroni and A. Ronveaux (eds), *Polynomes Orthogonaux et Applications: Proceedings of the Laguerre Symposium at Bar-le-Duc*, number 1171 in *Lecture Notes in Mathematics*, Springer-Verlag, New York, pp. 399–407. This article is in English; employs unusual implicit time-marching which is exact (instead of second order) for time integration of the two slowest-decaying diffusion eigenmodes.

Deville, M. and Mund, E.: 1985, Chebyshev pseudospectral solution of second-order elliptic equations with finite element pre-conditioning, *Journal of Computational Physics* **60**, 517–533.

Deville, M. and Mund, E.: 1991, Finite element preconditioning of collocation schemes for advection-diffusion equations, *in* R. Beauwens and P. de Groen (eds), *Proceedings of the IMACS International Symposium on Iterative Methods in Linear Algebra*, IMACS, North-Holland, Amsterdam, pp. 181–190.

Deville, M., Haldenwang, P. and Labrosse, G.: 1981, Comparison of time integration (finite difference and spectral)) for the nonlinear Burgers' equation, *in* H. Viviond (ed.), *Proceedings of the 4th GAMMConference on Nuemrical Methods in Fluid Mechanics*, Vieweg, Braunschweig.

Deville, M., Kleiser, L. and Montigny-Rannou, F.: 1984, Pressure and time treatment for Chebyshev spectral solution of a Stokes problem, *Internat. J. Numer. Meth. Fluids* **4**, 1149–1163. Backward Euler is raised by Richardson extrapolation to a second order time-marching. Four different schemes including penalty method, splitting, influence matrix and Morchoisne's space-time pseudospectral scheme. Good discussion of compatibility conditions on the initial condition.

Deville, M. O. and Mund, E. H.: 1990, Finite-element preconditioning for pseudospectral solutions of elliptic problems, *SIAM Journal of Scientific and Statistical Computing* **12**, 311–342.

Deville, M. O. and Mund, E. H.: 1992, Fourier analysis of finite element preconditioned collocation schemes, *SIAM Journal of Scientific and Statistical Computing* **13**(2), 596–610.

Deville, M. O., Mund, E. H. and Van Kemenade, V.: 1994, Preconditioned Chebyshev collocation methods and triangular finite elements, *in* C. Bernardi and Y. Maday (eds), *Analysis, Algorithms and Applications of Spectral and High Order Methods for Partial Differential Equations*, Selected Papers from the International Conference on Spectral and High Order Methods (ICOSAHOM '92), Le Corum, Montpellier, France, 22-26 June 1992, North-Holland, Amsterdam, pp. 193–200.

Devulder, C. and Marion, M.: 1992, A class of numerical algorithms for large time integration: the nonlinear Galerkin methods, *SIAM Journal of Numerical Analysis* **29**(2), 462–483. Many theorems.

Dickinson, B. A.: 1933, Energy levels of H_2^+ molecular ion, *J. Chem. Phys.* **1**, 317. Illustration of the Rayleigh variational principle in quantum chemistry;.

Dimitropoulus, C. and Beris, A. N.: 1997, An efficient and robust spectral solver for non-separable elliptic equations, *Journal of Computational Physics* **133**, 186–191. Biconjugate gradient outer iteration with a Concus-Golub inner iteration in which a separable Helmholtz equation is solved at each iteration.

Dimitropoulus, C. and Beris, A. N.: 1998, Efficient pseudospectral flow simulations in moderately complex geometries, *Journal of Computational Physics* **144**(2), 517–549. Pseudoconformal orthogonal curvilinear coordinates in two dimensions, Fourier in one coordinate and Chebyshev in the other.

Don, W. S. and Gottlieb, D.: 1990, Spectral simulation of unsteady flow past a cylinder, *in* C. Canuto and A. Quarteroni (eds), *Spectral and High Order Methods for Partial Differential Equations: Proceedings of the ICOSAHOM '89 Conference in Como, Italy*, North-Holland/Elsevier, Amsterdam, pp. 39–58. Also in Comput. Meths. Appl. Mech. Engrg., vol. 80, with the same page numbers.

Don, W. S. and Gottlieb, D.: 1994a, The Chebyshev-Legendre method: implementing Legendre methods on Chebyshev points, *SIAM Journal of Numerical Analysis* **31**, 1519–1534.

Don, W. S. and Gottlieb, D.: 1994b, Spectral simulation of supersonic reactive flows, *SIAM Journal of Numerical Analysis* **35**(6), 2370–2384. Shock waves with the use of the Gegenbauer regularization method for defeating Gibbs' Phenomenon and the Kosloff/Tal-Ezer map to allow a relatively long time step.

Don, W. S. and Solomonoff, A.: 1995, Accuracy and speed in computing the Chebyshev collocation derivative, *SIAM J. Sci. Comput.* **16**, 1253–1268.

Don, W. S. and Solomonoff, A.: 1997, Accuracy enhancement for higher derivatives using Chebyshev collocation and a mapping technique, *SIAM Journal of Scientific Computing* **18**(4), 1040–1055. The Kosloff/Tal-Ezer mapping is used to reduce roundoff error and allow a larger timestep.

Dongarra, J. J., Straughan, B. and Walker, D. W.: 1996, Chebyshev tau-QZ algorithm methods for calculating spectra of hydrodynamic stability problems, *Applied Numerical Mathematics* **22**, 399–434.

Doron, E., Hollingsworth, A., Hoskins, B. J. and Simmons, A. J.: 1974, A comparison of grid-point and spectral methods in a meteorological problem, *Quarterly Journal of the Royal Meteorological Society* **100**, 371–383.

Douglas, J. and Russell, T. F.: 1982, Numerical methods for convection-dominated diffusion problems based on combining the method of characteristics with finite element or finite difference procedures, *SIAM Journal of Numerical Analysis* **19**, 871–885. Not spectral; invention of a semi-Lagrangian scheme.

Drake, J., Foster, I., Michalakes, J., Toonen, B. and Worley, P.: 1995, Design and performance of a scalable parallel community climate model, *Parallel Comput.* **21**(10), 1571–1591. Parallel version, PCCM2, of the CCM2 spherical harmonics/vertical finite difference climate model. Performance on the IBM SP2 and Intel Paragon.

Driscoll, J., Healy, Jr., D. M. and Rockmore, D.: 1997, Fast discrete polynomial transform with applications to data analysis on distance transitive graphs, *SIAM J. Comput.* **26**, 1066–1099. $O(N \log_2(N))$ transform which works for any set of orthogonal polynomials; more efficient than Orszag(1986).

Driscoll, T. A. and Fornberg, B.: 1998, A block pseudospectral method for Maxwell's equations. I. One-dimensional case, *Journal of Computational Physics* **140**(1), 47–65. Employ a domain decomposition scheme in which fictitious points beyond the domain walls are used, in a sort of grid overlapping scheme, to allow a much more uniform separation between grid points than in a standard pseudospectral algorithm. This allows a relatively long time step at the cost of much increased domain-to-domain communication compared to the standard domain decomposition method. They generalize the scheme so that it works well even when there are discontinuous changes in material properties at domain walls.

Dubois, T., Jauberteau, F. and Témam, R.: 1990, The nonlinear Galerkin method for the two and three dimensional Navier-Stokes equations, *in* K. W. Morton (ed.), *Proceedings of the Twelfth International Conference on Numerical Methods in Fluid Dynamics*, Springer-Verlag, New York, pp. 117–120.

Dubois, T., Jauberteau, F. and Témam, R.: 1998, Incremental unknowns, multilevel methods and the numerical simulation of turbulence, *Comput. Meth. Appl. M.* **159**, 123–189.

Durran, D. R.: 1991, The third-order Adams-Bashforth method: an attractive alternative to leapfrog time-differencing, *Monthly Weather Review* **119**, 702–720.

Dutt, A. and Rokhlin, V.: 1993, Fast Fourier Transforms for nonequispaced data, *SIAM J Comput.* **14**, 1368–1393.

Dutt, A. and Rokhlin, V.: 1995, Fast Fourier Transforms for nonequispaced data, II, *Applied and Computational Harmonic Analysis* **2**, 85–110.

D'yakonov, E. G.: 1961, An iteration method for solving systems of finite difference equations, *Doklady Akademiia Nauk SSSR* **138**, 522–525. Not spectral; iteration preconditioned by separable PDE.

Dym, H. and McKean, H. P.: 1972, *Fourier Series and Integrals*, Academic Press, New York. 129 pp.

Eggert, N., Jarratt, M. and Lund, J.: 1987, Sinc function computation of Sturm-Liouville problems, *Journal of Computational Physics* **69**, 209–229.

Ehrenstein, U. and Peyret, R.: 1989, A Chebyshev collocation method for the Navier-Stokes equations with application to double-diffusive convection, *International Journal for Numerical Methods in Fluids* **9**, 427–452. Semi-implicit time integration with influence matrix method.

Eisen, H. and Heinrichs, W.: 1992, A new method of stabilization for singular perturbation problems with spectral methods, *SIAM Journal of Numerical Analysis* **29**, 107–122. Shows that basis functions which vanish at boundaries are much better conditioned than the Chebyshev polynomials from which these basis functions are formed.

Eisen, H., Heinrichs, W. and Witch, K.: 1991, Spectral collocation methods and polar coordinate singularities, *Journal of Computational Physics* **96**, 241–257.

Eisenstat, S. C., Elman, H. C. and Schultz, M. H.: 1983, Variational iterative methods for nonsymmetric systems of linear equations, *SIAM J. Numer. Anal.* **20**, 345–357.

El-Daou, M. K. and Ortiz, E. L.: 1993, Error analysis of the tau method: dependence of the error on the degree and on the length of the interval of approximation, *Computers Math. Applic.* **25**(7), 33–45.

El-Daou, M. K. and Ortiz, E. L.: 1994a, A recursive formulation of collocation in terms of canonical polynomials, *Computing* **52**, 177–202.

El-Daou, M. K. and Ortiz, E. L.: 1994b, The weighting subspaces of collocation and the Tau method, *in* J. D. Brown, M. T. Chu, D. C. Ellison and R. J. Plemmons (eds), *Proceedings of the Cornelius Lanczos International Centenary Conference*, Society for Industrial and Applied Mathematics, Society for Industrial and Applied Mathematics, Philadelphia, PA.

El-Daou, M. K. and Ortiz, E. L.: 1997, The uniform convergence of the Tau method for singularly perturbed problems, *Applied Mathematics Letters* 10(2), 91–94. Existence and stability of the algorithm is proved, independent of the perturbation parameter ϵ.

El-Daou, M. K., Ortiz, E. L. and Samara, H.: 1993, A unified approach to the tau method and Chebyshev series expansion techniques, *Computers Math. Applic.* 25(3), 73–82.

Eliassen, E. and Machenhauer, B.: 1974, On spectral representation of the vertical variation of the meteorological fields in numerical integration of a primitive equation model, *GARP WGNE Report 7*, World Meteorological Organization, Geneva, Switzerland. Legendre polynomials in the vertical.

Elliott, D. and Stenger, F.: 1984, Sinc method of solution of singular integral equations, *in* A. Gerasoulis and R. Vichnevetsky (eds), *Numerical Solution of Singular Integral Equations*, IMACS.

Ellsaesser, H. W.: 1966, Evaluation of spectral versus grid point methods of hemispheric numerical weather prediction, *J. Appl. Meteor.* 5, 246–262. Fig. 12 is a good illustration of late onset of spectral blocking due to violation of the CFL criterion after the advecting flow has intensified from its initial maximum.

Engelmann, F., Feix, M., Minardi, E. and Oxenius, J.: 1963, Nonlinear effects from vlasov's equation, *The Physics of Fluids* 6(2), 266–275. Low order Hermite series ($N = 2, 3$) for velocity coordinate; such low truncations are found to suppress instabilities that occur in some parameter regions for the full equations.

Erlebacher, G., Zang, T. A. and Hussaini, M. Y.: 1987, Spectral multigrid methods for the numerical simulation of turbulence, *in* S. McCormick and K. Stuben (eds), *Multigrid Methods*, Marcel Dekker, New York, pp. 177–194.

Errico, R. M.: 1984, The dynamic balance of a general circulation model, *Monthly Weather Review* 112, 2439–2454. Not spectral, but evaluation of the usefulness of slow manifold concept in a complicated model.

Errico, R. M.: 1989, The degree of Machenauer balance in a climate model, *Monthly Weather Review* 112, 2723–2733. Test of slow manifold ideas of Machenauer's NG(1) approximation in a global hydrodynamics-with-physics model.

Falqués, A. and Iranzo, V.: 1992, Edge waves on a longshore shear flow, *Physics of Fluids* pp. 2169–2190. Rational Chebyshev and Laguerre on semi-infinite domain.

Finlayson, B. A.: 1973, *The Method of Weighted Residuals and Variational Principles*, Academic, New York. 412 pp. Many good examples of low order pseudospectral methods, dubbed "orthogonal collocation" here, and mostly drawn from chemical engineering and fluid dynamics. Mostly Legendre and Gegenbauer polynomials rather than Chebyshev and no mention of the FFT.

Fischer, P. F.: 1990, Analysis and application of a parallel spectral element method for the solution of the Navier-Stokes equations, *Computer Methods in Applied Mechanics and Engineering* 80(1–3), 483–491.

Fischer, P. F.: 1994a, Domain decomposition methods for large scale parallel Navier-Stokes calculations, *in* A. Quarteroni (ed.), *Proceedings of the Sixth International Conference on Domain Decomposition Methods for Partial Differential Equations, Como, Italy*, AMS, Providence.

Fischer, P. F.: 1994b, Parallel domain decomposition for incompressible fluid dynamics, *Contemp. Math.* **157**, 313.

Fischer, P. F.: 1997, An overlapping Schwarz method for spectral element solution of the incompressible Navier-Stokes equations, *J. Comput. Phys.* **133**, 84–101.

Fischer, P. F.: 1998, Projection techniques for iterative solution of ax=b with successive right-hand sides, *Comput. Meth. Appl. M.* **163**(1–4), 193–204. Not spectral, but useful for semi-implicit time marching.

Fischer, P. F. and Gottlieb, D.: 1997, On the optimal number of subdomains for hyperbolic problems on parallel computers, *International Journal of Supercomputing and Appl. High Performance Computing* **11**, 65–76. Spectral elements.

Fischer, P. F. and Patera, A. T.: 1989, Parallel spectral element methods for the incompressible Navier-Stokes equations, *in* J. H. Kane and A. D. Carlson (eds), *Solution of Super Large Problems in Computational Mechanics*, Plenum, New York.

Fischer, P. F. and Patera, A. T.: 1991, Parallel spectral element solution of the Stokes problem, *Journal of Computational Physics* **92**(2), 380–421.

Fischer, P. F. and Patera, A. T.: 1992, Parallel spectral element solutions of eddy-promoter channel flow, *Proceedings of the European Research Community on Flow Turbulence and Combustion Workshop, Laussane, Switzerland*, Cambridge University Press, Cambridge.

Fischer, P. F. and Patera, A. T.: 1994, Parallel simulation of viscous incompressible flows, *Annual Reviews of Fluid Mechanics* **26**, 483–527. REVIEW.

Fischer, P. F. and Rønquist, E. M.: 1994, Spectral element methods for large scale parallel Navier-Stokes calculations, *in* C. Bernardi and Y. Maday (eds), *Analysis, Algorithms and Applications of Spectral and High Order Methods for Partial Differential Equations*, Selected Papers from the International Conference on Spectral and High Order Methods (ICOSAHOM '92), Le Corum, Montpellier, France, 22-26 June 1992, North-Holland, Amsterdam, pp. 69–76. Also in Comput. Methods. Appl. Mech. Engrg., vol. 116.

Fischer, P. F., Ho, L.-W., Karniadakis, G. E., Rønquist, E. M. and Patera, A. T.: 1988a, Recent advances in parallel spectral element simulation of unsteady incompressible flows, *Computers and Structures* **30**, 217–231.

Fischer, P. F., Rønquist, E. M. and Patera, A. T.: 1989, Parallel spectral element methods for viscous flow, *in* G. Carey (ed.), *Parallel Supercomputing: Methods, Algorithms and Applications*, Wiley, New York, pp. 223–238.

Fischer, P., Rønquist, E. M., Dewey, D. and Patera, A. T.: 1988b, Spectral element methods: Algorithms and architectures, *in* R. Glowinski, G. Golub, G. Meurant and J. Periaux (eds), *Proceedings of the First International Conference on Domain Decomposition Methods for Partial Differential Equations*, SIAM, SIAM, Philadelphia, pp. 173–197.

Fjørtoft, R.: 1952, On a numerical method of integrating the barotropic vorticity equation, *Tellus* **4**, 179–194. Not spectral; early use of Lagrangian coordinates in numerical meteorology.

Fjørtoft, R.: 1955, On a numerical method of integrating the barotropic vorticity equation, *Tellus* **7**, 462–480. Not spectral; early use of Lagrangian coordinates in numerical meteorology.

Flå, T.: 1992, A numerical energy conserving method for the DNLS equation, *Journal of Computational Physics* **101**, 71–79. Fourier pseudospectral scheme for the Derivative-Nonlinear Schroedinger equation.

Flatau, P., Boyd, J. P. and Cotton, W. R.: 1987, Symbolic algebra in applied mathematics and geophysical fluid dynamics — REDUCE examples, *Technical report*, Colorado State University, Department of Atmospheric Science, Fort Collins, CO 80523. Some examples of spectral methods in REDUCE.

Flyer, N.: 1998, Asymptotic upper bounds for the coefficients in the Chebyshev series expansion for a general order integral of a function, *Math. Comput.* **67**(224), 1601–1616.

Foias, C., Jolly, M. S., Kevrekidis, I. G. and Titi, E. S.: 1991, Dissipativity of numerical schemes, *Nonlinearity* **4**, 591–613. Blow-up of various schemes including Nonlinear Galerkin.

Foias, C., Jolly, M. S., Kevrekidis, I. G. and Titi, E. S.: 1994, On some dissipative fully discrete nonlinear Galerkin schemes for the Kuramoto-Sivashinsky equation, *Physics Letters A* **186**, 87–96.

Foias, C., Manley, O. and Témam, R.: 1988, Modelling of the interaction of small and large eddies in two dimensional turbulent flows, *Rairo-Math. Mod. Numer. Anal.* **22**(1), 93–118. First paper on approximate inertial manifolds, which led to rediscovery of Nonlinear Galerkin method.

Forbes, J. M. and Garrett, H. B.: 1976, Solar diurnal tide in the thermosphere, *Journal of the Atmospheric Sciences* **33**, 2226–2241.

Fornberg, B.: 1975, On a Fourier method for the integration of hyperbolic equations, *SIAM Journal of Numerical Analysis* **12**, 509–528.

Fornberg, B.: 1977, A numerical study of 2-d turbulence, *Journal of Computer Physics* **25**, 1–31.

Fornberg, B.: 1978, Pseudospectral calculations on 2-d turbulence and nonlinear waves, *SIAM-AMS Proceedings* **11**, 1–18.

Fornberg, B.: 1987, The pseudospectral method: Comparisons with finite differences for the elastic wave equation, *Geophysics* **52**, 483–501.

Fornberg, B.: 1988a, The pseudospectral method: accurate representation of interfaces in elastic wave calculations, *Geophysics* **53**, 625–637.

Fornberg, B.: 1988b, Generation of finite difference formulas on arbitrarily spaced grids, *Mathematics of Computation* **51**, 699–706. Simple recursion to compute finite difference weights. This is helpful for finite difference preconditioning of spectral methods; the finite difference problem must be solved on the same unevenly spaced grid as used by the Chebyshev or Legendre pseudospectral scheme.

Fornberg, B.: 1990a, High order finite differences and the pseudospectral method on staggered grids, *SIAM Journal of Numerical Analysis* **27**, 904–918.

Fornberg, B.: 1990b, An improved pseudospectral method for initial boundary value problems, *Journal of Computational Physics* **91**, 381–397. Uses additional boundary conditions, derived from the differential equation, to greatly reduce the largest eigenvalues of Chebyshev differentiation matrices, greatly reducing the "stiffness" of time-dependent problems and allowing a longer time step.

Fornberg, B.: 1992, Fast generation of weights in finite difference formulas, *in* G. D. Byrne and W. E. Schiesser (eds), *Recent Developments in Numerical Methods and Software for QDEs/DAEs/PDEs*, World Scientific, Singapore, pp. 97–123.

Fornberg, B.: 1995, A pseudospectral approach for polar and spherical geometries, *SIAM Journal of Scientific Computing* **16**, 1071–1081. Double Fourier series on latitude and longitude grid with strong filtering of large zonal wavenumbers near the poles to avoid CFL instability.

Fornberg, B.: 1996, *A Practical Guide to Pseudospectral Methods*, Cambridge University Press, New York.

Fornberg, B.: 1998, Calculation of weights in finite difference formulas, *SIAM Rev.* **40**(3), 685–691. Arbitrary order finite differences, which in the limit of order equal to the number of grid points gives a pseudospectral approximation.

Fornberg, B. and Merrill, D.: 1997, Comparison of finite difference and pseudospectral methods for convective flow over a sphere, *Geophys. Res. Lett.* **24**, 3245–3248. Second and fourth order finite differences and double Fourier series, all on a latitude-longitude grid with strong high latitude filtering of large zonal wavenumbers, are compared with spherical harmonics for a problem whose solution is very smooth. Equal accuracy for both spectral methods, but the double Fourier algorithm is faster.

Fornberg, B. and Sloan, D.: 1994, A review of pseudospectral methods for solving partial differential equations, *in* A. Iserles (ed.), *Acta Numerica*, Cambridge University Press, New York, pp. 203–267.

Fornberg, B. and Whitham, G. B.: 1978, A numerical and theoretical study of certain nonlinear wave phenomena, *Philosophical Transactions of the Royal Society of London* **289**, 373–404. Develops efficient Fourier pseudospectral method for Korteweg–deVries and related wave equations; the linear terms, which have constant coefficient, are integrated in time exactly.

Foster, I. T. and Worley, P. H.: 1997, Parallel algorithms for the spectral transform method, *SIAM J. Sci. Comput.* **18**(3), 806–837. Comparison of methods for spherical harmonics transforms for the nonlinear shallow water wave equations on various parallel architectures.

Fox, D. G. and Orszag, S. A.: 1973, Pseudospectral approximation to two-dimensional turbulence, *Journal of Computational Physics* **11**, 612–619.

Fox, L.: 1962, Chebyshev methods for ordinary differential equations, *Computer Journal* **4**, 318–331.

Fox, L. and Parker, I. B.: 1968, *Chebyshev Polynomials in Numerical Analysis*, Oxford University Press, London. Very readable and still good for general background, but the Clenshaw-type recursive algorithms for solving differential and integral equations are no longer popular except for special applications.

Fox, L., Hayes, L. and Mayers, D. F.: 1973, The double eigenvalue problem, *in* J. C. P. Miller (ed.), *Numerical Analysis*, Academic Press, pp. 93–112. Chebyshev solution of a 2d order ODE which has three boundary conditions imposed. Solutions exist because the problem has two independent eigenparameters.

Francis, P. E.: 1972, The possible use of Laguerre polynomials for representing the vertical structure of numerical models of the atmosphere, *Quart. J. Roy. Met. Soc.* **98**, 662–667. See also the correspondence by B. J. Hoskins and Francis, QJRMS 99, 571–572. The conclusion is that Laguerre polynomials demand a very short timestep, and are therefore useless in this context.

Francken, P., Deville, M. O. and Mund, E. H.: 1990, On the spectrum of the iteration operator associated to the finite element preconditioning of Chebyshev collocation calculations, *Computer Methods in Applied Mechanics and Engineering* **80**, 295–304.

Fraser, W. and Wilson, M. W.: 1966, Remarks on the Clenshaw-Curtis quadrature scheme, *SIAM Review* **8**, 322–327.

Froes Bunchaft, M. E.: 1997, Some extensions of the Lanczos-Ortiz theory of canonical polynomials in the Tau method, *Mathematics of Computation* **66**(218), 609–621. Extends and simplifies the canonical polynomial tau method; no numerical illustrations.

Frutos, J. and Sanz-Serna, J. M.: 1992, An easily implementable fourth-order method for the time integration of wave problems, *Journal of Computational Physics* **103**, 160–168.

Frutos, J. et al.: 1990, A Hamiltonian explicit algorithm with spectral accuracy for the 'good' Boussinesq equation, *Computer Methods in Applied Mechanics and Engineering* **80**, 417–423.

Funaro, D.: 1986, A multidomain spectral approximation of elliptic equations, *Methods for Partial Differential Equations* **2**, 187–205.

Funaro, D.: 1987a, Some results about the spectrum of the Chebyshev differencing operator, *in* E. L. Ortiz (ed.), *Numerical Approximations of P. D. E., Part III*, North-Holland, Amsterdam, pp. 271–284.

Funaro, D.: 1987b, A preconditioning matrix for the Chebyshev differencing operator, *SIAM Journal of Numerical Analysis* **24**, 1024–1031.

Funaro, D.: 1988a, Computing the inverse of the Chebyshev collocation derivative matrix, *SIAM Journal of Scientific and Statistical Computing* **9**, 1050–1058.

Funaro, D.: 1988b, Domain decomposition methods for pseudo spectral approximations. Part I. Second order equations in one dimension, *Numerische Mathematik* **52**, 329–344.

Funaro, D.: 1990a, Computational aspects of pseudospectral Laguerre approximations, *Appl. Numer. Math.* **6**(6), 447–457.

Funaro, D.: 1990b, Convergence analysis for pseudospectral multidomain approximations of linear advection equations, *IMA J. Numer. Anal.* **10**(1), 63–74.

Funaro, D.: 1990c, A variational formulation for the Chebyshev pseudospectral approximation of Neumann problems, *SIAM J. Numer. Anal.* **27**(3), 695–703.

Funaro, D.: 1991, Pseudospectral approximation of a PDE defined on a triangle, *Appl. Math. Comput.* **42**, 121–138. Subdivides each triangle into three quadrilaterals.

Funaro, D.: 1992a, Approximation by the Legendre collocation method of a model problem in electrophysiology, *Journal of Computational and Applied Mathematics* **43**, 261–271.

Funaro, D.: 1992b, *Polynomial Approximation of Differential Equations*, Springer-Verlag, New York. 313 pp.

Funaro, D.: 1993a, A new scheme for the approximation of advection-diffusion equations by collocation, *SIAM Journal of Numerical Analysis* **30**(6), 1664–1676.

Funaro, D.: 1993b, FORTRAN routines for spectral methods, *Report 891*, I. A. N.-C. N. R., Pavia, Italy. The 82 FORTRAN routines plus manual are available at the Web address ftp:/ftp.ian.pv.cnr.it/pub/splib.

Funaro, D.: 1994a, A fast solver for elliptic boundary-value problems in the square, *in* C. Bernardi and Y. Maday (eds), *Analysis, Algorithms and Applications of Spectral and High Order Methods for Partial Differential Equations*, Selected Papers from the International Conference on Spectral and High Order Methods (ICOSAHOM '92), Le Corum, Montpellier, France, 22-26 June 1992, North-Holland, Amsterdam, pp. 253–256. Also in Comp. Meths. Appl. Mech. Engrg., 116.

Funaro, D.: 1994b, Spectral elements in the approximation of boundary-value-problems in complex geometries, *Applied Numerical Mathematics* **15**(2), 201–205.

Funaro, D.: 1997a, Some remarks about the collocation method on a modified Legendre grid, *Computers and Mathematics with Applications* **33**, 95–103.

Funaro, D.: 1997b, *Spectral Elements for Transport-Dominated Equations*, Vol. 1 of *Lecture Notes in Computational Science and Engineering*, Springer-Verlag, Heidelberg. 200 pp.

Funaro, D. and Gottlieb, D.: 1988, A new method of imposing boundary conditions in pseudospectral approximations of hyperbolic equations, *Mathematics of Computation* **51**, 599–613.

Funaro, D. and Gottlieb, D.: 1991, Convergence results for pseudospectral approximations of hyperbolic systems by a penalty-type boundary treatment, *Mathematics of Computation* **57**, 585–596.

Funaro, D. and Heinrichs, W.: 1990, Some results about the pseudospectral approximation of one dimensional fourth order problems, *Numerische Mathematik* **58**, 399–418.

Funaro, D. and Kavian, O.: 1991, Approximation of some diffusion evolutions equations in unbounded domains by Hermite functions, *Mathematics of Computation* **57**, 597–619.

Funaro, D. and Rothman, E.: 1989, Preconditioning matrices for the pseudospectral approximations of first-order operators, *in* T. J. Chung and G. R. Karr (eds), *Finite Elements Analysis in Fluids*, UAH Press, Huntsville, Alabama, pp. 1458–1463.

Funaro, D. and Russo, A.: 1993, Approximation of advection-diffusion problems by a modified Legendre grid, *in* K. Morgan, E. Onate, J. Periaux, J. Peraire and O. C. Zienkiewicz (eds), *Finite Elements in Fluids, New Trends and Applications*, Pineridge Press, Battersea, England, pp. 1311–1318.

Funaro, D., Quarteroni, A. and Zanolli, P.: 1988, An iterative procedure with interface relaxation for domain decomposition methods, *SIAM Journal of Numerical Analysis* **25**, 1213–1236.

Funaro, O. C. D. and Kavian, O.: 1990, Laguerre spectral approximations of elliptic problems in exterior domains, *Computer Methods in Applied Mechanics and Engineering* **80**, 451–458.

Gad-el-Hak, M., Davis, S. H., McMurray, J. T. and Orszag, S. A.: 1984, On the stability of the decelerating laminar boundary layer, *Journal of Fluid Mechanics* **138**, 297–323.

Garba, A.: 1998, A mixed spectral/wavelet method for the solution of Stokes problem, *J. Comput. Phys.* **145**(1), 297–315. Chebyshev polynomials in the non-periodic coordinate, Daubechies wavelets in the spatially periodic coordinate.

García-Archilla, B.: 1995, Some practical experience with the time integration of dissipative equations, *Journal of Computational Physics* **122**, 25–29. Kuramoto-Sivashinsky equation is used to compare Nonlinear Galerkin and standard Galerkin methods; finds that standard stiff ODE methods are very effective, more so than Nonlinear Galerkin schemes.

García-Archilla, B.: 1996, A spectral method for the equal width equation, *Journal of Computational Physics* **125**, 395–402.

García-Archilla, B. and de Frutos, J.: 1995, Time integration of the non-linear Galerkin method, *IMA Journal of Numerical Analysis* **15**, 221–244. "The results show that for these problems [Kuramoto-Sivashinsky and reaction-diffusion equations], the non-linear Galerkin method is not competitive with either pure spectral Galerkin or pseudospectral discretizations", from the abstract.

Gardner, D. R., Trogdon, S. A. and Douglass, R. W.: 1989, A modified tau spectral method that eliminates spurious eigenvalues, *Journal of Computational Physics* **80**, 137–167.

Gary, J. and Helgason, R.: 1970, A matrix method for ordinary differential equation eigenvalue problems, *Journal of Computational Physics* **5**, 169–187. Shows that the QR and QZ matrix eigensolvers make high order discretizations especially advantageous.

Gautheir, S., Guillard, H., Lumpp, T., Malé, J., Peyret, R. and Renaud, F.: 1996, A spectral domain decomposition technique with moving interfaces for viscous incompressible flows, *ECCOMAS 96*, John Wiley, New York.

Gelb, A.: 1997, The resolution of the Gibbs phenomenon for spherical harmonics, *Math. Comput.* **66**, 699–717. Applies the Gegenbauer polynomial sequence acceleration of Gottlieb and Shu to discontinuous functions on the sphere.

Gelb, A. and Gottlieb, D.: 1998, Recovering grid-point values without Gibbs oscillations in two dimensional domains on the sphere, *J. Comput. Phys.* Submitted. Gegenbauer polynomial regularization for rectangular regions or a union of rectancles on the sphere with meteorological applications.

Gentleman, W. M.: 1972a, Implementing Clenshaw-Curtis quadrature, *Communications of the ACM* **15**(5), 353–355. Table with complete FORTRAN code including the necesary cosine-FFT.

Gentleman, W. M.: 1972b, Implementing Clenshaw-Curtis quadrature: I. Methodology and experience, *Communications of the ACM* **15**(5), 337–342. Careful analysis of performance on fifty test integrals.

Gentleman, W. M.: 1972c, Implementing Clenshaw-Curtis quadrature: II. Computing the cosine transformation, *Communications of the ACM* **15**(5), 343–346.

Ghaddar, N. K., Korczak, K. Z., Mikic, B. B. and Patera, A. T.: 1986a, Numerical investigation of incompressible flow in grooved channels, Part 1: Stability and self-sustained oscillations, *Journal of Fluid Mechanics* **163**, 99–127.

Ghaddar, N. K., Korczak, K. Z., Mikic, B. B. and Patera, A. T.: 1986b, Numerical investigation of incompressible flow in grooved channels, Part 2: Resonance and oscillatory heat transfer, *Journal of Fluid Mechanics* **168**, 541–567.

Ghosh, S., Hossain, M. and Matthaeus, W. H.: 1993, The application of spectral methods in simulating compressible fluid and magnetofluid turbulence, *Computer Physics Communications* **74**, 18–40.

Gill, A. W. and Sneddon, G. E.: 1995, Complex mapped matrix methods in hydrodynamic stability problems, *Journal of Computational Physics* **122**, 13–24. Derive analytic formulas for optimizing Boyd's mappings to detour around singularities in the complex plane to calculate eigenvalues for modes with a singularity on or near the real axis, as occur in linearized hydrodynamic stability and Sturm-Liouville eigenproblems of the Fourth Kind; illustrated with experiments.

Gill, A. W. and Sneddon, G. E.: 1996, Pseudospectral methods and composite complex maps for near-boundary critical latitudes, *Journal of Computational Physics* **129**(1), 1–7. Simple formulas for optimizing a change-of-coordinate to resolve a differential equation with singularities very close to the boundary; Chebyshev pseudospectral solution of eigenproblems.

Giraldo, F. X.: 1998, The Lagrange-Galerkin spectral element method on unstructured quadrilateral grids, *J. Comput. Phys.* **147**(1), 114–146. Semi-Lagrangian spectral elements for fluids.

Givi, P. and Madnia, C. K.: 1993, Spectral methods in combustion, *in* T. J. Chung (ed.), *Numerical Modeling in Combustion*, Hemisphere, Taylor and Francis, Washington, pp. 409–452.

Glatzmaier, G. A.: 1984, Numerical simulations of stellar convective dynamos. Part I. The model and method, *Journal of Computational Physics* **55**, 461–484. Spherical harmonics in latitude and longitude, Chebyshev polynomials in radius. This model, with refinements, has been used in more than forty successive articles on mantle convection and stellar fluid dynamics by Glatzmaier and collaborators.

Glatzmaier, G. A.: 1988, Numerical simulations of mantle convection: Time-dependent ,three-dimensional, compressible spherical shell, *Geophysical and Astrophysical Fluid Dynamics* **43**, 223–264.

Glatzmaier, G. A. and Roberts, P. H.: 1997, Simulating the geodynamo, *Contemp. Phys.* **38**(4), 269–288. C.

Glowinski, R., Keller, H. B. and Reinhart, L.: 1985, Continuation conjugate gradient methods for the least squares solution of nonlinear boundary value problems, *SIAM Journal of Scientific and Statistical Computing* **6**, 793–832.

Godon, P.: 1995, The propagation of acoustic waves and quasi-periodic oscillations in accretion disc boundary layers, *Monthly Notices of the Royal Astronomical Society* **274**, 61–74. Polar coordinates in a two-dimensional plane; Fourier-Chebyshev with 4th order Runge-Kutta. Application.

Godon, P.: 1996a, Accretion disc boundary layers around pre-main-sequence stars, *Monthly Notices of the Royal Astronomical Society* **279**(4), 1071–1082. Polar coordinates in a two-dimensional plane; Fourier-Chebyshev with 4th order Runge-Kutta. Application.

Godon, P.: 1996b, Accretion disk boundary layers in classical t tauri stars, *Astrophysical Journal* **463**(2), 674–680. Polar coordinates in a two-dimensional plane; Fourier-Chebyshev with 4th order Runge-Kutta. Kosloff/Tal-Ezer mapping.

Godon, P.: 1997a, Advection in accretion disk boundary layers, *Astrophysical Journal* **483**(1), 882–886. Polar coordinates in a two-dimensional plane through the center of an accretion disk around a star: Chebyshev in radius and Fourier in the polar angle.

Godon, P.: 1997b, Numerical modeling of tidal effects in polytropic accretion disks, *Astrophysical Journal* **480**(1), 329–343. Polar coordinates in a two-dimensional plane through the center of an accretion disk around a star: Chebyshev in radius and Fourier in the polar angle. Kosloff-Tal-Ezer mapping.

Godon, P., , Regev, O. and Shaviv, G.: 1995, One-dimensional time-dependent numerical modeling of accretion disc boundary-layers, *Monthly Notices of the Royal Astronomical Society* **275**(4), 1093–1101. Chebyshev-Fourier polar coordinate model with 4th order Runge-Kutta time-marching.

Godon, P. and Shaviv, G.: 1993, A two-dimensional time-dependent Chebyshev method of collocation for the study of astrophysical flows, *Comput. Meths. Appl. Mech. Engin.* **110**(1–2), 171–194. Chebyshev-Fourier polar coordinate model with 4th order Runge-Kutta time-marching.

Godon, P. and Shaviv, G.: 1995, The dynamics of two-dimensional local and finite perturbations in envelopes of rotating dwarf stars, *Astrophysical Journal* **447**, 797–806. Chebyshev-Fourier polar coordinate model with 4th order Runge-Kutta time-marching.

Goldhirsch, I., Orszag, S. A. and Maulik, B. K.: 1987, An efficient method for computing leading eigenvalues and eigenvectors of large asymmetric matrices, *Journal of Scientific Computing* **2**, 33–58.

Gordon, C. T. and Stern, W. F.: 1982, A description of the GFDL global spectral model, *Monthly Weather Review* **110**, 625–644.

Gottlieb, D.: 1981, The stability of pseudospectral Chebyshev methods, *Mathematics of Computation* **36**, 107–118.

Gottlieb, D.: 1984, Spectral methods for compressible flow problems, *in* Soubbaramayer and J. P. Boujot (eds), *Proceedings of the 9th International Conference on Numerical Methods in Fluid Dynamics, Saclay, France*, number 218 in *Lecture Notes in Physics*, Springer-Verlag, New York, pp. 48–61.

Gottlieb, D. and Hirsh, R. S.: 1989, Parallel pseudospectral domain decomposition techniques, *Journal of Scientific Computing* **4**(1), 309–326.

Gottlieb, D. and Lustman, L.: 1983a, The Dufort-Frankel Chebyshev method for parabolic initial value problems, *Computers and Fluids* **11**, 107–120.

Gottlieb, D. and Lustman, L.: 1983b, The spectrum of the Chebyshev collocation operator for the heat equation, *SIAM Journal of Numerical Analysis* **20**, 909–921. Global and subdomain cases are analyzed with numerical results for the potential equation of airfoil theory.

Gottlieb, D. and Orszag, S. A.: 1977, *Numerical Analysis of Spectral Methods*, SIAM, Philadelphia, PA. 200 pp.

Gottlieb, D. and Orszag, S. A.: 1980, High resolution spectral calculations of inviscid compressible flows, *in* (ed.), *Approximation Methods for Navier-Stokes Problems*, Springer-Verlag, New York, pp. 381–398.

Gottlieb, D. and Shu, C.-W.: 1994, Resolution properties of the Fourier method for discontinuous waves, *in* C. Bernardi and Y. Maday (eds), *Analysis, Algorithms and Applications of Spectral and High Order Methods for Partial Differential Equations*, Selected Papers from the International Conference on Spectral and High Order Methods (ICOSAHOM '92), Le Corum, Montpellier, France, 22-26 June 1992, North-Holland, Amsterdam, pp. 27–38. Also in Comput. Meths. Appl. Mech. Engrg., vol. 116.

Gottlieb, D. and Shu, C.-W.: 1995a, On the Gibbs phenomenon IV: Recovering exponential accuracy in a subinterval from a Gegenbauer partial sum of a piecewise analytic function, *Mathematics of Computation* **64**, 1081–1095.

Gottlieb, D. and Shu, C.-W.: 1995b, On the Gibbs phenomenon V: Recovering exponential accuracy from collocation point values of a piecewise analytic function, *Numerische Mathematik* **71**, 511–526.

Gottlieb, D. and Streett, C. L.: 1990, Quadrature imposition of compatibility conditions in Chebyshev methods, *Journal of Scientific Computing* **5**(3), 223–240.

Gottlieb, D. and Tadmor, E.: 1985, Recovering pointwise values of discontinuous data within spectral accuracy, *in* E. M. Murman and S. S. Abarbanel (eds), *Progress and Supercomputing in Computational Fluid Dynamics*, Birkhäuser, Boston, pp. 357–375.

Gottlieb, D. and Tadmor, E.: 1991, The CFL condition for spectral approximation to hyperbolic BVPs, *Mathematics of Computation* **56**, 565–588.

Gottlieb, D. and Témam, R.: 1993, Implementation of the Nonlinear Galerkin method with pseudospectral (collocation) discretizations, *Applied Numerical Mathematics* **12**, 119–134.

Gottlieb, D. and Turkel, E.: 1980, On time discretization for spectral methods, *Studies in Applied Mathematics* **63**, 67–86.

Gottlieb, D. and Turkel, E.: 1985, Topics in spectral methods for time dependent problems, *in* F. Brezzi (ed.), *Numerical Methods In Fluid Dynamics*, Springer-Verlag, New York, pp. 115–155.

Gottlieb, D., Hussaini, M. Y. and Orszag, S. A.: 1984a, Theory and application of spectral methods, *in* R. G. Voigt, D. Gottlieb and M. Y. Hussaini (eds), *Spectral Methods for Partial Differential Equations*, SIAM, Philadelphia, pp. 1–54.

Gottlieb, D., Lustman, L. and Orszag, S. A.: 1981a, Spectral calculations of one-dimensional inviscid compressible flow, *SIAM Journal of Scientific and Statistical Computing* **2**, 296–310.

Gottlieb, D., Lustman, L. and Streett, C. L.: 1984b, Spectral methods for two-dimensional shocks, *in* R. G. Voigt, D. Gottlieb and M. Y. Hussaini (eds), *Spectral Methods for Partial Differential Equations*, SIAM, Philadelphia, pp. 79–96.

Gottlieb, D., Lustman, L. and Tadmor, E.: 1987a, Stability analysis of spectral methods for hyperbolic initial-value problems, *SIAM Journal of Numerical Analysis* **24**, 241–258.

Gottlieb, D., Lustman, L. and Tadmor, E.: 1987b, Convergence of spectral methods for hyperbolic initial-value problems, *SIAM Journal of Numerical Analysis* **24**, 532–537.

Gottlieb, D., Orszag, S. A. and Turkel, E.: 1981b, Stability of pseudospectral and finite difference methods for variable coefficient problems, *Mathematics of Computation* **37**, 293–305.

Gottlieb, D., Shu, C.-W., Solomonoff, A. and Vandeven, H.: 1992, On the Gibbs phenomenon I: recovering exponential accuracy from the Fourier partial sum of a nonperiodic analytic function, *Journal of Computational and Applied Mathematics* **43**, 81–98.

Grant, F. C. and Feix, M. R.: 1967, Fourier-Hermite solutions of the Vlasov equations in the linearized limit, *The Physics of Fluids* **10**(4), 696–702. Damping term is added to improve the otherwise slow convergence of the Hermite series for the velocity coordinate.

Gravel, S. and Staniforth, A.: 1994, A mass-conserving semi-Lagrangian scheme for shallow-water equations, *Monthly Weather Review* **122**, 243–248.

Gravel, S., Staniforth, A. and Côté, J.: 1993, A stability analysis of a family of baroclinic semi-Lagrangian forecast models, *Monthly Weather Review* **121**, 815–824.

Greengard, L. and Strain, J.: 1991, The fast Gauss transform, *SIAM J. Scient. Stat. Comput.* **12**, 79–94. Sum of N Gaussians at N points in space; not a generalized FFT but a close cousin.

Gresho, P. M., Gartling, D. K., Torczynski, J. R., Cliffe, K. A., Winters, K. H., Garratt, T. J., Spence, A. and Goodrich, J. W.: 1993, Is the steady visous incompressible two-dimensional flow over a backward- facing step at Re=800 stable?, *International Journal for Numerical Methods in Fluids* **17**, 501–541. Shows, by applying several different algorithms to a classical benchmark, that a published spectral element calculation was in error because of insufficient convergence tests. See also Kaiktsis, Karniadakis and Orszag(1991, 1996).

Grosch, C. E. and Orszag, S. A.: 1977, Numerical solution of problems in unbounded regions: coordinate transforms, *Journal of Computational Physics* **25**, 273–296.

Guillard, H. and Desideri, J. A.: 1990, Iterative methods with spectral preconditioning for elliptic equations, *Comput. Meths. Appl. Mech. Engrg.* **80**(1–3), 305–312.

Guillard, H. and Peyret, R.: 1988, On the use of spectral methods for the numerical solution of stiff problems, *Computer Methods in Applied Mechanics and Engineering* **66**, 17–43. Flame propagation problems and Burgers' equation.

Guo, B.-Y.: 1998, *Spectral Methods and Their Applications*, World Scientific, Singapore. 360 pp. Emphasizes proofs rather than numerical examples.

Guo, B.-Y. and Manoranjan, V. S.: 1985, A spectral method for solving the RLW equation, *IMA Journal of Numerical Analysis* **5**, 307–318.

Gustafsson, N. and McDonald, A.: 1996, A comparison of the HIRLAM gridpoint and spectral semi-Lagrangian models, *Monthly Weather Review* **124**(9), 2008–2022. Both models work well and at comparable cost; semi-Lagrangian advection is more accurate than the alternatives.

Gwynllyw, D. R., Davies, A. R. and Phillips, T. N.: 1996, A moving spectral element approach to the dynamically loaded journal bearing problem, *Journal of Computational Physics* **123**, 476–494.

Haidvogel, D. B.: 1977, Quasigeostrophic regional and general circulation modelling: an efficient pseudospectral aproximation technique, *in* R. P. Shaw (ed.), *Computing Methods in Geophysical Mechanics, Volume 25*, ASME, New York. REVIEW.

Haidvogel, D. B.: 1983, Periodic and regional models, *in* A. P. Robinson (ed.), *Eddies in Marine Science*, Springer-Verlag, New York, pp. 404–437. REVIEW.

Haidvogel, D. B. and Zang, T. A.: 1979, The accurate solution of Poisson's equation by expansion in Chebyshev polynomials, *Journal of Computational Physics* **30**, 167–180. Classic paper on separable BVP PDE's.

Haidvogel, D. B., Curchitser, E. N., Iskandarani, M., Hughes, R. and Taylor, M.: 1997, Global modeling of the ocean and atmosphere using the spectral element method, *Atmosphere-Ocean* **35**(1), 505–531.

Haj, A., Phillips, C. and Delves, L. M.: 1980, The global element method for stationary advective problems, *International Journal of Numerical Methods in Engineering* **15**, 167–175.

Hald, O. H.: 1981, Convergence of Fourier methods for Navier-Stokes equations, *Journal of Computational Physics* **40**, 305–317. Proofs.

Haldenwang, P., Labrosse, G., Abboudi, S. and Deville, M.: 1984, Chebyshev 3-d spectral and 2-d pseudospectral solvers for the Helmholtz equation, *Journal of Computational Physics* **55**, 115–128. Comparison of three methods for constant coefficient PDE: diagonalization in two dimensions plus solution of tridiagonal systems in the third (Haidvogel and Zang, 1979), diagonalization in all three dimensions, and an iteration preconditioned by a fast direct finite difference solver for a separable system of cost $O(N^3 \log_2(N))$. The HZ algorithm was fastest, but loses about five digits due to roundoff for N as small as 64.

Haltiner, G. J. and Williams, R. T.: 1980, *Numerical Prediction and Dynamic Meteorology*, second edn, John Wiley. Chapter on spherical harmonics.

Han, H. C., Schultz, W. W., Boyd, J. P. and Schumack, M. R.: 1999, The flow in a elliptical journal bearing by the spectral element method for a rotating and translating shaft. To appear. The movement of the spinning central shaft was calculated simultaneously with the flow.

Hardiker, V.: 1997, A global numerical weather prediction model with variable resolution, *Monthly Weather Review* **125**(1), 59–73. Conformal mapping is used to make at T-83 spherical harmonics model as effective as a T-170 uniform resolution code for modelling hurricanes.

Harding, R. C.: 1968, Response of a one-dimensional Vlasov plasma to external electric fields, *Physics of Fluids* **11**(10), 2233–2240. Fourier for space coordinate, Hermite for velocity coordinate, finite difference for time. Little about numerics except that several hundred Hermite modes were needed because of problems resolved by later improvements (Holloway, 1996a,b).

Hargittai, I. and Hargittai, M.: 1994, *Symmetry: A Unifying Concept*, Shelter Publications, Bolinas, California. Good popularization of symmetry in art and nature.

Haugen, J. E. and Machenhauer, B.: 1993, A spectral limited-area model formulation with time-dependent boundary conditions applied to the shallow-water equations, *Monthly Weather Review* **121**, 2618–2630.

Haupt, S. E. and Boyd, J. P.: 1988, Modeling nonlinear resonance: A modification to Stokes' perturbation expansion, *Wave Motion* **10**, 83–98. Fourier basis for nonlinear eigenproblem. Analysis of the relationship between perturbation theory and the Galerkin numerical algorithm.

Haupt, S. E. and Boyd, J. P.: 1991, Double cnoidal waves of the Korteweg-deVries equation: The boundary value approach, *Physica D* **50**, 117–134. Two-dimensional Fourier basis for a nonlinear problem with two eigenparameters.

Haurwitz, B.: 1940, The motion of atmospheric disturbances on the spherical earth, *J. Marine Res.* **3**, 254–267. Shows spherical harmonics are travelling wave solutions of the shallow water equation, now known as "Rossby-Haurwitz" waves.

Healy, D. M., Rockmore, D. N., Kostelec, P. and Moore, S. S. B.: 1999, FFTs for the 2-sphere — Improvements and variations, *Adv. Appl. Math.* Submitted. Fast spherical harmonic transforms.

Heikes, R. and Randall, D. A.: 1995a, Numerical integration of the shallow-water equations on a twisted icosahedral grid. Part I: Basic design and result of tests., *Monthly Weather Review* **123**, 1862–1880. Non-spectral alternative to spherical harmonics.

Heikes, R. and Randall, D. A.: 1995b, Numerical integration of the shallow-water equations on a twisted icosahedral grid. Part II: A detailed description of the grid and an analysis of numerical accuracy., *Monthly Weather Review* **123**, 1881–1887. Non-spectral alternative to spherical harmonics.

Heinrichs, W.: 1987, *Kollokationsverfahren und Mehrgittermethoden bei elliptschen Randwertaufgaben*, Vol. 168 of *GMD-Bericht Nr.*, Oldenbourg-Verlag, Oldenbourg. Doctoral thesis.

Heinrichs, W.: 1988a, Line relaxation for spectral multigrid, *Journal of Computational Physics* **77**, 166–182.

Heinrichs, W.: 1988b, Multigrid methods for combined finite difference and Fourier problems, *Journal of Computational Physics* **78**, 424–436.

Heinrichs, W.: 1988c, Collocation and full multigrid methods, *Applied Mathematics and Computation* **28**, 35–45.

Heinrichs, W.: 1989a, Improved condition number for spectral methods, *Mathematics of Computation* **53**, 103–119.

Heinrichs, W.: 1989b, Spectral methods with sparse matrices, *Numerische Mathematik* **56**, 25–41.

Heinrichs, W.: 1989c, Konvergenzaussagen für Kollokationsverfahren bei elliptschen Randwertaufgaben, *Numerische Mathematik* **54**, 619–637.

Heinrichs, W.: 1990, Algebraic spectral multigrid methods, *Computer Methods in Applied Mechanics and Engineering* **80**, 281–289.

Heinrichs, W.: 1991a, A 3D spectral multigrid method, *Applied Mathematics and Computation* **41**, 117–128.

Heinrichs, W.: 1991b, Stabilization techniques for spectral methods, *Journal of Scientific Computing* **6**(1), 1–19. Basis functions that satisfy homogeneous boundary conditions reduce condition number.

Heinrichs, W.: 1991c, A stabilized treatment of the biharmonic operator with spectral methods, *SIAM Journal of Scientific and Statistical Computing* **12**, 1162–1172.

Heinrichs, W.: 1992a, A spectral multigrid method for the Stokes problem in streamfunction formulation, *Journal of Computational Physics* **102**, 310–318.

Heinrichs, W.: 1992b, A stabilized multidomain approach for singular perturbation methods, *Journal of Scientific Computing* **7**, 95–127.

Heinrichs, W.: 1992c, Strong convergence estimates for pseudospectral methods, *Applications of Mathematics* **37**(6), 401–417.

Heinrichs, W.: 1992d, Spectral projective Newton-methods for quasilinear elliptic boundary value problems, *Calcolo* **29**, 33–48.

Heinrichs, W.: 1993a, Distributive relaxations for the spectral Stokes operator, *Journal of Scientific Computing* **8**, 389–398.

Heinrichs, W.: 1993b, Splitting techniques for the pseudospectral approximation of the unsteady Stokes equation, *SIAM Journal of Numerical Analysis* **30**, 19–39.

Heinrichs, W.: 1993c, Spectral multi-grid techniques for the Navier-Stokes equations, *Computer Methods in Applied Mechanics and Engineering* **106**, 297–314.

Heinrichs, W.: 1993d, Spectral multigrid methods for the reformulated Stokes equations, *Journal of Computational Physics* **107**(2), 213–224.

Heinrichs, W.: 1993e, Domain decomposition for fourth order problems, *SIAM Journal of Numerical Analysis* **30**(2), 435–453.

Heinrichs, W.: 1993f, Finite element preconditioning for spectral multigrid methods, *Applied Mathematics and Computation* **59**(1), 19–40.

Heinrichs, W.: 1993g, Spectral multigrid methods for domain decomposition problems using patching techniques, *Applied Mathematics and Computation* **59**(2), 165–176.

Heinrichs, W.: 1993h, Defect correction for convection dominated flow, *in* F.-K. Hebeker, R. Rannacher and G. Wittum (eds), *Proceedings of the International Workshop on Numerical Methods for the Navier-Stokes Equations*, number 47 in *Notes on Numerical Fluid Mechanics*, Heidelberg, pp. 111–121.

Heinrichs, W.: 1994a, Spectral methods for singular perturbation problems, *Applications of Mathematics* **39**(3), 161–188.

Heinrichs, W.: 1994b, *Efficient iterative solution of spectral systems for the Navier-Stokes equations*, Vol. 2 of *Wissenschaftliche Schriftenreihe Mathematik*, Verlag Dr. Koster, Berlin. Habilitationshrift, 109 pp.

Heinrichs, W.: 1994c, Spectral viscosity for convection dominated flow, *Journal of Scientific Computing* **9**(2), 137–147.

Heinrichs, W.: 1994d, Defection correction for the advection-diffusion equation, *Computer Methods in Applied Mechanics and Engineering* **119**, 191–197.

Heinrichs, W.: 1996, Defect correction for convection dominated flow, *SIAM Journal of Scientific Computing* **17**(5), 1082–1091.

Heinrichs, W.: 1998a, Spectral collocation on triangular elements, *J. Comput. Phys.* **145**(2), 743–757.

Heinrichs, W.: 1998b, Splitting techniques for the unsteady Stokes equations, *SIAM J. Numer. Anal.* **35**(4), 1646–1662. Third order time accurate Uzawa scheme with a single, unstaggered grid for in a pseudospectral approach.

Held, I. M. and Suarez, M. J.: 1994, A proposal for the intercomparison of the dynamical cores of atmospheric general circulation models, *Bull. Amer. Meteor. Soc.* **75**, 1825–1830. Comparison of 20-level models:, leapfrog time-stepping $T63$ spherical harmonics versus 2d order finite difference with $G72$[144 grid points around the equator]. There "is an impressive degree of agreement between the two models" but "the climate of both models are sensitive to resolution and have not yet converged at the resolution presented here". This article is unusual in claiming similar results for low order and high order methods for complex flows at the same resolution.

Hendry, J. A. and Delves, L. M.: 1979, The global element method applied to a harmonic mixed boundary value problem, *Journal of Computational Physics* **33**, 33–44.

Hendry, J. A., Delves, L. M. and Mohamed, J.: 1982, Iterative solution of the global element equations, *Computer Methods in Applied Mechanics and Engineering* **35**, 271–283.

Herbst, B. M. and Ablowitz, M.: 1992, Numerical homoclinic instabilities in the sine-Gordon equation, *Quaestiones Mathematicae* **15**, 345–363.

Herbst, B. M. and Ablowitz, M.: 1993, Numerical chaos, symplectic integrators, and exponentially small splitting distances, *Journal of Computational Physics* **105**, 122–132.

Herring, J. R., Orszag, S. A., Kraichnan, R. H. and Fox, D. G.: 1974, Decay of two-dimensional homogeneous turbulence, *Journal of Fluid Mechanics* **66**, 417–444.

Hesthaven, J. S.: 1998a, From electrostatics to almost optimal nodal sets for polynomial interpolation in a simplex, *SIAM Journal of Numerical Analysis* **35**(2), 655–676. Good distribution of points for a non-tensor product grid in a triangular subdomain. An alternative is given by Taylor&Wingate(1999).

Hesthaven, J. S.: 1998b, Integration preconditioning of pseudospectral operators. I. Basic linear operators, *SIAM J. Numer. Anal.* **35**(2), 1571–1593.

Hesthaven, J. S.: 1999, A stable penalty method for the compressible Navier-Stokes equations. III. Multidimensional domain decomposition schemes, *SIAM Journal of Scientific Computing* **20**(1), 62–93.

Hesthaven, J. S. and Gottlieb, D.: 1996, A stable penalty method for the compressible Navier-Stokes equations. I. Open boundary conditions., *SIAM Journal of Scientific Computing* **17**(3), 579–612.

Hesthaven, J. S. and Gottlieb, D.: 1999, Stable spectral methods for conservation laws on triangles with unstructured grids, *Comput. Meths. Appl. Mech. Engrg.* **175**(3–4), 361–382.

Hesthaven, J. S., Dinesen, P. G. and Lynov, J. P.: 1999, Spectral collocation time-domain modeling of diffractive optical elements, *J. Comput. Phys.* **155**(2), 287–306. Pseudospectral multidomain with absorbing boundary layers. The Kosloff/Tal-Ezer map with $\beta = 1 - \cos(1/2)$ is used to double the timestep versus what would be stable with the usual unmapped Chebyshev grid.

Higgins, J. R.: 1977, *Completeness and Basis Properties of Sets of Special Functions*, Cambridge University Press, New York. This slim book is a very readable introduction to basis sets, but has no numerical content. Pp. 59-64 are a good discussion of complex-valued rational Chebyshev functions.

Hille, E.: 1939, Contributions to the theory of Hermitian series, *Duke Math. Journal* **5**, 875–936. This and the next three papers prove Hermite function convergence theorems.

Hille, E.: 1940a, Contributions to the theory of Hermitian series. II. The representation problem, *Transaction of the American Mathematical Society* **47**, 80–94.

Hille, E.: 1940b, A class of differential operators of infinite order, I, *Duke Mathematical Journal* **7**, 458–495.

Hille, E.: 1961, Sur les fonctions analytiques définies par des séries d'hermite, *Journal of Mathematiques Pures Appliques* **40**, 335–342.

Ho, L.-W. and Patera, A. T.: 1990, A Legendre spectral element method for simulation of unsteady incompressible viscous free-surface flows, *in* C. Canuto and A. Quarteroni (eds), *Spectral and High Order Methods for Partial Differential Equations: Proceedings of the ICOSAHOM '89 Conference in Como, Italy*, North-Holland/Elsevier, Amsterdam, pp. 355–366. Also in Comput. Meths. Appl. Mech. Engrg., vol. 80, with the same page numbers.

Ho, L. W. and Patera, A. T.: 1991, Variational formulation of three-dimensional viscous free-surface flows: natural imposition of surface tension boundary conditions, *International Journal for Numerical Methods in Fluids* **13**, 691–698.

Ho, L.-W. and Rønquist, E. M.: 1994, Spectral element solution of steady incompressible viscous free-surface flows, *in* C. Bernardi and Y. Maday (eds), *Analysis, Algorithms and Applications of Spectral and High Order Methods for Partial Differential Equations*, Selected Papers from the International Conference on Spectral and High Order Methods (ICOSAHOM '92), Le Corum, Montpellier, France, 22-26 June 1992, North-Holland, Amsterdam, pp. 347–368. Also in Finite Elements in Analysis and Design, vol. 16, pp. 207-229.

Ho, L.-W., Maday, Y., Patera, A. T. and Rønquist, E. M.: 1990, A high-order Lagrangian-decoupling method for the incompressible Navier-Stokes equations, *in* C. Canuto and A. Quarteroni (eds), *Spectral and High Order Methods for Partial Differential Equations: Proceedings of the ICOSAHOM '89 Conference in Como, Italy*, North-Holland/Elsevier, Amsterdam, pp. 65–90.

Hogan, T. F. and Rosmond, T. E.: 1991, The description of the Navy operational global atmospheric prediction systems's spectral forecast model, *Monthly Weather Review* **119**, 1786–1815.

Holloway, J. P.: 1996a, Spectral velocity discretizations for the Vlasov-Maxwell equations, *Transport Theory and Statistical Physics* **25**, 1–32.

Holloway, J. P.: 1996b, Hamiltonian spectral methods, *Journal of Computational Physics* **129**(1), 121–133. Fourier-Hermite spectral methods.

Holly, Jr., F. M. and Preissman, A.: 1977, Accurate evaluation of transport in two dimensions, *Journal of Hydraulics Division of the ASCE* **98**, 1259–1277. Semi-Lagrangian advection.

Holvorcem, P. R.: 1992, Asymptotic summation of Hermite series, *J. Phys. A* **25**(4), 909–924.

Holvorcem, P. R. and Vianna, M. L.: 1992, Integral-equation approach to tropical ocean dynamics. 2. Rossby-wave scattering from the equatorial Atlantic western boundary, *J. Marine Res.* **50**(1), 33–61. Boundary-element algorithm constructed through ingenious summation of slowly converging Hermite series.

Hortal, M. and Simmons, A. J.: 1991, Use of reduced Gaussian grids in spectral models, *Monthly Weather Review* **119**, 1057–1074.

Hoskins, B. J.: 1973, Comments on the possible use of Laguerre polynomials for representing the vertical structure of numerical models of the atmosphere, *Quart. J. Roy. Met. Soc.* **99**, 571–572.

Hoskins, B. J.: 1980, Representation of the earth topography using spherical harmonics, *Monthly Weather Review* **108**, 111–115. Exponential filter.

Hoskins, B. J. and Simmons, A. J.: 1975, A multi-layer spectral model and the semi-implicit method, *Quart. J. Roy. Met. Soc.* **101**, 637–655. Comparison of spherical harmonics/vertical finite differences model with a three-dimensional finite difference code.

Hua, B. L.: 1987, Periodic quasi-geostrophic models, *in* J. J. O'Brien (ed.), *Advanced Physical Oceanographic Numerical Modelling*, D. Reidel, Dordrect, Holland, pp. 233–254. REVIEW.

Hua, B. L. and Haidvogel, D. B.: 1986, Numerical simulations of the vertical structure of quasi-geostrophic turbulence, *Journal of the Atmospheric Sciences* **43**(23), 2923–2936. Fourier pseudospectral in the horizontal, normal modes in ocean depth for both constant and variable stratification.; detailed description of the numerical algorithms.

Huang, W.-Z. and Sloan, D. M.: 1993a, A new pseudospectral method with upwind features, *IMA Journal of Numerical Analysis* **13**, 413–430. The authors device is to approximate $u(x)$ by different polynomials in the same equation. Diffusive terms are treated by the usual Chebyshev-Lobatto approximation. The advective terms are approximated by a polynomial which omits one boundary point, keeping only the boundary value on the upwind side. When the shock layer is adequately resolved, upwinding is irrelevant. When the shock is underresolved. i. e., a boundary layer of thickness of ϵ where $\epsilon \ll 1/N$, they show their scheme is much better behaved than the standard pseudospectral method. There is an $O(1)$ error around the shock, but the pollution of areas away from the shock is greatly reduced. They show many analogies between their scheme and upwinded finite differences.

Huang, W. Z. and Sloan, D. M.: 1993b, Pole condition for singular problems, *Journal of Computational Physics* **107**, 254. Spectral methods for polar and spherical coordinates.

Huang, W.-Z. and Sloan, D. M.: 1994, The pseudospectral method for solving differential eigenvalue equations, *Journal of Computational Physics* **111**, 399–409.

Hussaini, M. Y. and Zang, T. A.: 1984, Iterative spectral methods and spectral solution to compressible flows, *in* D. G. R. Voigt and M. Y. Hussaini (eds), *Spectral Methods for PDEs*, SIAM, Philadelphia.

Hussaini, M. Y., Kopriva, D. A. and Patera, A. T.: 1989, Spectral collocation methods, *Applied Numerical Mathematics* **5**, 177–208.

Hussaini, M. Y., Kopriva, D. A., Salas, M. D. and Zang, T. A.: 1985a, Spectral methods for the Euler equations: Part I — Fourier methods and shock-capturing, *AIAA J.* **23**, 64–70. Shows that spectral shock-capturing, that is, employing a smoothing technique to resolve a shock whose location is not directly calculated, fails for an astrophysical problem.

Hussaini, M. Y., Kopriva, D. A., Salas, M. D. and Zang, T. A.: 1985b, Spectral methods for the Euler equations: Part II – Chebyshev methods and shock-fitting, *AIAA J.* **23**, 234–240. Shock-fitting explicitly computes the location of the shock and then maps the computational domain so that the shock lies along a grid line in the new coordinate. This is much more effective than shock capturing for strong shocks.

If, F., Berg, P., Christiansen, P. L. and Skovgaard, O.: 1987, Split-step spectral method for nonlinear Schrodinger equation with absorbing boundaries, *Journal of Computational Physics* **72**, 501–503. Solve NLS equation with periodic boundary conditions on large period L to imitate the infinite interval. An artificial damping $\gamma(x)$ is added so that waves propagating left and right out of the computational zone do not return. The NLS equation is modified to $i\,u_t + \frac{1}{2}u_{xx} + |u|^2 u = -i\,\gamma(x)\,u$, $\gamma(x) = \gamma_0\{\mathrm{sech}^2(\alpha[x - L/2]) + \mathrm{sech}^2(\alpha[x + L/2])\}$ that is, the artificial damping is large at the ends of the interval, but exponentially small in the middle where the solution should resemble that of the undamped equation.

Ioakimidis, N. I.: 1987, Quadrature methods for the determination of zeros of transcendental functions — a review, *in* P. Keast and G. Fairweather (eds), *Numerical Integration: Recent Developments, Software and Applications*, D. Reidel, Dordrecht, Holland, pp. 61–82.

Ioakimidis, N. I. and Anastasselou, E. G.: 1986, An elementary noniterative quadrature-type method for the numerical solution of a nonlinear equation, *Computing* **37**, 269–275. Gives an explicit, exponentially convergent formula for the root of a function $f(x)$ within an interval by combining Chebyshev-Lobatto and Chebyshev-Gauss quadrature.

Iranzo, V. and Falques, A.: 1992, Some spectral approximations for differential equations in unbounded domains, *Computer Methods in Applied Mechanics and Engineering* **98**, 105–126. Compares Laguerre-tau, Laguerre pseudospectral, and rational Chebyshev pseudospectral for $x \in [0, \infty]$.

Iskandarani, M., Haidvogel, D. B. and Boyd, J. P.: 1995, A staggered spectral finite element method for the shallow water equations, *Int. J. Num. Meths. Fluids* **20**, 393–414.

Ito, K. and Teglas, R.: 1986, Legendre-Tau approximations for functional differential equations, *SIAM Journal for Control and Optimization* **24**, 737–759.

Ito, K. and Teglas, R.: 1987, Legendre-Tau approximations for functional differential equations, part 2: The linear quadratic optimal control problem, *SIAM Journal for Control and Optimization*.

Jackson, E., She, Z.-S. and Orszag, S. A.: 1991, A case study in parallel computing. I. Homogeneous turbulence on a hypercube, *Journal of Scientific Computing* **6**(1), 27–46.

Jacobs, S. J.: 1990, A variable order pseudospectral method for two-point boundary value problems, *Journal of Computational Physics* **88**, 169–182.

Jacobs, S. J.: 1995, An accurate split step scheme for viscous incompressible fluid flow, *Journal of Computational Physics* **119**, 26–33.

Jakob-Chien, R. and Alpert, B. K.: 1997, A fast spherical filter with uniform resolution, *Journal of Computational Physics* **136**(2), 580–584.

Jakob-Chien, R., Hack, J. J. and Williamson, D. L.: 1995, Spectral transform solutions to the shallow water test set, *Journal of Computational Physics* **119**, 164–187.

Jakob, R.: 1993, *Fast and Parallel Spectral Transform Algorithms for Global Shallow Water Models*, PhD dissertation, University of Colorado, Department of Electrical and Computer Engineering. Also available as NCAR Cooperative Thesis CT-144 from the National Center for Atmospheric Research, Publications Office, P. O. Box 3000,Boulder, CO 80307; author later known as R. Jakob-Chien.

Jarratt, M., Lund, J. and Bowers, K. L.: 1990, Galerkin schemes and the sinc-Galerkin method for singular Sturm-Liouville problems, *Journal of Computational Physics* **89**, 41–62.

Jarraud, M. and Baede, A. P. M.: 1985, The use of spectral techniques in numerical weather prediction, *in* B. Enquist, S. Osher and R. Somerville (eds), *Large Scale Computations in Fluid Mechanics*, number 22 in *Lectures in Applied Mathematics*, American Mathematical Society, Washington, pp. 1–41. REVIEW.

Jauberteau, F., Rosier, C. and Témam, R.: 1990a, A nonlinear Galerkin method for the Navier-Stokes equations, *Computer Methods in Applied Mechanics and Engineering* **80**, 245–260.

Jauberteau, F., Rosier, C. and Témam, R.: 1990b, The nonlinear Galerkin method in computational fluid dynamics, *Applied Numerical Mathematics* **6**, 361–370.

Jensen, T. G. and Kopriva, D. A.: 1989, Comparison of finite difference and spectral collocation reduced gravity ocean model, *in* A. M. Davies (ed.), *Modeling Marine Systems*, number II, CRC Press, Boca Raton, Florida, pp. 25–39.

Jolly, M. S. and Xiong, C.: 1995, On computing the long-time solution of the two-dimensional Navier-Stokes equations, *Theoretical and Computational Fluid Dynamics* **7**, 261–278. Comparison of several variants of the Nonlinear Galerkin method with standard two-dimensional Fourier method.

Jolly, M. S., Kevrekidis, I. G. and Titi, E. S.: 1990, Approximate inertial manifolds for the Kuramoto-Sivashinsky equation: Analysis and computations, *Physica D* **44**, 38–60. Low order Fourier basis.

Jolly, M. S., Kevrekidis, I. G. and Titi, E. S.: 1991, Preserving dissipation in approximate inertial forms for the Kuramoto-Sivashinsky equation, *Dynamics and Differential Equations* **3**(2), 179–197. Several variations of Nonlinear Galerkin methods.

Jones, D. A., Margolin, L. G. and Titi, E. S.: 1995, On the effectiveness of the approximate inertial manifold — a computational study, *Theoretical and Computational Fluid Dynamics* **7**, 243–260. Comparison of Nonlinear Galerkin with standard spectral Galerkin; mostly theory with numerical illustrations using forced Burgers equation and forced Kuramoto-Sivashinsky equation.

Jones, W. B. and O'Brien, J. J.: 1996, Pseudo-spectral methods and linear instabilities in reaction-diffusion fronts, *Chaos* **6**(2), 219–228. Comparisons of two-dimensional Fourier scheme with second order finite differences; Fourier is superior for both accuracy and preserving symmetries.

Joyce, G., Knorr, G. and Meier, H. K.: 1971, Numerical integration methods of the Vlasov equation, *Journal of Computational Physics* **8**, 53–63. Reviews Fourier ("characteristic function") and Hermite series and introduces a method of moments for the velocity coordinate. The poorly-converging coefficients of the moments expansion are truncated by polynomial extrapolation in degree n. (The method of moments, which is badly conditioned at high order, is *not* recommended.

Jun, S.-R., Kwon, Y.-H. and Kang, S.: 1998, A variational spectral method for the two-dimensional Stokes problem, *Computers Math. Applic.* **35**(4), 1–17.

Kaiktsis, L., Karniadakis, G. E. and Orszag, S. A.: 1991, Onset of three-dimensionality, equilibria, and early transition in flow over a backward facing step, *Journal of Fluid Mechanics* **231**, 501–528.

Kaiktsis, L., Karniadakis, G. E. and Orszag, S. A.: 1996, Unsteadiness and convective instabilities in two-dimensional flow over a backward facing step, *Journal of Fluid Mechanics* **321**, 157–187. Spectral elements. Attempt to understand why the authors' early paper gave unsteady flow whereas all other studies found only steady solutions.

Kalnay de Rivas, E.: 1972, On the use of nonuniform grids in finite-difference equations, *Journal of Computational Physics* **10**, 202–210. Coordinate mappings to resolve boundary layers. Not spectral, but her cosine mapping gives the usual Chebyshev pseudospectral grid.

Kanamitsu, M., Tada, K., Kudo, T., Sato, N. and Isa, S.: 1983, Description of the JMA operational spectral model, *Journal of the Meteorological Society of Japan* **61**, 812–828.

Karageorghis, A.: 1988a, A note on the Chebyshev coefficients of the general order derivative of an infinitely differentiable function, *Journal of Computational and Applied Mathematics* **21**, 129–132.

Karageorghis, A.: 1988b, A note on the Chebyshev coefficients of the moments of the general order derivative of an infinitely differentiable function, *Journal of Computational and Applied Mathematics* **21**, 383–386.

Karageorghis, A.: 1993, On the equivalence between basis recombination and boundary bordering formulations for spectral collocation methods in rectangular domains, *Mathematics and Computers in Simulation* **35**, 113–123.

Karageorghis, A.: 1994a, Satisfaction of boundary conditions for Chebyshev collocation methods in cuboidal domains, *Computers and Mathematics with Applications* **27**, 85–90.

Karageorghis, A.: 1994b, A conforming spectral technique for biharmonic-type problems in rectangular domains, *Journal of Computational and Applied Mathematics* **51**, 275–278.

Karageorghis, A.: 1994c, Conforming spectral methods for Poisson problems in cuboidal domains, *Mathematics and Computers in Simulation* **9**, 341–350.

Karageorghis, A.: 1995, A fully conforming spectral collocation scheme for second and fourth order problems, *Computer Methods in Applied Mechanics and Engineering* **126**, 305–314.

Karageorghis, A. and Phillips, T. N.: 1991, Conforming Chebyshev spectral collocation methods for the solution of laminar flow in a constricted channel, *IMA Journal of Numerical Analysis* **11**, 33–55.

Karageorghis, A. and Phillips, T. N.: 1992, On the coefficients of differentiated expansions of ultraspherical polynomials, *Applied Numerical Mathematics* **9**, 133–141.

Karagheorghis, A. and Paprzycki, M.: 1996, An efficient direct method for fully conforming spectral collocation schemes, *Numerical Mathematics* **12**, 309–319. Solution of linear systems arising from fourth and second order problems through multidomain spectral methods.

Karagheorghis, A. and Paprzycki, M.: 1998, Direct methods for spectral approximations in nonconforming domain decompositions, *Comput. Math. Appl.* **35**(11), 75–82.

Karniadakis, G. E. and Orszag, S. A.: 1993a, Some novel aspects of spectral methods, *in* M. Y. Hussaini, A. Kumar and M. D. Salas (eds), *Algorithmic Trends in Computational Fluid Dynamics*, Springer-Verlag, New York, p. 245.

Karniadakis, G. E. and Orszag, S. A.: 1993b, Nodes, modes and flow codes, *Physics Today* **46**, 32–42. REVIEW.

Karniadakis, G. E. and Sherwin, S. J.: 1995a, A triangular and tetrahedral basis for high-order finite elements: applications to the incompressible Navier-Stokes equations, *International Journal for Numerical Methods in Engineering* **38**, 3775–.

Karniadakis, G. E. and Sherwin, S. J.: 1995b, Atriangular spectral element method: applications to the incompressible Navier-Stokes equations, *Computer Methods in Applied Mechanics and Engineering* **123**, 189–229.

Karniadakis, G. E. and Sherwin, S. J.: 1999, *Spectral/hp Element Methods for CFD*, Oxford University Press, Oxford. 448 pp., 140 illustrations. Detailed analysis of both quadrilateral and triangular elements with many numerical examples.

Karniadakis, G. E., Bullister, E. T. and Patera, A. T.: 1986, A spectral element method for solution of two-and three-dimensional incompresssible Navier-Stokes equations, *in* P. Bergan and K. J. Bathe (eds), *Proceedings of the Europe-U.S. Conference on Finite Element Methods for Nonlinear Problems*, Wunderlich/Springer-Verlag, New York, pp. 803–817.

Karniadakis, G. E., Israeli, M. and Orszag, S. A.: 1991, High-order time-accurate splitting methods for incompressible Navier-Stokes equations, *Journal of Computational Physics* **97**, 414–443.

Kasahara, A.: 1977, Numerical integration of the global barotropic primitive equations with Hough harmonic expansions, *Journal of the Atmospheric Sciences* **34**, 687–701.

Kasahara, A.: 1978, Further studies on a spectral model of the global barotropic primitive equations with hough harmonic expansions, *Journal of the Atmospheric Sciences* **35**, 2043–2051.

Kasahara, A. and Puri, K.: 1981, Spectral representation of three-dimensional global data by expansion in normal modes, *Mon. Wea. Rev.* **109**, 37–51. Hough functions in latitude and longitude, eigenfunctions of the "vertical structure" equation in height.

Katopodes, N., Sanders, B. F. and Boyd, J. P.: 1998, Short wave behavior of long wave equations, *Waterways, Coastal and Ocean Engineering Journal of ASCE* **124**(5), 238–247. Application of Fourier basis; methods described in Sanders, Katopodes and Boyd (1998).

Keller, H. B.: 1977, Numerical solution of bifurcation and nonlinear eigenvalue problems, *in* P. Rabinowitz (ed.), *Applications of Bifurcation Theory*, Academic Press, New York, pp. 359–384.

Keller, H. B.: 1992, *Numerical Methods for Two-Point Boundary-Value Problems*, Dover, New York. Reprints of good research papers on solving nonlinear algebraic equations as well as differential equations.

Kermode, M., McKerrell, A. and Delves, L. M.: 1985, The calculation of singular coefficients, *Computer Methods in Applied Mechanics and Engineering* **50**, 205–215. Corner singularities, global elements.

Khabibrakhmanov, I. K. and Summers, D.: 1998, The use of generalized Laguerre polynomials in spectral methods for nonlinear differential equations, *Comput. Math. Appl.* **36**(2), 65–70. Galerkin method, implemented by recurrence, and illustrated by the Blasius equation on the semi-infinite domain.

Khajah, H. G. and Ortiz, E. L.: 1992, Numerical approximation of solutions of functional equations using the Tau method, *Applied Numerical Mathematics* **9**, 461–474.

Khajah, H. G. and Ortiz, E. L.: 1993, Rational approximations: a tau method approach, *in* T. M. Rassias, H. M. Srivastava and A. Yanushauskas (eds), *Topics in polynomials of one and several variables and their applications*, Volume dedicated to the memory of P. L. Chebyshev, World Scientific, Singapore, pp. 323–333.

Khorrami, M. R. and Malik, M. R.: 1993, Efficient computation of spatial eigenvalues for hydrodynamic stability analysis, *Journal of Computational Physics* **104**(1), 267–272. Show that by neglecting terms quadratic in the eigenvalue, one can halve the size of the matrix given the QZ algorithm, and then safely refine the eigenvalues by iterative correction with the quadratic term reinserted.

Kida, S.: 1985, Three-dimensional periodic flows with high-symmetry, *Journal of the Physical Society of Japan* **54**, 2132–2136. Three-dimensional Fourier basis, greatly reduced by symmetry.

Kida, S. and Orszag, S. A.: 1990, Energy and spectral dynamics in forced compressible turbulence, *Journal of Scientific Computing* **5**(2), 85–126. Fourier 64^3 with Runge-Kutta-Gill timestepping.

Kikuchi, N.: 1986, *Finite Element Methods in Mechanics*, Cambridge University Press, Cambridge. Very readable, elementary introduction to finite elements and skyline matrix solvers.

Klimas, A. J.: 1987, A method for overcoming the velocity space filamentation problem in collisionless plasma model solutions, *Journal of Computational Physics* **68**, 202–226. Fourier-Galerkin for both velocity and space coordinates; introduces a filter which is quite effective in controlling the dissipation of small-scale structure in the velocity coordinate.

Klimas, A. J. and Farrell, W. M.: 1994, A splitting algorithm for Vlasov simulation with filamentation filtration, *Journal of Computational Physics* **110**, 150–163. Fourier-Galerkin for both velocity coordinate and space coordinate; shows a particular filter greatly improves convergence and reduces filamentation in the velocity.

Kopriva, D. A.: 1986, A spectral multidomain method for the solution of hyperbolic systems, *Applied Numerical Mathematics* **2**, 221–241.

Kopriva, D. A.: 1987, A practical assessment of spectral accuracy for hyperbolic problems with discontinuity, *Journal of Scientific Computing* **2**, 249–262.

Kopriva, D. A.: 1988, A multidomain spectral collocation computation of the sound generated by a shock-vortex interaction, *in* M. Schultz, D. Lee and R. Sternberg (eds), *Computational Acoustics and Wave Propagation*, North-Holland, Amsterdam.

Kopriva, D. A.: 1989a, Computation of hyperbolic equations on complicated domains with patched and overset Chebyshev grids, *SIAM Journal of Scientific and Statistical Computing* **10**, 120–132.

Kopriva, D. A.: 1989b, Domain decomposition with both spectral and finite difference methods for the accurate computation of flows with shocks, *Applied Numerical Mathematics* **6**, 141–151.

Kopriva, D. A.: 1991, Multidomain spectral solution of the Euler gas dynamics equations, *Journal of Computational Physics* **96**, 428–450.

Kopriva, D. A.: 1992, Spectral solution of inviscid supersonic flows over wedges and axisymmetric cones, *Computers and Fluids* **21**, 247–266.

Kopriva, D. A.: 1993a, Spectral solution of the viscous blunt-body problem, *AIAA Journal* **31**, 1235–1242.

Kopriva, D. A.: 1993b, Spectral solutions of high-speed flows over blunt cones, *AIAA Journal* **31**, 2227–2231.

Kopriva, D. A.: 1993c, Multidomain spectral solution of compressible viscous flows, *AIAA Journal* **31**, 3376–3384.

Kopriva, D. A.: 1994, Multidomain spectral solution of compressible viscous flows, *Journal of Computational Physics* **115**(1), 184–199.

Kopriva, D. A.: 1996a, Spectral solution of the viscous blunt body problem. II. Multidomain approximation, *AIAA Journal* **34**(3), 560–564.

Kopriva, D. A.: 1996b, A conservative, staggered-grid multidomain method for the Euler gas-dynamics equations, *in* A. V. Ilin and L. R. Scott (eds), *Proceedings of the Third International Conference on Spectral and High Order Methods*, Houston Journal of Mathematics, Houston, Texas, pp. 457–468.

Kopriva, D. A.: 1998, A staggered-grid multidomain spectral method for the compressible Navier-Stokes equations, *Journal of Computational Physics* **143**(1), 125–158. Nonconforming interfaces; complex geometry.

Kopriva, D. A. and Hussaini, M. Y.: 1989, Multidomain spectral solution of shock-turbulence interactions, *in* T. F. Chan, R. Glowinski, J. Periaux and O. B. Widlund (eds), *Domain Decomposition Methods*, SIAM, Philadelphia, pp. 340–350.

Kopriva, D. A. and Kolias, J. H.: 1996, A conservative staggered-grid Chebyshev multido-main method for compressible flows, *Journal of Computational Physics* 125, 244–261.

Kopriva, D. A. and Panchang, V. G.: 1989, Pseudospectral solution of two-dimensional water-wave propagation, *Mathematical and Computer Modelling* 12, 625–640.

Kopriva, D. A., Zang, T. A. and Hussaini, M. Y.: 1991, Spectral methods for the Euler equations: The blunt body revisited, *AIAA Journal* 29(9), 1458–1462.

Kopriva, D. A., Zang, T. A., Salas, M. D. and Hussaini, M. Y.: 1984, Pseudospectral solution of two-dimensional gas-dynamics problems, *in* M. Pandolfi and R. Piva (eds), *Proceedings 5th GAMM Conf. Numerical Methods on Fluid Mechanics*, Vieweg, Braunschweig, pp. 185–192.

Korczak, K. Z. and Patera, A. T.: 1986, An isoparametric spectral element method for solution of the Navier-Stokes equations in complex geometry, *Journal of Computational Physics* 62, 361–382.

Kosloff, D., Reshef, M. and Loewenthal, D.: 1984, Elastic wave calculations by the Fourier method, *Bulletin of the Seismological Society of America* 74, 875–891.

Kosloff, R. and Tal-Ezer, H.: 1993, A modified Chebyshev pseudospectral method with an $O(1/N)$ time step restriction, *Journal of Computational Physics* 104, 457–469. The key idea is to employ a mapping which almost converts the Chebyshev polynomials back into cosines. The interpolahon grid is not uniform, but the roots of the mapped polynomials are not as concentrated near the endpoints as those of the Chebyshev polynomials. When the stretching parameter is chosen so as to give an $O(1/N)$ timstep, the asymptotic rate of convergence is reduced from geometric to algebraic, however.

Koures, V. G.: 1996, Solving the Coulomb Shrödinger equation in $d = 2 + 1$ via sinc collocation, *Journal of Computational Physics* 128(1), 1–5.

Kreiss, H.-O.: 1991, Problems with different time scales, *Acta Numerica* 1, 101–139. Numerical algorithms for problems with a "slow manifold".

Krishnamurti, T. N.: 1962, Numerical integration of primitive equations by a quasi-Lagrangian advection scheme, *Journal of Applied Meteorology* 1, 508–521. Not spectral; early paper on semi-Lagrangian methods.

Kuo, H.-C. and Williams, R. T.: 1990, Semi-Lagrangian solutions to the inviscid Burgers equation, *Monthly Weather Review* 118, 1278–1288. Elementary, but good introduction to semi-Lagrangian schemes.

Lacour, C. and Maday, Y.: 1997, Two different approaches for matching nonconforming grids: The mortar element method and the FETI method, *BIT* 37(3), 720–738.

Lambiotte, J., Bokhari, S., Hussaini, M. Y. and Orszag, S. A.: 1982, Navier-Stokes solutions on the Cyber-203 by a pseudospectral technique, *10th IMACS World Congress on System Simulation and Scientific Computation*, Montreal. Splitting with mixed finite difference/spectral spatial discretization.

Lanczos, C.: 1938, Trigonometric interpolation of empirical and analytical functions, *Journal of Mathematics and Physics* 17, 123–199. The origin of both the pseudospectral method and the tau method. Lanczos is to spectral methods what Newton was to calculus.

Lanczos, C.: 1956, *Applied Analysis*, Prentice-Hall, Englewood Cliffs, New Jersey. 400 pp.

Lanczos, C.: 1966, *Discourse on Fourier Series*, Oliver and Boyd, Edinburgh.

Lanczos, C.: 1973, Legendre versus Chebyshev polynomials, *in* J. C. P. Miller (ed.), *Numerical Analysis*, Academic Press, New York, pp. 191–201. Shows Legendre is "superconvergent" and much better than Chebyshev at $x = \pm 1$.

Lander, J. and Hoskins, B. J.: 1997, Believable scales and parameterizations in a spectral transform model, *Mon. Weath. Rev.* **125**, 292–303.

Laprise, R.: 1992, The resolution of global spectral methods, *Bulletin of the American Meteorological Society* **73**, 1453–1454.

Launay, J., Bouchet, S., Randriamampianina, A., Bontoux, P. and Gibart, P.: 1994, Modeling and experiments on epitaxial growth on a GaAs hemisphere substrate at 1 g and under hypergravity, *in* L. L. Regel and W. R. Wilcox (eds), *Materials Processing in High Gravity*, Plenum Press, New York, pp. 139–160. Two-dimensional (r, z) crystal in cylindrical coordinates using a Chebyshev tau method with the Haidvogel-Zang diagonalization method and time integration by the software package LSODA.

Le Quéré , P. and Pécheux, J.: 1989, Multiple transitions in axisymmetric annulus convection, *Journal of Fluid Mechanics* **206**, 517–544. Chebyshev-tau spectral method.

Le Quéré , P. and Pécheux, J.: 1990, A three-dimensional pseudo-spectral algorithm for the computation of convection in a rotating annulus, *Computer Methods in Applied Mechanics and Engineering* **80**, 261–271.

Lee, N. Y., Schultz, W. W. and Boyd, J. P.: 1989, Stability of fluid in a rectangular enclosure by spectral method, *International Journal of Heat and Mass Transfer* **32**, 513–520.

Leith, C. E.: 1980, Nonlinear normal mode initialization and quasi-geostrophic theory, *Journal of the Atmospheric Sciences* **37**, 958–968.

Leovy, C. B.: 1964, Simple models of thermally driven mesospheric circulations, *Journal of the Atmospheric Sciences* **21**, 327–341. Spectral eigensolution of Laplace's Tidal equation.

Leslie, L. M. and Purser, R. J.: 1991, High order numerics in an unstaggered three-dimensional time-split semi-Lagrangian forecast model, *Monthly Weather Review* **119**, 1612–1623. Not spectral, but very high order spatial differences.

Levin, J. G., Iskandarani, M. and Haidvogel, D. B.: 1997, A spectral filtering procedure for eddy-resolving simulations with a spectral element ocean model, *J. Comput. Phys.* **137**(1), 130–154.

Lewis, D. L., Lund, J. and Bowers, K. L.: 1987, The space-time sinc-Galerkin method for parabolic problems, *International Journal for Numerical Methods in Engineering* **24**, 1629–1644. Sinc basis in *both* space and time.

Lewis, H. R. and Bellan, P. M.: 1990, Physical constraints on the coefficients of Fourier expansions in cylindrical coordinates, *Journal of Mathematical Physics* **31**, 2592–. Analysis of the symmetry conditions across $r = 0$ for both scalars and vectors.

Li, C. W. and Qin, M. Z.: 1988, A symplectic difference scheme for the infinite dimensional Hamiltonian system, *Journal of Computer and Applied Mathematics* **6**, 164–174.

Li, T.-Y.: 1987, Solving polynomial systems, *Math. Intelligencer* **9**, 33–39. Probability-one homotopy.

Lie, I.: 1993, Using implicit ODE methods with iterative linear equation solvers in spectral methods, *SIAM Journal of Scientific Computing* **14**, 1194–1213.

Liffman, K.: 1996, Comments on a collocation spectral solver for the Helmholtz equation, *Journal of Computational Physics* **128**(1), 254–258. Extends the linear, elliptic boundary value solver of Ehrenstein and Peyret(1986), which uses Chebyshev polynomials, to inhomogeneous Robin boundary conditions.

Lin, S.-H. and Pierrehumbert, R. T.: 1988, Does Ekman friction suppress baroclinic instability?, *Journal of the Atmospheric Sciences* **45**, 2920–2933. Solves a two-dimensional eigenvalue problem on the half-space ($[-\infty, \infty] \otimes [0, \infty]$) via a tensor product basis of rational Chebyshev functions, $TB_m(y)\,TL_n(z)$.

Lin, S. J. and Rood, R. B.: 1996, Multidimensional flux-form semi-Lagrangian transport schemes, *Mon. Weath. Rev.* **124**(9), 2046–2070. Non-spectral but popular alternative to spherical harmonic algorithms for weather forecasting and climate modeling.

Lindberg, C. and Broccoli, A. J.: 1996, Representation of topography in spectral climate models and its effect on simulated precipitation, *J. Climate* **9**, 2641–. Smoothing of spherical harmonic series for topography through non-uniform spherical spline.

Lindzen, R. S.: 1970, Internal equatorial planetary-scale waves in shear flow, *Journal of the Atmospheric Sciences* **27**, 394–407. Mixed elliptic-hyperbolic PDE.

Liu, K. M. and Ortiz, E. L.: 1982, Eigenvalue problems for singularly perturbed differential equations, *in* J. J. H. Miller (ed.), *Computational Methods for Boundary and Interior Layers, Vol. II*, Boole Press, Dublin, pp. 324–329.

Liu, K. M. and Ortiz, E. L.: 1986, Numerical solution of eigenvalue problems for partial differential equations with tau-lines method, *Computer Mathematics and Applications* **12B**, 1153–1168.

Liu, K. M. and Ortiz, E. L.: 1987a, Tau method approximate solution of high-order differential eigenvalue problems defined in the complex plane with an application of the Orr-Sommerfeld stability equation, *Comm. Appl. Numer. Meths.* **3**, 187–194.

Liu, K. M. and Ortiz, E. L.: 1987b, Tau method approximation of differential eigenvalue problems where the spectral parameter enters nonlinearly, *Journal of Computational Physics* **72**(2), 299–310.

Liu, K. M. and Ortiz, E. L.: 1989, Numerical solution of ordinary and partial functional-differential eigenvalue problems with the tau method, *Computing* **41**, 205–217.

Liu, K. M., Ortiz, E. L. and Pun, K.-S.: 1984, Numerical solution of Steklov's partial differential equation eigenvalue problem, *in* J. J. H. Miller (ed.), *Computational Methods for Boundary and Interior Layers, Vol. II*, Boole Press, Dublin, pp. 244–249.

Liu, Y., Liu, L. and Tang, T.: 1994, The numerical computation of connecting orbits in dynamical systems: A rational spectral approach, *Journal of Computational Physics* **111**, 373–380. TB_n basis for computing heteroclinic and homoclinic solutions (i. e., shock-like and soliton-like) solutions to nonlinear boundary value problems on an infinite interval.

Lomtev, I., Quillen, C. B. and Karniadakis, G. E.: 1998, Spectral/hp methods for viscous compressible flows on unstructured 2D meshes, *J. Comput. Phys.* **144**(2), 325–357. Spectral elements on triangles using a discontinuous Galerkin formulation.

Longuet-Higgins, M. S.: 1968, The eigenfunctions of Laplace's tidal equation over a sphere, *Phil. Trans. Royal Society of London, Series A* **262**, 511–607. Spherical harmonics; tridiagonal Galerkin matrices.

Lopez, J. M. and Shen, J.: 1998, An efficient spectral-projection method for the Navier-Stokes equations in cylindrical coordinates, *Journal of Computational Physics* **139**, 308–326.

Lorenz, E. N. and Krishnamurthy, V.: 1987, On the nonexistence of a slow manifold, *Journal of the Atmospheric Sciences* **44**, 2940–2950. Weakly non-local in time; nothing on spectral methods except that the five-mode system is derived by a Galerkin method.

Lund, J.: 1986, Symmetrization of the sinc-Galerkin method for boundary value problems, *Mathematics of Computation* **47**, 571–588.

Lund, J. and Bowers, K. L.: 1992, *Sinc Methods For Quadrature and Differential Equations*, Society for Industrial and Applied Mathematics, Philadelphia. 304 pp. Restricted to sinc functions only.

Lund, J. and Vogel, C. R.: 1990, A fully-Galerkin method for the numerical solution of an inverse problem in a parabolic partial differential equation, *Inverse Problems* **6**, 205–217. Sinc basis.

Lund, J., Bowers, K. L. and Carlson, T. S.: 1991, Fully sinc-Galerkin computation for boundary feedback stabilization, *Journal of Mathematical Systems, Estimation and Control* **1**, 165–182.

Lund, J., Bowers, K. L. and McArthur, K. M.: 1989, Symmetrization of the sinc-Galerkin method with block techniques for elliptic equations, *IMA Journal of Numerical Analysis* **9**, 29–46.

Lund, J. R. and Riley, B. B.: 1984, A sinc-collocation method for the computation of the eigenvalues of the radial Schroedinger equation, *IMA Journal of Numerical Analysis* **4**, 83–98.

Lundin, L.: 1980, A cardinal function method of solution of the equation $\Delta u = u - u^3$, *Mathematics of Computation* **35**, 747–756. Sinc (Whittaker cardinal) basis.

Lynch, P.: 1992, Richardson's barotropic forecast: A reappraisal, *Bulletin of the American Meteorological Society* **73**(1), 35–47.

Lynch, P. and Huang, X.: 1992, Initialization of the HIRLAM model using a digital filter, *Monthly Weather Review* **120**, 1019–1034. Not spectral; slow manifold initialization scheme.

Ma, H.: 1992, The equatorial basin response to a Rossby wave packet: The effects of nonlinear mechanism, *Journal of Marine Research* **50**, 567–609. Spectral element ocean model.

Ma, H.: 1993a, Trapped internal gravity waves in a geostrophic boundary current, *Journal of Fluid Mechanics* **247**, 205–229. Spectral element ocean model.

Ma, H.: 1993b, A spectral element basin model for the shallow water equations, *Journal of Computational Physics* **109**, 133–149. Spectral element ocean model.

Ma, H.: 1995, Parallel computation with the spectral element method, *in* A. Ecer, J. Periaux, N. Satofuka and S. Taylor (eds), *Parallel Computational Fluid Dynamics: Implementations and Results Using Parallel Computers*, Elsevier, Amsterdam, pp. 239–246.

Ma, H.: 1996a, Baroclinic wave motions in an equatorial ocean current and related temperature fluctuations, *J. Marine Res.* **54**(6), 1073–1096. Spectral elements.

Ma, H.: 1996b, The dynamics of North Brazil Current retroflection eddies, *Journal of Marine Research* **54**, 35–53. Spectral elements.

Ma, H.-P. and Guo, B.-Y.: 1986, The Fourier pseudo-spectral method with a restrain operator for the Korteweg-de Vries equation, *Journal of Computational Physics* **65**, 120–137.

Ma, H.-P. and Guo, B.-Y.: 1987, The Fourier pseudo-spectral method for solving two-dimensional vorticity equations, *IMA Journal of Numerical Analysis* **5**, 47–60.

Machenhauer, B.: 1977, On the dynamics of gravity oscillations in a shallow water model, with application to normal mode initialization, *Beitr. Phys. Atmos.* **50**, 253–271.

Machenhauer, B. and Daley, R.: 1972, A baroclinic primitive equation model with a spectral representation in three dimensions, *Report 4*, Institut for Teoretisch Meteorologi, Copenhagen University, Copenhagen, Denmark. 66 pp. Legendre series in the vertical, spherical harmonics in latitude and longitude.

Machenhauer, B. and Daley, R.: 1974, Hemispheric spectral model, number 14 in *GARP Publication Series*, World Meteorological Organization, Geneva, Switzerland, pp. 226–251. Weather forecasting model with Legendre polynomials in height and spherical harmonics in latitude and longitude.

Maday, Y. and Patera, A. T.: 1987, Spectral element methods for the incompressible Navier Stokes equations, *in* A. K. Noor and J. T. Oden (eds), *State of the Art Surveys on Computational Mechanics*, ASME, New York, pp. 71–143.

Maday, Y. and Rønquist, E. M.: 1990, Optimal error analysis of spectral methods with emphasis on non-constant coefficients and deformed geometries, *in* C. Canuto and A. Quarteroni (eds), *Spectral and High Order Methods for Partial Differential Equations: Proceedings of the ICOSAHOM '89 Conference in Como, Italy*, North-Holland/Elsevier, Amsterdam, pp. 91–115. Also in Comput. Meths. Appl. Mech. Engrg., vol. 80, with the same page numbers.

Maday, Y., Mavriplis, C. A. and Patera, A. T.: 1989, Nonconforming mortar element method: application to spectral discretizations, *in* T. F. Chan, R. Glowinski, J. Periaux and O. B. Widlund (eds), *Domain Decomposition Methods*, SIAM, Society for Industrial and Applied Mathematics SIAM, Philadelphia.

Maday, Y., Meiron, D., Patera, A. T. and Rønquist, E. M.: 1993, Analysis of iterative methods for the steady and unsteady Stokes problem: Application to spectral element discretizations, *SIAM Journal of Scientific Computing* **14**, 310–337.

Maday, Y., Patera, A. T. and Rønquist, E. M.: 1990, An operator integration-factor splitting method for time-dependent problems: Application to incompressible fluid flow, *Journal of Scientific Computing* **5**, 263–292. Generalizes the usual two-level and three-level schemes, which are of at most second order accuracy in time, to arbitrary order.

Maday, Y., Patera, A. T. and Rønquist, E. M.: 1992, The $P_N \times P_{N-2}$ method for the approximation of the Stokes problem, *Laboratoire d'Analyse Numérique, Paris VI* **11**, 4.

Maday, Y., Pernaud-Thomas, B. and Vandeven, H.: 1985, Reappraisal of Laguerre type spectral methods, *Rech. Aerosp.* **1985-6**, 13–35.

Makar, P. A. and Karpik, S. R.: 1996, Basis-spline interpolation on the sphere: application to semi-Lagrangian advection, *Monthly Weather Review* **124**, 182. Comparisons with spectral methods.

Malek, A. and Phillips, T. N.: 1995, Pseudospectral collocation methods for fourth-order differential equations, *IMA Journal of Numerical Analysis* **15**, 523–553.

Malik, M. R.: 1990, Numerical methods for hypersonic boundary layer stability, *Journal of Computational Physics* **86**(2), 376–413. Linear eigenvalue problems solved by single domain and multi-domain Chebyshev methods and compact 4th order differences; inverse Rayleigh iteration.

Malik, M. R. and Orszag, S. A.: 1987, Linear stability analysis of three-dimensional compressible boundary layers, *Journal of Scientific Computing* **2**, 77–98.

Mamun, C. K. and Tuckerman, L. S.: 1995, Asymmetry and Hopf bifurcation in spherical Couette flow, *Physics of Fluids* **7**(1), 80–91. Axisymmetric incompressible flow between two spheres using vorticity/streamfunction and influence matrix method. Good discussion of how, by modifying only twenty lines in the time-dependent code, steady states could be computed directly by a nonlinear Richardson's iteration with preconditioning by Stokes flow (that is, by inverting the operator which is treated implicitly in the semi-implicit time-marching algorithm).

Mansell, G., Merryfield, W., Shizgal, B. and Weinert, U.: 1993, A comparison of differential quadrature methods for the solution of partial differential equations, *Computer Methods in Applied Mechanics and Engineering* **104**, 295–316. Differential quadrature is the same or almost the same as the pseudospectral method.

Marchuk, G. I.: 1974, *Numerical Methods in Numerical Weather Prediction*, Academic Press, New York. 277 pp. Splitting and fractional steps time integration.

Marcus, P. A.: 1984a, Simulation of Taylor-Couette flow, Part 1. Numerical methods and comparison with experiment, *Journal of Fluid Mechanics* **146**, 45–64. Fourier-Fourier-Chebyshev computations of flow in an annulus in cylindrical coordinates, assumed periodic in z as well as θ. "Shift-and-reflect" symmetry halves the number of Fourier modes in the basis.

Marcus, P. A.: 1984b, Simulation of Taylor-Couette flow, Part 2. Numerical results for wavy vortex flow with one travelling wave, *Journal of Fluid Mechanics* **146**, 65–113.

Marcus, P. A. and Tuckerman, L. S.: 1987a, Simulation of flow between concentric rotating spheres. Part 1. Steady states, *J. Fluid Mech.* **185**, 1–30. Axisymmetric flows with splitting errors removed by a Green's function [influence matrix] method.

Marcus, P. A. and Tuckerman, L. S.: 1987b, Simulation of flow between concentric rotating spheres. Part 2. Transitions, *J. Fluid Mech.* **185**, 31–65.

Marcus, P. A., Orszag, S. A. and Patera, A. T.: 1983, Simulation of cylindrical Couette flow, *in* E. Krause (ed.), *8th International Conf. on Numerical Methods in Fluid Dynamics*, number 170 in *Lecture Notes in Physics*, Springer-Verlag, New York, pp. 371–376.

Marcus, P. J.: 1990, Vortex dynamics in a shearing zonal flow, *Journal of Fluid Mechanics* **215**, 393–430. Fourier-Chebyshev algorithm for annular flow from Marcus(1984a). Dealiasing using Orszag Two-Thirds Rule because it helped here, contrary to the experiments of Marcus(1984b). Compared hyperviscosity with filtering of high order modes, and preferred the latter because it was cheaper, not requiring additional boundary conditions.

Margolin, L. G. and Jones, D. A.: 1992, An approximate inertial manifold for computing Burgers' equation, *Physica D* **60**, 175–184.

Marion, M. and Témam, R.: 19, Nonlinear Galerkin methods, *SIAM Journal of Numerical Analysis* **26**, 1139–1157.

Marshall, H. G. and Boyd, J. P.: 1987, Solitons in a continuously stratified equatorial ocean, *Journal of Physical Oceanography* **17**, 1016–1031. Hermite function application.

Mason, J. C.: 1967, Chebyshev polynomial approximations for the L-shaped membrane eigenvalue problem, *SIAM Journal of Applied Mathematics* **15**, 172–186. Maps the L-shaped domain into a polygonal figure and then solves the problem on the square containing the polygon. This completely eliminates the corner singularities; his results are superb.

Matsushima, T. and Marcus, P. S.: 1995, A spectral method for polar coordinates, *Journal of Computational Physics* **120**, 365–374. One-side Jacobi polynomials in radius.

Matsushima, T. and Marcus, P. S.: 1997, A spectral method for unbounded domains, *Journal of Computational Physics* **137**(2), 321–345. Polar coordinates with Fourier basis in angle and a radial basis of rational functions which are the images of associated Legendre functions under an algebraic mapping; applications to hydrodynamic vortices.

Mavriplis, C.: 1989, Laguerre polynomials for infinite-domain spectral elements, *Journal of Computational Physics* **80**(2), 480–.

Mavriplis, C.: 1994, Adaptive mesh strategies for the spectral element method, *in* C. Bernardi and Y. Maday (eds), *Analysis, Algorithms and Applications of Spectral and High Order Methods for Partial Differential Equations*, Selected Papers from the International Conference on Spectral and High Order Methods (ICOSAHOM '92), Le Corum, Montpellier, France, 22-26 June 1992, North-Holland, Amsterdam, pp. 77–86. Also in Comput. Meths. Appl. Mech. Engr., vol. 116.

Mayer, E. W. and Powell, K. G.: 1992, Viscous and inviscid instabilities of a trailing vortex, *Journal of Fluid Mechanics* **245**, 91–114. Chebyshev polynomial/QR one-dimensional eigenvalue problem in radius r in polar coordinates with domain truncation at large but finite r. Integration on a contour deformed into the complex plane is used to resolve nearly-neutral modes. The infinite domain is truncated to an annulus for some runs to calculate "ring modes" which are concentrated at intermediate radius.

McArthur, K. M., Bowers, K. L. and Lund, J.: 1987a, Numerical implementation of the sinc-Galerkin method for second-order hyperbolic equations, *Numerical Methods for Partial Differential Equations* **3**, 169–185.

McArthur, K. M., Bowers, K. L. and Lund, J.: 1987b, The sinc method in multiple space dimensions: Model problems, *Numerische Mathematik* **56**, 789–816.

McCalpin, J. D.: 1988, A quantitative analysis of the dissipation inherent in Semi-Lagrangian advection, *Monthly Weather Review* **116**, 2330–2336. Not spectral, but good analysis of the damping which is implicit in off-grid interpolation.

McCrory, R. L. and Orszag, S. A.: 1980, Spectral methods for multi-dimensional diffusion problems, *Journal of Computational Physics* **37**, 93–112.

McDonald, A.: 1984, Accuracy of multiply-upstream semi-Lagrangian advective schemes I., *Monthly Weather Review* **112**, 1267–1275. Not spectral.

McDonald, A.: 1986, A semi-Lagrangian and semi-implicit two-time-level integration scheme, *Monthly Weather Review* **114**, 824–830. Not spectral; only first order accurate in time.

McDonald, A.: 1987, Accuracy of multiply-upstream semi-Lagrangian advection schemes II, *Monthly Weather Review* **115**, 1446–1450. Not spectral.

McDonald, A. and Bates, J. R.: 1987, Improving the estimate of the departure point position in a two-time-level semi-Lagrangian and semi-implicit model, *Monthly Weather Review* **115**, 737–739. Not spectral.

McDonald, A. and Bates, J. R.: 1989, Semi-Lagrangian integration of a gridpoint shallow-water model on the sphere, *Monthly Weather Review* **117**(1), 130–137. Not spectral.

McFadden, G. B., Murray, B. T. and Boisvert, R. F.: 1990, Elimination of spurious eigenvalues in the Chebyshev tau spectral method, *Journal of Computational Physics* **91**, 228–239.

McGregor, J. L.: 1993, Economical determination of departure points for semi-Lagrangian models, *Monthly Weather Review* **121**, 221–230. Not spectral per se; shows his $O(\tau^2)$ formula D_2 costs only 5 additions and 7 multiplications versus 30 and 60 for bicubic interpolation (in two space dimensions).

McKerrell, A.: 1988, The global element applied to fluid flow problems, *Computers and Fluids* **16**, 41–46. Corner singularities, domain decomposition algorithms, two-dimensional mappings.

McKerrell, A. and Delves, L. M.: 1984, Solution of the global element equations on the ICL DAP, *ICL Technical Journal* **4**, 50–58. Massively parallel computation with spectral domain decomposition.

McKerrell, A., Phillips, C. and Delves, L. M.: 1981, Chebyshev expansion methods for the solution of elliptic partial differential equations, *Journal of Computational Physics* **40**, 444–452.

McLaughlin, J. B. and Orszag, S. A.: 1982, Transition from periodic to chaotic thermal convection, *Journal of Fluid Mechanics* **122**, 123–142.

Mead, K. O. and Delves, L. M.: 1973, On the convergence rate of generalized Fourier expansions, *Journal of the Institute for Mathematics and Its Applicaitons* **12**, 247–259.

Mean, J. L. and Renaut, R. A.: 1999, Optimal Runge-Kutta methods for first order pseudospectral operators, *J. Comput. Phys.* **152**(1), 404–419.

Meiron, D. I., Baker, G. R. and Orszag, S. A.: 1982, Analytic structure of vortex sheet dynamics. Part 1. Kelvin-Helmholtz instability, *J. Fluid Mech.* **114**, 283–298. Singularities estimated from the slopes of the Fourier coefficients versus degree at different times.

Meiron, D. I., Orszag, S. A. and Israeli, M.: 1981, Applications of numerical conformal mapping, *Journal of Computational Physics* **40**, 345–360.

Mercier, B.: 1989, *An Introduction to the Numerical Analysis of Spectral Methods*, Vol. 318 of *Lecture Notes in Physics*, Springer-Verlag, New York. 200 pp. Author's translation, without updating, of an earlier monograph in French.

Mercier, B. and Raugel, G.: 1982, Resolution d'un problème aux limites dans un ouvert axisymétrique par éléments finis en r, z et séries de Fourier en théta, *R. A. I. R. O. Anal. Numer.* **16**, 67–100.

Merilees, P. E.: 1973a, An alternative scheme for the summation of a series of spherical harmonics, *J. Appl. Meteor.* **12**, 224–227. Instability of standard recurrence for the spherical harmonics for large zonal wavenumber; ill-conditioning of "Robert" basis.

Merilees, P. E.: 1973b, The pseudospectral approximation applied to the shallow water wave equations on a sphere, *Atmosphere* **11**, 13–20.

Merilees, P. E.: 1974, Numerical experiments with the pseudospectral method in spherical coordinates, *Atmosphere* **12**, 77–96.

Merzbacher, E.: 1970, *Quantum Mechanics*, 2 edn, Wiley, New York. 400 pp. Hermite funcs.; Galerkin meths. in quantum mechanics.

Metcalfe, R. W., Orszag, S. A., Brachet, M. E., Menon, S. and Riley, J.: 1987, Secondary instability of a temporally growing mixing layer, *Journal of Fluid Mechanics* **184**, 207–243.

Mills, R. D.: 1987, Using a small algebraic manipulation system to solve differential and integral equations by variational and approximation techniques, *J. Symbolic Comp.* **3**, 291–301.

Minion, M. L. and Brown, D. L.: 1997, Performance of under-resolved two-dimensional incompressible flow simulations, II, *J. Comput. Phys.* **138**(2), 734–765. Comparison of several numerical methods including a pseudospectral method; generation of spurious vortices whose size is much larger than twice the grid spacing through instability, excited by truncation error, of underresolved flows.

Mittal, R. and Balachandar, S.: 1996, Direct numerical simulation of flow past elliptic cylinders, *Journal of Computational Physics* **124**, 351–367. Chebyshev-Fourier time-dependent calculations using a single computational domain in elliptic coordinates. Good, careful discussion of both inflow and outflow boundary conditions, parabolization of the outflow, smooth blending of outflow into inflow.

Mizzi, A., Tribbia, J. and Curry, J.: 1995, Vertical spectral representation in primitive equation models of the atmosphere, *Monthly Weather Review* **123**(8), 2426–2446. Removed most obstacles to use of spectral methods in z, but their series of vertical normal modes converged rather slowly.

Moore, D. W. and Philander, S. G. H.: 1977, Modelling of the tropical oceanic circulation, *in* E. D. Goldberg (ed.), *The Sea, Volume 6*, Wiley, New York, pp. 319–361. Hermite functions.

Moore, S., Healy, Jr, D. M. and Rockmore, D.: 1993, Symmetry stabilization for fast discrete monomial transforms and polynomial evaluations, *Lin. Alg. Appl.* **192**, 249–299. $O(N \log_2(N))$ discrete Fourier transform on nonuniform grids.

Morf, R., Orszag, S. A. and Frisch, U.: 1980, Spontaneous singularity in three-dimensional, inviscid compressible flow, *Phys. Rev. Lett.* **44**, 572–575. The conclusion that singularities can form was (mostly) repudiated in Brachet *et al.*(1983).

Morgan, A. P.: 1987, *Solving Polynomial Systems Using Continuation for Scientific and Engineering Problems*, Prentice-Hall, Englewood Cliffs, New Jersey. Not spectral.

Morse, P. M. and Feshbach, H.: 1953, *Methods of Theoretical Physics*, McGraw-Hill, New York. 2000 pp, (in two volumes) Good treatise on solving linear partial differential equations through the method of separation of variables.

Morton, K. W.: 1985, Generalised Galerkin methods for hyperbolic problems, *Computer Methods in Applied Mechanics and Engineering* **52**, 847–871. Not spectral; independent invention of a semi-Lagrangian method.

Mulholland, L. and Sloan, D.: 1991, The effect of filtering on the pseudospectral solution of evolutionary partial differential equations, *Journal of Computational Physics* **96**, 369–390.

Mulholland, L. S. and Sloan, D. M.: 1992, The role of preconditioning in the solution of evolutionary partial differential equations by implicit Fourier pseudospectral methods, *Journal of Computational and Applied Mathematics* **42**, 157–174.

Mulholland, L. S., Huang, W.-Z. and Sloan, D. M.: 1998, Pseudospectral solution of near-singular problems using numerical coordinate transformations based on adaptivity, *SIAM J. Sci. Comput.* **19**(4), 1261–1289.

Mullholland, L. S., Qiu, Y. and Sloan, D. M.: 1997, Solution of evolutionary PDEs using adaptive finite differences with pseudospectral post-processing, *J. Comput. Phys.* **131**(2), 280–298.

Nakamura, S.: 1996, *Numerical Analysis and Graphic Visualization with MATLAB*, Prentice-Hall, Upper Saddle River, New Jersey. Not spectral.

Namasivayam, S. and Ortiz, E. L.: 1981, Best approximation and the numerical solution of partial differential equations with the Tau method, *Portug. Math.* **40**, 97–119.

Namasivayam, S. and Ortiz, E. L.: 1993, Error analysis of the Tau method: dependence of the approximation error on the choice of perturbation term, *Computers Math. applic.* **25**(1), 89–104.

Navarra, A.: 1987, An application of the Arnoldi's method to a geophysical fluid dynamics problem, *Journal of Computational Physics* **69**, 143–162. Mixed finite difference/spherical harmonics Galerkin treatment of boundary value problem; Arnoldi's method allows computations of unstable modes with as many as 13,000 unknowns.

Navarra, A., Stern, W. F. and Miyakoda, K.: 1994, Reduction of the Gibbs oscillation in spectral model simulations, *J. Climate* **7**, 1169–1183. Apply several different smoothers to spherical harmonics series with sucess. Note that the usual second order finite differences can be equivalently derived by applying Lanczos smoothing to the spectral sum for a derivative.

Navon, I. M.: 1987, PENT: A periodic pentadiagonal systems solver, *Comm. Appl. Numer. Meths.* **3**, 63–69. Not spectral, but useful matrix-solver.

Nicholls, D. P.: 1998, Travelling water waves: Spectral continuation methods with parallel implementation, *J. Comput. Phys.* **143**(1), 224–240. Fourier pseudospectral solution to a nonlinear eigenvalue problem.

Nosenchuck, D. M., Krist, S. E. and Zang, T. A.: 1987, Multigrid methods for the Navier-Stokes Computer, *in* S. McCormick and K. Stuben (eds), *Multigrid Methods*, Marcel Dekker, New York, pp. 491–516.

Nouri, F. Z. and Sloan, D. M.: 1989, A comparison of Fourier pseudospectral methods for the solution of the Korteweg-deVries equation, *Journal of Computational Physics* **83**, 324–344.

O'Connor, W. P.: 1995, The complex wavenumber eigenvalues of Laplace's tidal equations for oceans bounded by meridians, *Proceedings of the Royal Society of London A* **449**, 51–64. Associated Legendre functions through a recurrence-derived Galerkin methods, which is susceptible to errors [as confirmed by Corrigendum: O'Connor(1996).].

O'Connor, W. P.: 1996, The complex wavenumber eigenvalues of Laplace's tidal equations for oceans bounded by meridians: Corrigendum, *Proceedings of the Royal Society of London A* **452**, 1185–1187.

Orszag, S. A.: 1970, Transform method for calculation of vector coupled sums: Application to the spectral form of the vorticity equation, *Journal of the Atmospheric Sciences* **27**, 890–895. Implementing spectral methods in spherical harmonics.

Orszag, S. A.: 1971a, On the elimination of aliasing in finite difference schemes by filtering high-wavenumber components, *Journal of the Atmospheric Sciences* **28**, 1074. A two-paragraph classic.

Orszag, S. A.: 1971b, Accurate solution of the Orr-Sommerfeld equation, *Journal of Fluid Mechanics* **50**, 689–703. Combines Chebyshev method with QR or QZ matrix eigensolver to solve a linear stability problem.

Orszag, S. A.: 1971c, Numerical simulations of incompressible flows within simple boundaries: accuracy, *Journal of Fluid Mechanics* **49**, 75–112.

Orszag, S. A.: 1971d, Numerical simulations of incompressible flows within simple boundaries: Galerkin (spectral) representations, *Studies in Applied Mathematics* **50**, 293–327.

Orszag, S. A.: 1971e, Galerkin approximations to flows within slabs, spheres and cylinders, *Physical Review Letters* **26**, 1100–1103.

Orszag, S. A.: 1972, Comparison of pseudospectral and spectral approximations, *Studies in Applied Mathematics* **51**, 253–259.

Orszag, S. A.: 1974, Fourier series on spheres, *Monthly Weather Review* **102**, 56–75.

Orszag, S. A.: 1976, Turbulence and transition: A progress report, *in* A. I. vander Vooren and Zandbergen (eds), *Proceedings of the Fifth International Conference on Numerical Fluid Dynamics*, number 59 in *Lecture Notes in Physics*, Springer-Verlag, New York, pp. 39–51.

Orszag, S. A.: 1979, Spectral methods for problems in complex geometries, *in* S. V. Parter (ed.), *Numerical Methods for Partial Differential Equations*, Academic Press, New York. Shorter, preliminary version of Orszag (1980).

Orszag, S. A.: 1980, Spectral methods for problems in complex geometries, *Journal of Computational Physics* **37**, 70–92. Independent invention of finite difference preconditioning for solving boundary value problems by means of Chebyshev polynomials and iteration.

Orszag, S. A.: 1986, Fast eigenfunction transforms, *in* G. C. Rota (ed.), *Science and Computers*, Academic Press, New York, pp. 23–30. A generalization of the Fast Fourier Transform to functions that satisfy three-term recurrence relations.

Orszag, S. A. and Israeli, M.: 1974, Numerical simulation of incompressible flow, *Ann. Revs. Fluid Mech.* **6**, 281–318. REVIEW.

Orszag, S. A. and Kells, L. C.: 1980, Transition to turbulence in plane Poiseuille flow and plane Couette flow, *Journal of Fluid Mechanics* **96**, 159–205.

Orszag, S. A. and Patera, A. T.: 1980, Subcritical transition to turbulence in plane channel flows, *Physical Review Letters* **45**, 989–993.

Orszag, S. A. and Patera, A. T.: 1981, Subcritical transition to turbulence in planar shear flow, *in* R. E. Meyer (ed.), *Transition and Turbulence*, Academic Press, New York, pp. 127–146.

Orszag, S. A. and Patera, A. T.: 1983, Secondary instability of wall bounded shear flows, *Journal of Fluid Mechanics* **128**, 347–385.

Orszag, S. A. and Paterson, Jr., G. S.: 1972a, *Statistical Models of Turbulence*, Springer-Verlag, New York, chapter Numerical simulation of turbulence, pp. 127–147.

Orszag, S. A. and Paterson, Jr., G. S.: 1972b, Numerical simulation of three dimensional homogeneous isotropic turbulence, *Phys. Rev. Lett.* **28**, 76–79.

Orszag, S. A. and Tang, C. M.: 1979, Small-scale structure of two-dimensional magnetohydrodynamic turbulence, *Journal of Fluid Mechanics* **90**, 129–143.

Orszag, S. A., Israeli, M. and Deville, M.: 1986, Boundary conditions for incompressible flows, *Journal of Scientific Computing* **1**, 75–111. Thorough discussion of high order artificial boundary conditions for fractional step (splitting) time-marching so that the computed flows are both nondivergent and no-slip.

Ortiz, E. L.: 1969, The tau method, *SIAM Journal of Numerical Analysis* **6**, 480–492.

Ortiz, E. L.: 1987, Recent progress in the numerical treatment of singular problems for partial differential equations with techniques based on the Tau method, *in* E. L. Ortiz (ed.), *Numerical Approximations of P. D. E., Part III*, North-Holland, Amsterdam, pp. 83–98.

Ortiz, E. L. and Pham Ngoc Dinh, A.: 1987, Linear recursive schemes associated with some nonlinear partial differential equations in one dimension and the tau method, *SIAM J. Math. Anal.* **18**(2), 452–464.

Ortiz, E. L. and Pun, K.-S.: 1985, Numerical solution of nonlinear partial differential equations with the tau method, *Journal of Computational and Applied Mathematics* **12 and 13**, 511–516.

Ortiz, E. L. and Samara, H.: 1981, An operational approach to the Tau method for the numerical solution of nonlinear differential equations, *Computing (Wien)* **27**, 15–25.

Ortiz, E. L. and Samara, H.: 1983, Numerical solution of differential eigenvalue problems with an operational approach to the tau method, *Computing (Wien)* **31**, 95–103.

Owens, R. G.: 1998, Spectral approximations on the triangle, *Proc. R. Soc. Lond. A* **454**, 857–872. Derives a new polynomial basis for the triangle through a Sturm-Liouville eigenproblem in two dimensions; the eigenfunctions are products of Legendre polynomials in one transformed coordinate with hypergeometric polynomials in the other. A spectral rate of convergence is proved. The polynomials are used to generate a new cubature formula for the triangle.

Owens, R. G. and Phillips, T. N.: 1991, A spectral domain decomposition method for the planar non-Newtonian stick-slip problem, *Journal of Non-Newtonian Fluid Mechanics* **41**, 43–79.

Owens, R. G. and Phillips, T. N.: 1996, Steady viscoelastic flow past a sphere using spectral elements, *International Journal of Numerical Methods in Engineering* **39**(9), 1517–1534. Flow past sphere embedded in a cylindrical tube.

Panchang, V. G. and Kopriva, D. A.: 1989, Solution of two-dimensional water-wave propagation problems by Chebyshev collocation, *Mathematical and Computer Modelling* **12**, 625–640.

Pascal, F. and Basdevant, C.: 1992, Nonlinear Galerkin method and subgrid-scale model for two-dimensional turbulent flows, *Theoretical and Computational Fluid Dynamics* **3**, 267–284.

Patera, A. T.: 1984, A spectral element method for fluid dynamics: laminar flow in a channel expansion, *Journal of Computational Physics* **54**, 468–488.

Patera, A. T.: 1986, Fast direct Poisson solvers for high-order finite element discretizations in rectangularly decomposable domains, *Journal of Computational Physics* **65**, 474–480. Combines the eigenfunction scheme of Haidvogel and Zang(1979) with static condensation to create a direct method of $O(N^{5/2})$ cost.

Patera, A. T. and Orszag, S. A.: 1980, Transition and turbulence in plane channel flows, *in* R. W. MacCormack and W. C. Reynolds (eds), *Proceedings of the 7th International Conference on Numerical Methods in Fluid Dynamics*, Springer-Verlag, New York, pp. 329–335.

Patera, A. T. and Orszag, S. A.: 1981, Finite-amplitude stability of axisymmetric pipe flow, *Journal of Fluid Mechanics* **112**, 467.

Patera, A. T. and Orszag, S. A.: 1986, Instability of pipe flow, *in* A. R. Bishop, D. K. Campbell and B. Nicolaenko (eds), *Nonlinear Problems: Present and Future*, North Holland, Amsterdam, pp. 367–377.

Patterson Jr., G. S. and Orszag, S. A.: 1971, Spectral calculation of isotropic turbulence: Efficient removal of aliasing interaction, *Physics of Fluids* **14**, 2538–2541.

Pellerin, P., Laprise, R. and Zawadzki, I.: 1995, The performance of a semi-Lagrangian transport scheme for the advection-condensation problem, *Monthly Weather Review* **123**, 3318.

Pelz, R. B.: 1993, Parallel compact FFTs for real sequences, *SIAM J. Sci.Comput.* **14**(4), 914–935.

Phillips, N. J.: 1956, The general circulation of the atmosphere: A numerical experiment, *Quart. J. Roy. Met. Soc.* **82**, 123–164. The empirical discovery of aliasing instability.

Phillips, N. J.: 1959, An example of nonlinear computational instability, *The Atmosphere and the Sea in Motion*, Rockefeller Institute Press, New York, pp. 501–504. The theoretical discovery of aliasing instability.

Phillips, T. N.: 1988, On the Legendre coefficients of a general-order derivative of an infinitely differentiable function, *IMA Journal of Numerical Analysis* **8**, 455–459.

Phillips, T. N. and Davies, A. R.: 1988, On semi-infinite spectral elements for poisson problems with re-entrant boundary singularities, *Journal of Computational and Applied Mathematics* **21**, 173–188.

Phillips, T. N. and Karageorghis, A.: 1990, On the coefficients of integrated expansions of ultraspherical polynomials, *SIAM Journal on Numerical Analysis* **27**, 823–830.

Phillips, T. N. and Owens, R. G.: 1997, A mass conserving multi-domain spectral collocation method for the Stokes problem, *Comput. Fluids* **26**(8), 825–840.

Phillips, T. N., Zang, T. A. and Hussaini, M. Y.: 1986, Preconditioners for the spectral multigrid method, *IMA Journal of Numerical Analysis* **6**, 273–292.

Pinelli, A., Benocci, C. and Deville, M.: 1992, A preconditioing technique for Chebyshev collocated advection diffusion operators, *in* C. Hirsh (ed.), *First European CFD Conference Proceedings*, Elsevier, Amsterdam, p. 109.

Pinelli, A., Benocci, C. and Deville, M.: 1994, A Chebyshev collocation algorithm for the solution of advection-diffusion equations, *in* C. Bernardi and Y. Maday (eds), *Analysis, Algorithms and Applications of Spectral and High Order Methods for Partial Differential Equations*, Selected Papers from the International Conference on Spectral and High Order Methods (ICOSAHOM '92), Le Corum, Montpellier, France, 22-26 June 1992, North-Holland, Amsterdam, pp. 201–210. Also in Comput. Meths. Appl. Mech. Engrg., vol. 116.

Pinelli, A., Couzy, W., Deville, M. O. and Benocci, C.: 1996, An efficient iterative solution method for the Chebyshev collocation of advection-dominated transport problems, *SIAM Journal of Scientific Computing* **17**(3), 647–657.

Pironneau, O.: 1982, On the transport-diffusion algorithm and its application to the Navier-Stokes equations, *Numerische Mathematik* **38**, 309–322. Not spectral; invention of a semi-Lagrangian method.

Press, W. H., Flannery, B. H., Teukolsky, S. A. and Vetterling, W. T.: 1986, *Numerical Recipes: The Art of Scientific Computing*, Cambridge University Press, New York.

Priestley, A.: 1993, A quasi-conservative version of the semi-Lagrangian advection scheme, *Monthly Weather Review* **121**, 621. Not spectral.

Priymak, V. G.: 1995, Pseudospectral algorithms for Navier-Stokes simulation of turbulent flows in cylindrical geometry with coordinate singularities, *Journal of Computational Physics* **118**, 366–379.

Priymak, V. G. and Miyazaki, T.: 1998, Accurate Navier-Stokes investigation of transitional and turbulent flows in a circular pipe, *Journal of Computational Physics* **142**, 370–411. Chebyshev in radius, Fourier in the other two coordinates.

Promislow, K. and Témam, R.: 1991, Approximate interaction laws for small and large waves in the Ginzburg-Landau equation, *IMA Journal of Applied Mathematics* **46**, 121–136. Nonlinear Galerkin method and theory.

Pudykiewicz, J., Benoit, R. and Staniforth, A.: 1985, Preliminary results from a partial LRTAP model based on an existing meteorological forecast model, *Atmospheres and Oceans* **23**, 267–303. Not spectral; convergence of fixed point iteration for semi-Lagrangian methods.

Pulicani, J. P. and Ouazzani, J.: 1991, A Fourier-Chebyshev pseudospectral method for solving steady 3-D Navier-Stokes and heat equations in cylindrical cavities, *Computers and Fluids* **20**(2), 93–109.

Purnell, D. K.: 1976, Solution of the advective equation by upstream interpolation with a cubic spline, *Monthly Weather Review* **104**, 42–48.

Purser, R. J. and Leslie, L. M.: 1988, A semi-implicit semi-Lagrangian finite-difference scheme using high-order spatial differencing on a nonstaggered grid, *Monthly Weather Review* **116**, 2069–2080. Not spectral, but very high order spatial differences.

Purser, R. J. and Leslie, L. M.: 1991, An efficient interpolation procedure for high order three-dimensional semi-Lagrangian models, *Monthly Weather Review* **119**, 2492–2498. Not spectral, but high order spatial differences.

Purser, R. J. and Leslie, L. M.: 1994, An efficient semi-Lagrangian scheme using third-order semi-implicit time integration and forward trajectories, *Monthly Weather Review* **122**, 745–756.

Qin, M. Z. and Zhang, M. Q.: 1990, Multi-stage symplectic schems of two kinds of Hamiltonian systems for wave equations, *Computer Mathematics and Applications* **19**, 51–62.

Rand, R. H.: 1984, *Computer Algebra in Applied Mathematics: An Introduction to MACSYMA*, Pitman Research Notes in Mathematics, Pitman, New York.

Rand, R. H. and Armbruster, J.: 1987, *Perturbation Methods, Bifurcation Theory and Computer Algebra*, Springer-Verlag, New York. Not spectral, but many of the perturbation series are spectral series, and the order-by-order equations are solved by Galerkin methods.

Randriamampianina, A.: 1994, On the use of vorticity-vector-potential with a spectral tau method in rotating annular domains, *in* C. Bernardi and Y. Maday (eds), *Analysis, Algorithms and Applications of Spectral and High Order Methods for Partial Differential Equations*, Selected Papers from the International Conference on Spectral and High Order Methods (ICOSAHOM '92), Le Corum, Montpellier, France, 22-26 June 1992, North-Holland, Amsterdam, pp. 439–448. Three-dimensional vector potential for multiply connected domain with Fourier in aximuthal coordinate and Chebyshev in the radial and axial coordinates with time discretization by a mixed second order Adams-Bashforth/second order Backwards Euler [Backward Differentiation]. Also published in Finite Elements in Analysis and Design, vol. 16, pp. 299-307.

Randriamampianina, A., Bontoux, P. and Roux, B.: 1987, Ecoulements induits par la force gravique dans une cavité cylindrique en rotation, *International Journal of Heat and Mass Transfer* **30**, 1275–1292. Spectral tau-method study of the competition between buoyancy and rotational forces in a confined cavity.

Rančić, M., Purser, R. J. and Mesinger, F.: 1996, A global shallow water model using an expanded spherical cube: Gnomonic versus conformal coordinates, *Quart. J. Roy. Meteor. Soc.* **122**, 959–982. Finite difference model: alternative to spherical harmonics method.

Rasch, P. and Williamson, D.: 1990a, On shape-preserving interpolation and semi-Lagrangian transport, *SIAM Journal of Scientific and Statistical Computing* **11**, 656–687.

Rasch, P. and Williamson, D.: 1990b, Computational aspects of moisture transport in global models of the atmosphere, *Quarterly Journal of the Royal Meteorological Society* **116**, 1071–1090.

Rasch, P. and Williamson, D.: 1991, The sensitivity of a general circulation model climate to the moisture transport formulation, *Journal of Geophysical Research* **96**, 13123–13137.

Rashid, A., Cao, W.-M. and Guo, B.-Y.: 1993, Three level Fourier pseudospectral method for fluid flow with low Mach number, *Journal of SUST* **16**, 223–237.

Rashid, A., Cao, W.-M. and Guo, B.-Y.: 1994a, Three level Fourier spectral approximations for fluid flow with low Mach number, *AMC* **63**, 131–150.

Rashid, A., Cao, W.-M. and Guo, B.-Y.: 1994b, The Fourier pseudospectral method for fluid flow with low Mach number, *Journal of Applied Science* **12**, 1–12.

Raspo, I., Ouazzani, J. and Peyret, R.: 1994, A direct Chebyshev multidomain method for flow computation with application to rotating systems, *Contemporary Mathematics* **180**, 533–538. 2D axisymmetric Navier-Stokes flow in a cylinder in vorticity/stream function form with influence matrix solution; two domains with a singularity in the corner of one domain wall where the boundary conditions change; fast algebraic rate of convergence.

Raspo, I., Ouazzani, J. and Peyret, R.: 1996, A spectral multidomain technique for the computation of the Czochralski melt configuration, *International Journal for Numerical Methods in Fluid Mechanics* **6**, 31–58. Similar to previous article, but different engineering application.

Reid, J. K. and Walsh, J. E.: 1965, An elliptic eigenvalue problem for a re-entrant region, *SIAM Journal of Applied Mathematics* **13**, 837–850. Devised a global mapping employed by Mason (1967); not spectral.

Renaud, F. and Gauthier, S.: 1997, A dynamical pseudo-spectral domain decomposition technique: Application to viscous compressible flows, *Journal of Computational Physics* **131**(1), 89–108. Fourier-Chebyshev algorithm with adaptive mapping of the subdomains.

Renaut, R. and Su, Y.: 1997, Evaluation of chebyshev pseudospectral methods for third order differential equations, *Numerical Algorithms* **16**(3–4), 255–281.

Rennich, S. C. and Lele, S. K.: 1997, Numerical method for incompressible vortical flows with two unbounded directions, *J. Comput. Phys.* **137**, 101–129. Three-dimensional Fourier method with domain truncation, using analytical conditions on the irrotational part of the velocity to greatly reduce the size of the domain.

Richardson, L. F.: 1910, The approximate arithmetical solution by finite differences of physical problems involving differential equations with an application to stresses in a masonry dam, *Phil. Trans. Royal Society of London* **A210**, 307–357. Invention of Richardson's iteration.

Richardson, L. F.: 1922, *Weather Prediction by Numerical Processes*, Cambridge University Press, New York. Ur-text of computational weather forecasting.

Riley, J. J., Metcalfe, R. W. and Orszag, S. A.: 1986, Direct numerical simulations of chemically reacting turbulent mixing layers, *Physics of Fluids* **29**, 406–422.

Ritchie, H.: 1986, Eliminating the interpolation associated with the semi-Lagrangian scheme, *Monthly Weather Review* **114**(1), 135–146. Not spectral, but the time-marching algorithm was applied to spectral models in several of the author's later articles.

Ritchie, H.: 1987, Semi-Lagrangian advection on a Gaussian grid, *Monthly Weather Review* **115**, 608–619. Advection by a steady wind in spherical geometry.

Ritchie, H.: 1988, Application of the semi-Lagrangian method to a spectral model of the shallow-water equations, *Monthly Weather Review* **116**, 1587–1598. Surface of a sphere.

Ritchie, H.: 1991, Application of the semi-Lagrangian method to a multilevel spectral primitive equations model, *Quart. J. Roy. Met. Soc.* **117**, 91–106.

Ritchie, H. and Tanguay, M.: 1996, A comparison of spatially averaged Eulerian and semi-Lagrangian treatment of mountains, *Monthly Weather Review* **124**, 167.

Robert, A.: 1981, A stable numerical integration scheme for the primitive meteorological equations, *Atmosphere—Oceans* **19**, 35–46.

Robert, A.: 1982, A semi-Lagrangian and semi-implicit numerical integration scheme for the primitive meteorological equations, *Journal of the Meteorological Society of Japan* **60**, 319–324.

Robert, A. J.: 1966, The integration of a low order spectral form of the primitive meteorological equations, *J. Meteor. Soc.* **44**, 237–244. Alternatives to spherical harmonics. First use of what is now called the "Asselin" time filter to damp high temporal frequencies.

Robert, A., Yee, T. L. and Ritchie, H.: 1985, A semi-Lagrangian and semi-implicit numerical integration scheme for multilevel atmospheric models, *Monthly Weather Review* **113**, 388–394.

Robinson, A. C. and Saffman, P. G.: 1984, Stability and structure of stretched vortices, *Studies in Applied Mathematics* **68**, 163–181. Sinc series: 2D, nonlinear eigenvalue problem on an unbounded domain.

Robson, R. E. and Prytz, A.: 1993, The discrete ordinate/pseudo-spectral method: Review and application from a physicist's perspective, *Australian J. Physics* **46**, 465–495. REVIEW.

Rogallo, R. S.: 1977, An ILLIAC program for the numerical simulation of homogeneous, incompressible turbulence, *Technical Report TM-73203*, NASA.

Rogallo, R. S.: 1981, Numerical experiments in homogeneous turbulence, *Technical Report TM-81315*, NASA.

Rogallo, R. S. and Moin, P.: 1984, Numerical simulation of turbulent flows, *Annual Reviews of Fluid Mechanics* **16**, 99–137.

Ronchi, C., Iacono, R. and Paolucci, P. S.: 1996, The "Cubed Sphere": A new method for the solution of partial differential equations in spherical geometry, *J. Comput. Phys.* **124**(1), 93–114. Comparison of new finite difference method with spherical harmonics.

Rønquist, E. M.: 1988, *Optimal spectral element methods for the unsteady three-dimensional incompressible Navier-Stokes equations*, PhD dissertation, MIT, Department of Mechanical Engineering.

Rønquist, E. M.: 1996, Convection treatment using spectral elements of different order, *Int. J. Numer. Meth. Fluids* **22**(4), 241–264.

Rønquist, E. M. and Patera, A. T.: 1987a, A Legendre spectral element method for the Stefan problem, *Int. J. Num. Methods. Eng.* **24**, 2273–2299.

Rønquist, E. M. and Patera, A. T.: 1987b, Spectral element multigrid. I. Formulation and numerical results, *Journal of Scientific Computing* **2**, 389–406.

Runge, C.: 1903, *Zeit. f. Math. u. Phys.* **48**, 443–456.

Runge, C.: 1905, *Zeit. f. Math. u. Phys.* **53**, 117–123.

Runge, C.: 1913, *Berlauterung des Rechnungsformulars, u. s. w., Braunschweig*. These three papers describe Runge's independent discovery of what is now called the Fast Fourier Transform.

Saff, E. B. and Kuijlaars, A. B. J.: 1997, Distributing many points on a sphere, *Mathematical Intelligencer* **19**(1), 5–11. Semi-popular review of progress and open problems in generating nearly uniform distributions of points, such as collocation points, on a sphere.

Saffman, P. G. and Szeto, R.: 1981, Structure of a linear array of uniform vortices, *Studies in Applied Mathematics* **65**, 223–248. Application of sinc spectral method.

Salupere, A., Maugin, G. A., Engelbrecht, J. and Kalda, J.: 1996, On the KdV soliton formation and discrete spectral analysis, *Wave Motion* **23**, 49–66. High resolution Fourier solutions in one space dimension.

Sanders, B. F., Katopodes, N. and Boyd, J. P.: 1998, Unified pseudo-spectral solution to water wave equations, *Journal of Hydraulic Eng. ASCE* **124**, 2–12. Fourier pseudospectral calculations of a wide variety of one-dimensional nonlinear wave equations with careful comparisons with exact soliton solutions.

Sanugi, B. B. and Evans, D. J.: 1988, A Fourier series method for the numerical solution of the nonlinear advection problem, *Applied Mathematics Letters* **1**, 385–389.

Sanz-Serna, J. M. and Calvo, M. P.: 1994, *Numerical Hamiltonian Methods*, Chapman and Hall, London.

Sardeshmukh, P. D. and Hoskins, B. J.: 1984, Spatial smoothing on the sphere, *Monthly Weather Review* **112**, 2524–2529. If a filter applied to a spherical harmonics series depends only on the subscript n of the spherical harmonics, it is equivalent to an isotropic spatial filter whose effect depends only on distance from the point of observation.

Sawyer, J. S.: 1963, A semi-Lagrangian method of solving the vorticity equation, *Tellus* **15**(4), 336–342. Not spectral; coined the label "semi-Lagrangian".

Schaffer, S. and Stenger, F.: 1986, Multigrid-sinc methods, *Applied Mathematics and Computation* **19**, 311–319.

Schatz, M. F., Tagg, R. P., Swinney, H. L., Fischer, P. F. and Patera, A. T.: 1991, Supercritical transition in plane channel flow with spatially periodic perturbations, *Physical Review Letters*.

Schmidt, F.: 1977, Variable fine mesh in the spectral global models, *Beitr. Phys. Atmos.* **50**, 211–217. Conformal mapping to give high resolution in a chosen region.

Schmidt, F.: 1982, Cyclone tracing, *Beitr. Phys. Atmos.* **55**, 335–357. Conformal mapping is used to track cyclones by increasing resolution locally around the track of the storm; method of Schmidt(1977).

Schnack, D. D., Baxter, D. C. and Caramana, E. J.: 1984, A pseudospectral algorithm for three-dimensional magnetohydrodynamic simulation, *Journal of Computational Physics* **55**, 485–514.

Schneidesch, C., Deville, M. and Demaret, P.: 1990, Steady-state solution of a convection benchmark problem by multidomain Chebyshev collocation, *in* B. Roux (ed.), *Numerical Simulation of Oscillatory Convection in Low-Pr Fluids*, Friedr. Vieweg & Sohn Verlagsgesellschaft mbH, Braunschweig, part 4, pp. 256–261.

Schneidesch, C. R. and Deville, M. O.: 1993, Chebyshev collocation method and multidomain decomposition for Navier-Stokes equations in complex curved geometries, *Journal of Computational Physics* **106**, 234–257.

Schneidesch, C. R. and Deville, M. O.: 1994, Multidomain decomposition of curved geometries in the Chebyshev collocation method for thermal problems, *in* C. Bernardi and Y. Maday (eds), *Analysis, Algorithms and Applications of Spectral and High Order Methods for Partial Differential Equations*, Selected Papers from the International Conference on Spectral and High Order Methods (ICOSAHOM '92), Le Corum, Montpellier, France, 22-26 June 1992, North-Holland, Amsterdam, pp. 87–94. Also in Comput. Meths. Appl. Mech. Engrg., 116.

Schoeberl, M. R. and Geller, M. A.: 1977, A calculation of the structure of stationary planetary waves in winter, *Journal of the Atmospheric Sciences* **34**, 1235–1255. Application of finite differences and block-tridiagonal Gaussian elimination to a PDE of mixed elliptic-hyperbolic type.

Schultz, W. W., Huh, J. and Griffin, O. M.: 1994, Potential-energy in steep and breaking waves, *J. Fluid Mech.* **278**, 201–228. Spectral algorithm for two-dimensional potential flow.

Schultz, W. W., Lee, N.-Y. and Boyd, J. P.: 1989, Chebyshev pseudospectral method of viscous flows with corner singularities, *Journal of Scientific Computing* **4**, 1–24.

Schultz, W. W., Vanden-Broeck, J., Jiang, L. and Perlin, M.: 1998, Highly nonlinear standing water waves with small capillary effect, *J. Fluid Mech.* **369**, 253–272. Spectral boundary integral method is combined with Newton iteration to compute spatially and temporally periodic standing waves. A non-uniform node distribution (change-of-coordinate) is used to get accurate approximations near the limiting wave.

Schumack, M. R.: 1996, Application of the pseudospectral method to thermodynamic lubrication, *Int. J. Numer. Meth. Fluids* **23**(11), 1145–1161. Preprint.

Schumack, M. R., Schultz, W. W. and Boyd, J. P.: 1991, Spectral method solution of the Stokes equations on nonstaggered grids, *Journal of Computational Physics* **94**(1), 30–58.

Schumer, J. W. and Holloway, J. P.: 1998, Vlasov simulations using velocity-scaled Hermite representations, *J. Comput. Phys.* **144**(2), 626–661. Hermite function calculations in plasma physics.

Schuster, A.: 1903, On some definite integrals and a new method of reducing a function of spherical coordinates to a series of spherical harmonics, *Phil. Trans. Roy. Soc. London A* **200**, 181–223. Two-step transform in which data for a given longitudinal wavenumber is expanded first as a trigonometric series, and then this is converted to spherical harmonics at a savings of roughly a factor of two over direct computation of spherical harmonic coefficients.

Schwartz, C.: 1963, Estimating convergence rates of variational calculations, *Methods in Computational Physics* **2**, 241–266. How singularities affect the rate of convergence of Laguerre function series.

Sela, J.: 1980, Spectral modelling at the National Meteorological Center, *Monthly Weather Review* **108**, 1279–1292.

Sela, J. G.: 1995, Weather forecasting on parallel architectures, *Parallel Comput.* **21**(10), 1639–1654. Spherical harmonic modelling at the National Meteorological Center.

She, Z.-S., Jackson, E. and Orszag, S. A.: 1990, Vortex structure and dynamics in turbulence, *in* C. Canuto and A. Quarteroni (eds), *Spectral and High Order Methods for Partial Differential Equations: Proceedings of the ICOSAHOM '89 Conference in Como, Italy*, North-Holland/Elsevier, Amsterdam, pp. 173–184. Also in Comput. Meths. Appl. Mech. Engrg., vol. 80, with the same page numbers. Spectral computations of turbulence.

Shen, J.: 1994, Efficient spectral-Galerkin method I. Direct solvers of second- and fourth-order equations using Legendre polynomials, *SIAM Journal of Scientific Computing* **15**(6), 1489–1505.

Shen, J.: 1995a, Efficient spectral-Galerkin method II. Direct solvers of second- and fourth-order equations using Chebyshev polynomials, *SIAM Journal of Scientific Computing* **16**(6), 74–87.

Shen, J.: 1995b, On fast Poisson solver, inf-sup constant and iterative Stokes solver by Legendre Galerkin method, *Journal of Computational Physics* **116**, 184–188.

Shen, J.: 1997, Efficient spectral-Galerkin method III. Polar and cylindrical geometries, *SIAM J. Sci. Comput.* Fast methods for inverting the Laplace and other almost constant coefficient operators using orthogonal polynomials.

Shen, J.: 1999, Efficient spectral-Galerkin methods IV. Spherical geometries, *SIAM J. Scientific Comput.* **20**(4), 1438–1455. Double Fourier series method; comparisons with spherical harmonics and finite difference.

Shen, J. and Wang, S. H.: 1999, A fast and accurate numerical scheme for the primitive equations of the atmosphere, *SIAM J. Numer. Anal.* **36**(3), 719–737. Vector spherical harmonics with either Legendre or Chebyshev polynomials for the vertical coordinate; semi-implicit time marching with fast direct solution of the only non-trivial boundary value problems, which are one-dimensional in height only, one such ODE per spherical harmonic.

Sherwin, S. J.: 1997, Hierarchical *hp* finite elements in hybrid domains, *Finite Elements in Analysis and Design* **27**, 109–119.

Sherwin, S. J. and Karniadakis, G. E.: 1996, Tetrahedral *hp* finite elements: algorithms and flow simulations, *Journal of Computational Physics* **124**, 14–45. Tensor product spectral elements in three dimensions.

Sherwin, S. J., Karniadakis, G. E. and Orszag, S. A.: 1994, Numerical simulation of the ion etching process, *Journal of Computational Physics* **110**, 373.

Shoucri, M. M. and Gagné, R. R. J.: 1977, Numerical solution of a two-dimensional Vlasov equation, *Journal of Computational Physics* **25**, 94–103. Tensor product Hermite basis for the two velocity coordinates, finite differences for the two space dimensions.

Simmons, A. J. and Hoskins, B. J.: 1975, A comparison of spectral and finite-difference simulations of a growing baroclinic wave, *Quarterly Journal of the Royal Meteorological Society* **101**, 551–565.

Siyyam, H. I. and Syam, M. I.: 1997, An accurate solution of the Poisson equation by the Chebyshev-Tau method, *J. Comput. App. Math.* **85**, 1–10. Alternative to Haidvogel-Zang algorithm for separable PDEs.

Sloan, D. M.: 1991, Fourier pseudospectral solution of the Regularised Long Wave equation, *Journal of Computational and Applied Mathematics* **36**, 159–179.

Sloan, I. H.: 1995, Polynomial interpolation and hyperinterpolation over general regions, *Journal of Approximation Theory* **83**, 238–254. Computation of approximations on the surface of sphere as a truncated spherical harmonic series with coefficients evaluated by numerical quadrature. "Hyperinterpolation" is a coefficient computed by a quadrature with more points than unknowns in the truncated basis. For the sphere, interpolation is not possible but hyperinterpolation is successful.

Smith, N. R.: 1988, A truncated oceanic spectral model for equatorial thermodynamic studies, *Dyn. Atmos. Oceans* **12**, 313–337. Hermite spectral series.

Smith, R. C. and Bowers, K. L.: 1991, A fully Galerkin method for the recovery of stiffness and damping parameters in Euler-Bernoulli beam models, *Computation and Control II, Proceedings of the Bozeman Conference 1990*, number 11 in *Progress in Systems and Control Theory*, Birkhäuser, Boston, pp. 289–306.

Smith, R. C., Bogar, G. A., Bowers, K. L. and Lund, J.: 1991a, The sinc-Galerkin method for fourth-order differential equations, *SIAM Journal of Numerical Analysis* **28**, 760–788.

Smith, R. C., Bowers, K. L. and Lund, J.: 1991b, Efficient numerical solution of fourth-order problems in the modeling of flexible structures, *Computation and Control, Proceedings of the Bozeman Conference 1988*, number 1 in *Progress in Systems and Control Theory*, Birkhäuser, Boston, pp. 289–306.

Smith, R. C., Bowers, K. L. and Lund, J.: 1992, A fully sinc-Galerkin method for Euler-Bernoulli beam models, *Numerical Methods for Partial Differential Equations* **8**, 171–202.

Smolarkiewicz, P. and Rasch, P.: 1990, Monotone advection on the sphere: An Eulerian versus a semi-Lagrangian approach, *Journal of the Atmospheric Sciences* **48**, 793–810.

Sneddon, G. E.: 1996, Second-order spectral differentiation matrices, *SIAM Journal of Numerical Analysis* **33**(6), 2468–2487. Shows that different discretization methods (pseudospectral versus Galerkin versus tau) are equivalent, after splitting the matrix into two subproblems through parity, to a rank-one update to the symmetric differentiation matrix and a different rank-one update to the differentiation matrix for functions antisymmetric with respect to x. Claims that one can choose an update which makes the eigenvalues $O(N^2)$ at largest instead of $O(N^4)$ without drastic loss of accuracy in differentiation.

Solomonoff, A.: 1992, A fast algorithm for spectral differentiation, *Journal of Computational Physics* **98**, 174–177. Independent invention of the Parity Matrix Multiplication.

Spalart, P. R.: 1984, A spectral method for external viscous flows, *Contemporary Mathematics* **28**, 315–335. Divergence-free basis functions.

Spotz, W. F. and Swarztrauber, P. N.: 2000, A performance comparison of Associated Legendre projections, *Journal of Computational Physics*. Submitted. Comparisons of four algorithms for the Legendre transforms in latitude. The "weighted orthogonal complement" method (Swarztrauber and Spotz, 2000) is moderately faster than the alternatives for $N < 200$ including the Fast Multipole Method.

Spotz, W. F., Taylor, M. A. and Swarztrauber, P. N.: 1998, Fast shallow-water equation solvers in latitude-longitude coordinates, *Journal of Computational Physics* **145**(1), 432–444. Double Fourier series in both longitude and latitude, which would normally have serious pole troubles, are stabilized by using a filter. One filter is an improvement of a Fourier filter developed by Merilees. Another is the fast multipole-like filter develped by Jakob-Chien and Alpert and improved by Yarvin and Rokhlin. Both filters work well and eliminate most of the speed penalty associated with Legendre transforms in a pure spherical harmonics method.

Staniforth, A. and Côté, J.: 1991, Semi-Lagrangian integration schemes for atmospheric models — a review, *Monthly Weather Review* **119**, 2206–2223.

Stenger, F.: 1979, A "Sinc-Galerkin" method of solution of boundary value problems, *Mathematics of Computation* **33**, 85–109.

Stenger, F.: 1981, Numerical methods based on Whittaker cardinal or Sinc functions, *SIAM Review* **23**, 165–224.

Stenger, F.: 1993, *Sinc Methods*, Springer-Verlag, New York. 500 pp. Restricted to sinc functions; nothing on Chebyshev or Fourier algorithms.

Strain, J.: 1994, Fast spectrally-accurate solution of variable-coefficient elliptic problems, *Proceedings of the American Mathematical Society* **122**, 843–850. Iteration for non-separable BVPs in which a separable elliptic problem is solved by a fast direct spectral method at each iteration.

Strain, J.: 1995, Spectral methods for nonlinear parabolic systems, *Journal of Computational Physics* **122**, 1–12.

Straka, J. M., Wilhelmson, R. B., Wicker, L. J., Anderson, J. R. and Droegemeier, K. K.: 1993, Numerical solutions of a non-linear density current: A benchmark solution and comparisons, *International Journal for Numerical Methods in Fluids* **17**, 1–22.

Strang, G.: 1968, On the construction and comparison of difference schemes, *SIAM Journal of Numerical Analysis* **5**(3), 506–517. Independent invention of fractional steps time-marching, which is often called "Strang splitting" in the U. S.

Strang, G. and Fix, G. J.: 1973, *The Analysis of the Finite Element Method*, Prentice-Hall, Englewood Cliffs, New Jersey. 305 pp.

Straughan, B. and Walker, D. W.: 1996, Two very accurate and efficient methods for computing eigenvalues and eigenfunctions in porous convection problems, *Journal of Computational Physics* **127**, 128–141. Compound matrix and Chebyshev-tau algorithms.

Streett, C. L. and Zang, T. A.: 1984, Spectral methods for the solution of the boundary-layer equations, *Paper 84-0170*, AIAA. Nonlinear Richardson's iteration.

Streett, C. L., Zang, T. A. and Hussaini, M. Y.: 1985, Spectral multigrid methods with applications to transonic potential flow, *Journal of Computational Physics* **57**, 43–76.

Stuhne, G. R. and Peltier, W. R.: 1999, New icosahedral grid-point discretizations of the shallow water equations on the sphere, *J. Comput. Phys.* **148**(1), 23–58. Non-spectral alternative to spherical harmonics.

Su, Y. Y. and Khomami, B.: 1992, Numerical solution of eigenvalue problems using spectral techniques, *Journal of Computational Physics* **100**, 297–305.

Sulem, P. L., Sulem, C. and Patera, A.: 1985, A numerical simulation of singular solutions to the two-dimensional Cubic Schrodinger equation, *Communications in Pure and Applied Mathematics* **37**, 755–778.

Süli, E. and Ware, A.: 1991, A spectral method of characteristics for hyperbolic problems, *SIAM Journal of Numerical Analysis* **28**, 423–445.

Süli, E. and Ware, A. F.: 1992, Analysis of the spectral Lagrange-Galerkin method for the Navier-Stokes equations, *in* J. G. Heywood, K. Masuda, R. Rautmann and V. A. Solonnikov (eds), *The Navier-Stokes Equations II-Theory and Numerical Methods: Proceedings of a Conference in Oberwolfach, Germany*, Springer-Verlag, pp. 184–195.

Swarztrauber, P. N.: 1977, On the spectral approximation of discrete scalar and vector functions on the sphere, *SIAM Journal of Numerical Analysis* **16**, 934–949.

Swarztrauber, P. N.: 1981, The approximation of vector functions and their derivatives on the sphere, *SIAM Journal of Numerical Analysis* **18**, 191–210.

Swarztrauber, P. N.: 1982, Vectorizing the FFTs, *in* G. Rodrique (ed.), *Parallel Computations*, Academic Press, New York, pp. 51–83.

Swarztrauber, P. N.: 1986, Symmetric FFTs, *Mathematics of Computation* **47**, 323–346. Efficient methods of computing cosine and sine transforms.

Swarztrauber, P. N.: 1987, Multiprocessor FFTs, *Parallel Computing* **7**, 197–210.

Swarztrauber, P. N.: 1993, The vector harmonic transform method for solving partial differential equations in spherical geometry, *Monthly Weather Review* **121**, 3415–3437.

Swarztrauber, P. N.: 1996, Spectral transform methods for solving the shallow-water equations on the sphere, *Monthly Weather Review* **124**(4), 730–744.

Swarztrauber, P. N. and Kasahara, A.: 1985, The vector harmonic analysis of Laplace's tidal equations, *SIAM Journal of Scientific and Statistical Computing* **6**, 464–491. This and earlier individual papers by the two authors are very good works on the use of vector basis functions for geophysical flows.

Swarztrauber, P. N. and Spotz, W. F.: 2000, Generalized discrete spherical harmonic transform, *Journal of Computational Physics* **159**(2), 213–230. Introduces a new "weighted
   orthogonal complement" method for computing Associated Legendre function transforms in latitude; this is somewhat faster than alternatives (Spotz and Swarztrauber,
   2000) and reduces storage requirements by an order of magnitude.

Swarztrauber, P. N., Williamson, D. L. and Drake, J. B.: 1997, The Cartesian method for solving partial differential equations in spherical geometry, *Dyn. Atmos. Oceans* **27**, 679–
   706. Ingenious non-spectral scheme on an icosahedral grid.

Tan, B. and Boyd, J. P.: 1997, Dynamics of the Flierl-Petviashvili monopoles in a barotropic
   model with topographic forcing, *Wave Motion* **26**, 239–252. Double Fourier series for
   nonlinear initial value problem.

Tan, B. and Boyd, J. P.: 1998, Davydov soliton collisions, *Physics Letters A* **240**, 282–286.
   Double Fourier series for nonlinear initial value problem.

Tan, B. and Boyd, J. P.: 1999, Coupled-mode envelope solitary waves in a pair of cubic
   Schroedinger equations with cross modulation: Analytical solution and collisions,
   Chaos, Solitons and Fractals. in proof.

Tan, C. S.: 1985, Accurate solution of three-dimensional Poisson's equation in cylindrical
   coordinates by expansion in Chebyshev polynomials, *Journal of Computational Physics*
   59, 81–95.

Tang, T.: 1993, The Hermite spectral method for Gaussian-type functions, *SIAM Journal of
   Scientific Computing* **14**, 594–606.

Taylor, M.: 1995, Cubature for the sphere and the discrete spherical harmonic transform,
   SIAM Journal of Numerical Analysis **32**, 667–670.

Taylor, M. A. and Wingate, B. A.: 1999, The Fekete collocation points for triangular spectral
   elements, *SIAM Journal of Numerical Analysis.* Submitted.

Taylor, M., Tribbia, J. and Iskandarani, M.: 1997, The spectral element method for the shallow water equations on the sphere, *Journal of Computational Physics* **130**(1), 92–108.

Taylor, T. D., Hirsh, R. S. and Nadworny, M. M.: 1981, FFT versus conjugate gradient
   method for solutions of flow equations by pseudospectral methods, *in* H. Viviand
   (ed.), *Proceedings of the 4th GAMM Conference on Numerical Methods in Fluid Mechanics,*
   Vieweg, Braunschweig, pp. 311–325.

Taylor, T. D., Hirsh, R. S. and Nadworny, M. M.: 1984, Comparison of FFT, direct inversion,
   and conjugate gradient methods for use in pseudospectral methods, *Computers and
   Fluids* **12**, 1–9. Examines the penalty for using a nonstandard grid such that neither the
   FFT nor matrix multiplication transform are applicable, and one must invert matrices
   to calculate coefficients of $f(x)$ from grid point values of $f(x)$.

Témam, R.: 1989, Attractors for the Navier-Stokes equations: localization and approximation, *Journal of the Faculty of Science of the University of Tokyo, Sec. Ia, Mathematics*
   36, 629–647.

Témam, R.: 1991a, New emerging methods in numerical analysis; applications to fluid
   mechanics, *in* M. Gunzberger and N. Nicolaides (eds), *Computational Fluid Dynamics
   — Trend and Advances,* Cambridge University Press, Cambridge.

Témam, R.: 1991b, Stability analysis of the nonlinear Galerkin method, *Mathematics of Computation* **57**, 477–503.

Témam, R.: 1992, General methods for approximating inertial manifolds. applications to computing, *in* D. S. Broomhead and A. Iserles (eds), *The Dynamics of Numerics and the Numerics of Dynamics*, Clarendon Press, Oxford, Oxford, pp. 1–21.

Temperton, C.: 1975, Algorithms for the solutions of cyclic tridiagonal systems, *Journal of Computational Physics* **19**, 317–323.

Temperton, C.: 1983a, Fast mixed-radix real Fourier transforms, *Journal of Computational Physics* **52**, 340–350.

Temperton, C.: 1983b, Self-sorting mixed-radix fast Fourier transforms, *Journal of Computational Physics* **52**, 1–23.

Temperton, C.: 1985, Implementation of a self-sorting in-place prime factor FFT algorithm, *Journal of Computational Physics* **58**, 283–299.

Temperton, C.: 1992, A generalized prime factor FFT algorithm for any N=(2**p) (3**q) (5**r), *SIAM J. Sci. Stat. Comput.* **13**, 676–686.

Temperton, C. and Staniforth, A.: 1987, An efficient two-time-level semi-Lagrangian semi-implicit integration scheme, *Quarterly Journal of the Royal Meteorological Society* **113**, 1025–1039.

Thompson, J., Warsi, Z. and Mastin, C.: 1985, *Numerical Grid Generation*, North-Holland, New York. Not spectral.

Thompson, W. D.: 1917, *Growth and Form*, Cambridge University Press, Cambridge. Not spectral, 600 pp.

Thuburn, J.: 1997, A PV-based shallow-water model on a hexagonal-icosahedral grid, *Mon. Wea. Rev.* **125**, 2328–2347. Non-spectral alternative for flows on the surface of a sphere.

Trefethen, L. N. and Bau, III, D.: 1997, *Numerical Linear Algebra*, Society for Industrial and Applied Mathematics (SIAM), Philadelphia.

Tribbia, J. J.: 1984a, Modons in spherical geometry, *Geophys. Astrophys. Fluid Dyn.* **30**, 131–168. Spherical harmonics in a rotated coordinate system are used to construct nonlinear Rossby waves.

Tribbia, J. J.: 1984b, A simple scheme for high-order nonlinear normal initialization, *Monthly Weather Review* **112**, 278–284. Iterative scheme for initialization onto the slow manifold, illustrated by Hough-Hermite spectral basis.

Tse, K. L. and Chasnov, J. R.: 1998, A Fourier-Hermite pseudospectral method for penetrative convection, *Journal of Computational Physics* **142**(2), 489–505. Two Fourier coordinates plus Hermite functions in the unbounded vertical; motion is confined to an unstable layer far from any boundaries.

Tuckerman, L.: 1989, Divergence-free velocity field in nonperiodic geometries, *Journal of Computational Physics* **80**, 403–441. Very careful treatment of influence matrix method and spurious pressure modes in computing incompressible flow.

Vallis, G. K.: 1985, On the spectral integration of the quasi-geostrophic equations for doubly-periodic and channel flow, *Journal of the Atmospheric Sciences* **42**, 95–99.

Van Kemenade, V. and Deville, M.: 1994a, Preconditioned Chebyshev collocation for non-Newtonian fluid flows, *in* C. Bernardi and Y. Maday (eds), *Analysis, Algorithms and Applications of Spectral and High Order Methods for Partial Differential Equations*, Selected Papers from the International Conference on Spectral and High Order Methods (ICOSAHOM '92), Le Corum, Montpellier, France, 22-26 June 1992, North-Holland, Amsterdam, pp. 377–386. Also in Finite Elements in Analysis and Design, vo. 16, pp. 237-245.

Van Kemenade, V. and Deville, M. O.: 1994b, Application of spectral elements to viscoelastic creeping flows, *Journal of Non-Newtonian Fluid Mechanics* **51**, 277–308. Plane and cylindrical geometries; Maxwell-B fluid; comparisons with 4×4 SUPG finite elements.

van Loan, C.: 1992, *Computational Frameworks for the Fast Fourier Transform*, SIAM, Philadelphia.

Vandeven, H.: 1991, Family of spectral filters for discontinuous problems, *SIAM Journal of Scientific Computing* **6**, 159–192.

Verkley, W. T. M.: 1997a, A pseudo-spectral model for two-dimensional incompressible flow in a circular basin. I. Mathematical formulation, *Journal of Computational Physics* **136**(1), 100–114. Employs the basis set $r^{|m|} P_k^{0,|m|} (2r^2 - 1) \exp(im\theta)$ where $P_k^{\alpha,\beta}$ is the usual Jacobi polynomial; shows and exploits the fact that the inverse of Laplace operator has a banded Galerkin representation with this basis.

Verkley, W. T. M.: 1997b, A pseudo-spectral model for two-dimensional incompressible flow in a circular basin. II. Numerical examples, *Journal of Computational Physics* **136**(1), 115–131.

Vermer, J. G. and van Loon, M.: 1994, An evaluation of explicit pseudo-steady-state approximation schemes to stiff ODE systems from chemical kinetics, *Journal of Computational Physics* **113**, 347–352. The PSSA strategy, widely used in chemical kinetics, is equivalent to the lowest order Nonlinear Galerkin method: the ODEs for the fast components are replaced by algebraic relations.

Vianna, M. L. and Holvorcem, P. R.: 1992, Integral-equation approach to tropical ocean dynamics. 1. Theory and computational methods, *J. Marine Res.* **50**(1), 1–31. Boundary-element algorithm constructed through ingenious summation of slowly converging Hermite series.

Vinograde, B.: 1967, *Linear and Matrix Algebra*, D. C. Heath, Boston. Not spectral.

Voigt, R. G., Gottlieb, D. and Hussaini, M. Y. (eds): 1995, *Spectral Methods for Partial Differential Equations*, SIAM, Philadelphia.

Vozovoi, L., Israeli, M. and Averbuch, A.: 1996, Analysis and application of the Fourier-Gegenbauer method to stiff differential equations, *SIAM J. Numer. Anal.* **33**(5), 1844–1863.

Vozovoi, L., Weill, A. and Israeli, M.: 1997, Spectrally accurate solution of non-periodic differential equations by the Fourier-Gegenbauer method, *SIAM J. Numer. Anal.* **34**(4), 1451–1471.

Wahba, G.: 1990, *Spline Models for Observational Data*, Vol. 59 of *CBMS-NSF Regional Conference Series in Applied Mathematics*, SIAM, Philadelphia. 169 pp. Fourier and spherical harmonic splines.

Wang, D.: 1991, Semi-discrete Fourier spectral approximations of infinite dimensional Hamiltonian systems and conservation laws, *Computers Mathematics and Applications* **21**, 63–75.

Ware, A. F.: 1991, *A spectral Lagrange-Galerkin method for convection-dominated diffusion problems*, Ph.D. dissertation, Oxford University, Wolfson College. 106 pp.

Ware, A. F.: 1994, A spectral Lagrange-Galerkin method for convection-dominated diffusion problems, *in* C. Bernardi and Y. Maday (eds), *Analysis, Algorithms and Applications of Spectral and High Order Methods for Partial Differential Equations*, Selected Papers from the International Conference on Spectral and High Order Methods (ICOSAHOM '92), Le Corum, Montpellier, France, 22-26 June 1992, North-Holland, Amsterdam, pp. 227–234. Also in Comp. Meths. Appl. Mech. Engrg., 116.

Ware, A. F.: 1998, Fast approximate Fourier transforms for irregularly spaced data, *SIAM Rev.* **40**(4), 838–856.

Weideman, J. A. C.: 1992, The eigenvalues of Hermite and rational spectral differentiation matrices, *Numerische Mathematik* **61**, 409–431.

Weideman, J. A. C.: 1994a, Computation of the complex error function, *SIAM Journal of Numerical Analysis* **31**, 1497–1518. Errata: 1995, 32, 330–331]. These series of rational functions are useful for complex-valued z.

Weideman, J. A. C.: 1994b, Computing integrals of the complex error function, *Proceedings of Symposia in Applied Mathematics* **48**, 403–407. Short version of Weideman(1994a).

Weideman, J. A. C.: 1995a, Errata: computation of the complex error function, *SIAM Journal of Numerical Analysis* **32**, 330–331.

Weideman, J. A. C.: 1995b, Computing the Hilbert Transform on the real line, *Mathematics of Computation* **64**(210), 745–762. The Hilbert transform of a function f is defined by $H\{f\}(y) \equiv \frac{1}{\pi} PV \int_{-\infty}^{\infty} \frac{f(x)}{x-y} dx$ where PV is the principal value of the (singular) integral. Weideman shows that a set of complex-valued orthogonal functions, $\rho_n(x) = (l + ix)^n/(l - ix)^{n+l}, n = 0, \pm 1, \pm 2, \ldots$, are eigenfunctions of the transform. Weideman's algorithm using these functions is generally more efficient than the Fourier transform algorithm, $H\{f\}(y) = F^{-1}\{i\text{sign}(k)F\{f\}(k)\}$. The Hilbert transform is important in many physical applications, notably the Benjamin-Ono model for solitary waves in deep stratified fluid.

Weideman, J. A. C. and Cloot, A.: 1990, Spectral methods and mappings for evolution equations on the infinite line, *Computer Methods in Applied Mechanics and Engineering* **80**, 467–481.

Weideman, J. A. C. and James, R. L.: 1992, Pseudospectral methods for the Benjamin-Ono equation, *Advances in Computer Methods for Partial Differential Equations VII* pp. 371–377. Both Fourier series and rational functions.

Welander, P.: 1955, Studies on the general development of motion in a two dimensional fluid, *Tellus* **7**, 141–156. Not spectral; classic article that shows (i) a square blob of fluid becomes greatly stretched and deformed even by very smooth flow — becoming "chaotically" mixed and bounded by a "fractal" curve, though these terms were not yet invented which (ii) shows the unfeasibility of a fully Lagrangian (as opposed to semi-Lagrangian) time-marching for fluids.

Weyl, H.: 1952, *Symmetry*, Princeton University Press, Princeton, New Jersey. Popular but rigorous discussion of symmetry in art and nature and its mathematical embodiment in group theory.

White, Jr., A. B.: 1982, On the numerical solution of initial/boundary value problems in one space dimension, *SIAM Journal of Numerical Analysis* **19**(4), 683–697. Shocks in Burgers' equation are resolved by transforming the spatial coordinate to arclength, which requires simultaneously integrating an extra equation.

Whittaker, E. T.: 1915, On the functions which are represented by the expansions of the interpolation theory, *Proceedings of the Royal Society Edinburgh* **35**, 181–194. Pioneering paper on sinc functions.

Wiin-Nielsen, A.: 1959, On the application of trajectory methods in numerical forecasting, *Tellus* **11**(2), 180–196. Not spectral; earliest application of semi-Lagrangian time-marching in which the choice of the particles is reinitialized at each time step to those arriving at the points of the regular Eulerian-coordinates grid at that time level.

Williamson, D.: 1990, Semi-Lagrangian transport in the NMC spectral model, *Tellus A* **42**, 413–428.

Williamson, D. L. and Rasch, P. J.: 1989, Two-dimensional semi-Lagrangian transport with shape-preserving interpolation, *Monthly Weather Review* **117**, 102–129. Both Cartesian and Gaussian (spherical) grid; only pure advection is computed.

Wingate, B. A. and Boyd, J. P.: 1996, Spectral element methods on triangles for geophysical fluid dynamics problems, *in* A. V. Ilin and L. R. Scott (eds), *Proceedings of the Third International Conference on Spectral and High Order Methods*, Houston Journal of Mathematics, Houston, Texas, pp. 305–314.

Wingate, B. A. and Taylor, M. A.: 1999a, A Fekete point triangular spectral element method; application to the shallow water equations, *Journal of Computational Physics*. Submitted.

Wingate, B. A. and Taylor, M. A.: 1999b, The natural function space for triangular spectral elements, *SIAM Journal of Numerical Analysis*. Submitted.

Wragg, A.: 1966, The use of Lanczos-τ methods in the numerical solution of a Stefan problem, *Computer Journal* **9**, 106–109.

Wunsch, C., Haidvogel, D. B., Iskandarani, M. and Hughes, R.: 1997, Dynamics of the long-period tides, *Progress in Oceanography* **40**(1–4), 81–108. Spectral elements.

Xu, C. and Maday, Y.: 1998, A spectral element method for the time-dependent two-dimensional Euler equations: applications to flow simulations, *J. Comput. Appl. Math.* **91**, 63–85. Show that a spectral approximation with polynomials of equal degree for pressure and velocity is well-posed. (Not true for the Navier-Stokes equation.).

Yakhot, A., Orszag, S. A., Yakhot, V. and Israeli, M.: 1989, Renormalization group formulation of large-eddy simulations, *Journal of Scientific Computing* **4**(2), 139–158.

Yanenko, N. N.: 1971, *The Method of Fractional Steps*, Springer-Verlag, New York. 160 pages.

Yang, G., Belleudy, P. and Temperville, A.: 1991, A higher-order Eulerian scheme for coupled advection-diffusion transport, *International Journal for Numerical Methods in Fluids* **12**, 43–58.

Yang, J. and Hsu, E.: 1991, On the use of the reach-back characteristics method for the calculation of dispersion, *International Journal for Numerical Methods in Fluids* **12**, 225–235.

Yang, T.-S. and Akylas, T. R.: 1996, Weakly nonlocal gravity-capillary solitary waves, *Physics of Fluids* **8**(6), 1506–1514. Rational Chebyshev functions (in somewhat unorthodox notation) and a special radiation basis function are combined with Newton's iteration to compute oscillatory tails even when the tails are smaller than 10^{-13} by combining collocation with Newton's iteration.

Yang, Z.-H. and Keller, H. B.: 1986, Multiple laminar flows through curved pipes, *Applied Numerical Mathematics* **2**, 257–271. The Dean problem of viscous steady flow through a coiled circular pipe is studied using Fourier expansion in polar angle and finite differences; good discussion of pseudoarclength continuation to trace an intricate structure of solution branches.

Yarvin, N. and Rohklin, V.: 1998, A generalized one-dimensional fast multipole method with application to filtering of spherical harmonics, *J. Comput. Phys.* **147**(2), 594–609.

Yee, S. Y. K.: 1981, Solution of Poisson's equation on a sphere by double Fourier series, *Monthly Weather Review* **109**, 501–505.

Yin, G.: 1994, Sinc-collocation method with orthogonalization for singular Poisson-like problems, *Mathematics of Computation* **62**, 21–40.

Young, D. M. and Gregory, R. T.: 1972, *A Survey of Numerical Mathematics, Vol. 1*, Addison-Wesley, Reading, MA. pg. 329–339; not spectral.

Zampieri, E. and Tagliani, A.: 1997, Numerical approximation of elastic waves by implicit spectral methods, *Comput. Meth. Appl. M.* **144**(1–2), 33–50. Legendre polynomials in single-domain spectral element (variational) algorithm combined with implicit time-marching; absorbing boundaries are used to imitate an infinite domain.

Zang, T.: 1991, On the rotation and skew-symmetric forms for incompressible flow simulations, *Applied Numerical Mathematics* **7**, 27–40.

Zang, T. A. and Hussaini, M. Y.: 1986, On spectral multigrid methods for the time-dependent Navier-Stokes equations, *Applied Mathematics and Computation* **19**, 359–372.

Zang, T. A., Krist, S. E. and Hussaini, M. Y.: 1989, Resolution requirements for numerical simulations of transition, *Journal of Scientific Computing* **4**(2), 197–219.

Zang, T. A., Wong, Y. S. and Hussaini, M. Y.: 1984, Spectral multigrid methods for elliptic equations II, *Journal of Computational Physics* **54**, 489–507.

Zebib, A.: 1984, A Chebyshev method for the solution of boundary value problems, *Journal of Computational Physics* **53**, 443–455. Assumes expansion for highest derivative of unknown rather than u itself.

Zebib, A.: 1987a, Stability of viscous flow past a circular cylinder, *Journal of Engineering Mathematics* **21**, 155–165.

Zebib, A.: 1987b, Removal of spurious modes encountered in solving stability problems by spectral methods, *Journal of Computational Physics* **70**(2), 521–525.

Zhao, S. and Yedlin, M. J.: 1994, A new iterative Chebyshev spectral method for solving the elliptic equation $\nabla \otimes (\sigma \nabla u) = f$, *Journal of Computational Physics* **113**, 215–223. The Haidvogel-Zang fast direct method for solving separable PDEs through Chebyshev series is the key operation in a fast iterative method for non-separable elliptic equations.

Zheng, J.-D., Zhang, R.-F. and Guo, B.-Y.: 1989, The Fourier pseudo-spectral method for the SRLW equation, *Applied Mathematics and Mechanics* **10**, 843–852.

Zrahia, U., Orszag, S. A. and Israeli, M.: 1996, Mixed analytical/numerical-spectral element algorithm for efficient solution of problems with interior or boundary layers, *in* A. V. Ilin and L. R. Scott (eds), *Proceedings of the Third International Conference on Spectral and High Order Methods*, Houston Journal of Mathematics, Houston, Texas, pp. 315–326. Analytical solution for boundary layer combined with spectral element solution for the outer flow.